Science and Mathematics for Engineering

Why is knowledge of science and mathematics important in engineering?

A career in any engineering field will require both basic and advanced mathematics and science. Without mathematics and science to determine principles, calculate dimensions and limits, explore variations, prove concepts, and so on, there would be no mobile telephones, televisions, stereo systems, video games, microwave ovens, computers, or virtually anything electronic. There would be no bridges, tunnels, roads, skyscrapers, automobiles, ships, planes, rockets or most things mechanical. There would be no metals beyond the common ones, such as iron and copper, no plastics, no synthetics. In fact, society would most certainly be less advanced without the use of mathematics and science throughout the centuries and into the future.

Electrical engineers require mathematics and science to design, develop, test, or supervise the manufacturing and installation of electrical equipment, components, or systems for commercial, industrial, military, or scientific use.

Mechanical engineers require mathematics and science to perform engineering duties in planning and designing tools, engines, machines, and other mechanically functioning equipment; they oversee installation, operation, maintenance, and repair of such equipment as centralised heat, gas, water, and steam systems.

Aerospace engineers require mathematics and science to perform a variety of engineering work in designing, constructing, and testing aircraft, missiles, and spacecraft; they conduct basic and applied research to evaluate adaptability of materials and equipment to aircraft design and manufacture and recommend improvements in testing equipment and techniques.

Nuclear engineers require mathematics and science to conduct research on nuclear engineering problems or apply principles and theory of nuclear science to problems concerned with release, control, and utilisation of nuclear energy and nuclear waste disposal.

Petroleum engineers require mathematics and science to devise methods to improve oil and gas well production and determine the need for new or modified tool designs; they oversee drilling and offer technical advice to achieve economical and satisfactory progress.

Industrial engineers require mathematics and science to design, develop, test, and evaluate integrated systems for managing industrial production processes, including human work factors, quality control, inventory control, logistics and material flow, cost analysis, and production coordination.

Environmental engineers require mathematics and science to design, plan, or perform engineering duties in the prevention, control, and remediation of environmental health hazards, using various engineering disciplines; their work may include waste treatment, site remediation, or pollution control technology.

Civil engineers require mathematics and science in all levels in civil engineering – structural engineering, hydraulics and geotechnical engineering are all fields that employ mathematical tools such as differential equations, tensor analysis, field theory, numerical methods and operations research.

Knowledge of mathematics and science is therefore needed by each of the engineering disciplines listed above.

It is intended that this text – *Science and Mathematics for Engineering* – will provide a step by step approach to learning fundamental mathematics and science needed for your engineering studies.

John Bird is the former Head of Applied Electronics in the Faculty of Technology at Highbury College, Portsmouth, U.K. More recently, he has combined freelance lecturing at the University of Portsmouth, with Examiner responsibilities for Advanced Mathematics with City and Guilds and examining for International Baccalaureate. He has some 45 years experience of successfully teaching, lecturing, instructing, training, educating and planning of trainee engineers study programmes. He is the author of 135 textbooks on engineering and mathematical subjects with worldwide sales of over one million copies. He is a chartered engineer, a chartered mathematician, a chartered scientist and a Fellow of three professional institutions. He is currently lecturing at the Defence College of Marine Engineering in the Defence College of Technical Training at H.M.S. Sultan, Gosport, Hampshire, U.K, one of the largest technical training establishments in Europe.

Science and Mathematics for Engineering

Sixth Edition

John Bird BSc(Hons), CEng, CSci, CMath, FIET, FIMA, FCollT

Routledge
Taylor & Francis Group

LONDON AND NEW YORK

Sixth edition published 2020
by Routledge
2 Park Square, Milton Park, Abingdon, Oxon, OX14 4RN

and by Routledge
52 Vanderbilt Avenue, New York, NY 10017

Routledge is an imprint of the Taylor & Francis Group, an informa business

© 2020 John Bird

The right of John Bird to be identified as author of this work has been asserted by him in accordance with sections 77 and 78 of the Copyright, Designs and Patents Act 1988.

All rights reserved. No part of this book may be reprinted or reproduced or utilised in any form or by any electronic, mechanical, or other means, now known or hereafter invented, including photocopying and recording, or in any information storage or retrieval system, without permission in writing from the publishers.

Trademark notice: Product or corporate names may be trademarks or registered trademarks, and are used only for identification and explanation without intent to infringe.

First edition published by Elsevier 1995
Fifth edition published by Routledge 2015

British Library Cataloguing-in-Publication Data
A catalogue record for this book is available from the British Library

Library of Congress Cataloging-in-Publication Data
Names: Bird, J. O., author.
Title: Science and mathematics for engineering / John Bird.
Other titles: Science for engineering
Description: Sixth edition. | Boca Raton : Taylor & Francis, a CRC title, part of the Taylor & Francis imprint, a member of the Taylor & Francis Group, the academic division of T&F Informa, plc, 2020. | Includes index. | Revised edition : Science for engineering. 5th ed. London ; New York : Routledge, 2015.
Identifiers: LCCN 2019020085| ISBN 9780367204747 (pbk) | ISBN 9780367204754 (hbk) | ISBN 9780429261701 (ebk)
Subjects: LCSH: Engineering. | Science. | Engineering mathematics–Examinations, questions, etc.
Classification: LCC TA145 .B53 2020 | DDC 500.2024/62–dc23
LC record available at https://lccn.loc.gov/2019020085

ISBN: 978-0-367-20475-4 (hbk)
ISBN: 978-0-367-20474-7 (pbk)
ISBN: 978-0-429-26170-1 (ebk)

Typeset in Times
by Servis Filmsetting Ltd, Stockport, Cheshire

Visit the companion website: www.routledge.com/cw/bird

To Sue

Contents

Science and Mathematics for Engineering. 978-0-367-20475-4, © John Bird. Published by Taylor & Francis. All rights reserved.

Preface

'Science and Mathematics for Engineering, 6th Edition', aims to develop in the reader an understanding of fundamental scientific and applied mathematical principles which will enable the solution of elementary engineering problems. The aims are to describe engineering systems in terms of basic scientific laws and principles, to investigate the behaviour of simple linear systems in engineering, to calculate the response of engineering systems to changes in variables, and to determine the response of such engineering systems to changes in parameters. In particular, the aim is to develop an understanding of applied mathematics, static's, dynamics, electrical principles, energy and engineering systems.

In this sixth edition of *Science and Mathematics for Engineering*, (formerly entitled *Science for Engineering*), additional material on metric to imperial conversions, logarithms and exponential functions, heat exchangers, resistor colour coding, lithium-ion and glass batteries, solar energy, rectification, global climate change and the generation of electricity have all been added to the text.

The website **http://www.routledge.com/cw/bird** contains full solutions to all of the problems in the text, together with much more – see page xiv.

The text covers the following courses of study:

  (i) **Mathematics for Engineering Technicians** (BTEC First Certificate/Diploma, level 2)

 (ii) **Applied Electrical and Mechanical Science for Engineering** (BTEC National Certificate/Diploma, level 2)

(iii) **Mathematics for Engineering Technicians** (BTEC National Certificate/Diploma, level 3)

 (iv) **Any introductory/access/foundation course** involving Engineering Science and basic Mathematics

This sixth edition of *Science and Mathematics for Engineering* is arranged in four sections.

Section 1, Applied Mathematics, chapters 1 to 14, provides the basic mathematical tools needed to effectively understand the science applications in sections II, III and IV. Basic arithmetic, fractions, decimals, percentages, indices, units, prefixes and engineering notation, calculations and evaluation of formulae, algebra, simple equations, transposition of formulae, simultaneous equations, logarithms and exponential functions, straight line graphs, trigonometry, areas of common shapes, the circle, and volumes of common solids are covered in this first section.

Section II, Mechanical Applications, chapters, 15 to 33, covers SI units, density, atomic structure of matter, speed and velocity, acceleration, forces acting at a point, work, energy and power, simply supported beams, linear and angular motion, friction, simple machines, the effects of forces on materials, linear momentum and impulse, torque, pressure in fluids, heat energy and transfer, thermal expansion, ideal gas laws and the measurement of temperature.

Section III, Electrical Applications, chapters 34 to 44, covers an introduction to electric circuits, resistance variation, batteries and alternative sources of energy, series and parallel networks, Kirchhoff's laws, magnetism and electromagnetism, electromagnetic induction, alternating voltages and currents, capacitors and inductors, electrical measuring instruments and measurements, and global climate change and future electricity production.

Section IV, Engineering systems, chapter 45, covers an overview of the principles of electronic and mechanical engineering systems, forming a basis for further studies.

Each topic in the text is presented in a way that assumes in the reader little previous knowledge of that topic. Theory is introduced in each chapter by an outline of essential information, definitions, formulae, laws and procedures. The theory is kept to a minimum, for problem solving is extensively used to establish and exemplify the theory. It is intended that readers will gain real understanding through seeing problems solved and then through solving similar problems themselves.

Science and Mathematics for Engineering, 6th Edition contains some **650 worked problems**, together with **430 multiple-choice questions,** and over **1440 further**

Science and Mathematics for Engineering. 978-0-367-20475-4, © John Bird. Published by Taylor & Francis. All rights reserved.

questions, arranged in **219 Exercises**, all with answers at the back of the book; the Exercises appear at regular intervals – every 2 or 3 pages – throughout the text. Also included are **432 short answer questions**, the answers for which can be determined from the preceding material in that particular chapter. **463 line diagrams** further enhance the understanding of the theory. All of the problems – multiple-choice, short answer and further questions – mirror where possible practical situations in science and engineering.

Full of solutions to the further problems and a **PowerPoint presentation of all the illustrations** contained in the text is available on the website – see below.

At regular intervals throughout the text are fifteen **Revision Tests** to check understanding. For example, Revision Test 1 covers material contained in chapters 1 and 2, Revision Test 2 covers the material contained in chapters 3 and 4, and so on. These Revision Tests do not have answers given since it is envisaged that lecturers/instructors could set the tests for students to attempt as part of their course structure. Lecturers' may obtain solutions of the Revision Tests on the website – see below.

A list of the **main formulae** is included at the end of the book for easy reference, together with a comprehensive **glossary of terms**.

'Learning by Example' is at the heart of **'Science and Mathematics for Engineering 6ᵗʰ Edition'**.

JOHN BIRD
**Defence School of Marine Engineering,
HMS Sultan, formerly University of Portsmouth
and Highbury College, Portsmouth**

The following support material is available from http://www.routledge.com/cw/bird

For Students:

1. **Full worked solutions to over 1440 further questions contained in the 219 Practice Exercises**

2. **A list of Essential Formulae**

3. **A full glossary of terms**

4. **430 multiple-choice questions**

5. **Information on 38 Famous Engineers and Scientists mentioned in the text**

For Lecturers/Instructors:

1–5. As per students 1–5 above

6. **Full solutions and marking scheme for each of the 15 Revision Tests; also, each test may be downloaded for distribution to students.**

7. **All 463 illustrations used in the text may be downloaded for use in PowerPoint presentations**

Section I

Applied mathematics

Chapter 1

Basic arithmetic

Why it is important to understand: **Basic arithmetic**

Being numerate, i.e. having an ability to add, subtract, multiply and divide whole numbers with some confidence, goes a long way towards helping you become competent at mathematics. Of course electronic calculators are a marvellous aid to the quite complicated calculations often required in engineering; however, having a feel for numbers 'in our head' can be invaluable when estimating. Do not spend too much time on this chapter because we deal with the calculator later; however, try to have some idea how to do quick calculations in the absence of a calculator. You will feel more confident in dealing with numbers and calculations if you can do this.

At the end of this chapter, you should be able to:

- understand positive and negative integers
- add and subtract whole numbers
- multiply and divide two whole numbers
- multiply numbers up to 12×12 by rote
- determine the highest common factor from a set of numbers
- determine the lowest common factor from a set of numbers
- appreciate the order of operation when evaluating expressions
- understand the use of brackets in expressions
- evaluate expressions containing $+$, $-$, $\times$, $\div$ and brackets

1.1 Introduction

Whole Numbers

Whole Numbers are simply the numbers **0, 1, 2, 3, 4, 5, . . .**

Counting Numbers

Counting Numbers are whole numbers, but **without the zero**, i.e. **1, 2, 3, 4, 5, . . .**

Natural Numbers

Natural Numbers can mean either counting numbers or whole numbers.

Integers

Integers are like whole numbers, but they **also include negative numbers**

Examples of integers include $. . . -5, -4, -3, -2, -1, 0, 1, 2, 3, 4, 5, . . .$

Science and Mathematics for Engineering. 978-0-367-20475-4, © John Bird. Published by Taylor & Francis. All rights reserved.

Arithmetic Operators

The four basic arithmetic operators are: add (+), subtract (−), multiply (×) and divide (÷).

It is assumed that adding, subtracting, multiplying and dividing reasonably small numbers can be achieved without a calculator. However, if revision of this area is needed then some worked problems are included in the following sections.

When **unlike signs** occur together in a calculation, the overall sign is **negative**.

For example,

$$3 + (-4) = 3 + -4 = 3 - 4 = -1$$

and

$$(+5) \times (-2) = -10$$

Like signs together give an overall **positive sign**. For example,

$$3 - (-4) = 3 - -4 = 3 + 4 = 7$$

and

$$(-6) \times (-4) = +24$$

Prime Numbers

A prime number can be divided, without a remainder, only by itself and by 1. For example, 17 can be divided only by 17 and by 1. Other examples of prime numbers are 2, 3, 5, 7, 11, 13, 19 and 23.

1.2 Revision of addition and subtraction

You can probably already add two or more numbers together and subtract one number from another. However, if you need revision of this, then the following worked problems should be helpful.

Problem 1. Determine $735 + 167$

```
  H T U
    7 3 5
  + 1 6 7
  ───────
    9 0 2
  ───────
    1 1
```

(i) $5 + 7 = 12$. Place the 2 in the units (U) column. Carry the 1 in the tens (T) column.

(ii) $3 + 6 + 1$ (carried) $= 10$. Place the 0 in the tens column. Carry the 1 in the hundreds (H) column.

(iii) $7 + 1 + 1$ (carried) $= 9$. Place the 9 in the hundreds column.

Hence, $\mathbf{735 + 167 = 902}$

Problem 2. Determine $632 - 369$

```
  H T U
    6 3 2
  − 3 6 9
  ───────
    2 6 3
```

(i) $2 - 9$ is not possible; therefore 'borrow' 1 from the tens column (leaving 2 in the tens column). In the units column, this gives us $12 - 9 = 3$

(ii) Place 3 in the units column.

(iii) $2 - 6$ is not possible; therefore 'borrow' 1 from the hundreds column (leaving 5 in the hundreds column). In the tens column, this gives us $12 - 6 = 6$

(iv) Place the 6 in the tens column.

(v) $5 - 3 = 2$

(vi) Place the 2 in the hundreds column.

Hence, $\mathbf{632 - 369 = 263}$

Problem 3. Add 27, −74, 81 and −19

This problem is written as $27 - 74 + 81 - 19$.

Adding the positive integers:	27
	81
Sum of positive integers is:	108
Adding the negative integers:	74
	19
Sum of negative integers is:	93
Taking the sum of the negative integers from the sum of the positive integers gives:	108
	−93
	15

Thus, $\mathbf{27 - 74 + 81 - 19 = 15}$

Problem 4.   Subtract −74 from 377

This problem is written as $377 − −74$. Like signs together give an overall positive sign, hence

$$377 − −74 = 377 + 74$$

$$\begin{array}{r} 3\,7\,7 \\ +\ \ 7\,4 \\ \hline 4\,5\,1 \end{array}$$

Thus, **377 − −74 = 451**

Now try the following Practice Exercise

Practice Exercise 1   Further problems on addition and subtraction (answers on page 531)

In problems 1–11, determine the values of the expressions given, without using a calculator:

1. $67\,\text{kg} − 82\,\text{kg} + 34\,\text{kg}$

2. $851\,\text{mm} − 372\,\text{mm}$

3. $124 − 273 + 481 − 398$

4. $£927 − £114 + £182 − £183 − £247$

5. $647 − 872$

6. $2417 − 487 + 2424 − 1778 − 4712$

7. $£2715 − £18{,}250 + £11{,}471 − £1509 + £113{,}274$

8. $47 + (−74) − (−23)$

9. $813 − (−674)$

10. $−23{,}148 − 47{,}724$

11. $\$53{,}774 − \$38{,}441$

1.3   Revision of multiplication and division

You can probably already multiply two numbers together and divide one number by another. However, if you need to revise this then the following worked problems should be helpful.

Problem 5.   Determine $86 × 7$

$$\begin{array}{r} \text{H T U} \\ 8\,6 \\ ×\ \ \ \ 7 \\ \hline 6\,0\,2 \\ \hline 4 \end{array}$$

(i) $7 × 6 = 42$. Place the 2 in the units (U) column and 'carry' the 4 into the tens (T) column.

(ii) $7 × 8 = 56; 56 + 4\ (\text{carried}) = 60$. Place the 0 in the tens column and the 6 in the hundreds (H) column.

Hence, **86 × 7 = 602**

A good grasp of **multiplication tables** is needed when multiplying such numbers; a reminder of the multiplication table up to $12 × 12$ is shown on page 6. Confidence in handling numbers will be greatly improved if this table is memorised.

Problem 6.   Determine $764 × 38$

$$\begin{array}{r} 7\,6\,4 \\ ×\ \ \ 3\,8 \\ \hline 6\,1\,1\,2 \\ 2\,2\,9\,2\,0 \\ \hline 2\,9\,0\,3\,2 \end{array}$$

(i) $8 × 4 = 32$. Place the 2 in the units column and carry 3 into the tens column.

(ii) $8 × 6 = 48; 48 + 3\ (\text{carried}) = 51$. Place the 1 in the tens column and carry the 5 into the hundreds column.

(iii) $8 × 7 = 56; 56 + 5\ (\text{carried}) = 61$. Place the 1 in the hundreds column and the 6 in the thousands column.

(iv) Place 0 in the units column under the 2

(v) $3 × 4 = 12$. Place the 2 in the tens column and carry the 1 into the hundreds column.

(vi) $3 × 6 = 18; 18 + 1\ (\text{carried}) = 19$. Place the 9 in the hundreds column and carry the 1 into the thousands column.

(vii) $3 × 7 = 21; 21 + 1\ (\text{carried}) = 22$. Place a 2 in the thousands column and the other 2 in the ten thousands column.

(viii) $6112 + 22{,}920 = 29{,}032$

Multiplication table

×	2	3	4	5	6	7	8	9	10	11	12
2	**4**	6	8	10	12	14	16	18	20	22	24
3	6	**9**	12	15	18	21	24	27	30	33	36
4	8	12	**16**	20	24	28	32	36	40	44	48
5	10	15	20	**25**	30	35	40	45	50	55	60
6	12	18	24	30	**36**	42	48	54	60	66	72
7	14	21	28	35	42	**49**	56	63	70	77	84
8	16	24	32	40	48	56	**64**	72	80	88	96
9	18	27	36	45	54	63	72	**81**	90	99	108
10	20	30	40	50	60	70	80	90	**100**	110	120
11	22	33	44	55	66	77	88	99	110	**121**	132
12	24	36	48	60	72	84	96	108	120	132	**144**

Hence, **764 × 38 = 29,032**

Again, knowing multiplication tables is rather important when multiplying such numbers.

It is appreciated, of course, that such a multiplication can, and probably will, be performed using a **calculator**. However, there are times when a calculator may not be available and it is then useful to be able to calculate the 'long way'.

Problem 7. Determine $1834 \div 7$

$$\begin{array}{r} 262 \\ 7\overline{)1834} \end{array}$$

(i) 7 into 18 goes 2, remainder 4. Place the 2 above the 8 of 1834 and carry the 4 remainder to the next digit on the right, making it 43.

(ii) 7 into 43 goes 6, remainder 1. Place the 6 above the 3 of 1834 and carry the 1 remainder to the next digit on the right, making it 14.

(iii) 7 into 14 goes 2, remainder 0. Place 2 above the 4 of 1834.

Hence, $1834 \div 7 = 1834/7 = \dfrac{1834}{7} = 262$

The method shown is called **short division**.

Problem 8. Determine $5796 \div 12$

$$\begin{array}{r} 483 \\ 12\overline{)5796} \\ \underline{48} \\ 99 \\ \underline{96} \\ 36 \\ \underline{36} \\ 00 \end{array}$$

(i) 12 into 5 won't go. 12 into 57 goes 4; place 4 above the 7 of 5796

(ii) $4 \times 12 = 48$; place the 48 below the 57 of 5796

(iii) $57 - 48 = 9$

(iv) Bring down the 9 of 5796 to give 99

(v) 12 into 99 goes 8; place 8 above the 9 of 5796

(vi) $8 \times 12 = 96$; place 96 below the 99

(vii) $99 - 96 = 3$

(viii) Bring down the 6 of 5796 to give 36

(ix) 12 into 36 goes 3 exactly.

(x) Place the 3 above the final 6

(xi) $3 \times 12 = 36$; place the 36 below the 36

(xii) $36 - 36 = 0$

Hence, $5796 \div 12 = 5796/12 = \dfrac{5796}{12} = 483$

The method shown is called **long division**.

Now try the following Practice Exercise

Practice Exercise 2   Further problems on multiplication and division (answers on page 531)

Determine the values of the expressions given in problems 1–7, without using a calculator:

1. (a) 78×6 (b) 124×7

2. (a) £261×7 (b) £462×9

3. (a) $783 \text{ kg} \times 11$ (b) $73 \text{ kg} \times 8$

4. (a) $27 \text{ mm} \times 13$ (b) $77 \text{ mm} \times 12$

5. (a) $288 \text{ m} \div 6$ (b) $979 \text{ m} \div 11$

6. (a) $\dfrac{1813}{7}$ (b) $\dfrac{896}{16}$

7. (a) $\dfrac{88737}{11}$ (b) $46858 \div 14$

8. A screw has a mass of 15 grams. Calculate, in kilograms, the mass of 1200 such screws ($1 \text{ kg} = 1000 \text{ g}$).

9. Holes are drilled 36 mm apart in a metal plate. If a row of 26 holes is drilled, determine the distance, in centimetres, between the centres of the first and last holes.

10. A builder needs to clear a site of bricks and top soil. The total weight to be removed is 696 tonnes. Trucks can carry a maximum load of 24 tonnes. Determine the number of truck loads needed to clear the site.

1.4   Highest common factors and lowest common multiples

When two or more numbers are multiplied together, the individual numbers are called **factors**. Thus a factor is a number which divides into another number exactly. The **highest common factor (HCF)** is the largest number which divides into two or more numbers exactly.

For example, consider the numbers 12 and 15.

The factors of 12 are 1, 2, 3, 4, 6 and 12 (i.e. all the numbers that divide into 12).

The factors of 15 are 1, 3, 5 and 15 (i.e. all the numbers that divide into 15).

1 and 3 are the only **common factors**, i.e. numbers which are factors of **both** 12 and 15.

Hence, **the HCF of 12 and 15 is 3** since 3 is the highest number which divides into **both** 12 and 15.

A **multiple** is a number which contains another number an exact number of times. The smallest number which is exactly divisible by each of two or more numbers is called the **lowest common multiple (LCM)**.

For example, the multiples of 12 are 12, 24, 36, 48, 60, 72, . . .

and the multiples of 15 are 15, 30, 45, 60, 75, . . .

60 is a common multiple (i.e. a multiple of **both** 12 and 15) and there are no lower common multiples.

Hence, **the LCM of 12 and 15 is 60** since 60 is the lowest number that both 12 and 15 divide into.

Here are some further problems involving the determination of HCFs and LCMs.

Problem 9.   Determine the HCF of the numbers 12, 30 and 42

Probably the simplest way of determining an HCF is to express each number in terms of its lowest factors. This is achieved by repeatedly dividing by the prime numbers 2, 3, 5, 7, 11, 13, . . . (where possible) in turn. Thus

$$
\begin{aligned}
12 &= 2 \times 2 \times 3 \\
30 &= 2 \qquad\quad \times 3 \times 5 \\
42 &= 2 \qquad\quad \times 3 \times 7
\end{aligned}
$$

The factors which are common to each of the numbers are 2 in column 1 and 3 in column 3, shown by the dotted lines. Hence, the **HCF is 2×3**, i.e. **6**. That is, 6 is the largest number which will divide into 12, 30 and 42.

Problem 10.   Determine the LCM of the numbers 12, 42 and 90

The LCM is obtained by finding the lowest factors of each of the numbers, as shown in Problem 9 above, and then selecting the largest group of any of the factors present. Thus

$$12 = \boxed{2 \times 2} \times 3$$

$$42 = 2 \quad \times 3 \qquad \times \boxed{7}$$

$$90 = 2 \quad \times \boxed{3 \times 3} \times \boxed{5}$$

The largest group of any of the factors present are shown by the dotted lines and are 2×2 in 12, 3×3 in 90, 5 in 90 and 7 in 42.

Hence, **the LCM is $2 \times 2 \times 3 \times 3 \times 5 \times 7 = 1260$**, and is the smallest number which 12, 42 and 90 will all divide into exactly.

Now try the following Practice Exercise

Practice Exercise 3   Further problems on highest common factors and lowest common multiples (answers on page 531)

Find (a) the HCF and (b) the LCM of the following groups of numbers:

1. 8, 12
2. 60, 72
3. 50, 70
4. 270, 900
5. 6, 10, 14
6. 12, 30, 45
7. 10, 15, 70, 105
8. 90, 105, 300

1.5   Order of operation and brackets

1.5.1   Order of operation

Sometimes addition, subtraction, multiplication, division, powers and brackets may all be involved in a calculation. For example,

$$5 - 3 \times 4 + 24 \div (3 + 5) - 3^2$$

This is an extreme example but will demonstrate the order that is necessary when evaluating.

When we read, we read from left to right. However with mathematics there is a definite order of precedence which we need to adhere to.

The order is as follows:

Brackets
Order (or p**O**wer)
Division
Multiplication
Addition
Subtraction

Notice that the first letters of each word spell **BODMAS**, a handy *aide-mémoire*.

Order means p**O**wer. For example, $4^2 = 4 \times 4 = 16$

$$5 - 3 \times 4 + 24 \div (3 + 5) - 3^2$$

is evaluated as follows:

$5 - 3 \times 4 + 24 \div (3 + 5) - 3^2$
$= 5 - 3 \times 4 + 24 \div 8 - 3^2$ (**B**racket is removed and $3 + 5$ replaced with 8)
$= 5 - 3 \times 4 + 24 \div 8 - 9$ (**O**rder means p**O**wer – in this case $3^2 = 3 \times 3 = 9$)
$= 5 - 3 \times 4 + 3 - 9$ (**D**ivision $24 \div 8 = 3$)
$= 5 - 12 + 3 - 9$ (**M**ultiplication $-3 \times 4 = -12$)
$= 8 - 12 - 9$ (**A**ddition $5 + 3 = 8$)
$= \mathbf{-13}$ (**S**ubtraction $8 - 12 - 9 = -13$)

In practice, **it does not matter if multiplication is performed before division or if subtraction is performed before addition**. What is important is that **the process of multiplication and division must be completed before addition and subtraction**.

1.5.2   Brackets and operators

The basic laws governing the **use of brackets and operators** are shown by the following examples:

(a) $2 + 3 = 3 + 2$, i.e. the order of numbers when adding does not matter;

(b) $2 \times 3 = 3 \times 2$, i.e. the order of numbers when multiplying does not matter;

(c) $2 + (3 + 4) = (2 + 3) + 4$, i.e. the use of brackets when adding does not affect the result;

(d) $2 \times (3 \times 4) = (2 \times 3) \times 4$, i.e. the use of brackets when multiplying does not affect the result;

(e) $2 \times (3 + 4) = 2(3 + 4) = 2 \times 3 + 2 \times 4$, i.e. a number placed outside of a bracket indicates that the whole contents of the bracket must be multiplied by that number;

(f) $(2 + 3)(4 + 5) = (5)(9) = 5 \times 9 = 45$, i.e. adjacent brackets indicate multiplication;

(g) $2[3 + (4 \times 5)] = 2[3 + 20] = 2 \times 23 = 46$, i.e. when an expression contains inner and outer brackets, **the inner brackets are removed first**.

Here are some further problems where BODMAS needs to be used.

Problem 11. Find the value of $6 + 4 \div (5 - 3)$

The order of precedence of operations is remembered by using the mnemonic BODMAS. Thus

$$6 + 4 \div (5 - 3) = 6 + 4 \div 2 \quad \text{(\textbf{B}rackets)}$$
$$= 6 + 2 \quad \text{(\textbf{D}ivision)}$$
$$= \mathbf{8} \quad \text{(\textbf{A}ddition)}$$

Problem 12. Determine the value of $13 - 2 \times 3 + 14 \div (2 + 5)$

$$13 - 2 \times 3 + 14 \div (2 + 5) = 13 - 2 \times 3 + 14 \div 7 \quad \text{(\textbf{B})}$$
$$= 13 - 2 \times 3 + 2 \quad \text{(\textbf{D})}$$
$$= 13 - 6 + 2 \quad \text{(\textbf{M})}$$
$$= 15 - 6 \quad \text{(\textbf{A})}$$
$$= \mathbf{9} \quad \text{(\textbf{S})}$$

Problem 13. Evaluate $16 \div (2 + 6) + 18[3 + (4 \times 6) - 21]$

$$16 \div (2 + 6) + 18[3 + (4 \times 6) - 21]$$
$$= 16 \div (2 + 6) + 18[3 + 24 - 21] \quad \text{(\textbf{B}, inner bracket is determined first)}$$

$$= 16 \div 8 + 18 \times 6 \quad \text{(\textbf{B})}$$
$$= 2 + 18 \times 6 \quad \text{(\textbf{D})}$$
$$= 2 + 108 \quad \text{(\textbf{M})}$$
$$= \mathbf{110} \quad \text{(\textbf{A})}$$

Note that a number outside of a bracket multiplies all that is inside the brackets. In this case,

$$18[3 + 24 - 21] = 18[6] \text{ which means } 18 \times 6 = 108$$

Now try the following Practice Exercise

Practice Exercise 4   Further problems on order of precedence and brackets (answers on page 531)

Evaluate the following expressions:

1. $14 + 3 \times 15$
2. $17 - 12 \div 4$
3. $86 + 24 \div (14 - 2)$
4. $7(23 - 18) \div (12 - 5)$
5. $63 - 8(14 \div 2) + 26$
6. $\dfrac{40}{5} - 42 \div 6 + (3 \times 7)$
7. $\dfrac{(50 - 14)}{3} + 7(16 - 7) - 7$
8. $\dfrac{(7 - 3)(1 - 6)}{4(11 - 6) \div (3 - 8)}$

For fully worked solutions to each of the problems in Exercises 1 to 4 in this chapter, go to the website:
www.routledge.com/cw/bird

Chapter 2

Fractions, decimals and percentages

Why it is important to understand: **Fractions, decimals and percentages**

Engineers use fractions all the time, examples including stress to strain ratios in mechanical engineering, chemical concentration ratios and reaction rates, and ratios in electrical equations to solve for current and voltage. Fractions are also used everywhere in science, from radioactive decay rates to statistical analysis. Calculators are able to handle calculations with fractions. However, there will be times when a quick calculation involving addition, subtraction, multiplication and division of fractions is needed.

Real life applications of ratio and proportion are numerous. When you prepare recipes, paint your house, or repair gears in a large machine or in a car transmission, you use ratios and proportions. Trusses must have the correct ratio of pitch to support weight of roof and snow, cement must be the correct mixture to be sturdy, and doctors are always calculating ratios as they determine medications. Almost every job uses ratios one way or another: ratios are used in building & construction, model making, art & crafts, land surveying, die and tool making, food and cooking, chemical mixing, in automobile manufacturing and in airplane and parts making. Engineers use ratios to test structural and mechanical systems for capacity and safety issues. Millwrights use ratio to solve pulley rotation and gear problems. Operating engineers apply ratios to ensure the correct equipment is used to safely move heavy materials such as steel on worksites. It is therefore important that we have some working understanding of ratio and proportion.

Engineers and scientists also use decimal numbers all the time in calculations. Calculators are able to handle calculations with decimals; however, there will be times when a quick calculation involving addition, subtraction, multiplication and division of decimals is needed.

Engineers and scientists also use percentages all the time in calculations; calculators are able to handle calculations with percentages. For example, percentage change is commonly used in engineering, statistics, physics, finance, chemistry, and economics.

When you feel able to do calculations with basic arithmetic, fractions, ratios, decimals and percentages, all with or without the aid of a calculator, then suddenly mathematics doesn't seem quite so difficult.

Science and Mathematics for Engineering. 978-0-367-20475-4. © John Bird. Published by Taylor & Francis. All rights reserved.

At the end of this chapter, you should be able to:

- understand the terminology numerator, denominator, proper and improper fractions and mixed numbers
- add and subtract fractions
- multiply and divide two fractions
- appreciate the order of operation when evaluating expressions involving fractions
- define ratio
- perform calculations with ratios
- define direct proportion
- perform calculations with direct proportion
- define inverse proportion
- convert a decimal number to a fraction and vice-versa
- understand and use significant figures and decimal places in calculations
- add and subtract decimal numbers
- multiply and divide decimal numbers
- understand the term 'percentage'
- convert decimals to percentages and vice versa
- calculate the percentage of a quantity
- express one quantity as a percentage of another quantity
- calculate percentage error and percentage change

2.1   Fractions

A mark of 9 out of 14 in an examination may be written as $\dfrac{9}{14}$ or 9/14. $\dfrac{9}{14}$ is an example of a fraction.

The number above the line, i.e. 9, is called the **numerator**. The number below the line, i.e. 14, is called the **denominator**. When the value of the numerator is less than the value of the denominator, the fraction is called a **proper fraction**. $\dfrac{9}{14}$ is an example of a proper fraction.

When the value of the numerator is greater than the value of the denominator, the fraction is called an **improper fraction**. $\dfrac{5}{2}$ is an example of a improper fraction.

A **mixed number** is a combination of a whole number and a fraction. $2\dfrac{1}{2}$ is an example of a mixed number.

In fact, $\dfrac{5}{2} = 2\dfrac{1}{2}$

There are a number of everyday examples where fractions are readily referred to. A supermarket advertises a $\dfrac{1}{5}$ off a six-pack of beer; if the beer normally costs £2 then it will now cost £1.60

$\dfrac{3}{4}$ of the employees of a company are women; if the company has 48 employees, then 36 are women.

Calculators are able to handle calculations with fractions. However, to understand a little more about fractions we will in this chapter show how to add, subtract, multiply and divide with fractions without the use of a calculator.

Problem 1.   Change the following improper fractions into mixed numbers:

$$\text{(a) } \frac{9}{2} \quad \text{(b) } \frac{13}{4}$$

(a) $\dfrac{9}{2}$ means 9 halves and $\dfrac{9}{2}$ means $9 \div 2$, and $9 \div 2 = 4$ and one half, i.e.

$$\frac{9}{2} = 4\frac{1}{2}$$

(b) $\dfrac{13}{4}$ means 13 quarters and $\dfrac{13}{4} = 13 \div 4$, and $13 \div 4 = 3$ and 1 quarter, i.e.

$$\frac{13}{4} = 3\frac{1}{4}$$

Problem 2.   Change the following mixed numbers into improper fractions:

$$\text{(a) } 5\frac{3}{4} \quad \text{(b) } 1\frac{7}{9}$$

(a) $5\frac{3}{4}$ means $5 + \frac{3}{4}$

5 contains $5 \times 4 = 20$ quarters. Thus, $5\frac{3}{4}$ contains $20 + 3 = 23$ quarters, i.e.

$$5\frac{3}{4} = \frac{\mathbf{23}}{\mathbf{4}}$$

The quick way to change $5\frac{3}{4}$ into an improper fraction is: $\frac{4 \times 5 + 3}{4} = \frac{\mathbf{23}}{\mathbf{4}}$

(b) $1\frac{7}{9} = \frac{9 \times 1 + 7}{9} = \frac{\mathbf{16}}{\mathbf{9}}$

2.1.1 Adding and subtracting fractions

When the denominators of two (or more) fractions to be added are the same, the fractions can be added 'on sight'.

For example, $\frac{2}{9} + \frac{5}{9} = \frac{7}{9}$ and $\frac{3}{8} + \frac{1}{8} = \frac{4}{8}$

In the latter example, dividing both the 4 and the 8 by 4 gives $\frac{4}{8} = \frac{1}{2}$ which is the simplified answer.

This is called **cancelling**.

Addition and subtraction of fractions is demonstrated in the following worked examples.

Problem 3. Simplify $\frac{1}{3} + \frac{1}{2}$

(i) Make the denominators the same for each fraction. The lowest number that both denominators divide into is called the **lowest common multiple** or **LCM** (see chapter 1, page 7). In this example, the LCM of 3 and 2 is 6.

(ii) 3 divides into 6 twice. Multiplying both numerator and denominator of $\frac{1}{3}$ by 2 gives:

$$\frac{1}{3} = \frac{2}{6}$$

(iii) 2 divides into 6, three times. Multiplying both numerator and denominator of $\frac{1}{2}$ by 3 gives:

$$\frac{1}{2} = \frac{3}{6}$$

(iv) Hence,

$$\frac{1}{3} + \frac{1}{2} = \frac{2}{6} + \frac{3}{6} = \frac{5}{6}$$

Problem 4. Simplify $\frac{3}{4} - \frac{7}{16}$

(i) Make the denominators the same for each fraction. The lowest common multiple (LCM) of 4 and 16 is 16

(ii) 4 divides into 16, 4 times. Multiplying both numerator and denominator of $\frac{3}{4}$ by 4 gives:

$$\frac{3}{4} = \frac{12}{16}$$

(iii) $\frac{7}{16}$ already has a denominator of 16

(iv) Hence,

$$\frac{3}{4} - \frac{7}{16} = \frac{12}{16} - \frac{7}{16} = \frac{\mathbf{5}}{\mathbf{16}}$$

Problem 5. Simplify $4\frac{2}{3} - 1\frac{1}{6}$

$4\frac{2}{3} - 1\frac{1}{6}$ is the same as $\left(4\frac{2}{3}\right) - \left(1\frac{1}{6}\right)$, which is the same as $\left(4 + \frac{2}{3}\right) - \left(1 + \frac{1}{6}\right)$, which is the same as $4 + \frac{2}{3} - 1 - \frac{1}{6}$, which is the same as: $3 + \frac{2}{3} - \frac{1}{6}$, which is the same as: $3 + \frac{4}{6} - \frac{1}{6} = 3 + \frac{3}{6} = 3 + \frac{1}{2}$

Thus, $4\frac{2}{3} - 1\frac{1}{6} = \mathbf{3\frac{1}{2}}$

Now try the following Practice Exercise

Practice Exercise 5 Introduction to fractions (answers on page 531)

1. Change the improper fraction $\frac{15}{7}$ into a mixed number

2. Change the mixed number $2\frac{4}{9}$ into an improper fraction

3. A box contains 165 paper clips. 60 clips are removed from the box. Express this as a fraction in its simplest form

4. Order the following fractions from the smallest to the largest: $\frac{4}{9}, \frac{5}{8}, \frac{3}{7}, \frac{1}{2}, \frac{3}{5}$

Evaluate, in fraction form, the expressions given in problems 5–14.

5. $\frac{1}{3} + \frac{2}{5}$ 6. $\frac{5}{6} - \frac{4}{15}$

7. $\frac{1}{2} + \frac{2}{5}$ 8. $\frac{7}{16} - \frac{1}{4}$

9. $\frac{2}{7} + \frac{3}{11}$ 10. $\frac{2}{9} - \frac{1}{7} + \frac{2}{3}$

11. $3\frac{2}{5} - 2\frac{1}{3}$ 12. $\frac{7}{27} - \frac{2}{3} + \frac{5}{9}$

13. $5\frac{3}{13} + 3\frac{3}{4}$ 14. $4\frac{5}{8} - 3\frac{2}{5}$

2.1.2 Multiplication and division of fractions

Multiplication

To multiply two or more fractions together, the numerators are first multiplied to give a single number, and this becomes the new numerator of the combined fraction. The denominators are then multiplied together to give the new denominator of the combined fraction.

For example, $\frac{2}{3} \times \frac{4}{7} = \frac{2 \times 4}{3 \times 7} = \frac{8}{21}$

Problem 6. Find the value of $\frac{3}{7} \times \frac{14}{15}$

Dividing numerator and denominator by 3 gives:

$$\frac{3}{7} \times \frac{14}{15} = \frac{1}{7} \times \frac{14}{5} = \frac{1 \times 14}{7 \times 5}$$

Dividing numerator and denominator by 7 gives:

$$\frac{1 \times 14}{7 \times 5} = \frac{1 \times 2}{1 \times 5} = \frac{2}{5}$$

This process of dividing both the numerator and denominator of a fraction by the same factor(s) is called **cancelling**.

Problem 7. Evaluate $1\frac{3}{5} \times 2\frac{1}{3} \times 3\frac{3}{7}$

Mixed numbers **must** be expressed as improper fractions before multiplication can be performed. Thus,

$$1\frac{3}{5} \times 2\frac{1}{3} \times 3\frac{3}{7} = \left(\frac{5}{5} + \frac{3}{5}\right) \times \left(\frac{6}{3} + \frac{1}{3}\right) \times \left(\frac{21}{7} + \frac{3}{7}\right)$$

$$= \frac{8}{5} \times \frac{7}{3} \times \frac{24}{7} = \frac{8 \times 1 \times 8}{5 \times 1 \times 1} = \frac{64}{5}$$

$$= \mathbf{12\frac{4}{5}}$$

Division

The simple rule for division is: '**change the division sign into a multiplication sign and invert the second fraction**'.

For example, $\frac{2}{3} \div \frac{3}{4} = \frac{2}{3} \times \frac{4}{3} = \frac{8}{9}$

Problem 8. Simplify $\frac{3}{7} \div \frac{8}{21}$

$$\frac{3}{7} \div \frac{8}{21} = \frac{3}{7} \times \frac{21}{8} = \frac{3}{1} \times \frac{3}{8} \text{ by cancelling}$$

$$= \frac{3 \times 3}{1 \times 8} = \frac{9}{8} = \mathbf{1\frac{1}{8}}$$

Problem 9. Simplify $3\frac{2}{3} \times 1\frac{3}{4} \div 2\frac{3}{4}$

Mixed numbers must be expressed as improper fractions before multiplication and division can be performed:

$$3\frac{2}{3} \times 1\frac{3}{4} \div 2\frac{3}{4} = \frac{11}{3} \times \frac{7}{4} \div \frac{11}{4} = \frac{11}{3} \times \frac{7}{4} \times \frac{4}{11}$$

$$= \frac{1 \times 7 \times 1}{3 \times 1 \times 1} \text{ by cancelling}$$

$$= \frac{7}{3} = \mathbf{2\frac{1}{3}}$$

Now try the following Practice Exercise

Practice Exercise 6   Multiplying and dividing fractions (answers on page 531)

Evaluate the following:

1. $\dfrac{2}{5} \times \dfrac{4}{7}$

2. $\dfrac{3}{4} \times \dfrac{8}{11}$

3. $\dfrac{3}{4} \times \dfrac{5}{9}$

4. $\dfrac{17}{35} \times \dfrac{15}{68}$

5. $\dfrac{3}{5} \times \dfrac{7}{9} \times 1\dfrac{2}{7}$

6. $\dfrac{1}{4} \times \dfrac{3}{11} \times 1\dfrac{5}{39}$

7. $\dfrac{2}{9} \div \dfrac{4}{27}$

8. $\dfrac{3}{8} \div \dfrac{45}{64}$

9. $\dfrac{3}{8} \div \dfrac{5}{32}$

10. $2\dfrac{1}{4} \times 1\dfrac{2}{3}$

11. $1\dfrac{1}{3} \div 2\dfrac{5}{9}$

12. $2\dfrac{3}{4} \div 3\dfrac{2}{3}$

13. $\dfrac{1}{9} \times \dfrac{3}{4} \times 1\dfrac{1}{3}$

14. $3\dfrac{1}{4} \times 1\dfrac{3}{5} \div \dfrac{2}{5}$

15. If a storage tank is holding 450 litres when it is three-quarters full, how much will it contain when it is two-thirds full?

16. A tank contains 24,000 litres of oil. Initially, $\dfrac{7}{10}$ of the contents are removed, then $\dfrac{3}{5}$ of the remainder is removed. How much oil is left in the tank?

2.1.3   Order of operation with fractions

As stated in chapter 1, sometimes addition, subtraction, multiplication, division, powers and brackets can all be involved in a calculation. A definite order of precedence must be adhered to. The order is:

Brackets
Order (or p**O**wer)
Division
Multiplication
Addition
Subtraction
This is demonstrated in the following worked problems.

Problem 10.   Simplify $\dfrac{1}{4} - 2\dfrac{1}{5} \times \dfrac{5}{8} + \dfrac{9}{10}$

$\dfrac{1}{4} - 2\dfrac{1}{5} \times \dfrac{5}{8} + \dfrac{9}{10} = \dfrac{1}{4} - \dfrac{11}{5} \times \dfrac{5}{8} + \dfrac{9}{10}$

$\qquad = \dfrac{1}{4} - \dfrac{11}{1} \times \dfrac{1}{8} + \dfrac{9}{10}$ by cancelling

$\qquad = \dfrac{1}{4} - \dfrac{11}{8} + \dfrac{9}{10}$ **(M)**

$\qquad = \dfrac{1 \times 10}{4 \times 10} - \dfrac{11 \times 5}{8 \times 5} + \dfrac{9 \times 4}{10 \times 4}$

(since the LCM of 4, 8 and 10 is 40)

$\qquad = \dfrac{10}{40} - \dfrac{55}{40} + \dfrac{36}{40}$

$\qquad = \dfrac{10 - 55 + 36}{40}$ **(A/S)**

$\qquad = -\dfrac{9}{40}$

Problem 11.   Evaluate

$\dfrac{1}{3}$ of $\left(5\dfrac{1}{2} - 3\dfrac{3}{4}\right) + 3\dfrac{1}{5} \div \dfrac{4}{5} - \dfrac{1}{2}$

$\dfrac{1}{3}$ of $\left(5\dfrac{1}{2} - 3\dfrac{3}{4}\right) + 3\dfrac{1}{5} \div \dfrac{4}{5} - \dfrac{1}{2}$

$\qquad = \dfrac{1}{3}$ of $1\dfrac{3}{4} + 3\dfrac{1}{5} \div \dfrac{4}{5} - \dfrac{1}{2}$ **(B)**

$\qquad = \dfrac{1}{3} \times \dfrac{7}{4} + \dfrac{16}{5} \div \dfrac{4}{5} - \dfrac{1}{2}$ **(O)**

(Note that the 'of' is replaced with a multiplication sign.)

$\qquad = \dfrac{1}{3} \times \dfrac{7}{4} + \dfrac{16}{5} \times \dfrac{5}{4} - \dfrac{1}{2}$ **(D)**

$\qquad = \dfrac{1}{3} \times \dfrac{7}{4} + \dfrac{4}{1} \times \dfrac{1}{1} - \dfrac{1}{2}$ by cancelling

$\qquad = \dfrac{7}{12} + \dfrac{4}{1} - \dfrac{1}{2}$ **(M)**

$\qquad = \dfrac{7}{12} + \dfrac{48}{12} - \dfrac{6}{12}$ **(A/S)**

$\qquad = \dfrac{49}{12}$

$\qquad = 4\dfrac{1}{12}$

Now try the following Practice Exercise

Practice Exercise 7   Order of precedence with fractions (answers on page 531)

Evaluate the following:

1. $2\dfrac{1}{2} - \dfrac{3}{5} \times \dfrac{20}{27}$

2. $\dfrac{1}{3} - \dfrac{3}{4} \times \dfrac{16}{27}$

3. $\dfrac{1}{2} + \dfrac{3}{5} \div \dfrac{9}{15} - \dfrac{1}{3}$

4. $\dfrac{1}{5} + 2\dfrac{2}{3} \div \dfrac{5}{9} - \dfrac{1}{4}$

5. $\dfrac{4}{5} \times \dfrac{1}{2} - \dfrac{1}{6} \div \dfrac{2}{5} + \dfrac{2}{3}$

6. $\dfrac{3}{5} - \left(\dfrac{2}{3} - \dfrac{1}{2}\right) \div \left(\dfrac{5}{6} \times \dfrac{3}{2}\right)$

7. $\dfrac{1}{2}$ of $\left(4\dfrac{2}{5} - 3\dfrac{7}{10}\right) + \left(3\dfrac{1}{3} \div \dfrac{2}{3}\right) - \dfrac{2}{5}$

8. $\dfrac{6\dfrac{2}{3} \times 1\dfrac{2}{5} - \dfrac{1}{3}}{6\dfrac{3}{4} \div 1\dfrac{1}{2}}$

2.2   Ratio and proportion

Ratio is a way of comparing amounts of something; it shows how much bigger one thing is than the other. Some practical examples include mixing paint, or sand and cement, or screen wash. Gears, map scales, food recipes, scale drawings and metal alloy constituents all use ratios.

Two quantities are in **direct proportion** when they increase or decrease in the **same ratio**. There arc several practical engineering laws which rely on direct proportion. Also, calculating currency exchange rates and converting imperial to metric units rely on direct proportion.

Sometimes, as one quantity increases at a particular rate, another quantity decreases at the same rate; this is called **inverse proportion**. For example, the time taken to do a job is inversely proportional to the number of people in a team: double the people, half the time.

Here are some worked examples to help us understand more about ratios.

Problem 12. In a class, the ratio of female to male students is 6:27. Reduce the ratio to its simplest form

Both 6 and 27 can be divided by 3.

Thus, 6:27 is the same as **2:9**.

6:27 and 2:9 are called **equivalent ratios**.

It is normal to express ratios in their lowest, or simplest, form. In this example, the simplest form is **2:9**, which means for every 2 females in the class there are 9 male students.

Problem 13. A gear wheel having 128 teeth is in mesh with a 48 tooth gear. What is the gear ratio?

Gear ratio = 128:48

A ratio can be simplified by finding common factors.

128 and 48 can both be divided by 2, i.e. 128:48 is the same as 64:24

64 and 24 can both be divided by 8, i.e. 64:24 is the same as 8:3

There is no number that divides completely into both 8 and 3, so 8:3 is the simplest ratio, i.e. **the gear ratio is 8:3**

128:48 is equivalent to 64:24 which is equivalent to 8:3

8:3 is the simplest form.

Problem 14. A wooden pole is 2.08 m long. Divide it in the ratio of 7 to 19

Since the ratio is 7:19, the total number of parts is $7 + 19 = 26$ parts.

26 parts corresponds to 2.08m = 208 cm, hence, 1 part corresponds to $\dfrac{208}{26} = 8$

Thus, 7 parts corresponds to $7 \times 8 = $ **56 cm,** and 19 parts corresponds to $19 \times 8 = $ **152 cm.**

Hence, **2.08 m divides in the ratio of 7:19 as 56 cm to 152 cm**.

(Check: 56 + 152 must add up to 208, otherwise an error would have been made.)

Problem 15. A map scale is 1:30,000. On the map the distance between two schools is 6 cm. Determine the actual distance between the schools, giving the answer in kilometres.

Actual distance between schools $= 6 \times 30,000$ cm

$$= 180,000 \text{ cm}$$

$$= \dfrac{180,000}{100} \text{m} = 1800 \text{ m} = \dfrac{1800}{1000} \text{m} = \mathbf{1.80 \text{ km}}$$

(1 mile $\approx$ 1.6 km, hence the schools are just over 1 mile apart.)

Now try the following Practice Exercise

Practice Exercise 8   Ratios (answers on page 532)

1. In a box of 333 paper clips, 9 are defective. Express the non-defective paper clips as a ratio of the defective paper clips, in its simplest form.

2. A gear wheel having 84 teeth is in mesh with a 24 tooth gear. Determine the gear ratio in its simplest form.

3. In a box of 2000 nails, 120 are defective. Express the non-defective nails as a ratio of the defective ones, in its simplest form.

4. A metal pipe 3.36 m long is to be cut into two in the ratio 6 to 15. Calculate the length of each piece.

5. On the instructions for cooking a turkey it says that it needs to be cooked 45 minutes for every kilogram. How long will it take to cook a 7 kg turkey?

6. In a will, £6440 is to be divided between three beneficiaries in the ratio 4:2:1. Calculate the amount each receives.

7. A local map has a scale of 1:22500. The distance between two motorways is 2.7 km. How far are they apart on the map?

8. A machine produces 320 bolts in a day. Calculate the number of bolts produced by 4 machines in 7 days.

2.3   Decimals

The **decimal system of numbers is based on the digits 0 to 9**.

There are a number of everyday occurrences where we use decimal numbers. For example, a radio is, say, tuned to 107.5 MHz FM. **107.5 is an example of a decimal number**. In a shop, a pair of trainers cost, say, £57.95. **57.95 is another example of a decimal number**. 57.95 is a decimal fraction where a decimal point separates the integer, i.e. 57, from the fractional part, i.e. 0.95

57.95 actually means:

$$(5 \times 10) + (7 \times 1) + \left(9 \times \frac{1}{10}\right) + \left(5 \times \frac{1}{100}\right)$$

2.3.1   Significant figures and decimal places

A number which can be expressed exactly as a decimal fraction is called a **terminating decimal**.
For example,

$$3\frac{3}{16} = 3.1825 \text{ is a terminating decimal.}$$

A number which cannot be expressed exactly as a decimal fraction is called a **non-terminating decimal**.
For example,

$$1\frac{5}{7} = 1.7142857\ldots \text{ is a non-terminating decimal.}$$

A non-terminating decimal may be expressed in two ways, depending on the accuracy required:

(a) correct to a number of **significant figures**, or

(b) correct to a number of **decimal places**, i.e. the number of figures after the decimal point.

The last digit in is unaltered if the next digit on the right is in the group of numbers 0, 1, 2, 3, or 4.
For example,

1.7142857 … = 1.713 correct to 3 significant figures
= 1.714 correct to 3 decimal places

since the next digit on the right in this example is 2.
The last digit in the answer is increased by 1 if the next digit on the right is in the group of numbers 5, 6, 7, 8 or 9.
For example,

1.7142857 … = 1.7144 correct to 4 significant figures

= 1.7143 correct to 4 decimal places

since the next digit on the right in this example is 8.

Problem 16.   Express 15.36815 correct to:
(a) 2 decimal places,   (b) 3 significant figures,
(c) 3 decimal places,   (d) 6 significant figures

(a)   15.36815 = **15.37** correct to 2 decimal places.

(b)   15.36815 = **15.4** correct to 3 significant figures.

(c)   15.36815 = **15.368** correct to 3 decimal places.

(d)   15.36815 = **15.3682** correct to 6 significant figures.

Problem 17.   Express 0.004369 correct to (a) 4 decimal places, (b) 3 significant figures

(a) $0.004369 = \mathbf{0.0044}$ correct to 4 decimal places.

(b) $0.004369 = \mathbf{0.00437}$ correct to 3 significant figures.

(Note that the zeros to the right of the decimal point do not count as significant figures.)

Now try the following Practice Exercise

Practice Exercise 9   Significant figures and decimal places (answers on page 532)

1. Express 14.1794 correct to 2 decimal places.

2. Express 2.7846 correct to 4 significant figures.

3. Express 65.3792 correct to 2 decimal places.

4. Express 43.2746 correct to 4 significant figures.

5. Express 1.2973 correct to 3 decimal places.

6. Express 0.0005279 correct to 3 significant figures.

2.3.2   Adding and subtracting decimal numbers

When adding or subtracting decimal numbers, care needs to be taken to ensure that the decimal points are beneath each other. This is demonstrated in the following worked examples.

Problem 18.   Evaluate $46.8 + 3.06 + 2.4 + 0.09$ and give the answer correct to 3 significant figures

The decimal points are placed under each other as shown. Each column is added, starting from the right.

$$\begin{array}{r} 46.8 \\ 3.06 \\ 2.4 \\ +\ 0.09 \\ \hline 52.35 \\ \hline 11\ 1 \end{array}$$

(i) $6 + 9 = 15$. Place 5 in the hundredths column. Carry 1 in the tenths column.

(ii) $8 + 0 + 4 + 0 + 1$ (carried) $= 13$. Place the 3 in the tenths column. Carry the 1 in the units column.

(iii) $6 + 3 + 2 + 0 + 1$ (carried) $= 12$. Place the 2 in the units column. Carry the 1 in the tens column.

(iv) $4 + 1$ (carried) $= 5$. Place the 5 in the hundreds column.

Hence,

$\mathbf{46.8 + 3.06 + 2.4 + 0.09 = 52.35}$
$\qquad\qquad \mathbf{= 52.4, \ correct \ to \ 3}$
$\qquad\qquad\qquad \mathbf{significant \ figures}.$

Problem 19.   Evaluate $64.46 - 28.77$ and give the answer correct to 1 decimal place

As with addition, the decimal points are placed under each other as shown.

$$\begin{array}{r} 64.46 \\ -\ 28.77 \\ \hline 35.69 \end{array}$$

(i) $6 - 7$ is not possible; therefore 'borrow' 1 from the tenths column. This gives $16 - 7 = 9$. Place 9 in hundredths column.

(ii) $3 - 7$ is not possible; therefore 'borrow' 1 from the units column. This gives $13 - 7 = 6$. Place the 6 in the tenths column.

(iii) $3 - 8$ is not possible; therefore 'borrow' from the hundreds column. This gives $13 - 8 = 5$. Place the 5 in the units column.

(iv) $5 - 2 = 3$. Place the 3 in the hundreds column.

Hence,

$\mathbf{64.46 - 28.77 = 35.69}$
$\qquad\qquad \mathbf{= 35.7 \ correct \ to \ 1 \ decimal \ place}.$

Now try the following Practice Exercise

Practice Exercise 10   Adding and subtracting decimal numbers (answers on page 532)

Determine the following without using a calculator

1. Evaluate $37.69 + 42.6$, correct to 3 significant figures.

2. Evaluate $378.1 - 48.85$, correct to 1 decimal place.

3. Evaluate $68.92 + 34.84 - 31.223$, correct to 4 significant figures.

4. Evaluate $67.841 - 249.55 + 56.883$, correct to 2 decimal places.

5. Evaluate $483.24 - 120.44 - 67.49$, correct to 4 significant figures.

2.3.3 Multiplying and dividing decimal numbers

When multiplying decimal fractions:

(a) the numbers are multiplied as if they were integers, and

(b) the position of the decimal point in the answer is such that there are as many digits to the right of it as the sum of the digits to the right of the decimal points of the two numbers being multiplied together.

This is demonstrated in the following worked examples.

Problem 20. Evaluate 37.6×5.4

$$
\begin{array}{r}
376 \\
\times\ 54 \\
\hline
1504 \\
18800 \\
\hline
20304 \\
\hline
\end{array}
$$

(a) $376 \times 54 = 20304$

(b) As there are $1 + 1 = 2$ digits to the right of the decimal points of the two numbers being multiplied together, $37.\underline{6} \times 5.\underline{4}$, then

$$37.6 \times 5.4 = 203.04$$

Problem 21. Evaluate $44.25 \div 1.2$, correct to (a) 3 significant figures, and (b) 2 decimal places

$44.25 \div 1.2 = \dfrac{44.25}{1.2}$ The denominator is multiplied by 10 to change it into an integer. The numerator is also multiplied by 10 to keep the fraction the same.

Thus, $\dfrac{44.25}{1.2} = \dfrac{44.25 \times 10}{1.2 \times 10} = \dfrac{442.5}{12}$

The long division is similar to the long division of integers and the steps are as shown.

$$
\begin{array}{r}
36.875 \\
12\overline{)442.500} \\
\underline{36} \\
82 \\
\underline{72} \\
105 \\
\underline{96} \\
90 \\
\underline{84} \\
60 \\
\underline{60} \\
0 \\
\end{array}
$$

(i) 12 into 44 goes 3; place the 3 above the second 4 of 442.500

(ii) $3 \times 12 = 36$; place the 36 below the 44 of 442.500

(iii) $44 - 36 = 8$

(iv) Bring down the 2 to give 82

(v) 12 into 82 goes 6; place the 6 above the 2 of 442.500

(vi) $6 \times 12 = 72$; place 72 below the 82

(vii) $82 - 72 = 10$

(viii) Bring down the 5 to give 105

(ix) 12 into 105 goes 8: place the 8 above the 5 of 442.500

(x) $8 \times 12 = 96$; place 96 below the 105

(xi) $105 - 96 = 9$

(xii) Bring down the 0 to give 90

(xiii) 12 into 90 goes 7: place the 7 above the first zero of 442.500

(xiv) $7 \times 12 = 84$; place the 84 below the 90

(xv) $90 - 84 = 6$

(xvi) Bring down the 0 to give 60

(xvii) 12 into 60 gives 5 exactly; place the 5 above the second zero of 442.500

(xviii) Hence, $44.25 \div 1.2 = \dfrac{442.5}{12} = 36.875$

So

(a) $44.25 \div 1.2 = 36.9$, **correct to 3 significant figures**.

(b) $44.25 \div 1.2 = 36.88$, **correct to 2 decimal places**.

Problem 22. Express $7\frac{2}{3}$ as a decimal fraction, correct to 4 significant figures

Dividing 2 by 3 gives $\frac{2}{3} = 0.666666\ldots$

and $\qquad\qquad 7\frac{2}{3} = 7.666666\ldots$

Hence, $7\frac{2}{3} = \mathbf{7.667}$ **correct to 4 significant figures**.
Note that 7.6666... is called **7.6 recurring** and is written as $7.\dot{6}$

Now try the following Practice Exercise

Practice Exercise 11   Multiplying and dividing decimal numbers (answers on page 532)

In problems 1–5, evaluate without using a calculator.

1. Evaluate 3.57×1.4

2. Evaluate 67.92×0.7

3. Evaluate $548.28 \div 1.2$

4. Evaluate $478.3 \div 1.1$, correct to 5 significant figures.

5. Evaluate $563.48 \div 0.9$, correct to 4 significant figures.

In problems 6–9, express as decimal fractions to the accuracy stated:

6. $\frac{4}{9}$, correct to 3 significant figures.

7. $\frac{17}{27}$, correct to 5 decimal places.

8. $1\frac{9}{16}$, correct to 4 significant figures.

9. $13\frac{31}{37}$, correct to 2 decimal places.

10. Evaluate $421.8 \div 17$, (a) correct to 4 significant figures and (b) correct to 3 decimal places

11. Evaluate $\dfrac{0.0147}{2.3}$, (a) correct to 5 decimal places and (b) correct to 2 significant figures

12. Evaluate (a) $\dfrac{12.\dot{6}}{1.5}$ (b) $5.\dot{2} \times 12$

2.4   Percentages

Percentages are used to give a common standard. The use of percentages is very common in many aspects of commercial life, as well as in engineering. Interest rates, sale reductions, pay rises, exams and VAT are all examples where percentages are used.

We are familiar with the symbol for percentage, i.e. %. Here are some examples:

- A pair of trainers in a shop cost £60. They are advertised in a sale as **20% off**. How much will you pay?
- If you earn £20,000 p.a. and you receive a **2.5% pay rise**, how much extra will you have to spend in the next year?
- A book costing £18 can be purchased on the internet for 30% less. What will be its cost?

The easiest way to understand percentages is to go through some worked examples.

Problem 23. Express 0.015 as a percentage

To express a decimal number as a percentage, merely multiply by 100, i.e.

$$0.015 = 0.015 \times 100\%$$

$$= \mathbf{1.5\%}$$

Multiplying a decimal number by 100 means moving the decimal point 2 places **to the right**.

Problem 24. Express 6.5% as a decimal number

$$6.5\% = \frac{6.5}{100} = \mathbf{0.065}$$

Dividing by 100 means moving the decimal point 2 places **to the left**.

Problem 25. Express $\frac{5}{8}$ as a percentage

$$\frac{5}{8} = \frac{5}{8} \times 100\%$$

$$= \frac{500}{8}\% = \mathbf{62.5\%}$$

Problem 26. In two successive tests a student gains marks of 57/79 and 49/67. Is the second mark better or worse than the first?

$$57/79 = \frac{57}{79} = \frac{57}{79} \times 100\% = \frac{5700}{79}\%$$
$$= \mathbf{72.15\%} \text{ correct to 2 decimal places.}$$
$$49/67 = \frac{49}{67} = \frac{49}{67} \times 100\% = \frac{4900}{67}\%$$
$$= \mathbf{73.13\%} \text{correct to 2 decimal places.}$$

Hence, **the second test is marginally better than the first test**.

This question demonstrates how much easier it is to compare two fractions when they are expressed as percentages.

Problem 27. Express 75% as a fraction

$$75\% = \frac{75}{100}$$
$$= \frac{\mathbf{3}}{\mathbf{4}}$$

The fraction $\frac{75}{100}$ is reduced to its simplest form by cancelling, i.e. dividing numerator and denominator by 25.

Problem 28. Find 27% of £65

$$27\% \text{ of £65} = \frac{27}{100} \times 65$$
$$= \mathbf{£17.55} \text{ by calculator}$$

Problem 29. Express 23 cm as a percentage of 72 cm, correct to the nearest 1%

$$23 \text{ cm as a percentage of } 72\,\text{cm} = \frac{23}{72} \times 100\%$$
$$= 31.94444\ldots\%$$
$$= \mathbf{32\%} \text{ correct to}$$
$$\text{the nearest } 1\%.$$

Problem 30. Express 47 minutes as a percentage of 2 hours, correct to 1 decimal place

Note that it is essential that the two quantities are in the **same units**.

$$\text{Working in minute units, 2 hours} = 2 \times 60$$
$$= 120 \text{ minutes}$$
$$47 \text{ minutes as a percentage of 120 minutes}$$
$$= \frac{47}{120} \times 100\%$$
$$= \mathbf{39.2\%} \text{ correct to 1 decimal place.}$$

Problem 31. A box of resistors increases in price from £45 to £52. Calculate the percentage change in cost, correct to 3 significant figures

$$\% \text{ change} = \frac{\text{new value – original value}}{\text{original value}} \times 100\%$$
$$= \frac{52 - 45}{45} \times 100\% = \frac{7}{45} \times 100\%$$
$$= \mathbf{15.6\%} = \textbf{percentage change in cost}.$$

Problem 32. A drilling speed should be set to 400 rev/min. The nearest speed available on the machine is 412 rev/min. Calculate the percentage over-speed

$$\% \text{over-speed} = \frac{\text{available speed – correct speed}}{\text{correct speed}} \times 100\%$$
$$= \frac{412 - 400}{400} \times 100\% = \frac{12}{400} \times 100\%$$
$$= \mathbf{3\%}$$

Now try the following Practice Exercise

Practice Exercise 12   Percentages (answers on page 532)

In problems 1–3, express the given numbers as percentages.

1. 0.0032   2. 1.734   3. 0.057

4. Express 20% as a decimal number.

5. Express 1.25% as a decimal number.

6. Express $\frac{11}{16}$ as a percentage.

7. Express as percentages, correct to 3 significant figures:
    (a) $\frac{7}{33}$ (b) $\frac{19}{24}$ (c) $1\frac{11}{16}$

8. Place the following in order of size, the smallest first, expressing each as percentages, correct to 1 decimal place: (a) $\frac{12}{21}$ (b) $\frac{9}{17}$
    (c) $\frac{5}{9}$ (d) $\frac{6}{11}$

9. Express 31.25% as a fraction in its simplest form.

10. Express 56.25% as a fraction in its simplest form.

11. Calculate 43.6% of 50 kg.

12. Determine 36% of 27 m.

13. Calculate correct to 4 significant figures:
     (a) 18% of 2758 t   (b) 47% of 18.42 g
     (c) 147% of 14.1 s

14. Express: (a) 140 kg as a percentage of 1 t
     (b) 47 s as a percentage of 5 min (c) 13.4 cm
     as a percentage of 2.5 m

15. Express 325 mm as a percentage of 867 mm,
     correct to 2 decimal places.

16. Express 408 g as a percentage of 2.40 kg

17. When signing a new contract, a Premier-
     ship footballer's pay increases from £15,500
     to £21,500 per week. Calculate the percent-
     age pay increase, correct to 3 significant
     figures.

18. A metal rod 1.80 m long is heated and its
     length expands by 48.6 mm. Calculate the
     percentage increase in length.

19. A machine part has a length of 36 mm. The
     length is incorrectly measured as 36.9 mm.
     Determine the percentage error in the
     measurement.

20. A resistor has a value of 820 Ω ± 5%.
     Determine the range of resistance values
     expected.

21. For each of the following resistors, determine
     the (i) minimum value, (ii) maximum value:
     (a) 680 Ω ± 20%   (b) 47 kΩ ± 5%

22. An engine speed is 2400 rev/min. The speed
     is increased by 8%. Calculate the new speed.

**For fully worked solutions to each of the problems in Exercises 5 to 12 in this chapter,
go to the website:**
www.routledge.com/cw/bird

Revision Test 1: Arithmetic, fractions, decimals and percentages

This assignment covers the material contained in chapters 1 and 2. *The marks for each question are shown in brackets at the end of each question.*

1. Evaluate
    $1009 \, \text{cm} - 356 \, \text{cm} - 742 \, \text{cm} + 94 \, \text{cm}$. (3)

2. Determine £284 × 9 (3)

3. Evaluate:
    (a) $-11239 - (-4732) + 9639$
    (b) -164×-12
    (c) 367×-19 (8)

4. Calculate: (a) $\$153 \div 9$ (b) $1397 \, \text{g} \div 11$ (6)

5. A small component has a mass of 27 grams. Calculate the mass, in kilograms, of 750 such components. (3)

6. Find (a) the highest common factor, and (b) the lowest common multiple of the following numbers: 15 40 75 120 (7)

Evaluate the expressions in questions 7 to 12.

7. $7 + 20 \div (9 - 5)$ (3)

8. $147 - 21(24 \div 3) + 31$ (3)

9. $40 \div (1 + 4) + 7[8 + (3 \times 8) - 27]$ (5)

10. $\dfrac{(7 - 3)(2 - 5)}{3(9 - 5) \div (2 - 6)}$ (4)

11. $\dfrac{(7 + 4 \times 5) \div 3 + 6 \div 2}{2 \times 4 + (5 - 8) - 2^2 + 3}$ (5)

12. $\dfrac{\left(4^2 \times 5 - 8\right) \div 3 + 9 \times 8}{4 \times 3^2 - 20 \div 5}$ (5)

13. Simplify:
    (a) $\dfrac{3}{4} - \dfrac{7}{15}$
    (b) $1\dfrac{5}{8} - 2\dfrac{1}{3} + 3\dfrac{5}{6}$ (8)

14. A training college has 375 students of whom 120 are girls. Express this as a fraction in its simplest form. (2)

15. A tank contains 30,000 litres of oil. Initially, $\dfrac{7}{10}$ of the contents are removed, then $\dfrac{4}{9}$ of the remainder is removed. How much oil is left in the tank? (4)

16. Evaluate:
    (a) $1\dfrac{7}{9} \times \dfrac{3}{8} \times 3\dfrac{3}{5}$
    (b) $6\dfrac{2}{3} \div 1\dfrac{1}{3}$
    (c) $1\dfrac{1}{3} \times 2\dfrac{1}{5} \div \dfrac{2}{5}$ (10)

17. Calculate:
    (a) $\dfrac{1}{4} \times \dfrac{2}{5} - \dfrac{1}{5} \div \dfrac{2}{3} + \dfrac{4}{15}$
    (b) $\dfrac{\dfrac{2}{3} + 3\dfrac{1}{5} \times 2\dfrac{1}{2} + 1\dfrac{1}{3}}{8\dfrac{1}{3} \div 3\dfrac{1}{3}}$ (8)

18. Simplify: $\left\{ \dfrac{1}{13} \text{ of } \left(2\dfrac{9}{10} - 1\dfrac{3}{5} \right) \right\} + \left(2\dfrac{1}{3} \div \dfrac{2}{3} \right) - \dfrac{3}{4}$ (8)

19. Convert 0.048 to a proper fraction. (2)

20. Convert 6.4375 to a mixed number. (3)

21. Express $\dfrac{9}{32}$ as a decimal fraction. (2)

22. Express 0.0784 correct to 2 decimal places. (2)

23. Express 0.0572953 correct to 4 significant figures. (2)

24. Evaluate:
    (a) $46.7 + 2.085 + 6.4 + 0.07$
    (b) $68.51 - 136.34$ (4)

25. Determine 2.37×1.2 (3)

26. Evaluate $250.46 \div 1.1$, correct to 1 decimal place. (3)

27. Evaluate $5.\dot{2} \times 15$ (2)

28. Express 56.25% as a fraction in its simplest form.   (3)

29. 12.5% of a length of wood is 70 cm. What is the full length?   (3)

30. A metal rod, 1.20 m long, is heated and its length expands by 42 mm. Calculate the percentage increase in length.   (2)

31. In a box of 2000 nails, 120 are defective. Express the non-defective nails as a ratio of the defective ones, in its simplest form.   (3)

32. Prize money in a lottery totals £3801 and is shared among three winners in the ratio 4:2:1. How much does the first prize winner receive?   (3)

33. A simple machine has an effort:load ratio of 3:37. Determine the effort, in newtons, to lift a load of 5.55 kN   (3)

34. If 16 cans of lager weigh 8.32 kg, what will 28 cans weigh?   (3)

35. Hooke's law states that stress is directly proportional to strain within the elastic limit of a material. When for brass the stress is 21 MPa, the strain is 250×10^{-6}. Determine the stress when the strain is 350×10^{-6}.   (4)

36. If 12 inches $= 30.48$ cm, find the number of millimetres in 23 inches.   (3)

37. If x is inversely proportional to y and $x = 12$ when $y = 0.4$, determine: (a) the value of x when y is 3, and (b) the value of y when $x = 2$ (5)

For lecturers/instructors/teachers, fully worked solutions to each of the problems in Revision Test 1, together with a full marking scheme, are available at the website:

www.routledge.com/cw/bird

Chapter 3

Indices, units, prefixes and engineering notation

Why it is important to understand: **Indices, units, prefixes and engineering notation**

Powers and roots are used extensively in mathematics and engineering, so it is important to get a good grasp of what they are and how, and why, they are used. Being able to multiply powers together by adding their indices is particularly useful for disciplines like engineering and electronics, where quantities are often expressed as a value multiplied by some power of ten. In the field of electrical engineering, for example, the relationship between electric current, voltage and resistance in an electrical system is critically important, and yet the typical unit values for these properties can differ by several orders of magnitude. Studying, or working, in an engineering discipline, you very quickly become familiar with powers and roots and laws of indices.

In engineering there are many different quantities to get used to, and hence many units to become familiar with. For example, force is measured in newtons, electric current is measured in amperes and pressure is measured in pascals. Sometimes the units of these quantities are either very large or very small and hence prefixes are used. For example, 1000 pascals may be written as 10^3 Pa which is written as 1 kPa in prefix form, the k being accepted as a symbol to represent 1000 or 10^3. Studying, or working, in an engineering discipline, you very quickly become familiar with the standard units of measurement, the prefixes used and engineering notation. An electronic calculator is extremely helpful with engineering notation.

At the end of this chapter, you should be able to:

- understand the terms base, index and power
- understand square roots
- perform calculations with powers and roots
- state the laws of indices
- state the seven SI units
- understand derived units
- recognise common engineering units
- understand common prefixes used in engineering
- express decimal numbers in standard form
- use engineering notation and prefix form with engineering units

Science and Mathematics for Engineering. 978-0-367-20475-4, © John Bird. Published by Taylor & Francis. All rights reserved.

3.1 Powers and roots

3.1.1 Indices

The number 16 is the same as $2 \times 2 \times 2 \times 2$ and $2 \times 2 \times 2 \times 2$ can be abbreviated to 2^4. When written as 2^4, 2 is called the **base** and the 4 is called the **index** or **power**. 2^4 is read as '**two to the power of four**'. Similarly, 3^5 is read as '**three to the power of 5**'.

When the indices are 2 and 3 they are given special names, i.e. 2 is called 'squared' and 3 is called 'cubed'. Thus, 4^2 is called '**four squared**' rather than '4 to the power of 2' and 5^3 is called '**five cubed**' rather than '5 to the power of 3'.

When no index is shown, the power is 1. For example, 2 means 2^1.

Problem 1.   Evaluate (a) 2^6 (b) 3^4

(a) 2^6 means $2 \times 2 \times 2 \times 2 \times 2 \times 2$ (i.e. 2 multiplied by itself 6 times)
and $2 \times 2 \times 2 \times 2 \times 2 \times 2 = \mathbf{64}$
i.e. $2^6 = \mathbf{64}$

(b) 3^4 means $3 \times 3 \times 3 \times 3$ (i.e. 3 multiplied by itself 4 times)
and $3 \times 3 \times 3 \times 3 = \mathbf{81}$
i.e. $3^4 = \mathbf{81}$

Problem 2.   Evaluate $3^3 \times 2^2$

$$3^3 \times 2^2 = 3 \times 3 \times 3 \times 2 \times 2$$
$$= 27 \times 4 = \mathbf{108}$$

3.1.2 Square roots

When a number is multiplied by itself the product is called a square. For example, the square of 3 is $3 \times 3 = 3^2 = 9$

A square root is the reverse process, i.e. the value of the base which when multiplied by itself gives the number, i.e. the square root of 9 is 3

The symbol $\sqrt{}$ is used to denote a square root. Thus, $\sqrt{9} = 3$. Similarly, $\sqrt{4} = 2$ and $\sqrt{25} = 5$

Because $-3 \times -3 = 9$ then $\sqrt{9}$ also equals -3. Thus $\sqrt{9} = +3$ or -3 which is usually written as $\sqrt{9} = \pm 3$. Similarly, $\sqrt{16} = \pm 4$ and $\sqrt{36} = \pm 6$

The square root of, say, 9 may also be written in index form as $9^{\frac{1}{2}}$
$$9^{\frac{1}{2}} \equiv \sqrt{9} = \pm 3$$

Problem 3.   Evaluate $\dfrac{3^2 \times 2^3 \times \sqrt{36}}{\sqrt{16} \times 4}$ taking only positive square roots

$$\frac{3^2 \times 2^3 \times \sqrt{36}}{\sqrt{16} \times 4} = \frac{3 \times 3 \times 2 \times 2 \times 2 \times 6}{4 \times 4}$$
$$= \frac{9 \times 8 \times 6}{16} = \frac{9 \times 1 \times 6}{2}$$
$$= \frac{9 \times 1 \times 3}{1} \quad \text{by cancelling}$$
$$= \mathbf{27}$$

Problem 4.   Evaluate $\dfrac{10^4 \times \sqrt{100}}{10^3}$ taking the positive square root only

$$\frac{10^4 \times \sqrt{100}}{10^3} = \frac{10 \times 10 \times 10 \times 10 \times 10}{10 \times 10 \times 10}$$
$$= \frac{1 \times 1 \times 1 \times 10 \times 10}{1 \times 1 \times 1} \quad \text{by cancelling}$$
$$= \frac{100}{1} = \mathbf{100}$$

Now try the following Practice Exercise

Practice Exercise 13   Powers and roots (answers on page 532)

Evaluate the following without the aid of a calculator.

1.   Evaluate 3^3

2.   Evaluate 2^7

3.   Evaluate 10^5

4.   Evaluate $2^4 \times 3^2 \times 2 \div 3$

5.   Evaluate $25^{\frac{1}{2}}$

6.   Evaluate $\dfrac{10^5}{10^3}$

7. Evaluate $\dfrac{10^2 \times 10^3}{10^5}$

8. Evaluate $\dfrac{2^5 \times 64^{\frac{1}{2}} \times 3^2}{\sqrt{144} \times 3}$ taking positive square roots only.

3.2 Laws of indices

(i) From earlier, $2^2 \times 2^3 = (2 \times 2) \times (2 \times 2 \times 2)$
$$= 32$$
$$= 2^5$$
Hence, $2^2 \times 2^3 = 2^5$
or $2^2 \times 2^3 = 2^{2+3}$

This is the first law of indices, which demonstrates that **when multiplying two or more numbers having the same base, the indices are added**.

(ii) $\dfrac{2^5}{2^3} = \dfrac{2 \times 2 \times 2 \times 2 \times 2}{2 \times 2 \times 2} = \dfrac{1 \times 1 \times 1 \times 2 \times 2}{1 \times 1 \times 1}$
$$= \dfrac{2 \times 2}{1} = 4 = 2^2$$
Hence, $\dfrac{2^5}{2^3} = 2^2$ or $\dfrac{2^5}{2^3} = 2^{5-3}$

This is the second law of indices, which demonstrates that **when dividing two numbers having the same base, the index in the denominator is subtracted from the index in the numerator**.

(iii) $(3^5)^2 = 3^{5 \times 2} = 3^{10}$ and $\left(2^2\right)^3 = 2^{2 \times 3} = 2^6$

This is the third law of indices, which demonstrates that **when a number which is raised to a power is raised to a further power, the indices are multiplied**.

(iv) $3^0 = 1$ and $17^0 = 1$

This is the fourth law of indices, which demonstrates that **when a number has an index of 0, its value is 1**.

(v) $3^{-4} = \dfrac{1}{3^4}$ and $\dfrac{1}{2^{-3}} = 2^3$

This is the fifth law of indices, which demonstrates that **a number raised to a negative power is the reciprocal of that number raised to a positive power**.

(vi) $8^{2/3} = \sqrt[3]{8^2} = (2)^2 = 4$ and
$$25^{1/2} = \sqrt[2]{25^1} = \sqrt{25^1} = \pm 5$$
(Note that $\sqrt{} \equiv \sqrt[2]{}$)

This is the sixth law of indices, which demonstrates that **when a number is raised to a fractional power the denominator of the fraction is the root of the number and the numerator is the power**.

Here are some worked examples using the laws of indices.

Problem 5. Evaluate in index form $5^3 \times 5 \times 5^2$

$5^3 \times 5 \times 5^2 = 5^3 \times 5^1 \times 5^2$ (Note that 5 means 5^1)
$$= 5^{3+1+2} \qquad \text{from law (i)}$$
$$= \mathbf{5^6}$$

Problem 6. Evaluate $\dfrac{3^5}{3^4}$

$$\dfrac{3^5}{3^4} = 3^{5-4} \qquad \text{from law (ii)}$$
$$= 3^1$$
$$= \mathbf{3}$$

Problem 7. Evaluate $\dfrac{2^4}{2^4}$

$$\dfrac{2^4}{2^4} = 2^{4-4} \qquad \text{from law (ii)}$$
$$= 2^0$$

But $\dfrac{2^4}{2^4} = \dfrac{2 \times 2 \times 2 \times 2}{2 \times 2 \times 2 \times 2} = \dfrac{16}{16} = 1$

Hence, $\mathbf{2^0 = 1}$ from law (iv)

Any number raised to the power of zero equals 1.
For example, $6^0 = 1$, $128^0 = 1$ and $13742^0 = 1$ and so on.

Problem 8. Evaluate $\dfrac{10^3 \times 10^2}{10^8}$

$$\frac{10^3 \times 10^2}{10^8} = \frac{10^{3+2}}{10^8} = \frac{10^5}{10^8} \qquad \text{from law (i)}$$

$$= 10^{5-8} = 10^{-3} \qquad \text{from law (ii)}$$

$$= \frac{1}{10^{+3}} = \frac{1}{1000} \qquad \text{from law (v)}$$

Hence, $\dfrac{10^3 \times 10^2}{10^8} = 10^{-3} = \dfrac{1}{1000} = \mathbf{0.001}$

Understanding powers of 10 is important, especially when dealing with prefixes later in the chapter. For example,

$$10^2 = 100,\ 10^3 = 1000,\ 10^4 = 10,000$$

$$10^5 = 100,000,\ 10^6 = 1,000,000$$

$$10^{-1} = \frac{1}{10} = 0.1,\ 10^{-2} = \frac{1}{10^2} = \frac{1}{100} = 0.01$$

and so on

Problem 9. Evaluate: (a) $5^2 \times 5^3 \div 5^4$
(b) $(3 \times 3^5) \div (3^2 \times 3^3)$

From laws (i) and (ii):

(a) $\quad 5^2 \times 5^3 \div 5^4 = \dfrac{5^2 \times 5^3}{5^4} = \dfrac{5^{(2+3)}}{5^4}$

$$= \frac{5^5}{5^4} = 5^{(5-4)} = 5^1 = \mathbf{5}$$

(b) $\quad (3 \times 3^5) \div (3^2 \times 3^3) = \dfrac{3 \times 3^5}{3^2 \times 3^3} = \dfrac{3^{(1+5)}}{3^{(2+3)}}$

$$= \frac{3^6}{3^5} = 3^{6-5} = 3^1 = \mathbf{3}$$

Problem 10. Simplify: (a) $(2^3)^4$　(b) $(3^2)^5$ expressing the answers in index form

From law (iii):

(a) $\quad (2^3)^4 = 2^{3 \times 4} = \mathbf{2^{12}}$

(b) $\quad (3^2)^5 = 3^{2 \times 5} = \mathbf{3^{10}}$

Problem 11. Evaluate (a) $4^{1/2}$ (b) $16^{3/4}$ (c) $27^{2/3}$
(d) $9^{-1/2}$

(a) $\quad 4^{1/2} = \sqrt{4} = \mathbf{\pm 2}$

(b) $\quad 16^{3/4} = \sqrt[4]{16^3} = (2)^3 = \mathbf{8}$

(Note that it does not matter whether the 4th root of 16 is found first or whether 16 cubed is found first – the same answer will result.)

(c) $\quad 27^{2/3} = \sqrt[3]{27^2} = (3)^2 = \mathbf{9}$

(d) $\quad 9^{-1/2} = \dfrac{1}{9^{1/2}} = \dfrac{1}{\sqrt{9}} = \dfrac{1}{\pm 3} = \mathbf{\pm \dfrac{1}{3}}$

Problem 12. Evaluate $\dfrac{3^3 \times 5^7}{5^3 \times 3^4}$

The laws of indices only apply to terms **having the same base**. Grouping terms having the same base, and then applying the laws of indices to each of the groups independently gives:

$$\frac{3^3 \times 5^7}{5^3 \times 3^4} = \frac{3^3}{3^4} \times \frac{5^7}{5^3} = 3^{(3-4)} \times 5^{(7-3)}$$

$$= 3^{-1} \times 5^4 = \frac{5^4}{3^1} = \frac{625}{3} = \mathbf{208\dfrac{1}{3}}$$

Problem 13. FInd the value of $\dfrac{2^3 \times 3^5 \times (7^2)^2}{7^4 \times 2^4 \times 3^3}$

$$\frac{2^3 \times 3^5 \times (7^2)^2}{7^4 \times 2^4 \times 3^3} = 2^{3-4} \times 3^{5-3} \times 7^{2 \times 2 - 4}$$

$$= 2^{-1} \times 3^2 \times 7^0$$

$$= \frac{1}{2} \times 3^2 \times 1 = \frac{9}{2} = \mathbf{4\dfrac{1}{2}}$$

Problem 14. Evaluate: $\dfrac{4^{1.5} \times 8^{1/3}}{2^2 \times 32^{-2/5}}$

$$4^{1.5} = 4^{3/2} = \sqrt{4^3} = 2^3 = 8, \quad 8^{1/3} = \sqrt[3]{8} = 2,$$

$$2^2 = 4, \quad 32^{-2/5} = \frac{1}{32^{2/5}} = \frac{1}{\sqrt[5]{32^2}} = \frac{1}{2^2} = \frac{1}{4}$$

Hence, $\dfrac{4^{1.5} \times 8^{1/3}}{2^2 \times 32^{-2/5}} = \dfrac{8 \times 2}{4 \times \dfrac{1}{4}} = \dfrac{16}{1} = \mathbf{16}$

Now try the following Practice Exercise

Practice Exercise 14 Laws of indices (answers on page 532)

Determine the following without the aid of a calculator:

1. Evaluate $2^2 \times 2 \times 2^4$

2. Evaluate $3^5 \times 3^3 \times 3$ in index form

3. Evaluate $\dfrac{2^7}{2^3}$

4. Evaluate $\dfrac{3^3}{3^5}$

5. Evaluate 7^0

6. Evaluate $\dfrac{2^3 \times 2 \times 2^6}{2^7}$

7. Evaluate $\dfrac{10 \times 10^6}{10^5}$

8. Evaluate $10^4 \div 10$

9. Evaluate $\dfrac{10^3 \times 10^4}{10^9}$

10. Evaluate $5^6 \times 5^2 \div 5^7$

11. Evaluate $(7^2)^3$ in index form

12. Evaluate $(3^3)^2$

13. Evaluate $\dfrac{3^2 \times 3^{-4}}{3^3}$

14. Evaluate $\dfrac{7^2 \times 7^{-3}}{7 \times 7^{-4}}$

15. Evaluate $\dfrac{2^3 \times 2^{-4} \times 2^5}{2 \times 2^{-2} \times 2^6}$

16. Evaluate $\dfrac{5^{-7} \times 5^2}{5^{-8} \times 5^3}$

In problems 17–19, simplify the expressions given, expressing the answers in index form and with positive indices.

17. $\dfrac{3^3 \times 5^2}{5^4 \times 3^4}$

18. $\dfrac{7^{-2} \times 3^{-2}}{3^5 \times 7^4 \times 7^{-3}}$

19. $\dfrac{4^2 \times 9^3}{8^3 \times 3^4}$

In problems 24–28, evaluate the expressions given.

20. $\left(\dfrac{1}{3^2}\right)^{-1}$

21. $81^{0.25}$

22. $16^{-\frac{1}{4}}$

23. $\left(\dfrac{4}{9}\right)^{1/2}$

3.3 Introduction to engineering units

Of considerable importance in engineering is a knowledge of units of engineering quantities, the prefixes used with units, and engineering notation. We need to know, for example, that

$$80\,\text{kV} = 80 \times 10^3\,\text{V} \text{ which means } 80{,}000 \text{ volts}$$

and $25\,\text{mA} = 25 \times 10^{-3}\,\text{A}$ which means 0.025 amperes

and $50\,\text{nF} = 50 \times 10^{-9}\,\text{F}$ which means 0.000000050 farads

This is explained in the following sections.

3.4 SI units

The system of units used in engineering and science is the *Système Internationale d'Unités* (**International system of units**) usually abbreviated to SI units, and is based on the metric system. This was introduced in 1960 and is now adopted by the majority of countries as the official system of measurement.

There are, of course, many other units than the seven shown above. These other units are called **derived units** and are defined in terms of the standard units listed above. For example, speed is measured in metres per second, therefore using two of the standard units, i.e. length and time.

Some derived units are given **special names**. For example, force = mass × acceleration, has units of kilogram metre per second squared, which uses three of the base units, i.e. kilograms, metres and seconds. The unit of kg m/s^2 is given the special name of a **newton***.

The basic seven units used in the SI system are listed below with their symbols

Quantity	Unit	Symbol
Length	metre	m (1 m = 100 cm = 1000 mm)
Mass	kilogram	kg (1 kg = 1000 g)
Time	second	s
Electric current	ampere	A
Thermodynamic temperature	kelvin	K (K = °C + 273)
Luminous intensity	candela	cd
Amount of substance	mole	mol

*Who was Newton? – **Sir Isaac Newton** (25 December 1642–20 March 1727) was an English polymath. Newton showed that the motions of objects are governed by the same set of natural laws, by demonstrating the consistency between Kepler's laws of planetary motion and his theory of gravitation. The SI unit of force is the newton, named in his honour. To find out more about Newton go to www.routledge.com/cw/bird

Below is a list of some quantities and their units that are common in engineering

Quantity	Unit	Symbol
Length	metre	m
Area	square metre	m^2
Volume	cubic metre	m^3
Mass	kilogram	kg
Time	second	s
Electric current	ampere	A
Speed, velocity	metre per second	m/s
Acceleration	metre per second squared	m/s^2
Density	kilogram per cubic metre	kg/m^3
Temperature	kelvin or Celsius	K or °C
Angle	radian or degree	rad or °
Angular velocity	radian per second	rad/s
Frequency	hertz	Hz
Force	newton	N
Pressure	pascal	Pa
Energy, work	joule	J
Power	watt	W
Charge, quantity of electricity	coulomb	C
Electric potential	volt	V
Capacitance	farad	F
Electrical resistance	ohm	Ω
Inductance	henry	H
Moment of force	newton metre	N m

3.5 Common prefixes

SI units may be made larger or smaller by using prefixes which denote multiplication or division by a particular amount. The most common multiples are listed on page 31. A knowledge of indices is needed since all of the prefixes are powers of 10 with indices that are a multiple of 3.

Here are some examples of prefixes used with engineering units.

A **frequency of 15 GHz** means 15×10^9 Hz, which is 15,000,000,000 hertz, i.e. 15 gigahertz is written as 15 GHz and is equal to 15 thousand million hertz*. Instead of writing 15,000,000,000 hertz, it is much neater, takes up less space and prevents errors caused by having so many zeros, to write the frequency as 15 GHz.

A **voltage of 40 MV** means 40×10^6 V, which is 40,000,000 volts, i.e. 40 megavolts is written as 40 MV and is equal to 40 million volts.

An **inductance of 12 mH** means 12×10^{-3} H or $\dfrac{12}{10^3}$ H or $\dfrac{12}{1000}$ H, which is 0.012 H, i.e. 12 millihenries is written as 12 mH and is equal to 12 thousandths of a henry*.

A **time of 150 ns** means 150×10^{-9} s or $\dfrac{150}{10^9}$ s, which is 0.000 000 150 s, i.e. 150 nanoseconds is written as 150 ns and is equal to 150 thousand millionths of a second.

A **force of 20 kN** means 20×10^3 N, which is 20,000 newtons, i.e. 20 kilonewtons is written as 20 kN and is equal to 20 thousand newtons.

Now try the following Practice Exercise

Practice Exercise 15   SI units and common prefixes (answers on page 532)

1. State the SI unit of volume

2. State the SI unit of capacitance

3. State the SI unit of area

4. State the SI unit of velocity

5. State the SI unit of density

6. State the SI unit of energy

7. State the SI unit of charge

8. State the SI unit of power

9. State the SI unit of electric potential

*Who was Hertz? – **Heinrich Rudolf Hertz** (22 February 1857–1 January 1894) was the first person to conclusively prove the existence of electromagnetic waves. To find out more go to www.routledge.com/cw/bird

*Who was Henry? – **Joseph Henry** (17 December 1797–13 May 1878) was an American scientist who discovered the electromagnetic phenomenon of self-inductance. To find out more go to www.routledge.com/cw/bird

Prefix	Name	Meaning	
G	giga	multiply by 10^9	i.e. $\times 1,000,000,000$
M	mega	multiply by 10^6	i.e. $\times 1,000,000$
k	kilo	multiply by 10^3	i.e. $\times 1000$
m	milli	multiply by 10^{-3}	i.e. $\times \dfrac{1}{10^3} = \dfrac{1}{1000} = 0.001$
μ	micro	multiply by 10^{-6}	i.e. $\times \dfrac{1}{10^6} = \dfrac{1}{1,000,000} = 0.000\,001$
n	nano	multiply by 10^{-9}	i.e. $\times \dfrac{1}{10^9} = \dfrac{1}{1,000,000,000} = 0.000\,000\,001$
p	pico	multiply by 10^{-12}	i.e. $\times \dfrac{1}{10^{12}} = \dfrac{1}{1,000,000,000,000} = 0.000\,000\,000\,001$

10. State which quantity has the unit kg

11. State which quantity has the unit symbol Ω

12. State which quantity has the unit Hz

13. State which quantity has the unit m/s^2

14. State which quantity has the unit symbol A

15. State which quantity has the unit symbol H

16. State which quantity has the unit symbol m

17. State which quantity has the unit symbol K

18. State which quantity has the unit rad/s

19. What does the prefix 'G' mean?

20. What is the symbol and meaning of the prefix 'milli'?

21. What does the prefix 'p' mean?

22. What is the symbol and meaning of the prefix 'mega'?

3.6 Standard form

A number written with one digit to the left of the decimal point and multiplied by 10 raised to some power is said to be written in **standard form**.
For example, $43,645 = 4.3645 \times 10^4$ in standard form and $0.0534 = 5.34 \times 10^{-2}$ in standard form.

Problem 15.   Express in standard form: (a) 38.71 (b) 3746 (c) 0.0124

For a number to be in standard form, it is expressed with only one digit to the left of the decimal point. Thus:

(a) 38.71 must be divided by 10 to achieve one digit to the left of the decimal point and it must also be multiplied by 10 to maintain the equality, i.e.

$$38.71 = \frac{38.71}{10} \times 10 = \mathbf{3.871 \times 10} \text{ in standard form}$$

(b) $3746 = \dfrac{3746}{1000} \times 1000 = \mathbf{3.746 \times 10^3}$ in standard form

(c) $0.0124 = 0.0124 \times \dfrac{100}{100} = \dfrac{1.24}{100} = \mathbf{1.24 \times 10^{-2}}$ in standard form.

Problem 16.   Express the following numbers, which are in standard form, as decimal numbers: (a) 1.725×10^{-2} (b) 5.491×10^4 (c) 9.84×10^0

(a) $1.725 \times 10^{-2} = \dfrac{1.725}{100} = \mathbf{0.01725}$ (i.e. move the decimal point 2 places to the left)

(b) $5.491 \times 10^4 = 5.491 \times 10,000 = \mathbf{54910}$ (i.e. move the decimal point 4 places to the right)

(c) $9.84 \times 10^0 = 9.84 \times 1 = \mathbf{9.84}$ (since $10^0 = 1$).

Problem 17. Express in standard form, correct to 3 significant figures: (a) $\frac{3}{8}$ (b) $19\frac{2}{3}$ (c) $741\frac{9}{16}$

(a) $\frac{3}{8} = 0.375$, and expressing it in standard form gives:

$$0.375 = \mathbf{3.75 \times 10^{-1}}$$

(b) $19\frac{2}{3} = 19.\dot{6} = \mathbf{1.97 \times 10}$ in standard form, correct to 3 significant figures.

(c) $741\frac{9}{16} = 741.5625 = \mathbf{7.42 \times 10^2}$ in standard form, correct to 3 significant figures.

Problem 18. Express the following numbers, given in standard form, as fractions or mixed numbers: (a) 2.5×10^{-1} (b) 6.25×10^{-2} (c) 1.354×10^2

(a) $2.5 \times 10^{-1} = \frac{2.5}{10} = \frac{25}{100} = \mathbf{\frac{1}{4}}$

(b) $6.25 \times 10^{-2} = \frac{6.25}{100} = \frac{625}{10,000} = \mathbf{\frac{1}{16}}$

(c) $1.354 \times 10^2 = 135.4 = 135\frac{4}{10} = \mathbf{135\frac{2}{5}}$

Problem 19. Evaluate (a) $(3.75 \times 10^3)(6 \times 10^4)$
(b) $\dfrac{3.5 \times 10^5}{7 \times 10^2}$ expressing answers in standard form

(a) $(3.75 \times 10^3)(6 \times 10^4) = (3.75 \times 6)(10^{3+4})$
$$= 22.50 \times 10^7$$
$$= \mathbf{2.25 \times 10^8}$$

(b) $\dfrac{3.5 \times 10^5}{7 \times 10^2} = \dfrac{3.5}{7} \times 10^{5-2} = 0.5 \times 10^3 = \mathbf{5 \times 10^2}$

Now try the following Practice Exercise

Practice Exercise 16 Standard form
(answers on page 533)

In problems 1–5, express in standard form:

1. (a) 73.9 (b) 28.4 (c) 197.62

2. (a) 2748 (b) 33,170 (c) 274,218

3. (a) 0.2401 (b) 0.0174 (c) 0.00923

4. (a) 1702.3 (b) 10.04 (c) 0.0109

5. (a) $\frac{1}{2}$ (b) $11\frac{7}{8}$
   (c) $130\frac{3}{5}$ (d) $\frac{1}{32}$

In problems 6 and 7, express the numbers given as integers or decimal fractions.

6. (a) 1.01×10^3 (b) 9.327×10^2
   (c) 5.41×10^4 (d) 7×10^0

7. (a) 3.89×10^{-2} (b) 6.741×10^{-1}
   (c) 8×10^{-3}

In problems 8 and 9, evaluate the given expressions, stating the answers in standard form.

8. (a) $(4.5 \times 10^{-2})(3 \times 10^3)$
   (b) $2 \times (5.5 \times 10^4)$

9. (a) $\dfrac{6 \times 10^{-3}}{3 \times 10^{-5}}$

   (b) $\dfrac{(2.4 \times 10^3)(3 \times 10^{-2})}{(4.8 \times 10^4)}$

10. Write the following statements in standard form.

    (a) The density of aluminium is $2710 \, \text{kg m}^{-3}$

    (b) Poisson's ratio for gold is 0.44

    (c) The impedance of free space is $376.73 \, \Omega$

    (d) The electron rest energy is $0.511 \, \text{MeV}$

    (e) Proton charge–mass ratio is 95,789,700 C kg^{-1}

    (f) The normal volume of a perfect gas is $0.02241 \, \text{m}^3 \, \text{mol}^{-1}$

3.7 Engineering notation

In engineering, standard form is not as important as engineering notation.
Engineering notation is similar to standard form except that the power of 10 **is always a multiple of 3**. For example, $43,645 = 43.645 \times 10^3$ in engineering

notation and $0.0534 = 53.4 \times 10^{-3}$ in engineering notation.

In the list of engineering prefixes on page 31 it is apparent that all prefixes involve powers of 10 that are multiples of 3. For example, a force of 43,645 N can re-written as 43.645×10^3 N and from the list of prefixes can then be expressed as 43.645 kN.

Thus, **43,645 N $\equiv$ 43.645 kN**

To help further, there is an 'ENG' button on your calculator. Enter the number 43,645 into your calculator and then press '='. Now press the 'ENG' button and the answer is 43.645×10^3. We then have to appreciate that 10^3 is the prefix 'kilo' giving **43,645 N $\equiv$ 43.645 kN**. In another example, let a current be 0.0745 A. Enter 0.0745 into your calculator. Press '='. Now press 'ENG' and the answer is 74.5×10^{-3}. We then have to appreciate that 10^{-3} is the prefix 'milli' giving **0.0745 A $\equiv$ 74.5 mA**

Problem 20. Express the following in engineering notation and in prefix form:
(a) 300,000 W (b) 0.000068 H

(a) Enter 300,000 into the calculator. Press '='
Now press 'ENG' and the answer is 300×10^3
From the table of prefixes on page 31, 10^3 corresponds to kilo.

Hence, $300,000 \text{ W} = 300 \times 10^3$ W in engineering notation
$= \textbf{300 kW}$ in prefix form.

(b) Enter 0.000068 into the calculator. Press '='
Now press 'ENG' and the answer is 68×10^{-6}
From the table of prefixes on page 31, 10^{-6} corresponds to micro.

Hence, $0.000068 \text{ H} = 68 \times 10^{-6}$ H in engineering notation
$= \textbf{68}\mu\textbf{H}$ in prefix form.

Problem 21. Rewrite (a) 63×10^4V in kV
(b) 3100 pF in nF

(a) Enter 63×10^4 into the calculator. Press '='
Now press 'ENG' and the answer is 630×10^3
From the table of prefixes on page 31, 10^3 corresponds to kilo.

Hence, $63 \times 10^4 \text{V} = 630 \times 10^3 \text{V} = \textbf{630 kV}$.

(b) Enter 3100×10^{-12} into the calculator. Press '='
Now press 'ENG' and the answer is 3.1×10^{-9}
From the table of prefixes on page 31, 10^{-9} corresponds to nano.

Hence, $3100 \text{ pF} = 3100 \times 10^{-12}\text{F} = 3.1 \times 10^{-9}\text{F}$
$= \textbf{3.1 nF}$

Problem 22. Rewrite (a) 14,700 mm in metres, (b) 276 cm in metres, (c) 3.375 kg in grams

(a) 1 m = 1000 mm hence,
$1 \text{ mm} = \dfrac{1}{1000} = \dfrac{1}{10^3} = 10^{-3}$ m
Hence, $14,700 \text{ mm} = 14700 \times 10^{-3}\text{m} = \textbf{14.7 m}$

(b) 1 m = 100 cm hence $1 \text{ cm} = \dfrac{1}{100} = \dfrac{1}{10^2} = 10^{-2}$ m
Hence, $276 \text{ cm} = 276 \times 10^{-2}\text{m} = \textbf{2.76 m}$

(c) $1 \text{ kg} = 1000 \text{ g} = 10^3$ g
Hence, $3.375 \text{ kg} = 3.375 \times 10^3\text{g} = \textbf{3375 g}$

Now try the following Practice Exercise

Practice Exercise 17 Engineering notation (answers on page 533)

In problems 1–12, express in engineering notation in prefix form:

1. 60,000 Pa
2. 0.00015 W
3. 5×10^7 V
4. 5.5×10^{-8} F
5. 100,000 W
6. 0.00054 A
7. $15 \times 10^5 \, \Omega$
8. 225×10^{-4} V
9. 35,000,000,000 Hz
10. 1.5×10^{-11} F
11. 0.000017 A
12. 46200 Ω
13. Rewrite 0.003 mA in μA

14. Rewrite 2025 kHz as MHz

15. Rewrite 6250 cm in metres.

16. Rewrite 34.6 g in kg

In problems 17 and 18, use a calculator to evaluate in engineering notation:

17. $4.5 \times 10^{-7} \times 3 \times 10^{4}$

18. $\dfrac{\left(1.6 \times 10^{-5}\right)\left(25 \times 10^{3}\right)}{\left(100 \times 10^{-6}\right)}$

19. The distance from Earth to the moon is around 3.8×10^{8} m. State the distance in kilometres.

20. The radius of a hydrogen atom is 0.53×10^{-10} m. State the radius in nanometres.

21. The tensile stress acting on a rod is 5600000 Pa. Write this value in engineering notation.

22. The expansion of a rod is 0.0043 m. Write this in engineering notation.

3.8 Metric conversions

Length in metric units

$$\mathbf{1\ m = 100\ cm = 1000\ mm}$$

$$\mathbf{1\ cm = \frac{1}{100}\ m = \frac{1}{10^{2}}\ m = 10^{-2}\ m}$$

$$\mathbf{1\ mm = \frac{1}{1000}\ m = \frac{1}{10^{3}}\ m = 10^{-3}\ m}$$

Problem 23. Rewrite 14,700 mm in metres

$1\ m = 1000\ mm$ hence, $1\ mm = \dfrac{1}{1000} = \dfrac{1}{10^{3}} = 10^{-3}\ m$

Hence, $14,700\ mm = 14,700 \times 10^{-3}$ m $= \mathbf{14.7\ m}$

Problem 24. Rewrite 276 cm in metres

$1\ m = 100\ cm$ hence, $1\ cm = \dfrac{1}{100} = \dfrac{1}{10^{2}} = 10^{-2}\ m$

Hence, $276\ cm = 276 \times 10^{-2}$ m $= \mathbf{2.76\ m}$

Now try the following Practice Exercise

Practice Exercise 18   Length in metric units (Answers on page 533)

1. State 2.45 m in millimetres

2. State 1.675 m in centimetres

3. State the number of millimetres in 65.8 cm

4. Rewrite 25,400 mm in metres

5. Rewrite 5632 cm in metres

6. State the number of millimetres in 4.356 m

7. How many centimetres are there in 0.875 m?

8. State a length of 465 cm in (a) mm (b) m

9. State a length of 5040 mm in (a) cm (b) m

10. A machine part is measured as 15.0 cm ± 1%. Between what two values would the measurement be? Give the answer in millimetres.

Areas in metric units

Area is a measure of the size or extent of a plane surface. Area is measured in **square units** such as mm², cm² and m².

1 m = 100 cm

1 m = 100 cm

The area of the above square is 1 m²

$$\mathbf{1\ m^{2} = 100\ cm \times 100\ cm = 10000\ cm^{2} = 10^{4}\ cm^{2}}$$

i.e. to change from square metres to square centimetres, multiply by 10^{4}

Hence, $\mathbf{2.5\ m^{2} = 2.5 \times 10^{4}\ cm^{2}}$

and $\quad \mathbf{0.75\ m^{2} = 0.75 \times 10^{4}\ cm^{2}}$

Since $\mathbf{1\ m^{2} = 10^{4}\ cm^{2}}$ then $1\ cm^{2} = \frac{1}{10^{4}}\ m^{2} = 10^{-4}\ m^{2}$

i.e. to change from square centimetres to square metres, multiply by 10^{-4}

Hence, $\mathbf{52\ cm^{2} = 52 \times 10^{-4}\ m^{2}}$

and $\quad \mathbf{643\ cm^{2} = 643 \times 10^{-4}\ m^{2}}$

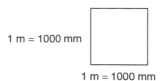

1 m = 1000 mm

1 m = 1000 mm

The area of the above square is $1 m^2$

$$1 m^2 = 1000 mm \times 1000 mm = 1000000 mm^2 = 10^6 mm^2$$

i.e. to change from square metres to square millimetres, multiply by 10^6

Hence, $7.5 m^2 = 7.5 \times 10^6 mm^2$

and $0.63 m^2 = 0.63 \times 10^6 mm^2$

Since $1 m^2 = 10^6 mm^2$

then $1 mm^2 = \frac{1}{10^6} m^2 = 10^{-6} m^2$

i.e. to change from square millimetres to square metres, multiply by 10^{-6}

Hence, $235 mm^2 = 235 \times 10^{-6} m^2$

and $47 mm^2 = 47 \times 10^{-6} m^2$

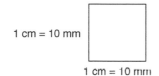

1 cm = 10 mm

1 cm = 10 mm

The area of the above square is $1 cm^2$

$$1 cm^2 = 10 mm \times 10 mm = 100 mm^2 = 10^2 mm^2$$

i.e. to change from square centimetres to square millimetres, multiply by 100 or 10^2

Hence, $3.5 cm^2 = 3.5 \times 10^2 mm^2 = 350 mm^2$

and $0.75 cm^2 = 0.75 \times 10^2 mm^2 = 75 mm^2$

Since $1 cm^2 = 10^2 mm^2$

then $1 mm^2 = \frac{1}{10^2} cm^2 = 10^{-2} cm^2$

i.e. to change from square millimetres to square centimetres, multiply by 10^{-2}

Hence, $250 mm^2 = 250 \times 10^{-2} cm^2 = 2.5 cm^2$

and $85 mm^2 = 85 \times 10^{-2} cm^2 = 0.85 cm^2$

Problem 25. Rewrite $12 m^2$ in square centimetres

$1 m^2 = 10^4 cm^2$ hence, $\mathbf{12\,m^2 = 12 \times 10^4 cm^2}$

Problem 26. Rewrite $50 cm^2$ in square metres

$1 cm^2 = 10^{-4} m^2$ hence, $\mathbf{50\,cm^2 = 50 \times 10^{-4} m^2}$

Problem 27. Rewrite $2.4 m^2$ in square millimetres

$1 m^2 = 10^6 mm^2$ hence, $\mathbf{2.4\,m^2 = 2.4 \times 10^6 mm^2}$

Problem 28. Rewrite $147 mm^2$ in square metres

$1 mm^2 = 10^{-6} m^2$ hence, $\mathbf{147\,mm^2 = 147 \times 10^{-6} m^2}$

Problem 29. Rewrite $34.5 cm^2$ in square millimetres

$1 cm^2 = 10^2 mm^2$
hence, $\mathbf{34.5\,cm^2 = 34.5 \times 10^2 mm^2 = 3450\,mm^2}$

Problem 30. Rewrite $400 mm^2$ in square centimetres

$1 mm^2 = 10^{-2} cm^2$c
hence, $\mathbf{400\,mm^2 = 400 \times 10^{-2} cm^2 = 4\,cm^2}$

Problem 31. The top of a small rectangular table is 800 mm long and 500 mm wide. Determine its area in (a) mm^2 (b) cm^2 (c) m^2

(a) **Area of rectangular table top** $= l \times b = 800 \times 500$
$$= \mathbf{400,000\ mm^2}$$

(b) Since 1 cm = 10 mm then $1 cm^2 = 1 cm \times 1 cm = 10 mm \times 10 mm = 100 mm^2$

or $1 mm^2 = \dfrac{1}{100} = 0.01\ cm^2$

Hence, $\mathbf{400,000\ mm^2} = 400,000 \times 0.01\ cm^2$
$$= \mathbf{4000\ cm^2}$$

(c) $1 cm^2 = 10^{-4} m^2$
hence, $\mathbf{4000\ cm^2} = 4000 \times 10^{-4} m^2 = \mathbf{0.4\ m^2}$

Now try the following Practice Exercise

Practice Exercise 19   Areas in metric units
(Answers on page 533)

1. Rewrite $8 m^2$ in square centimetres

2. Rewrite $240 cm^2$ in square metres

3. Rewrite $3.6 m^2$ in square millimetres

4. Rewrite $350 mm^2$ in square metres

5. Rewrite $50 cm^2$ in square millimetres

6. Rewrite $250 mm^2$ in square centimetres

7. A rectangular piece of metal is 720 mm long and 400 mm wide. Determine its area in

    (a) mm^2 (b) cm^2 (c) m^2

Volumes in metric units

The **volume** of any solid is a measure of the space occupied by the solid.

Volume is measured in **cubic units** such as mm^3, cm^3 and m^3.

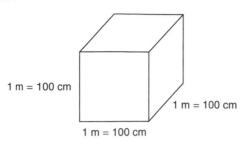

The volume of the cube shown is 1 m^3

$$1\,m^3 = 100\,cm \times 100\,cm \times 100\,cm$$

$$= 1000000\,cm^2 = 10^6 cm^2$$

$$1\,litre = 1000\,cm^3$$

i.e. to change from cubic metres to cubic centimetres, multiply by 10^6

Hence, $3.2\,m^3 = 3.2 \times 10^6 cm^3$

and $0.43\,m^3 = 0.43 \times 10^6 cm^3$

Since $1\,m^3 = 10^6 cm^3$ then $1\,cm^3 = \dfrac{1}{10^6}m^3 = 10^{-6}\,m^3$

i.e. to change from cubic centimetres to cubic metres, multiply by 10^{-6}

Hence, $140\,cm^3 = 140 \times 10^{-6}m^3$

and $2500\,cm^3 = 2500 \times 10^{-6}m^3$

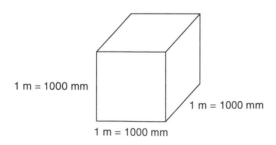

The volume of the cube shown is 1 m^3

$$1\,m^3 = 1000\,mm \times 1000\,mm \times 1000\,mm$$

$$= 1000000000\,mm^3 = 10^9 mm^3$$

i.e. to change from cubic metres to cubic millimetres, multiply by 10^9

Hence, $4.5\,m^3 = 4.5 \times 10^9 mm^3$

and $0.25\,m^3 = 0.25 \times 10^9 mm^3$

Since $1\,m^3 = 10^9 mm^2$

then $1\,mm^3 = \dfrac{1}{10^9}m^3 = 10^{-9}\,m^3$

i.e. to change from cubic millimetres to cubic metres, multiply by 10^{-9}

Hence, $\textbf{500}\,\textbf{mm}^3 = \textbf{500} \times \textbf{10}^{-9}\textbf{m}^3$

and $4675\,mm^3 = 4675 \times 10^{-9}m^3$ or $4.675 \times 10^{-6}m^3$

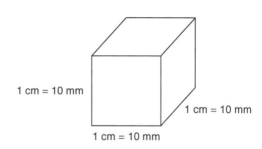

The volume of the cube shown is 1 cm^3

$$1\,cm^3 = 10\,mm \times 10\,mm \times 10\,mm$$

$$= 1000\,mm^3 = 10^3 mm^3$$

i.e. to change from cubic centimetres to cubic millimetres, multiply by 1000 or 10^3

Hence, $5\,cm^3 = 5 \times 10^3 mm^3 = 5000\,mm^3$

and $0.35\,cm^3 = 0.35 \times 10^3 mm^3 = 350\,mm^3$

Since $1\,cm^3 = 10^3 mm^3$

then $1\,mm^3 = \dfrac{1}{10^3}cm^3 = 10^{-3}\,cm^3$

i.e. to change from cubic millimetres to cubic centimetres, multiply by 10^{-3}

Hence, $650\,mm^3 = 650 \times 10^{-3}cm^3 = 0.65\,cm^3$

and $75\,mm^3 = 75 \times 10^{-3}cm^3 = 0.075\,cm^3$

Problem 32.   Rewrite 1.5 m^3 in cubic centimetres

$1\,m^3 = 10^6 cm^3$ hence, $\textbf{1.5}\,\textbf{m}^3 = \textbf{1.5} \times \textbf{10}^6\textbf{cm}^3$

Problem 33.   Rewrite 300 cm^3 in cubic metres

$1\,cm^3 = 10^{-6}m^3$ hence, $\textbf{300}\,\textbf{cm}^3 = \textbf{300} \times \textbf{10}^{-6}\textbf{m}^3$

Problem 34.   Rewrite 0.56 m^3 in cubic millimetres

$1\,m^3 = 10^9 mm^3$

hence, $\textbf{0.56}\,\textbf{m}^3 = \textbf{0.56} \times \textbf{10}^9\textbf{mm}^3$ or $\textbf{560} \times \textbf{10}^6\textbf{mm}^3$

Problem 35. Rewrite 1250 mm³ in cubic metres

$1 \text{ mm}^3 = 10^{-9}\text{m}^3$
hence, $\textbf{1250 mm}^3 = \textbf{1250} \times \textbf{10}^{-9}\textbf{m}^3$ or $\textbf{1.25} \times \textbf{10}^{-6}\textbf{m}^3$

Problem 36. Rewrite 8 cm³ in cubic millimetres

$1 \text{ cm}^3 = 10^3\text{mm}^3$
hence, $\textbf{8 cm}^3 = \textbf{8} \times \textbf{10}^3\textbf{mm}^3 = \textbf{8000 mm}^3$

Problem 37. Rewrite 600 mm³ in cubic centimetres

$1 \text{ mm}^3 = 10^{-3}\text{cm}^3$
hence, $\textbf{600 mm}^3 = \textbf{600} \times \textbf{10}^{-3}\textbf{cm}^3 = \textbf{0.6 cm}^3$

Problem 38. A water tank is in the shape of a rectangular prism having length 1.2 m, breadth 50 cm and height 250 mm. Determine the capacity of the tank (a) m³ (b) cm³ (c) litres

Capacity means volume. When dealing with liquids, the word capacity is usually used.

(a) Capacity of water tank = l × b × h
where l = 1.2 m, b = 50 cm and h = 250 mm. To use this formula, all dimensions **must** be in the same units.
Thus, l = 1.2 m, b = 0.50 m and h = 0.25 m (since 1 m = 100 cm = 1000 mm)
Hence, **capacity of tank** = 1.2 × 0.50 × 0.25
= **0.15 m³**

(b) $1 \text{ m}^3 = 10^6\text{cm}^3$
Hence, **capacity** = 0.15 m³ = 0.15 × 10⁶ cm³
= **150,000 cm³**

(c) 1 litre = 1000 cm³
Hence, $\textbf{150,000 cm}^3 = \dfrac{150,000}{1000} = \textbf{150 litres}$

Now try the following Practice Exercise

Practice Exercise 20   Volumes in metric units (Answers on page 533)

1. Rewrite 2.5 m³ in cubic centimetres.
2. Rewrite 400 cm³ in cubic metres.
3. Rewrite 0.87 m³ in cubic millimetres.
4. Change a volume of 2,400,000 cm³ to cubic metres.
5. Rewrite 1500 mm³ in cubic metres.
6. Rewrite 400 mm³ in cubic centimetres.
7. Rewrite 6.4 cm³ in cubic millimetres.
8. Change a volume of 7500 mm³ to cubic centimetres.
9. An oil tank is in the shape of a rectangular prism having length 1.5 m, breadth 60 cm and height 200 mm. Determine the capacity of the tank in (a) m³ (b) cm³ (c) litres.

3.9 Metric – US/Imperial conversions

The Imperial System (which uses yards, feet, inches, etc. to measure length) was developed over hundreds of years in the UK, then the French developed the Metric System (metres) in 1670, which soon spread through Europe, even to England itself in 1960. But the USA and a few other countries still prefer feet and inches. When converting from metric to imperial units, or vice versa, one of the following tables (3.1 to 3.8) should help.

Table 3.1 Metric to imperial length

Metric	US or Imperial
1 millimetre, mm	0.03937 inch
1 centimetre, cm = 10 mm	0.3937 inch
1 metre, m = 100 cm	1.0936 yard
1 kilometre, km = 1000 m	0.6214 mile

Problem 39. Calculate the number of inches in 350 mm, correct to 2 decimal places

350 mm = 350 × 0.03937 inches = **13.78 inches** from Table 3.1

Problem 40. Calculate the number of inches in 52 cm, correct to 4 significant figures

52 cm = 52 × 0.3937 inches = **20.47 inches** from Table 3.1

Problem 41. Calculate the number of yards in 74 m, correct to 2 decimal places

74 m = 74 × 1.0936 yards = **80.93 yds** from Table 3.1

Problem 42. Calculate the number of miles in 12.5 km, correct to 3 significant figures

12.5 km = 12.5 × 0.6214 miles = **7.77 miles** from Table 3.1

Table 3.2 Imperial to metric length

US or Imperial	US or Imperial
1 inch, in	2.54 cm
1 foot, ft = 12 in	0.3048 m
1 yard, yd = 3 ft	0.9144 m
1 mile = 1760 yd	1.6093 km
1 nautical mile = 2025.4 yd	1.853 km

Problem 43. Calculate the number of centimetres in 35 inches, correct to 1 decimal places

35 inches = 35 × 2.54 cm = **88.9 cm** from Table 3.2

Problem 44. Calculate the number of metres in 66 inches, correct to 2 decimal places

66 inches = $\frac{66}{12}$ feet = $\frac{66}{12}$ × 0.3048 m = **1.68 m** from Table 3.2

Problem 45. Calculate the number of metres in 50 yards, correct to 2 decimal places

50 yards = 50 × 0.9144 m = **45.72 m** from Table 3.2

Problem 46. Calculate the number of kilometres in 7.2 miles, correct to 2 decimal places

7.2 miles = 7.2 × 1.6093 km = **11.59 km** from Table 3.2

Problem 47. Calculate the number of (a) yards (b) kilometres in 5.2 nautical miles

(a) 5.2 nautical miles = 5.2 × 2025.4 yards
= **10532 yards** from Table 3.2
(b) 5.2 nautical miles = 5.2 × 1.853 km
= **9.636 km** from Table 3.2

Table 3.3 Metric to imperial area

Metric	US or Imperial
1 cm^2 = 100 mm^2	0.1550 in^2
1 m^2 = 10, 000 cm^2	1.1960 yd^2
1 hectare, ha = 10, 000 m^2	2.4711 acres
1 km^2= 100 ha	0.3861 mile^2

Problem 48. Calculate the number of square inches in 47 cm^2, correct to 4 significant figures

47 cm^2 = 47 × 0.1550 in^2 = **7.285 in^2** from Table 3.3

Problem 49. Calculate the number of square yards in 20 m^2, correct to 2 decimal places

20 m^2= 20 × 1.1960 yd^2 = **23.92 yd^2** from Table 3.3

Problem 50. Calculate the number of acres in 23 hectares of land, correct to 2 decimal places

23 hectares = 23 × 2.4711 acres = **56.84 acres** from Table 3.3

Problem 51. Calculate the number of square miles in a field of 15 km^2 area, correct to 2 decimal places

15 km^2= 15 × 0.3861 mile^2 = **5.79 mile^2** from Table 3.3

Table 3.4 Imperial to metric area

US or Imperial	Metric
1 in^2	6.4516 cm^2
1 ft^2= 144 in^2	0.0929 m^2
1 yd^2= 9 ft^2	0.8361 m^2
1 acre = 4840 yd^2	4046.9 m^2
1 mile^2= 640 acres	2.59 km^2

Problem 52. Calculate the number of square centimetres in 17.5 in^2, correct to the nearest square centimetre

17.5 in^2 = 17.5 × 6.4516 cm^2 = **113 cm^2** from Table 3.4

Problem 53. Calculate the number of square metres in 205 ft^2, correct to 2 decimal places

205 ft^2 = 205 × 0.0929 m^2 = **19.04 m^2** from Table 3.4

Problem 54. Calculate the number of square metres in 11.2 acres, correct to the nearest square metre

11.2 acres = 11.2 × 4046.9 m^2 = **45325 m^2** from Table 3.4

Problem 55. Calculate the number of square kilometres in 12.6 $mile^2$, correct to 2 decimal places

12.6 $mile^2$ = 12.6 × 2.59 km^2 = **32.63 km^2** from Table 3.4

Table 3.5 Metric to imperial volume/capacity

Metric	US or Imperial
1 cm^3	0.0610 in^3
1 dm^3 = 1000 cm^3	0.0353 ft^3
1 m^3 = 1000 dm^3	1.3080 yd^3
1 litre = 1 dm^3 = 1000 cm^3	2.113 fluid pt = 1.7598 pt

Problem 56. Calculate the number of cubic inches in 123.5 cm^3, correct to 2 decimal places

123.5 cm^3 = 123.5 × 0.0610 cm^3 = **7.53 cm^3** from Table 3.5

Problem 57. Calculate the number of cubic feet in 144 dm^3, correct to 3 decimal places

144 dm^3 = 144 × 0.0353 ft^3 = **5.083 ft^3** from Table 3.5

Problem 58. Calculate the number of cubic yards in 5.75 m^3, correct to 4 significant figures

5.75 m^3 = 5.75 × 1.3080 yd^3 = **7.521 yd^3** from Table 3.5

Problem 59. Calculate the number of US fluid pints in 6.34 litres of oil, correct to 1 decimal place

6.34 litre = 6.34 × 2.113 US fluid pints = **13.4 US fluid pints** from Table 3.5

Problem 60. Calculate the number of cubic centimetres in 3.75 in^3, correct to 2 decimal places

3.75 in^3 = 3.75 × 16.387 cm^3 = **61.45 cm^3** from Table 3.6

Table 3.6 Imperial to metric volume/capacity

US or Imperial	Metric
1 in^3	16.387 cm^3
1 ft^3	0.02832 m^3
1 US fl oz = 1.0408 UK fl oz	0.0296 litres
1 US pint (16 fl oz) = 0.8327 UK pt	0.4732 litres
1 US gal (231 in^3) = 0.8327 UK gal	3.7854 litres

Problem 61. Calculate the number of cubic metres in 210 ft^3, correct to 3 significant figures

210 ft^3 = 210 × 0.02832 m^3 = **5.95 m^3** from Table 3.6

Problem 62. Calculate the number of litres in 4.32 US pints, correct to 3 decimal places

4.32 US pints = 4.32 × 0.4732 litres = **2.044 litres** from Table 3.6

Problem 63. Calculate the number of litres in 8.62 US gallons, correct to 2 decimal places

8.62 US gallons = 8.62 × 3.7854 litre = **32.63 litres** from Table 3.6

Table 3.7 Metric to imperial mass

Metric	US or Imperial
1 g = 1000 mg	0.0353 oz
1 kg = 1000 g	2.2046 lb
1 tonne, t, = 1000 kg	1.1023 short ton
1 tonne, t, = 1000 kg	0.9842 long ton

The British ton is the long ton, which is 2240 pounds, and the US ton is the short ton which is 2000 pounds.

Problem 64. Calculate the number of ounces in a mass of 1346 g, correct to 2 decimal places

1346 g = 1346 × 0.0353 oz = **47.51 oz** from Table 3.7

Problem 65. Calculate the mass, in pounds, in a 210.4 kg mass, correct to 4 significant figures

210.4 kg = 210.4 × 2.2046 lb = **463.8 lb** from Table 3.7

Problem 66. Calculate the number of short tons in 5000 kg, correct to 2 decimal places

5000 kg = 5 t = 5 × 1.1023 short tons = **5.51 short tons** from Table 3.7

Table 3.8 Imperial to metric mass

US or Imperial	Metric
1 oz = 437.5 grain	28.35 g
1 lb = 16 oz	0.4536 kg
1 stone = 14 lb	6.3503 kg
1 hundredweight, cwt = 112 lb	50.802 kg
1 short ton	0.9072 tonne
1 long ton	1.0160 tonne

Problem 67. Calculate the number of grams in 5.63 oz, correct to 4 significant figures

5.63 oz = 5.63 × 28.35 g = **159.6 g** from Table 3.8

Problem 68. Calculate the number of kilograms in 75 oz, correct to 3 decimal places

$75 \text{ oz} = \frac{75}{16} \text{lb} = \frac{75}{16} \times 0.4536 \text{ kg} = \textbf{2.126 kg}$ from Table 3.8

Problem 69. Convert 3.25 cwt into (a) pounds (b) kilograms

(a) 3.25 cwt = 3.25 × 112 lb = **364 lb** from Table 3.8

(b) 3.25 cwt = 3.25 × 50.802 kg = **165.1 kg** from Table 3.8

Temperature

To convert from Celsius to Fahrenheit, first multiply by 9/5, then add 32.
To convert from Fahrenheit to Celsius, first subtract 32, then multiply by 5/9

Problem 70. Convert 35° C to degrees Fahrenheit

$F = \frac{9}{5}C + 32$ hence $35° C = \frac{9}{5}(35) + 32 = 63 + 32$

$$= \textbf{95° F}$$

Problem 71. Convert 113° F to degrees Celsius

$C = \frac{5}{9}(F - 32)$ hence $113° F = \frac{5}{9}(113 - 32) = \frac{5}{9}(81)$

$$= \textbf{45° C}$$

Now try the following Practice Exercise

Practice Exercise 21   Metric/Imperial conversions (Answers on page 533)

In the following Problems, use the metric/imperial conversions in Tables 3.1 to 3.8

1. Calculate the number of inches in 476 mm, correct to 2 decimal places.

2. Calculate the number of inches in 209 cm, correct to 4 significant figures.

3. Calculate the number of yards in 34.7 m, correct to 2 decimal places.

4. Calculate the number of miles in 29.55 km, correct to 2 decimal places.

5. Calculate the number of centimetres in 16.4 inches, correct to 2 decimal places.

6. Calculate the number of metres in 78 inches, correct to 2 decimal places.

7. Calculate the number of metres in 15.7 yards, correct to 2 decimal places.

8. Calculate the number of kilometres in 3.67 miles, correct to 2 decimal places.

9. Calculate the number of (a) yards (b) kilometres in 11.23 nautical miles.

10. Calculate the number of square inches in 62.5 cm^2, correct to 4 significant figures.

11. Calculate the number of square yards in 15.2 m^2, correct to 2 decimal places.

12. Calculate the number of acres in 12.5 hectares, correct to 2 decimal places.

13. Calculate the number of square miles in 56.7 km^2, correct to 2 decimal places.

14. Calculate the number of square centimetres in 6.37 in^2, correct to the nearest square centimetre.

15. Calculate the number of square metres in 308.6 ft^2, correct to 2 decimal places.

16. Calculate the number of square metres in 2.5 acres, correct to the nearest square metre.

17. Calculate the number of square kilometres in 21.3 mile^2, correct to 2 decimal places.

18. Calculate the number of cubic inches in 200.7 cm^3, correct to 2 decimal places.

19. Calculate the number of cubic feet in 214.5 dm^3, correct to 3 decimal places.

20. Calculate the number of cubic yards in 13.45 m^3, correct to 4 significant figures.

21. Calculate the number of US fluid pints in 15 litres, correct to 1 decimal place.

22. Calculate the number of cubic centimetres in 2.15 in^3, correct to 2 decimal places.

23. Calculate the number of cubic metres in 175 ft^3, correct to 4 significant figures.

24. Calculate the number of litres in 7.75 US pints, correct to 3 decimal places.

25. Calculate the number of litres in 12.5 US gallons, correct to 2 decimal places.

26. Calculate the number of ounces in 980 g, correct to 2 decimal places.

27. Calculate the mass, in pounds, in 55 kg, correct to 4 significant figures.

28. Calculate the number of short tons in 4000 kg, correct to 3 decimal places.

29. Calculate the number of grams in 7.78 oz, correct to 4 significant figures.

30. Calculate the number of kilograms in 57.5 oz, correct to 3 decimal places.

31. Convert 2.5 cwt into (a) pounds (b) kilograms.

32. Convert 55°C to degrees Fahrenheit.

33. Convert 167°F to degrees Celsius.

For fully worked solutions to each of the problems in Exercises 13 to 21 in this chapter, go to the website:
www.routledge.com/cw/bird

Chapter 4

Calculations and evaluation of formulae

Why it is important to understand: **Calculations and evaluation of formulae**

The availability of electronic pocket calculators, at prices which all can afford, has had a considerable impact on engineering education. Engineers and student engineers now use calculators all the time since calculators are able to handle a very wide range of calculations. You will feel more confident to deal with all aspects of engineering studies if you are able to correctly use a calculator.

At the end of this chapter, you should be able to:

- use a calculator to add, subtract, multiply and divide decimal numbers
- use a calculator to evaluate square, cube, reciprocal, power, root and $\times 10^x$ functions
- use a calculator to evaluate expressions containing fractions and trigonometric functions
- use a calculator to evaluate expressions containing π and e^x functions
- evaluate formulae, given values

4.1 Introduction

In engineering, calculations often need to be performed. For simple numbers it is useful to be able to use mental arithmetic. However, when numbers are larger an electronic calculator needs to be used.

There are several calculators on the market, many of which will be satisfactory for our needs. It is essential to have a **scientific notation calculator** which will have all the necessary functions needed, and more. This chapter assumes you have a **CASIO fx-83ES** or **fx-991ES calculator**, or similar.

4.2 Use of a calculator

Use your calculator to check the following worked problems.

Science and Mathematics for Engineering. 978-0-367-20475-4, © John Bird. Published by Taylor & Francis. All rights reserved.

Problem 1.   Evaluate $\dfrac{12.47 \times 31.59}{70.45 \times 0.052}$ correct to 4 significant figures

(i)   Type in 12.47

(ii)   Press $\times$

(iii)   Type in 31.59

(iv)   Press $\div$

(v)   The denominator must have brackets, i.e. press (

(vi)   Type in 70.45×0.052 and complete the bracket, i.e.)

(vii)   Press = and the answer 107.530518... appears

Hence, $\dfrac{12.47 \times 31.59}{70.45 \times 0.052} = 107.5$ **correct to 4 significant figures**.

Problem 2.   Evaluate 0.17^2 in engineering form

(i)   Type in 0.17

(ii)   Press x^2 and 0.17^2 appears on the screen

(iii)   Press shift and = and the answer 0.0289 appears

(iv)   Press the ENG function and the answer changes to 28.9×10^{-3} which is engineering form

Hence, $0.17^2 = 28.9 \times 10^{-3}$ **in engineering form**. The ENG function is extremely important in engineering calculations.

Problem 3.   Change 0.0000538 into engineering form

(i)   Type in 0.0000538

(ii)   Press = then ENG

Hence, $0.0000538 = 53.8 \times 10^{-6}$ **in engineering form**.

Problem 4.   Evaluate 1.4^3

(i)   Type in 1.4

(ii)   Press x^3 and 1.4^3 appears on the screen

(iii)   Press = and the answer $\dfrac{343}{125}$ appears

(iv)   Press the S $\Leftrightarrow$ D function and the fraction changes to a decimal 2.744

Thus, $1.4^3 = 2.744$

Problem 5.   Evaluate $\dfrac{1}{3.2}$

(i)   Type in 3.2

(ii)   Press x^{-1} and 3.2^{-1} appears on the screen

(iii)   Press = and the answer $\dfrac{5}{16}$ appears

(iv)   Press the S $\Leftrightarrow$ D function and the fraction changes to a decimal 0.3125

Thus, $\dfrac{1}{3.2} = 0.3125$

Problem 6.   Evaluate 1.5^5 correct to 4 significant figures

(i)   Type in 1.5

(ii)   Press $x^{\square}$ and $1.5^{\square}$ appears on the screen

(iii)   Press 5 and 1.5^5 appears on the screen

(iv)   Press shift and = and the answer 7.59375 appears

Thus, $1.5^5 = 7.594$ **correct to 4 significant figures**.

Problem 7.   Evaluate $\sqrt{361}$

(i)   Press the $\sqrt{\square}$ function

(ii)   Type in 361 and $\sqrt{361}$ appears on the screen

(iii)   Press = and the answer 19 appears

Thus, $\sqrt{361} = 19$

Problem 8.   Evaluate $\sqrt[4]{81}$

(i)   Press the $\sqrt[\square]{\square}$ function

(ii)   Type in 4 and $\sqrt[4]{\square}$ appears on the screen

(iii)   Press $\rightarrow$ to move the cursor and then type in 81 and $\sqrt[4]{81}$ appears on the screen

(iv)   Press = and the answer 3 appears

Thus, $\sqrt[4]{81} = 3$

Now try the following Practice Exercise

Practice Exercise 22   Using a calculator
(answers on page 533)

1. Evaluate $\dfrac{17.35 \times 34.27}{41.53 \div 3.76}$ correct to 3 decimal places

2. Evaluate $\dfrac{(4.527 + 3.63)}{(452.51 \div 34.75)} + 0.468$ correct to 5 significant figures

3. Evaluate $52.34 - \dfrac{(912.5 \div 41.46)}{(24.6 - 13.652)}$ correct to 3 decimal places

4. Evaluate 3.5^2

5. Evaluate $(0.036)^2$ in engineering form

6. Evaluate 1.563^2 correct to 5 significant figures

7. Evaluate 3.14^3 correct to 4 significant figures

8. Evaluate $(0.38)^3$ correct to 4 decimal places

9. Evaluate $\dfrac{1}{1.75}$ correct to 3 decimal places

10. Evaluate $\dfrac{1}{0.0250}$

11. Evaluate $\dfrac{1}{0.00725}$ correct to 1 decimal place

12. Evaluate $\dfrac{1}{0.065} - \dfrac{1}{2.341}$ correct to 4 significant figures

13. Evaluate 2.1^4

14. Evaluate $(0.22)^5$ correct to 5 significant figures in engineering form

15. Evaluate $(1.012)^7$ correct to 4 decimal places

16. Evaluate $1.1^3 + 2.9^4 - 4.4^2$ correct to 4 significant figures

17. Evaluate $\sqrt{123.7}$ correct to 5 significant figures

18. Evaluate $\sqrt{0.69}$ correct to 4 significant figures

19. Evaluate $\sqrt[3]{17}$ correct to 3 decimal places

20. Evaluate $\sqrt[5]{3.12}$ correct to 4 decimal places

21. Evaluate $\sqrt[6]{2451} - \sqrt[4]{46}$ correct to 3 decimal places

Express the answers to questions 22–25 in engineering form.

22. Evaluate $5 \times 10^{-3} \times 7 \times 10^8$

23. Evaluate $\dfrac{3 \times 10^{-4}}{8 \times 10^{-9}}$

24. Evaluate $\dfrac{6 \times 10^3 \times 14 \times 10^{-4}}{2 \times 10^6}$

25. Evaluate $\dfrac{99 \times 10^5 \times 6.7 \times 10^{-3}}{36.2 \times 10^{-4}}$ correct to 4 significant figures

Use your calculator to check the following worked problems.

Problem 9. Evaluate $\dfrac{1}{4} + \dfrac{2}{3}$

(i) Press the $\dfrac{\square}{\square}$ function

(ii) Type in 1

(iii) Press $\downarrow$ on the cursor key and type in 4

(iv) $\dfrac{1}{4}$ appears on the screen

(v) Press $\rightarrow$ on the cursor key and type in $+$

(vi) Press the $\dfrac{\square}{\square}$ function

(vii) Type in 2

(viii) Press $\downarrow$ on the cursor key and type in 3

(ix) Press $\rightarrow$ on the cursor key

(x) Press $=$ and the answer $\dfrac{11}{12}$ appears

(xi) Press the $S \Leftrightarrow D$ function and the fraction changes to a decimal $0.9166666\ldots$

Thus, $\dfrac{1}{4} + \dfrac{2}{3} = \dfrac{11}{12} = \mathbf{0.9167}$ as a decimal, correct to 4 decimal places.

Problem 10. Evaluate $5\dfrac{1}{5} - 3\dfrac{3}{4}$

(i) Press Shift then the $\dfrac{\square}{\square}$ function and $\square\dfrac{\square}{\square}$ appears on the screen

(ii)   Type in 5 then → on the cursor key

(iii)   Type in 1 and ↓ on the cursor key

(iv)   Type in 5 and $5\dfrac{1}{5}$ appears on the screen

(v)   Press → on the cursor key

(vi)   Type in − and then press Shift then the $\dfrac{\square}{\square}$ function and $5\dfrac{1}{5} - \square\dfrac{\square}{\square}$ appears on the screen

(vii)   Type in 3 then → on the cursor key

(viii)   Type in 3 and ↓ on the cursor key

(ix)   Type in 4 and $5\dfrac{1}{5} - 3\dfrac{3}{4}$ appears on the screen

(x)   Press = and the answer $\dfrac{29}{20}$ appears

(xi)   Press the S ⇔ D function and the fraction changes to a decimal 1.45

Thus, $5\dfrac{1}{5} - 3\dfrac{3}{4} = \dfrac{29}{20} = 1\dfrac{9}{20} = 1.45$ as a decimal.

Problem 11.   Evaluate sin 38°

(i)   Make sure your calculator is in degrees mode

(ii)   Press the sin function and sin(appears on the screen

(iii)   Type in 38 and close the bracket with) and sin(38) appears on the screen

(iv)   Press = and the answer 0.615661475… appears

Thus, **sin 38° = 0.6157, correct to 4 decimal places**.

Problem 12.   Evaluate 5.3 tan(2.23 rad)

(i)   Make sure your calculator is in radian mode by pressing Shift then Setup then 4 and a small R appears at the top of the screen

(ii)   Type in 5.3 then press the tan function and 5.3tan(appears on the screen

(iii)   Type in 2.23 and close the bracket with) and 5.3tan(2.23) appears on the screen

(iv)   Press = and the answer − 6.84021262… appears

Thus, **5.3 tan(2.23 rad) = −6.8402, correct to 4 decimal places**.

Problem 13.   Evaluate 3.57 π

(i)   Enter 3.57

(ii)   Press Shift and the × 10^x key and 3.57 π appears on the screen

(iii)   Either press shift and = (or = and S ⇔ D) and the value of 3.57 π appears in decimal form as 11.2154857…

Hence,   **3.57π = 11.22 correct to 4 significant figures**.

Problem 14.   Evaluate $e^{2.37}$

(i)   Press Shift and then press the ln function key and $e^{\square}$ appears on the screen

(ii)   Enter 2.37 and $e^{2.37}$ appears on the screen

(iii)   Either press shift and = (or = and S ⇔ D) and the value of $e^{2.37}$ appears in decimal as 10.6973922…

Hence, $e^{2.37} = 10.70$ **correct to 4 significant figures**.

Now try the following Practice Exercise

Practice Exercise 23   Using a calculator (answers on page 533)

1.   Evaluate $\dfrac{2}{3} - \dfrac{1}{6} + \dfrac{3}{7}$ as a fraction

2.   Evaluate $2\dfrac{5}{6} + 1\dfrac{5}{8}$ as a decimal, correct to 4 significant figures

3.   Evaluate $\dfrac{1}{3} - \dfrac{3}{4} \times \dfrac{8}{21}$ as a fraction

4.   Evaluate $8\dfrac{8}{9} \div 2\dfrac{2}{3}$ as a mixed number

5.   Evaluate $\dfrac{\left(4\dfrac{1}{5} - 1\dfrac{2}{3}\right)}{\left(3\dfrac{1}{4} \times 2\dfrac{3}{5}\right)} - \dfrac{2}{9}$ as a decimal, correct to 3 significant figures

Evaluate problems 6–12, each correct to 4 decimal places.

6.   Evaluate sin 15.78°

7. Evaluate cos 63.74°

8. Evaluate tan 39.55° − sin 52.53°

9. Evaluate sin(0.437 rad)

10. Evaluate cos(1.42 rad)

11. Evaluate tan(5.673 rad)

12. Evaluate $\dfrac{(\sin 42.6°)(\tan 83.2°)}{\cos 13.8°}$

Evaluate problems 13–18, each correct to 4 significant figures.

13. $1.59\,\pi$

14. $\pi^2\left(\sqrt{13}-1\right)$

15. $3e^{(2\pi-1)}$

16. $2\pi e^{\frac{\pi}{3}}$

17. $\sqrt{\left[\dfrac{5.52\pi}{2e^{-2}\times\sqrt{26.73}}\right]}$

18. $\sqrt{\left[\dfrac{e^{(2-\sqrt{3})}}{\pi\times\sqrt{8.57}}\right]}$

4.3 Evaluation of formulae

The statement $y = mx + c$ is called a **formula** for y in terms of m, x and c.

y, m, x and c are called **symbols** or **variables**.

When given values of m, x and c we can evaluate y. There are a large number of formulae used in engineering and in this section we will insert numbers in place of symbols to evaluate engineering quantities. Here are some practical examples. Check with your calculator that you agree with the working and answers.

Problem 15. The surface area A of a hollow cone is given by $A = \pi r l$. Determine, correct to 1 decimal place, the surface area when $r = 3.0$ cm and $l = 8.5$ cm

$$A = \pi r l = \pi(3.0)(8.5)\ \text{cm}^2$$

Hence, **surface area $A = 80.1\,\text{cm}^2$, correct to 1 decimal place**.

Problem 16. Velocity v is given by $v = u + at$. If $u = 9.54$ m/s, $a = 3.67$ m/s^2 and $t = 7.82$ s, find v, correct to 3 significant figures

$$v = u + at = 9.54 + 3.67 \times 7.82$$
$$= 9.54 + 28.6994 = 38.2394$$

Hence, **velocity v = 38.2 m/s, correct to 3 significant figures**.

Problem 17. The area A of a circle is given by $A = \pi r^2$. Determine the area correct to 2 decimal places, given radius $r = 5.23$ m

$$A = \pi r^2 = \pi(5.23)^2 = \pi(27.3529)$$

Hence, **area, A = 85.93 m^2, correct to 2 decimal places**.

Problem 18. The volume V cm^3 of a right circular cone is given by $V = \dfrac{1}{3}\pi r^2 h$. Given that radius $r = 2.45$ cm and height $h = 18.7$ cm, find the volume, correct to 4 significant figures

$$V = \frac{1}{3}\pi r^2 h = \frac{1}{3}\pi(2.45)^2(18.7) = \frac{1}{3}\times\pi\times 2.45^2\times 18.7$$
$$= 117.544521\ldots$$

Hence, **volume, V = 117.5 cm^3, correct to 4 significant figures**.

Problem 19. Force F newtons is given by the formula $F = \dfrac{Gm_1 m_2}{d^2}$, where m_1 and m_2 are masses, d their distance apart and G is a constant. Find the value of the force given that $G = 6.67 \times 10^{-11}$, $m_1 = 7.36$, $m_2 = 15.5$ and $d = 22.6$. Express the answer in standard form, correct to 3 significant figures

$$F = \frac{Gm_1 m_2}{d^2} = \frac{(6.67\times 10^{-11})(7.36)(15.5)}{(22.6)^2}$$
$$= \frac{(6.67)(7.36)(15.5)}{(10^{11})(510.76)} = \frac{1.490}{10^{11}}$$

Hence, **force $F = 1.49 \times 10^{-11}$ newtons, correct to 3 significant figures**.

Problem 20. The time of swing, t seconds, of a simple pendulum is given by: $t = 2\pi\sqrt{\dfrac{l}{g}}$

Determine the time, correct to 3 decimal places, given that $l = 12.9$ and $g = 9.81$

$$t = 2\pi\sqrt{\frac{l}{g}} = (2\pi)\sqrt{\frac{12.9}{9.81}} = 7.20510343\ldots$$

Hence, **time t = 7.205 s, correct to 3 decimal places.**

Now try the following Practice Exercise

Practice Exercise 24   Evaluation of formulae
(answers on page 534)

1. The circumference C of a circle is given by the formula $C = 2\pi r$. Determine the circumference given $r = 8.40\,\text{mm}$

2. A formula used in connection with gases is $R = \dfrac{PV}{T}$. Evaluate R when $P = 1500$, $V = 5$ and $T = 200$

3. The velocity of a body is given by $v = u + at$. The initial velocity u is measured when time t is 15 s and found to be 12 m/s. If the acceleration a is 9.81 m/s² calculate the final velocity v

4. Find the distance s, given that $s = \dfrac{1}{2}gt^2$. Time $t = 0.032$ seconds and acceleration due to gravity $g = 9.81\,\text{m/s}^2$. Give the answer in millimetres

5. The energy stored in a capacitor is given by $E = \dfrac{1}{2}CV^2$ joules. Determine the energy when capacitance $C = 5 \times 10^{-6}$ farads and voltage $V = 240\,\text{V}$

6. Resistance R_2 is given by $R_2 = R_1(1 + \alpha t)$. Find R_2, correct to 4 significant figures, when $R_1 = 220$, $\alpha = 0.00027$ and $t = 75.6$

7. Density $= \dfrac{\text{mass}}{\text{volume}}$. Find the density when the mass is 2.462 kg and the volume is 173 cm³

Give the answer in units of kg/m³ (Note that $1\text{cm}^3 = 10^{-6}\text{m}^3$)

8. Velocity $=$ frequency $\times$ wavelength. Find the velocity when the frequency is 1825 Hz and the wavelength is 0.154 m

9. Evaluate resistance R_T, given $\dfrac{1}{R_T} = \dfrac{1}{R_1} + \dfrac{1}{R_2} + \dfrac{1}{R_3}$ when $R_1 = 5.5\,\Omega$, $R_2 = 7.42\,\Omega$ and $R_3 = 12.6\,\Omega$

10. Power $= \dfrac{\text{force} \times \text{distance}}{\text{time}}$. Find the power when a force of 3760 N raises an object a distance of 4.73 m in 35 s

11. The potential difference, V volts, available at battery terminals is given by $V = E - Ir$. Evaluate V when $E = 5.62$, $I = 0.70$ and $R = 4.30$

12. Given force $F = \dfrac{1}{2}m(v^2 - u^2)$, find F when $m = 18.3$, $v = 12.7$ and $u = 8.24$

13. Energy, E joules, is given by the formula $E = \dfrac{1}{2}LI^2$. Evaluate the energy when $L = 5.5$ and $I = 1.2$

14. The current I amperes in an a.c. circuit is given by $I = \dfrac{V}{\sqrt{(R^2 + X^2)}}$. Evaluate the current when $V = 250$, $R = 11.0$ and $X = 16.2$

15. Distance s metres is given by the formula $s = ut + \dfrac{1}{2}at^2$. If $u = 9.50$, $t = 4.60$ and $a = -2.50$, evaluate the distance.

16. The area A of any triangle is given by $A = \sqrt{[s(s - a)(s - b)(s - c)]}$ where $s = \dfrac{a + b + c}{2}$. Evaluate the area, given $a = 3.60\,\text{cm}$, $b = 4.00\,\text{cm}$ and $c = 5.20\,\text{cm}$

For fully worked solutions to each of the problems in Exercises 22 to 24 in this chapter,
go to the website: www.routledge.com/cw/bird

This assignment covers the material contained in chapters 3 and 4. *The marks for each question are shown in brackets at the end of each question.*

1. Evaluate: (a) $3 \times 2^3 \times 2^2$ (b) $49^{\frac{1}{2}}$ (4)

2. Evaluate $\dfrac{3^2 \times \sqrt{36} \times 2^2}{3 \times 81^{\frac{1}{2}}}$ taking positive square roots only.   (3)

3. Evaluate $6^4 \times 6 \times 6^2$ in index form.   (3)

4. Evaluate: (a) $\dfrac{2^7}{2^2}$ (b) $\dfrac{10^4 \times 10 \times 10^5}{10^6 \times 10^2}$ (4)

5. Evaluate: (a) $\dfrac{2^3 \times 2 \times 2^2}{2^4}$ (b) $\dfrac{(2^3 \times 16)^2}{(8 \times 2)^3}$

    (c) $\left(\dfrac{1}{4^2}\right)^{-1}$ (7)

6. Evaluate: (a) $(27)^{-\frac{1}{3}}$ (b) $\dfrac{\left(\dfrac{3}{2}\right)^{-2} - \dfrac{2}{9}}{\left(\dfrac{2}{3}\right)^2}$ (5)

7. State the SI unit of: (a) capacitance (b) electrical potential (c) work.   (3)

8. State the quantity that has an SI unit of: (a) kilograms (b) henry (c) hertz (d) m^3.   (4)

9. Express the following in engineering notation in prefix form: (a) $250{,}000\,\text{J}$ (b) $0.05\,\text{H}$ (c) $2 \times 10^8\,\text{W}$ (d) $750 \times 10^{-8}\,\text{F}$.   (4)

10. Rewrite: (a) $0.0067\,\text{mA}$ in μA (b) $40 \times 10^4\text{kV}$ as MV.   (2)

11. Evaluate the following, correct to 4 significant figures: $3.3^2 - 2.7^3 + 1.8^4$ (3)

12. Evaluate $\sqrt{6.72} - \sqrt[3]{2.54}$ correct to 3 decimal places.   (3)

13. Evaluate $\dfrac{1}{0.0071} - \dfrac{1}{0.065}$ correct to 4 significant figures.   (2)

14. The potential difference, V volts, available at battery terminals is given by: $V = E - Ir$. Evaluate V when $E = 7.23$, $I = 1.37$ and $r = 3.60$ (3)

15. Evaluate $\dfrac{4}{9} + \dfrac{1}{5} - \dfrac{3}{8}$ as a decimal, correct to 3 significant figures.   (3)

16. Evaluate $\dfrac{16 \times 10^{-6} \times 5 \times 10^9}{2 \times 10^7}$ in engineering form.   (2)

17. Evaluate resistance R, given $\dfrac{1}{R} = \dfrac{1}{R_1} + \dfrac{1}{R_2} + \dfrac{1}{R_3}$ when $R_1 = 3.6\,\text{k}\Omega$, $R_2 = 7.2\ \text{k}\Omega$ and $R_3 = 13.6\,\text{k}\Omega$.   (3)

18. Evaluate $6\dfrac{2}{7} - 4\dfrac{5}{9}$ as a mixed number and as a decimal, correct to 3 decimal places.   (3)

19. Evaluate, correct to 3 decimal places: $\sqrt{\left[\dfrac{2e^{1.7} \times 3.67^3}{4.61 \times \sqrt{3\pi}}\right]}$ (3)

20. If $a = 0.270$, $b = 15.85$, $c = 0.038$, $d = 28.7$ and $e = 0.680$, evaluate v correct to 3 significant figures, given that: $v = \sqrt{\left(\dfrac{ab}{c} - \dfrac{d}{e}\right)}$ (4)

21. Evaluate the following, each correct to 2 decimal places: (a) $\left(\dfrac{36.2^2 \times 0.561}{27.8 \times 12.83}\right)^3$

 (b) $\sqrt{\left(\dfrac{14.69^2}{\sqrt{17.42} \times 37.98}\right)}$ (4)

22. If $1.6\,\text{km} = 1$ mile, determine the speed of 45 miles/hour in kilometres per hour.   (2)

23. The area A of a circle is given by $A = \pi r^2$. Find the area of a circle of radius $r = 3.73\,\text{cm}$, correct to 2 decimal places.   (3)

24. Evaluate B, correct to 3 significant figures, when $W = 7.20$, $v = 10.0$ and $g = 9.81$, given that $B = \dfrac{Wv^2}{2g}$ (3)

25. Rewrite $32\ \text{cm}^2$ in square millimetres.   (1)

26. A rectangular tabletop is 1500 mm long and 800 mm wide. Determine its area in
    (a) mm^2 (b) cm^2 (c) m^2 (3)

27. Rewrite 0.065 m^3 in cubic millimetres.   (1)

28. Rewrite 20000 mm^3 in cubic metres.   (1)

29. Rewrite 8.3 cm^3 in cubic millimetres.   (1)

30. A petrol tank is in the shape of a rectangular prism having length 1.0 m, breadth 75 cm and height 120 mm. Determine the capacity of the tank in
    (a) m^3 (b) cm^3 (c) litres   (3)

For lecturers/instructors/teachers, fully worked solutions to each of the problems in Revision Test 2, together with a full marking scheme, are available at the website:

www.routledge.com/cw/bird

Chapter 5

Basic algebra

Why it is important to understand: **Basic algebra**

Algebra is one of the most fundamental tools for engineers because it allows them to determine the value of something (length, material constant, temperature, mass, and so on) given values that they do know (possibly other length, material properties, mass). Although the types of problems that mechanical, chemical, civil, environmental, electrical engineers deal with vary, all engineers use algebra to solve problems. An example where algebra is frequently used is in simple electrical circuits, where the resistance is proportional to voltage. Using Ohm's Law, or V = IR, an engineer simply multiplies the current in a circuit by the resistance to determine the voltage across the circuit. Engineers and scientists use algebra in many ways, and so frequently that they don't even stop to think about it. Depending on what type of engineer you choose to be, you will use varying degrees of algebra, but in all instances algebra lays the foundation for the mathematics you will need to become an engineer. Algebra is a form of mathematics that allows you to work with unknowns. If you do not know what a number is, arithmetic does not allow you to use it in calculations. Algebra has variables. Variables are labels for numbers and measurements you do not yet know. Algebra lets you use these variables in equations and formulae. A basic form of mathematics, algebra is nevertheless among the most commonly used forms of mathematics in the workforce. Although relatively simple, algebra possesses a powerful problem-solving tool used in many fields of engineering. For example, in designing a rocket to go to the moon, an engineer must use algebra to solve for flight trajectory, how long to burn each thruster and at what intensity, and at what angle to lift off. An engineer uses mathematics all the time – and in particular, algebra. Becoming familiar with algebra will make all engineering mathematics and science studies so much easier.

At the end of this chapter, you should be able to:

- understand basic operations in algebra
- add, subtract, multiply and divide using letters instead of numbers
- state the laws of indices in letters instead of numbers
- simplify algebraic expressions using the laws of indices
- use brackets with basic operations in algebra
- understand factorisation
- factorise simple algebraic expressions
- use the laws of precedence to simplify algebraic expressions

Science and Mathematics for Engineering. 978-0-367-20475-4. © John Bird. Published by Taylor & Francis. All rights reserved.

5.1 Introduction

We are already familiar with evaluating formulae using a calculator from chapter 4. For example, if the length of a football pitch is L and its width is b, then the formula for the area A is given by:

$$A = L \times b$$

This is an **algebraic equation**.
If $L = 120$ m and $b = 60$ m, then the area $A = 120 \times 60 = 7200\,\text{m}^2$.
The temperature in Fahrenheit F is given by:

$$F = \frac{9}{5}C + 32$$

where C is the temperature in Celsius*. This is also an **algebraic equation**.
If $C = 100°C$, then $F = \frac{9}{5} \times 100 + 32$

$$= 180 + 32 = 212°F$$

*Who was Celsius? – **Anders Celsius** (27 November 1701–25 April 1744) was the Swedish astronomer that proposed the Celsius temperature scale in 1742 which takes his name. To find out more go to www.routledge.com/cw/bird

5.2 Basic operations

Algebra merely uses letters to represent numbers. If, say, a, b, c and d represent any four numbers, then in algebra:

(i) $a + a + a + a = 4a$ For example, if $a = 2$, then $2 + 2 + 2 + 2 = 4 \times 2 = 8$

(ii) **$5b$ means $5 \times b$** For example, if $b = 4$, then $5b = 5 \times 4 = 20$

(iii) **$2a + 3b + a - 2b = 2a + a + 3b - 2b = 3a + b$**
Only similar terms can be combined in algebra. The $2a$ and the $+ a$ can be combined to give $3a$ and the $3b$ and $-2b$ can be combined to give $1b$, which is written as b. Also, with terms separated by $+$ and $-$ signs, the order in which they are written does not matter. In this example, $2a + 3b + a - 2b$ is the same as $2a + a + 3b - 2b$ which is the same as $3b + a + 2a - 2b$ and so on. (Note that the first term, i.e. $2a$, means $+2a$)

(iv) **$4abcd = 4 \times a \times b \times c \times d$**
For example, if $a = 3$, $b = -2$, $c = 1$ and $d = -5$, then $4abcd = 4 \times 3 \times -2 \times 1 \times -5 = 120$
(Note that $\times - = +$)

(v) **$(a)(c)(d)$ means $a \times c \times d$**
Brackets are often used instead of multiplication signs. For example,
$(2)(5)(3)$ means $2 \times 5 \times 3 = 30$

(vi) **$ab = ba$**
If $a = 2$ and $b = 3$ then 2×3 is exactly the same as 3×2, i.e. 6

(vii) **$b^2 = b \times b$** For example, if $b = 3$, then $3^2 = 3 \times 3 = 9$

(viii) **$a^3 = a \times a \times a$** For example, if $a = 2$, then $2^3 = 2 \times 2 \times 2 = 8$

Here are some worked examples to help get a feel for basic operations in this introduction to algebra.

5.2.1 Addition and subtraction

Problem 1. Find the sum of $4x$, $3x$, $-2x$, $-x$

$$4x + 3x + -2x + -x = 4x + 3x - 2x - x$$

(Note that $+ \times - = -$)

$$= 4x$$

Problem 2. Find the sum of $5x$, $3y$, z, $-3x$, $-4y$ and $6z$

$$5x + 3y + z + -3x + -4y + 6z$$
$$= 5x + 3y + z - 3x - 4y + 6z$$
$$= 5x - 3x + 3y - 4y + z + 6z$$
$$= 2x - y + 7z$$

Note that the order can be changed when terms are separated by $+$ and $-$ signs. Only similar terms can be combined.

Problem 3. Simplify $4x^2 - x - 2y + 5x + 3y$

$$4x^2 - x - 2y + 5x + 3y = 4x^2 + 5x - x + 3y - 2y$$
$$= 4x^2 + 4x + y$$

5.2.2 Multiplication and division

Problem 4. Simplify $bc \times abc$

$$bc \times abc = a \times b \times b \times c \times c$$
$$= a \times b^2 \times c^2$$
$$= ab^2c^2$$

Problem 5. Simplify $-2p \times -3p$

$$- \times - = + \quad \text{hence, } -2p \times -3p = 6p^2$$

Problem 6. Simplify $ab \times b^2c \times a$

$$ab \times b^2c \times a = a \times a \times b \times b \times b \times c$$
$$= a^2 \times b^3 \times c$$
$$= a^2b^3c$$

Problem 7. Evaluate $3ab + 4bc - abc$ when $a = 3$, $b = 2$ and $c = 5$

$$3ab + 4bc - abc = 3 \times a \times b + 4 \times b \times c - a \times b \times c$$
$$= 3 \times 3 \times 2 + 4 \times 2 \times 5 - 3 \times 2 \times 5$$
$$= 18 + 40 - 30$$
$$= 28$$

Problem 8. Multiply $2a + 3b$ by $a + b$

Each term in the first expression is multiplied by a, then each term in the first expression is multiplied by b, and the two results are added. The usual layout is shown below.

$$
\begin{array}{r}
2a + 3b \\
a + b \\
\hline
\end{array}
$$

Multiplying by a gives: $\quad 2a^2 + 3ab$
Multiplying by b gives $\qquad\qquad 2ab + 3b^2$

Adding gives: $\qquad\qquad \overline{2a^2 + 5ab + 3b^2}$

Thus, $(2a + 3b)(a + b) = 2a^2 + 5ab + 3b^2$

Problem 9. Simplify $2x \div 8xy$

$$2x \div 8xy \text{ means } \frac{2x}{8xy}$$

$$\frac{2x}{8xy} = \frac{2 \times x}{8 \times x \times y}$$
$$= \frac{1 \times 1}{4 \times 1 \times y} \text{ by cancelling}$$
$$= \frac{1}{4y}$$

Now try the following Practice Exercise

Practice Exercise 25   Basic operations in algebra (answers on page 534)

1. Find the sum of $4a$, $-2a$, $3a$, $-8a$

2. Find the sum of $2a$, $5b$, $-3c$, $-a$, $-3b$ and $7c$

3. Simplify $5ab - 4a + ab + a$

4. Simplify $2x - 3y + 5z - x - 2y + 3z + 5x$

5. Add $x - 2y + 3$ to $3x + 4y - 1$

6. Subtract $a - 2b$ from $4a + 3b$

7. From $a + b - 2c$ take $3a + 2b - 4c$

8. Simplify $pq \times pq^2r$

9. Simplify $-4a \times -2a$

10. Simplify $3 \times -2q \times -q$

11. Evaluate $3pq - 5qr - pqr$ when $p = 3$, $q = -2$ and $r = 4$

12. If $x = 5$ and $y = 6$, evaluate: $\dfrac{23(x - y)}{y + xy + 2x}$

13. If $a = 4, b = 3, c = 5$ and $d = 6$, evaluate $\dfrac{3a + 2b}{3c - 2d}$

14. Simplify $2x \div 14xy$

15. Multiply $3a - b$ by $a + b$

16. Simplify $3a \div 9ab$

5.3   Laws of indices

The laws of indices with numbers were covered in chapter 3; the laws of indices in algebraic terms are as follows:

(i) $a^m \times a^n = a^{m+n}$

For example, $a^3 \times a^4 = a^{3+4} = a^7$

(ii) $\dfrac{a^m}{a^n} = a^{m-n}$ For example, $\dfrac{c^5}{c^2} = c^{5-2} = c^3$

(iii) $(a^m)^n = a^{mn}$ For example, $\left(d^2\right)^3 = d^{2 \times 3} = d^6$

(iv) $a^{\frac{m}{n}} = \sqrt[n]{a^m}$ For example, $x^{\frac{4}{3}} = \sqrt[3]{x^4}$

(v) $a^{-n} = \dfrac{1}{a^n}$ For example, $3^{-2} = \dfrac{1}{3^2} = \dfrac{1}{9}$

(vi) $a^0 = 1$ For example, $17^0 = 1$

Here are some worked examples to demonstrate these laws of indices.

Problem 10.   Simplify $a^2 b^3 c \times ab^2 c^5$

$$a^2 b^3 c \times ab^2 c^5 = a^2 \times b^3 \times c \times a \times b^2 \times c^5$$
$$= a^2 \times b^3 \times c^1 \times a^1 \times b^2 \times c^5$$

Grouping together like terms gives:

$$a^2 \times a^1 \times b^3 \times b^2 \times c^1 \times c^5$$

Using law (i) of indices gives:

$$a^{2+1} \times b^{3+2} \times c^{1+5} = a^3 \times b^5 \times c^6$$

i.e. $$a^2 b^3 c \times ab^2 c^5 = a^3 b^5 c^6$$

Problem 11.   Simplify $\dfrac{x^5 y^2 z}{x^2 y z^3}$

$$\frac{x^5 y^2 z}{x^2 y z^3} = \frac{x^5 \times y^2 \times z}{x^2 \times y \times z^3}$$
$$= \frac{x^5}{x^2} \times \frac{y^2}{y^1} \times \frac{z}{z^3}$$
$$= x^{5-2} \times y^{2-1} \times z^{1-3} \quad \text{by law (ii) of indices}$$
$$= x^3 \times y^1 \times z^{-2}$$
$$= x^3 y z^{-2} \text{ or } \frac{x^3 y}{z^2}$$

Problem 12.   Simplify: $\dfrac{a^3 b^2 c^4}{abc^{-2}}$ and evaluate when $a = 3, b = \dfrac{1}{4}$ and $c = 2$

Using law (ii) of indices:

$$\frac{a^3}{a} = a^{3-1} = a^2, \quad \frac{b^2}{b} = b^{2-1} = b \text{ and}$$

$$\frac{c^4}{c^{-2}} = c^{4--2} = c^6$$

Thus, $\dfrac{a^3 b^2 c^4}{abc^{-2}} = a^2 b c^6$

When $a = 3, b = \dfrac{1}{4}$ and $c = 2$,

$$a^2 b c^6 = (3)^2 \left(\frac{1}{4}\right)(2)^6 = (9)\left(\frac{1}{4}\right)(64) = \mathbf{144}$$

Problem 13.   Simplify: $(p^3)^2 (q^2)^4$

Using law (iii) of indices gives:

$$(p^3)^2 (q^2)^4 = p^{3 \times 2} \times q^{2 \times 4}$$
$$= p^6 q^8$$

Problem 14.   Simplify $\dfrac{(x^2 y^{1/2})(\sqrt{x} \sqrt[3]{y^2})}{(x^5 y^3)^{1/2}}$

Using laws (iii) and (iv) of indices gives:

$$\frac{(x^2 y^{1/2})(\sqrt{x} \sqrt[3]{y^2})}{(x^5 y^3)^{1/2}} = \frac{(x^2 y^{1/2})(x^{1/2} y^{2/3})}{x^{5/2} y^{3/2}}$$

Using laws (i) and (ii) of indices gives:

$$x^{2+\frac{1}{2}} \; \tfrac{5}{2}y^{\frac{1}{2}+\frac{2}{3}-\frac{3}{2}} = x^0y^{-\frac{1}{3}} = y^{-\frac{1}{3}} \text{ or } \frac{1}{y^{1/3}} \text{ or } \frac{1}{\sqrt[3]{y}}$$

from laws (v) and (vi) of indices.

Now try the following Practice Exercise

**Practice Exercise 26   Laws of indices
(answers on page 534)**

In problems 1–16, simplify the following, giving each answer as a power:

1. $z^2 \times z^6$
2. $a \times a^2 \times a^5$
3. $n^8 \times n^{-5}$
4. $b^4 \times b^7$
5. $b^2 \div b^5$
6. $c^5 \times c^3 \div c^4$
7. $\dfrac{m^5 \times m^6}{m^4 \times m^3}$
8. $\dfrac{(x^2)(x)}{x^6}$
9. $\left(x^3\right)^4$
10. $\left(y^2\right)^{-3}$
11. $\left(t \times t^3\right)^2$
12. $\left(c^{-7}\right)^{-2}$
13. $\left(\dfrac{a^2}{a^5}\right)^3$
14. $\left(\dfrac{1}{b^3}\right)^4$
15. $\left(\dfrac{b^2}{b^7}\right)^{-2}$
16. $\dfrac{1}{(s^3)^3}$

17. Simplify $\dfrac{a^5bc^3}{a^2b^3c^2}$ and evaluate when $a = \dfrac{3}{2}$, $b = \dfrac{1}{2}$ and $c = \dfrac{2}{3}$

In problems 18 and 19, simplify the given expressions:

18. $\dfrac{(abc)^2}{(a^2b^{-1}c^{-3})^3}$
19. $\dfrac{(a^3b^{1/2}c^{-1/2})(ab)^{1/3}}{(\sqrt{a^3}\sqrt{b}\,c)}$

5.4 Brackets

With algebra

(i) $2(a + b) = 2a + 2b$

(ii) $(a + b)(c + d) = a(c + d) + b(c + d)$
$$= ac + ad + bc + bd$$

Here are some worked examples to help understanding of brackets with algebra.

Problem 15.   Determine $2b(a - 5b)$

$$2b(a - 5b) = 2b \times a + 2b \times -5b$$
$$= 2ba - 10b^2$$
$$= \mathbf{2ab - 10b^2} \quad \text{(Note that } 2ba \text{ is the same as } 2ab\text{).}$$

Problem 16.   Determine $(3x + 4y)(x - y)$

$$(3x + 4y)(x - y) = 3x(x - y) + 4y(x - y)$$
$$= 3x^2 - 3xy + 4yx - 4y^2$$
$$= 3x^2 - 3xy + 4xy - 4y^2$$
(Note that $4yx$ is the same as $4xy$)
$$= \mathbf{3x^2 + xy - 4y^2}$$

Problem 17.   Simplify $3(2x - 3y) - (3x - y)$

$$3(2x - 3y) - (3x - y) = 3 \times 2x - 3 \times 3y - 3x - -y$$
(Note that $-(3x - y) = -1(3x - y)$ and the
-1 multiplies **both** terms in the bracket)
$$= 6x - 9y - 3x + y$$
(Note: $- \times - = +$)
$$= 6x - 3x + y - 9y$$
$$= \mathbf{3x - 8y}$$

Problem 18.   Simplify $(2x - 3y)^2$

$$(2x - 3y)^2 = (2x - 3y)(2x - 3y)$$
$$= 2x(2x - 3y) - 3y(2x - 3y)$$
$$= 2x \times 2x + 2x \times -3y$$
$$\qquad\quad -3y \times 2x - 3y \times -3y$$
$$= 4x^2 - 6xy - 6xy + 9y^2$$
(Note: $+ \times - = -$ and $- \times - = +$)
$$= \mathbf{4x^2 - 12xy + 9y^2}$$

Problem 19. Remove the brackets from the expression and simplify: $2\left[x^2 - 3x(y + x) + 4xy\right]$

$2\left[x^2 - 3x(y + x) + 4xy\right] = 2\left[x^2 - 3xy - 3x^2 + 4xy\right]$
(Whenever more than one type of brackets are involved, **always start with the inner brackets**)

$$= 2\left[-2x^2 + xy\right]$$
$$= -4x^2 + 2xy$$
$$= \mathbf{2xy - 4x^2}$$

Now try the following Practice Exercise

Practice Exercise 27   Brackets (answers on page 534)

Expand the brackets in problems 1–20:

1. $(x + 2)(x + 3)$ 2. $(x + 4)(2x + 1)$

3. $(2x + 3)^2$ 4. $(2j - 4)(j + 3)$

5. $(2x + 6)(2x + 5)$ 6. $(pq + r)(r + pq)$

7. $(x + 6)^2$ 8. $(5x + 3)^2$

9. $(2x - 6)^2$ 10. $(2x - 3)(2x + 3)$

11. $3a(b - 2a)$ 12. $2x(x - y)$

13. $(2a - 5b)(a + b)$

14. $3(3p - 2q) - (q - 4p)$

15. $(3x - 4y) + 3(y - z) - (z - 4x)$

16. $(2a + 5b)(2a - 5b)$

17. $(x - 2y)^2$ 18. $2x + [y - (2x + y)]$

19. $3a + 2[a - (3a - 2)]$

20. $4\left[a^2 - 3a(2b + a) + 7ab\right]$

5.5 Factorisation

The **factors** of 8 are 1, 2, 4 and 8 because 8 divides by 1, 2, 4 and 8. The factors of 24 are 1, 2, 3, 4, 6, 8, 12 and 24 because 24 divides by 1, 2, 3, 4, 6, 8, 12 and 24. The **common factors** of 8 and 24 are 1, 2, 4 and 8 since 1, 2, 4 and 8 are factors of both 8 and 24.
The **highest common factor (HCF)** is the largest number that divides into two or more terms.

Hence, the HCF of 8 and 24 is 8, as explained in chapter 1.
When two or more terms in an algebraic expression contain a common factor, then this factor can be shown outside of a bracket. For example, $df + dg = d(f + g)$, which is just the reverse of $d(f + g) = df + dg$. This process is called **factorisation**.
Here are some worked examples to help understanding of factorising in algebra.

Problem 20. Factorise $ab - 5ac$

a is common to both of the terms ab and $-5ac$. a is therefore taken outside of the bracket. What goes inside the bracket?

(i)   What multiplies a to make ab? Answer: b

(ii) What multiplies a to make $-5ac$? Answer: $-5c$

Hence, $b - 5c$ appears in the bracket.
Thus, $ab - 5ac = \mathbf{a(b - 5c)}$

Problem 21. Factorise $2x^2 + 14xy^3$

For the numbers 2 and 14, the highest common factor (HCF) is 2 (i.e. 2 is the largest number that divides into both 2 and 14). For the x terms, x^2 and x, the HCF is x. Thus, the HCF of $2x^2$ and $14xy^3$ is $2x$. $2x$ is therefore taken outside of the bracket. What goes inside the bracket?

(i)   What multiplies $2x$ to make $2x^2$? Answer: x

(ii) What multiplies $2x$ to make $14xy^3$? Answer: $7y^3$

Hence $x + 7y^3$ appears in the bracket.
Thus, $2x^2 + 14xy^3 = \mathbf{2x\left(x + 7y^3\right)}$

Problem 22. Factorise $3x^3y - 12xy^2 + 15xy$

For the numbers 3, 12 and 15, the highest common factor is 3 (i.e. 3 is the largest number that divides into 3, 12 and 15). For the x terms, x^3, x and x, the HCF is x. For the y terms, y, y^2 and y, the HCF is y.
Thus, the HCF of $3x^3y$ and $12xy^2$ and $15xy$ is $3xy$.
$3xy$ is therefore taken outside of the bracket. What goes inside the bracket?

(i)   What multiplies $3xy$ to make $3x^3y$? Answer: x^2

(ii) What multiplies $3xy$ to make $-12xy^2$? Answer: $-4y$

(iii) What multiplies $3xy$ to make $15xy$? Answer: 5

Hence, $x^2 - 4y + 5$ appears in the bracket.

Thus, $3x^3y - 12xy^2 + 15xy = \mathbf{3xy\,(x^2 - 4y + 5)}$

Now try the following Practice Exercise

Practice Exercise 28 Factorisation (answers on page 534)

Factorise and simplify the following:

1. $2x + 4$ 2. $2xy - 8xz$

3. $pb + 2pc$ 4. $2x + 4xy$

5. $4d^2 - 12df^5$ 6. $4x + 8x^2$

7. $2q^2 + 8qn$ 8. $rs + rp + rt$

9. $x + 3x^2 + 5x^3$ 10. $abc + b^3c$

11. $3x^2y^4 - 15xy^2 + 18xy$

12. $4p^3q^2 - 10pq^3$ 13. $21a^2b^2 - 28ab$

14. $2xy^2 + 6x^2y + 8x^3y$

15. $2x^2y - 4xy^3 + 8x^3y^4$

5.6   Laws of precedence

Sometimes addition, subtraction, multiplication, division, powers and brackets can all be involved in an algebraic expression. With mathematics there is a definite order of precedence (first met in chapter 1) which we need to adhere to. With the **laws of precedence** the order is:

Brackets

Order (or p**O**wer)

Division

Multiplication

Addition

Subtraction

The first letter of each word spells **BODMAS**.
Here are some examples to help understanding of BODMAS with algebra.

Problem 23. Simplify $2x + 3x \times 4x - x$

$$2x + 3x \times 4x - x = 2x + 12x^2 - x \quad \text{(M)}$$

$$= 2x - x + 12x^2$$

$$= \mathbf{x + 12x^2} \quad \text{(S)}$$

$$\text{or } \mathbf{x(1 + 12x)} \quad \text{by factorising.}$$

Problem 24. Simplify $(y + 4y) \times 3y - 5y$

$$(y + 4y) \times 3y - 5y = 5y \times 3y - 5y \quad \text{(B)}$$

$$= \mathbf{15y^2 - 5y} \quad \text{(M)}$$

$$\text{or } \mathbf{5y(3y - 1)} \quad \text{by factorising.}$$

Problem 25. Simplify $t \div 2t + 3t - 5t$

$$t \div 2t + 3t - 5t = \frac{t}{2t} + 3t - 5t \quad \text{D}$$

$$= \frac{1}{2} + 3t - 5t \quad \text{by cancelling}$$

$$= \mathbf{\frac{1}{2} - 2t} \quad \text{S}$$

Problem 26. Simplify $x \div (4x + x) - 3x$

$$x \div (4x + x) - 3x = x \div 5x - 3x \quad \text{(B)}$$

$$= \frac{x}{5x} - 3x \quad \text{(D)}$$

$$= \mathbf{\frac{1}{5} - 3x} \quad \text{by cancelling.}$$

Problem 27. Simplify

$$5a + 3a \times 2a + a \div 2a - 7a$$

$$5a + 3a \times 2a + a \div 2a - 7a$$

$$= 5a + 3a \times 2a + \frac{a}{2a} - 7a \quad \text{(D)}$$

$$= 5a + 3a \times 2a + \frac{1}{2} - 7a \quad \text{by cancelling}$$

$$= 5a + 6a^2 + \frac{1}{2} - 7a \quad \text{(M)}$$

$$= -2a + 6a^2 + \frac{1}{2} \quad \text{(S)}$$

$$= \mathbf{6a^2 - 2a + \frac{1}{2}}$$

Now try the following Practice Exercise

Practice Exercise 29   Laws of precedence
(answers on page 534)

Simplify the following:

1. $3x + 2x \times 4x - x$

2. $(2y + y) \times 4y - 3y$

3. $4b + 3b \times (b - 6b)$

4. $8a \div 2a + 6a - 3a$

5. $6x \div (3x + x) - 4x$

6. $4t \div (5t - 3t + 2t)$

7. $3y + 2y \times 5y + 2y \div 8y - 6y$

8. $(x + 2x)3x + 2x \div 6x - 4x$

9. $5a + 2a \times 3a + a \div (2a - 9a)$

10. $(3t + 2t)(5t + t) \div (t - 3t)$

**For fully worked solutions to each of the problems in Exercises 25 to 29 in this chapter,
go to the website:**
www.routledge.com/cw/bird

Chapter 6

Solving simple equations

Why it is important to understand: Solving simple equations

In mathematics, engineering and science, formulae are used to relate physical quantities to each other. They provide rules so that if we know the values of certain quantities, we can calculate the values of others. Equations occur in all branches of engineering. Simple equations always involve one unknown quantity which we try to find when we solve the equation. In reality, we all solve simple equations in our heads all the time without even noticing it. If, for example, you have bought two CDs, each for the same price, and a DVD, and know that you spent £25 in total and that the DVD was £11, then you actually solve the linear equation $2x + 11 = 25$ to find out that the price of each CD was £7. It is probably true to say that there is no branch of engineering, physics, economics, chemistry and computer science which does not require the solution of simple equations. The ability to solve simple equations is another stepping stone on the way to having confidence to handle engineering mathematics and science.

At the end of this chapter, you should be able to:

- distinguish between an algebraic expression and an algebraic equation
- maintain the equality of a given equation whilst applying arithmetic operations
- solve linear equations in one unknown including those involving brackets and fractions
- form and solve linear equations involved with practical situations
- evaluate formulae by substitution of data

6.1 Introduction

$3x - 4$ is an example of an **algebraic expression**.
$3x - 4 = 2$ is an example of an **algebraic equation** (i.e. it contains an '=' sign).
An equation is simply a statement that two expressions are equal.

Hence, $A = \pi r^2$ (where A is the area of a circle of radius r)

$$F = \frac{9}{5}C + 32$$ (which relates Fahrenheit and Celsius temperatures)

and $y = 3x + 2$ (which is the equation of a straight line graph)

are all examples of equations.

Science and Mathematics for Engineering. 978-0-367-20475-4, © John Bird. Published by Taylor & Francis. All rights reserved.

6.2   Solving equations

To '**solve an equation**' means '**to find the value of the unknown**'. For example, to solve $3x - 4 = 2$ means that the value of x is required. In this example, $x = 2$. The purpose of this chapter is to show how to solve such equations.

Many equations occur in engineering and it is essential that we can solve them when needed. Here are some examples to demonstrate how simple equations are solved.

Problem 1.   Solve the equation: $4x = 20$

Dividing each side of the equation by 4 gives:

$$\frac{4x}{4} = \frac{20}{4}$$

i.e. $x = 5$ by cancelling which is the solution to the equation $4x = 20$

The same operation **must** be applied to both sides of an equation so that the equality is maintained. We can do anything we like to an equation, **as long as we do the same to both sides**. This is, in fact, the only rule to remember when solving simple equations (and also when transposing formulae, which we do in the next chapter).

Problem 2.   Solve the equation: $\dfrac{2x}{5} = 6$

Multiplying both sides by 5 gives: $5\left(\dfrac{2x}{5}\right) = 5(6)$

Cancelling and removing brackets gives: $2x = 30$

Dividing both sides of the equation by 2 gives:

$$\frac{2x}{2} = \frac{30}{2}$$

Cancelling gives: $x = 15$

which is the solution of the equation $\dfrac{2x}{5} = 6$

Problem 3.   Solve the equation: $a - 5 = 8$

Adding 5 to both sides of the equation gives:

$$a - 5 + 5 = 8 + 5$$

i.e. $a = 8 + 5$

i.e. $a = 13$

which is the solution of the equation $a - 5 = 8$

Note that adding 5 to both sides of the above equation results in the '-5' moving from the left hand side (LHS) to the right hand side (RHS), but the sign is changed to '$+$'.

Problem 4.   Solve the equation: $x + 3 = 7$

Subtracting 3 from both sides gives: $x + 3 - 3 = 7 - 3$

i.e. $x = 7 - 3$

i.e. $x = 4$

which is the solution of the equation $x + 3 = 7$

Note that subtracting 3 from both sides of the above equation results in the '$+3$' moving from the LHS to the RHS, but the sign is changed to '$-$'. So we can move straight from $x + 3 = 7$ to $x = 7 - 3$

Thus a term can be moved from one side of an equation to the other **as long as a change in sign is made**.

Problem 5.   Solve the equation: $6x + 1 = 2x + 9$

In such equations the terms containing x are grouped on one side of the equation and the remaining terms grouped on the other side of the equation. As in Problems 3 and 4, changing from one side of an equation to the other must be accompanied by a change of sign.

Since $6x + 1 = 2x + 9$

then $6x - 2x = 9 - 1$

i.e. $4x = 8$

Dividing both sides by 4 gives: $\dfrac{4x}{4} = \dfrac{8}{4}$

Cancelling gives: $x = 2$

which is the solution of the equation $6x + 1 = 2x + 9$.

In the above examples, the solutions can be checked. Thus, in Problem 5, where $6x + 1 = 2x + 9$, if $x = 2$ then:

$$\text{LHS of equation} = 6(2) + 1 = 13$$

$$\text{RHS of equation} = 2(2) + 9 = 13$$

Since the LHS equals the RHS, then $x = 2$ must be the correct solution of the equation.

When solving simple equations, always check your answers by substituting your solution back into the original equation.

Problem 6.   Solve the equation: $4 - 3p = 2p - 11$

In order to keep the p term positive the terms in p are moved to the RHS and the constant terms to the LHS.

Similar to the previous Problem, if $4 - 3p = 2p - 11$
then
$$4 + 11 = 2p + 3p$$
i.e.
$$15 = 5p$$

Dividing both sides by 5 gives:
$$\frac{15}{5} = \frac{5p}{5}$$

Cancelling gives: $\mathbf{3 = p}$ or $\mathbf{p = 3}$
which is the solution of the equation $4 - 3p = 2p - 11$.
By substituting $p = 3$ into the original equation, the
solution may be checked.

$$\text{LHS} = 4 - 3(3) = 4 - 9 = -5$$

$$\text{RHS} = 2(3) - 11 = 6 - 11 = -5$$

Since LHS = RHS, the solution $p = 3$ must be correct.
If, in this example, the unknown quantities had been
grouped initially on the LHS instead of the RHS
then:
$$-3p - 2p = -11 - 4$$
i.e.
$$-5p = -15$$
from which,
$$\frac{-5p}{-5} = \frac{-15}{-5}$$
and
$$p = 3,$$
as before.
It is often easier, however, to work with positive values
where possible.

Problem 7.   Solve the equation: $3(x - 2) = 9$

Removing the bracket gives: $3x - 6 = 9$
Rearranging gives: $3x = 9 + 6$
i.e. $3x = 15$
Dividing both sides by 3 gives: $x = 5$
which is the solution of the equation $3(x - 2) = 9$.
The equation may be checked by substituting $x = 5$ back
into the original equation.

Problem 8.   Solve the equation:
$4(2r - 3) - 2(r - 4) = 3(r - 3) - 1$

Removing brackets gives:
$$8r - 12 - 2r + 8 = 3r - 9 - 1$$
Rearranging gives: $8r - 2r - 3r = -9 - 1 + 12 - 8$
i.e. $3r = -6$
Dividing both sides by 3 gives: $r = \dfrac{-6}{3} = -2$
which is the solution of the equation
$4(2r - 3) - 2(r - 4) = 3(r - 3) - 1$

The solution may be checked by substituting $r = -2$
back into the original equation.

$$\text{LHS} = 4(-4 - 3) - 2(-2 - 4) = -28 + 12 = -16$$

$$\text{RHS} = 3(-2 - 3) - 1 = -15 - 1 = -16$$

Since LHS = RHS then $r = -2$ is the correct solution.

Now try the following Practice Exercise

Practice Exercise 30   Solving simple
equations (answers on page 534)

Solve the following equations:

1. $2x + 5 = 7$

2. $8 - 3t = 2$

3. $\dfrac{2}{3}c - 1 = 3$

4. $2x - 1 = 5x + 11$

5. $7 - 4p = 2p - 5$

6. $2a + 6 - 5a = 0$

7. $3x - 2 - 5x = 2x - 4$

8. $20d - 3 + 3d = 11d + 5 - 8$

9. $2(x - 1) = 4$

10. $16 = 4(t + 2)$

11. $5(f - 2) - 3(2f + 5) + 15 = 0$

12. $2x = 4(x - 3)$

13. $6(2 - 3y) - 42 = -2(y - 1)$

14. $2(3g - 5) - 5 = 0$

15. $4(3x + 1) = 7(x + 4) - 2(x + 5)$

Here are some further worked examples of solving
simple equations.

Problem 9.   Solve the equation: $\dfrac{4}{x} = \dfrac{2}{5}$

The lowest common multiple (LCM) of the denominators, i.e. the lowest algebraic expression that both x and
5 will divide into, is $5x$.

Multiplying both sides by $5x$ gives:

$$5x\left(\frac{4}{x}\right) = 5x\left(\frac{2}{5}\right)$$

Cancelling gives: $\quad 5(4) = x(2)$

i.e. $\quad\quad\quad\quad\quad 20 = 2x \quad\quad\quad\quad\quad (1)$

Dividing both sides by 2 gives: $\quad \dfrac{20}{2} = \dfrac{2x}{2}$

Cancelling gives: $\quad\quad\quad$ **$10 = x$ or $x = 10$**

which is the solution of the equation $\dfrac{4}{x} = \dfrac{2}{5}$

When there is just one fraction on each side of the equation, as in this example, there is a quick way to arrive at equation (1) without needing to find the LCM of the denominators.

We can move from $\dfrac{4}{x} = \dfrac{2}{5}$ to: $4 \times 5 = 2 \times x$ by what is called '**cross-multiplication**'.

In general, if $\dfrac{a}{b} = \dfrac{c}{d}$ then: $ad = bc$

We can use cross-multiplication when there is only one fraction on each side of the equation.

Problem 10. Solve the equation:
$$\frac{2y}{5} + \frac{3}{4} + 5 = \frac{1}{20} - \frac{3y}{2}$$

The LCM of the denominators is 20, i.e. the lowest number that 4, 5, 20 and 2 will divide into.
Multiplying each term by 20 gives:

$$20\left(\frac{2y}{5}\right) + 20\left(\frac{3}{4}\right) + 20(5) = 20\left(\frac{1}{20}\right) - 20\left(\frac{3y}{2}\right)$$

Cancelling gives: $\quad 4(2y) + 5(3) + 100 = 1 - 10(3y)$

i.e. $\quad\quad\quad\quad 8y + 15 + 100 = 1 - 30y$

Rearranging gives: $\quad\quad\quad 8y + 30y = 1 - 15 - 100$

i.e. $\quad\quad\quad\quad\quad\quad\quad 38y = -114$

Dividing both sides by 38 gives: $\dfrac{38y}{38} = \dfrac{-114}{38}$

Cancelling gives: $\quad\quad\quad\quad\quad$ **$y = -3$**

which is the solution of the equation

$$\frac{2y}{5} + \frac{3}{4} + 5 = \frac{1}{20} - \frac{3y}{2}$$

Problem 11. Solve the equation: $\sqrt{x} = 2$

Whenever square root signs are involved in an equation, both sides of the equation must be squared.

Squaring both sides gives: $\quad\quad \left(\sqrt{x}\right)^2 = (2)^2$

i.e. $\quad\quad\quad\quad\quad\quad\quad\quad\quad$ **$x = 4$**

which is the solution of the equation $\sqrt{x} = 2$

Problem 12. Solve the equation: $2\sqrt{d} = 8$

Whenever square roots are involved in an equation, the square root term needs to be isolated on its own before squaring both sides.

Cross-multiplying gives: $\quad\quad \sqrt{d} = \dfrac{8}{2}$

Cancelling gives: $\quad\quad\quad\quad \sqrt{d} = 4$

Squaring both sides gives: $\quad\quad \left(\sqrt{d}\right)^2 = (4)^2$

i.e. $\quad\quad\quad\quad\quad\quad\quad\quad\quad$ **$d = 16$**

which is the solution of the equation $2\sqrt{d} = 8$

Problem 13. Solve the equation: $x^2 = 25$

Whenever a square term is involved, the square root of both sides of the equation must be taken.
Taking the square root of both sides gives: $\sqrt{x^2} = \sqrt{25}$

i.e. $\quad\quad\quad\quad\quad\quad\quad\quad\quad$ **$x = \pm 5$**

which is the solution of the equation $x^2 = 25$

Problem 14. Solve the equation: $\dfrac{15}{4t^2} = \dfrac{2}{3}$

We need to rearrange the equation to get the t^2 term on its own.

Cross-multiplying gives: $\quad\quad 15(3) = 2(4t^2)$

i.e. $\quad\quad\quad\quad\quad\quad\quad\quad 45 = 8t^2$

Dividing both sides by 8 gives: $\quad \dfrac{45}{8} = \dfrac{8t^2}{8}$

By cancelling: $\quad\quad\quad\quad\quad 5.625 = t^2$

or $\quad\quad\quad\quad\quad\quad\quad\quad t^2 = 5.625$

Taking the square root of both sides gives:

$$\sqrt{t^2} = \sqrt{5.625}$$

i.e. $t = \pm 2.372$, correct to 4 significant figures
which is the solution of the equation $\dfrac{15}{4t^2} = \dfrac{2}{3}$

Now try the following Practice Exercise

Practice Exercise 31   Solving simple equations (answers on page 534)

Solve the following equations:

1. $\dfrac{1}{5}d + 3 = 4$

2. $2 + \dfrac{3}{4}y = 1 + \dfrac{2}{3}y + \dfrac{5}{6}$

3. $\dfrac{1}{4}(2x - 1) + 3 = \dfrac{1}{2}$

4. $\dfrac{1}{5}(2f - 3) + \dfrac{1}{6}(f - 4) + \dfrac{2}{15} = 0$

5. $\dfrac{x}{3} - \dfrac{x}{5} = 2$

6. $1 - \dfrac{y}{3} = 3 + \dfrac{y}{3} - \dfrac{y}{6}$

7. $\dfrac{2}{a} = \dfrac{3}{8}$

8. $\dfrac{1}{3n} + \dfrac{1}{4n} = \dfrac{7}{24}$

9. $\dfrac{x + 3}{4} = \dfrac{x - 3}{5} + 2$

10. $\dfrac{2}{a - 3} = \dfrac{3}{2a + 1}$

11. $\dfrac{x}{4} - \dfrac{x + 6}{5} = \dfrac{x + 3}{2}$

12. $3\sqrt{t} = 9$

13. $2\sqrt{y} = 5$

14. $4 = \sqrt{\left(\dfrac{3}{a}\right)} + 3$

15. $10 = 5\sqrt{\left(\dfrac{x}{2} - 1\right)}$

16. $16 = \dfrac{t^2}{9}$

6.3 Practical problems involving simple equations

There are many practical situations in engineering where solving equations is needed.

Here are some worked examples to demonstrate typical practical situations.

Problem 15. Applying the principle of moments to a beam results in the following equation:

$$F \times 3 = (7.5 - F) \times 2$$

where F is the force in newtons. Determine the value of F

Removing brackets gives:	$3F = 15 - 2F$
Rearranging gives:	$3F + 2F = 15$
i.e.	$5F = 15$
Dividing both sides by 5 gives:	$\dfrac{5F}{5} = \dfrac{15}{5}$

from which, **force $F = 3$ N**

Problem 16. $PV = mRT$ is the characteristic gas equation. Find the value of gas constant R when pressure $P = 3 \times 10^6$ Pa, volume $V = 0.90m^3$, mass $m = 2.81$ kg and temperature $T = 231$ K

Dividing both sides of $PV = mRT$ by mT gives:

$$\dfrac{PV}{mT} = \dfrac{mRT}{mT}$$

Cancelling gives: $\dfrac{PV}{mT} = R$

Substituting values gives: $R = \dfrac{\left(3 \times 10^6\right)(0.90)}{(2.81)(231)}$

Using a calculator, **gas constant $R = 4160$ J/(kg K)**, correct to 4 significant figures.

Problem 17. The temperature coefficient of resistance α may be calculated from the formula $R_t = R_0(1 + \alpha t)$. Find α given $R_t = 0.928$, $R_0 = 0.80$ and $t = 40$

Since $R_t = R_0(1 + \alpha t)$ then:

$$0.928 = 0.80[1 + \alpha(40)]$$
$$0.928 = 0.80 + (0.8)(\alpha)(40)$$
$$0.928 - 0.80 = 32\alpha$$
$$0.128 = 32\alpha$$

Hence, $\alpha = \dfrac{0.128}{32} = \mathbf{0.004}$

Problem 18. The distance s metres travelled in time t seconds is given by the formula $s = ut + \dfrac{1}{2}at^2$, where u is the initial velocity in m/s and a is the acceleration in m/s^2. Find the acceleration of the body if it travels 168 m in 6 s, with an initial velocity of 10 m/s

$s = ut + \dfrac{1}{2}at^2$, and $s = 168, u = 10$ and $t = 6$

Hence,
$$168 = (10)(6) + \frac{1}{2}a(6)^2$$
$$168 = 60 + 18a$$
$$168 - 60 = 18a$$
$$108 = 18a$$
$$a = \frac{108}{18} = 6$$

Hence, the acceleration of the body is 6 m/s^2

Problem 19. When three resistors in an electrical circuit are connected in parallel the total resistance R_T is given by: $\dfrac{1}{R_T} = \dfrac{1}{R_1} + \dfrac{1}{R_2} + \dfrac{1}{R_3}$. Find the total resistance when $R_1 = 5\,\Omega, R_2 = 10\,\Omega$ and $R_3 = 30\,\Omega$

$$\frac{1}{R_T} = \frac{1}{5} + \frac{1}{10} + \frac{1}{30} = \frac{6 + 3 + 1}{30} = \frac{10}{30} = \frac{1}{3}$$

Taking the reciprocal of both sides gives: $\boldsymbol{R_T = 3\,\Omega}$.

Alternatively, if $\dfrac{1}{R_T} = \dfrac{1}{5} + \dfrac{1}{10} + \dfrac{1}{30}$ the LCM of the denominators is $30\,R_T$.

Hence
$$30R_T\left(\frac{1}{R_T}\right) = 30R_T\left(\frac{1}{5}\right) + 30R_T\left(\frac{1}{10}\right) + 30R_T\left(\frac{1}{30}\right)$$

Cancelling gives: $30 = 6R_T + 3R_T + R_T$

i.e. $30 = 10R_T$

and $\boldsymbol{R_T} = \dfrac{30}{10} = \boldsymbol{3\,\Omega}$, as above.

___

Now try the following Practice Exercise

Practice Exercise 32   Practical problems involving simple equations (answers on page 535)

1. A formula used for calculating resistance of a cable is $R = \dfrac{\rho L}{a}$. Given $R = 1.25, L = 2500$ and $a = 2 \times 10^{-4}$ find the value of ρ

2. Force F newtons is given by $F = ma$, where m is the mass in kilograms and a is the acceleration in metres per second squared. Find the acceleration when a force of 4 kN is applied to a mass of 500 kg

3. $PV = mRT$ is the characteristic gas equation. Find the value of m when $P = 100 \times 10^3$, $V = 3.00, R = 288$ and $T = 300$

4. When three resistors R_1, R_2 and R_3 are connected in parallel the total resistance R_T is determined from $\dfrac{1}{R_T} = \dfrac{1}{R_1} + \dfrac{1}{R_2} + \dfrac{1}{R_3}$

   (a) Find the total resistance when $R_1 = 3\,\Omega$, $R_2 = 6\,\Omega$ and $R_3 = 18\,\Omega$

   (b) Find the value of R_3 given that $R_T = 3\,\Omega$, $R_1 = 5\,\Omega$ and $R_2 = 10\,\Omega$

5. Ohm's law may be represented by $I = V/R$, where I is the current in amperes, V is the voltage in volts and R is the resistance in ohms. A soldering iron takes a current of 0.30 A from a 240 V supply. Find the resistance of the element

6. The stress, σ pascals, acting on the reinforcing rod in a concrete column is given in the following equation: $500 \times 10^{-6}\sigma + 2.67 \times 10^5 = 3.55 \times 10^5$. Find the value of the stress in MPa.

___

Here are some further worked examples on solving simple equations in practical situations.

Problem 20. Power in a d.c. circuit is given by $P = \dfrac{V^2}{R}$ where V is the supply voltage and R is the circuit resistance. Find the supply voltage if the circuit resistance is 1.25 Ω and the power measured is 320 W

Since $P = \dfrac{V^2}{R}$ then $320 = \dfrac{V^2}{1.25}$

$$(320)(1.25) = V^2$$

i.e. $V^2 = 400$

Supply voltage $V = \sqrt{400} = \pm 20\,V$

Problem 21.   The stress f in a material of a thick cylinder can be obtained from: $\dfrac{D}{d} = \sqrt{\left(\dfrac{f+p}{f-p}\right)}$

Calculate the stress, given that $D = 21.5$, $d = 10.75$ and $p = 1800$

Since $\dfrac{D}{d} = \sqrt{\left(\dfrac{f+p}{f-p}\right)}$ then $\dfrac{21.5}{10.75} = \sqrt{\left(\dfrac{f+1800}{f-1800}\right)}$

i.e. $\qquad\qquad\qquad 2 = \sqrt{\left(\dfrac{f+1800}{f-1800}\right)}$

Squaring both sides gives: $\qquad 4 = \dfrac{f+1800}{f-1800}$

Cross-multiplying gives:

$$4(f - 1800) = f + 1800$$

$$4f - 7200 = f + 1800$$

$$4f - f = 1800 + 7200$$

$$3f = 9000$$

$$f = \dfrac{9000}{3} = 3000$$

Hence, $\qquad\qquad$ **stress, $f = 3000$**

Now try the following Practice Exercise

Practice Exercise 33   Practical problems involving simple equations (answers on page 535)

1. Given $R_2 = R_1(1 + \alpha t)$, find α given $R_1 = 5.0$, $R_2 = 6.03$ and $t = 51.5$

2. If $v^2 = u^2 + 2as$, find u given $v = 24$, $a = -40$ and $s = 4.05$

3. The relationship between the temperature on a Fahrenheit scale and that on a Celsius scale is given by $F = \dfrac{9}{5}C + 32$. Express $113°F$ in degrees Celsius

4. If $t = 2\pi\sqrt{\dfrac{w}{Sg}}$, find the value of S given $w = 1.219$, $g = 9.81$ and $t = 0.3132$

5. A rectangular laboratory has a length equal to one and a half times its width and a perimeter of $40\,m$. Find its length and width

6. Applying the principle of moments to a beam results in the following equation:

$$F \times 3 = (5 - F) \times 7$$

    where F is the force in newtons. Determine the value of F

For fully worked solutions to each of the problems in Exercises 30 to 33 in this chapter, go to the website:
www.routledge.com/cw/bird

This assignment covers the material contained in chapters 5 and 6. *The marks for each question are shown in brackets at the end of each question.*

1. Evaluate $3pqr^3 - 2p^2qr + pqr$
   when $p = \dfrac{1}{2}$, $q = -2$ and $r = 1$ (3)

   In problems 2–7, simplify the expressions.

2. $\dfrac{9p^2qr^3}{3pq^2r}$ (3)

3. $2(3x - 2y) - (4y - 3x)$ (3)

4. $(x - 2y)(2x + y)$ (3)

5. $p^2q^{-3}r^4 \times pq^2r^{-3}$ (3)

6. $(3a - 2b)^2$ (3)

7. $\dfrac{a^4b^2c}{ab^3c^2}$ (3)

8. Factorise:

   (a) $2x^2y^3 - 10xy^2$

   (b) $21ab^2c^3 - 7a^2bc^2 + 28a^3bc^4$ (5)

9. Factorise and simplify
   $\dfrac{2x^2y + 6xy^2}{x + 3y} - \dfrac{x^3y^2}{x^2y}$ (5)

10. Remove the brackets and simplify
    $10a - [3(2a - b) - 4(b - a) + 5b]$ (4)

11. Simplify: $x \div 5x - x + (2x - 3x)x$ (4)

12. Simplify: $3a + 2a \times 5a + 4a \div 2a - 6a$ (4)

13. Solve the equations:

    (a) $3a = 39$

    (b) $2x - 4 = 9$ (3)

14. Solve the equations:

    (a) $\dfrac{4}{9}y = 8$

    (b) $6x - 1 = 4x + 5$ (4)

15. Solve the equation:
    $5(t - 2) - 3(4 - t) = 2(t + 3) - 40$ (4)

16. Solve the equations:

    (a) $\dfrac{3}{2x + 1} = \dfrac{1}{4x - 3}$

    (b) $2x^2 = 162$ (7)

17. Kinetic energy is given by the formula, $E_k = \dfrac{1}{2}mv^2$ joules, where m is the mass in kilograms and v is the velocity in metres per second. Evaluate the velocity when $E_k = 576 \times 10^{-3}$ J and the mass is 5 kg. (4)

18. An approximate relationship between the number of teeth T on a milling cutter, the diameter of the cutter D and the depth of cut d is given by: $T = \dfrac{12.5\,D}{D + 4d}$ Evaluate d when $T = 10$ and $D = 32$ (5)

19. The modulus of elasticity E is given by the formula: $E = \dfrac{FL}{xA}$ where F is force in newtons, L is the length in metres, x is the extension in metres and A the cross-sectional area in square metres. Evaluate A, in square centimetres, when $E = 80 \times 10^9$N/m^2, $x = 2$ mm, $F = 100 \times 10^3$N and $L = 2.0$ m (5)

For lecturers/instructors/teachers, fully worked solutions to each of the problems in Revision Test 3, together with a full marking scheme, are available at the website:

www.routledge.com/cw/bird

Transposing formulae

Why it is important to understand: **Transposing equations**

As was mentioned in the last chapter, formulae are used frequently in almost all aspects of engineering in order to relate a physical quantity to one or more others. Many well known physical laws are described using formulae – for example, Ohm's law, $V = I \times R$, or Newton's second law of motion, $F = m \times a$. In an everyday situation, imagine you buy 5 identical items for £20. How much did each item cost? If you divide £20 by 5 to get an answer of £4, you are actually applying transposition of a formula. Transposing formulae is a basic skill required in all aspects of engineering. The ability to transpose formulae is yet another stepping stone on the way to having confidence to handle engineering mathematics and science.

At the end of this chapter, you should be able to:

- define 'subject of the formula'
- transpose equations whose terms are connected by plus and/or minus signs
- transpose equations that involve fractions
- transpose equations that contain a root or power
- transpose equations in which the subject appears in more than one term

7.1   Introduction

In the formula $I = \frac{V}{R}$, I is called the **subject of the formula**. Similarly, in the formula $y = mx + c$, y is the subject of the formula.

When a symbol other than the subject is required to be the subject, then the formula needs to be rearranged to make a new subject. This rearranging process is called **transposing the formula** or **transposition**.

7.2   Transposing formulae

There are no new rules for transposing formulae. The same rules as were used for simple equations in chapter 6 are used, i.e. **the balance of an equation must be maintained**. Whatever is done to one side of an equation must be done to the other.

Here are some worked examples to help understanding of transposing formulae.

Science and Mathematics for Engineering. 978-0-367-20475-4, © John Bird. Published by Taylor & Francis. All rights reserved.

Problem 1. Transpose $p = q + r + s$ to make r the subject

The object is to obtain r on its own on the left hand side (LHS) of the equation. Changing the equation around so that r is on the LHS gives:

$$q + r + s = p \qquad (1)$$

From the previous chapter on simple equations, a term can be moved from one side of an equation to the other side as long as the sign is changed.
Rearranging gives: $r = p - q - s$
Mathematically, we have subtracted $q + s$ from both sides of equation (1).

Problem 2. If $a + b = w - x + y$, express x as the subject

As stated in the previous problem, a term can be moved from one side of an equation to the other side but with a change of sign.
Hence, rearranging gives: $x = w + y - a - b$

Problem 3. Transpose $v = f\lambda$ to make λ the subject

$v = f\lambda$ relates velocity v, frequency f and wavelength λ.
Rearranging gives: $\qquad\qquad f\lambda = v$
Dividing both sides by f gives: $\quad \dfrac{f\lambda}{f} = \dfrac{v}{f}$
Cancelling gives: $\qquad\qquad\qquad \lambda = \dfrac{v}{f}$

Problem 4. When a body falls freely through a height h, the velocity v is given by $v^2 = 2gh$. Express this formula with h as the subject

Rearranging gives: $\qquad\qquad 2gh = v^2$
Dividing both sides by $2g$ gives: $\dfrac{2gh}{2g} = \dfrac{v^2}{2g}$
Cancelling gives: $\qquad\qquad\qquad h = \dfrac{v^2}{2g}$

Problem 5. If $I = \dfrac{V}{R}$, rearrange to make V the subject

$I = \dfrac{V}{R}$ is Ohm's law, where I is the current, V is the voltage and R is the resistance.

Rearranging gives: $\qquad\qquad \dfrac{V}{R} = I$

Multiplying both sides by R gives: $R\left(\dfrac{V}{R}\right) = R(I)$

Cancelling gives: $\qquad\qquad\qquad V = IR$

Problem 6. Transpose $a = \dfrac{F}{m}$ for m

$a = \dfrac{F}{m}$ relates acceleration a, force F and mass m.
Rearranging gives: $\qquad\qquad \dfrac{F}{m} = a$

Multiplying both sides by m gives: $m\left(\dfrac{F}{m}\right) = m(a)$

Cancelling gives: $\qquad\qquad\qquad F = ma$
Rearranging gives: $\qquad\qquad ma = F$

Dividing both sides by a gives: $\qquad \dfrac{ma}{a} = \dfrac{F}{a}$

i.e. $\qquad\qquad\qquad\qquad m = \dfrac{F}{a}$

Problem 7. Rearrange the formula $R = \dfrac{\rho L}{A}$ to make (i) A the subject, and (ii) L the subject

$R = \dfrac{\rho L}{A}$ relates resistance R of a conductor, resistivity ρ, conductor length L and conductor cross-sectional area A.

(i) Rearranging gives: $\qquad\qquad \dfrac{\rho L}{A} = R$
  Multiplying both sides by A gives:
  $$A\left(\dfrac{\rho L}{A}\right) = A(R)$$
  Cancelling gives: $\qquad\qquad \rho L = AR$
  Rearranging gives: $\qquad\qquad AR = \rho l$
  Dividing both sides by R gives: $\dfrac{AR}{R} = \dfrac{\rho L}{R}$
  Cancelling gives: $\qquad\qquad A = \dfrac{\rho L}{R}$

(ii) Multiplying both sides of $\dfrac{\rho L}{A} = R$ by A gives:
  $$\rho L = AR$$
  Dividing both sides by ρ gives: $\dfrac{\rho L}{\rho} = \dfrac{AR}{\rho}$
  Cancelling gives: $\qquad\qquad L = \dfrac{AR}{\rho}$

Now try the following Practice Exercise

Practice Exercise 34   Transposing formulae
(answers on page 535)

Make the symbol indicated in brackets the subject of each of the formulae shown and express each in its simplest form.

1. $a + b = c - d - e$ (d)

2. $y = 7x$ (x)

3. $pv = c$ (v)

4. $v = u + at$ (a)

5. $x + 3y = t$ (y)

6. $c = 2\pi r$ (r)

7. $y = mx + c$ (x)

8. $I = PRT$ (T)

9. $X_L = 2\pi f L$ (L)

10. $Q = mc\Delta T$ (c)

11. $I = \dfrac{E}{R}$ (R)

12. $y = \dfrac{x}{a} + 3$ (x)

13. $F = \dfrac{9}{5}C + 32$ (C)

14. $pV = mRT$ (R)

7.3 Further transposing of formulae

Here are some more transposition examples to help us further understand how more difficult formulae are transposed.

Problem 8.   Transpose the formula $v = u + \dfrac{Ft}{m}$, to make F the subject

$v = u + \dfrac{Ft}{m}$ relates final velocity v, initial velocity u, force F, mass m and time t $\left(\dfrac{F}{m}\text{ is acceleration } a\right)$

Rearranging gives: $u + \dfrac{Ft}{m} = v$

and $\dfrac{Ft}{m} = v - u$

Multiplying each side by m gives:

$$m\left(\dfrac{Ft}{m}\right) = m(v - u)$$

Cancelling gives: $Ft = m(v - u)$

Dividing both sides by t gives: $\dfrac{Ft}{t} = \dfrac{m(v - u)}{t}$

Cancelling gives: $F = \dfrac{m(v - u)}{t}$ or $F = \dfrac{m}{t}(v - u)$

This shows two ways of expressing the answer. There is often more than one way of expressing a transposed answer. In this case, both of these equations for F are equivalent; neither one is more correct than the other.

Problem 9.   The final length L_2 of a piece of wire heated to $\theta°C$ is given by the formula $L_2 = L_1(1 + \alpha\theta)$ where L_1 is the original length. Make the coefficient of expansion α the subject

Rearranging gives: $L_1(1 + \alpha\theta) = L_2$

Removing the brackets gives: $L_1 + L_1\alpha\theta = L_2$

Rearranging gives: $L_1\alpha\theta = L_2 - L_1$

Dividing both sides by $L_1\theta$ gives: $\dfrac{L_1\alpha\theta}{L_1\theta} = \dfrac{L_2 - L_1}{L_1\theta}$

Cancelling gives: $\alpha = \dfrac{L_2 - L_1}{L_1\theta}$

An alternative method of transposing $L_2 = L_1(1 + \alpha\theta)$ for α is:

Dividing both sides by L_1 gives: $\dfrac{L_2}{L_1} = 1 + \alpha\theta$

Subtracting 1 from both sides gives: $\dfrac{L_2}{L_1} - 1 = \alpha\theta$

or $\alpha\theta = \dfrac{L_2}{L_1} - 1$

Dividing both sides by θ gives: $\alpha = \dfrac{\dfrac{L_2}{L_1} - 1}{\theta}$

The two answers $\alpha = \dfrac{L_2 - L_1}{L_1\,\theta}$ and $\alpha = \dfrac{\dfrac{L_2}{L_1} - 1}{\theta}$ look quite different. They are, however, equivalent. The first answer looks tidier but is no more correct than the second answer.

Problem 10.   A formula for the distance s moved by a body is given by: $s = \dfrac{1}{2}(v + u)t$. Rearrange the formula to make u the subject

Rearranging gives: $\dfrac{1}{2}(v+u)t = s$

Multiplying both sides by 2 gives: $(v+u)t = 2s$

Dividing both sides by t gives: $\dfrac{(v+u)t}{t} = \dfrac{2s}{t}$

Cancelling gives: $v+u = \dfrac{2s}{t}$

Rearranging gives: $u = \dfrac{2s}{t} - v$ or $u = \dfrac{2s - vt}{t}$

Problem 11. A formula for kinetic energy is $k = \dfrac{1}{2}mv^2$. Transpose the formula to make v the subject

Rearranging gives: $\dfrac{1}{2}mv^2 = k$

Whenever the prospective new subject is a squared term, that term is isolated on the LHS, and then the square root of both sides of the equation is taken.

Multiplying both sides by 2 gives: $mv^2 = 2k$

Dividing both sides by m gives: $\dfrac{mv^2}{m} = \dfrac{2k}{m}$

Cancelling gives: $v^2 = \dfrac{2k}{m}$

Taking the square root of both sides gives:

$$\sqrt{v^2} = \sqrt{\left(\dfrac{2k}{m}\right)}$$

i.e. $$v = \sqrt{\left(\dfrac{2k}{m}\right)}$$

Problem 12. Given $t = 2\pi\sqrt{\dfrac{l}{g}}$, find g in terms of t, l and π

Whenever the prospective new subject is within a square root sign, it is best to isolate that term on the LHS and then to square both sides of the equation.

Rearranging gives: $2\pi\sqrt{\dfrac{l}{g}} = t$

Dividing both sides by 2π gives: $\sqrt{\dfrac{l}{g}} = \dfrac{t}{2\pi}$

Squaring both sides gives: $\dfrac{l}{g} = \left(\dfrac{t}{2\pi}\right)^2 = \dfrac{t^2}{4\pi^2}$

Cross-multiplying (i.e. multiplying each term by $4\pi^2 g$), gives: $4\pi^2 l = gt^2$

or $gt^2 = 4\pi^2 l$

Dividing both sides by t^2 gives: $\dfrac{gt^2}{t^2} = \dfrac{4\pi^2 l}{t^2}$

Cancelling gives: $g = \dfrac{4\pi^2 l}{t^2}$

Now try the following Practice Exercise

Practice Exercise 35   Further transposing formulae (answers on page 535)

Make the symbol indicated in brackets the subject of each of the formulae shown, and express each in its simplest form.

1. $S = \dfrac{a}{1-r}$ (r)

2. $y = \dfrac{\lambda(x-d)}{d}$ (x)

3. $A = \dfrac{3(F-f)}{L}$ (f)

4. $y = \dfrac{AB^2}{5CD}$ (D)

5. $R = R_0(1 + \alpha t)$ (t)

6. $\dfrac{1}{R} = \dfrac{1}{R_1} + \dfrac{1}{R_2}$ (R_2)

7. $I = \dfrac{E-e}{R+r}$ (R)

8. $y = 4ab^2c^2$ (b)

9. $\dfrac{P_1 V_1}{T_1} = \dfrac{P_2 V_2}{T_2}$ (V_2)

10. $t = 2\pi\sqrt{\dfrac{L}{g}}$ (L)

11. $v^2 = u^2 + 2as$ (u)

12. $N = \sqrt{\left(\dfrac{a+x}{y}\right)}$ (a)

13. The lift force, L, on an aircraft is given by: $L = \dfrac{1}{2}\rho v^2 a c$ where ρ is the density, v is the velocity, a is the area and c is the lift coefficient. Transpose the equation to make the velocity the subject.

14. $\dfrac{P_1 V_1}{T_1} = \dfrac{P_2 V_2}{T_2}$ (T_2)

7.4 More difficult transposing of formulae

Here are some more transposition examples to help us further understand how more difficult formulae are transposed.

Problem 13. (a) Transpose $S = \sqrt{\dfrac{3d(L-d)}{8}}$ to make L the subject. (b) Evaluate L when $d = 1.65$ and $S = 0.82$

The formula $S = \sqrt{\dfrac{3d(L-d)}{8}}$ represents the sag S at the centre of a wire.

(a) Squaring both sides gives: $S^2 = \dfrac{3d(L-d)}{8}$

Multiplying both sides by 8 gives:
$$8S^2 = 3d(L-d)$$

Dividing both sides by $3d$ gives:
$$\frac{8S^2}{3d} = L - d$$

Rearranging gives: $\quad L = d + \dfrac{8S^2}{3d}$

(b) When $d = 1.65$ and $S = 0.82$,
$$L = d + \frac{8S^2}{3d} = 1.65 + \frac{8 \times 0.82^2}{3 \times 1.65} = \textbf{2.737}$$

Problem 14. Make b the subject of the formula $a = \dfrac{x-y}{\sqrt{bd+be}}$

Rearranging gives: $\quad \dfrac{x-y}{\sqrt{bd+be}} = a$

Multiplying both sides by $\sqrt{bd+be}$ gives:
$$x - y = a\sqrt{bd+be}$$

or $\quad\quad a\sqrt{bd+be} = x - y$

Dividing both sides by a gives:
$$\sqrt{bd+be} = \frac{x-y}{a}$$

Squaring both sides gives: $\quad bd + be = \left(\dfrac{x-y}{a}\right)^2$

Factorising the LHS gives: $b(d+e) = \left(\dfrac{x-y}{a}\right)^2$

Dividing both sides by $(d+e)$ gives:
$$b = \frac{\left(\dfrac{x-y}{a}\right)^2}{(d+e)} \quad \text{or} \quad b = \frac{(x-y)^2}{a^2(d+e)}$$

Problem 15. If $a = \dfrac{b}{1+b}$ make b the subject of the formula

Rearranging gives: $\quad\quad \dfrac{b}{1+b} = a$

Multiplying both sides by $(1+b)$ gives:
$$b = a(1+b)$$

Removing the brackets gives: $\quad\quad b = a + ab$

Rearranging to obtain terms in b on the LHS gives:
$$b - ab = a$$

Factorising the LHS gives: $\quad\quad b(1-a) = a$

Dividing both sides by $(1-a)$ gives: $\quad \boldsymbol{b = \dfrac{a}{1-a}}$

Problem 16. Transpose the formula $V = \dfrac{Er}{R+r}$ to make r the subject

Rearranging gives: $\quad\quad \dfrac{Er}{R+r} = V$

Multiplying both sides by $(R+r)$ gives:
$$Er = V(R+r)$$

Removing the brackets gives: $\quad\quad Er = VR + Vr$

Rearranging to obtain terms in r on the LHS gives:
$$Er - Vr = VR$$

Factorising gives: $\quad\quad r(E-V) = VR$

Dividing both sides by $(E-V)$ gives: $\quad \boldsymbol{r = \dfrac{VR}{E-V}}$

Now try the following Practice Exercise

Practice Exercise 36   Further transposing formulae (answers on page 535)

Make the symbol indicated in brackets the subject of each of the formulae shown in problems 1–6, and express each in its simplest form.

1. $y = \dfrac{a^2 m - a^2 n}{x}$ (a)

2. $M = \pi(R^4 - r^4)$ (R)

3. $x + y = \dfrac{r}{3 + r}$ (r)

4. $m = \dfrac{\mu L}{L + rCR}$ (L)

5. $a^2 = \dfrac{b^2 - c^2}{b^2}$ (b)

6. $\dfrac{x}{y} = \dfrac{1 + r^2}{1 - r^2}$ (r)

7. A formula for the focal length f of a convex lens is: $\dfrac{1}{f} = \dfrac{1}{u} + \dfrac{1}{v}$. Transpose the formula to make v the subject and evaluate v when $f = 5$ and $u = 6$

8. The quantity of heat Q is given by the formula $Q = mc(t_2 - t_1)$. Make t_2 the subject of the formula and evaluate t_2 when $m = 10$, $t_1 = 15$, $c = 4$ and $Q = 1600$

9. The velocity v of water in a pipe appears in the formula $h = \dfrac{0.03Lv^2}{2dg}$. Express v as the subject of the formula and evaluate v when $h = 0.712$, $L = 150$, $d = 0.30$ and $g = 9.81$

10. The sag S at the centre of a wire is given by the formula: $S = \sqrt{\left(\dfrac{3d(l - d)}{8}\right)}$ Make l the subject of the formula and evaluate l when $d = 1.75$ and $S = 0.80$

11. An approximate relationship between the number of teeth T on a milling cutter, the diameter of cutter D and the depth of cut d is given by: $T = \dfrac{12.5\,D}{D + 4d}$. Determine the value of D when $T = 10$ and $d = 4$ mm

12. A simply supported beam of length L has a centrally applied load F and a uniformly distributed load of w per metre length of beam. The reaction at the beam support is given by:

$$R = \frac{1}{2}(F + wL)$$

Rearrange the equation to make w the subject. Hence determine the value of w when $L = 4\,\text{m}$, $F = 8\,\text{kN}$ and $R = 10\,\text{kN}$

13. The rate of heat conduction through a slab of material, Q, is given by the formula $Q = \dfrac{kA(t_1 - t_2)}{d}$ where t_1 and t_2 are the temperatures of each side of the material, A is the area of the slab, d is the thickness of the slab, and k is the thermal conductivity of the material. Rearrange the formula to obtain an expression for t_2

14. The slip, s, of a vehicle is given by: $s = \left(1 - \dfrac{r\,\omega}{v}\right) \times 100\%$ where r is the tyre radius, ω is the angular velocity and v the velocity. Transpose to make r the subject of the formula.

15. The critical load, F newtons, of a steel column may be determined from the formula $L\sqrt{\dfrac{F}{E\,I}} = n\pi$ where L is the length, EI is the flexural rigidity, and n is a positive integer. Transpose for F and hence determine the value of F when $n = 1$, $E = 0.25 \times 10^{12}\,\text{N/m}^2$, $I = 6.92 \times 10^{-6}\,\text{m}^4$ and $L = 1.12m$

16. The flow of slurry along a pipe on a coal processing plant is given by: $V = \dfrac{\pi p r^4}{8\eta\ell}$ Transpose the equation for r

For fully worked solutions to each of the problems in Exercises 34 to 36 in this chapter, go to the website:
www.routledge.com/cw/bird

Chapter 8

Solving simultaneous equations

Why it is important to understand: **Solving simultaneous equations**

Simultaneous equations arise a great deal in engineering and science, some applications including theory of structures, data analysis, electrical circuit analysis and air traffic control. Systems that consist of a small number of equations can be solved analytically using standard methods from algebra (as explained in this chapter). Systems of large numbers of equations require the use of numerical methods and computers. Matrices are generally used to solve simultaneous equations. Solving simultaneous equations is an important skill required in all aspects of engineering.

At the end of this chapter, you should be able to:

- solve simultaneous equations in two unknowns by substitution
- solve simultaneous equations in two unknowns by elimination
- solve simultaneous equations involving practical situations

8.1 Introduction

Only one equation is necessary when finding the value of a **single unknown quantity** (as with simple equations in chapter 6). However, when an equation contains **two unknown quantities** it has an infinite number of solutions. When two equations are available connecting the same two unknown values then a unique solution is possible. Similarly, for three unknown quantities it is necessary to have three equations in order to solve for a particular value of each of the unknown quantities, and so on.

Equations which have to be solved together to find the unique values of the unknown quantities, which are true for each of the equations, are called **simultaneous equations**.

Science and Mathematics for Engineering. 978-0-367-20475-4, © John Bird. Published by Taylor & Francis. All rights reserved.

Two methods of solving simultaneous equations analytically are: (a) by **substitution**, and (b) by **elimination**.

8.2 Solving simultaneous equations in two unknowns

The method of solving simultaneous equations is demonstrated in the following worked problems.

Problem 1. Solve the following equations for x and y, (a) by substitution, and (b) by elimination:

$$x + 2y = -1 \qquad (1)$$
$$4x - 3y = 18 \qquad (2)$$

(a) **By substitution**

From equation (1): $x = -1 - 2y$
Substituting this expression for x into equation (2) gives:

$$4(-1 - 2y) - 3y = 18$$

This is now a simple equation in y.
Removing the brackets gives:

$$-4 - 8y - 3y = 18$$
$$-11y = 18 + 4 = 22$$
$$y = \frac{22}{-11} = -2$$

Substituting $y = -2$ into equation (1) gives:

$$x + 2(-2) = -1$$
$$x - 4 = -1$$
$$x = -1 + 4 = 3$$

Thus, $x = 3$ and $y = -2$ is the solution to the simultaneous equations.
(Check: In equation (2), since $x = 3$ and $y = -2$,

$$\text{LHS} = 4(3) - 3(-2) = 12 + 6 = 18 = \text{RHS.})$$

(b) **By elimination**

$$x + 2y = -1 \qquad (1)$$
$$4x - 3y = 18 \qquad (2)$$

If equation (1) is multiplied throughout by 4 the coefficient of x will be the same as in equation (2), giving:

$$4x + 8y = -4 \qquad (3)$$

Subtracting equation (3) from equation (2) gives:

$$4x - 3y = 18 \qquad (2)$$
$$\underline{4x + 8y = -4} \qquad (3)$$
$$0 - 11y = 22$$

Hence, $\quad y = \dfrac{22}{-11} = -2$

(Note, in the above subtraction,

$$18 - -4 = 18 + 4 = 22)$$

Substituting $y = -2$ into either equation (1) or equation (2) will give $x = 3$ as in method (a). The solution $x = 3$, $y = -2$ is the only pair of values that satisfies both of the original equations.

Problem 2. Solve, by a substitution method, the simultaneous equations

$$3x - 2y = 12 \qquad (1)$$
$$x + 3y = -7 \qquad (2)$$

From equation (2), $x = -7 - 3y$
Substituting for x in equation (1) gives:

$$3(-7 - 3y) - 2y = 12$$
i.e. $\qquad -21 - 9y - 2y = 12$
$$-11y = 12 + 21 = 33$$

Hence, $y = \dfrac{33}{-11} = -3$
Substituting $y = -3$ in equation (2) gives:

$$x + 3(-3) = -7$$
i.e. $\qquad x - 9 = -7$
Hence $\qquad x = -7 + 9 = 2$

Thus, $x = 2, y = -3$ is the solution of the simultaneous equations.
(Such solutions should always be checked by substituting values into each of the original two equations.)

Problem 3.   Use an elimination method to solve the simultaneous equations

$$3x + 4y = 5 \qquad (1)$$

$$2x - 5y = -12 \qquad (2)$$

If equation (1) is multiplied throughout by 2 and equation (2) by 3, then the coefficient of x will be the same in the newly formed equations. Thus:

$$2 \times \text{equation (1) gives :} \qquad 6x + 8y = 10 \qquad (3)$$

$$3 \times \text{equation (2) gives :} \qquad 6x - 15y = -36 \qquad (4)$$

Equation (3) – equation (4) gives:

$$0 + 23y = 46$$

i.e. $$y = \frac{46}{23} = 2$$

(Note $+ 8y - -15y = 8y + 15y = 23y$ and $10 - -36 = 10 + 36 = 46$)

Substituting $y = 2$ in equation (1) gives:

$$3x + 4(2) = 5$$

from which $$3x = 5 - 8 = -3$$

and $$x = -1$$

Checking, by substituting $x = -1$ and $y = 2$ in equation (2), gives:

$$\text{LHS} = 2(-1) - 5(2) = -2 - 10 = -12 = \text{RHS}$$

Hence, $x = -1$ and $y = 2$ is the solution of the simultaneous equations.

The elimination method is the most common method of solving simultaneous equations.

Problem 4.   Solve

$$7x - 2y = 26 \qquad (1)$$

$$6x + 5y = 29 \qquad (2)$$

When equation (1) is multiplied by 5 and equation (2) by 2 the coefficients of y in each equation are numerically the same, i.e. 10, but are of opposite sign.

$$5 \times \text{equation (1) gives :} \quad 35x - 10y = 130 \quad (3)$$

$$2 \times \text{equation (2) gives :} \quad 12x + 10y = 58 \quad (4)$$

Adding equations (3) and (4) gives: $$47x + 0 = 188$$

Hence, $$x = \frac{188}{47} = 4$$

[Note that when the signs of common coefficients are **different** the two equations are **added**, and when the signs of common coefficients are the **same** the two equations are **subtracted** (as in Problems 1 and 3)]

Substituting $x = 4$ in equation (1) gives:

$$7(4) - 2y = 26$$

$$28 - 2y = 26$$

$$28 - 26 = 2y$$

$$2 = 2y$$

Hence, $$y = 1$$

Checking, by substituting $x = 4$ and $y = 1$ in equation (2), gives:

$$\text{LHS} = 6(4) + 5(1) = 24 + 5 = 29 = \text{RHS}$$

Thus, **the solution is $x = 4$, $y = 1$**

Now try the following Practice Exercise

Practice Exercise 37   Solving simultaneous equations (answers on page 535)

Solve the following simultaneous equations and verify the results.

1. $2x - y = 6$
    $x + y = 6$

2. $2x - y = 2$
    $x - 3y = -9$

3. $x - 4y = -4$
    $5x - 2y = 7$

4. $3x - 2y = 10$
    $5x + y = 21$

5. $2x - 7y = -8$
    $3x + 4y = 17$

6. $a + 2b = 8$
    $b - 3a = -3$

7. $a + b = 7$
    $a - b = 3$

8. $2x + 5y = 7$
    $x + 3y = 4$

9. $3s + 2t = 12$
    $4s - t = 5$

10. $3x - 2y = 13$
    $2x + 5y = -4$

11. $5m - 3n = 11$ 12. $8a - 3b = 51$

   $3m + n = 8$ $3a + 4b = 14$

8.3 Further solving of simultaneous equations

Here are some further worked problems on solving simultaneous equations.

Problem 5.   Solve

$$3p = 2q \tag{1}$$
$$4p + q + 11 = 0 \tag{2}$$

Rearranging gives:

$$3p - 2q = 0 \tag{3}$$
$$4p + q = -11 \tag{4}$$

Multiplying equation (4) by 2 gives:

$$8p + 2q = -22 \tag{5}$$

Adding equations (3) and (5) gives:

$$11p + 0 = -22$$
$$p = \frac{-22}{11} = -2$$

Substituting $p = -2$ into equation (1) gives:

$$3(-2) = 2q$$
$$-6 = 2q$$
$$q = \frac{-6}{2} = -3$$

Checking, by substituting $p = -2$ and $q = -3$ into equation (2) gives:

LHS $= 4(-2) + (-3) + 11 = -8 - 3 + 11 = 0 =$ RHS

Hence, **the solution is $p = -2$, $q = -3$**

Problem 6.   Solve

$$\frac{x}{8} + \frac{5}{2} = y \tag{1}$$
$$13 - \frac{y}{3} = 3x \tag{2}$$

Whenever fractions are involved in simultaneous equations it is often easier to first remove them. Thus, multiplying equation (1) by 8 gives:

$$8\left(\frac{x}{8}\right) + 8\left(\frac{5}{2}\right) = 8y$$

i.e. $x + 20 = 8y \tag{3}$

Multiplying equation (2) by 3 gives:

$$39 - y = 9x \tag{4}$$

Rearranging equations (3) and (4) gives:

$$x - 8y = -20 \tag{5}$$
$$9x + y = 39 \tag{6}$$

Multiplying equation (6) by 8 gives:

$$72x + 8y = 312 \tag{7}$$

Adding equations (5) and (7) gives:

$$73x + 0 = 292$$
$$x = \frac{292}{73} = \mathbf{4}$$

Substituting $x = 4$ into equation (5) gives:

$$4 - 8y = -20$$
$$4 + 20 = 8y$$
$$24 = 8y$$
$$y = \frac{24}{8} = \mathbf{3}$$

Checking: substituting $x = 4$, $y = 3$ in the original equations, gives:

Equation (1): LHS $= \dfrac{4}{8} + \dfrac{5}{2} = \dfrac{1}{2} + 2\dfrac{1}{2} = 3 = y =$ RHS

Equation (2): LHS $= 13 - \dfrac{3}{3} = 13 - 1 = 12$

RHS $= 3x = 3(4) = 12$

Hence, **the solution is $x = 4$, $y = 3$**

Now try the following Practice Exercise

Practice Exercise 38   Solving simultaneous equations (answers on page 535)

Solve the following simultaneous equations and verify the results.

1. $7p + 11 + 2q = 0$
   $-1 = 3q - 5p$

2. $\dfrac{x}{2} + \dfrac{y}{3} = 4$
   $\dfrac{x}{6} - \dfrac{y}{9} = 0$

3. $\dfrac{a}{2} - 7 = -2b$
   $12 = 5a + \dfrac{2}{3}b$

4. $\dfrac{3}{2}s - 2t = 8$
   $\dfrac{s}{4} + 3t = -2$

5. $\dfrac{x}{5} + \dfrac{2y}{3} = \dfrac{49}{15}$
   $\dfrac{3x}{7} - \dfrac{y}{2} + \dfrac{5}{7} = 0$

6. $v - 1 = \dfrac{u}{12}$
   $u + \dfrac{v}{4} - \dfrac{25}{2} = 0$

8.4   Practical problems involving simultaneous equations

There are a number of situations in engineering and science where the solution of simultaneous equations is required. Some are demonstrated in the following worked problems.

Problem 7.   The law connecting friction F and load L for an experiment is of the form $F = aL + b$, where a and b are constants. When $F = 5.6\,\text{N}$, $L = 8.0$ N, and when $F = 4.4\,\text{N}$, $L = 2.0\,\text{N}$. Find the values of a and b and the value of F when $L = 6.5\,\text{N}$

Substituting $F = 5.6$, $L = 8.0$ into $F = aL + b$ gives:

$$5.6 = 8.0a + b \qquad (1)$$

Substituting $F = 4.4$, $L = 2.0$ into $F = aL + b$ gives:

$$4.4 = 2.0a + b \qquad (2)$$

Subtracting equation (2) from equation (1) gives:

$$1.2 = 6.0a$$

$$a = \frac{1.2}{6.0} = \frac{1}{5} \text{ or } \mathbf{0.2}$$

Substituting $a = \dfrac{1}{5}$ into equation (1) gives:

$$5.6 = 8.0\left(\frac{1}{5}\right) + b$$

$$5.6 = 1.6 + b$$

$$5.6 - 1.6 = b$$

i.e. $\qquad\qquad\qquad \mathbf{b = 4}$

Checking, substituting $a = \dfrac{1}{5}$ and $b = 4$ in equation (2), gives:

$$\text{RHS} = 2.0\left(\frac{1}{5}\right) + 4 = 0.4 + 4 = 4.4 = \text{LHS}$$

Hence, $\boldsymbol{a = \dfrac{1}{5}}$ **and** $\boldsymbol{b = 4}$

When $L = 6.5$, $F = aL + b = \dfrac{1}{5}(6.5) + 4 = 1.3 + 4$, i.e. $F = 5.30\,\text{N}$

Problem 8.   The distance s metres from a fixed point of a vehicle travelling in a straight line with constant acceleration, $a\,\text{m/s}^2$, is given by $s = ut + \dfrac{1}{2}at^2$, where u is the initial velocity in m/s and t the time in seconds. Determine the initial velocity and the acceleration given that $s = 42\,\text{m}$ when $t = 2\,\text{s}$, and $s = 144\,\text{m}$ when $t = 4\,\text{s}$. Find also the distance travelled after 3 s

Substituting $s = 42$, $t = 2$ into $s = ut + \dfrac{1}{2}at^2$ gives:

$$42 = 2u + \frac{1}{2}a(2)^2$$

i.e. $\qquad\qquad 42 = 2u + 2a \qquad (1)$

Substituting $s = 144$, $t = 4$ into $s = ut + \dfrac{1}{2}at^2$ gives:

$$144 = 4u + \frac{1}{2}a(4)^2$$

i.e. $\qquad\qquad 144 = 4u + 8a \qquad (2)$

Multiplying equation (1) by 2 gives:

$$84 = 4u + 4a \qquad (3)$$

Subtracting equation (3) from equation (2) gives:

$$60 = 0 + 4a$$

and $$a = \frac{60}{4} = 15$$

Substituting $a = 15$ into equation (1) gives:

$$42 = 2u + 2(15)$$

$$42 - 30 = 2u$$

$$u = \frac{12}{2} = 6$$

Substituting $a = 15$, $u = 6$ in equation (2) gives:

RHS $= 4(6) + 8(15) = 24 + 120 = 144 =$ LHS

Hence, **the initial velocity, $u = 6$ m/s and the acceleration, $a = 15$ m/s^2**

Distance travelled after 3 s is given by $s = ut + \frac{1}{2}at^2$, where $t = 3$, $u = 6$ and $a = 15$

Hence, $s = (6)(3) + \frac{1}{2}(15)(3)^2 = 18 + 67.5$,

i.e. **distance travelled after 3 s = 85.5 m**

Problem 9. The molar heat capacity of a solid compound is given by the equation $c = a + bT$, where a and b are constants. When $c = 52$, $T = 100$, and when $c = 172$, $T = 400$. Determine the values of a and b

When $c = 52$, $T = 100$, hence

$$52 = a + 100b \qquad (1)$$

When $c = 172$, $T = 400$, hence

$$172 = a + 400b \qquad (2)$$

Equation (2) – equation (1) gives:

$$120 = 300b$$

from which, $$b = \frac{120}{300} = 0.4$$

Substituting $b = 0.4$ in equation (1) gives:

$$52 = a + 100(0.4)$$

$$a = 52 - 40 = 12$$

Hence, **$a = 12$ and $b = 0.4$**

Now try the following Practice Exercise

Practice Exercise 39   Practical problems involving simultaneous equations (answers on page 535)

1. In a system of pulleys, the effort P required to raise a load W is given by $P = aW + b$, where a and b are constants. If $W = 40$ when $P = 12$ and $W = 90$ when $P = 22$, find the values of a and b

2. Applying Kirchhoff's laws to an electrical circuit produces the following equations:

$$5 = 0.2I_1 + 2(I_1 - I_2)$$
$$12 = 3I_2 + 0.4I_2 - 2(I_1 - I_2)$$

Determine the values of currents I_1 and I_2

3. Velocity v is given by the formula $v = u + at$. If $v = 20$ when $t = 2$ and $v = 40$ when $t = 7$ find the values of u and a. Hence find the velocity when $t = 3.5$

4. $y = mx + c$ is the equation of a straight line of slope m and y-axis intercept c. If the line passes through the point where $x = 2$ and $y = 2$, and also through the point where $x = 5$ and $y = 0.5$, find the slope and y-axis intercept of the straight line.

5. The molar heat capacity of a solid compound is given by the equation $c = a + bT$. When $c = 52$, $T = 100$, and when $c = 172$, $T = 400$. Find the values of a and b

6. In a system of forces, the relationship between two forces F_1 and F_2 is given by:

$$5F_1 + 3F_2 + 6 = 0$$
$$3F_1 + 5F_2 + 18 = 0$$

Solve for F_1 and F_2

7. For a balanced beam, the equilibrium of forces is given by: $R_1 + R_2 = 12.0$ kN As a result of taking moments: $0.2R_1 + 7 \times 0.3 + 3 \times 0.6 = 0.8R_2$ Determine the values of the reaction forces R_1 and R_2

For fully worked solutions to each of the problems in Exercises 37 to 39 in this chapter, go to the website: www.routledge.com/cw/bird

Chapter 9

Logarithms and exponential functions

Why it is important to understand: **Logarithms and exponential functions**

All types of engineers use natural and common logarithms. Chemical engineers use them to measure radioactive decay and pH solutions, both of which are measured on a logarithmic scale. The Richter scale which measures earthquake intensity is a logarithmic scale. Biomedical engineers use logarithms to measure cell decay and growth, and also to measure light intensity for bone mineral density measurements. In electrical engineering, a dB (decibel) scale is very useful for expressing attenuations in radio propagation and circuit gains, and logarithms are used for implementing arithmetic operations in digital circuits.

Exponential functions are used in engineering, physics, biology and economics. There are many quantities that grow exponentially; some examples are population, compound interest and charge in a capacitor. With exponential growth, the rate of growth increases as time increases. We also have exponential decay; some examples are radioactive decay, atmospheric pressure, Newton's law of cooling and linear expansion. Understanding and using logarithms and exponential functions is therefore important in many branches of science and engineering.

At the end of this chapter, you should be able to:

- define base, power, exponent, index and logarithm
- distinguish between common and Napierian (i.e. hyperbolic or natural) logarithms
- state the laws of logarithms
- simplify logarithmic expressions
- solve equations involving logarithms
- solve indicial equations
- sketch graphs of $\log_{10} x$ and $\log_e x$
- evaluate exponential functions using a calculator
- plot graphs of exponential functions
- evaluate Napierian logarithms using a calculator
- solve equations involving Napierian logarithms
- appreciate the many examples of laws of growth and decay in engineering and science
- perform calculations involving the laws of growth and decay

Science and Mathematics for Engineering. 978-0-367-20475-4, © John Bird. Published by Taylor & Francis. All rights reserved.

9.1 Introduction to logarithms

From chapter 3, we know that: $16 = 2^4$
The number 4 is called the **power** or the **exponent** or the **index**. In the expression 2^4, the number 2 is called the **base**.
In another example, we know that: $64 = 8^2$
In this example, 2 is the power, or exponent, or index. The number 8 is the base.

What is a logarithm?

Consider the expression $16 = 2^4$
An alternative, yet equivalent, way of writing this expression is: $\log_2 16 = 4$
This is stated as 'log to the base 2 of 16 equals 4'.
We see that the logarithm is the same as the power or index in the original expression. It is the base in the original expression which becomes the base of the logarithm.

The two statements: $16 = 2^4$ and

$\log_2 16 = 4$ are equivalent.

If we write either of them, we are automatically implying the other.
In general, if a number y can be written in the form a^x, then the index x is called the 'logarithm of y to the base of a',

i.e.   **if $y = a^x$ then $x = \log_a y$**

If we write down that $64 = 8^2$ then the equivalent statement using logarithms is: $\log_8 64 = 2$
In another example, if we write down that: $\log_3 27 = 3$ then the equivalent statement using powers is: $3^3 = 27$
So the two sets of statements, one involving powers and one involving logarithms, are equivalent.

Common logarithms

From above, if we write down that: $1000 = 10^3$, then $3 = \log_{10} 1000$
This may be checked using the 'log' button on your calculator.
Logarithms having a base of 10 are called **common logarithms** and $\log_{10}$ is usually abbreviated to lg.
The following values may be checked by using a calculator:

$\lg 27.5 = 1.4393\ldots,\quad \lg 378.1 = 2.5776\ldots$ and
$\lg 0.0204 = -1.6903\ldots$

Napierian logarithms

Logarithms having a base of e (where 'e' is a mathematical constant approximately equal to 2.7183) are called **hyperbolic, Napierian** or **natural logarithms**, and $\log_e$ is usually abbreviated to ln.
The following values may be checked by using a calculator:

$\ln 3.65 = 1.2947\ldots,\quad \ln 417.3 = 6.0338\ldots$ and
$\ln 0.182 = -1.7037\ldots$

More on Napierian logarithms is explained in Section 9.7.
Logarithms to any base may be evaluated using a scientific notation calculator. In science and engineering, evaluating common logarithms (i.e. 'log' on the calculator) and Napierian logarithms (i.e. 'ln' on the calculator) are the most commonly used.

9.2 Laws of logarithms

There are three laws of logarithms, which apply to any base:

(i)   To multiply two numbers:
  $\log (A \times B) = \log A + \log B$

(ii) To divide two numbers:
  $\log\left(\dfrac{A}{B}\right) = \log A - \log B$

(iii) To raise a number to a power: $\log A^n = n \log A$

Here are some worked problems to help understanding of the laws of logarithms.

Problem 1. Write $\log 4 + \log 7$ as the logarithm of a single number

$\log 4 + \log 7 = \log(7\times4)$ by the first law of logarithms
  $= \log 28$

Problem 2. Write $\log 16 - \log 2$ as the logarithm of a single number

$\log 16 - \log 2 = \log\left(\dfrac{16}{2}\right)$ by the second law
  of logarithms
  $= \log 8$

Problem 3. Write $2\log 3$ as the logarithm of a single number

$2\log 3 = \log 3^2$ by the third law of logarithms

$\qquad = \mathbf{log\ 9}$

Problem 4. Write $\frac{1}{2}\log 25$ as the logarithm of a single number

$\frac{1}{2}\log 25 = \log 25^{\frac{1}{2}}$ by the third law of logarithms

$\qquad = \log \sqrt{25} = \mathbf{log\ 5}$

Problem 5. Simplify: $\log 64 - \log 128 + \log 32$

$64 = 2^6$, $128 = 2^7$ and $32 = 2^5$

Hence, $\log 64 - \log 128 + \log 32$

$= \log 2^6 - \log 2^7 + \log 2^5$

$= 6\log 2 - 7\log 2 + 5\log 2$ by the third law

$\qquad\qquad\qquad\qquad\qquad$ of logarithms

$= \mathbf{4\ log\ 2}$

Problem 6. Solve the equation:
$\log(x-1) + \log(x+8) = 2\log(x+2)$

$\text{LHS} = \log(x-1) + \log(x+8) = \log(x-1)(x+8)$

$\qquad\qquad\qquad$ from the first law of logarithms

$\qquad = \log(x^2 + 7x - 8)$

$\text{RHS} = 2\log(x+2) = \log(x+2)^2$

$\qquad\qquad\qquad$ from the third law of logarithms

$\qquad = \log(x^2 + 4x + 4)$

Hence, $\log(x^2 + 7x - 8) = \log(x^2 + 4x + 4)$

from which, $x^2 + 7x - 8 = x^2 + 4x + 4$

i.e. $\qquad\qquad 7x - 8 = 4x + 4$

i.e. $\qquad\qquad 3x = 12$

and $\qquad\qquad \mathbf{x = 4}$

Problem 7. Solve the equation: $\frac{1}{2}\log 4 = \log x$

$\frac{1}{2}\log 4 = \log 4^{\frac{1}{2}}$ from the third law of logarithms

$\qquad = \log \sqrt{4}$ from the laws of indices

Hence, $\quad \frac{1}{2}\log 4 = \log x$

becomes $\log \sqrt{4} = \log x$

i.e. $\qquad \log 2 = \log x$

from which, $\quad 2 = x$

i.e. the **solution of the equation is: x = 2**

Now try the following Practice Exercise

**Practice Exercise 40 Laws of logarithms
(Answers on page 536)**

In Problems 1 to 10, write as the logarithm of a single number:

1. $\log 2 + \log 3$

2. $\log 3 + \log 5$

3. $\log 3 + \log 4 - \log 6$

4. $\log 7 + \log 21 - \log 49$

5. $2\log 2 + \log 3$

6. $2\log 2 + 3\log 5$

7. $2\log 2 + \log 5 - \log 10$

8. $\log 27 - \log 9 + \log 81$

9. $\log 64 + \log 32 - \log 128$

10. $\log 8 - \log 4 + \log 32$

Solve the equations given in Problems 11 to 15:

11. $\log x^4 - \log x^3 = \log 5x - \log 2x$

12. $\log 2t^3 - \log t = \log 16 + \log t$

13. $2\log b^2 - 3\log b = \log 8b - \log 4b$

14. $\log(x+1) + \log(x-1) = \log 3$

15. $\frac{1}{3}\log 27 = \log(0.5a)$

9.3 Indicial equations

The laws of logarithms may be used to solve certain equations involving powers – called **indicial equations**. For example, to solve, say, $3^x = 27$, logarithms to a base of 10 are taken of both sides,

i.e. $\log_{10} 3^x = \log_{10} 27$

and $x \log_{10} 3 = \log_{10} 27$ by the third law of logarithms

Rearranging gives: $x = \dfrac{\log_{10} 27}{\log_{10} 3} = \dfrac{1.43136\ldots}{0.47712\ldots} = 3$

which may be readily checked.

(Note, $\dfrac{\log 27}{\log 3}$ is **not** equal to $\log \dfrac{27}{3}$)

Problem 8. Solve the equation: $2^x = 5$, correct to 4 significant figures

Taking logarithms to base 10 of both sides of $2^x = 5$ gives:

$$\log_{10} 2^x = \log_{10} 5$$

i.e. $x \log_{10} 2 = \log_{10} 5$ by the third law of logarithms

Rearranging gives: $\mathbf{x} = \dfrac{\log_{10} 5}{\log_{10} 2} = \dfrac{0.6989700\ldots}{0.3010299\ldots} =$

$\quad$ **2.322**, correct to 4 significant figures.

Problem 9. Solve the equation: $x^{2.7} = 34.68$, correct to 4 significant figures

Taking logarithms to base 10 of both sides gives:

$$\log_{10} x^{2.7} = \log_{10} 34.68$$

$$2.7 \log_{10} x = \log_{10} 34.68$$

Hence, $\log_{10} x = \dfrac{\log_{10} 34.68}{2.7} = 0.57040$

Thus, $\mathbf{x} = $ antilog $0.57040 = 10^{0.57040} = \mathbf{3.719}$, correct to 4 significant figures.

Problem 10. A gas follows the polytropic law $PV^{1.25} = C$. Determine the new volume of the gas, given that its original pressure and volume are 101 kPa and 0.35m^3, respectively, and its final pressure is 1.18 MPa

If $PV^{1.25} = C$ then $P_1 V_1{}^{1.25} = P_2 V_2{}^{1.25}$

$P_1 = 101$ kPa, $P_2 = 1.18$ MPa and $V_1 = 0.35$m^3

$$P_1 V_1{}^{1.25} = P_2 V_2{}^{1.25}$$

i.e. $(101 \times 10^3)(0.35)^{1.25} = (1.18 \times 10^6) V_2{}^{1.25}$

from which, $V_2{}^{1.25} = \dfrac{(101 \times 10^3)(0.35)^{1.25}}{(1.18 \times 10^6)}$

$$= 0.02304$$

Taking logarithms of both sides of the equation gives:

$$\log_{10} V_2{}^{1.25} = \log_{10} 0.02304$$

i.e. $1.25 \log_{10} V_2 = \log_{10} 0.02304$ from the third law of logarithms

and $\qquad \log_{10} V_2 = \dfrac{\log_{10} 0.02304}{1.25} = -1.3100$

from which, **volume, $V_2 = 10^{-1.3100} = \mathbf{0.049}$ m^3**

Now try the following Practice Exercise

Practice Exercise 41 Indicial equations (Answers on page 536)

In problems 1 to 6, solve the indicial equations for x, each correct to 4 significant figures:

1. $3^x = 6.4$
2. $2^x = 9$
3. $x^{1.5} = 14.91$
4. $25.28 = 4.2^x$
5. $x^{-0.25} = 0.792$
6. $0.027^x = 3.26$
7. The decibel gain n of an amplifier is given by: $n = 10 \log_{10}\left(\dfrac{P_2}{P_1}\right)$ where P_1 is the power input and P_2 is the power output. Find the power gain $\dfrac{P_2}{P_1}$ when n = 25 decibels.
8. A gas follows the polytropic law $PV^{1.26} = C$. Determine the new volume of the gas, given that its original pressure and volume are 101 kPa and 0.42m^3, respectively, and its final pressure is 1.25 MPa.

9. With measuring earthquakes, the Richter local magnitude, ML is given by:

$$ML = \log_{10}(A) + 2.56\log_{10}(D) - 1.67$$

where A is the measured ground motion, in micrometres, and D is the distance from the event, in kilometres. Calculate ML, given $A = 5\mu m$ and $D = 400,000$ m.

9.4 Graphs of logarithmic functions

A graph of $y = \log_{10} x$ is shown in Figure 9.1 and a graph of $y = \log_e x$ is shown in Figure 9.2. Both are seen to be of similar shape; in fact, the same general shape occurs for a logarithm to any base.

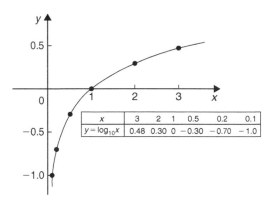

x	3	2	1	0.5	0.2	0.1
$y = \log_{10}x$	0.48	0.30	0	−0.30	−0.70	−1.0

Figure 9.1

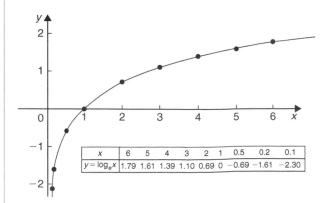

x	6	5	4	3	2	1	0.5	0.2	0.1
$y = \log_e x$	1.79	1.61	1.39	1.10	0.69	0	−0.69	−1.61	−2.30

Figure 9.2

9.5 Exponential functions

An exponential function is one which contains e^x, e being a constant called the exponent and having an approximate value of 2.7183. The exponent arises from the natural laws of growth and decay and is used as a base for natural or Napierian logarithms.

The most common method of evaluating an exponential function is by using a scientific notation **calculator**. Use your calculator to check the following values:

$e^1 = 2.7182818$, correct to 8 significant figures,

$e^{-1.618} = 0.1982949$, each correct to 7 significant figures,

$e^{0.12} = 1.1275$, correct to 5 significant figures,

$e^{-2.785} = 0.0617291$, correct to 7 decimal places.

Problem 11. Evaluate the following correct to 4 decimal places, using a calculator:

$$0.0256(e^{5.21} - e^{2.49})$$

$0.0256(e^{5.21} - e^{2.49})$

$$= 0.0256(183.094058... - 12.0612761...)$$

$$= \textbf{4.3784}, \text{ correct to 4 decimal places.}$$

Problem 12. Evaluate the following correct to 4 decimal places, using a calculator:

$$5\left(\frac{e^{0.25} - e^{-0.25}}{e^{0.25} + e^{-0.25}}\right)$$

$5\left(\dfrac{e^{0.25} - e^{-0.25}}{e^{0.25} + e^{-0.25}}\right)$

$$= 5\left(\frac{1.28402541... - 0.77880078...}{1.28402541... + 0.77880078...}\right)$$

$$= 5\left(\frac{0.5052246}{2.0628262}\right)$$

$$= \textbf{1.2246,} \text{ correct to 4 decimal places.}$$

Problem 13. The instantaneous voltage v in a capacitive circuit is related to time t by the equation: $v = Ve^{-t/CR}$ where V, C and R are constants. Determine v, correct to 4 significant figures, when $t = 50$ ms, $C = 10\ \mu F$, $R = 47$ kΩ and $V = 300$ volts

$$v = Ve^{-t/CR} = 300e^{(-50\times10^{-3})/(10\times10^{-6}\times47\times10^{3})}$$

Using a calculator, $v = 300e^{-0.1063829...}$

$$= 300(0.89908025...) = \textbf{269.7 volts}$$

Now try the following Practice Exercise

Practice Exercise 42 Evaluating exponential functions (Answers on page 536)

1. Evaluate the following, correct to 4 significant figures:

   (a) $e^{-1.8}$ (b) $e^{-0.78}$ (c) e^{10}

2. Evaluate the following, correct to 5 significant figures:

   (a) $e^{1.629}$ (b) $e^{-2.7483}$ (c) $0.62e^{4.178}$

In Problems 3 and 4, evaluate correct to 5 decimal places:

3. (a) $\frac{1}{7}e^{3.4629}$ (b) $8.52e^{-1.2651}$ (c) $\frac{5e^{2.6921}}{3e^{1.1171}}$

4. (a) $\frac{5.6823}{e^{-2.1347}}$ (b) $\frac{e^{2.1127} - e^{-2.1127}}{2}$

   (c) $\frac{4(e^{-1.7295} - 1)}{e^{3.6817}}$

5. The length of a bar, l, at a temperature θ is given by $l = l_0 e^{\alpha\theta}$, where l_0 and α are constants.

   Evaluate l, correct to 4 significant figures, where $l_0 = 2.587$, $\theta = 321.7$ and $\alpha = 1.771 \times 10^{-4}$.

6. When a chain of length 2L is suspended from two points, 2D metres apart, on the same horizontal level: $D = k\left\{\ln\left(\frac{L + \sqrt{L^2 + k^2}}{k}\right)\right\}$
   Evaluate D when $k = 75$ m and $L = 180$ m.

9.6 Graphs of exponential functions

Values of e^x and e^{-x} obtained from a calculator, correct to 2 decimal places, over a range $x = -3$ to $x = 3$, are shown in the following table.

x	−3.0	−2.5	−2.0	−1.5	−1.0	−0.5	0
e^x	0.05	0.08	0.14	0.22	0.37	0.61	1.00
e^{-x}	20.09	12.18	7.39	4.48	2.72	1.65	1.00

x	0.5	1.0	1.5	2.0	2.5	3.0
e^x	1.65	2.72	4.48	7.39	12.18	20.09
e^{-x}	0.61	0.37	0.22	0.14	0.08	0.05

Figure 9.3 shows graphs of $y = e^x$ and $y = e^{-x}$

Problem 14. Plot a graph of $y = 2e^{0.3x}$ over a range of $x = -2$ to $x = 3$. Hence determine the value of y when $x = 2.2$ and the value of x when $y = 1.6$

A table of values is drawn up as shown below.

x	−3	−2	−1	0	1	2	3
$2e^{0.3x}$	0.81	1.10	1.48	2.00	2.70	3.64	4.92

A graph of $y = 2e^{0.3x}$ is shown plotted in Figure 9.4.

From the graph, **when x = 2.2, y = 3.87** and **when y = 1.6, x = −0.74**

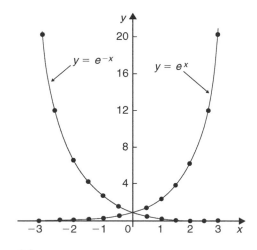

Figure 9.3

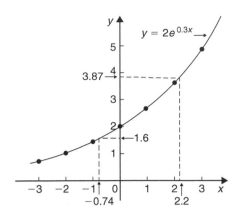

Figure 9.4

Problem 15. The decay of voltage, v volts, across a capacitor at time t seconds is given by $v = 250e^{-t/3}$ Draw a graph showing the natural decay curve over the first 6 seconds. From the graph, find (a) the voltage after 3.4 s, and (b) the time when the voltage is 150 V.

A table of values is drawn up as shown below.

t	0	1	2	3
$e^{-t/3}$	1.00	0.7165	0.5134	0.3679
$v = 250e^{-t/3}$	250.0	179.1	128.4	91.97

t	4	5	6
$e^{-t/3}$	0.2636	0.1889	0.1353
$v = 250e^{-t/3}$	65.90	47.22	33.83

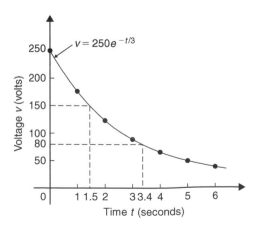

Figure 9.5

The natural decay curve of $v = 250e^{-t/3}$ is shown in Figure 9.5.   **Figure 9.5**

From the graph:

(a) **when time t = 3.4 s, voltage v = 80 volts**

and (b) **when voltage v = 150 volts, time t = 1.5 seconds**

Now try the following Practice Exercise

Practice Exercise 43 Exponential graphs (Answers on page 536)

1. Plot a graph of y = $3e^{0.2x}$ over the range $x = -3$ to $x = 3$. Hence determine the value of y when $x = 1.4$ and the value of x when y = 4.5

2. Plot a graph of $y = \frac{1}{2}e^{-1.5x}$ over a range $x = -1.5$ to $x = 1.5$ and hence determine the value of y when x = -0.8 and the value of x when y = 3.5

3. In a chemical reaction the amount of starting material C cm^3 left after t minutes is given by: $C = 40e^{-0.006t}$. Plot a graph of C against t and determine (a) the concentration C after 1 hour, and (b) the time taken for the concentration to decrease by half.

4. The rate at which a body cools is given by $\theta = 250e^{-0.05t}$ where the excess of temperature of a body above its surroundings at time t minutes is $\theta°C$. Plot a graph showing the natural decay curve for the first hour of cooling. Hence determine (a) the temperature after 25 minutes, and (b) the time when the temperature is 195°C.

9.7   Napierian logarithms

As mentioned in Section 9.1, logarithms having a base of 'e' are called **hyperbolic, Napierian** or **natural logarithms** and the Napierian logarithm of x is written as $\log_e x$, or more commonly as ln x. Logarithms were invented by John Napier, a Scotsman (1550–1617).

The most common method of evaluating a Napierian logarithm is by a scientific notation **calculator**. Use your calculator to check the following values:

ln 4.328 = 1.46510554 ... = 1.4651, correct to 4 decimal places

$\ln 1 = 0$

$\ln 527 = 6.2672$, correct to 5 significant figures

$\ln 0.17 = -1.772$, correct to 4 significant figures

$\ln e^3 = 3$

$\ln e^1 = 1$

From the last two examples we can conclude that:

$$\log_e e^x = x$$

This is useful when **solving equations involving exponential functions**.

For example, to solve $e^{3x} = 7$, take Napierian logarithms of both sides, which gives:

$$\ln e^{3x} = \ln 7$$

i.e. $3x = \ln 7$

from which $x = \dfrac{1}{3}\ln 7 = \mathbf{0.6486}$,

correct to 4 decimal places.

Problem 16.   Evaluate the following, each correct to 5 significant figures:

(a) $\dfrac{1}{2}\ln 4.7291$

(b) $\dfrac{\ln 7.8693}{7.8693}$

(c) $\dfrac{3.17\ln 24.07}{e^{-0.1762}}$

(a) $\dfrac{1}{2}\ln 4.7291 = \dfrac{1}{2}(1.5537349\ldots) = \mathbf{0.77687}$,

correct to 5 significant figures

(b) $\dfrac{\ln 7.8693}{7.8693} = \dfrac{2.06296911\ldots}{7.8693} = \mathbf{0.26215}$, correct

to 5 significant figures

(c) $\dfrac{3.17\ln 24.07}{e^{-0.1762}} = \dfrac{3.17(3.18096625\ldots)}{0.83845027\ldots} = \mathbf{12.027}$,

correct to 5 significant figures.

Problem 17.   Solve the equation: $9 = 4e^{-3x}$ to find x, correct to 4 significant figures

---

Rearranging $9 = 4e^{-3x}$ gives: $\dfrac{9}{4} = e^{-3x}$

Taking the reciprocal of both sides gives:

$$\frac{4}{9} = \frac{1}{e^{-3x}} = e^{3x}$$

Taking Napierian logarithms of both sides gives:

$$\ln\left(\frac{4}{9}\right) = \ln(e^{3x})$$

Since $\log_e e^{\alpha} = \alpha$, then $\ln\left(\dfrac{4}{9}\right) = 3x$

Hence, $\mathbf{x} = \dfrac{1}{3}\ln\left(\dfrac{4}{9}\right) = \dfrac{1}{3}(-0.81093) = \mathbf{-0.2703}$,

correct to 4 significant figures.

Problem 18.   Given $32 = 70(1 - e^{-\frac{t}{2}})$ determine the value of t, correct to 3 significant figures

---

Rearranging $32 = 70(1 - e^{-\frac{t}{2}})$ gives:

$$\frac{32}{70} = 1 - e^{-\frac{t}{2}}$$

and $e^{-\frac{t}{2}} = 1 - \dfrac{32}{70} = \dfrac{38}{70}$

Taking the reciprocal of both sides gives:

$$e^{\frac{t}{2}} = \frac{70}{38}$$

Taking Napierian logarithms of both sides gives:

$$\ln e^{\frac{t}{2}} = \ln\left(\frac{70}{38}\right)$$

i.e. $\dfrac{t}{2} = \ln\left(\dfrac{70}{38}\right)$

from which, $\mathbf{t} = 2\ln\left(\dfrac{70}{38}\right) = \mathbf{1.22}$, correct to 3 significant figures.

Problem 19.   Solve $\dfrac{7}{4} = e^{3x}$ correct to 4 significant figures

---

Take natural logs of both sides gives:

$$\ln\frac{7}{4} = \ln e^{3x}$$

$$\ln\frac{7}{4} = 3x\ln e$$

Since, $\ln\dfrac{7}{4} = 3x$

i.e. $0.55962 = 3x$

i.e. $\mathbf{x = 0.1865}$,

correct to 4 significant figures

Now try the following Practice Exercise

Practice Exercise 44 Evaluating Napierian logarithms (Answers on page 536)

In Problems 1 and 2, evaluate correct to 5 significant figures:

1.   (a) $\dfrac{1}{3}\ln 5.2932$ (b) $\dfrac{\ln 82.473}{4.829}$

    (c) $\dfrac{5.62\ln 321.62}{e^{1.2942}}$

2.   (a) $\dfrac{1.786\ln e^{1.76}}{\lg 10^{1.41}}$ (b) $\dfrac{5e^{-0.1629}}{2\ln 0.00165}$

    (c) $\dfrac{\ln 4.8629 - \ln 2.4711}{5.173}$

In Problems 3 to 11 solve the given equations, each correct to 4 significant figures.

3. $1.5 = 4e^{2t}$

4. $7.83 = 2.91e^{-1.7x}$

5. $16 = 24(1 - e^{-\frac{t}{2}})$

6. $5.17 = \ln\left(\dfrac{x}{4.64}\right)$

7. $\ln x = 2.40$

8. $24 + e^{2x} = 45$

9. $5 = e^{x+1} - 7$

10. $5 = 8\left(1 - e^{\frac{-x}{2}}\right)$

11. $e^{(x+1)} = 3e^{(2x-5))}$

12.   If $\dfrac{P}{Q} = 10\log_{10}\left(\dfrac{R_1}{R_2}\right)$ find the value of R_1 when $P = 160$, $Q = 8$ and $R_2 = 5$

13.   The velocity v_2 of a rocket is given by: $v_2 = v_1 + C\ln\left(\dfrac{m_1}{m_2}\right)$ where v_1 is the initial rocket velocity, C is the velocity of the jet exhaust gases, m_1 is the mass of the rocket before the jet engine is fired, and m_2 is the mass of the rocket after the jet engine is switched off. Calculate the velocity of the rocket given $v_1 = 600$ m/s, $C = 3500$ m/s, $m_1 = 8.50 \times 10^4$ kg and $m_2 = 7.60 \times 10^4$ kg

14.   The work done in an isothermal expansion of a gas from pressure p_1 to p_2 is given by:

$$w = w_0\ln\left(\dfrac{p_1}{p_2}\right)$$

If the initial pressure $p_1 = 7.0$ kPa, calculate the final pressure p_2 if $w = 3w_0$

9.8 Laws of growth and decay

Laws of exponential growth and decay are of the form $y = Ae^{-kx}$ and $y = A(1 - e^{-kx})$, where A and k are constants. When plotted, the form of these equations is as shown in Figure 9.6.

The laws occur frequently in engineering and science and examples of quantities related by a natural law include:

(i)   Linear expansion $l = l_0 e^{\alpha\theta}$

(ii)   Change in electrical resistance with temperature $R_0 = R_0 e^{\alpha\theta}$

(iii)   Tension in belts $T_1 = T_0 e^{\mu\theta}$

(iv)   Newton's law of cooling $\theta = \theta_0 e^{-kt}$

(v)   Biological growth $y = y_0 e^{kt}$

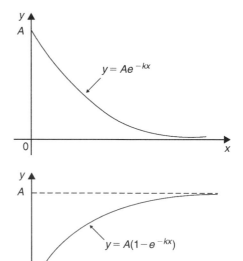

Figure 9.6

(vi) Discharge of a capacitor $q = Qe^{-\frac{t}{CR}}$

(vii) Atmospheric pressure $p = p_0 e^{-h/c}$

(viii) Radioactive decay $N = N_0 e^{-\lambda t}$

(ix) Decay of current in an inductive circuit $i = Ie^{-\frac{Rt}{L}}$

(x) Growth of current in a capacitive circuit
$i = I(1 - e^{-\frac{t}{CR}})$

Here are some worked problems to demonstrate the laws of growth and decay.

Problem 20. The resistance R of an electrical conductor at temperature $\theta°C$ is given by $R = R_0 e^{\alpha\theta}$, where α is a constant and $R_0 = 5k\Omega$. Determine the value of α correct to 4 significant figures, when $R = 6\ k\Omega$ and $\theta = 1500°C$. Also, find the temperature, correct to the nearest degree, when the resistance R is 5.4 kΩ

Transposing $R = R_0 e^{\alpha\theta}$ gives:
$$\frac{R}{R_0} = e^{\alpha\theta}$$

Taking Napierian logarithms of both sides gives:
$$\ln\frac{R}{R_0} = \ln e^{\alpha\theta} = \alpha\theta$$

Hence, $\alpha = \frac{1}{\theta}\ln\frac{R}{R_0} = \frac{1}{1500}\ln\left(\frac{6\times10^3}{5\times10^3}\right)$

$$= \frac{1}{1500}(0.1823215\ldots) = 1.215477.. \times 10^{-4}$$

Hence, $\alpha = \mathbf{1.215 \times 10^{-4}}$ correct to 4 significant figures.

From above, $\ln\frac{R}{R_0} = \alpha\theta$ hence $\theta = \frac{1}{\alpha}\ln\frac{R}{R_0}$

When $R = 5.4 \times 10^3$, $\alpha = 1.215477.. \times 10^{-4}$ and $R_0 = 5 \times 10^3$

$$\theta = \frac{1}{1.215477.. \times 10^{-4}}\ln\left(\frac{5.4\times10^3}{5\times10^3}\right)$$

$$= \frac{10^4}{1.215477..}(7.696104.. \times 10^{-2}) = \mathbf{633°C}$$ correct to the nearest degree.

Problem 21. In an experiment involving Newton's law of cooling, the temperature $\theta(°C)$ is given by: $\theta = \theta_0 e^{-kt}$. Find the value of constant k when $\theta_0 = 56.6°C$, $\theta = 16.5°C$ and $t = 79.0$ seconds

Transposing $\theta = \theta_0 e^{-kt}$ gives:
$$\frac{\theta}{\theta_0} = e^{-kt} \text{ from which, } \frac{\theta_0}{\theta} = \frac{1}{e^{-kt}} = e^{kt}$$

Taking Napierian logarithms of both sides gives:
$$\ln\frac{\theta_0}{\theta} = kt$$

from which, $k = \frac{1}{t}\ln\frac{\theta_0}{\theta}$

$$= \frac{1}{79.0}\ln\left(\frac{56.6}{16.5}\right) = \frac{1}{79.0}(1.2326486..)$$

Hence, $\mathbf{k = 0.01560}$ or $\mathbf{15.60\times10^{-3}}$

Problem 22. The current i amperes flowing in a capacitor at time t seconds is given by:
$i = 8.0(1 - e^{-\frac{t}{CR}})$, where the circuit resistance R is 25 kΩ and capacitance C is 16 μF. Determine (a) the current i after 0.5 seconds and (b) the time, to the nearest millisecond, for the current to reach 6.0 A. Sketch the graph of current against time

(a) Current $i = 8.0(1 - e^{-\frac{t}{CR}})$

$$= 8.0[1 - e^{-0.5/(16\times10^{-6}\times25\times10^3)}]$$

$$= 8.0(1 - e^{-1.25})$$

$$= 8.0(1 - 0.2865047..)$$

$$= 8.0(0.7134952..)$$

$$= \mathbf{5.71\ amperes}$$

(b) Transposing $i = 8.0(1 - e - \frac{t}{CR})$ gives:
$$\frac{i}{8.0} = 1 - e^{-\frac{t}{CR}}$$

from which, $e^{-\frac{t}{CR}} = 1 - \frac{i}{8.0} = \frac{8.0-i}{8.0}$

Taking the reciprocal of both sides gives: $e^{\frac{t}{CR}} = \frac{8.0}{8.0-i}$

Taking Napierian logarithms of both sides gives:
$$\frac{t}{CR} = \ln\left(\frac{8.0}{8.0-i}\right)$$

Hence,
$$t = CR\ln\left(\frac{8.0}{8.0-i}\right)$$

When $i = 6.0$ A,
$$t = (16\times10^{-6})(25\times10^3)\ln\left(\frac{8.0}{8.0-6.0}\right)$$

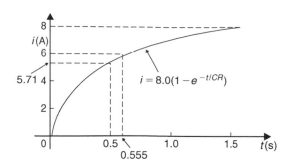

Figure 9.7

i.e.

$$t = \frac{400}{10^3} \ln \left(\frac{8.0}{2.0} \right) = 0.4 \ln 4.0$$

$$= 0.4(1.3862943..) = 0.5545s$$

$$= \textbf{555 ms} \text{ correct to the nearest ms.}$$

A graph of current against time is shown in Figure 9.7.

Now try the following Practice Exercise

Practice Exercise 45 Laws of growth and decay (Answers on page 536)

1. The temperature, $T°C$, of a cooling object varies with time, t minutes, according to the equation: $T = 150e^{-0.04t}$. Determine the temperature when (a) $t = 0$, (b) $t = 10$ minutes.

2. The pressure p pascals at height h metres above ground level is given by: $p = p_0 e^{-h/C}$, where p_0 is the pressure at ground level and C is a constant. Find pressure p when $p_0 = 1.012 \times 10^5$ Pa, height h = 1420 m and C = 71500.

3. The voltage drop, v volts, across an inductor L henrys at time t seconds is given by: $v = 200e^{-\frac{Rt}{L}}$, where R = 150 Ω and $L = 12.5 \times 10^{-3}$ H. Determine (a) the voltage when $t = 160 \times 10^{-6}$ s, and (b) the time for the voltage to reach 85 V.

4. The length l metres of a metal bar at temperature t°C is given by $l = l_0 e^{\alpha t}$, where l_0 and α are constants. Determine (a) the value of l when $l_0 = 1.894$, $\alpha = 2.038 \times 10^{-4}$ and t = 250°C, and (b) the value of l_0 when l = 2.416, t = 310°C and $\alpha = 1.682 \times 10^{-4}$

5. The temperature $\theta_2°C$ of an electrical conductor at time t seconds is given by: $\theta_2 = \theta_1(1 - e^{-t/T})$, where θ_1 is the initial temperature and T seconds is a constant. Determine (a) θ_2 when $\theta_1 = 159.9°C$, $t = 30$ s and T = 80 s, and (b) the time t for θ_2 to fall to half the value of θ_1 if T remains at 80 s

6. A belt is in contact with a pulley for a sector of $\theta = 1.12$ radians and the coefficient of friction between these two surfaces is $\mu = 0.26$. Determine the tension on the taut side of the belt, T Newtons, when tension on the slack side is given by $T_0 = 22.7$ Newtons, given that these quantities are related by the law $T = T_0 e^{\mu\theta}$

7. The instantaneous current i at time t is given by: $i = 10e^{-t/CR}$ when a capacitor is being charged. The capacitance C is 7×10^{-6} farads and the resistance R is 0.3×10^6 ohms. Determine:

   (a) the instantaneous current when t is 2.5 seconds, and

   (b) the time for the instantaneous current to fall to 5 amperes.

   Sketch a curve of current against time from $t = 0$ to $t = 6$ seconds.

8. The amount of product x (in mol/cm³) found in a chemical reaction starting with 2.5 mol/cm³ of reactant is given by: $x = 2.5 (1 - e^{-4t})$ where t is the time, in minutes, to form product x. Plot a graph at 30 second intervals up to 2.5 minutes and determine x after 1 minute.

9. The current i flowing in a capacitor at time t is given by: $i = 12.5(1 - e^{-t/CR})$ where resistance R is 30 kΩ and the capacitance C is 20 μF. Determine (a) the current flowing after 0.5 seconds, and (b) the time for the current to reach 10 amperes.

10. The percentage concentration C of the starting material in a chemical reaction varies with time t according to the equation $C = 100e^{-0.004t}$. Determine the concentration when (a) $t = 0$, (b) $t = 100$ s, (c) $t = 1000$ s

11. The current i flowing through a diode at room temperature is given by: $i = i_S \left(e^{40V} - 1\right)$ amperes. Calculate the current flowing in a silicon diode when the reverse saturation current $i_S = 50$ nA and the forward voltage $V = 0.27$ V

12. A formula for chemical decomposition is given by: $C = A\left(1 - e^{-\frac{t}{10}}\right)$ where t is the time in seconds. Calculate the time, in milliseconds, for a compound to decompose to a value of $C = 0.12$ given $A = 8.5$

13. The mass, m, of pollutant in a water reservoir decreases according to the law $m = m_0\, e^{-0.1t}$ where t is the time in days and m_0 is the initial mass. Calculate the percentage decrease in the mass after 60 days, correct to 3 decimal places.

14. A metal bar is cooled with water. Its temperature, in °C, is given by: $\theta = 15 + 1300\, e^{-0.2t}$ where t is the time in minutes. Calculate how long it will take for the temperature, θ, to decrease to 36°C, correct to the nearest second.

For fully worked solutions to each of the problems in Exercises 40 to 45 in this chapter, go to the website:
www.routledge.com/cw/bird

This assignment covers the material contained in chapters 7 to 9. *The marks for each question are shown in brackets at the end of each question.*

1. Transpose $p - q + r = a - b$ for b (2)

2. Make π the subject of the formula $r = \dfrac{c}{2\pi}$ (2)

3. Transpose $V = \dfrac{1}{3}\pi r^2 h$ for h (2)

4. Transpose $I = \dfrac{E - e}{R + r}$ for E (2)

5. Transpose $k = \dfrac{b}{ad - 1}$ for d (4)

6. Make g the subject of the formula
$t = 2\pi \sqrt{\dfrac{L}{g}}$ (3)

7. Transpose $A = \dfrac{\pi R^2 \theta}{360}$ for R (2)

8. Make r the subject of the formula:
$x + y = \dfrac{r}{3 + r}$ (5)

9. Make L the subject of the formula:
$m = \dfrac{\mu L}{L + rCR}$ (5)

10. The surface area A of a rectangular prism is given by the formula: $A = 2(bh + hl + lb)$. Evaluate b when $A = 11{,}750\,\text{mm}^2$, $h = 25\,\text{mm}$ and $l = 75\,\text{mm}$.   (4)

11. The velocity v of water in a pipe appears in the formula: $h = \dfrac{0.03\,L\,v^2}{2\,dg}$. Evaluate v when $h = 0.384, d = 0.20, L = 80$ and $g = 10$ (3)

12. A formula for the focal length f of a convex lens is: $\dfrac{1}{f} = \dfrac{1}{u} + \dfrac{1}{v}$. Evaluate v when $f = 4$ and $u = 20$ (4)

In problems 13 and 14, solve the simultaneous equations.

13. (a) $2x + y = 6$ (b) $4x - 3y = 11$
        $5x - y = 22$ $3x + 5y = 30$ (9)

14. (a) $3a - 8 + \dfrac{b}{8} = 0$
    $b + \dfrac{a}{2} = \dfrac{21}{4}$

   (b) $\dfrac{2p + 1}{5} - \dfrac{1 - 4q}{2} = \dfrac{5}{2}$
    $\dfrac{1 - 3p}{7} + \dfrac{2q - 3}{5} + \dfrac{32}{35} = 0$ (18)

15. In an engineering process two variables x and y are related by the equation: $y = ax + \dfrac{b}{x}$ where a and b are constants. Evaluate a and b if $y = 15$ when $x = 1$ and $y = 13$ when $x = 3$.   (5)

16. Evaluate the following, correct to 4 significant figures.
    (a) $3.2 \ln 4.92 - 5 \lg 17.9$ (b) $\dfrac{5(1 - e^{-2.65})}{e^{1.73}}$ (4)

17. Solve the following equations.
    (a) $5^x = 2$ (b) $3e^{2x} = 4.2$ (7)

18. Evaluate the following, each correct to 3 decimal places.

    (a) $\ln 462.9$

    (b) $\ln 0.0753$

    (c) $\dfrac{\ln 3.68 - \ln 2.91}{4.63}$ (3)

19. Evaluate v given that $v = E\left(1 - e^{-\frac{t}{CR}}\right)$ volts when $E = 100\,\text{V}$, $C = 15\,\mu\,\text{F}$, $R = 50\,k\Omega$ and $t = 1.5\text{s}$. Also, determine the time when the voltage is 60V.   (8)

20. Plot a graph of $y = \frac{1}{2}e^{-1.2x}$ over the range $x = -2$ to $x = +1$ and hence determine, correct to 1 decimal place,

    (a) the value of y when $x = -0.75$, and

    (b) the value of x when $y = 4.0$ (8)

For lecturers/instructors/teachers, fully worked solutions to each of the problems in Revision Test 4, together with a full marking scheme, are available at the website:
www.routledge.com/cw/bird

Chapter 10

Straight line graphs

Why it is important to understand: **Straight line graphs**

Graphs have a wide range of applications in engineering and in physical sciences because of their inherent simplicity. A graph can be used to represent almost any physical situation involving discrete objects and the relationship between them. If two quantities are directly proportional and one is plotted against the other, a straight line is produced. Examples include an applied force on the end of a spring plotted against spring extension, the speed of a flywheel plotted against time, and strain in a wire plotted against stress (Hooke's law). In engineering, the straight line graph is the most basic and important graph to draw and evaluate.

At the end of this chapter, you should be able to:

- understand rectangular axes, scales and co-ordinates
- plot given co-ordinates and draw the best straight line graph
- determine the gradient of a straight line graph
- estimate the vertical-axis intercept
- state the equation of a straight line graph
- plot straight line graphs involving practical engineering examples

10.1 Introduction to graphs

A graph is a visual representation of information, showing how one quantity varies with another related quantity.

We often see graphs in newspapers or in business reports, in travel brochures and government publications. For example, a graph of the share price (in pence) over a six-month period for a drinks company Fizzy Pops is given in Figure 10.1. Generally, we see that the share price increases to a high of 400p in June, but dips down to around 280p in August before recovering slightly in September.

A graph should convey information more quickly to the reader than if the same information was explained in words.

When this chapter is completed you should be able to draw up a table of values, plot co-ordinates, determine the gradient and state the equation of a straight line graph. Some typical practical examples are included where straight lines are used.

Science and Mathematics for Engineering. 978-0-367-20475-4, © John Bird. Published by Taylor & Francis. All rights reserved.

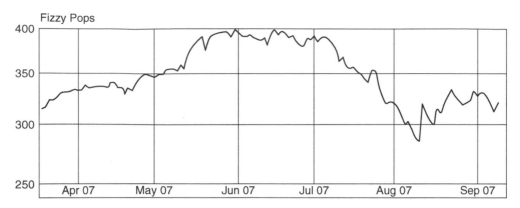

Figure 10.1

10.2 Axes, scales and co-ordinates

We are probably all familiar with reading a map to locate a town, or reading a local map to locate a particular street. For example, a street map of central Portsmouth is shown in Figure 10.2. Notice the squares drawn horizontally and vertically on the map; this is called a **grid** and enables us to locate a place of interest or a particular road. Most maps contain such a grid.

We locate places of interest on the map by stating a letter and a number – this is called the **grid reference**. For example, on the map, the Portsmouth & Southsea station is in square D2, King's Theatre is in square E5, HMS Warrior is in square A2, Gunwharf Quays is in square B3 and High Street is in square B4. Portsmouth & Southsea station is located by moving horizontally along the bottom of the map until reaching the square labelled D, and then moving vertically upwards until square 2 is met.

The letter/number, D2, is referred to as **co-ordinates**, i.e. co-ordinates are used to locate the position of a point on a map. If you are familiar with using a map in this way then you should have no difficulties with graphs, because similar co-ordinates are used with graphs.

As stated earlier, a **graph** is a visual representation of information, showing how one quantity varies with another related quantity.

The most common method of showing a relationship between two sets of data is to use a pair of reference axes – these are two lines drawn at right angles to each other, (often called **Cartesian**, named after **Descartes***, or **rectangular axes**), as shown in Figure 10.3.

The horizontal axis is labelled the *x*-axis, and the vertical axis is labelled the *y*-axis. The point where *x* is 0 and *y* is 0 is called the **origin**. *x* values have **scales** that are positive to the right of the origin and negative to the left.

y values have scales that are positive up from the origin and negative down from the origin.

Co-ordinates are written with brackets and a comma in between two numbers. For example, point A is shown

*Who was Descartes? – **René Descartes** (31 March 1596–11 February 1650) was a French philosopher, mathematician and writer. He wrote many influential texts including *Meditations on First Philosophy*. Descartes is best known for the philosophical statement '*Cogito ergo sum*' (I think, therefore I am), found in part IV of *Discourse on the Method*. To find out more go to www.routledge.com/cw/bird

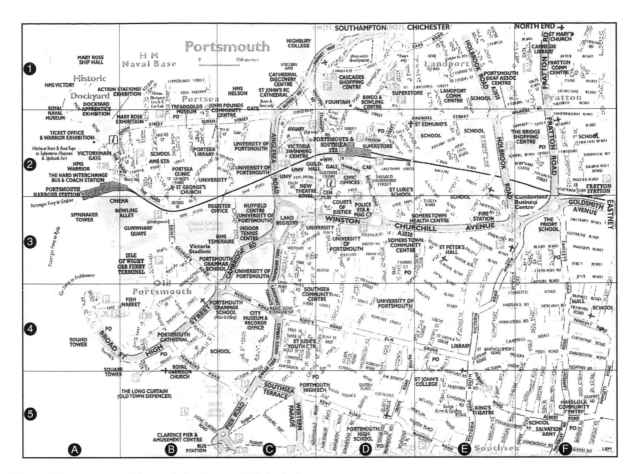

Figure 10.2 Reprinted with permission from AA Media Ltd.

with co-ordinates (3, 2) and is located by starting at the origin and moving 3 units in the positive *x* direction (i.e. to the right) and then 2 units in the positive *y* direction (i.e. up).

When co-ordinates are stated the first number is always the *x* value, and the second number is always the *y* value. Also in Figure 10.3, point B has co-ordinates (−4, 3) and point C has co-ordinates (−3, −2).

10.3 Straight line graphs

The distances travelled by a car in certain periods of time are shown in the following table of values:

Time (s)	10	20	30	40	50	60
Distance travelled (m)	50	100	150	200	250	300

We will plot time on the horizontal (or *x*) axis with a scale of 1 cm = 10 s. We will plot distance on the vertical

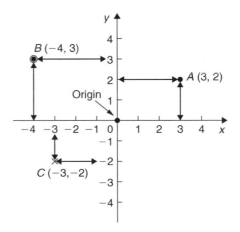

Figure 10.3

(or *y*) axis with a scale of 1 cm = 50 m. (When choosing scales it is better to choose scales such as 1 cm = 1 unit or 1 cm = 2 units or 1 cm = 10 units because it makes reading values between these values easier.)

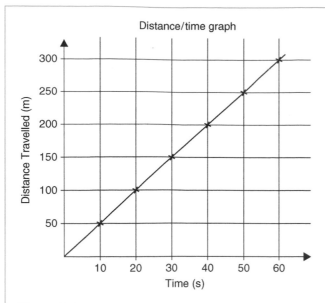

Figure 10.4

With the above data, the (x, y) co-ordinates become (time, distance) co-ordinates, i.e. the co-ordinates are (10, 50), (20, 100), (30, 150) and so on.

The co-ordinates are shown plotted in Figure 10.4 using crosses. (Alternatively, a dot or a dot and circle may be used as shown in Figure 10.3.)

A straight line is drawn through the plotted co-ordinates as shown in Figure 10.4.

Student task

The following table gives the force F newtons which, when applied to a lifting machine, overcomes a corresponding load of L newtons.

F (newtons)	19	35	50	93	125	147
L (newtons)	40	120	230	410	540	680

1. Plot L horizontally and F vertically.

2. Scales are normally chosen such that the graph occupies as much space as possible on the graph paper. So in this case, the following scales are chosen:

    Horizontal axis (i.e. L): 1 cm = 50 N

    Vertical axis (i.e. F): 1 cm = 10 N.

3. Draw the axes and label them L (newtons) for the horizontal axis and F (newtons) for the vertical axis.

4. Label the origin as 0.

5. Write on the horizontal scaling at 100, 200, 300 and so on, every 2 cm.

6. Write on the vertical scaling at 10, 20, 30 and so on, every 1 cm.

7. Plot on the graph the co-ordinates (40, 19), (120, 35), (230, 50), (410, 93), (540, 125) and (680, 147) marking each with a cross or a dot.

8. Using a ruler, draw the best straight line through the points. You will notice that not all of the points lie exactly on a straight line. This is quite normal with experimental values. In a practical situation it would be surprising if all of the points lay exactly on a straight line.

9. Extend the straight line at each end.

10. From the graph, determine the force applied when the load is 325 N. It should be close to 75 N. This process of finding an equivalent value within the given data is called **interpolation**. Similarly, determine the load that a force of 45 N will overcome. It should be close to 170 N

11. From the graph, determine the force needed to overcome a 750 N load. It should be close to 161 N. This process of finding an equivalent value outside the given data is called **extrapolation**. To extrapolate we need to have extended the straight line drawn. Similarly, determine the force applied when the load is zero. It should be close to 11 N. Where the straight line crosses the vertical axis is called the **vertical-axis intercept**. So in this case, the vertical-axis intercept = 11 N at co-ordinates (0, 11)

The graph you have drawn should look something like Figure 10.5.

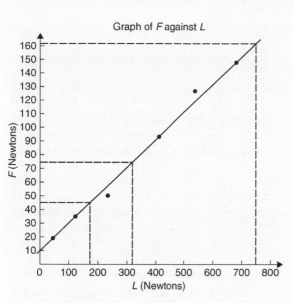

Figure 10.5

In another example, let the relationship between two variables x and y be $y = 3x + 2$

When $x = 0$, $y = 0 + 2 = 2$

When $x = 1$, $y = 3 + 2 = 5$

When $x = 2$, $y = 6 + 2 = 8$, and so on.

The co-ordinates $(0, 2)$, $(1, 5)$ and $(2, 8)$ have been produced and are plotted, with others, as shown in Figure 10.6.
When the points are joined together **a straight line graph results**, i.e. $y = 3x + 2$ is a straight line graph.

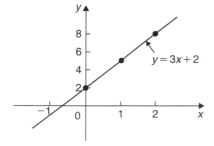

Figure 10.6

10.3.1 Summary of general rules to be applied when drawing graphs

(a) Give the graph a title clearly explaining what is being illustrated.

(b) Choose scales such that the graph occupies as much space as possible on the graph paper being used.

(c) Choose scales so that interpolation is made as easy as possible. Usually scales such as 1 cm = 1 unit, or 1 cm = 2 units, or 1 cm = 10 units are used. Awkward scales such as 1 cm = 3 units or 1 cm = 7 units should not be used.

(d) The scales need not start at zero, particularly when starting at zero produces an accumulation of points within a small area of the graph paper.

(e) The co-ordinates, or points, should be clearly marked. This is achieved either by a cross, or a dot and circle, or just by a dot (see Figure 10.3).

(f) A statement should be made next to each axis explaining the numbers represented with their appropriate units.

(g) Sufficient numbers should be written next to each axis without cramping.

Problem 1. Plot the graph $y = 4x + 3$ in the range $x = -3$ to $x = +4$. From the graph, find (a) the value of y when $x = 2.2$, and (b) the value of x when $y = -3$

Whenever an equation is given and a graph is required, a table giving corresponding values of the variable is necessary. The table is achieved as follows:

When $x = -3$, $y = 4x + 3 = 4(-3) + 3$
                   $= -12 + 3 = -9$
When $x = -2$, $y = 4(-2) + 3$
                   $= -8 + 3 = -5$, and so on.

Such a table is shown below:

x	-3	-2	-1	0	1	2	3	4
y	-9	-5	-1	3	7	11	15	19

The co-ordinates $(-3, -9)$, $(-2, -5)$, $(-1, -1)$ and so on, are plotted and joined together to produce the straight line shown in Figure 10.7. (Note that the scales used on the x- and y-axes do not have to be the same.) From the graph:

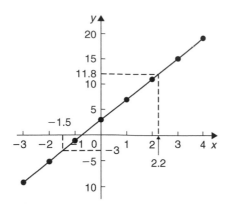

Figure 10.7

(a) when $x = 2.2$, **$y = 11.8$**, and

(b) when $y = -3$, **$x = -1.5$**

Now try the following Practice Exercise

Practice Exercise 46   Straight line graphs (answers on page 536)

1. Assuming graph paper measuring 20 cm by 20 cm is available, suggest suitable scales for the following ranges of values:
    (a) Horizontal axis: 3 V to 55 V Vertical axis: 10 Ω to 180 Ω
    (b) Horizontal axis: 7 m to 86 m Vertical axis: 0.3 V to 1.69 V
    (c) Horizontal axis: 5 N to 150 N Vertical axis: 0.6 mm to 3.4 mm

2. Corresponding values obtained experimentally for two quantities are:

x	−5	−3	−1	0	2	4
y	−13	−9	−5	−3	1	5

    Plot a graph of y (vertically) against x (horizontally) to scales of 2 cm = 1 for the horizontal x-axis and 1 cm = 1 for the vertical y-axis. (This graph will need the whole of the graph paper with the origin somewhere in the centre of the paper.)

    From the graph find:
    (a) the value of y when $x = 1$
    (b) the value of y when $x = -2.5$
    (c) the value of x when $y = -6$
    (d) the value of x when $y = 7$

3. Corresponding values obtained experimentally for two quantities are:

x	−2.0	−0.5	0	1.0	2.5	3.0	5.0
y	−13.0	−5.5	−3.0	2.0	9.5	12.0	22.0

    Use a horizontal scale for x of 1 cm = $\frac{1}{2}$ unit and a vertical scale for y of 1 cm = 2 units and draw a graph of x against y. Label the graph and each of its axes. By interpolation, find from the graph the value of y when x is 3.5

4. Draw a graph of $y - 3x + 5 = 0$ over a range of $x = -2$ to $x = 4$. Hence determine (a) the value of y when $x = 1.3$ and (b) the value of x when $y = -9.2$

5. The speed n rev/min of a motor changes when the voltage V across the armature is varied. The results are shown in the following table:

n (rev/min)	560	720	900	1010	1240	1410
V (volts)	80	100	120	140	160	180

    It is suspected that one of the readings taken of the speed is inaccurate. Plot a graph of speed (horizontally) against voltage (vertically) and find this value. Find also

    (a) the speed at a voltage of 132 V, and

    (b) the voltage at a speed of 1300 rev/min.

10.4   Gradients, intercepts and equation of a graph

10.4.1   Gradients

The **gradient** or **slope** of a straight line is the ratio of the change in the value of y to the change in the value of x between any two points on the line. If, as x increases ($\rightarrow$), y also increases ($\uparrow$), then the gradient is positive. In Figure 10.8(a), a straight line graph $y = 2x + 1$ is shown. To find the gradient of this straight line, choose two points on the straight line graph, such as A and C. Then construct a right-angled triangle, such as ABC, where BC is vertical and AB is horizontal.

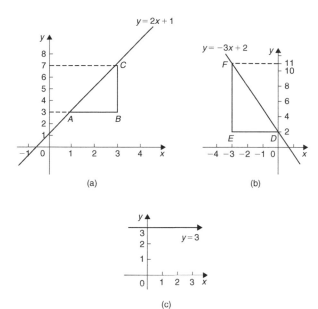

Figure 10.8

Then, **gradient of AC** $= \dfrac{\text{change in } y}{\text{change in } x} = \dfrac{CB}{BA}$

$$= \frac{7-3}{3-1} = \frac{4}{2} = 2$$

In Figure 10.8(b), a straight line graph $y = -3x + 2$ is shown. To find the gradient of this straight line, choose two points on the straight line graph, such as D and F. Then construct a right-angled triangle, such as DEF, where EF is vertical and DE is horizontal.

Then, **gradient of DF** $= \dfrac{\text{change in } y}{\text{change in } x} = \dfrac{FE}{ED}$

$$= \frac{11-2}{-3-0} = \frac{9}{-3} = -3$$

Figure 10.8(c) shows a straight line graph $y = 3$. **Since the straight line is horizontal the gradient is zero**.

10.4.2   y-axis intercept

The value of y when $x = 0$ is called the **y-axis intercept**. In Figure 10.8(a) the y-axis intercept is 1 and in Figure 10.8(b) the y-axis intercept is 2.

10.4.3   Equation of a straight line graph

The general equation of a straight line graph is:

$$y = mx + c$$

where m is the gradient and c is the y-axis intercept. Thus, as we have found in Figure 10.8(a), $y = 2x + 1$ represents a straight line of gradient 2 and y-axis intercept 1. So, given an equation $y = 2x + 1$, we are able to state, on sight, that the gradient $= 2$ and the y-axis intercept $=1$, without the need for any analysis. Similarly, in Figure 10.8(b), $y = -3x + 2$ represents a straight line of gradient -3 and y-axis intercept 2. In Figure 10.8(c), $y = 3$ may be re-written as $y = 0x + 3$ and therefore represents a straight line of gradient 0 and y-axis intercept 3.

Here are some worked problems to help understanding of gradients, intercepts and equations of graphs.

Problem 2.   Plot the following graphs on the same axes between the range $x = -4$ to $x = +4$, and determine the gradient of each.
(a) $y = x$ (b) $y = x + 2$
(c) $y = x + 5$ (d) $y = x - 3$

A table of co-ordinates is produced for each graph.

(a) $y = x$

x	−4	−3	−2	−1	0	1	2	3	4
y	−4	−3	−2	−1	0	1	2	3	4

(b) $y = x + 2$

x	−4	−3	−2	−1	0	1	2	3	4
y	−2	−1	0	1	2	3	4	5	6

(c) $y = x + 5$

x	−4	−3	−2	−1	0	1	2	3	4
y	1	2	3	4	5	6	7	8	9

(d) $y = x - 3$

x	−4	−3	−2	−1	0	1	2	3	4
y	−7	−6	−5	−4	−3	−2	−1	0	1

The co-ordinates are plotted and joined for each graph. The results are shown in Figure 10.9. Each of the straight lines produced are parallel to each other, i.e. the slope or gradient is the same for each.

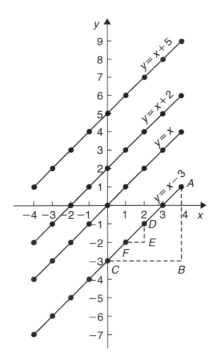

Figure 10.9

To find the gradient of any straight line, say, $y = x - 3$, a horizontal and vertical component needs to be constructed. In Figure 10.9, AB is constructed vertically at $x = 4$ and BC constructed horizontally at $y = -3$

The gradient of $AC = \dfrac{AB}{BC} = \dfrac{1 - (-3)}{4 - 0} = \dfrac{4}{4} = 1$

i.e. the gradient of the straight line $y = x - 3$ is 1, which could have been deduced 'on sight' since $y = 1x - 3$ represents a straight line graph with gradient 1 and y-axis intercept of -3.

The actual positioning of AB and BC is unimportant for the gradient is also given by:

$$\frac{DE}{EF} = \frac{-1 - (-2)}{2 - 1} = \frac{1}{1} = 1$$

The slope or gradient of each of the straight lines in Figure 10.9 is thus 1, since they are all parallel to each other.

Problem 3.   Plot the following graphs on the same axes between the values $x = -3$ to $x = +3$ and determine the gradient and y-axis intercept of each.
(a) $y = 3x$ (b) $y = 3x + 7$
(c) $y = -4x + 4$ (d) $y = -4x - 5$

A table of co-ordinates is drawn up for each equation.

(a) $y = 3x$

x	-3	-2	-1	0	1	2	3
y	-9	-6	-3	0	3	6	9

(b) $y = 3x + 7$

x	-3	-2	-1	0	1	2	3
y	-2	1	4	7	10	13	16

(c) $y = -4x + 4$

x	-3	-2	-1	0	1	2	3
y	16	12	8	4	0	-4	-8

(d) $y = -4x - 5$

x	-3	-2	-1	0	1	2	3
y	7	3	-1	-5	-9	-13	-17

Each of the graphs is plotted as shown in Figure 10.10, and each is a straight line. $y = 3x$ and $y = 3x + 7$ are parallel to each other and thus have the same gradient. The gradient of AC is given by:

$$\frac{CB}{BA} = \frac{16 - 7}{3 - 0} = \frac{9}{3} = 3$$

Hence, the gradient of both $y = 3x$ and $y = 3x + 7$ is 3, which could have been deduced 'on sight'.

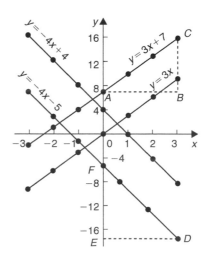

Figure 10.10

$y = -4x + 4$ and $y = -4x - 5$ are parallel to each other and thus have the same gradient. The gradient of DF is given by:

$$\frac{FE}{ED} = \frac{-5 - (-17)}{0 - 3} = \frac{12}{-3} = -4$$

Hence, the gradient of both $y = -4x + 4$ and $y = -4x - 5$ is -4, which, again, could have been deduced 'on sight'.

The y-axis intercept means the value of y where the straight line cuts the y-axis. From Figure 10.10,

$$y = 3x \text{ cuts the } y\text{-axis at } y = 0$$

$$y = 3x + 7 \text{ cuts the } y\text{-axis at } y = +7$$

$$y = -4x + 4 \text{ cuts the } y\text{-axis at } y = +4$$

$$\text{and } y = -4x - 5 \text{ cuts the } y\text{-axis at } y = -5$$

Some general conclusions can be drawn from the graphs shown in Figures 10.9 and 10.10.
When an equation is of the form $y = mx + c$, where m and c are constants, then

(i)   a graph of y against x produces a straight line,

(ii) m represents the slope or gradient of the line, and

(iii) c represents the y-axis intercept.

Thus, given an equation such as $y = 3x + 7$, it may be deduced 'on sight' that its gradient is $+3$ and its y-axis intercept is $+7$, as shown in Figure 10.10. Similarly, if $y = -4x - 5$, then the gradient is -4 and the y-axis intercept is -5, as shown in Figure 10.10.
When plotting a graph of the form $y = mx + c$, only two co-ordinates need be determined. When the co-ordinates are plotted a straight line is drawn between the two points. Normally, three co-ordinates are determined, the third one acting as a check.

Problem 4.   Plot the graph $3x + y + 1 = 0$ and $2y - 5 = x$ on the same axes and find their point of intersection

---

Rearranging $3x + y + 1 = 0$ gives: $y = -3x - 1$
Rearranging $2y - 5 = x$ gives: $2y = x + 5$ and $y = \frac{1}{2}x + 2\frac{1}{2}$
Since both equations are of the form $y = mx + c$ both are straight lines. Knowing an equation is a straight line means that only two co-ordinates need to be plotted and a straight line drawn through them. A third co-ordinate

is usually determined to act as a check. A table of values is produced for each equation as shown below.

x	1	0	-1
$-3x - 1$	-4	-1	2

x	2	0	-3
$\frac{1}{2}x + 2\frac{1}{2}$	$3\frac{1}{2}$	$2\frac{1}{2}$	1

The graphs are plotted as shown in Figure 10.11.
The two straight lines are seen to intersect at $(-1, 2)$

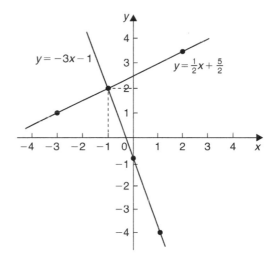

Figure 10.11

Problem 5.   Determine the gradient of the straight line graph passing through the co-ordinates
(a) $(-2, 5)$ and $(3, 4)$ (b) $(-2, -3)$ and $(-1, 3)$

From Figure 10.12, a straight line graph passing through co-ordinates (x_1, y_1) and (x_2, y_2) has a gradient given by:

$$m = \frac{y_2 - y_1}{x_2 - x_1}$$

(a)   A straight line passes through $(-2, 5)$ and $(3, 4)$, hence $x_1 = -2$, $y_1 = 5$, $x_2 = 3$ and $y_2 = 4$, hence

$$\textbf{gradient, } m = \frac{y_2 - y_1}{x_2 - x_1} = \frac{4 - 5}{3 - (-2)} = -\frac{1}{5}$$

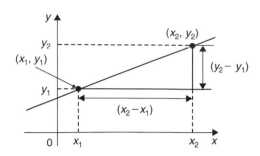

Figure 10.12

(b)   A straight line passes through $(-2, -3)$ and $(-1, 3)$, hence $x_1 = -2$, $y_1 = -3$, $x_2 = -1$ and $y_2 = 3$, hence **gradient**,

$$m = \frac{y_2 - y_1}{x_2 - x_1} = \frac{3 - (-3)}{-1 - (-2)} = \frac{3 + 3}{-1 + 2} = \frac{6}{1} = \mathbf{6}$$

Now try the following Practice Exercise

Practice Exercise 47   Gradients, intercepts and equation of a graph (answers on page 536)

1.   The equation of a line is $4y = 2x + 5$. A table of corresponding values is produced and is shown below. Complete the table and plot a graph of y against x. Find the gradient of the graph.

x	−4	−3	−2	−1	0	1	2	3	4
y		−0.25			1.25			3.25	

2.   Determine the gradient and intercept on the y-axis for each of the following equations:
   (a) $y = 4x - 2$ (b) $y = -x$
   (c) $y = -3x - 4$ (d) $y = 4$

3.   Determine the gradient and y-axis intercept for each of the following equations and sketch the graphs.
   (a) $y = 6x - 3$ (b) $y = -2x + 4$
   (c) $y = 3x$ (d) $y = 7$

4.   Determine the gradient of the straight line graphs passing through the co-ordinates:

(a) $(2, 7)$ and $(-3, 4)$

(b) $(-4, -1)$ and $(-5, 3)$

(c) $\left(\dfrac{1}{4}, -\dfrac{3}{4}\right)$ and $\left(-\dfrac{1}{2}, \dfrac{5}{8}\right)$

5.   State which of the following equations will produce graphs which are parallel to one another:
   (a) $y - 4 = 2x$ (b) $4x = -(y + 1)$
   (c) $x = \dfrac{1}{2}(y + 5)$ (d) $1 + \dfrac{1}{2}y = \dfrac{3}{2}x$
   (e) $2x = \dfrac{1}{2}(7 - y)$

6.   Draw on the same axes the graphs of $y = 3x - 5$ and $3y + 2x = 7$. Find the co-ordinates of the point of intersection. Check the result obtained by solving the two simultaneous equations algebraically

7.   A piece of elastic is tied to a support so that it hangs vertically, and a pan, on which weights can be placed, is attached to the free end. The length of the elastic is measured as various weights are added to the pan and the results obtained are as follows:

Load W (N)	5	10	15	20	25
Length l (cm)	60	72	84	96	108

Plot a graph of load (horizontally) against length (vertically) and determine: (a) the value of the length when the load is 17 N, (b) the value of load when the length is 74 cm, (c) its gradient, and (d) the equation of the graph

10.5   Practical problems involving straight line graphs

When a set of co-ordinate values are given or are obtained experimentally and it is believed that they follow a law of the form $y = mx + c$, then if a straight line can be drawn reasonably close to most of the co-ordinate values when plotted, this verifies that a law of the form $y = mx + c$ exists. From the graph, constants m (i.e. gradient) and c (i.e. y-axis intercept) can be determined.

Here are some worked problems where practical situations are featured.

Problem 6. The temperature in degrees Celsius* and the corresponding values in degrees Fahrenheit are shown in the table below. Construct rectangular axes, choose a suitable scale and plot a graph of degrees Celsius (on the horizontal axis) against degrees Fahrenheit (on the vertical scale)

°C	10	20	40	60	80	100
°F	50	68	104	140	176	212

From the graph find (a) the temperature in degrees Fahrenheit at 55°C, (b) the temperature in degrees Celsius at 167°F, (c) the Fahrenheit temperature at 0°C, and (d) the Celsius temperature at 230°F

The co-ordinates (10, 50), (20, 68), (40, 104) and so on are plotted as shown in Figure 10.13. When the co-ordinates are joined, a straight line is produced. Since a straight line results there is a linear relationship between degrees Celsius and degrees Fahrenheit.

(a) To find the Fahrenheit temperature at 55°C a vertical line AB is constructed from the horizontal axis to meet the straight line at B. The point where the horizontal line BD meets the vertical axis indicates the equivalent Fahrenheit temperature.
Hence, 55°C is equivalent to 131°F
This process of finding an equivalent value in between the given information in the above table is called **interpolation**.

(b) To find the Celsius temperature at 167°F, a horizontal line EF is constructed as shown in Figure 10.13. The point where the vertical line FG cuts the horizontal axis indicates the equivalent Celsius temperature.
Hence, 167°F is equivalent to 75°C

(c) If the graph is assumed to be linear even outside of the given data, then the graph may be extended at both ends (shown by broken lines in Figure 10.13). From Figure 10.13, **0°C corresponds to 32°F**

(d) **230°F is seen to correspond to 110°C**
The process of finding equivalent values outside of the given range is called **extrapolation**.

Problem 7. In an experiment demonstrating Hooke's law*, the strain in an aluminium wire was measured for various stresses. The results were:

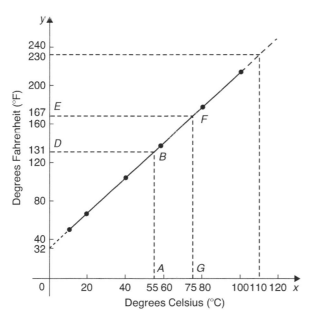

Figure 10.13

*Who was Hooke? – **Robert Hooke** (28 July 1635–3 March 1703) was an English natural philosopher, architect and polymath who, amongst other things, discovered the law of elasticity. To find out more about Hooke go to www.routledge.com/cw/bird

*Who was Celsius go to **www.routledge.com/cw/bird** (For image refer to page 43)

Stress N/mm^2	4.9	8.7	15.0
Strain	0.00007	0.00013	0.00021

Stress N/mm^2	18.4	24.2	27.3
Strain	0.00027	0.00034	0.00039

Plot a graph of stress (vertically) against strain (horizontally). Find: (a) Young's Modulus of Elasticity* for aluminium which is given by the gradient of the graph, (b) the value of the strain at a stress of 20 N/mm^2, and (c) the value of the stress when the strain is 0.00020

The co-ordinates (0.00007, 4.9), (0.00013, 8.7) and so on, are plotted as shown in Figure 10.14. The graph produced is the best straight line which can be drawn corresponding to these points. (With experimental

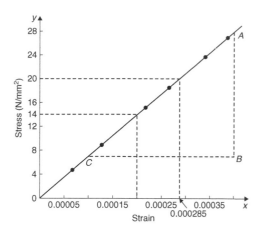

Figure 10.14

results it is unlikely that all the points will lie exactly on a straight line.) The graph, and each of its axes, are labelled. Since the straight line passes through the origin, then stress is directly proportional to strain for the given range of values.

(a) The gradient of the straight line AC is given by

$$\frac{AB}{BC} = \frac{28 - 7}{0.00040 - 0.00010} = \frac{21}{0.00030}$$

$$= \frac{21}{3 \times 10^{-4}} = \frac{7}{10^{-4}} = 7 \times 10^4$$

$$= 70,000 \, \text{N/mm}^2$$

Thus, **Young's Modulus of Elasticity for aluminium is 70,000 N/mm^2**.
Since $1 \, \text{m}^2 = 10^6 \, \text{mm}^2$, 70,000 N/mm^2 is equivalent to $70,000 \times 10^6$ N/m^2, i.e. **70 × 10^9 N/m^2 (or pascals)**.

From Figure 10.14:

(b) the value of the strain at a stress of 20 N/mm^2 is **0.000285**, and

(c) the value of the stress when the strain is 0.00020 is **14 N/mm^2**.

Problem 8. The following values of resistance R ohms and corresponding voltage V volts are obtained from a test on a filament lamp.

R ohms	30	48.5	73	107	128
V volts	16	29	52	76	94

*Who was Young? – **Thomas Young** (13 June 1773–10 May 1829) was an English polymath. He is perhaps best known for his work on Egyptian hieroglyphics and the Rosetta Stone, but Young also made notable scientific contributions to the fields of vision, light, solid mechanics, energy, physiology, language and musical harmony. *Young's modulus* relates the stress in a body to its associated strain. To find out more about Young go to www.routledge.com/cw/bird

Choose suitable scales and plot a graph with R representing the vertical axis and V the horizontal axis. Determine (a) the gradient of the graph, (b) the R-axis intercept value, (c) the equation of the graph, (d) the value of resistance when the voltage is 60 V, and (e) the value of the voltage when the resistance is 40 ohms. (f) If the graph were to continue in the same manner, what value of resistance would be obtained at 110 V?

The co-ordinates (16, 30), (29, 48.5) and so on, are shown plotted in Figure 10.15, where the best straight line is drawn through the points.

(a) The slope or gradient of the straight line AC is given by:

$$\frac{AB}{BC} = \frac{135 - 10}{100 - 0} = \frac{125}{100} = 1.25$$

(Note that the vertical line AB and the horizontal line BC may be constructed anywhere along the length of the straight line. However, calculations are made easier if the horizontal line BC is carefully chosen, in this case, 100)

(b) The R-axis intercept is at $R = 10$ ohms (by extrapolation).

(c) The equation of a straight line is $y = mx + c$, when y is plotted on the vertical axis and x on the horizontal axis. m represents the gradient and c the y-axis intercept. In this case, R corresponds to y, V corresponds to x, $m = 1.25$ and $c = 10$. Hence the equation of the graph is:

$$R = (1.25V + 10)\ \Omega$$

From Figure 10.15,

(d) when the voltage is 60 V, the resistance is **85 Ω**

(e) when the resistance is 40 Ω, the voltage is **24 V**, and

(f) by extrapolation, when the voltage is 110 V, the resistance is **147 Ω**

Now try the following Practice Exercise

Practice Exercise 48   Practical problems involving straight line graphs (answers on page 537)

1. The resistance R ohms of a copper winding is measured at various temperatures $t°C$ and the results are as follows:

R ohms	112	120	126	131	134
$t°C$	20	36	48	58	64

Plot a graph of R (vertically) against t (horizontally) and find from it (a) the temperature when the resistance is 122 Ω and (b) the resistance when the temperature is 52°C

2. The speed of a motor varies with armature voltage as shown by the following experimental results:

n(rev/min)	285	517	615	750	917	1050
V volts	60	95	110	130	155	175

Plot a graph of speed (horizontally) against voltage (vertically) and draw the best straight line through the points. Find from the graph (a) the speed at a voltage of 145 V, and (b) the voltage at a speed of 400 rev/min

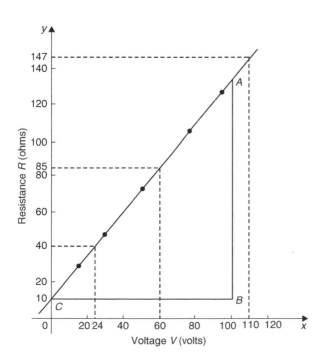

Figure 10.15

3. The following table gives the force F newtons which, when applied to a lifting machine, overcomes a corresponding load of L newtons.

Force F newtons	25	47	64	120	149	187
Load L newtons	50	140	210	430	550	700

Choose suitable scales and plot a graph of F (vertically) against L (horizontally). Draw the best straight line through the points. Determine from the graph (a) the gradient, (b) the F-axis intercept, (c) the equation of the graph, (d) the force applied when the load is 310 N, and (e) the load that a force of 160 N will overcome. (f) If the graph were to continue in the same manner, what value of force will be needed to overcome a 800 N load?

4. The velocity v of a body after varying time intervals t was measured as follows:

t (seconds)	2	5	8	11	15	18
v (m/s)	16.9	19.0	21.1	23.2	26.0	28.1

Plot v vertically and t horizontally and draw a graph of velocity against time. Determine from the graph (a) the velocity after 10 s, (b) the time at 20 m/s and (c) the equation of the graph

5. The mass m of a steel joist varies with length L as follows:

Mass m (kg)	80	100	120	140	160
Length L (m)	3.00	3.74	4.48	5.23	5.97

Plot a graph of mass (vertically) against length (horizontally). Determine the equation of the graph

6. An experiment with a set of pulley blocks gave the following results:

Effort E (newtons)	9.0	11.0	13.6	17.4	20.8	23.6
Load L (newtons)	15	25	38	57	74	88

Plot a graph of effort (vertically) against load (horizontally) and determine (a) the gradient, (b) the vertical axis intercept, (c) the law of the graph, (d) the effort when the load is 30 N and (e) the load when the effort is 19 N

7. The variation of pressure p in a vessel with temperature T is believed to follow a law of the form $p = aT + b$, where a and b are constants. Verify this law for the results given below and determine the approximate values of a and b. Hence determine the pressures at temperatures of 285 K and 310 K and the temperature at a pressure of 250 kPa.

Pressure p kPa	244	247	252	258	262	267
Temperature T K	273	277	282	289	294	300

For fully worked solutions to each of the problems in Exercises 46 to 48 in this chapter,
go to the website:
www.routledge.com/cw/bird

Chapter 11

Introduction to trigonometry

Why it is important to understand: **Introduction to trigonometry**

Knowledge of angles and triangles is very important in engineering. Trigonometry is needed in surveying and architecture, for building structures/systems, designing bridges and solving scientific problems. Trigonometry is also used in electrical engineering: the functions that relate angles and side lengths in right angled triangles are useful in expressing how a.c. electric current varies with time. Engineers use triangles to determine how much force it will take to move along an incline, GPS satellite receivers use triangles to determine exactly where they are in relation to satellites orbiting hundreds of miles away. Whether you want to build a skateboard ramp, a stairway, or a bridge, you can't escape trigonometry.

At the end of this chapter, you should be able to:

- state the theorem of Pythagoras and use it to find the unknown side of a right angled triangle
- define sine, cosine and tangent of an angle in a right angled triangle
- evaluate trigonometric ratios of angles
- solve right angled triangles
- sketch sine, cosine and tangent waveforms
- state and use the sine rule
- state and use the cosine rule
- use various formulae to determine the area of any triangle
- apply the sine and cosine rules to solving practical trigonometric problems

11.1 Introduction

Trigonometry is a subject that involves the measurement of sides and angles of triangles and their relationship with each other. There are many applications in engineering and science where a knowledge of trigonometry is needed.

11.2 The theorem of Pythagoras

The **theorem of Pythagoras*** states:
'In any right-angled triangle, the square on the hypotenuse is equal to the sum of the squares on the other two sides.'

Science and Mathematics for Engineering. 978-0-367-20475-4, © John Bird. Published by Taylor & Francis. All rights reserved.

In the right-angled triangle *ABC* shown in Figure 11.1, this means:

$$b^2 = a^2 + c^2 \qquad (1)$$

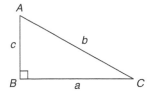

Figure 11.1

If the lengths of any two sides of a right-angled triangle are known, then the length of the third side may be calculated by Pythagoras's theorem.

From equation (1): $b = \sqrt{a^2 + c^2}$

Transposing equation (1) for *a* gives: $a^2 = b^2 - c^2$ from which, $a = \sqrt{b^2 - c^2}$

*Who was Pythagoras? – **Pythagoras of Samos** (born about 570 BC and died about 495 BC) was an Ionian Greek philosopher and mathematician. He is best known for the Pythagorean theorem, which states that in a right-angled triangle $a^2 + b^2 = c^2$. To find out more go to www.routledge.com/cw/bird

Transposing equation (1) for *c* gives: $c^2 = b^2 - a^2$ from which, $c = \sqrt{b^2 - a^2}$

Here are some worked problems to demonstrate the theorem of Pythagoras.

Problem 1. In Figure 11.2, find the length of *BC*

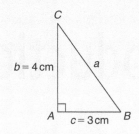

Figure 11.2

From Pythagoras, $a^2 = b^2 + c^2$

i.e. $a^2 = 4^2 + 3^2$

$= 16 + 9 = 25$

Hence, $a = \sqrt{25} = 5\,\text{cm}$

$\sqrt{25} = \pm 5$ but in a practical example like this an answer of $a = -5\,\text{cm}$ has no meaning; so we take only the positive answer.

Thus $a = BC = 5\,\text{cm}$

ABC is a **3, 4, 5 triangle**. There are not many right-angled triangles which have integer values (i.e. whole numbers) for all three sides.

Problem 2. In Figure 11.3, find the length of *EF*

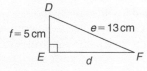

Figure 11.3

By Pythagoras's theorem: $e^2 = d^2 + f^2$

Hence, $13^2 = d^2 + 5^2$

$169 = d^2 + 25$

$d^2 = 169 - 25 = 144$

Thus, $d = \sqrt{144} = 12\,\text{cm}$

i.e. $d = EF = 12\,\text{cm}$

DEF **is a 5, 12, 13 triangle**, another right-angled triangle which has integer values for all three sides.

> **Problem 3.** Two aircraft leave an airfield at the same time. One travels due north at an average speed of 300 km/h and the other due west at an average speed of 220 km/h. Calculate their distance apart after 4 hours

After 4 hours, the first aircraft has travelled

$$4 \times 300 = 1200 \text{ km, due north,}$$

and the second aircraft has travelled

$$4 \times 220 = 880 \text{ km due west,}$$

as shown in Figure 11.4. The distance apart after 4 hours is *BC*
From Pythagoras's theorem:

$$BC^2 = 1200^2 + 880^2$$

$$= 1,440,000 + 774,400 = 2,214,400$$

and $\quad BC = \sqrt{2,214,400} = 1488 \text{ km}$

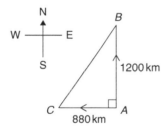

Figure 11.4

Hence, distance apart after 4 hours = 1488 km.

Now try the following Practice Exercise

Practice Exercise 49   Theorem of Pythagoras (answers on page 537)

1. Find the length of side *x* in Figure 11.5(a)

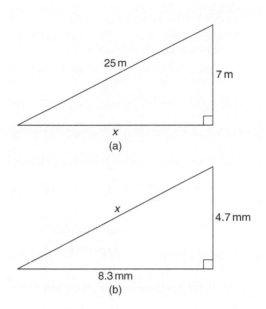

(a)

(b)

Figure 11.5

2. Find the length of side *x* in Figure 11.5(b), correct to 3 significant figures

3. In a triangle *ABC*, *AB* = 17 cm, *BC* = 12 cm and ∠*ABC* = 90°. Determine the length of *AC*, correct to 2 decimal places

4. A tent peg is 4.0 m away from a 6.0 m high tent. What length of rope, correct to the nearest centimetre, runs from the top of the tent to the peg?

5. In a triangle *ABC*, ∠*B* is a right angle, *AB* = 6.92 cm and *BC* = 8.78 cm. Find the length of the hypotenuse

6. In a triangle *CDE*, *D* = 90°, *CD* = 14.83 mm and *CE* = 28.31 mm. Determine the length of *DE*

7. A man cycles 24 km due south and then 20 km due east. Another man, starting at the same time as the first man, cycles 32 km due east and then 7 km due south. Find the distance between the two men

8. A ladder 3.5 m long is placed against a perpendicular wall with its foot 1.0 m from the wall. How far up the wall (to the nearest centimetre) does the ladder reach? If the foot of the ladder is now moved 30 cm further away from the wall, how far does the top of the ladder fall?

9. Two ships leave a port at the same time. One travels due west at 18.4 knots and the other due south at 27.6 knots. If 1 knot = 1 nautical mile per hour, calculate how far apart the two ships are after 4 hours

11.3 Sines, cosines and tangents

With reference to angle θ in the right-angled triangle ABC shown in Figure 11.6:

$$\textbf{sine } \boldsymbol{\theta} = \frac{\textbf{opposite side}}{\textbf{hypotenuse}}$$

'Sine' is abbreviated to 'sin', thus $\sin \theta = \dfrac{BC}{AC}$

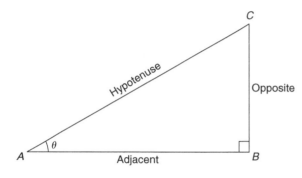

Figure 11.6

Also, $\textbf{cosine } \boldsymbol{\theta} = \dfrac{\textbf{adjacent side}}{\textbf{hypotenuse}}$

'Cosine' is abbreviated to 'cos', thus $\cos \theta = \dfrac{AB}{AC}$

Finally, $\textbf{tangent } \boldsymbol{\theta} = \dfrac{\textbf{opposite side}}{\textbf{adjacent side}}$

'Tangent' is abbreviated to 'tan', thus $\tan \theta = \dfrac{BC}{AB}$

These three trigonometric ratios only apply to right-angled triangles.
Remembering these three equations is very important.
SOH CAH TOA is one way of remembering them.

SOH indicates $\textbf{sin} = \textbf{opposite} \div \textbf{hypotenuse}$

CAH indicates $\textbf{cos} = \textbf{adjacent} \div \textbf{hypotenuse}$

TOA indicates $\textbf{tan} = \textbf{opposite} \div \textbf{adjacent}$

Here are some worked problems to help familiarise ourselves with trigonometric ratios.

Problem 4. In triangle PQR shown in Figure 11.7, determine $\sin \theta$, $\cos \theta$ and $\tan \theta$

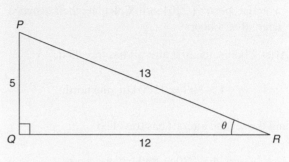

Figure 11.7

$$\sin \theta = \frac{\text{opposite side}}{\text{hypotenuse}} = \frac{PQ}{PR} = \frac{5}{13} = \textbf{0.3846}$$

$$\cos \theta = \frac{\text{adjacent side}}{\text{hypotenuse}} = \frac{QR}{PR} = \frac{12}{13} = \textbf{0.9231}$$

$$\tan \theta = \frac{\text{opposite side}}{\text{adjacent side}} = \frac{PQ}{QR} = \frac{5}{12} = \textbf{0.4167}$$

Problem 5. In triangle ABC of Figure 11.8, determine length AC, $\sin C$, $\cos C$, $\tan C$, $\sin A$, $\cos A$ and $\tan A$

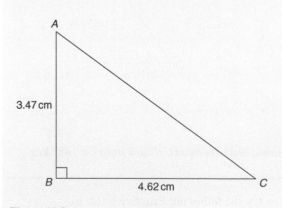

Figure 11.8

By Pythagoras, $AC^2 = AB^2 + BC^2$

i.e. $AC^2 = 3.47^2 + 4.62^2$

from which, $AC = \sqrt{3.47^2 + 4.62^2} = \textbf{5.778 cm}$

$$\sin C = \frac{\text{opposite side}}{\text{hypotenuse}} = \frac{AB}{AC} = \frac{3.47}{5.778} = \mathbf{0.6006}$$

$$\cos C = \frac{\text{adjacent side}}{\text{hypotenuse}} = \frac{BC}{AC} = \frac{4.62}{5.778} = \mathbf{0.7996}$$

$$\tan C = \frac{\text{opposite side}}{\text{adjacent side}} = \frac{AB}{BC} = \frac{3.47}{4.62} = \mathbf{0.7511}$$

$$\sin A = \frac{\text{opposite side}}{\text{hypotenuse}} = \frac{BC}{AC} = \frac{4.62}{5.778} = \mathbf{0.7996}$$

$$\cos A = \frac{\text{adjacent side}}{\text{hypotenuse}} = \frac{AB}{AC} = \frac{3.47}{5.778} = \mathbf{0.6006}$$

$$\tan A = \frac{\text{opposite side}}{\text{adjacent side}} = \frac{BC}{AB} = \frac{4.62}{3.47} = \mathbf{1.3314}$$

Problem 6.   If $\tan B = \dfrac{8}{15}$, determine the value of $\sin B$, $\cos B$, $\sin A$ and $\tan A$

A right-angled triangle ABC is shown in Figure 11.9. If $\tan B = \dfrac{8}{15}$, then $AC = 8$ and $BC = 15$.

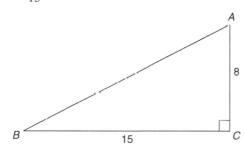

Figure 11.9

By Pythagoras, $AB^2 = AC^2 + BC^2$

i.e. $AB^2 = 8^2 + 15^2$

from which, $AB = \sqrt{8^2 + 15^2} = 17$

$$\mathbf{\sin B} = \frac{AC}{AB} = \frac{8}{17} \text{ or } \mathbf{0.4706}$$

$$\mathbf{\cos B} = \frac{BC}{AB} = \frac{15}{17} \text{ or } \mathbf{0.8824}$$

$$\mathbf{\sin A} = \frac{BC}{AB} = \frac{15}{17} \text{ or } \mathbf{0.8824}$$

$$\mathbf{\tan A} = \frac{BC}{AC} = \frac{15}{8} \text{ or } \mathbf{1.8750}$$

Problem 7.   Point A lies at co-ordinate (2, 3) and point B at (8, 7). Determine the distance AB

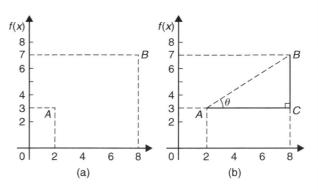

Figure 11.10

Points A and B are shown in Figure 11.10(a). In Figure 11.10(b), the horizontal and vertical lines AC and BC are constructed.

Since ABC is a right-angled triangle, and $AC = (8 - 2) = 6$ and $BC = (7 - 3) = 4$, then by Pythagoras's theorem,

$$AB^2 = AC^2 + BC^2 = 6^2 + 4^2$$

and $$\mathbf{AB} = \sqrt{6^2 + 4^2} = \sqrt{52}$$

$$= \mathbf{7.211} \text{ correct to 3}$$
decimal places.

Now try the following Practice Exercise

Practice Exercise 50   Trigonometric ratios (answers on page 537)

1. Sketch a triangle XYZ such that $\angle Y = 90°$, $XY = 9\,\text{cm}$ and $YZ = 40\,\text{cm}$. Determine $\sin Z$, $\cos Z$, $\tan X$ and $\cos X$

2. In triangle ABC shown in Figure 11.11, find $\sin A$, $\cos A$, $\tan A$, $\sin B$, $\cos B$ and $\tan B$

Figure 11.11

3. If $\cos A = \dfrac{15}{17}$ find $\sin A$ and $\tan A$, in fraction form

4. For the right-angled triangle shown in Figure 11.12, find (a) $\sin \alpha$, (b) $\cos \theta$ and (c) $\tan \theta$

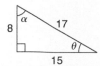

Figure 11.12

5. If $\tan \theta = \dfrac{7}{24}$, find $\sin \theta$ and $\cos \theta$ in fraction form

6. Point P lies at co-ordinate $(-3, 1)$ and point Q at $(5, -4)$. Determine the distance PQ

11.4   Evaluating trigonometric ratios of acute angles

The easiest way to evaluate trigonometric ratios of any angle is to use a calculator.
Use a calculator to check the following (each correct to 4 decimal places):

$$\sin 29^\circ = \mathbf{0.4848} \quad \sin 53.62^\circ = \mathbf{0.8051}$$

$$\cos 67^\circ = \mathbf{0.3907} \quad \cos 83.57^\circ = \mathbf{0.1120}$$

$$\tan 34^\circ = \mathbf{0.6745} \quad \tan 67.83^\circ = \mathbf{2.4541}$$

$$\sin 67^\circ 43' = \sin 67\frac{43}{60}^\circ = \sin 67.7166666\ldots^\circ = \mathbf{0.9253}$$

$$\cos 13^\circ 28' = \cos 13\frac{28}{60}^\circ = \cos 13.466666\ldots^\circ = \mathbf{0.9725}$$

$$\tan 56^\circ 54' = \tan 56\frac{54}{60}^\circ = \tan 56.90^\circ = \mathbf{1.5340}$$

If we know the value of a trigonometric ratio and need to find the angle we use the **inverse function** on our calculators. For example, using shift and sin on our calculator gives $\sin^{-1}$.
If, for example, we know the sine of an angle is 0.5 then the value of the angle is given by:

$$\sin^{-1} 0.5 = 30^\circ \text{ (check that } \sin 30^\circ = 0.5)$$

Similarly, if

$$\cos \theta = 0.4371 \text{ then } \theta = \cos^{-1} 0.4371 = \mathbf{64.08^\circ}$$

and if

$$\tan A = 3.5984 \text{ then } A = \tan^{-1} 3.5984 = \mathbf{74.47^\circ},$$

each correct to 2 decimal places.

Use your calculator to check the following worked problems.

Problem 8.   Determine, correct to 4 decimal places, $\sin 43^\circ\ 39'$

$$\sin 43^\circ 39' = \sin 43\frac{39}{60}^\circ = \sin 43.65^\circ = \mathbf{0.6903}$$

This answer can be obtained using the **calculator** as follows:

1. Press sin   2. Enter 43   3. Press ° '''
4. Enter 39   5. Press ° '''   6. Press)
7. Press =   Answer = **0.6902512....**
                    = **0.6903**, correct to 4 decimal places

Problem 9.   Determine, correct to 3 decimal places, $6 \cos 62^\circ\ 12'$

$$6 \cos 62^\circ 12' = 6 \cos 62\frac{12}{60}^\circ = 6 \cos 62.20^\circ = \mathbf{2.798}$$

This answer can be obtained using the **calculator** as follows:

1. Enter 6   2. Press cos   3. Enter 62
4. Press ° '''   5. Enter 12   6. Press ° '''
7. Press)   8. Press =   Answer = **2.798319....**
                    = **2.798**, correct to 3 decimal places

Problem 10.   Evaluate sin 1.481, correct to 4 significant figures

sin 1.481 means the sine of 1.481 **radians**. (If there is no degree sign (°) then radians are assumed). Therefore the calculator needs to be on the radian function.
Hence, $\sin 1.481 = \mathbf{0.9960}$

Problem 11.   Evaluate tan 2.93, correct to 4 significant figures

Again, since there is no degree sign, then 2.93 means 2.93 radians.
Hence, $\mathbf{\tan 2.93 = -0.2148}$
It is important to know when to have your calculator on either degrees mode or radian mode. A lot of mistakes can arise from this if we are not careful.

Problem 12.   Find the acute angle $\sin^{-1} 0.4128$ in degrees, correct to 2 decimal places

$\sin^{-1} 0.4128$ means 'the angle whose sine is 0.4128'
Using a calculator:

1. Press shift   2. Press sin   3. Enter 0.4128

5. Press)   6. Press =

The answer 24.380848... is displayed.
Hence, **$\sin^{-1} 0.4128 = 24.38°$** correct to 2 decimal places.

Problem 13.   Find the acute angle $\cos^{-1} 0.2437$ in degrees and minutes

$\cos^{-1} 0.2437$ means 'the angle whose cosine is 0.2437'
Using a calculator:

1. Press shift   2. Press cos   3. Enter 0.2437

4. Press)   5. Press =

The answer 75.894979... is displayed

6. Press ° ''' and 75° 53' 41.93" is displayed

Hence, **$\cos^{-1} 0.2437 = 75.89° = 77°54'$** correct to the nearest minute.

Problem 14.   Find the acute angle $\tan^{-1} 7.4523$ in degrees and minutes

$\tan^{-1} 7.4523$ means 'the angle whose tangent is 7.4523'
Using a calculator:

1. Press shift   2. Press tan   3. Enter 7.4523
4. Press)   5. Press =

The answer 82.357318... is displayed

6. Press ° ''' and 82° 21' 26.35" is displayed

Hence, **$\tan^{-1} 7.4523 = 82.36° = 82°21'$** correct to the nearest minute.

Problem 15.   In triangle *EFG* in Figure 11.13, calculate angle *G*

Figure 11.13

With reference to $\angle G$, the two sides of the triangle given are the opposite side *EF* and the hypotenuse *EG*; hence, sine is used,

i.e. $\sin G = \dfrac{2.30}{8.71} = 0.26406429...$

from which, $G = \sin^{-1} 0.26406429...$

i.e. $G = 15.311360...$

Hence, $\angle G = $ **$15.31°$** or **$15°19'$**.

Now try the following Practice Exercise

Practice Exercise 51   Evaluating trigonometric ratios (answers on page 537)

1. Determine, correct to 4 decimal places, $3 \sin 66° \, 41'$

2. Determine, correct to 3 decimal places, $5 \cos 14° 15'$

3. Determine, correct to 4 significant figures, $7 \tan 79° \, 9'$

4. Determine (a) $\cos 1.681$ (b) $\tan 3.672$

5. Find the acute angle $\sin^{-1} 0.6734$ in degrees, correct to 2 decimal places

6. Find the acute angle $\cos^{-1} 0.9648$ in degrees, correct to 2 decimal places

7. Find the acute angle $\tan^{-1} 3.4385$ in degrees, correct to 2 decimal places

8. Find the acute angle $\sin^{-1} 0.1381$ in degrees and minutes

9. Find the acute angle $\cos^{-1} 0.8539$ in degrees and minutes

10. Find the acute angle $\tan^{-1} 0.8971$ in degrees and minutes

11. In the triangle shown in Figure 11.14, determine angle θ, correct to 2 decimal places

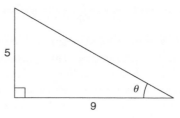

Figure 11.14

12. In the triangle shown in Figure 11.15, determine angle θ in degrees and minutes

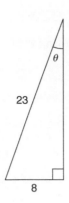

Figure 11.15

13. For the supported beam AB shown in Figure 11.16, determine (a) the angle the supporting stay CD makes with the beam, i.e. θ, correct to the nearest degree, (b) the length of the stay, CD, correct to the nearest centimetre.

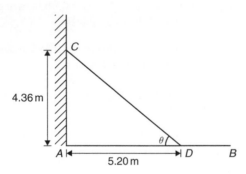

Figure 11.16

11.5 Solving right-angled triangles

'Solving a right-angled triangle' means 'finding the unknown sides and angles'. This is achieved using (i) the theorem of Pythagoras and/or (ii) trigonometric ratios. Six pieces of information describe a triangle completely, i.e. three sides and three angles. As long as at least three facts are known, the other three can usually be calculated.

Here are some worked problems to demonstrate the solution of right-angled triangles.

Problem 16. In the triangle ABC shown in Figure 11.17, find the lengths AC and AB

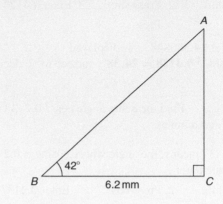

Figure 11.17

There is usually more than one way to solve such a triangle.

In triangle ABC,

$$\tan 42° = \frac{AC}{BC} = \frac{AC}{6.2}$$

(Remember SOH CAH TOA)

Transposing gives:

$$AC = 6.2 \tan 42° = \mathbf{5.583\,mm}$$

$$\cos 42° = \frac{BC}{AB} = \frac{6.2}{AB} \quad \text{from which,}$$

$$AB = \frac{6.2}{\cos 42°} = \mathbf{8.343\,mm}$$

Alternatively, by Pythagoras, $AB^2 = AC^2 + BC^2$
from which, $AB = \sqrt{AC^2 + BC^2} = \sqrt{5.583^2 + 6.2^2}$

$$= \sqrt{69.609889} = \mathbf{8.343\,mm}$$

Problem 17. Sketch a right-angled triangle ABC such that $B = 90°$, $AB = 5$ cm and $BC = 12$ cm. Determine the length of AC and hence evaluate $\sin A$, $\cos C$ and $\tan A$

Triangle ABC is shown in Figure 11.18.

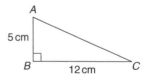

Figure 11.18

By Pythagorus's theorem, $AC = \sqrt{5^2 + 12^2} = 13$

By definition: $\sin A = \dfrac{\text{opposite side}}{\text{hypotenuse}} = \dfrac{12}{13}$ or $\mathbf{0.9231}$

(Remember SOH CAH TOA)

$$\cos C = \frac{\text{adjacent side}}{\text{hypotenuse}} = \frac{12}{13} \text{ or } \mathbf{0.9231}$$

and $\quad \tan A = \dfrac{\text{opposite side}}{\text{adjacent side}} = \dfrac{12}{5} \text{ or } \mathbf{2.400}$

Problem 18. In triangle PQR shown in Figure 11.19, find the lengths of PQ and PR

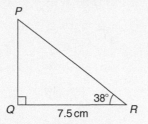

Figure 11.19

$\tan 38° = \dfrac{PQ}{QR} = \dfrac{PQ}{7.5}$ hence,

$\qquad PQ = 7.5 \tan 38° = 7.5(0.7813) = \mathbf{5.860\,cm}$

$\cos 38° = \dfrac{QR}{PR} = \dfrac{7.5}{PR}$ hence,

$$PR = \frac{7.5}{\cos 38°} = \frac{7.5}{0.7880} = \mathbf{9.518\,cm}$$

[Check: Using Pythagoras's theorem:
$(7.5)^2 + (5.860)^2 = 90.59 = (9.518)^2$]

Problem 19. Solve the triangle ABC shown in Figure 11.20

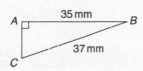

Figure 11.20

To 'solve triangle ABC' means 'to find the length AC and angles B and C'.

$$\sin C = \frac{35}{37} = 0.94595 \text{ hence,}$$

$$C = \sin^{-1} 0.94595 = \mathbf{71.08°}$$

$B = 180° - 90° - 71.08° = \mathbf{18.92°}$ (since angles in a triangle add up to $180°$)

$\sin B = \dfrac{AC}{37}$ hence,

$\quad AC = 37 \sin 18.92° = 37(0.3242) = \mathbf{12.0\,mm}$

or, using Pythagoras's theorem: $37^2 = 35^2 + AC^2$ from which, $AC = \sqrt{(37^2 - 35^2)} = \mathbf{12.0\,mm}$.

Problem 20. An electricity pylon stands on horizontal ground. At a point 80 m from the base of the pylon, the angle of elevation of the top of the pylon is $23°$. Calculate the height of the pylon to the nearest metre

Figure 11.21 shows the pylon AB and the angle of elevation of A from point C is $23°$.

$$\tan 23° = \frac{AB}{BC} = \frac{AB}{80}$$

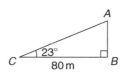

Figure 11.21

Hence, height of pylon

$$AB = 80 \tan 23° = 80(0.4245) = 33.96\,m$$

$$= \mathbf{34\,m\ to\ the\ nearest\ metre}.$$

Now try the following Practice Exercise

Practice Exercise 52   Solving right-angled triangles (answers on page 537)

1. Calculate the dimensions shown as x in Figures 11.22(a) to (f), each correct to 4 significant figures

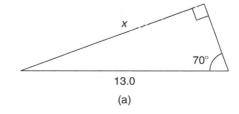

(a)

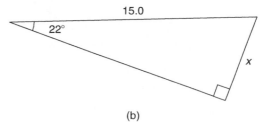

(b)

Figure 11.22

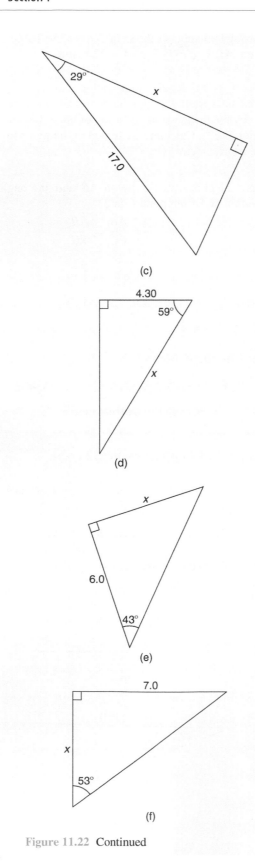

(c)

(d)

(e)

(f)

Figure 11.22 Continued

2. Find the unknown sides and angles in the right-angled triangles shown in Figure 11.23. The dimensions shown are in centimetres

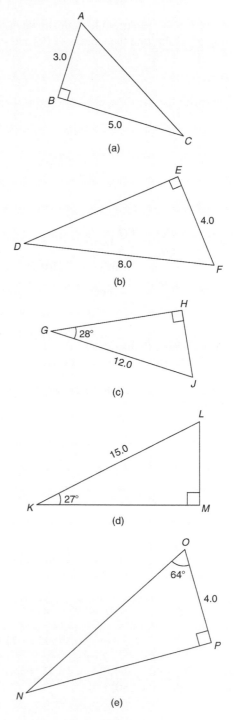

(a)

(b)

(c)

(d)

(e)

Figure 11.23

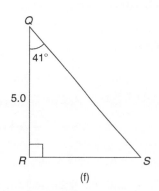

Figure 11.23 Continued

3. A ladder rests against the top of the perpendicular wall of a building and makes an angle of 73° with the ground. If the foot of the ladder is 2 m from the wall, calculate the height of the building

4. Determine the length x in Figure 11.24

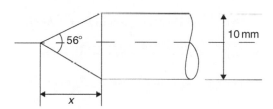

Figure 11.24

5. A vertical tower stands on level ground. At a point 105 m from the foot of the tower the angle of elevation of the top is 19°. Find the height of the tower

11.6 Graphs of trigonometric functions

By drawing up tables of values from 0° to 360°, graphs of $y = \sin A$, $y = \cos A$ and $y = \tan A$ may be plotted. Values obtained with a calculator (correct to 3 decimal places – which is more than sufficient for plotting graphs), using 30° intervals, are shown below, with the respective graphs shown in Figure 11.25.

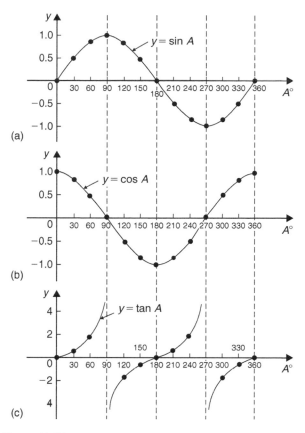

Figure 11.25

(a) $y = \sin A$

A	0	30°	60°	90°	120°	150°	180°
$\sin A$	0	0.500	0.866	1.000	0.866	0.500	0

A	210°	240°	270°	300°	330°	360°
$\sin A$	−0.500	−0.866	−1.000	−0.866	−0.500	0

(b) $y = \cos A$

A	0	30°	60°	90°	120°	150°	180°
$\cos A$	1.000	0.866	0.500	0	−0.500	−0.866	−1.000

A	210°	240°	270°	300°	330°	360°
$\cos A$	−0.866	−0.500	0	0.500	0.866	1.000

(c) $y = \tan A$

A	0	30°	60°	90°	120°	150°	180°
$\tan A$	0	0.577	1.732	∞	−1.732	−0.577	0

A	210°	240°	270°	300°	330°	360°
tan A	0.577	1.732	∞	−1.732	−0.577	0

From Figure 11.25 it is seen that:

(i) Sine and cosine graphs oscillate between peak values of ±1

(ii) The cosine curve is the same shape as the sine curve but displaced by 90°

(iii) The sine and cosine curves are continuous and they repeat at intervals of 360°; the tangent curve appears to be discontinuous and repeats at intervals of 180°

11.7 Sine and cosine rules

To **'solve a triangle'** means 'to find the values of unknown sides and angles'.
If a triangle is **right-angled**, trigonometric ratios and the theorem of Pythagoras may be used for its solution, as shown earlier. However, for a **non-right-angled triangle**, trigonometric ratios and Pythagoras's theorem **cannot** be used. Instead, two rules, called the **sine rule** and the **cosine rule**, are used.

11.7.1 Sine rule

With reference to triangle ABC of Figure 11.26, the **sine rule** states:

$$\frac{a}{\sin A} = \frac{b}{\sin B} = \frac{c}{\sin C}$$

Figure 11.26

The rule may be used only when:

(i) 1 side and any 2 angles are initially given, or

(ii) 2 sides and an angle (not the included angle) are initially given.

11.7.2 Cosine rule

With reference to triangle ABC of Figure 11.26, the **cosine rule** states:

$$a^2 = b^2 + c^2 - 2bc\cos A$$

or

$$b^2 = a^2 + c^2 - 2ac\cos B$$

or

$$c^2 = a^2 + b^2 - 2ab\cos C$$

The rule may be used only when:

(i) 2 sides and the included angle are initially given, or

(ii) 3 sides are initially given.

11.8 Area of any triangle

The **area of any triangle** such as ABC of Figure 11.26 is given by:

(i) $\dfrac{1}{2} \times$ **base** $\times$ **perpendicular height**

or (ii) $\dfrac{1}{2}ab\sin C$ or $\dfrac{1}{2}ac\sin B$ or $\dfrac{1}{2}bc\sin A$

or (iii) $\sqrt{[s(s-1)(s-b)(s-c)]}$ where

$$s = \frac{a+b+c}{2}$$

11.9 Worked problems on the solution of triangles and their areas

Problem 21. In a triangle XYZ, $\angle X = 51°$, $\angle Y = 67°$ and $YZ = 15.2$ cm. Solve the triangle and find its area

The triangle XYZ is shown in Figure 11.27. Solving the triangle means finding $\angle Z$ and sides XZ and XY.

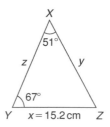

Figure 11.27

Since the angles in a triangle add up to $180°$, then: $Z = 180° - 51° - 67° = \mathbf{62°}$.

Applying the sine rule: $\dfrac{15.2}{\sin 51°} = \dfrac{y}{\sin 67°} = \dfrac{z}{\sin 62°}$

Using $\dfrac{15.2}{\sin 51°} = \dfrac{y}{\sin 67°}$

and transposing gives: $y = \dfrac{15.2 \sin 67°}{\sin 51°}$

$= \mathbf{18.00\,cm = XZ}$

Using $\dfrac{15.2}{\sin 51°} = \dfrac{z}{\sin 62°}$

and transposing gives: $z = \dfrac{15.2 \sin 62°}{\sin 51°}$

$= \mathbf{17.27\,cm = XY}$

Area of triangle $XYZ = \dfrac{1}{2}xy \sin Z$

$= \dfrac{1}{2}(15.2)(18.00) \sin 62° = \mathbf{120.8\,cm^2}$

(or area $= \dfrac{1}{2}xz \sin Y = \dfrac{1}{2}(15.2)(17.27) \sin 67°$

$= \mathbf{120.8\,cm^2}$)

It is always worth checking with triangle problems that the longest side is opposite the largest angle, and vice-versa. In this problem, Y is the largest angle and XZ is the longest of the three sides.

Problem 22.   Solve the triangle ABC given $B = 78°51'$, $AC = 22.31$ mm and $AB = 17.92$ mm. Find also its area

Triangle ABC is shown in Figure 11.28. Solving the triangle means finding angles A and C and side BC.

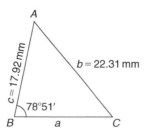

Figure 11.28

Applying the sine rule: $\dfrac{22.31}{\sin 78°51'} = \dfrac{17.92}{\sin C}$

from which, $\sin C = \dfrac{17.92 \sin 78°51'}{22.31} = 0.7881$

Hence, $C = \sin^{-1} 0.7881 = 52°0'$ or $128°0'$

Since $B = 78°51'$, C cannot be $128°0'$, since $128°0' + 78°51'$ is greater than $180°$. Thus only $\mathbf{C = 52°0'}$ is valid. Angle $A = 180° - 78°51' - 52°0' = \mathbf{49°9'}$

Applying the sine rule: $\dfrac{a}{\sin 49°9'} = \dfrac{22.31}{\sin 78°51'}$

from which, $a = \dfrac{22.31 \sin 49°9'}{\sin 78°51'} = \mathbf{17.20\,mm}$

Hence, $\mathbf{A = 49°9'}$, $\mathbf{C = 52°0'}$ and $\mathbf{BC = 17.20\,mm}$

Area of triangle $ABC = \dfrac{1}{2}ac \sin B$

$= \dfrac{1}{2}(17.20)(17.92) \sin 78°51'$

$= \mathbf{151.2\,mm^2}$

Problem 23.   A triangle ABC has sides $a = 9.0$ cm, $b = 7.5$ cm and $c = 6.5$ cm. Determine its three angles and its area

Triangle ABC is shown in Figure 11.29. It is usual first to calculate the largest angle to determine whether the triangle is acute or obtuse. In this case the largest angle is A (i.e. opposite the longest side).

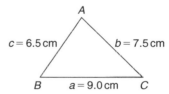

Figure 11.29

Applying the cosine rule: $a^2 = b^2 + c^2 - 2bc \cos A$

from which, $2bc \cos A = b^2 + c^2 - a^2$

and $\cos A = \dfrac{b^2 + c^2 - a^2}{2bc} = \dfrac{7.5^2 + 6.5^2 - 9.0^2}{2(7.5)(6.5)}$

$= 0.1795$

Hence, $A = \cos^{-1} 0.1795 = \mathbf{79.67°}$

(or $280.33°$, which is clearly impossible)

The triangle is thus acute-angled since $\cos A$ is positive. (If $\cos A$ had been negative, angle A would be obtuse, i.e. lie between $90°$ and $180°$.)

Applying the sine rule:

$$\frac{9.0}{\sin 79.67°} = \frac{7.5}{\sin B}$$

from which, $\qquad \sin B = \dfrac{7.5 \sin 79.67°}{9.0} = 0.8198$

Hence, $\qquad\qquad B = \sin^{-1} 0.8198 = \mathbf{55.07°}$

and $\qquad C = 180° - 79.67° - 55.07° = \mathbf{45.26°}$

Area $= \sqrt{[s(s-a)(s-b)(s-c)]}$, where

$s = \dfrac{a+b+c}{2} = \dfrac{9.0+7.5+6.5}{2} = 11.5\,\text{cm}$

Hence,

$$\textbf{area} = \sqrt{[11.5(11.5-9.0)(11.5-7.5)(11.5-6.5)]}$$

$$= \sqrt{[11.5(2.5)(4.0)(5.0)]} = \mathbf{23.98\,cm^2}$$

Alternatively, $\textbf{area} = \dfrac{1}{2} ac \sin B$

$$= \frac{1}{2}(9.0)(6.5) \sin 55.07° = \mathbf{23.98\,cm^2}.$$

Now try the following Practice Exercise

Practice Exercise 53   Solution of triangles and their areas (answers on page 537)

In problems 1 and 2, use the sine rule to solve the triangles *ABC* and find their areas.

1. $A = 29°, B = 68°, b = 27\,\text{mm}$

2. $B = 71°26', C = 56°32', b = 8.60\,\text{cm}$

In problems 3 and 4, use the sine rule to solve the triangles *DEF* and find their areas.

3. $d = 17\,\text{cm}, f = 22\,\text{cm}, F = 26°$

4. $d = 32.6\,\text{mm}, e = 25.4\,\text{mm}, D = 104°22'$

In problems 5 and 6, use the cosine and sine rules to solve the triangles *PQR* and find their areas.

5. $q = 12\,\text{cm}, r = 16\,\text{cm}, P = 54°$

6. $q = 3.25\,\text{m}, r = 4.42\,\text{m}, P = 105°$

In problems 7 and 8, use the cosine and sine rules to solve the triangles *XYZ* and find their areas.

7. $x = 10.0\,\text{cm}, y = 8.0\,\text{cm}, z = 7.0\,\text{cm}$

8. $x = 21\,\text{mm}, y = 34\,\text{mm}, z = 42\,\text{mm}$

11.10   Practical situations involving trigonometry

There are a number of **practical situations** where the use of trigonometry is needed to find unknown sides and angles of triangles. This is demonstrated in the following worked problems.

Problem 24. A room 8.0 m wide has a span roof which slopes at 33° on one side and 40° on the other. Find the length of the roof slopes, correct to the nearest centimetre

A section of the roof is shown in Figure 11.30.

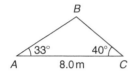

Figure 11.30

Angle at ridge, $B = 180° - 33° - 40° = 107°$.

From the sine rule: $\qquad \dfrac{8.0}{\sin 107°} = \dfrac{a}{\sin 33°}$

from which, $\quad a = \dfrac{8.0 \sin 33°}{\sin 107°} = \mathbf{4.556\,m} = BC$

Also from the sine rule: $\quad \dfrac{8.0}{\sin 107°} = \dfrac{c}{\sin 40°}$

from which, $\quad c = \dfrac{8.0 \sin 40°}{\sin 107°} = \mathbf{5.377\,m} = AB$

Hence, the roof slopes are 4.56 m and 5.38 m, correct to the nearest centimetre.

Problem 25. Two voltage phasors are shown in Figure 11.31. If $V_1 = 40\,\text{V}$ and $V_2 = 100\,\text{V}$ determine the value of their resultant (i.e. length *OA*) and the angle the resultant makes with V_1

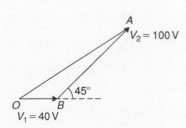

Figure 11.31

Angle $OBA = 180° - 45° = 135°$.
Applying the cosine rule:

$$OA^2 = V_1^2 + V_2^2 - 2V_1V_2 \cos OBA$$

$$= 40^2 + 100^2 - \{2(40)(100)\cos 135°\}$$

$$= 1600 + 10{,}000 - \{-5657\}$$

$$= 1600 + 10{,}000 + 5657 = 17{,}257$$

Thus, **resultant, OA** $= \sqrt{17{,}257} = $ **131.4 V**

Applying the sine rule: $\dfrac{131.4}{\sin 135°} = \dfrac{100}{\sin AOB}$

from which, $\sin AOB = \dfrac{100\sin 135°}{131.4} = 0.5381$

Hence, angle $AOB = \sin^{-1} 0.5381 = 32.55°$ (or $147.45°$, which is not possible).
Hence, the resultant voltage is 131.4 volts at 32.55° to V_1

Problem 26. In Figure 11.32, PR represents the inclined jib of a crane and is 10.0 m long. PQ is 4.0 m long. Determine the inclination of the jib to the vertical and the length of tie QR

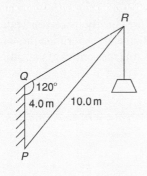

Figure 11.32

Applying the sine rule: $\dfrac{PR}{\sin 120°} = \dfrac{PQ}{\sin R}$

from which, $\sin R = \dfrac{PQ\sin 120°}{PR} = \dfrac{(4.0)\sin 120°}{10.0}$

$$= 0.3464$$

Hence, $\angle R = \sin^{-1} 0.3464 = 20.27°$ (or $159.73°$, which is not possible).
$\angle P = 180° - 120° - 20.27° = $ **39.73°, which is the inclination of the jib to the vertical**.

Applying the sine rule: $\dfrac{10.0}{\sin 120°} = \dfrac{QR}{\sin 39.73°}$

from which,   **length of tie, QR** $= \dfrac{10.0\sin 39.73°}{\sin 120°}$

$$= \textbf{7.38 m}.$$

Problem 27. The area of a field is in the form of a quadrilateral $ABCD$ as shown in Figure 11.33. Determine its area

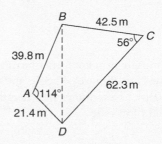

Figure 11.33

A diagonal drawn from B to D divides the quadrilateral into two triangles.
Area of quadrilateral $ABCD =$ area of triangle $ABD +$ area of triangle BCD.

$$= \frac{1}{2}(39.8)(21.4)\sin 114°$$

$$+ \frac{1}{2}(42.5)(62.3)\sin 56°$$

$$= 389.04 + 1097.5 = \textbf{1487 m}^2.$$

Now try the following Practice Exercise

Practice Exercise 54   Practical situations involving trigonometry (answers on page 537)

1. A ship P sails at a steady speed of 45 km/h in a direction of W 32° N (i.e. a bearing of 302°) from a port. At the same time another ship Q leaves the port at a steady speed of 35 km/h in a direction N 15° E (i.e. a bearing of 015°). Determine their distance apart after 4 hours

2. A jib crane is shown in Figure 11.34. If the tie rod PR is 8.0 m long and PQ is 4.5 m long determine (a) the length of jib RQ, and (b) the angle between the jib and the tie rod

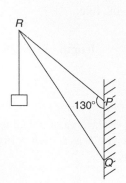

Figure 11.34

3. A building site is in the form of a quadrilateral as shown in Figure 11.35, and its area is 1510 m². Determine the length of the perimeter of the site

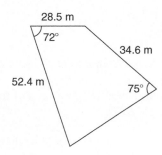

Figure 11.35

4. Determine the length of members *BF* and *EB* in the roof truss shown in Figure 11.36

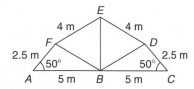

Figure 11.36

5. A laboratory 9.0 m wide has a span roof which slopes at 36° on one side and 44° on the other. Determine the lengths of the roof slopes

6. *PQ* and *QR* are the phasors representing the alternating currents in two branches of a circuit. Phasor *PQ* is 20.0 A and is horizontal. Phasor *QR* (which is joined to the end of *PQ* to form triangle *PQR*) is 14.0 A and is at an angle of 35° to the horizontal. Determine the resultant phasor *PR* and the angle it makes with phasor *PQ*

7. Calculate, correct to 3 significant figures, the co-ordinates *x* and *y* to locate the hole centre at *P* shown in Figure 11.37

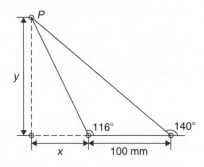

Figure 11.37

8. 16 holes are equally spaced on a pitch circle of 70 mm diameter. Determine the length of the chord joining the centres of two adjacent holes

For fully worked solutions to each of the problems in Exercises 49 to 54 in this chapter, go to the website:
www.routledge.com/cw/bird

Revision Test 5: Straight line graphs and trigonometry

This assignment covers the material contained in chapters 10 and 11. *The marks for each question are shown in brackets at the end of each question.*

1. Determine the value of P in the following table of values

x	0	1	4
$y = 3x - 5$	-5	-2	P

    (2)

2. Corresponding values obtained experimentally for two quantities are:

x	-5	-3	-1	0	2	4
y	-17	-11	-5	-2	4	10

    Plot a graph of y (vertically) against x (horizontally) to scales of $1 \text{ cm} = 1$ for the horizontal x-axis and $1 \text{ cm} = 2$ for the vertical y-axis.
    From the graph find:

    (a) the value of y when $x = 3$

    (b) the value of y when $x = -4$

    (c) the value of x when $y = 1$

    (d) the value of x when $y = -20$ (8)

3. If graphs of y against x were to be plotted for each of the following, state (i) the gradient, and (ii) the y-axis intercept.
    (a) $y = -5x + 3$ (b) $y = 7x$ (c) $2y + 4 = 5x$
    (d) $5x + 2y = 6$ (e) $2x - \dfrac{y}{3} = \dfrac{7}{6}$ (10)

4. The resistance R ohms of a copper winding is measured at various temperatures $t°C$ and the results are as follows:

R (Ω)	38	47	55	62	72
t (°C)	16	34	50	64	84

    Plot a graph of R (vertically) against t (horizontally) and find from it (a) the temperature when the resistance is 50 Ω (b) the resistance when the temperature is 72°C (c) the gradient (d) the equation of the graph.   (10)

5. In triangle JKL in Figure RT5.1, find (a) length KJ correct to 3 significant figures, (b) sin L and tan K, each correct to 3 decimal places.   (4)

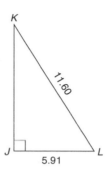

Figure RT5.1

6. Two ships leave a port at the same time. Ship X travels due west at 30 km/h and ship Y travels due north. After 4 hours the two ships are 130 km apart. Calculate the velocity of ship Y.   (4)

7. If $\sin A = \dfrac{12}{37}$ find tan A in fraction form.   (3)

8. Evaluate $5 \tan 62°11'$ correct to 3 significant figures.   (2)

9. Determine the acute angle $\cos^{-1} 0.3649$ in degrees and minutes.   (2)

10. In triangle PQR in Figure RT5.2, find angle P in decimal form, correct to 2 decimal places.   (2)

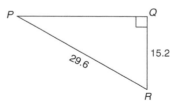

Figure RT5.2

11. Evaluate, correct to 3 significant figures:
$3 \tan 81.27° − 5 \cos 7.32° − 6 \sin 54.81°$ (2)

12. In triangle ABC in Figure RT5.3, find lengths AB and AC, correct to 2 decimal places. (4)

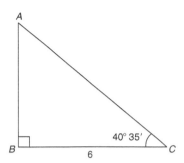

Figure RT5.3

13. From a point P, the angle of elevation of a 40 m high electricity pylon is 20°. How far is point P from the base of the pylon, correct to the nearest metre? (3)

14. A triangular plot of land ABC is shown in Figure RT5.4. Solve the triangle and determine its area. (10)

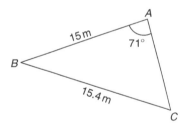

Figure RT5.4

15. A car is travelling 20 m above sea level. It then travels 500 m up a steady slope of 17°. Determine, correct to the nearest metre, how high the car is now above sea level. (3)

16. Figure RT5.5 shows a roof truss PQR with rafter $PQ = 3$ m. Calculate the length of (a) the roof rise PP', (b) rafter PR, and (c) the roof span QR. Find also (d) the cross-sectional area of the roof truss. (11)

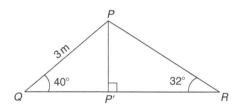

Figure RT5.5

17. Solve triangle ABC given $b = 10$ cm, $c = 15$ cm and $\angle A = 60°$. (10)

For lecturers/instructors/teachers, fully worked solutions to each of the problems in Revision Test 5, together with a full marking scheme, are available at the website:
www.routledge.com/cw/bird

Chapter 12

Areas of common shapes

Why it is important to understand: **Areas of common shapes**

To paint, wallpaper or panel a wall, you must know the total area of the wall so you can buy the appropriate amount of finish. When designing a new building, or seeking planning permission, it is often necessary to specify the total floor area of the building. In construction, calculating the area of a gable end of a building is important when determining the amount of bricks and mortar to order. When using a bolt, the most important thing is that it is long enough for your particular application and it may also be necessary to calculate the shear area of the bolt connection. Ridge vents allow a home to properly vent, while disallowing rain or other forms of precipitation to leak into the attic or crawlspace underneath the roof. Equal amounts of cool air and warm air flowing through the vents is paramount for proper heat exchange. Calculating how much surface area is available on the roof aids in determining how long the ridge vent should run. Arches are everywhere, from sculptures and monuments to pieces of architecture and strings on musical instruments; finding the height of an arch or its cross-sectional area is often required. Determining the cross-sectional areas of beam structures is vitally important in design engineering. There are thus a large number of situations in engineering where determining area is important.

At the end of this chapter, you should be able to:

- state the SI unit of area
- identify common polygons – triangle, quadrilateral, pentagon, hexagon, heptagon and octagon
- identify common quadrilaterals – rectangle, square, parallelogram, rhombus and trapezium
- calculate areas of common shapes
- appreciate that areas of similar shapes are proportional to the squares of the corresponding linear dimensions

12.1   Introduction

Area is a measure of the size or extent of a plane surface. It is measured in **square units** such as mm², cm² and m². This chapter deals with finding areas of common shapes.

In engineering it is often important to be able to calculate simple areas of various shapes. In everyday life it is important to be able to measure area to, say, lay a carpet,

Science and Mathematics for Engineering. 978-0-367-20475-4. © John Bird. Published by Taylor & Francis. All rights reserved.

or to order sufficient paint for a decorating job, or to order sufficient bricks for a new wall.

On completing this chapter you will be able to recognise common shapes and be able to find the areas of rectangles, squares, parallelograms, triangles, trapeziums and circles.

12.2   Common shapes

12.2.1   Polygons

A polygon is a closed plane figure bounded by straight lines. A polygon which has:

(a)   3 sides is called a **triangle** – see Figure 12.1(a)

(b)   4 sides is called a **quadrilateral** – see Figure 12.1(b)

(c)   5 sides is called a **pentagon** – see Figure 12.1(c)

(d)   6 sides is called a **hexagon** – see Figure 12.1(d)

(e)   7 sides is called a **heptagon** – see Figure 12.1(e)

(f)   8 sides is called an **octagon** – see Figure 12.1(f)

12.2.2   Quadrilaterals

There are five types of quadrilateral, these being rectangle, square, parallelogram, rhombus and trapezium.

If the opposite corners of any quadrilateral are joined by a straight line, two triangles are produced. Since the sum of the angles of a triangle is 180°, the sum of the angles of a quadrilateral is 360°.

Rectangle

In the rectangle *ABCD* shown in Figure 12.2:

(i)   all four angles are right angles,

(ii)   opposite sides are parallel and equal in length, and

(iii)   diagonals *AC* and *BD* are equal in length and bisect one another.

Square

In the square *PQRS* shown in Figure 12.3:

(i)   all four angles are right angles,

(ii)   opposite sides are parallel,

(iii)   all four sides are equal in length, and

(iv)   diagonals *PR* and *QS* are equal in length and bisect one another at right angles.

Parallelogram

In the parallelogram *WXYZ* shown in Figure 12.4:

(i)   opposite angles are equal,

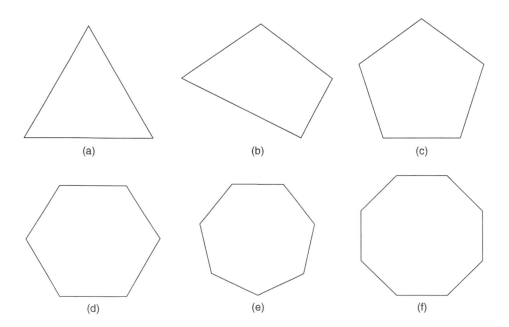

(a)   (b)   (c)

(d)   (e)   (f)

Figure 12.1

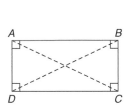

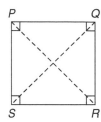

Figure 12.2

Figure 12.3

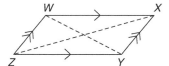

Figure 12.4

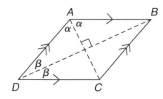

Figure 12.5

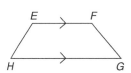

Figure 12.6

(ii)   opposite sides are parallel and equal in length, and

(iii)   diagonals *WY* and *XZ* bisect one another.

Rhombus

In the rhombus *ABCD* shown in Figure 12.5:

 (i)   opposite angles are equal,

 (ii)   opposite angles are bisected by a diagonal,

 (iii)   opposite sides are parallel,

 (iv)   all four sides are equal in length, and

 (v)   diagonals *AC* and *BD* bisect one another at right angles.

Trapezium

In the trapezium *EFGH* shown in Figure 12.6:

(i)   only one pair of sides is parallel.

Problem 1.   State the types of quadrilateral shown in Figure 12.7 and determine the angles marked *a* to *l*

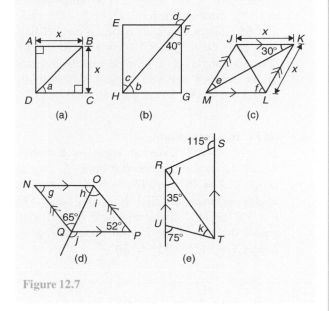

Figure 12.7

(a)   **ABCD is a square**
The diagonals of a square bisect each of the right angles, hence

$$a = \frac{90°}{2} = 45°$$

(b)   **EFGH is a rectangle**
In triangle *FGH*, $40° + 90° + b = 180°$, since angles in a triangle add up to $180°$, from which, **b = 50°**. Also, **c = 40°** (alternate angles between parallel lines *EF* and *HG*). (Alternatively, *b* and *c* are complementary, i.e. add up to $90°$) $d = 90° + c$ (external angle of a triangle equals the sum of the interior opposite angles), hence **d = 90° + 40° = 130°** (or ∠EFH = 50° and **d = 180° − 50° = 130°**)

(c)   **JKLM is a rhombus**
The diagonals of a rhombus bisect the interior angles and opposite internal angles are equal. Thus, ∠JKM = ∠MKL = ∠JMK = ∠LMK = 30°, hence **e = 30°**. In triangle *KLM*, 30° + ∠KLM + 30° = 180° (angles in a triangle add up to 180°), hence, ∠KLM = 120°. The diagonal *JL* bisects ∠KLM, hence, $f = \frac{120°}{2} = 60°$

(d) *NOPQ* **is a parallelogram**

$g = 52°$ since opposite interior angles of a parallelogram are equal. In triangle *NOQ*, $g + h + 65° = 180°$ (angles in a triangle add up to 180°), from which, $h = 180° - 65° - 52° = 63°$. $i = 65°$ (alternate angles between parallel lines *NQ* and *OP*), $j = 52° + i = 52° + 65° = 117°$ (external angle of a triangle equals the sum of the interior opposite angles). Alternatively, $\angle PQO = h = 63°$; hence, $j = 180° - 63° = 117°$

(e) *RSTU* **is a trapezium**

$35° + k = 75°$ (external angle of a triangle equals the sum of the interior opposite angles), hence, $k = 40°$. $\angle STR = 35°$ (alternate angles between parallel lines *RU* and *ST*). $l + 35° = 115°$ (external angle of a triangle equals the sum of the interior opposite angles), hence, $l = 115° - 35° = 80°$

Now try the following Practice Exercise

Practice Exercise 55   Common shapes (answers on page 537)

1. Find the angles p and q in Figure 12.8(a)
2. Find the angles r and s in Figure 12.8(b)
3. Find the angle t in Figure 12.8(c)

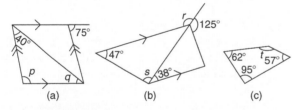

(a)   (b)   (c)

Figure 12.8

12.3   Calculating areas of common shapes

The formulae for the areas of common shapes are shown in Table 12.1.
Here are some worked problems to demonstrate how the formulae are used to determine the area of common shapes.

Problem 2.   Calculate the area and length of perimeter of the square shown in Figure 12.9

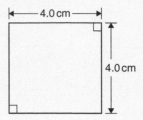

Figure 12.9

$$\textbf{Area of square} = x^2 = (4.0)^2 = 4.0\,\text{cm} \times 4.0\,\text{cm}$$
$$= \textbf{16.0}\,\textbf{cm}^2$$

(Note the unit of area is cm × cm = cm², i.e. square centimetres or centimetres squared.)

$$\textbf{Perimeter of square} = 4.0\,\text{cm} + 4.0\,\text{cm} + 4.0\,\text{cm}$$
$$+ 4.0\,\text{cm} = \textbf{16.0}\,\textbf{cm}^2$$

Problem 3.   Calculate the area and length of perimeter of the rectangle shown in Figure 12.10

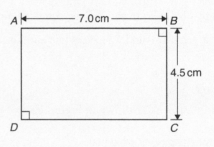

Figure 12.10

$$\textbf{Area of rectangle} = l \times b = 7.0 \times 4.5$$
$$= \textbf{31.5}\,\textbf{cm}^2$$

$$\textbf{Perimeter of rectangle} = 7.0\,\text{cm} + 4.5\,\text{cm}$$
$$+ 7.0\,\text{cm} + 4.5\,\text{cm}$$
$$= \textbf{23.0}\,\textbf{cm}$$

Table 12.1 **Formulae for the areas of common shapes**

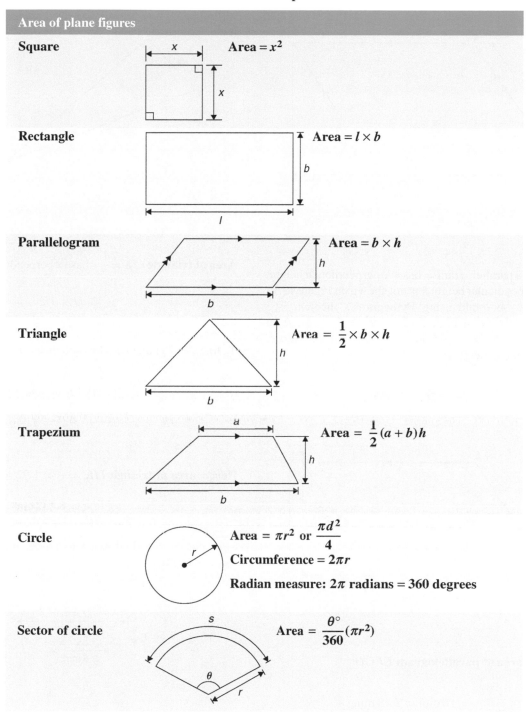

Area of plane figures	
Square	Area $= x^2$
Rectangle	Area $= l \times b$
Parallelogram	Area $= b \times h$
Triangle	Area $= \dfrac{1}{2} \times b \times h$
Trapezium	Area $= \dfrac{1}{2}(a+b)h$
Circle	Area $= \pi r^2$ or $\dfrac{\pi d^2}{4}$ Circumference $= 2\pi r$ Radian measure: 2π radians $= 360$ degrees
Sector of circle	Area $= \dfrac{\theta^\circ}{360}(\pi r^2)$

Problem 4. Calculate the area of the parallelogram shown in Figure 12.11

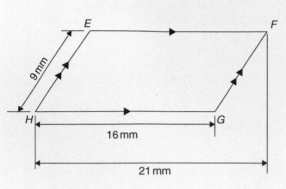

Figure 12.11

Area of a parallelogram = base × perpendicular height. The perpendicular height h is not shown on Figure 12.11 but may be found using Pythagoras's theorem (see chapter 10). From Figure 12.12, $9^2 = 5^2 + h^2$, from which, $h^2 = 9^2 - 5^2 = 81 - 25 = 56$. Hence, perpendicular height,

$$h = \sqrt{56} = 7.48 \, \text{mm}$$

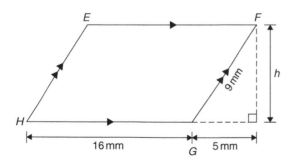

Figure 12.12

Hence, **area of parallelogram EFGH**

$$= 16 \, \text{mm} \times 7.48 \, \text{mm}$$

$$= \mathbf{120 \, mm^2}$$

Problem 5. Calculate the area of the triangle shown in Figure 12.13

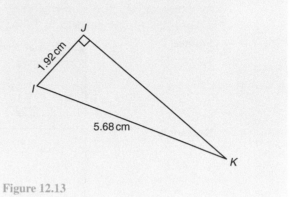

Figure 12.13

Area of triangle IJK $= \dfrac{1}{2} \times$ base × perpendicular height

$$= \dfrac{1}{2} \times IJ \times JK$$

To find JK, Pythagoras's theorem is used, i.e.

$$5.68^2 = 1.92^2 + JK^2 \text{ from which,}$$

$$JK = \sqrt{5.68^2 - 1.92^2} = 5.346 \, \text{cm}$$

Hence, **area of triangle IJK** $= \dfrac{1}{2} \times 1.92 \times 5.346$

$$= \mathbf{5.132 \, cm^2}$$

Problem 6. Calculate the area of the trapezium shown in Figure 12.14

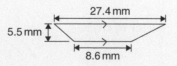

Figure 12.14

Area of a trapezium $= \dfrac{1}{2} \times$ (sum of parallel sides)

× (perpendicular distance

between the parallel sides).

Hence, **area of trapezium** *LMNO*

$$= \frac{1}{2} \times (27.4 + 8.6) \times 5.5$$

$$= \frac{1}{2} \times 36 \times 5.5 = \mathbf{99\,mm^2}$$

Problem 7. A rectangular tray is 820 mm long and 400 mm wide. Find its area in (a) mm², (b) cm², (c) m²

(a) **Area of tray** = length × width

$$= 820 \times 400$$

$$= \mathbf{328{,}000\,mm^2}$$

(b) Since 1 cm = 10 mm, then 1 cm² = 1 cm × 1 cm

$$= 10\,mm \times 10\,mm = 100\,mm^2$$

or $1\,mm^2 = \dfrac{1}{100}\,cm^2 = 0.01\,cm^2$

Hence, **328,000 mm²** = 328,000 × 0.01 cm²

$$= \mathbf{3280\,cm^2}$$

(c) Since 1 m = 100 cm, then 1 m² = 1 m × 1 m

$$= 100\,cm \times 100\,cm = 10{,}000\,cm^2$$

or $1\,cm^2 = \dfrac{1}{10{,}000}\,m^2 = 0.0001\,m^2$

Hence, **3280 cm²** = 3280 × 0.0001 m²

$$= \mathbf{0.3280\,m^2}$$

Problem 8. The outside measurements of a picture frame are 100 cm by 50 cm. If the frame is 4 cm wide, find the area of the wood used to make the frame

A sketch of the frame is shown shaded in Figure 12.15.

Area of wood = area of large rectangle

$$- \text{ area of small rectangle}$$

$$= (100 \times 50) - (92 \times 42)$$

$$= 5000 - 3864 = \mathbf{1136\,cm^2}$$

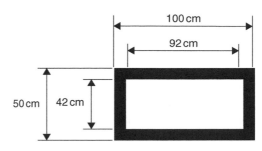

Figure 12.15

Problem 9. Find the cross-sectional area of the girder shown in Figure 12.16

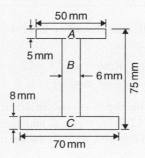

Figure 12.16

The girder may be divided into three separate rectangles as shown.

Area of rectangle $A = 50 \times 5 = 250\,mm^2$

Area of rectangle $B = (75 - 8 - 5) \times 6$

$$= 62 \times 6 = 372\,mm^2$$

Area of rectangle $C = 70 \times 8 = 560\,mm^2$

Total area of girder $= 250 + 372 + 560$

$$= \mathbf{1182\,mm^2} \text{ or } \mathbf{11.82\,cm^2}$$

Problem 10. Figure 12.17 shows the gable end of a building. Determine the area of brickwork in the gable end

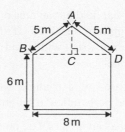

Figure 12.17

The shape is that of a rectangle and a triangle.

$$\text{Area of rectangle} = 6 \times 8 = 48\,\text{m}^2$$

$$\text{Area of triangle} = \frac{1}{2} \times \text{base} \times \text{height}$$

$CD = 4$ m, $AD = 5$ m, hence $AC = 3$ m (since it is a 3, 4, 5 triangle – or by Pythagoras).

Hence, area of triangle $ABD = \frac{1}{2} \times 8 \times 3 = 12\,\text{m}^2$.

Total area of brickwork $= 48 + 12$

$$= \textbf{60}\,\textbf{m}^\textbf{2}$$

Now try the following Practice Exercise

Practice Exercise 56   Areas of common shapes (answers on page 538)

1. Name the types of quadrilateral shown in Figure 12.18(i) to (iv), and determine for each (a) the area, and (b) the perimeter

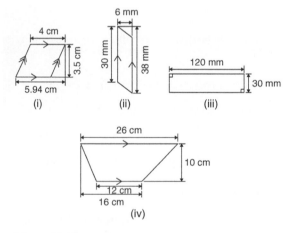

Figure 12.18

2. A rectangular plate is 85 mm long and 42 mm wide. Find its area in square centimetres

3. A rectangular field has an area of 1.2 hectares and a length of 150 m. If 1 hectare = 10,000 m² find (a) its width, and (b) the length of a diagonal

4. Find the area of a triangle whose base is 8.5 cm and perpendicular height 6.4 cm

5. A square has an area of 162 cm². Determine the length of a diagonal

6. A rectangular picture has an area of 0.96 m². If one of the sides has a length of 800 mm, calculate, in millimetres, the length of the other side

7. Determine the area of each of the angle iron sections shown in Figure 12.19

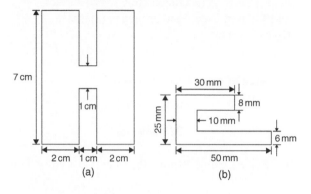

Figure 12.19

8. Figure 12.20 shows a 4 m wide path around the outside of a 41 m by 37 m garden. Calculate the area of the path

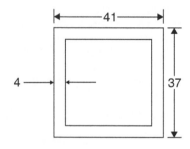

Figure 12.20

9. The area of a trapezium is 13.5 cm² and the perpendicular distance between its parallel sides is 3 cm. If the length of one of the parallel sides is 5.6 cm, find the length of the other parallel side

10. Calculate the area of the steel plate shown in Figure 12.21

11. Determine the area of an equilateral triangle of side 10.0 cm

12. If paving slabs are produced in 250 mm by 250 mm squares, determine the number of slabs required to cover an area of 2 m²

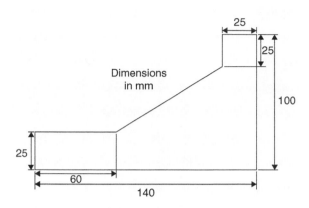

Figure 12.21

Here are some further worked problems on finding areas of common shapes, using the formulae in Table 12.1, page 127.

Problem 11. Find the area of the circle having a radius of 5 cm

$$\textbf{Area of circle} = \pi r^2 = \pi(5)^2 = 25\pi$$
$$= \textbf{78.54 cm}^2$$

Problem 12. Find the area of the circle having a diameter of 15 mm

$$\textbf{Area of circle} = \frac{\pi d^2}{4} = \frac{\pi(15)^2}{4} = \frac{225\pi}{4}$$
$$= \textbf{176.7 mm}^2$$

Problem 13. Find the area of the circle having a circumference of 70 mm

Circumference, $c = 2\pi r$, hence

$$\text{radius, } r = \frac{c}{2\pi} = \frac{70}{2\pi} = \frac{35}{\pi} \text{ mm.}$$
$$\textbf{Area of circle} = \pi r^2 = \pi \left(\frac{35}{\pi}\right)^2 = \frac{35^2}{\pi}$$
$$= \textbf{389.9 mm}^2 \text{ or } \textbf{3.899 cm}^2$$

Problem 14. Calculate the area of the sector of a circle having diameter 80 mm with angle subtended at the centre $107°42'$

If diameter $= 80$ mm, then radius, $r = 40$ mm, and

$$\text{Area of sector} = \frac{107°42'}{360}(\pi 40^2) = \frac{107\dfrac{42}{60}}{360}(\pi 40^2)$$
$$= \frac{107.7}{360}(\pi 40^2)$$
$$= \textbf{1504 mm}^2 \text{ or } \textbf{15.04 cm}^2$$

Problem 15. A hollow shaft has an outside diameter of 5.45 cm and an inside diameter of 2.25 cm. Calculate the cross-sectional area of the shaft

The cross-sectional area of the shaft is shown by the shaded part in Figure 12.22 (often called an **annulus**).

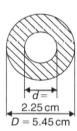

Figure 12.22

Area of shaded part = area of large circle
$$\quad - \text{ area of small circle}$$
$$= \frac{\pi D^2}{4} - \frac{\pi d^2}{4} = \frac{\pi}{4}(D^2 - d^2)$$
$$= \frac{\pi}{4}(5.45^2 - 2.25^2)$$
$$= \textbf{19.35 cm}^2$$

Now try the following Practice Exercise

Practice Exercise 57   Areas of common shapes (answers on page 538)

1. A rectangular garden measures 40 m by 15 m. A 1 m flower border is made round

the two shorter sides and one long side. A circular swimming pool of diameter 8 m is constructed in the middle of the garden. Find, correct to the nearest square metre, the area remaining

2. Determine the area of circles having (a) a radius of 4 cm (b) a diameter of 30 mm (c) a circumference of 200 mm

3. An annulus has an outside diameter of 60 mm and an inside diameter of 20 mm. Determine its area

4. If the area of a circle is 320 mm², find (a) its diameter, and (b) its circumference

5. Calculate the areas of the following sectors of circles:

   (a) radius 9 cm, angle subtended at centre 75°

   (b) diameter 35 mm, angle subtended at centre 48°37′

6. Determine the shaded area of the template shown in Figure 12.23

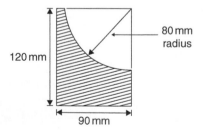

Figure 12.23

7. An archway consists of a rectangular opening topped by a semi-circular arch as shown in Figure 12.24. Determine the area of the opening if the width is 1 m and the greatest height is 2 m

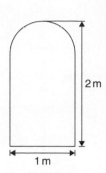

Figure 12.24

Here are some further worked problems involving common shapes.

Problem 16. Calculate the area of a regular octagon, if each side is 5 cm and the width across the flats is 12 cm

An octagon is an 8-sided polygon. If radii are drawn from the centre of the polygon to the vertices then 8 equal triangles are produced, as shown in Figure 12.25.

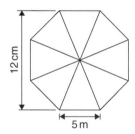

Figure 12.25

$$\text{Area of one triangle} = \frac{1}{2} \times \text{base} \times \text{height}$$

$$= \frac{1}{2} \times 5 \times \frac{12}{2} = 15 \, \text{cm}^2$$

Area of octagon $= 8 \times 15 = \mathbf{120 \, cm^2}$

Problem 17. Determine the area of a regular hexagon which has sides 8 cm long

A hexagon is a 6-sided polygon which may be divided into 6 equal triangles as shown in Figure 12.26.

The angle subtended at the centre of each triangle is $360°/6 = 60°$. The other two angles in the triangle add up to $120°$ and are equal to each other. Hence each of the triangles is equilateral with each angle $60°$ and each side 8 cm.

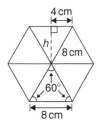

Figure 12.26

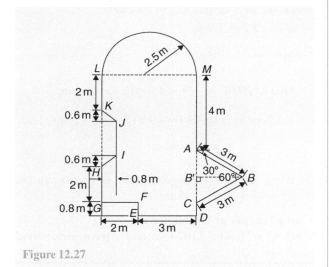

Figure 12.27

Area of one triangle

$$= \frac{1}{2} \times \text{ base } \times \text{ height}$$

$$= \frac{1}{2} \times 8 \times h$$

h is calculated using Pythagoras' theorem:

$$8^2 = h^2 + 4^2$$

from which,

$$h = \sqrt{8^2 - 4^2}$$

$$= 6.928 \, \text{cm}$$

Hence,

$$\text{Area of one triangle } = \frac{1}{2} \times 8 \times 6.928$$

$$= 27.71 \, \text{cm}^2$$

Area of hexagon $= 6 \times 27.71$

$$= \mathbf{166.3 \, cm^2}$$

Problem 18. Figure 12.27 shows a plan of a floor of a building which is to be carpeted. Calculate the area of the floor in square metres. Calculate the cost, correct to the nearest pound, of carpeting the floor with carpet costing £16.80 per m^2, assuming 30% extra carpet is required due to wastage in fitting

Area of floor plan

$$= \text{area of triangle } ABC$$

$$+ \text{ area of semicircle}$$

$$+ \text{ area of rectangle } CGLM$$

$$+ \text{ area of rectangle } CDEF$$

$$- \text{ area of trapezium } HIJK$$

Triangle ABC is equilateral since $AB = BC = 3$ m and hence angle $B'CB = 60°$.

i.e.
$$\sin B'CB = BB'/3$$
$$BB' = 3 \sin 60° = 2.598 \, \text{m}$$

$$\text{Area of triangle } ABC = \frac{1}{2}(AC)(BB')$$

$$= \frac{1}{2}(3)(2.598)$$

$$= 3.897 \, \text{m}^2$$

$$\text{Area of semicircle } = \frac{1}{2}\pi r^2 = \frac{1}{2}\pi(2.5)^2$$

$$= 9.817 \, \text{m}^2$$

$$\text{Area of } CGLM = 5 \times 7 = 35 \, \text{m}^2$$

$$\text{Area of } CDEF = 0.8 \times 3 = 2.4 \, \text{m}^2$$

$$\text{Area of } HIJK = \frac{1}{2}(KH + IJ)(0.8)$$

Since $MC = 7$ m then $LG = 7$ m, hence
$JI = 7 - 5.2 = 1.8$ m

Hence,

$$\text{Area of } HIJK = \frac{1}{2}(3 + 1.8)(0.8) = 1.92 \, \text{m}^2$$

Total floor area $= 3.897 + 9.817 + 35 + 2.4 - 1.92$

$$= 49.194 \, \text{m}^2$$

To allow for 30% wastage, amount of carpet required
$= 1.3 \times 49.194 = 63.95 \, \text{m}^2$

Cost of carpet at £16.80 per m²
$= 63.95 \times 16.80 = £1074$, correct to the nearest pound.

Now try the following Practice Exercise

Practice Exercise 58   Areas of common shapes (answers on page 538)

1. Calculate the area of a regular octagon if each side is 20 mm and the width across the flats is 48.3 mm

2. Determine the area of a regular hexagon which has sides 25 mm

3. A plot of land is in the shape shown in Figure 12.28. Determine

    (a)   its area in hectares ($1 \, \text{ha} = 10^4 \, \text{m}^2$), and

    (b)   the length of fencing required, to the nearest metre, to completely enclose the plot of land

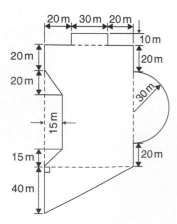

Figure 12.28

12.4   Areas of similar shapes

Figure 12.29 shows two squares, one of which has sides three times as long as the other.

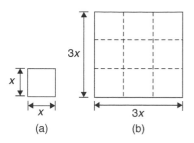

Figure 12.29

$$\text{Area of Figure 12.29(a)} = (x)(x) = x^2$$

$$\text{Area of Figure 12.29(b)} = (3x)(3x) = 9x^2$$

Hence, Figure 12.29(b) has an area $(3)^2$, i.e. 9 times the area of Figure 12.29(a).

Summarising, **the areas of similar shapes are proportional to the squares of corresponding linear dimensions**.

Problem 19.   A rectangular garage is shown on a building plan having dimensions 10 mm by 20 mm. If the plan is drawn to a scale of 1 to 250, determine the true area of the garage in square metres

$$\text{Area of garage on the plan} = 10 \, \text{mm} \times 20 \, \text{mm}$$

$$= 200 \, \text{mm}^2$$

Since the areas of similar shapes are proportional to the squares of corresponding dimensions then:

$$\text{True area of garage} = 200 \times (250)^2$$

$$= 12.5 \times 10^6 \, \text{mm}^2$$

$$= \frac{12.5 \times 10^6}{10^6} \, \text{m}^2$$

$$\text{since } 1 \, \text{m}^2 = 10^6 \, \text{mm}^2$$

$$= \mathbf{12.5 \, m^2}$$

Now try the following Practice Exercise

Practice Exercise 59 Areas of similar shapes (answers on page 538)

1. The area of a park on a map is 500 mm^2. If the scale of the map is 1 to 40,000 determine the true area of the park in hectares (1 hectare = 10^4 m^2)

2. A model of a boiler is made having an overall height of 75 mm corresponding to an overall height of the actual boiler of 6 m. If the area of metal required for the model is 12,500 mm^2, determine, in square metres, the area of metal required for the actual boiler

3. The scale of an Ordnance Survey map is 1:2500. A circular sports field has a diameter of 8 cm on the map. Calculate its area in hectares, giving your answer correct to 3 significant figures (1 hectare = 10^4 m^2)

For fully worked solutions to each of the problems in Exercises 55 to 59 in this chapter, go to the website:
www.routledge.com/cw/bird

Chapter 13

The circle

Why it is important to understand: **The circle**

A circle is one of the fundamental shapes of geometry; it consists of all the points that are equidistant from a central point. Knowledge of calculations involving circles is needed with crank mechanisms, with determinations of latitude and longitude, with pendulums, and even in the design of paper clips. The floodlit area at a football ground, the area an automatic garden sprayer sprays and the angle of lap of a belt drive all rely on calculations involving the arc of a circle. The ability to handle calculations involving circles and their properties is clearly essential in several branches of engineering design.

At the end of this chapter, you should be able to:

- define a circle
- state some properties of a circle – including radius, circumference, diameter, semicircle, quadrant, tangent, sector, chord, segment and arc
- appreciate the angle in a semicircle is a right angle
- define a radian, and change radians to degrees, and vice versa
- determine arc length, area of a circle and area of a sector of a circle

13.1 Introduction

A **circle** is a plain figure enclosed by a curved line, every point on which is equidistant from a point within, called the **centre**.

In chapter 11, worked problems on the area of circles and sectors was demonstrated. In this chapter, properties of circles are listed, arc lengths are calculated, together with more practical examples on areas of sectors of circles.

13.2 Properties of circles

(a) The distance from the centre to the curve is called the **radius** *r* of the circle (see *OP* in Figure 13.1).

(b) The boundary of a circle is called the **circumference** *c*.

(c) Any straight line passing through the centre and touching the circumference at each end is called

Science and Mathematics for Engineering. 978-0-367-20475-4, © John Bird. Published by Taylor & Francis. All rights reserved.

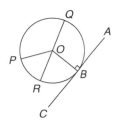

Figure 13.1

the **diameter** *d* (see *QR* in Figure 13.1). Thus, **d = 2r**.

(d) The ratio $\dfrac{\text{circumference}}{\text{diameter}}$ = a constant for any circle. This constant is denoted by the Greek letter π (pronounced 'pie'), where $\pi = 3.14159$, correct to 5 decimal places (check with your calculator). Hence, $\dfrac{c}{d} = \pi$, or $c = \pi d$, or $c = 2\pi r$

(e) A **semicircle** is one half of a whole circle.

(f) A **quadrant** is one quarter of a whole circle.

(g) A **tangent** to a circle is a straight line which meets the circle at one point only and does not cut the circle when produced. *AC* in Figure 13.1 is a tangent to the circle since it touches the curve at point *B* only. If radius *OB* is drawn, then **angle ABO is a right angle.**

(h) A **sector** of a circle is the part of a circle between radii (for example, the portion *OXY* of Figure 13.2 is a sector). If a sector is less than a semicircle it is called a **minor sector**, if greater than a semicircle it is called a **major sector**.

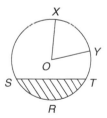

Figure 13.2

(i) A **chord** of a circle is any straight line which divides the circle into two parts and is terminated at each end by the circumference. *ST* in Figure 13.2 is a chord.

(j) A **segment** is the name given to the parts into which a circle is divided by a chord. If the segment is less than a semicircle it is called a **minor**

segment (see shaded area in Figure 13.2). If the segment is greater than a semicircle it is called a **major segment** (see the un-shaded area in Figure 13.2).

(k) An **arc** is a portion of the circumference of a circle. The distance *SRT* in Figure 13.2 is called a **minor arc** and the distance *SXYT* is called a **major arc**.

(l) The angle at the centre of a circle, subtended by an arc, is double the angle at the circumference subtended by the same arc. With reference to Figure 13.3,

Angle *AOC* = 2 × angle *ABC*

Figure 13.3

(m) The angle in a semicircle is a right angle (see angle *BQP* in Figure 13.3).

Problem 1. Find the circumference of a circle of radius 12.0 cm

Circumference, $c = 2 \times \pi \times \text{radius} = 2\pi r = 2\pi(12.0)$

$$= \mathbf{75.40\,cm}$$

Problem 2. If the diameter of a circle is 75 mm, find its circumference

Circumference, $c = \pi \times \text{diameter} = \pi d = \pi(75)$

$$= \mathbf{235.6\,mm}$$

Problem 3. Determine the radius of a circular pond if its perimeter is 112 m

Perimeter = circumference, $c = 2\pi r$

Hence, **radius of pond,** $r = \dfrac{c}{2\pi} = \dfrac{112}{2\pi} = \mathbf{17.83\,cm}$

Problem 4.   In Figure 13.4, *AB* is a tangent to the circle at *B*. If the circle radius is 40 mm and *AB* = 150 mm, calculate the length *AO*

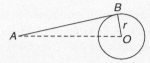

Figure 13.4

A tangent to a circle is at right angles to a radius drawn from the point of contact, i.e. **ABO = 90°**. Hence, using Pythagoras's theorem:

$$AO^2 = AB^2 + OB^2$$

from which, $AO = \sqrt{AB^2 + OB^2}$

$$= \sqrt{150^2 + 40^2} = \mathbf{155.2\,mm}$$

Now try the following Practice Exercise

Practice Exercise 60   Properties of a circle (answers on page 538)

1.   Calculate the length of the circumference of a circle of radius 7.2 cm

2.   If the diameter of a circle is 82.6 mm, calculate the circumference of the circle

3.   Determine the radius of a circle whose circumference is 16.52 cm

4.   Find the diameter of a circle whose perimeter is 149.8 cm

5.   A crank mechanism is shown in Figure 13.5, where *XY* is a tangent to the circle at point *X*. If the circle radius *OX* is 10 cm and length *OY* is 40 cm, determine the length of the connecting rod *XY*

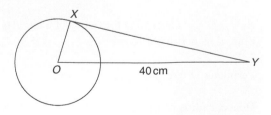

Figure 13.5

6.   If the circumference of the earth is 40,000 km at the equator, calculate its diameter

7.   Calculate the length of wire in the paper clip shown in Figure 13.6. The dimensions are in millimetres

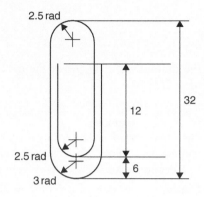

Figure 13.6

13.3   Radians and degrees

One **radian** is defined as the angle subtended at the centre of a circle by an arc equal in length to the radius.

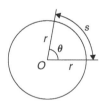

Figure 13.7

With reference to Figure 13.7, for arc length *s*,

$$\theta \text{ radians} = \frac{s}{r}$$

When *s* = whole circumference (= $2\pi r$) then

$$\theta = \frac{s}{r} = \frac{2\pi r}{r} = 2\pi$$

i.e.   **2π radians = 360°** or **π radians = 180°**

Thus, $\mathbf{1\,rad} = \dfrac{\mathbf{180°}}{\pi} = \mathbf{57.30°}$, correct to 2 decimal places

Since π rad $= 180°$, then $\dfrac{\pi}{2} = 90°$, $\dfrac{\pi}{3} = 60°$, $\dfrac{\pi}{4} = 45°$ and so on.

Problem 5.   Convert to radians: (a) 125°, (b) 69°47′

(a)   Since $180° = \pi$ rad, then $1° = \dfrac{\pi}{180}$ rad,

therefore $125° = 125\left(\dfrac{\pi}{180}\right)$ rad $=$ **2.182 radians**

(b) $69°47' = 69\dfrac{47°}{60} = 69.783°$ (or, with your calculator, enter 69°47′ using the ° ′ ″ function, press = and press ° ′ ″ again)

and $69.783° = 69.783\left(\dfrac{\pi}{180}\right)$ rad

$=$ **1.218 radians**.

Problem 6.   Convert to degrees and minutes: (a) 0.749 radians, (b) $3\pi/4$ radians

(a)   Since π rad $= 180°$, then 1 rad $= \dfrac{180°}{\pi}$

therefore 0.749 rad $= 0.749\left(\dfrac{180}{\pi}\right)° = 42.915°$

$0.915° = (0.915 \times 60)' = 55'$, correct to the nearest minute.

Hence, **0.749 radians $= 42°55'$**

(b)   Since 1 rad $= \left(\dfrac{180}{\pi}\right)°$ then

$\dfrac{3\pi}{4}$ rad $= \dfrac{3\pi}{4}\left(\dfrac{180}{\pi}\right)° = \dfrac{3}{4}(180)° =$ **135°**

Problem 7.   Express in radians, in terms of π, (a) 150°, (b) 270°, (c) 37.5°

Since $180° = \pi$ rad, then $1° = \dfrac{\pi}{180}$ rad

(a) $150° = 150\left(\dfrac{\pi}{180}\right)$ rad $= \dfrac{5\pi}{6}$ **rad**

(b) $270° = 270\left(\dfrac{\pi}{180}\right)$ rad $= \dfrac{3\pi}{2}$ **rad**

(c) $37.5° = 37.5\left(\dfrac{\pi}{180}\right)$ rad $= \dfrac{75\pi}{360}$ rad $= \dfrac{5\pi}{24}$ **rad**

Now try the following Practice Exercise

Practice Exercise 61   Radians and degrees (answers on page 538)

1.   Convert to radians in terms of π:
     (a)   30°,   (b)   75°,   (c)   225°

2.   Convert to radians, correct to 3 decimal places:
     (a)   48°,   (b)   84°51′,   (c)   232°15′

3.   Convert to degrees:
     (a) $\dfrac{7\pi}{6}$ rad,   (b) $\dfrac{4\pi}{9}$ rad,   (c) $\dfrac{7\pi}{12}$ rad

4.   Convert to degrees and minutes:
     (a)   0.0125 rad,   (b)   2.69 rad,
     (c)   7.241 rad

5.   A car engine speed is 1000 rev/min. Convert this speed into rad/s.

13.4   Arc length and area of circles and sectors

13.4.1   Arc length

From the definition of the radian in the previous section and Figure 13.7,

arc length, $s = r\theta$ where θ is in radians.

13.4.2   Area of circle

From chapter 12, for any circle, area $= \pi \times (\text{radius})^2$

i.e.   **area $= \pi r^2$**

Since, $r = \dfrac{d}{2}$, then   **area $= \pi r^2$ or $\dfrac{\pi d^2}{4}$**

13.4.3   Area of sector

Area of a sector $= \dfrac{\theta}{360}(\pi r^2)$ when θ is in degrees

$= \dfrac{\theta}{2\pi}(\pi r^2)$

$= \dfrac{1}{2}r^2\theta$ when θ is in radians.

Problem 8. A hockey pitch has a semicircle of radius 14.63 m around each goal net. Find the area enclosed by the semicircle, correct to the nearest square metre

Area of a semicircle $= \frac{1}{2}\pi r^2$

When $r = 14.63$ m, area $= \frac{1}{2}\pi(14.63)^2$

i.e. **area of semicircle $= 336$ m^2**

Problem 9. Find the area of a circular metal plate, correct to the nearest square millimetre, having a diameter of 35.0 mm

Area of a circle $= \pi r^2 = \frac{\pi d^2}{4}$

When $d = 35.0$ mm, area $= \frac{\pi(35.0)^2}{4}$

i.e. **area of circular plate $= 962$ mm^2**

Problem 10. Find the area of a circle having a circumference of 60.0 mm

Circumference $c = 2\pi r$,

from which, radius $r = \frac{c}{2\pi} = \frac{60.0}{2\pi} = \frac{30.0}{\pi}$

Area of a circle $= \pi r^2$

i.e. **area $= \pi \left(\frac{30.0}{\pi}\right)^2 = 286.5$ mm^2**

Problem 11. Find the length of arc of a circle of radius 5.5 cm when the angle subtended at the centre is 1.20 radians

Length of arc $s = r\theta$, where θ is in radians.
Hence, **arc length, $s = (5.5)(1.20) = 6.60$ cm**

Problem 12. Determine the diameter and circumference of a circle if an arc of length 4.75 cm subtends an angle of 0.91 radians

Since arc length $s = r\theta$ then radius

$r = \frac{s}{\theta} = \frac{4.75}{0.91} = 5.22$ cm

Diameter $= 2 \times$ radius $= 2 \times 5.22 = $ **10.44 cm**
Circumference $c = \pi d = \pi(10.44) = $ **32.80 cm**

Problem 13. If an angle of 125° is subtended by an arc of a circle of radius 8.4 cm, find the length of (a) the minor arc, and (b) the major arc, correct to 3 significant figures

Since $180° = \pi$ rad then $1° = \left(\frac{\pi}{180}\right)$ rad and

$125° = 125\left(\frac{\pi}{180}\right)$ rad

(a) Length of minor arc,

$s = r\theta = (8.4)(125)\left(\frac{\pi}{180}\right) = $ **18.3 cm**

correct to 3 significant figures.

(b) Length of major arc $=$ (circumference $-$ minor arc) $= 2\pi(8.4) - 18.3 = $ **34.5 cm**, correct to 3 significant figures.

(Alternatively, major arc $= r\theta$

$= 8.4(360 - 125)\left(\frac{\pi}{180}\right) = $ **34.5 cm**)

Problem 14. A football stadium floodlight can spread its illumination over an angle of 45° to a distance of 55 m. Determine the maximum area that is floodlit

Floodlit area $=$ area of sector $= \frac{1}{2}r^2\theta$

$= \frac{1}{2}(55)^2\left(45 \times \frac{\pi}{180}\right)$

$= $ **1188 m^2**

Problem 15. An automatic garden spray produces a spray to a distance of 1.8 m and revolves through an angle α which may be varied. If the desired spray catchment area is to be 2.5 m^2, to what should angle α be set, correct to the nearest degree?

Area of sector $= \frac{1}{2}r^2\theta$, hence $2.5 = \frac{1}{2}(1.8)^2\alpha$

from which, $\alpha = \frac{2.5 \times 2}{1.8^2} = 1.5432$ radians

1.5432 rad $= \left(1.5432 \times \frac{180}{\pi}\right)° = 88.42°$

Hence, **angle $\alpha = 88°$**, correct to the nearest degree.

Problem 16. The angle of a tapered groove is checked using a 20 mm diameter roller as shown in Figure 13.8. If the roller lies 2.12 mm below the top of the groove, determine the value of angle θ

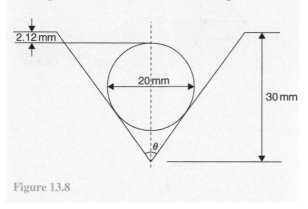

Figure 13.8

In Figure 13.9, triangle ABC is right-angled at C (see section 13.2 (g), page 137)

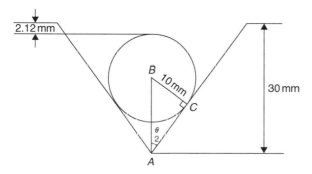

Figure 13.9

Length $BC = 10$ mm (i.e. the radius of the circle), and $AB = 30 - 10 - 2.12 = 17.88$ mm from Figure 13.9. Hence, $\sin \dfrac{\theta}{2} = \dfrac{10}{17.88}$ and $\dfrac{\theta}{2} = \sin^{-1}\left(\dfrac{10}{17.88}\right) = 34°$ and **angle $\theta = 68°$**

Now try the following Practice Exercise

Practice Exercise 62   Arc length and area of circles and sectors (answers on page 538)

1. Calculate the area of a circle of radius 6.0 cm, correct to the nearest square centimetre

2. The diameter of a circle is 55.0 mm. Determine its area, correct to the nearest square millimetre

3. The perimeter of a circle is 150 mm. Find its area, correct to the nearest square millimetre

4. Find the area of the sector, correct to the nearest square millimetre, of a circle having a radius of 35 mm, with angle subtended at centre of 75°

5. An annulus has an outside diameter of 49.0 mm and an inside diameter of 15.0 mm. Find its area correct to 4 significant figures

6. Find the area, correct to the nearest square metre, of a 2 m wide path surrounding a circular plot of land 200 m in diameter

7. A rectangular park measures 50 m by 40 m. A 3 m flower bed is made round the two longer sides and one short side. A circular fish pond of diameter 8.0 m is constructed in the centre of the park. It is planned to grass the remaining area. Find, correct to the nearest square metre, the area of grass

8. With reference to Figure 13.10, determine (a) the perimeter, and (b) the area

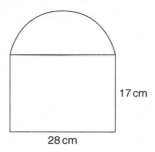

Figure 13.10

9. Find the area of the shaded portion of Figure 13.11

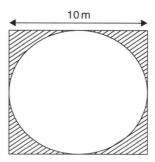

Figure 13.11

10. Find the length of an arc of a circle of radius 8.32 cm when the angle subtended at the centre is 2.14 radians. Calculate also the area of the minor sector formed

11. If the angle subtended at the centre of a circle of diameter 82 mm is 1.46 rad, find the lengths of the (a) minor arc, and (b) major arc

12. A pendulum of length 1.5 m swings through an angle of 10° in a single swing. Find, in centimetres, the length of the arc traced by the pendulum bob

13. Determine the angle of lap, in degrees and minutes, if 180 mm of a belt drive are in contact with a pulley of diameter 250 mm

14. Determine the number of complete revolutions a motorcycle wheel will make in travelling 2 km, if the wheel's diameter is 85.1 cm

15. The floodlights at a sports ground spread their illumination over an angle of 40° to a distance of 48 m. Determine (a) the angle in radians, and (b) the maximum area that is floodlit

16. Find the area swept in 50 minutes by the minute hand of a large floral clock, if the hand is 2 m long

17. Determine (a) the shaded area in Figure 13.12, (b) the percentage of the whole sector that the area of the shaded area represents

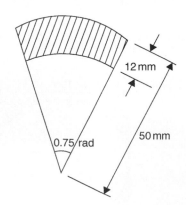

Figure 13.12

18. Determine the length of steel strip required to make the clip shown in Figure 13.13

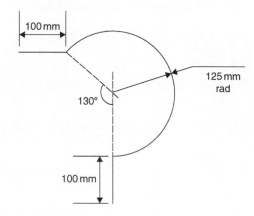

Figure 13.13

19. A 50° tapered hole is checked with a 40 mm diameter ball as shown in Figure 13.14. Determine the length shown as x

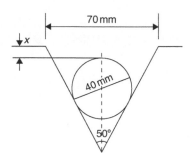

Figure 13.14

For fully worked solutions to each of the problems in Exercises 60 to 62 in this chapter, go to the website:

Chapter 14

Volumes of common solids

Why it is important to understand: **Volumes of common solids**

There are many practical applications where volumes and surface areas of common solids are required. Examples include determining capacities of oil, water, petrol and fish tanks, ventilation shafts and cooling towers, determining volumes of blocks of metal, ball-bearings, boilers and buoys, and calculating the cubic metres of concrete needed for a path. Finding the surface areas of loudspeaker diaphragms and lampshades provide further practical examples. Understanding these calculations is essential for the many practical applications in engineering, construction, architecture and science.

At the end of this chapter, you should be able to:

- state the SI unit of volume
- calculate the volumes and surface areas of cuboids, cylinders, prisms, pyramids, cones and spheres
- appreciate that volumes of similar bodies are proportional to the cubes of the corresponding linear dimensions

14.1 Introduction

The **volume** of any solid is a measure of the space occupied by the solid. Volume is measured in **cubic units** such as mm^3, cm^3 and m^3.

This chapter deals with finding volumes of common solids; in engineering it is often important to be able to calculate volume or capacity, to estimate, say, the amount of liquid, such as water, oil or petrol, in differently shaped containers.

A **prism** is a solid with a constant cross-section and with two ends parallel. The shape of the end is used to describe the prism. For example, there are rectangular prisms (called cuboids), triangular prisms and circular prisms (called cylinders).

On completing this chapter you will be able to calculate the volumes and surface areas of rectangular and other prisms, cylinders, pyramids, cones and spheres. Volumes of similar shapes are also considered.

14.2 Calculating volumes and surface areas of common solids

14.2.1 Cuboid or rectangular prism

A cuboid is a solid figure bounded by six rectangular faces; all angles are right angles and opposite faces are

Science and Mathematics for Engineering. 978-0-367-20475-4, © John Bird. Published by Taylor & Francis. All rights reserved.

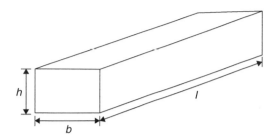

Figure 14.1

equal. A typical cuboid is shown in Figure 14.1 with length l, breadth b and height h.

$$\textbf{Volume of cuboid} = \textbf{\textit{l}} \times \textbf{\textit{b}} \times \textbf{\textit{h}}$$

and

$$\textbf{surface area} = \textbf{2 \textit{bh}} + \textbf{2 \textit{hl}} + \textbf{2 \textit{lb}} = \textbf{2 (\textit{bh} + \textit{hl} + \textit{lb})}$$

A **cube** is a square prism. If all the sides of a cube are x then

$$\textbf{Volume} = \textbf{\textit{x}}^3 \textbf{ and surface area} = \textbf{6\textit{x}}^2$$

Problem 1. A cuboid has dimensions of 12 cm by 4 cm by 3 cm. Determine: (a) its volume and (b) its total surface area

The cuboid is similar to Figure 14.1, with $l = 12$ cm, $b = 4$ cm and $h = 3$ cm.

(a) **Volume of cuboid** $= l \times b \times h = 12 \times 4 \times 3$
$$= \textbf{144 cm}^3$$

(b) **Surface area** $= 2(bh + hl + lb)$
$$= 2(4 \times 3 + 3 \times 12 + 12 \times 4)$$
$$= 2(12 + 36 + 48)$$
$$= 2 \times 96 = \textbf{192 cm}^2$$

Problem 2. An oil tank is the shape of a cube, each edge being of length 1.5 m. Determine: (a) the maximum capacity of the tank in m³ and litres, and (b) its total surface area ignoring input and output orifices

(a) **Volume of oil tank** $=$ volume of cube
$$= 1.5 \text{ m} \times 1.5 \text{ m} \times 1.5 \text{ m}$$
$$= 1.5^3 \text{ m}^3 = \textbf{3.375 m}^3$$

$$1 \text{ m}^3 = 100 \text{ cm} \times 100 \text{ cm} \times 100 \text{ cm} = 10^6 \text{ cm}^3$$

Hence,

$$\text{volume of tank} = 3.375 \times 10^6 \text{ cm}^3$$

1 litre $= 1000$ cm³ hence, **oil tank capacity**

$$= \frac{3.375 \times 10^6}{1000} \text{litres} = \textbf{3375 litres}.$$

(b) Surface area of one side $= 1.5 \text{ m} \times 1.5 \text{ m}$
$$= 2.25 \text{ m}^2$$

A cube has six identical sides, hence

$$\textbf{total surface area of oil tank} = \textbf{6} \times \textbf{2.25}$$
$$= \textbf{13.5 m}^2$$

Problem 3. A water tank is the shape of a rectangular prism having length 2 m, breadth 75 cm and height 500 mm. Determine the capacity of the tank in (a) m³, (b) cm³, (c) litres

Capacity means volume; when dealing with liquids, the word capacity is usually used.
The water tank is similar in shape to Figure 14.1, with $l = 2$ m, $b = 75$ cm and $h = 500$ mm.

(a) Capacity of water tank $= l \times b \times h$. To use this formula, all dimensions **must** be in the same units. Thus, $l = 2$ m, $b = 0.75$ m and $h = 0.5$ m (since 1 m $= 100$ cm $= 1000$ mm). Hence,

$$\textbf{capacity of tank} = 2 \times 0.75 \times 0.5 = \textbf{0.75 m}^3$$

(b) $1 \text{ m}^3 = 1 \text{ m} \times 1 \text{ m} \times 1 \text{ m}$
$$= 100 \text{ cm} \times 100 \text{ cm} \times 100 \text{ cm},$$
i.e. $\textbf{1 m}^3 = \textbf{1,000,000} = \textbf{10}^6 \textbf{ cm}^3$. Hence,

$$\textbf{capacity} = 0.75 \text{ m}^3 = 0.75 \times 10^6 \text{ cm}^3$$
$$= \textbf{750,000 cm}^3$$

(c) 1 litre $= 1000$ cm³. Hence,

$$\textbf{750,000 cm}^3 = \frac{750,000}{1000} = \textbf{750 litres}.$$

14.2.2 Cylinder

A cylinder is a circular prism. A cylinder of radius r and height h is shown in Figure 14.2.

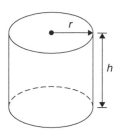

Figure 14.2

$$\text{Volume} = \pi r^2 h$$

$$\text{Curved surface area} = 2\pi rh$$

$$\text{Total surface area} = 2\pi rh + 2\pi r^2$$

Total surface area means the curved surface area plus the area of the two circular ends.

> **Problem 4.** A solid cylinder has a base diameter of 12 cm and a perpendicular height of 20 cm. Calculate (a) the volume, and (b) the total surface area

(a) $\text{Volume} = \pi r^2 h = \pi \times \left(\dfrac{12}{2}\right)^2 \times 20$

$$= 720\pi = \mathbf{2262\,cm^3}$$

(b) **Total surface area**

$$= 2\pi rh + 2\pi r^2$$

$$= (2 \times \pi \times 6 \times 20) + (2 \times \pi \times 6^2)$$

$$= 240\pi + 72\pi = 312\pi = \mathbf{980\,cm^2}$$

> **Problem 5.** A copper pipe has the dimensions shown in Figure 14.3. Calculate the volume of copper in the pipe, in cubic metres

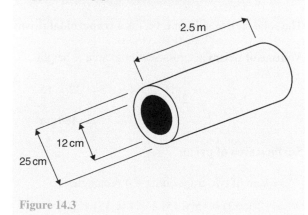

Figure 14.3

Outer diameter $D = 25\,\text{cm} = 0.25\,\text{m}$ and inner diameter $d = 12\,\text{cm} = 0.12\,\text{m}$

Area of cross-section of copper

$$= \frac{\pi D^2}{4} - \frac{\pi d^2}{4} = \frac{\pi\,(0.25)^2}{4} - \frac{\pi\,(0.12)^2}{4}$$

$$= 0.0491 - 0.0113 = 0.0378\,\text{m}^2$$

Hence, **volume of copper**

$$= (\text{cross-sectional area}) \times \text{length of pipe}$$

$$= 0.0378 \times 2.5 = \mathbf{0.0945\,m^3}$$

14.2.3 More prisms

A right-angled triangular prism is shown in Figure 14.4 with dimensions b, h and l.

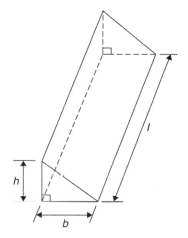

Figure 14.4

$$\text{Volume} = \frac{1}{2}\,bhl$$

and

$$\textbf{surface area} = \textbf{area of each end}$$

$$\textbf{+ area of three sides}$$

Notice that the volume is given by the area of the end (i.e. area of triangle $= \dfrac{1}{2}\,bh$) multiplied by the length l. In fact, the volume of any shaped prism is given by the area of an end multiplied by the length.

> **Problem 6.** Determine the volume (in cm³) of the shape shown in Figure 14.5

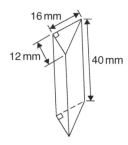

Figure 14.5

The solid shown in Figure 14.5 is a triangular prism. The volume V of any prism is given by: $V = Ah$, where A is the cross-sectional area and h is the perpendicular height. Hence,

$$\textbf{volume} = \frac{1}{2} \times 16 \times 12 \times 40 = 3840\,\text{mm}^3$$

$$= \textbf{3.840}\,\textbf{cm}^3$$

$$(\text{since } 1\,\text{cm}^3 = 1000\,\text{mm}^3)$$

Problem 7. Calculate the volume of the right-angled triangular prism shown in Figure 14.6. Also, determine its total surface area

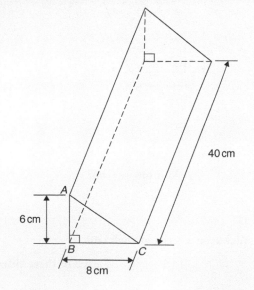

Figure 14.6

Volume of a right-angled triangular prism

$$= \frac{1}{2}bhl = \frac{1}{2} \times 8 \times 6 \times 40$$

i.e. **volume** $= \textbf{960}\,\textbf{cm}^3$

Total surface area $=$ area of each end

$$+ \text{ area of three sides.}$$

In triangle ABC, $AC^2 = AB^2 + BC^2$

from which, $AC = \sqrt{AB^2 + BC^2} = \sqrt{6^2 + 8^2}$

$$= 10\,\text{cm}$$

Hence, total surface area

$$= 2\left(\frac{1}{2}bh\right) + (AC \times 40) + (BC \times 40) + (AB \times 40)$$

$$= (8 \times 6) + (10 \times 40) + (8 \times 40) + (6 \times 40)$$

$$= 48 + 400 + 320 + 240$$

i.e. **total surface area** $= \textbf{1008}\,\textbf{cm}^2$

Problem 8. Calculate the volume and total surface area of the solid prism shown in Figure 14.7

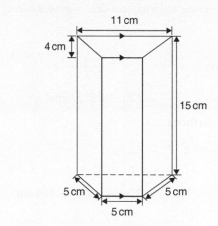

Figure 14.7

The solid shown in Figure 14.7 is a **trapezoidal prism**.

Volume of prism $=$ cross-sectional area $\times$ height

$$= \frac{1}{2}(11 + 5)4 \times 15 = 32 \times 15$$

$$= \textbf{480}\,\textbf{cm}^3$$

Surface area of prism

$$= \text{sum of two trapeziums} + 4 \text{ rectangles}$$

$$= (2 \times 32) + (5 \times 15) + (11 \times 15) + 2(5 \times 15)$$

$$= 64 + 75 + 165 + 150 = \textbf{454}\,\textbf{cm}^2$$

Now try the following Practice Exercise

Practice Exercise 63   Volumes and surface areas of common shapes (answers on page 538)

1. Change a volume of 1,200,000 cm³ to cubic metres

2. Change a volume of 5000 mm³ to cubic centimetres

3. A metal cube has a surface area of 24 cm². Determine its volume

4. A rectangular block of wood has dimensions of 40 mm by 12 mm by 8 mm. Determine

   (a) its volume, in cubic millimetres, and

   (b) its total surface area in square millimetres

5. Determine the capacity, in litres, of a fish tank measuring 90 cm by 60 cm by 1.8 m, given 1 litre = 1000 cm³

6. A rectangular block of metal has dimensions of 40 mm by 25 mm by 15 mm. Determine its volume in cm³. Find also its mass if the metal has a density of 9 g/cm³

7. Determine the maximum capacity, in litres, of a fish tank measuring 50 cm by 40 cm by 2.5 m (1 litre = 1000 cm³)

8. Determine how many cubic metres of concrete are required for a 120 m long path, 150 mm wide and 80 mm deep

9. A cylinder has a diameter 30 mm and height 50 mm. Calculate

   (a) its volume in cubic centimetres, correct to 1 decimal place, and

   (b) the total surface area in square centimetres, correct to 1 decimal place

10. Find (a) the volume, and (b) the total surface area of a right-angled triangular prism of length 80 cm and whose triangular end has a base of 12 cm and perpendicular height 5 cm

11. A steel ingot whose volume is 2 m² is rolled out into a plate which is 30 mm thick and 1.80 m wide. Calculate the length of the plate in metres

12. Calculate the volume of a metal tube whose outside diameter is 8 cm and whose inside diameter is 6 cm, if the length of the tube is 4 m

13. The volume of a cylinder is 400 cm³. If its radius is 5.20 cm, find its height. Determine also its curved surface area

14. A cylinder is cast from a rectangular piece of alloy 5 cm by 7 cm by 12 cm. If the length of the cylinder is to be 60 cm, find its diameter

15. Find the volume and the total surface area of a regular hexagonal bar of metal of length 3 m if each side of the hexagon is 6 cm

16. A block of lead 1.5 m by 90 cm by 750 mm is hammered out to make a square sheet 15 mm thick. Determine the dimensions of the square sheet, correct to the nearest centimetre

17. A cylinder is cast from a rectangular piece of alloy 5.20 cm by 6.50 cm by 19.33 cm. If the height of the cylinder is to be 52.0 cm, determine its diameter, correct to the nearest centimetre

18. How much concrete is required for the construction of the path shown in Figure 14.8, if the path is 12 cm thick?

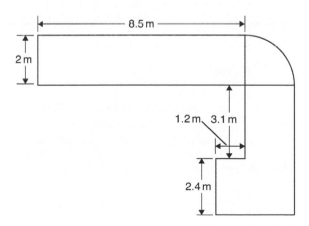

Figure 14.8

14.2.4 Pyramids

Volume of any pyramid

$$= \frac{1}{3} \times \text{area of base} \times \text{perpendicular height}$$

A square-based pyramid is shown in Figure 14.9 with base dimension x by x and perpendicular height h. For the square-base pyramid shown,

$$\text{volume} = \frac{1}{3}x^2h$$

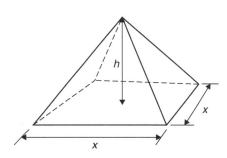

Figure 14.9

Problem 9. A square pyramid has a perpendicular height of 16 cm. If a side of the base is 6 cm, determine the volume of a pyramid

Volume of pyramid

$$= \frac{1}{3} \times \text{area of base} \times \text{perpendicular height}$$

$$= \frac{1}{3} \times (6 \times 6) \times 16$$

$$= 192 \text{ cm}^3$$

Problem 10. Determine the volume and the total surface area of the square pyramid shown in Figure 14.10 if its perpendicular height is 12 cm

Volume of pyramid

$$= \frac{1}{3} \times \text{area of base} \times \text{perpendicular height}$$

$$= \frac{1}{3}(5 \times 5) \times 12$$

$$= 100 \text{ cm}^3$$

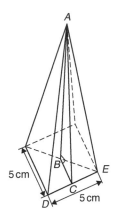

Figure 14.10

The total surface area consists of a square base and 4 equal triangles.

Area of triangle ADE

$$= \frac{1}{2} \times \text{base} \times \text{perpendicular height}$$

$$= \frac{1}{2} \times 5 \times AC$$

The length AC may be calculated using Pythagoras's theorem on triangle ABC, where $AB = 12$ cm, $BC = \frac{1}{2} \times 5 = 2.5$ cm

$$AC = \sqrt{AB^2 + BC^2} = \sqrt{12^2 + 2.5^2} = 12.26 \text{ cm}$$

Hence,

$$\text{area of triangle } ADE = \frac{1}{2} \times 5 \times 12.26 = 30.65 \text{ cm}^2$$

Total surface area of pyramid $= (5 \times 5) + 4(30.65)$

$$= 147.6 \text{ cm}^2$$

Problem 11. A rectangular prism of metal having dimensions of 5 cm by 6 cm by 18 cm is melted down and recast into a pyramid having a rectangular base measuring 6 cm by 10 cm. Calculate the perpendicular height of the pyramid, assuming no waste of metal

Volume of rectangular prism $= 5 \times 6 \times 18 = 540 \text{ cm}^3$.

Volume of pyramid

$$= \frac{1}{3} \times \text{area of base} \times \text{perpendicular height}.$$

Hence, $540 = \dfrac{1}{3} \times (6 \times 10) \times h$

from which, $h = \dfrac{3 \times 540}{6 \times 10} = 27$ cm

i.e. **perpendicular height of pyramid = 27 cm**.

14.2.5   Cones

A cone is a circular-based pyramid. A cone of base radius r and perpendicular height h is shown in Figure 14.11.

$$\text{Volume} = \dfrac{1}{3} \times \text{area of base} \times \text{perpendicular height}$$

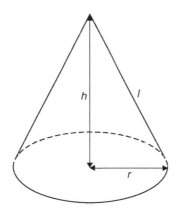

Figure 14.11

i.e. $$\textbf{Volume} = \dfrac{1}{3}\boldsymbol{\pi r^2 h}$$

$$\textbf{Curved surface area} = \boldsymbol{\pi r l}$$

$$\textbf{Total surface area} = \boldsymbol{\pi r l + \pi r^2}$$

> **Problem 12.** Calculate the volume, in cubic centimetres, of a cone of radius 30 mm and perpendicular height 80 mm

Volume of cone $= \dfrac{1}{3}\pi r^2 h = \dfrac{1}{3} \times \pi \times 30^2 \times 80$

$$= 75398.2236\ldots \text{mm}^3$$

1 cm = 10 mm and
$1\,\text{cm}^3 = 10\,\text{mm} \times 10\,\text{mm} \times 10\,\text{mm} = 10^3\,\text{mm}^3$ or

$$\textbf{1 mm}^3 = \textbf{10}^{-3}\,\textbf{cm}^3$$

Hence, $75398.2236\ldots\text{mm}^3$
$$= 75398.2236\ldots \times 10^{-3}\,\text{cm}^3$$
i.e.

$$\textbf{volume} = \textbf{75.40 cm}^3$$

Alternatively, from the question, $r = 30\,\text{mm} = 3\,\text{cm}$ and $h = 80\,\text{mm} = 8\,\text{cm}$. Hence,

$$\textbf{volume} = \dfrac{1}{3}\pi r^2 h = \dfrac{1}{3} \times \pi \times 3^2 \times 8 = \textbf{75.40 cm}^3$$

> **Problem 13.** Determine the volume and total surface area of a cone of radius 5 cm and perpendicular height 12 cm

The cone is shown in Figure 14.12.

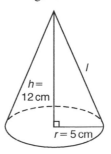

Figure 14.12

$$\textbf{Volume of cone} = \dfrac{1}{3}\pi r^2 h = \dfrac{1}{3} \times \pi \times 5^2 \times 12$$

$$= \textbf{314.2 cm}^3$$

Total surface area = curved surface area + area of base
$$= \pi r l + \pi r^2$$

From Figure 14.12, slant height l may be calculated using Pythagoras's theorem:

$$l = \sqrt{12^2 + 5^2} = 13\,\text{cm}$$

Hence, **total surface area** $= (\pi \times 5 \times 13) + (\pi \times 5^2)$
$$= \textbf{282.7 cm}^2$$

14.2.6   Spheres

For the sphere shown in Figure 14.13:

$$\textbf{Volume} = \dfrac{4}{3}\boldsymbol{\pi r^3} \quad \text{and} \quad \textbf{surface area} = \textbf{4}\boldsymbol{\pi r^2}$$

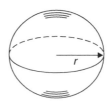

Figure 14.13

Problem 14. Find the volume and surface area of a sphere of diameter 10 cm

Since diameter $= 10$ cm, then radius $r = 5$ cm.

Volume of sphere $= \dfrac{4}{3}\pi r^3 = \dfrac{4}{3} \times \pi \times 5^3$

$$= \mathbf{523.6\,cm^3}$$

Surface area of sphere $= 4\pi r^2 = 4 \times \pi \times 5^2$

$$= \mathbf{314.2\,cm^2}$$

Problem 15. The surface area of a sphere is 201.1 cm². Find the diameter of the sphere, and hence its volume

Surface area of sphere $= 4\pi r^2$.

Hence, $201.1\,\text{cm}^2 = 4 \times \pi \times r^2$

from which, $\qquad r^2 = \dfrac{201.1}{4 \times \pi} = 16.0$

and $\qquad$ radius $r = \sqrt{16.0} = 4.0$ cm

from which, **diameter** $= 2 \times r = 2 \times 4.0 = \mathbf{8.0\,cm}$

$$\textbf{Volume of sphere} = \frac{4}{3}\pi r^3 = \frac{4}{3} \times \pi \times (4.0)^3$$

$$= \mathbf{268.1\,cm^3}$$

Now try the following Practice Exercise

Practice Exercise 64 Volumes and surface areas of common shapes (answers on page 538)

1. If a cone has a diameter of 80 mm and a perpendicular height of 120 mm, calculate its volume in cm³ and its curved surface area

2. A square pyramid has a perpendicular height of 4 cm. If a side of the base is 2.4 cm long find the volume and total surface area of the pyramid

3. A sphere has a diameter of 6 cm. Determine its volume and surface area

4. A pyramid having a square base has a perpendicular height of 25 cm and a volume of 75 cm³ Determine, in centimetres, the length of each side of the base

5. A cone has a base diameter of 16 mm and a perpendicular height of 40 mm. Find its volume correct to the nearest cubic millimetre

6. Determine (a) the volume, and (b) the surface area of a sphere of radius 40 mm

7. The volume of a sphere is 325 cm³. Determine its diameter

8. Given the radius of the earth is 6380 km, calculate, in engineering notation (a) its surface area in km² and (b) its volume in km³

9. An ingot whose volume is 1.5 m³ is to be made into ball bearings whose radii are 8.0 cm. How many bearings will be produced from the ingot, assuming 5% wastage?

10. A spherical chemical storage tank has an internal diameter of 5.6 m. Calculate the storage capacity of the tank, correct to the nearest cubic metre. If 1 litre $= 1000$ cm³, determine the tank capacity in litres

14.3 Summary of volumes and surface areas of common solids

A summary of volumes and surface areas of regular solids is given in Table 14.1.

14.4 Calculating more complex volumes and surface areas

Here are some worked problems involving more complex and composite solids.

Problem 16. A wooden section is shown in Figure 14.14 on page 152. Find (a) its volume, in m³, and (b) its total surface area

(a) The section of wood is a prism whose end comprises a rectangle and a semicircle. Since the radius of the semicircle is 8 cm, the diameter is 16 cm. Hence the rectangle has dimensions 12 cm by 16 cm.

Table 14.1 Volumes and surface areas of regular solids

Rectangular prism (or cuboid)

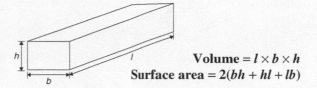

$$\text{Volume} = l \times b \times h$$
$$\text{Surface area} = 2(bh + hl + lb)$$

Cylinder

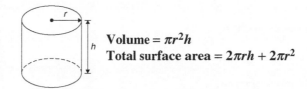

$$\text{Volume} = \pi r^2 h$$
$$\text{Total surface area} = 2\pi rh + 2\pi r^2$$

Triangular prism

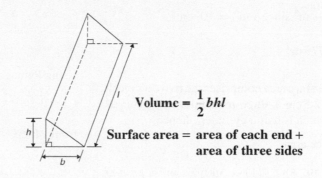

$$\text{Volume} = \frac{1}{2} bhl$$

$$\text{Surface area} = \text{area of each end} + \text{area of three sides}$$

Pyramid

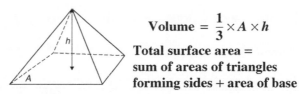

$$\text{Volume} = \frac{1}{3} \times A \times h$$

$$\text{Total surface area} = \text{sum of areas of triangles forming sides} + \text{area of base}$$

Cone

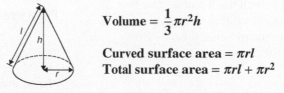

$$\text{Volume} = \frac{1}{3} \pi r^2 h$$

$$\text{Curved surface area} = \pi rl$$
$$\text{Total surface area} = \pi rl + \pi r^2$$

Sphere

$$\text{Volume} = \frac{4}{3} \pi r^3$$
$$\text{Surface area} = 4\pi r^2$$

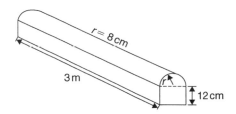

Figure 14.14

$$\text{Area of end} = (12 \times 16) + \frac{1}{2}\pi 8^2 = 292.5\,\text{cm}^2$$

Volume of wooden section

$$= \text{area of end} \times \text{perpendicular height}$$

$$= 292.5 \times 300 = \mathbf{87{,}750\,cm^3}$$

$$= \frac{87750}{10^6}\,\text{m}^3 \text{ since } 1\,\text{m}^3 = 10^6\,\text{cm}^3$$

$$= \mathbf{0.08775\,m^3}$$

(b) The total surface area comprises the two ends (each of area $292.5\,\text{cm}^2$), three rectangles and a curved surface (which is half a cylinder), hence,

total surface area

$$= (2 \times 292.5) + 2(12 \times 300)$$

$$+ (16 \times 300) + \frac{1}{2}(2\pi \times 8 \times 300)$$

$$= 585 + 7200 + 4800 + 2400\pi$$

$$= \mathbf{20{,}125\,cm^2} \text{ or } \mathbf{2.0125\,m^2}$$

> **Problem 17.** A pyramid has a rectangular base $3.60\,\text{cm}$ by $5.40\,\text{cm}$. Determine the volume and total surface area of the pyramid if each of its sloping edges is $15.0\,\text{cm}$

The pyramid is shown in Figure 14.15. To calculate the volume of the pyramid the perpendicular height EF is required. Diagonal BD is calculated using Pythagoras's theorem,

i.e. $BD = \sqrt{[3.60^2 + 5.40^2]} = 6.490\,\text{cm}$

Hence, $EB = \frac{1}{2}BD = \frac{6.490}{2} = 3.245\,\text{cm}$

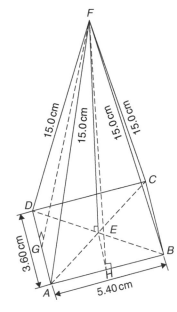

Figure 14.15

Using Pythagoras's theorem on triangle BEF gives

$$BF^2 = EB^2 + EF^2$$

from which, $EF = \sqrt{(BF^2 - EB^2)}$

$$= \sqrt{15.0^2 - 3.245^2} = 14.64\,\text{cm}$$

Volume of pyramid

$$= \frac{1}{3}\,(\text{area of base})(\text{perpendicular height})$$

$$= \frac{1}{3}(3.60 \times 5.40)(14.64) = \mathbf{94.87\,cm^3}$$

Area of triangle ADF (which equals triangle BCF) $= \frac{1}{2}(AD)(FG)$, where G is the midpoint of AD. Using Pythagoras's theorem on triangle FGA gives

$$FG = \sqrt{[15.0^2 - 1.80^2]} = 14.89\,\text{cm}$$

Hence, area of triangle $ADF = \frac{1}{2}(3.60)(14.89)$

$$= 26.80\,\text{cm}^2$$

Similarly, if H is the mid-point of AB, then

$$FH = \sqrt{15.0^2 - 2.70^2} = 14.75\,\text{cm},$$

hence, area of triangle ABF (which equals triangle CDF) $= \frac{1}{2}(5.40)(14.75) = 39.83\,\text{cm}^2$

Total surface area of pyramid

$$= 2(26.80) + 2(39.83) + (3.60)(5.40)$$

$$= 53.60 + 79.66 + 19.44$$

$$= \mathbf{152.7\,cm^2}$$

Problem 18.   Calculate the volume and total surface area of a hemisphere of diameter 5.0 cm

Volume of hemisphere $= \dfrac{1}{2}$ (volume of sphere)

$$= \frac{2}{3}\pi r^3 = \frac{2}{3}\pi \left(\frac{5.0}{2}\right)^3$$

$$= \mathbf{32.7\,cm^3}$$

Total surface area

$$= \text{curved surface area} + \text{area of circle}$$

$$= \frac{1}{2}(\text{surface area of sphere}) + \pi r^2$$

$$= \frac{1}{2}(4\pi r^2) + \pi r^2$$

$$= 2\pi r^2 + \pi r^2 = 3\pi r^2 = 3\pi \left(\frac{5.0}{2}\right)^2$$

$$= \mathbf{58.9\,cm^2}$$

Problem 19.   A rectangular piece of metal having dimensions 4 cm by 3 cm by 12 cm is melted down and recast into a pyramid having a rectangular base measuring 2.5 cm by 5 cm. Calculate the perpendicular height of the pyramid

Volume of rectangular prism of metal $= 4 \times 3 \times 12$

$$= 144\,cm^3$$

Volume of pyramid

$$= \frac{1}{3}(\text{area of base})(\text{perpendicular height}).$$

Assuming no waste of metal,

$$144 = \frac{1}{3}(2.5 \times 5)\,(\text{height})$$

i.e. **perpendicular height of pyramid** $= \dfrac{144 \times 3}{2.5 \times 5}$

$$= \mathbf{34.56\,cm}$$

Problem 20.   A rivet consists of a cylindrical head, of diameter 1 cm and depth 2 mm, and a shaft of diameter 2 mm and length 1.5 cm. Determine the volume of metal in 2000 such rivets

Radius of cylindrical head $= \dfrac{1}{2}$ cm $= 0.5$ cm and

height of cylindrical head $= 2$ mm $= 0.2$ cm.

Hence, volume of cylindrical head

$$= \pi r^2 h = \pi(0.5)^2(0.2) = 0.1571\,cm^3$$

Volume of cylindrical shaft

$$= \pi r^2 h = \pi \left(\frac{0.2}{2}\right)^2 (1.5) = 0.0471\,cm^3$$

Total volume of 1 rivet $= 0.1571 + 0.0471$

$$= 0.2042\,cm^3$$

Volume of metal in 2000 such rivets

$$= 2000 \times 0.2042 = \mathbf{408.4\,cm^3}$$

Problem 21.   A block of copper having a mass of 50 kg is drawn out to make 500 m of wire of uniform cross-section. Given that the density of copper is 8.91 g/cm³, calculate (a) the volume of copper, (b) the cross-sectional area of the wire, and (c) the diameter of the cross-section of the wire

(a)   A density of 8.91 g/cm³ means that 8.91 g of copper has a volume of 1 cm³, or 1 g of copper has a volume of (1/8.91) cm³.

$$\text{Density} = \frac{\text{mass}}{\text{volume}}$$

$$\text{from which,} \quad \text{volume} = \frac{\text{mass}}{\text{density}}$$

Hence, 50 kg, i.e. 50,000 g, has a

$$\mathbf{volume} = \frac{\text{mass}}{\text{density}} = \frac{50,000}{8.91}\,cm^3 = \mathbf{5612\,cm^3}$$

(b)   Volume of wire $=$ area of circular cross-section

$$\times \text{length of wire.}$$

Hence, 5612 cm³ $=$ area $\times (500 \times 100\,\text{cm})$,

$$\text{from which,} \quad \mathbf{area} = \frac{5612}{500 \times 100}\,cm^2$$

$$= \mathbf{0.1122\,cm^2}$$

(c) Area of circle $= \pi r^2$ or $\dfrac{\pi d^2}{4}$

hence, $0.1122 = \dfrac{\pi d^2}{4}$

from which, $d = \sqrt{\left(\dfrac{4 \times 0.1122}{\pi}\right)} = 0.3780\,\text{cm}$

i.e. **diameter of cross-section is 3.780 mm**

> **Problem 22.** A boiler consists of a cylindrical section of length 8 m and diameter 6 m, on one end of which is surmounted a hemispherical section of diameter 6 m, and on the other end a conical section of height 4 m and base diameter 6 m. Calculate the volume of the boiler and the total surface area

The boiler is shown in Figure 14.16.

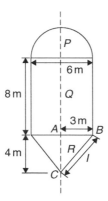

Figure 14.16

Volume of hemisphere $P = \dfrac{2}{3}\pi r^3$

$= \dfrac{2}{3} \times \pi \times 3^3 = 18\pi \text{ m}^3$

Volume of cylinder $Q = \pi r^2 h = \pi \times 3^2 \times 8$

$= 72\pi \text{ m}^3$

Volume of cone $R = \dfrac{1}{3}\pi r^2 h = \dfrac{1}{3} \times \pi \times 3^2 \times 4$

$= 12\pi \text{ m}^3$

Total volume of boiler $= 18\pi + 72\pi + 12\pi = 102\pi$

$= \textbf{320.4 m}^3$

Surface area of hemisphere $P = \dfrac{1}{2}(4\pi r^2)$

$= 2 \times \pi \times 3^2 = 18\pi \text{ m}^2$

Curved surface area of cylinder $Q = 2\pi rh$

$= 2 \times \pi \times 3 \times 8$

$= 48\pi \text{ m}^2$

The slant height of the cone l is obtained by Pythagoras's theorem on triangle ABC, i.e.

$l = \sqrt{(4^2 + 3^2)} = 5$

Curved surface area of cone R

$= \pi rl = \pi \times 3 \times 5 = 15\pi \text{ m}^2$

Total surface area of boiler $= 18\pi + 48\pi + 15\pi$

$= 81\pi = \textbf{254.5 m}^2$

Now try the following Practice Exercise

Practice Exercise 65   More complex volumes and surface areas (answers on page 538)

1. Find the total surface area of a hemisphere of diameter 50 mm

2. Find (a) the volume, and (b) the total surface area of a hemisphere of diameter 6 cm

3. Determine the mass of a hemispherical copper container whose external and internal radii are 12 cm and 10 cm, assuming that 1 cm³ of copper weighs 8.9 g

4. A metal plumb bob comprises a hemisphere surmounted by a cone. If the diameter of the hemisphere and cone are each 4 cm and the total length is 5 cm, find its total volume

5. A marquee is in the form of a cylinder surmounted by a cone. The total height is 6 m and the cylindrical portion has a height of 3.5 m, with a diameter of 15 m. Calculate the surface area of material needed to make the marquee assuming 12% of the material is wasted in the process

6. Determine (a) the volume and (b) the total surface area of the following solids:

    (i) a cone of radius 8.0 cm and perpendicular height 10 cm

    (ii) a sphere of diameter 7.0 cm

(iii) a hemisphere of radius 3.0 cm

(iv) a 2.5 cm by 2.5 cm square pyramid of perpendicular height 5.0 cm

(v) a 4.0 cm by 6.0 cm rectangular pyramid of perpendicular height 12.0 cm

(vi) a 4.2 cm by 4.2 cm square pyramid whose sloping edges are each 15.0 cm

(vii) a pyramid having an octagonal base of side 5.0 cm and perpendicular height 20 cm

7. A metal sphere weighing 24 kg is melted down and recast into a solid cone of base radius 8.0 cm. If the density of the metal is 8000 kg/m³ determine

(a) the diameter of the metal sphere and

(b) the perpendicular height of the cone, assuming that 15% of the metal is lost in the process

8. A buoy consists of a hemisphere surmounted by a cone. The diameter of the cone and hemisphere is 2.5 m and the slant height of the cone is 4.0 m. Determine the volume and surface area of the buoy

9. A petrol container is in the form of a central cylindrical portion 5.0 m long with a hemispherical section surmounted on each end. If the diameters of the hemisphere and cylinder are both 1.2 m determine the capacity of the tank in litres (1 litre = 1000 cm³)

10. Figure 14.17 shows a metal rod section. Determine its volume and total surface area

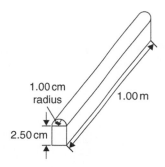

Figure 14.17

1.00 cm radius

1.00 m

2.50 cm

11. The cross-section of part of a circular ventilation shaft is shown in Figure 14.18, ends AB and CD being open. Calculate

(a) the volume of the air, correct to the nearest litre, contained in the part of the system shown, neglecting the sheet metal thickness, (given 1 litre = 1000 cm³),

(b) the cross-sectional area of the sheet metal used to make the system, in square metres, and

(c) the cost of the sheet metal if the material costs £11.50 per square metre, assuming that 25% extra metal is required due to wastage

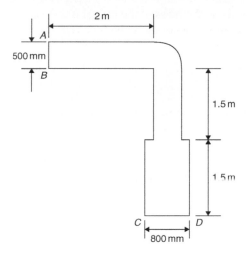

Figure 14.18

14.5 Volumes of similar shapes

Figure 14.19 shows two cubes, one of which has sides three times as long as those of the other.

Volume of Figure 14.19(a) $= (x)(x)(x) = x^3$

Volume of Figure 14.19(b) $= (3x)(3x)(3x) = 27x^3$

Hence, Figure 14.19(b) has a volume $(3)^3$, i.e. 27 times the volume of Figure 14.19(a).

Summarising, **the volumes of similar bodies are proportional to the cubes of corresponding linear dimensions**.

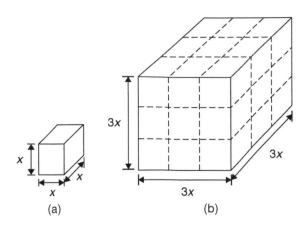

Figure 14.19

Mass = density × volume, and since both car and model are made of the same material then:

$$\frac{\text{Mass of model}}{\text{Mass of car}} = \left(\frac{1}{50}\right)^3$$

Hence, mass of model

$$= (\text{mass of car})\left(\frac{1}{50}\right)^3 = \frac{1000}{50^3} = \mathbf{0.008\,kg} \text{ or } \mathbf{8\,g}.$$

Now try the following Practice Exercise

Problem 23. A car has a mass of 1000 kg. A model of the car is made to a scale of 1 to 50. Determine the mass of the model if the car and its model are made of the same material

$$\frac{\text{Volume of model}}{\text{Volume of car}} = \left(\frac{1}{50}\right)^3$$

since the volume of similar bodies are proportional to the cube of corresponding dimensions.

Practice Exercise 66   Volumes of similar shapes (answers on page 538)

1. The diameter of two spherical bearings are in the ratio 2:5. What is the ratio of their volumes?

2. An engineering component has a mass of 400 g. If each of its dimensions are reduced by 30% determine its new mass

For fully worked solutions to each of the problems in Exercises 63 to 66 in this chapter, go to the website:
www.routledge.com/cw/bird

Revision Test 6: Areas and volumes

This assignment covers the material contained in chapters 12 to 14. *The marks for each question are shown in brackets at the end of each question.*

1. A rectangular metal plate has an area of $9600 \, \text{cm}^2$. If the length of the plate is $1.2 \, \text{m}$, calculate the width, in centimetres. (3)

2. Calculate the cross-sectional area of the angle iron section shown in Figure RT6.1, the dimensions being in millimetres. (4)

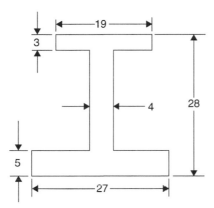

Figure RT6.1

3. Find the area of the trapezium *MNOP* shown in Figure RT6.2 when $a = 6.3 \, \text{cm}$, $b = 11.7 \, \text{cm}$ and $h = 5.5 \, \text{cm}$. (3)

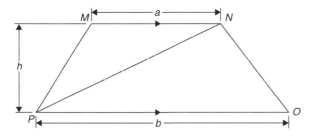

Figure RT6.2

4. Find the area of the triangle *DEF* in Figure RT6.3, correct to 2 decimal places. (4)

5. A rectangular park measures $150 \, \text{m}$ by $70 \, \text{m}$. A $2 \, \text{m}$ flower border is constructed round the two longer sides and one short side. A circular fish pond of diameter $15 \, \text{m}$ is in the centre of the park and the remainder of the park is grass. Calculate, correct to the nearest square metre, the area of (a) the fish pond, (b) the flower borders, and (c) the grass. (6)

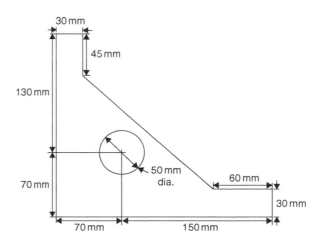

Figure RT6.3

6. A swimming pool is $55 \, \text{m}$ long and $10 \, \text{m}$ wide. The perpendicular depth at the deep end is $5 \, \text{m}$ and at the shallow end is $1.5 \, \text{m}$, the slope from one end to the other being uniform. The inside of the pool needs two coats of a protective paint before it is filled with water. Determine how many litres of paint will be needed if 1 litre covers $10 \, \text{m}^2$. (7)

7. A steel template is of the shape shown in Figure RT6.4, the circular area being removed. Determine the area of the template, in square centimetres, correct to 1 decimal place. (8)

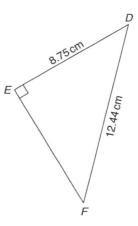

Figure RT6.4

8. Determine the shaded area in Figure RT6.5, correct to the nearest square centimetre. (3)

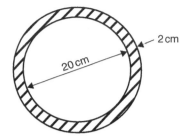

9. Determine the diameter of a circle, correct to the nearest millimetre, whose circumference is 178.4 cm. (2)

10. The circumference of a circle is 250 mm. Find its area, correct to the nearest square millimetre. (4)

11. Find the area of the sector of a circle having a radius of 50.0 mm, with angle subtended at centre of 120°. (3)

12. Determine the total area of the shape shown in Figure RT6.6, correct to 1 decimal place. (7)

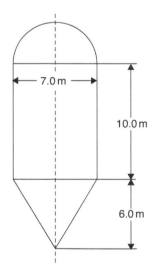

Figure RT6.6

13. The radius of a circular cricket ground is 75 m. The boundary is painted with white paint and 1 tin of paint will paint a line 22.5 m long. How many tins of paint are needed? (3)

14. Find the area of a 2 m wide path surrounding a circular plot of land 200 m in diameter. (3)

15. A cyclometer shows 2530 revolutions in a distance of 3.7 km. Find the diameter of the wheel in centimetres, correct to 2 decimal places. (4)

16. The minute hand of a wall clock is 10.5 cm long. How far does the tip travel in the course of 24 hours? (4)

17. Convert

   (a) 125°47′ to radians,

   (b) 1.724 radians to degrees and minutes. (4)

18. Calculate the length of metal strip needed to make the clip shown in Figure RT6.7. (7)

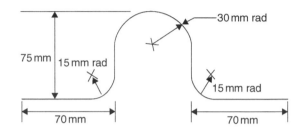

Figure RT6.7

19. A lorry has wheels of radius 50 cm. Calculate the number of complete revolutions a wheel makes (correct to the nearest revolution) when travelling 3 miles (assume 1 mile = 1.6 km). (4)

20. A rectangular block of alloy has dimensions of 60 mm by 30 mm by 12 mm. Calculate the volume of the alloy in cubic centimetres. (3)

21. Find the volume of a cylinder of radius 5.6 cm and height 15.5 cm. Give the answer correct to the nearest cubic centimetre. (3)

22. A garden roller is 0.35 m wide and has a diameter of 0.20 m. What area will it roll in making 40 revolutions? (4)

23. Find the volume of a cone of height 12.5 cm and base diameter 6.0 cm, correct to 1 decimal place. (3)

24. Find (a) the volume, and (b) the total surface area of the right-angled triangular prism shown in Figure RT6.8. (8)

25. A pyramid having a square base has a volume of 86.4 cm³. If the perpendicular height is 20 cm, determine the length of each side of the base. (4)

26. A copper pipe is 80 m long. It has a bore of 80 mm and an outside diameter of 100 mm. Calculate, in cubic metres, the volume of copper in the pipe. (4)

27. Find (a) the volume, and (b) the surface area of a sphere of diameter 25 mm. (4)

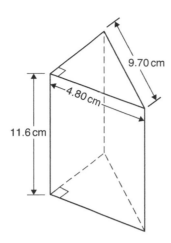

Figure RT6.8

28. A piece of alloy with dimensions 25 mm by 60 mm by 1.60 m is melted down and recast into a cylinder whose diameter is 150 mm. Assuming no wastage, calculate the height of the cylinder in centimetres, correct to 1 decimal place. (4)

29. A rectangular storage container has dimensions 3.2 m by 90 cm by 60 cm. Determine its volume in (a) m^3, (b) cm^3. (4)

30. Calculate (a) the volume, and (b) the total surface area of a 10 cm by 15 cm rectangular pyramid of height 20 cm. (8)

31. A water container is of the form of a central cylindrical part 3.0 m long and diameter 1.0 m, with a hemispherical section surmounted at each end as shown in Figure RT6.9. Determine the maximum capacity of the container, correct to the nearest litre (1 litre $= 1000 \, cm^3$). (5)

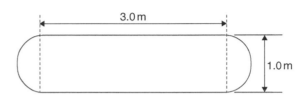

Figure RT6.9

32. A boat has a mass of 20,000 kg. A model of the boat is made to a scale of 1 to 80. If the model is made of the same material as the boat, determine the mass of the model (in grams). (3)

For lecturers/instructors/teachers, fully worked solutions to each of the problems in Revision Test 6, together with a full marking scheme, are available at the website:

www.routledge.com/cw/bird

Multiple-choice questions on applied mathematics

Exercise 67   (answers on page 539)

1. The relationship between the temperature in degrees Fahrenheit (F) and the temperature in degrees Celsius (C) is given by: $F = \dfrac{9}{5}C + 32$.

    135°F is equivalent to:

    (a) 43°C   (b) 57.2°C
    (c) 185.4°C   (d) 184°C

2. When $a = 2$, $b = 1$ and $c = -3$, the engineering expression $2ab - bc + abc$ is equal to:

    (a) 13   (b) −5   (c) 7   (d) 1

3. 11 mm expressed as a percentage of 41 mm is:

    (a) 2.68, correct to 3 significant figures

    (b) 26.83, correct to 2 decimal places

    (c) 2.6, correct to 2 significant figures

    (d) 0.2682, correct to 4 decimal places

4. When two resistors R_1 and R_2 are connected in parallel the formula $\dfrac{1}{R_T} = \dfrac{1}{R_1} + \dfrac{1}{R_2}$ is used to determine the total resistance R_T. If $R_1 = 470\,\Omega$ and $R_2 = 2.7k\Omega$, R_T (correct to 3 significant figures) is equal to:

    (a) 2.68 Ω   (b) 400 Ω
    (c) 473 Ω   (d) 3170 Ω

5. $1\dfrac{1}{3} + 1\dfrac{2}{3} \div 2\dfrac{2}{3} - \dfrac{1}{3}$ is equal to:

    (a) $1\dfrac{5}{8}$ (b) $\dfrac{19}{24}$ (c) $2\dfrac{1}{21}$ (d) $1\dfrac{2}{7}$

6. Four engineers can complete a task in 5 hours. Assuming the rate of work remains constant, six engineers will complete the task in:

    (a) 126 h   (b) 4 h 48 min

    (c) 3 h 20 min   (d) 7 h 30 min

7. In an engineering equation $\dfrac{3^4}{3^r} = \dfrac{1}{9}$. The value of r is:

    (a) −6   (b) 2   (c) 6   (d) −2

8. A graph of resistance against voltage for an electrical circuit is shown in Figure M1. The equation relating resistance R and voltage V is:

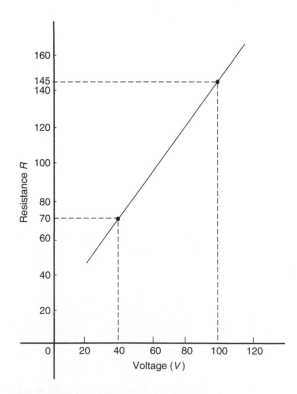

Figure M1

Science and Mathematics for Engineering. 978-0-367-20475-4, © John Bird. Published by Taylor & Francis. All rights reserved.

(a) $R = 1.45\,V + 40$ (b) $R = 0.8\,V + 20$
(c) $R = 1.45\,V + 20$ (d) $R = 1.25\,V + 20$

9. $\dfrac{1}{9.83} - \dfrac{1}{16.91}$ is equal to:

   (a) 0.0425, correct to 4 significant figures

   (b) 4.26×10^{-2}, correct to 2 decimal places

   (c) -0.141, correct to 3 decimal places

   (d) 0.042, correct to 2 significant figures

10. A formula for the focal length f of a convex lens is $\dfrac{1}{f} = \dfrac{1}{u} + \dfrac{1}{v}$. When $f = 4$ and $u = 6$, v is:

    (a) -2 (b) $\dfrac{1}{12}$ (c) 12 (d) $-\dfrac{1}{2}$

11. The engineering expression $\left(\dfrac{p + 2p}{p}\right) \times 3p - 2p$ simplifies to:

    (a) $7p$ (b) $6p$ (c) $9p^2 - 2p$ (d) $3p$

12. $(9.2 \times 10^2 + 1.1 \times 10^3)\,\text{cm}$ in standard form is equal to:

    (a) $10.3 \times 10^2\,\text{cm}$ (b) $20.2 \times 10^2\,\text{cm}$
    (c) $10.3 \times 10^3\,\text{cm}$ (d) $2.02 \times 10^3\,\text{cm}$

13. The current I in an a.c. circuit is given by: $I = \dfrac{V}{\sqrt{R^2 + X^2}}$ When $R = 4.8, X = 10.5$ and $I = 15$, the value of voltage V is:

    (a) 173.18 (b) 1.30
    (c) 0.98 (d) 229.50

14. In the engineering equation: $2^4 \times 2^t = 64$, the value of t is:

    (a) 1 (b) 2 (c) 3 (d) 4

15. The height s of a mass projected vertically upwards at time t is given by: $s = ut - \dfrac{1}{2}g\,t^2$. When $g = 10, t = 1.5$ and $s = 3.75$, the value of u is:

    (a) 10 (b) -5 (c) $+5$ (d) -10

16. In the right-angled triangle ABC shown in Figure M2, sine A is given by:

    (a) b/a (b) c/b (c) b/c (d) a/b

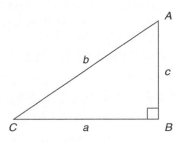

Figure M2

17. In the right-angled triangle ABC shown in Figure M2, cosine C is given by:

    (a) a/b (b) c/b (c) a/c (d) b/a

18. In the right-angled triangle shown in Figure M2, tangent A is given by:

    (a) b/c (b) a/c (c) a/b (d) c/a

19. Which of the straight lines shown in Figure M3 has the equation $y + 4 = 2x$?

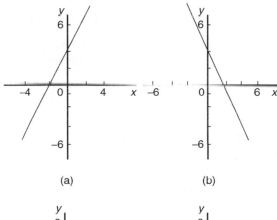

(a)   (b)

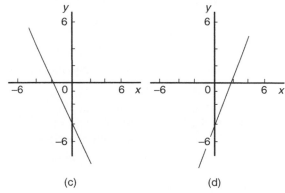

(c)   (d)

Figure M3

20. The quantity of heat Q is given by the formula $Q = mc(t_2 - t_1)$. When $m = 5, t_1 = 20, c = 8$ and $Q = 1200$, the value of t_2 is:

    (a) 10   (b) 1.5   (c) 21.5   (d) 50

21. When $p = 3$, $q = -\dfrac{1}{2}$ and $r = -2$, the engineering expression $2p^2q^3r^4$ is equal to:

    (a) -36 (b) 1296   (c) 36   (d) 18

22. Transposing $v = f\lambda$ to make wavelength λ the subject gives:

    (a) $\dfrac{v}{f}$ (b) $v + f$ (c) $f - v$ (d) $\dfrac{f}{v}$

23. The lowest common multiple (LCM) of the numbers 15, 105, 420 and 1155 is:

    (a) 4620   (b) 1155   (c) 15   (d) 1695

24. $\dfrac{3}{4} \div 1\dfrac{3}{4}$ is equal to:

    (a) $\dfrac{3}{7}$ (b) $1\dfrac{9}{16}$ (c) $1\dfrac{5}{16}$ (d) $2\dfrac{1}{2}$

25. The equation of a straight line graph is $2y = 5 - 4x$. The gradient of the straight line is:

    (a) -2 (b) $\dfrac{5}{2}$ (c) -4 (d) 5

26. If $(4^x)^2 = 16$, the value of x is:

    (a) 4   (b) 2   (c) 3   (d) 1

27. Transposing $I = \dfrac{V}{R}$ for resistance R gives:

    (a) $I - V$ (b) $\dfrac{V}{I}$ (c) $\dfrac{I}{V}$ (d) VI

28. 23 mm expressed as a percentage of 3.25 cm is:

    (a) 0.708, correct to 3 decimal places
    (b) 7.08, correct to 2 decimal places
    (c) 70.77, correct to 4 significant figures
    (d) 14.13, correct to 2 decimal places

29. $16^{-\frac{3}{4}}$ is equal to:

    (a) 8   (b) $-\dfrac{1}{2^3}$ (c) 4   (d) $\dfrac{1}{8}$

30. In the engineering equation $\dfrac{(3^2)^3}{(3)(9)^2} = 3^{2x}$, the value of x is:

    (a) 0   (b) $\dfrac{1}{2}$ (c) 1   (d) $1\dfrac{1}{2}$

31. If $x = \dfrac{57.06 \times 0.0711}{\sqrt{0.0635}}$ cm, which of the following statements is correct?

    (a) $x = 16$ cm, correct to 2 significant figures

    (b) $x = 16.09$ cm, correct to 4 significant figures

    (c) $x = 1.61 \times 10^1$ cm, correct to 3 decimal places

    (d) $x = 16.099$ cm, correct to 3 decimal places

32. Volume $= \dfrac{\text{mass}}{\text{density}}$. The density (in kg/m^3) when the mass is 2.532 kg and the volume is 162 cm^3 is:

    (a) 0.01563 kg/m^3 (b) 410.2 kg/m^3
    (c) 15630 kg/m^3 (d) 64.0 kg/m^3

33. $(5.5 \times 10^2)(2 \times 10^3)$ cm in standard form is equal to:

    (a) 11×10^6 cm   (b) 1.1×10^6 cm
    (c) 11×10^5 cm   (d) 1.1×10^5 cm

34. In the triangular template ABC shown in Figure M4, the length AC is:

    (a) 6.17 cm   (b) 11.17 cm
    (c) 9.22 cm   (d) 12.40 cm

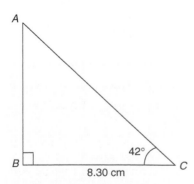

Figure M4

35. $3k + 2k \times 4k + k \div (6k - 2k)$ simplifies to:

    (a) $\dfrac{25k}{4}$ (b) $1 + 2k$

    (c) $\dfrac{1}{4} + k(3 + 8k)$ (d) $20k^2 + \dfrac{1}{4}$

36. Correct to 3 decimal places, $\sin(-2.6 \, \text{rad})$ is:

    (a) 0.516 (b) -0.045
    (c) -0.516 (d) 0.045

37. The engineering expression
    $a^3b^2c \times ab^{-4}c^3 \div a^2bc^{-2}$ is equivalent to:

    (a) $a^2b^5c^2$ (b) $\dfrac{a^2c^6}{b^3}$

    (c) a^6bc^2 (d) $a^{\frac{3}{2}}b^{-7}c^{-\frac{3}{2}}$

38. The highest common factor (HCF) of the numbers 12, 30 and 42 is:

    (a) 42 (b) 12 (c) 420 (d) 6

39. Resistance R ohms varies with temperature t according to the formula $R = R_0(1 + \alpha t)$. Given $R = 21 \, \Omega$, $\alpha = 0.004$ and $t = 100$, R_0 has a value of:

    (a) $21.4 \, \Omega$ (b) $29.4 \, \Omega$
    (c) $15 \, \Omega$ (d) $0.067 \, \Omega$

40. The solution of the simultaneous equations $3x - 2y = 13$ and $2x + 5y = -4$ is:

    (a) $x = -2, y = 3$ (b) $x = 1, y = -5$
    (c) $x = 3, y = -2$ (d) $x = -7, y = 2$

41. An engineering equation is $y = 2a^2b^3c^4$. When $a = 2$, $b = -\dfrac{1}{2}$ and $c = 2$, the value of y is:

    (a) -16 (b) -12 (c) 16 (d) 12

42. Electrical resistance $R = \dfrac{\rho L}{a}$; transposing this equation for L gives:

    (a) $\dfrac{Ra}{\rho}$ (b) $\dfrac{R}{a\rho}$ (c) $\dfrac{a}{R\rho}$ (d) $\dfrac{\rho a}{R}$

43. For the right-angled triangle PQR shown in Figure M5, angle R is equal to:

    (a) $41.41°$ (b) $48.59°$
    (c) $36.87°$ (d) $53.13°$

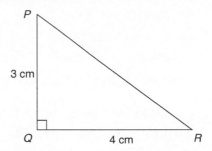

Figure M5

44. The value of $\dfrac{0.01372}{7.5}$ m is equal to:

    (a) 1.829 m, correct to 4 significant figures

    (b) 0.001829 m, correct to 6 decimal places

    (c) 0.0182 m, correct to 4 significant figures

    (d) 1.829×10^{-2} m, correct to 3 decimal places

45. A graph of y against x, two engineering quantities, produces a straight line. A table of values is shown below:

    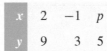

x	2	-1	p
y	9	3	5

    The value of p is:

    (a) $-\dfrac{1}{2}$ (b) -2 (c) 3 (d) 0

46. The engineering expression $\dfrac{(16 \times 4)^2}{(8 \times 2)^4}$ is equal to:

    (a) 4 (b) 2^{-4} (c) $\dfrac{1}{2^2}$ (d) 1

47. The area A of a triangular piece of land of sides a, b and c may be calculated using:

$$A = \sqrt{[s(s-a)(s-b)(s-c)]}$$

where $s = \dfrac{a+b+c}{2}$. When $a = 15\,\text{m}$, $b = 11\,\text{m}$ and $c = 8\,\text{m}$, the area, correct to the nearest square metre, is:

(a) $1836\,\text{m}^2$ (b) $648\,\text{m}^2$

(c) $445\,\text{m}^2$ (d) $43\,\text{m}^2$

48. $0.001010\,\text{cm}$ expressed in standard form is:

(a) $1010 \times 10^{-6}\,\text{cm}$ (b) $1.010 \times 10^{-3}\,\text{cm}$

(c) $0.101 \times 10^{-2}\,\text{cm}$ (d) $1.010 \times 10^3\,\text{cm}$

49. In a system of pulleys, the effort P required to raise a load W is given by $P = aW + b$, where a and b are constants. If $W = 40$ when $P = 12$ and $W = 90$ when $P = 22$, the values of a and b are:

(a) $a = 5, b = \dfrac{1}{4}$ (b) $a = 1, b = -28$

(c) $a = \dfrac{1}{3}, b = -8$ (d) $a = \dfrac{1}{5}, b = 4$

50. The value of $\dfrac{\ln 2}{e^2 \lg 2}$, correct to 3 significant figures, is:

(a) 0.0588 (b) 0.312

(c) 17.0 (d) 3.209

51. $(16^{-\frac{1}{4}} - 27^{-\frac{2}{3}})$ is equal to:

(a) $\dfrac{7}{18}$ (b) -7 (c) $1\dfrac{8}{9}$ (d) $-8\dfrac{1}{2}$

52. If $\cos A = \dfrac{12}{13}$, then $\sin A$ is equal to:

(a) $\dfrac{5}{13}$ (b) $\dfrac{13}{12}$ (c) $\dfrac{5}{12}$ (d) $\dfrac{12}{5}$

53. The area of triangle XYZ in Figure M6 is:

(a) $24.22\,\text{cm}^2$ (b) $19.35\,\text{cm}^2$

(c) $38.72\,\text{cm}^2$ (d) $32.16\,\text{cm}^2$

54. 3.810×10^{-3} is equal to:

(a) 0.003810 (b) 0.0381

(c) 3810 (d) 0.3810

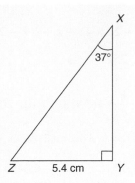

Figure M6

55. A graph relating effort E (plotted vertically) against load L (plotted horizontally) for a set of pulleys is given by $L + 30 = 6E$. The gradient of the graph is:

(a) $\dfrac{1}{6}$ (b) 5   (c) 6   (d) $\dfrac{1}{5}$

56. The value of $42 \div 3 - 2 \times 3 + 15 \div (1 + 4)$ is:

(a) 39   (b) 131   (c) $43\dfrac{1}{5}$ (d) 11

57. The value, correct to 3 decimal places, of $\cos\left(\dfrac{-3\pi}{4}\right)$ is:

(a) 0.999 (b) 0.707

(c) -0.999 (d) -0.707

58. $\dfrac{5^2 \times 5^{-3}}{5^{-4}}$ is equivalent to:

(a) 5^{-5} (b) 5^{24} (c) 5^3 (d) 5

59. The value of $24 \div \dfrac{8}{3} - 16[2 + (-3 \times 4) - 10]$ is:

(a) -311 (b) 329   (c) -256 (d) 384

60. The value of $\dfrac{3.67 \ln 21.28}{e^{-0.189}}$, correct to 4 valid figures, is:

(a) 9.289 (b) 13.56

(c) 13.5566 (d) -3.844×10^9

61. A triangle has sides $a = 9.0\,\text{cm}$, $b = 8.0\,\text{cm}$ and $c = 6.0\,\text{cm}$. Angle A is equal to:

(a) $82.42°$ (b) $56.49°$

(c) $78.58°$ (d) $79.87°$

62. The value of $\dfrac{2}{5}$ of $\left(4\dfrac{1}{2}-3\dfrac{1}{4}\right)+5\div\dfrac{5}{16}-\dfrac{1}{4}$ is:

    (a) $17\dfrac{7}{20}$ (b) $80\dfrac{1}{2}$ (c) $16\dfrac{1}{4}$ (d) 88

63. In Figure M7, *AB* is called:

    (a) the circumference of the circle
    (b) a sector of the circle
    (c) a tangent to the circle
    (d) a major arc of the circle

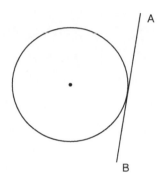

Figure M7

64. The surface area of a sphere of diameter 40 mm is:

    (a) $201.06\,\text{cm}^2$ (b) $33.51\,\text{cm}^2$
    (c) $268.08\,\text{cm}^2$ (d) $50.27\,\text{cm}^2$

65. A vehicle has a mass of 2000 kg. A model of the vehicle is made to a scale of 1 to 100. If the vehicle and model are made of the same material, the mass of the model is:

    (a) 2 g   (b) 20 kg   (c) 200 g   (d) 20 g

66. $\dfrac{3\pi}{4}$ radians is equivalent to:

    (a) $135°$ (b) $270°$ (c) $45°$ (d) $67.5°$

67. The area of the path shown shaded in Figure M8 is:

    (a) $300\,\text{m}^2$ (b) $234\,\text{m}^2$
    (c) $124\,\text{m}^2$ (d) $66\,\text{m}^2$

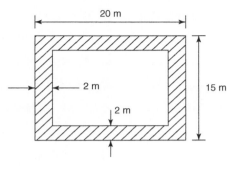

Figure M8

68. If the circumference of a circle is 100 mm, its area is:

    (a) $314.2\,\text{cm}^2$ (b) $7.96\,\text{cm}^2$
    (c) $31.83\,\text{mm}^2$ (d) $78.54\,\text{cm}^2$

69. A rectangular building is shown on a building plan having dimensions 20 mm by 10 mm. If the plan is drawn to a scale of 1 to 300, the true area of the building in m² is:

    (a) $60,000\,\text{m}^2$ (b) $18\,\text{m}^2$
    (c) $0.06\,\text{m}^2$ (d) $1800\,\text{m}^2$

70. A water tank is in the shape of a rectangular prism having length 1.5 m, breadth 60 cm and height 300 mm. If 1 litre = 1000 cm³, the capacity of the tank is:

    (a) 27 litres   (b) 2.7 litres
    (c) 2700 litres   (d) 270 litres

71. An arc of a circle of length 5.0 cm subtends an angle of 2 radians. The circumference of the circle is:

    (a) 2.5 cm   (b) 10.0 cm
    (c) 5.0 cm   (d) 15.7 cm

72. The total surface area of a cylinder of length 20 cm and diameter 6 cm is:

    (a) $433.54\,\text{cm}^2$ (b) $56.55\,\text{cm}^2$
    (c) $980.18\,\text{cm}^2$ (d) $226.19\,\text{cm}^2$

73. In the equation $5.0=3.0\ln\dfrac{2.9}{x}$ has a value correct to 3 significant figures of:

    (a) 1.59 (b) 0.392
    (c) 0.0625 (d) 0.548

For a copy of this multiple choice test, go to the website:

www.routledge.com/cw/bird

Mechanical applications

Chapter 15

SI units and density

Why it is important to understand: **SI units and density**

In engineering there are many different quantities to get used to, and hence many units to become familiar with. For example, force is measured in newtons, electric current is measured in amperes and pressure is measured in pascals. Sometimes the units of these quantities are either very large or very small and hence prefixes are used. For example, 1,000,000 newtons may be written as 10^6 N which is written as 1 MN in prefix form, the M being accepted as a symbol to represent 1,000,000 or 10^6. Studying, or working, in an engineering discipline, you very quickly become familiar with the standard units of measurement, the prefixes used and engineering notation. An electronic calculator is extremely helpful with engineering notation. Knowledge about density inspired engineers to build ships, surfboards and buoys; density is also an important property that engineers consider in the engines they design for cars and power plants.

At the end of this chapter, you should be able to:

- state the seven SI units
- understand common prefixes used in engineering
- use engineering notation and prefix form with engineering units
- define density and relative density
- perform simple calculations involving density

15.1 SI units

As discussed in chapter 3, the system of units used in engineering and science is the **Système Internationale d'Unités** (International system of units), usually abbreviated to SI units, and which is based on the metric system. It was introduced in 1960 and is now adopted by the majority of countries as the official system of measurement.

The basic units in the SI system are listed below with their symbols:

Quantity	Unit	Symbol	
Length	metre	m	(1m = 100 cm = 1000 mm)
Mass	kilogram	kg	(1 kg = 1000 g)
Time	second	s	

Science and Mathematics for Engineering. 978-0-367-20475-4, © John Bird. Published by Taylor & Francis. All rights reserved.

Quantity	Unit	Symbol	
Electric current	ampere	A	
Thermodynamic temperature	kelvin	K	$(K = {}^{\circ}C + 273)$
Luminous intensity	candela	cd	
Amount of substance	mole	mol	

SI units may be made larger or smaller by using **prefixes** which denote multiplication or division by a particular amount. The eight most common multiples, with their meaning, are listed below:

Prefix	Name	Meaning
T	tera	multiply by 10^{12} i.e. $\times 1,000,000,000,000$
G	giga	multiply by 10^9 i.e. $\times 1,000,000,000$
M	mega	multiply by 10^6 i.e. $\times 1,000,000$
k	kilo	multiply by 10^3 i.e. $\times 1000$
m	milli	multiply by 10^{-3} i.e. $\times \dfrac{1}{10^3} = \dfrac{1}{1000} = 0.001$
μ	micro	multiply by 10^{-6} i.e. $\times \dfrac{1}{10^6} = \dfrac{1}{1,000,000} = 0.000001$
n	nano	multiply by 10^{-9} i.e. $\times \dfrac{1}{10^9} = \dfrac{1}{1,000,000,000} = 0.000000001$
p	pico	multiply by 10^{-12} i.e. $\times \dfrac{1}{10^{12}} = \dfrac{1}{1,000,000,000,000} = 0.000000000001$

Length is the distance between two points. The standard unit of length is the **metre**, although the **centimetre (cm)**, **millimetre (mm)** and **kilometre (km)**, are often used.

$$1\,cm = 10\,mm, \ 1\,m = 100\,cm = 1000\,mm$$
$$\text{and } 1\,km = 1000\,m$$

Area is a measure of the size or extent of a plane surface and is measured by multiplying a length by a length.

If the lengths are in metres then the unit of area is the **square metre (m^2)**.

$$1\,m^2 = 1\,m \times 1\,m = 100\,cm \times 100\,cm$$
$$= 10,000\,cm^2 \text{ or } 10^4\,cm^2$$
$$= 1000\,mm \times 1000\,mm$$
$$= 1,000,000\,mm^2 \text{ or } 10^6\,mm^2$$

Conversely, $\mathbf{1\,cm^2 = 10^{-4}\,m^2}$ and $\mathbf{1\,mm^2 = 10^{-6}\,m^2}$.

Volume is a measure of the space occupied by a solid and is measured by multiplying a length by a length by a length. If the lengths are in metres then the unit of volume is in **cubic metres (m^3)**.

$$1\,m^3 = 1\,m \times 1\,m \times 1\,m$$
$$= 100\,cm \times 100\,cm \times 100\,cm = 10^6\,cm^3$$
$$= 1000\,mm \times 1000\,mm \times 1000\,mm = 10^9\,mm^3$$

Conversely, $\mathbf{1\,cm^3 = 10^{-6}\,m^3}$ and $\mathbf{1\,mm^3 = 10^{-9}\,m^3}$. Another unit used to measure volume, particularly with liquids, is the **litre (l)**, where 1 litre = 1000 cm^3.
Mass is the amount of matter in a body and is measured in **kilograms (kg)**.

$$1\,kg = 1000\,g \ (\text{or conversely, } 1\,g = 10^{-3}\,kg)$$

and 1 tonne (t) = 1000 kg.

Problem 1. Express (a) a length of 36 mm in metres, (b) 32,400 mm^2 in square metres, and (c) 8,540,000 mm^3 in cubic metres

(a) $1\,m = 10^3\,mm$ or $1\,mm = 10^{-3}\,m$
Hence, $36\,mm = 36 \times 10^{-3}\,m = \dfrac{36}{10^3}\,m$
$$= \dfrac{36}{1000}\,m = \mathbf{0.036\,m}$$

(b) $1\,m^2 = 10^6\,mm^2$ or $1\,mm^2 = 10^{-6}\,m^2$
Hence,
$$32,400\,mm^2 = 32,400 \times 10^{-6}\,m^2$$
$$= \dfrac{32,400}{1,000,000}\,m^2 = \mathbf{0.0324\,m^2}$$

(c) $1\,m^3 = 10^9\,mm^3$ or $1\,mm^3 = 10^{-9}\,m^3$

Hence,

$$8{,}540{,}000\,\text{mm}^3 = 8{,}540{,}000 \times 10^{-9}\text{m}^3$$

$$= \frac{8{,}540{,}000}{10^9}\text{m}^3$$

$$= \mathbf{8.54 \times 10^{-3}\,m^3}$$

$$\text{or } \mathbf{0.00854\,m^3}$$

Problem 2. Determine the area of a room 15 m long by 8 m wide in (a) m², (b) cm², (c) mm²

(a) Area of room $= 15\,\text{m} \times 8\,\text{m} = \mathbf{120\,m^2}$

(b) $120\,\text{m}^2 = 120 \times 10^4\,\text{cm}^2$, since $1\,\text{m}^2 = 10^4\,\text{cm}^2$

$$= \mathbf{1{,}200{,}000\,cm^2} \text{ or } \mathbf{1.2 \times 10^6\,cm^2}$$

(c) $120\,\text{m}^2 = 120 \times 10^6\,\text{mm}^2$, since $1\,\text{m}^2 = 10^6\,\text{mm}^2$

$$= \mathbf{120{,}000{,}000\,mm^2} \text{ or } \mathbf{0.12 \times 10^9\,mm^2}$$

(Note, it is usual to express the power of 10 as a multiple of 3, i.e. $\times 10^3$ or $\times 10^6$ or $\times 10^{-9}$ and so on.)

Problem 3. A cube has sides each of length 50 mm. Determine the volume of the cube in cubic metres

Volume of cube $= 50\,\text{mm} \times 50\,\text{mm} \times 50\,\text{mm}$

$$= 125{,}000\,\text{mm}$$

$1\,\text{mm}^3 = 10^{-9}\,\text{m}^3$, thus

volume $= 125{,}000 \times 10^{-9}\,\text{m}^3 = \mathbf{0.125 \times 10^{-3}\,m^3}$

Problem 4. A container has a capacity of 2.5 litres. Calculate its volume in (a) m³, (b) mm³

Since 1 litre $= 1000\,\text{cm}^3$, 2.5 litres $= 2.5 \times 1000\,\text{cm}^3$

$$= 2500\,\text{cm}^3$$

(a) $2500\,\text{cm}^3 = 2500 \times 10^{-6}\,\text{m}^3$

$$= \mathbf{2.5 \times 10^{-3}\,m^3} \text{ or } \mathbf{0.0025\,m^3}$$

(b) $2500\,\text{cm}^3 = 2500 \times 10^3\,\text{mm}^3$

$$= \mathbf{2{,}500{,}000\,mm^3} \text{ or } \mathbf{2.5 \times 10^6\,mm^3}$$

Now try the following Practice Exercise

Practice Exercise 68   Further problems on SI units (answers on page 539)

1. Express (a) a length of 52 mm in metres, (b) 20,000 mm² in square metres, (c) 10,000,000 mm³ in cubic metres

2. A garage measures 5 m by 2.5 m. Determine the area in (a) m², (b) mm²

3. The height of the garage in question 2 is 3 m. Determine the volume in (a) m³, (b) mm³

4. A bottle contains 6.3 litres of liquid. Determine the volume in (a) m³, (b) cm³, (c) mm³

15.2 Density

Density is the mass per unit volume of a substance. The symbol used for density is ρ (Greek letter rho) and its units are kg/m³.

$$\text{Density} = \frac{\text{mass}}{\text{volume}}$$

$$\text{i.e. } \boldsymbol{\rho = \frac{m}{V}} \quad \text{or} \quad \boldsymbol{m = \rho V} \quad \text{or} \quad \boldsymbol{V = \frac{m}{\rho}}$$

where m is the mass in kg, V is the volume in m³ and ρ is the density in kg/m³.

Some **typical values of densities** include:

Aluminium	2700 kg/m³	Steel	7800 kg/m³
Cast iron	7000 kg/m³	Petrol	700 kg/m³
Cork	250 kg/m³	Lead	11,400 kg/m³
Copper	8900 kg/m³	Water	1000 kg/m³

The **relative density** of a substance is the ratio of the density of the substance to the density of water, i.e.

$$\textbf{relative density} = \frac{\textbf{density of substance}}{\textbf{density of water}}$$

Relative density has no units, since it is the ratio of two similar quantities. Typical values of relative densities can be determined from the above (since water has a density of 1000 kg/m³), and include:

Aluminium	2.7	Steel	7.8
Cast iron	7.0	Petrol	0.7
Cork	0.25	Lead	11.4
Copper	8.9		

Anything with a density greater than 1000 sinks in pure water, and anything with a density less than 1000 floats. Not surprisingly, steel, with a density of 7800 kg/m³, sinks. Ice, however, with a density of 917 kg/m³, floats. Similarly, from the above list, cork floats in water.

So how can a ship, which is made of metal with a specific gravity very much greater than water, float? The ship floats because of the shape of its hull. The ship is mostly hollow and displaces water weighing more than the weight of the ship. Averaged throughout, the ship is less dense than the water overall, so the effective specific gravity or relative density of the ship is less than that of water.

The relative density of a liquid may be measured using a **hydrometer**, which is usually made of glass and consists of a cylindrical stem and a bulb weighted with a heavy material to make it float upright.

Another method of measuring the specific gravity or relative density is to use the **specific gravity bottle method**. The bottle is weighed, filled with the liquid whose specific gravity is to be found, and weighed again. The difference in weights is divided by the weight of an equal volume of water to give the specific gravity of the liquid. For gases a method essentially the same as the bottle method for liquids is used.

Problem 5. Determine the density of 50 cm³ of copper if its mass is 445 g

Volume = 50 cm³ = 50×10^{-6} m³ and
mass = 445 g = 445×10^{-3} kg.

$$\textbf{Density} = \frac{\text{mass}}{\text{volume}} = \frac{445 \times 10^{-3}\,\text{kg}}{50 \times 10^{-6}\,\text{m}^3} = \frac{445}{50} \times 10^3$$

$$= \textbf{8.9} \times \textbf{10}^3\,\textbf{kg/m}^3 \text{ or } \textbf{8900 kg/m}^3$$

Problem 6. The density of aluminium is 2700 kg/m³. Calculate the mass of a block of aluminium which has a volume of 100 cm³

Density, $\rho = 2700$ kg/m³ and
volume $V = 100$ cm³ $= 100 \times 10^{-6}$ m³.

Since density = mass/volume, then

$$\text{mass} = \text{density} \times \text{volume}.$$

Hence

$$\textbf{mass} = \rho V = 2700\,\text{kg/m} \times 100 \times 10^{-6}\,\text{m}^3$$

$$= \frac{2700 \times 100}{10^6}\,\text{kg} = \textbf{0.270 kg} \text{ or } \textbf{270 g}.$$

Problem 7. Determine the volume, in litres, of 20 kg of paraffin oil of density 800 kg/m³

$$\text{Density} = \frac{\text{mass}}{\text{volume}} \text{ hence, volume} = \frac{\text{mass}}{\text{density}}$$

$$\text{Thus, volume} = \frac{m}{\rho} = \frac{20\,\text{kg}}{800\,\text{kg/m}^3} = \frac{1}{40}\,\text{m}^3$$

$$= \frac{1}{40} \times 10^6\,\text{cm}^3 = 25{,}000\,\text{cm}^3$$

1 litre = 1000 cm³ hence 25,000 cm³ $= \dfrac{25{,}000}{1000}$

$$= \textbf{25 litres}.$$

Problem 8. Determine the relative density of a piece of steel of density 7850 kg/m³. Take the density of water as 1000 kg/m³

$$\textbf{Relative density} = \frac{\text{density of steel}}{\text{density of water}} = \frac{7850}{1000} = \textbf{7.85}$$

Problem 9. A piece of metal 200 mm long, 150 mm wide and 10 mm thick has a mass of 2700 g. What is the density of the metal?

$$\text{Volume of metal} = 200\,\text{mm} \times 150\,\text{mm} \times 10\,\text{mm}$$

$$= 300{,}000\,\text{mm}^3 = 3 \times 10^5\,\text{mm}^3$$

$$= \frac{3 \times 10^5}{10^9}\,\text{m}^3 = 3 \times 10^{-4}\,\text{m}^3$$

Mass = 2700 g = 2.7 kg.

$$\textbf{Density} = \frac{\text{mass}}{\text{volume}} = \frac{2.7\,\text{kg}}{3 \times 10^{-4}\,\text{m}^3} = 0.9 \times 10^4\,\text{kg/m}^3$$

$$= \textbf{9000 kg/m}^3$$

Problem 10. Cork has a relative density of 0.25. Calculate (a) the density of cork and (b) the volume in cubic centimetres of 50 g of cork. Take the density of water to be 1000 kg/m³

(a)   Relative density $= \dfrac{\text{density of cork}}{\text{density of water}}$

from which, density of cork

$\qquad = \text{relative density} \times \text{density of water},$

i.e. **density of cork, $\rho = 0.25 \times 1000$**

$$= \mathbf{250\,kg/m^3}$$

(b)   Density $= \dfrac{\text{mass}}{\text{volume}}$ from which,

$\quad$ volume $= \dfrac{\text{mass}}{\text{density}}$

Mass $m = 50\,\text{g} = 50 \times 10^{-3}\,\text{kg}$

Hence, **volume $V = \dfrac{m}{\rho} = \dfrac{50 \times 10^{-3}\,\text{kg}}{250\,\text{kg/m}^3} = \dfrac{0.05}{250}\,\text{m}^3$**

$$= \dfrac{0.05}{250} \times 10^6\,\text{cm}^3 = \mathbf{200\,cm^3}$$

Now try the following Practice Exercises

Practice Exercise 69   Further problems on density (answers on page 539)

1.   Determine the density of $200\,\text{cm}^3$ of lead which has a mass of $2280\,\text{g}$

2.   The density of iron is $7500\,\text{kg/m}^3$. If the volume of a piece of iron is $200\,\text{cm}^3$, determine its mass

3.   Determine the volume, in litres, of $14\,\text{kg}$ of petrol of density $700\,\text{kg/m}^3$

4.   The density of water is $1000\,\text{kg/m}^3$. Determine the relative density of a piece of copper of density $8900\,\text{kg/m}^3$

5.   A piece of metal $100\,\text{mm}$ long, $80\,\text{mm}$ wide and $20\,\text{mm}$ thick has a mass of $1280\,\text{g}$. Determine the density of the metal

6.   Some oil has a relative density of 0.80. Determine (a) the density of the oil, and (b) the volume of $2\,\text{kg}$ of oil. Take the density of water as $1000\,\text{kg/m}^3$

Practice Exercise 70   Short answer questions on SI units and density

1.   State the SI units for length, mass and time

2.   State the SI units for electric current and thermodynamic temperature

3.   What is the meaning of the following prefixes?
     (a) M   (b) m   (c) μ (d) k

In questions 4–8, complete the statements.

4. $1\,\text{m} = ...\,\text{mm}$ and $1\,\text{km} = ...\,\text{m}$

5. $1\,\text{m}^2 = ...\,\text{cm}^2$ and $1\,\text{cm}^2 = ...\,\text{mm}^2$

6. $1\,\text{l} = ...\,\text{cm}^3$ and $1\,\text{m}^3 = ...\,\text{mm}^3$

7. $1\,\text{kg} = ...\,\text{g}$ and $1\,\text{t} = ...\,\text{kg}$

8. $1\,\text{mm}^2 = ...\,\text{m}^2$ and $1\,\text{cm}^3 = ...\,\text{m}^3$

9.   Define density

10. What is meant by 'relative density'?

11. Relative density of liquids may be measured using a ....

12. How can a ship, which is made of metal with a specific gravity very much greater than water, float?

13. Briefly explain the specific gravity bottle method to measure relative density

Practice Exercise 71   Multiple-choice questions on SI units and density (answers on page 539)

1.   Which of the following statements is true? $1000\,\text{mm}^3$ is equivalent to:
     (a) $1\,\text{m}^3$ (b) $10^{-3}\,\text{m}^3$
     (c) $10^{-6}\,\text{m}^3$ (d) $10^{-9}\,\text{m}^3$

2.   Which of the following statements is true?
     (a) $1\,\text{mm}^2 = 10^{-4}\,\text{m}^2$ (b) $1\,\text{cm}^3 = 10^{-3}\,\text{m}^3$
     (c) $1\,\text{mm}^3 = 10^{-6}\,\text{m}^3$ (d) $1\,\text{km}^2 = 10^{10}\,\text{cm}^2$

3.   Which of the following statements is false? 1000 litres is equivalent to:
     (a) $10^3\,\text{m}^3$ (b) $10^6\,\text{cm}^3$
     (c) $10^9\,\text{mm}^3$ (d) $10^3\,\text{cm}^3$

4.   Let mass $= A$, volume $= B$ and density $= C$. Which of the following statements is false?
     (a) $C = A - B$ (b) $C = \dfrac{A}{B}$
     (c) $B = \dfrac{A}{C}$ (d) $A = BC$

5. The density of $100 \, cm^3$ of a material having a mass of $700 \, g$ is:
   (a) $70,000 \, kg/m^3$ (b) $7000 \, kg/m^3$
   (c) $7 \, kg/m^3$ (d) $70 \, kg/m^3$

6. An alloy has a relative density of 10. If the density of water is $1000 \, kg/m^3$, the density of the alloy is:
   (a) $100 \, kg/m^3$ (b) $0.01 \, kg/m^3$
   (c) $10,000 \, kg/m^3$ (d) $1010 \, kg/m^3$

7. Three fundamental physical quantities in the SI system are:
   (a) mass, velocity, length
   (b) time, length, mass
   (c) energy, time, length
   (d) velocity, length, mass

8. $60 \, \mu s$ is equivalent to:
   (a) $0.06 \, s$ (b) $0.00006 \, s$
   (c) 1000 minutes   (d) $0.6 \, s$

9. The density (in kg/m^3) when the mass is $2.532 \, kg$ and the volume is $162 \, cm^3$ is:
   (a) $0.01563 \, kg/m^3$ (b) $410.2 \, kg/m^3$
   (c) $15,630 \, kg/m^3$ (d) $64.0 \, kg/m^3$

10. Which of the following statements is true? $100 \, cm^3$ is equal to:
    (a) $10^{-4} \, m^3$ (b) $10^{-2} \, m^3$
    (c) $0.1 \, m^3$ (d) $10^{-6} \, m^3$

For fully worked solutions to each of the problems in Exercises 68 to 71 in this chapter, go to the website:

Chapter 16

Atomic structure of matter

Why it is important to understand: **Atomic structure of matter**

Originally, matter was defined as something that takes up space and has weight. It took a while for scientists to realise that air and other gases were also matter. The scientific definition of matter is that it is something that takes up space and has mass (instead of weight). On closer examination of matter it was found that matter was not continuous but was made up of tiny particles. This brought about the Molecular Theory of Matter and then the Atomic Theory of Matter, which stated that matter consisted of tiny particles called molecules and atoms. It was then thought that atoms were the smallest units of matter and were indivisible. However, further experiments showed that atoms were made up of even smaller subatomic particles – protons, neutrons, and electrons. After these were considered indivisible, it was found that protons and neutrons where made up of even smaller particles called quarks. Also, a number of other particles were discovered that were not part of the atom. This chapter gives a brief introduction to the atomic structure of matter and has important applications in materials science & engineering. Materials science is an interdisciplinary field involving the properties of matter and its applications to various areas of science and engineering. It includes elements of applied physics and chemistry, as well as chemical, mechanical, civil and electrical engineering.

At the end of this chapter, you should be able to:

- define elements, atoms, molecules and compounds
- define mixtures, solutions, suspensions and solubility
- understand crystals and crystallisation
- define polycrystalline substances and alloys
- appreciate how a metal is hardened and annealed

16.1 Elements, atoms, molecules and compounds

There are a very large number of different substances in existence, each substance containing one or more of a number of basic materials called elements.

An element is a substance which cannot be separated into anything simpler by chemical means.

There are 92 naturally occurring elements and 13 others that have been artificially produced.

Some examples of common elements with their symbols are: Hydrogen H, Helium He, Carbon C, Nitrogen N, Oxygen O, Sodium Na, Magnesium Mg, Aluminium

Science and Mathematics for Engineering. 978-0-367-20475-4, © John Bird. Published by Taylor & Francis. All rights reserved.

Al, Silicon Si, Phosphorus P, Sulphur S, Potassium K, Calcium Ca, Iron Fe, Nickel Ni, Copper Cu, Zinc Zn, Silver Ag, Tin Sn, Gold Au, Mercury Hg, Lead Pb and Uranium U.

16.1.1   Atoms

Elements are made up of very small parts called atoms.

An atom is the smallest part of an element which can take part in a chemical change and which retains the properties of the element.

Each of the elements has a unique type of atom.

In atomic theory, a model of an atom can be regarded as a miniature solar system. It consists of a central nucleus around which negatively charged particles called **electrons** orbit in certain fixed bands called **shells**. The nucleus contains positively charged particles called **protons** and particles having no electrical charge called **neutrons**.

An electron has a very small mass compared with protons and neutrons. An atom is electrically neutral, containing the same number of protons as electrons. The number of protons in an atom is called the **atomic number** of the element of which the atom is part. The arrangement of the elements in order of their atomic number is known as the **periodic table**.

The simplest atom is hydrogen, which has 1 electron orbiting the nucleus and 1 proton in the nucleus. The atomic number of hydrogen is thus 1. The hydrogen atom is shown diagrammatically in Figure 16.1(a). Helium has 2 electrons orbiting the nucleus, both of them occupying the same shell at the same distance from the nucleus, as shown in Figure 16.1(b).

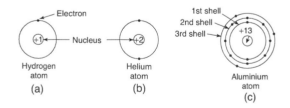

Figure 16.1

The first shell of an atom can have up to 2 electrons only, the second shell can have up to 8 electrons only and the third shell up to 18 electrons only. Thus an aluminium atom that has 13 electrons orbiting the nucleus is arranged as shown in Figure 16.1(c).

16.1.2   Molecules

When elements combine together, the atoms join to form a basic unit of new substance. This independent group of atoms bonded together is called a molecule.

A molecule is the smallest part of a substance which can have a separate stable existence.

All molecules of the same substance are identical. Atoms and molecules are the **basic building blocks** from which matter is constructed.

16.1.3   Compounds

When elements combine chemically their atoms inter-link to form molecules of a new substance called a compound.

A compound is a new substance containing two or more elements chemically combined so that their properties are changed.

For example, the elements hydrogen and oxygen are quite unlike water, which is the compound they produce when chemically combined.

The components of a compound are in fixed proportion and are difficult to separate. Examples include:

(i)   water (H_2O), where 1 molecule is formed by 2 hydrogen atoms combining with 1 oxygen atom,

(ii)   carbon dioxide (CO_2), where 1 molecule is formed by 1 carbon atom combining with 2 oxygen atoms,

(iii)   sodium chloride (NaCl) (common salt), where 1 molecule is formed by 1 sodium atom combining with 1 chlorine atom, and

(iv)   copper sulphate ($CuSO_4$), where 1 molecule is formed by 1 copper atom, 1 sulphur atom and 4 oxygen atoms combining.

16.2   Mixtures, solutions, suspensions and solubility

16.2.1   Mixtures

A mixture is a combination of substances which are not chemically joined together.

Mixtures have the same properties as their components. Also, the components of a mixture have no fixed proportion and are easy to separate. Examples include:

(i)   oil and water,

(ii)   sugar and salt,

(iii) air, which is a mixture of oxygen, nitrogen, carbon dioxide and other gases,

(iv) iron and sulphur,

(v) sand and water.

Mortar is an example of a mixture – consisting of lime, sand and water.

Compounds can be distinguished from mixtures in the following ways:

(i) The properties of a compound are different to its constituent components, whereas a mixture has the same properties as it constituent components.

(ii) The components of a compound are in fixed proportion, whereas the components of a mixture have no fixed proportion.

(iii) The atoms of a compound are joined, whereas the atoms of a mixture are free.

(iv) When a compound is formed, heat energy is produced or absorbed, whereas when a mixture is formed, little or no heat is produced or absorbed.

> **Problem 1.** State whether the following substances are elements, compounds or mixtures: (a) carbon, (b) salt, (c) mortar, (d) sugar, (e) copper

(a) carbon is an **element**.

(b) salt, i.e. sodium chloride, is a **compound** of sodium and chlorine.

(c) mortar is a **mixture** of lime, sand and water.

(d) sugar is a **compound** of carbon, hydrogen and oxygen.

(e) copper is an **element**.

16.2.2   Solutions

A solution is a mixture in which other substances are dissolved.

A solution is a mixture from which the two constituents may not be separated by leaving it to stand, or by filtration. For example, sugar dissolves in tea, salt dissolves in water and copper sulphate crystals dissolve in water leaving it a clear blue colour. The substance that is dissolved, which may be solid, liquid or gas, is called the **solute**, and the liquid in which it dissolves is called the **solvent**. Hence **solvent + solute = solution**.

A solution has a clear appearance and remains unchanged with time.

16.2.3   Suspensions

A suspension is a mixture of a liquid and particles of a solid which do not dissolve in the liquid.

The solid may be separated from the liquid by leaving the suspension to stand, or by filtration. Examples include:

(i) sand in water,

(ii) chalk in water,

(iii) petrol and water.

16.2.4   Solubility

If a material dissolves in a liquid the material is said to be **soluble**. For example, sugar and salt are both soluble in water. If, at a particular temperature, sugar is continually added to water and the mixture stirred there comes a point when no more sugar can dissolve. Such a solution is called saturated.

A solution is saturated if no more solute can be made to dissolve, with the temperature remaining constant.

Solubility is a measure of the maximum amount of a solute which can be dissolved in 0.1 kg of a solvent, at a given temperature.

For example, the solubility of potassium chloride at $20°C$ is 34 g per 0.1 kg of water, or, its percentage solubility is 34%.

(i) Solubility is dependent on temperature. When solids dissolve in liquids, as the temperature is increased, in most cases the amount of solid that will go into solution also increases. (More sugar is dissolved in a cup of hot tea than in the same amount of cold water.) There are exceptions to this, for the solubility of common salt in water remains almost constant and the solubility of calcium hydroxide decreases as the temperature increases.

(ii) Solubility is obtained more quickly when small particles of a substance are added to a liquid than when the same amount is added in large particles. For example, sugar lumps take longer to dissolve in tea than does granulated sugar.

(iii) A solid dissolves in a liquid more quickly if the mixture is stirred or shaken, i.e. solubility depends on the speed of agitation.

Problem 2. State whether the following mixtures are solutions or suspensions: (a) soda water, (b) chalk and water, (c) sea water, (d) petrol and water

(a) Soda water is a **solution** of carbon dioxide and water.

(b) Chalk and water is a **suspension**, the chalk sinking to the bottom when left to stand.

(c) Sea water is a **solution** of salt and water.

(d) Petrol and water is a **suspension**, the petrol floating on the top of the water when left to stand.

Problem 3. Determine the solubility and the percentage solubility of common salt (sodium chloride) if, at a particular temperature, 180 g dissolves in 500 g of water

The solubility is a measure of the maximum amount of sodium chloride that can be dissolved in 0.1 kg (i.e. 100 g) of water.

180 g of salt dissolves in 500 g of water. Hence, $\frac{180}{5}$ g of salt, i.e. 36 g, dissolves in 100 g of water.
Hence **the solubility of sodium chloride is 36 g in 0.1 kg of water**.

$$\textbf{Percentage solubility} = \frac{36}{100} \times 100\% = \textbf{36\%}$$

16.3 Crystals

A crystal is a regular, orderly arrangement of atoms or molecules forming a distinct pattern, i.e. an orderly packing of basic building blocks of matter. Most solids are crystalline in form and these include crystals such as common salt and sugar as well as the metals. Substances that are non-crystalline are called amorphous, examples including glass and wood. **Crystallisation** is the process of isolating solids from solution in a crystalline form. This may be carried out by adding a solute to a solvent until saturation is reached, raising the temperature, adding more solute, and repeating the process until a fairly strong solution is obtained, and then allowing the solution to cool, when crystals will separate. There are several examples of crystalline form that occur naturally, including graphite, quartz, diamond and common salt.

Crystals can vary in size but always have a regular geometric shape with flat faces, straight edges and specific angles between the sides. Two common shapes of crystals are shown in Figure 16.2. The angles between the faces of the common salt crystal (Figure 16.2(a)) are always 90° and those of a quartz crystal (Figure 16.2(b)) are always 60°. A particular material always produces exactly the same shape of crystal.

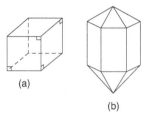

(a)

(b)

Figure 16.2

Figure 16.3 shows a crystal lattice of sodium chloride. This is always a cubic shaped crystal, being made up of 4 sodium atoms and 4 chlorine atoms. The sodium chloride crystals then join together as shown.

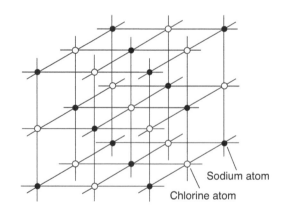

Sodium atom
Chlorine atom

Figure 16.3

16.4 Metals

Metals are **polycrystalline** substances. This means that they are made up of a large number of crystals joined at the boundaries, the greater the number of boundaries the stronger the material.

Every metal, in the solid state, has its own crystal structure. To form an **alloy**, different metals are mixed when molten, since in the molten state they do not have a crystal lattice. The molten solution is then left to cool and solidify. The solid formed is a mixture of different crystals, and an alloy is thus referred to as a **solid solution**. Examples include:

(i)   brass, which is a combination of copper and zinc,

(ii)   steel, which is mainly a combination of iron and carbon, and

(iii)   bronze, which is a combination of copper and tin.

Alloys are produced to enhance the properties of the metal, such as strength. For example, when a small proportion of nickel (say, 2–4%) is added to iron, the strength of the material is greatly increased. By controlling the percentage of nickel added, materials having different specifications may be produced.

A metal may be **hardened** by heating it to a high temperature, then cooling it very quickly. This produces a large number of crystals and therefore many boundaries. The greater the number of crystal boundaries, the stronger is the metal.

A metal is **annealed** by heating it to a high temperature, and then allowing it to cool very slowly. This causes larger crystals, thus less boundaries and hence a softer metal.

Now try the following Practice Exercises

Practice Exercise 72   Further problems on the atomic structure of matter (answers on page 539)

1.   State whether the following are elements, compounds or mixtures:

(a) town gas, (b) water, (c) oil and water, (d) aluminium

2.   The solubility of sodium chloride is 0.036 kg in 0.1 kg of water. Determine the amount of water required to dissolve 432 g of sodium chloride

3.   Describe, with appropriate sketches, a model depicting the structure of the atom

4.   Explain, with the aid of a sketch, what is meant by a crystal, and give two examples of materials with a crystalline structure

Practice Exercise 73   Short answer questions on the atomic structure of matter

1.   What is an element? State three examples

2.   Distinguish between atoms and molecules

3.   What is a compound? State three examples

4.   Distinguish between compounds and mixtures

5.   State three examples of a mixture

6.   Define a solution and state one example

7.   Define a suspension and state one example

8.   Define (a) solubility, (b) a saturated solution

9.   State three factors influencing the solubility of a solid in a liquid

10.   What is a crystal? Give three examples of crystals that occur naturally

11.   Briefly describe the process of crystallisation from a solution

12.   What does polycrystalline mean?

13.   How are alloys formed? State three examples of metallic alloys

Practice Exercise 74   Multiple-choice questions on the atomic structure of matter (answers on page 539)

1.   Which of the following statements is false?

(a)   The properties of a mixture are derived from the properties of its component parts

(b)   Components of a compound are in fixed proportion

(c)   In a mixture, the components may be present in any proportion

(d)   The properties of compounds are related to the properties of their component parts

2.   Which of the following is a compound?

(a) carbon (b) silver (c) salt (d) ink

3. Which of the following is a mixture?

   (a) air (b) water (c) lead (d) salt

4. Which of the following is a suspension?

   (a) soda water   (b) chalk and water
   (c) lemonade   (d) sea water

5. Which of the following is false?

   (a) carbon dioxide is a mixture
   (b) sand and water is a suspension
   (c) brass is an alloy
   (d) common salt is a compound

6. When sugar is completely dissolved in water the resulting clear liquid is termed a:

   (a) solvent   (b) solution

   (c) solute   (d) suspension

7. The solubility of potassium chloride is 34 g in 0.1 kg of water. The amount of water required to dissolve 510 g of potassium chloride is:

   (a) 173.4 g   (b) 1.5 kg
   (c) 340 g   (d) 17.34 g

8. Which of the following statements is false?

   (a) When two metals are combined to form an alloy, the strength of the resulting material is greater than that of either of the two original metals

   (b) An alloy may be termed a solid solution

   (c) In a solution, the substance that is dissolved is called the solvent

   (d) The atomic number of an element is given by the number of protons in the atom

For fully worked solutions to each of the problems in Exercises 72 to 74 in this chapter, go to the website:
www.routledge.com/cw/bird

Chapter 17

Speed and velocity

Why it is important to understand: **Speed and velocity**

Speed is a quantity that refers to 'how fast an object is moving'. Speed can be thought of as the rate at which an object covers distance. A fast-moving object has a high speed and covers a relatively large distance in a short amount of time. Contrast this to a slow-moving object that has a low speed; it covers a relatively small amount of distance in the same amount of time. An object with no movement at all has a zero speed. For human beings, an average walking speed is about 3 mph, a top athlete can run at 37km/h in a 200 m sprint, a cyclist can average around 12 mph, and a 747 airplane has an average speed of around 565 mph. Velocity is a quantity that refers to 'the rate at which an object changes its position'; as such, velocity is direction aware. When evaluating the velocity of an object, we need to also keep track of direction. It would not be enough to say that a car has a velocity of 65 km/h; we must also include direction information in order to fully describe the velocity of the object. For example, we may describe a car's velocity as being 65 km/h, due west. This is one of the essential differences between speed and velocity. Speed is a scalar quantity and does not keep track of direction; velocity is a vector quantity and is direction aware (for more on scalars and vectors, see chapter 19).

At the end of this chapter, you should be able to:

- define speed
- construct and perform calculations on distant/time graphs
- construct and perform calculations on speed/time graphs
- define velocity

17.1 Speed

Speed is the rate of covering distance and is given by:

$$\text{speed} = \frac{\text{distance travelled}}{\text{time taken}}$$

The usual units for speed are metres per second (m/s or m s^{-1}), or kilometres per hour (km/h or km h^{-1}). Thus if a person walks 5 kilometres in 1 hour, the speed of the person is $\frac{5}{1}$, that is, 5 kilometres per hour.

The symbol for the SI unit of speed (and velocity) is written as m s^{-1}, called the 'index notation'. However, engineers usually use the symbol m/s, called the

Science and Mathematics for Engineering. 978-0-367-20475-4, © John Bird. Published by Taylor & Francis. All rights reserved.

'oblique notation', and it is this notation which is largely used in this chapter and other chapters on mechanics. One of the exceptions is when labelling the axes of graphs, when two obliques occur, and in this case the index notation is used. Thus for speed or velocity, the axis markings are speed/m s^{-1} or velocity/m s^{-1}.

Problem 1. A man walks 600 metres in 5 minutes. Determine his speed in (a) metres per second and (b) kilometres per hour

(a) **Speed** $= \dfrac{\text{distance travelled}}{\text{time taken}} = \dfrac{600\,\text{m}}{5\,\text{min}}$

$\qquad = \dfrac{600\,\text{m}}{5\,\text{min}} \times \dfrac{1\,\text{min}}{60\,\text{s}} = \mathbf{2\,m/s}$

(b) $\mathbf{2\,m/s} = \dfrac{2\,\text{m}}{1\,\text{s}} \times \dfrac{1\,\text{km}}{1000\,\text{m}} \times \dfrac{3600\,\text{s}}{1\,\text{h}}$

$\qquad = 2 \times 3.6 = \mathbf{7.2\,km/h}$

(Note: to change from m/s to km/h, multiply by 3.6)

Problem 2. A car travels at 50 kilometres per hour for 24 minutes. Find the distance travelled in this time

Since speed $= \dfrac{\text{distance travelled}}{\text{time taken}}$ then

$\qquad$ distance travelled $=$ speed $\times$ time taken

Time $= 24$ minutes $= \dfrac{24}{60}$ hours, hence

$\qquad$ **distance travelled** $= 50\,\dfrac{\text{km}}{\text{h}} \times \dfrac{24}{60}\,\text{h} = \mathbf{20\,km}$

Problem 3. A train is travelling at a constant speed of 25 metres per second for 16 kilometres. Find the time taken to cover this distance

Since speed $= \dfrac{\text{distance travelled}}{\text{time taken}}$ then,

$\qquad$ time taken $= \dfrac{\text{distance travelled}}{\text{speed}}$

16 kilometres $= 16{,}000$ metres, hence,

$\qquad$ time taken $= \dfrac{16{,}000}{\dfrac{25\,\text{m}}{1\,\text{s}}} = 16{,}000\,\text{m} \times \dfrac{1\,\text{s}}{25\,\text{m}} = 640\,\text{s}$

and $640\,\text{s} = 640\,\text{s} \times \dfrac{1\,\text{min}}{60\,\text{s}} = 10\,\dfrac{2}{3}\,\text{min}$ or **10 min 40 s**

Now try the following Practice Exercise

Practice Exercise 75 Further problems on speed (answers on page 539)

1. A train covers a distance of 96 km in 1 h 20 min. Determine the average speed of the train (a) in km/h and (b) in m/s

2. A horse trots at an average speed of 12 km/h for 18 minutes; determine the distance covered by the horse in this time

3. A ship covers a distance of 1365 km at an average speed of 15 km/h. How long does it take to cover this distance?

17.2 Distance/time graph

One way of giving data on the motion of an object is graphically. A graph of distance travelled (the scale on the vertical axis of the graph) against time (the scale on the horizontal axis of the graph) is called a **distance/time graph**. Thus if an aeroplane travels 500 kilometres in its first hour of flight and 750 kilometres in its second hour of flight, then after 2 hours, the total distance travelled is (500 + 750) kilometres, that is, 1250 kilometres. The distance/time graph for this flight is shown in Figure 17.1.

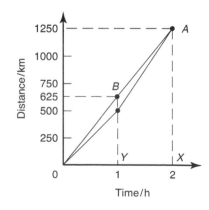

Figure 17.1

The **average speed** is given by:

$$\frac{\text{total distance travelled}}{\text{total time taken}}$$

Thus, the average speed of the aeroplane is

$$\frac{(500 + 750)\,\text{km}}{(1 + 1)\,\text{h}} = \frac{1250}{2} = 625\,\text{km/h}$$

If points 0 and A are joined in Figure 17.1, the slope of line $0A$ is defined as

$$\frac{\text{change in distance (vertical)}}{\text{change in time (horizontal)}}$$

for any two points on line $0A$.

For point A, the change in distance is AX, that is, 1250 kilometres, and the change in time is $0X$, that is, 2 hours. Hence the average speed is $\frac{1250}{2}$, i.e. 625 kilometres per hour.

Alternatively, for point B on line $0A$, the change in distance is BY, that is, 625 kilometres, and the change in time is $0Y$, that is, 1 hour, hence the average speed is $\frac{625}{1}$, i.e. 625 kilometres per hour.

In general, the average speed of an object travelling between points M and N is given by the slope of line MN on the distance/time graph.

Problem 4. A person travels from point 0 to A, then from A to B and finally from B to C. The distances of A, B and C from 0 and the times, measured from the start to reach points A, B and C are as shown:

	A	B	C
Distance (m)	100	200	250
Time (s)	40	60	100

Plot the distance/time graph and determine the speed of travel for each of the three parts of the journey

The vertical scale of the graph is distance travelled and the scale is selected to span 0 to 250 m, the total distance travelled from the start. The horizontal scale is time and spans 0 to 100 seconds, the total time taken to cover the whole journey. Co-ordinates corresponding to A, B and C are plotted and $0A$, AB and BC are joined by straight lines. The resulting distance/time graph is shown in Figure 17.2.

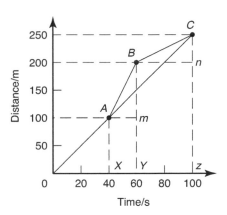

Figure 17.2

The speed is given by the slope of the distance/time graph.

Speed for part $0A$ of the journey = slope of $0A$

$$= \frac{AX}{0X} = \frac{100\,\text{m}}{40\,\text{s}} = \textbf{2.5 m/s}$$

Speed for part AB of the journey = slope of AB

$$= \frac{Bm}{Am} = \frac{(200 - 100)\,\text{m}}{(60 - 40)\,\text{s}} = \textbf{5 m/s}$$

Speed for part BC of the journey = slope of BC

$$= \frac{Cn}{Bn} = \frac{(250 - 200)\,\text{m}}{(100 - 60)\,\text{s}} = \textbf{1.25 m/s}$$

Problem 5. Determine the average speed (both in m/s and km/h) for the whole journey for the information given in Problem 4

$$\text{Average speed} = \frac{\text{total distance travelled}}{\text{total time taken}}$$

$$= \text{slope of line } 0C$$

From Figure 17.2, slope of line $0C$

$$= \frac{Cz}{0z} = \frac{250\,\text{m}}{100\,\text{s}} = \textbf{2.5 m/s}$$

$$2.5\,\text{m/s} = \frac{2.5\,\text{m}}{1\,\text{s}} \times \frac{1\,\text{km}}{1000\,\text{h}} \times \frac{3600\,\text{s}}{1\,\text{h}}$$

$$= (2.5 \times 3.6)\,\text{km/h} = \textbf{9 km/h}$$

Hence, **the average speed is 2.5 m/s or 9 km/h**

Problem 6. A coach travels from town A to town B, a distance of 40 kilometres at an average speed of 55 kilometres per hour. It then travels from town B to town C, a distance of 25 kilometres in 35 minutes. Finally, it travels from town C to town D at an average speed of 60 kilometres per hour in 45 minutes. Determine (a) the time taken to travel from A to B, (b) the average speed of the coach from B to C, (c) the distance from C to D, (d) the average speed of the whole journey from A to D

(a) From town A to town B:

Since speed $= \dfrac{\text{distance travelled}}{\text{time taken}}$ then,

time taken $= \dfrac{\text{distance travelled}}{\text{speed}}$

$= \dfrac{40\,\text{km}}{\dfrac{55\,\text{km}}{1\,\text{h}}} = 40\,\text{km} \times \dfrac{1\,\text{h}}{55\,\text{km}}$

$= 0.727\,\text{h or } \textbf{43.64 minutes}.$

(b) From town B to town C:

Since speed $= \dfrac{\text{distance travelled}}{\text{time taken}}$

and $35\,\text{min} = \dfrac{35}{60}\,\text{h}$ then,

speed $= \dfrac{25\,\text{km}}{\dfrac{35}{60}\,\text{h}} = \dfrac{25 \times 60}{35}\,\text{km/h} = \textbf{42.86 km/h}.$

(c) From town C to town D:

Since speed $= \dfrac{\text{distance travelled}}{\text{time taken}}$ then,

distance travelled $=$ speed $\times$ time taken,

and $45\,\text{min} = \dfrac{3}{4}\,\text{h}$ hence,

distance travelled $= 60\,\dfrac{\text{km}}{\text{h}} \times \dfrac{3}{4}\,\text{h} = \textbf{45 km}$

(d) From town A to town D:

average speed $= \dfrac{\text{total distance travelled}}{\text{total time taken}}$

$= \dfrac{(40 + 25 + 45)\,\text{km}}{\left(\dfrac{43.64}{60} + \dfrac{35}{60} + \dfrac{45}{60}\right)\,\text{h}} = \dfrac{110\,\text{km}}{\dfrac{123.64}{60}\,\text{h}}$

$= \dfrac{110 \times 60}{123.64}\,\text{km/h} = \textbf{53.38 km/h}$

Now try the following Practice Exercise

Practice Exercise 76   Further problems on distance/time graphs (answers on page 539)

1. Using the information given in the distance/time graph shown in Figure 17.3, determine the average speed when travelling from 0 to A, A to B, B to C, 0 to C and A to C

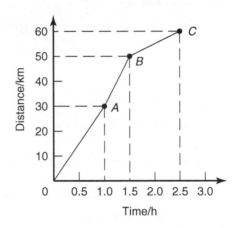

Figure 17.3

2. The distances travelled by an object from point 0 and the corresponding times taken to reach A, B, C and D, respectively, from the start are as shown:

Points	Start	A	B	C	D
Distance (m)	0	20	40	60	80
Time (s)	0	5	12	18	25

Draw the distance/time graph and hence determine the average speeds from 0 to A, A to B, B to C, C to D and 0 to D

3. A train leaves station A and travels via stations B and C to station D. The times the train passes the various stations are as shown:

Station	A	B	C	D
Times	10.55 am	11.40 am	12.15 pm	12.50 pm

The average speeds are: A to B, 56 km/h, B to C, 72 km/h, and C to D, 60 km/h. Calculate the total distance from A to D

4. A gun is fired 5 km north of an observer and the sound takes 15 s to reach him. Determine the average velocity of sound waves in air at this place

5. The light from a star takes 2.5 years to reach an observer. If the velocity of light is 330×10^6 m/s, determine the distance of the star from the observer in kilometres, based on a 365-day year

17.3   Speed/time graph

If a graph is plotted of speed against time, the area under the graph gives the distance travelled. This is demonstrated in Problem 7.

Problem 7. The motion of an object is described by the speed/time graph given in Figure 17.4. Determine the distance covered by the object when moving from 0 to B

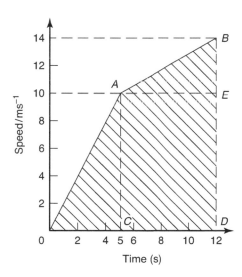

Figure 17.4

The distance travelled is given by the area beneath the speed/time graph, shown shaded in Figure 17.4.

Area of triangle $0AC$

$$= \frac{1}{2} \times \text{base} \times \text{perpendicular height}$$

$$= \frac{1}{2} \times 5\,\text{s} \times 10\,\frac{\text{m}}{\text{s}} = 25\,\text{m}$$

Area of rectangle $AEDC$

$$= \text{base} \times \text{height}$$

$$= (12 - 5)\,\text{s} \times (10 - 0)\,\frac{\text{m}}{\text{s}} = 70\,\text{m}$$

Area of triangle ABE

$$= \frac{1}{2} \times \text{base} \times \text{perpendicular height}$$

$$= \frac{1}{2} \times (12 - 5)\,\text{s} \times (14 - 10)\,\frac{\text{m}}{\text{s}}$$

$$= \frac{1}{2} \times 7\,\text{s} \times 4\,\frac{\text{m}}{\text{s}} = 14\,\text{m}$$

Hence, **the distance covered by the object moving from 0 to B** is:

$$(25 + 70 + 14)\,\text{m} -\ \textbf{109 m}$$

Now try the following Practice Exercise

Practice Exercise 77   Further problems on speed/time graphs (answers on page 539)

1. The speed/time graph for a car journey is shown in Figure 17.5. Determine the distance travelled by the car

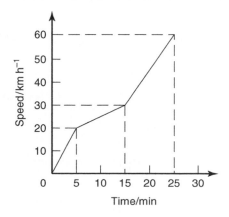

Figure 17.5

2. The motion of an object is as follows:

   A to B, distance 122 m, time 64 s

   B to C, distance 80 m at an average speed of 20 m/s

   C to D, time 7 s at an average speed of 14 m/s

   Determine the overall average speed of the object when travelling from A to D

17.4 Velocity

The **velocity** of an object is the speed of the object **in a specified direction**. Thus, if a plane is flying due south at 500 kilometres per hour, its speed is 500 kilometres per hour, but its velocity is 500 kilometres per hour due south. It follows that if the plane had flown in a circular path for one hour at a speed of 500 kilometres per hour, so that one hour after taking off it is again over the airport, its average velocity in the first hour of flight is zero. The average velocity is given by:

$$\frac{\text{distance travelled in a specific direction}}{\text{time taken}}.$$

If a plane flies from place 0 to place A, a distance of 300 kilometres in one hour, A being due north of 0, then $0A$ in Figure 17.6 represents the first hour of flight. It then flies from A to B, a distance of 400 kilometres during the second hour of flight, B being due east of A, thus AB in Figure 17.6 represents its second hour of flight.

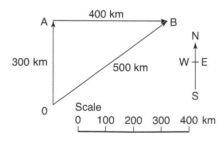

Figure 17.6

Its average velocity for the two hour flight is:

$$\frac{\text{distance } 0B}{2 \text{ hours}} = \frac{500 \text{ km}}{2 \text{ h}} = 250 \text{ km/h in direction } 0B$$

A graph of velocity (scale on the vertical axis) against time (scale on the horizontal axis) is called a **velocity/time graph**. The graph shown in Figure 17.7 represents a plane flying for 3 hours at a constant speed of 600 kilometres per hour in a specified direction. The shaded area represents velocity (vertically) multiplied by time (horizontally), and has units of

$$\frac{\text{kilometres}}{\text{hours}} \times \text{hours},$$

i.e. kilometres, and represents the distance travelled in a specific direction. In this case,

$$\text{distance} = 600 \, \frac{\text{km}}{\text{h}} \times 3 \, \text{h} = 1800 \, \text{km}$$

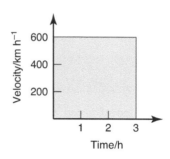

Figure 17.7

Another method of determining the distance travelled is from:

distance travelled = average velocity × time

Thus if a plane travels due south at 600 kilometres per hour for 20 minutes, the distance covered is

$$\frac{600 \text{ km}}{1 \text{ h}} \times \frac{20}{60} \text{ h} = 200 \text{ km}$$

Now try the following Practice Exercises

Practice Exercise 78   Short answer questions on speed and velocity (answers on page 539)

1. Speed is defined as ....

2. Speed is given by $\frac{\ldots\ldots}{\ldots\ldots}$

3. The usual units for speed are .... or ....

4. Average speed is given by $\frac{\ldots\ldots}{\ldots\ldots}$

5. The velocity of an object is ....

6. Average velocity is given by $\dfrac{\cdots}{\cdots}$

7. The area beneath a velocity/time graph represents the ....

8. Distance travelled = ... × ....

9. The slope of a distance/time graph gives the ....

10. The average speed can be determined from a distance/time graph from ......

Practice Exercise 79   Multiple-choice questions on speed and velocity (answers on page 540)

An object travels for 3 s at an average speed of 10 m/s and then for 5 s at an average speed of 15 m/s. In questions 1–3, select the correct answers from those given below:

(a) 105 m/s   (b) 3 m   (c) 30 m
(d) 13.125 m/s   (e) 3.33 m   (f) 0.3 m
(g) 75 m   (h) $\dfrac{1}{3}$ m   (i) 12.5 m/s

1. The distance travelled in the first 3 s

2. The distance travelled in the latter 5 s

3. The average speed over the 8 s period

4. Which of the following statements is false?

   (a) Speed is the rate of covering distance

   (b) Speed and velocity are both measured in m/s units

   (c) Speed is the velocity of travel in a specified direction

   (d) The area beneath the velocity/time graph gives distance travelled

In questions 5–7, use the table to obtain the quantities stated, selecting the correct answer from (a) to (i) of those given below.

Distance	Time	Speed
20 m	30 s	X
5 km	Y	20 km/h
Z	3 min	10 m/min

(a) 30 m   (b) $\dfrac{1}{4}$ h   (c) 600 m/s

(d) $3\dfrac{1}{3}$ m   (e) $\dfrac{2}{3}$ m/s   (f) $\dfrac{3}{10}$ m

(g) 4 h   (h) $1\dfrac{1}{4}$ m/s   (i) 100 h

5. Quantity X

6. Quantity Y

7. Quantity Z

Questions 8–10, refer to the distance/time graph shown in Figure 17.8.

8. The average speed when travelling from 0 to A is:

   (a) 3 m/s   (b) 1.5 m/s
   (c) 0.67 m/s   (d) 0.66 m/s

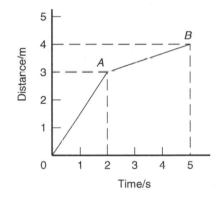

Figure 17.8

9. The average speed when travelling from A to B is:

   (a) 3 m/s   (b) 1.5 m/s
   (c) 0.67 m/s   (d) 0.33 m/s

10. The average overall speed when travelling from 0 to B is:

    (a) 0.8 m/s   (b) 1.2 m/s
    (c) 1.5 m/s   (d) 20 m/s

11. A car travels at 60 km/h for 25 minutes. The distance travelled in that time is:

    (a) 25 km   (b) 1500 km
    (c) 2.4 km   (d) 416.7 km

An object travels for 4 s at an average speed of 16 m/s, and for 6 s at an average speed of 24 m/s. Use these data to answer questions 12–14.

12.   The distance travelled in the first 4 s is:

    (a) 0.25 km   (b) 64 m

    (c) 4 m   (d) 0.64 km

13.   The distance travelled in the latter 6 s is:

    (a) 4 m   (b) 0.25 km

    (c) 144 m   (d) 14.4 km

14.   The average speed over the 10 s period is:

    (a) 20 m/s   (b) 50 m/s

    (c) 40 m/s   (d) 20.8 m/s

For fully worked solutions to each of the problems in Exercises 75 to 79 in this chapter, go to the website:
www.routledge.com/cw/bird

Chapter 18

Acceleration

Why it is important to understand: Acceleration

Acceleration may be defined as a 'change in velocity'. This change can be in the magnitude (speed) of the velocity or the direction of the velocity. In daily life we use *acceleration* as a term for the speeding up of objects and *decelerating* for the slowing down of objects. If there is a change in the velocity, whether it is slowing down or speeding up, or changing its direction, we say that the object is accelerating. If an object is moving at constant speed in a circular motion – such as a satellite orbiting the earth – it is said to be accelerating because change in direction of motion means its velocity is changing even if speed may be constant. This is called centripetal (directed towards the centre) acceleration. On the other hand, if the direction of motion of the object is not changing but its speed is, this is called tangential acceleration. If the direction of acceleration is in the same direction as that of velocity then the object is said to be speeding up or accelerating. If the acceleration and velocity are in opposite directions then the object is said to be slowing down or decelerating. An example of constant acceleration is the effect of the gravity of earth on an object in free fall. Measurement of the acceleration of a vehicle enables an evaluation of the overall vehicle performance and response. Detection of rapid negative acceleration of a vehicle is used to detect vehicle collision and deploy airbags. The measurement of acceleration is also used to measure seismic activity, inclination and machine vibration. Vibration monitoring is used in industries such as automotive manufacturing, machine tool applications, pharmaceutical production, power generation and power plants, pulp and paper, sugar mills, food and beverage production, water and wastewater, hydropower, petrochemical and steel manufacturing.

At the end of this chapter, you should be able to:

- define acceleration
- construct and perform calculations on velocity/time graphs
- define free-fall
- derive and use the equation of motion $v = u + at$

Science and Mathematics for Engineering. 978-0-367-20475-4, © John Bird. Published by Taylor & Francis. All rights reserved.

18.1 Introduction to acceleration

Acceleration is the rate of change of velocity with time. The average acceleration a is given by:

$$a = \frac{\text{change in velocity}}{\text{time taken}}$$

The usual units are metres per second squared (m/s² or m s⁻²). If u is the initial velocity of an object in metres per second, v is the final velocity in metres per second and t is the time in seconds elapsing between the velocities of u and v, then:

$$\textbf{average acceleration}, a = \frac{v - u}{t} \text{ m/s}^2$$

18.2 Velocity/time graph

A graph of velocity (scale on the vertical axis) against time (scale on the horizontal axis) is called a velocity/time graph, as introduced in chapter 17. For the velocity/time graph shown in Figure 18.1, the slope of line $0A$ is given by $\frac{AX}{0X}$. AX is the change in velocity from an initial velocity u of zero to a final velocity v of 4 metres per second. $0X$ is the time taken for this change in velocity, thus:

$$\frac{AX}{0X} = \frac{\text{change in velocity}}{\text{time taken}}$$

$$= \text{the acceleration in the first two seconds.}$$

From the graph: $\frac{AX}{0X} = \frac{4 \text{ m/s}}{2 \text{ s}} = 2 \text{ m/s}^2$, i.e. the acceleration is 2 m/s².

Similarly, the slope of line AB in Figure 18.1 is given by $\frac{BY}{AY}$, i.e. the acceleration between 2 s and 5 s is

$$\frac{8 - 4}{5 - 2} = \frac{4}{3} = 1\frac{1}{3} \text{ m/s}^2.$$

In general, the slope of a line on a velocity/time graph gives the acceleration.

The words 'velocity' and 'speed' are commonly interchanged in everyday language. Acceleration is a vector quantity and is correctly defined as the rate of change of velocity with respect to time. However, acceleration is also the rate of change of speed with respect to time in a certain specified direction.

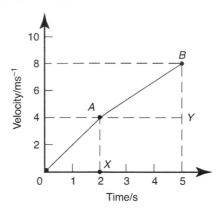

Figure 18.1

Problem 1. The speed of a car travelling along a straight road changes uniformly from zero to 50 km/h in 20 s. It then maintains this speed for 30 s and finally reduces speed uniformly to rest in 10 s. Draw the speed/time graph for this journey

The vertical scale of the speed/time graph is speed (km h⁻¹) and the horizontal scale is time (s). Since the car is initially at rest, then at time 0 s, the speed is 0 km/h. After 20 s, the speed is 50 km/h, which corresponds to point A on the speed/time graph shown in Figure 18.2. Since the change in speed is uniform, a straight line is drawn joining points 0 and A.

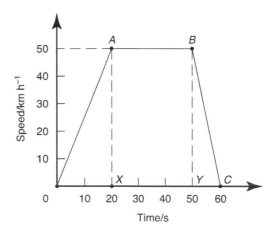

Figure 18.2

The speed is constant at 50 km/h for the next 30 s, hence, horizontal line AB is drawn in Figure 18.2 for the time period 20 s to 50 s. Finally, the speed falls from 50 km/h at 50 s to zero in 10 s, hence point C on the speed/time graph in Figure 18.2 corresponds to a speed of zero and a time of 60 s. Since the reduction in speed is uniform, a straight line is drawn joining BC. Thus, the speed/time graph for the journey is as shown in Figure 18.2.

Problem 2. For the speed/time graph shown in Figure 18.2, find the acceleration for each of the three stages of the journey

From above, the slope of line $0A$ gives the uniform acceleration for the first 20 s of the journey.

$$\text{Slope of } 0A = \frac{AX}{0X} = \frac{(50-0)\,\text{km/h}}{(20-0)\,\text{s}} = \frac{50\,\text{km/h}}{20\,\text{s}}$$

Expressing 50 km/h in metre/second units gives:

$$50\frac{\text{km}}{\text{h}} = \frac{50\,\text{km}}{1\,\text{h}} \times \frac{1000\,\text{m}}{1\,\text{km}} \times \frac{1\,\text{h}}{3600\,\text{s}} = \frac{50}{3.6}\text{m/s}$$

(Note: to change from km/h to m/s, divide by 3.6)

$$\text{Thus, } \frac{50\,\text{km/h}}{20\,\text{s}} = \frac{\frac{50}{3.6}\text{m/s}}{20\,\text{s}} = 0.694\ \text{m/s}^2$$

i.e. **the acceleration during the first 20 s is 0.694 m/s²**.

Acceleration is defined as

$$\frac{\text{change in velocity}}{\text{time taken}} \quad \text{or} \quad \frac{\text{change of speed}}{\text{time taken}}$$

since the car is travelling along a straight road. Since there is no change in speed for the next 30 s (line AB in Figure 18.2 is horizontal), **then the acceleration for this period is zero.**

From above, the slope of line BC gives the uniform deceleration for the final 10 s of the journey.

$$\text{Slope of } BC = \frac{BY}{YC} = \frac{50\,\text{km/h}}{10\,\text{s}}$$

$$= \frac{\frac{50}{3.6}\text{m/s}}{10\,\text{s}}$$

$$= 1.39\ \text{m/s}^2$$

i.e. **the deceleration during the final 10 s is 1.39 m/s²**. Alternatively, **the acceleration is – 1.39 m/s²**.

Now try the following Practice Exercise

Practice Exercise 80   Further problems on velocity/time graphs and acceleration (answers on page 540)

1. A coach increases velocity from 4 km/h to 40 km/h at an average acceleration of $0.2\ \text{m/s}^2$. Find the time taken for this increase in velocity

2. A ship changes velocity from 15 km/h to 20 km/h in 25 min. Determine the average acceleration in m/s^2 of the ship during this time

3. A cyclist travelling at 15 km/h changes velocity uniformly to 20 km/h in 1 min, maintains this velocity for 5 min and then comes to rest uniformly during the next 15 s. Draw a velocity/time graph and hence determine the accelerations in m/s^2: (a) during the first minute, (b) for the next 5 minutes, and (c) for the last 10 s

4. Assuming uniform accelerations between points draw the velocity/time graph for the data given below, and hence determine the accelerations from A to B, B to C and C to D:

Point	A	B	C	D
Speed (m/s)	25	5	30	15
Time (s)	15	25	35	45

18.3   Free-fall and equation of motion

If a dense object such as a stone is dropped from a height, called **free-fall**, it has a constant acceleration of approximately $9.8\ \text{m/s}^2$. In a vacuum, all objects have this same constant acceleration, vertically downwards, that is, a feather has the same acceleration as a stone. However, if free-fall takes place in air, dense objects have the constant acceleration of $9.8\ \text{m/s}^2$ over short distances, but objects that have a low density, such as feathers, have little or no acceleration.

For bodies moving with a constant acceleration, the average acceleration is the constant value of the acceleration, and since from section 18.1:

$$a = \frac{v-u}{t}$$

then

$$a \times t = v - u \text{ from which } \boldsymbol{v = u + at}$$

where u is the initial velocity in m/s,
v is the final velocity in m/s,
a is the constant acceleration in m/s^2 and
t is the time in s.

When symbol a has a negative value, it is called **deceleration** or **retardation**. The equation $v = u + at$ is called an **equation of motion**.

> **Problem 3.** A stone is dropped from an aeroplane. Determine (a) its velocity after 2 s and (b) the increase in velocity during the third second, in the absence of all forces except that due to gravity

The stone is free-falling and thus has an acceleration a of approximately 9.8 m/s^2 (taking downward motion as positive). From above:

$$\text{final velocity } v = u + at$$

(a) The initial downward velocity of the stone u is zero. The acceleration a is 9.8 m/s^2 downwards and the time during which the stone is accelerating is 2 s. Hence, final velocity, $v = u + at = 0 + 9.8 \times 2 = 19.6$ m/s, i.e. **the velocity of the stone after 2 s is approximately 19.6 m/s**

(b) From part (a), the velocity after two seconds u is 19.6 m/s. The velocity after 3 s, applying $v = u + at$, is $v = 19.6 + 9.8 \times 3 = 49$ m/s. Thus, **the change in velocity during the third second is $(49 - 19.6) = 29.4$ m/s**

(Since the value $a = 9.8$ m/s^2 is only an approximate value, then the answer is an approximate value.)

> **Problem 4.** Determine how long it takes an object, which is free-falling, to change its speed from 100 km/h to 150 km/h, assuming all other forces, except that due to gravity, are neglected

The initial velocity u is 100 km/h, i.e. $\dfrac{100}{3.6}$ m/s (see Problem 2). The final velocity v is 150 km/h, i.e. $\dfrac{150}{3.6}$ m/s. Since the object is free-falling, the acceleration a is approximately 9.8 m/s^2 downwards (i.e. in a positive direction).

From above, $v = u + at$, i.e. $\dfrac{150}{3.6} = \dfrac{100}{3.6} + 9.8 \times t$.

Transposing, gives $9.8 \times t = \dfrac{150 - 100}{3.6} = \dfrac{50}{3.6}$

Hence, $\qquad \text{time } t = \dfrac{50}{3.6 \times 9.8} = 1.42$ s

Since the value of a is only approximate, and rounding-off errors have occurred in calculations, then **the approximate time for the velocity to change from 100 km/h to 150 km/h is 1.42 s**

> **Problem 5.** A train travelling at 30 km/h accelerates uniformly to 50 km/h in 2 minutes. Determine the acceleration

$$30 \text{ km/h} = \frac{30}{3.6} \text{ m/s (see Problem 2)},$$

$$50 \text{ km/h} = \frac{50}{3.6} \text{ m/s}$$

and $\quad 2 \text{ min} = 2 \times 60 = 120$ s.

From above,

$$v = u + at, \quad \text{i.e. } \frac{50}{3.6} = \frac{30}{3.6} + a \times 120$$

Transposing, gives $\quad 120 \times a = \dfrac{50 - 30}{3.6}$

and $\qquad\qquad a = \dfrac{20}{3.6 \times 120}$

$$= 0.0463 \text{ m/s}^2$$

i.e. **the uniform acceleration of the train is 0.0463 m/s^2**

> **Problem 6.** A car travelling at 50 km/h applies its brakes for 6 s and decelerates uniformly at 0.5 m/s^2. Determine its velocity in km/h after the 6 s braking period

The initial velocity $u = 50$ km/h $= \dfrac{50}{3.6}$ m/s (see Problem 2).
From above, $v = u + at$. Since the car is decelerating, i.e. it has a negative acceleration, then $a = -0.5$ m/s^2 and t is 6 s.

Thus, final velocity $\quad v = \dfrac{50}{3.6} + (-0.5)(6)$

$$= 13.89 - 3 = 10.89 \text{ m/s}$$

$$10.89 \text{ m/s} = 10.89 \times 3.6 = 39.2 \text{ km/h}$$

(Note: to convert m/s to km/h, multiply by 3.6)
Thus, **the velocity after braking is 39.2 km/h**

Problem 7. A cyclist accelerates uniformly at $0.3\,\text{m/s}^2$ for 10 s, and his speed after accelerating is $20\,\text{km/h}$. Find his initial speed

The final speed $v = \dfrac{20}{3.6}\,\text{m/s}$, time $t = 10\,\text{s}$, and acceleration $a = 0.3\,\text{m/s}^2$.
From above, $v = u + at$, where u is the initial speed.

Hence, $\dfrac{20}{3.6} = u + 0.3 \times 10$

from which, $u = \dfrac{20}{3.6} - 3 = 2.56\,\text{m/s}$

$2.56\,\text{m/s} = 2.56 \times 3.6\,\text{km/h}$ (see Problem 6)

$= 9.2\,\text{km/h}$

i.e. **the initial speed of the cyclist is 9.2 km/h**

Now try the following Practice Exercises

Practice Exercise 81   Further problems on free-fall and equation of motion (answers on page 540)

1. An object is dropped from the third floor of a building. Find its approximate velocity 1.25 s later if all forces except that of gravity are neglected

2. During free fall, a ball is dropped from point A and is travelling at $100\,\text{m/s}$ when it passes point B. Calculate the time for the ball to travel from A to B if all forces except that of gravity are neglected

3. A piston moves at $10\,\text{m/s}$ at the centre of its motion and decelerates uniformly at $0.8\,\text{m/s}^2$. Determine its velocity 3 s after passing the centre of its motion

4. The final velocity of a train after applying its brakes for 1.2 min is $24\,\text{km/h}$. If its uniform retardation is $0.06\,\text{m/s}^2$, find its velocity before the brakes are applied

5. A plane in level flight at $400\,\text{km/h}$ starts to descend at a uniform acceleration of $0.6\,\text{m/s}^2$. It levels off when its velocity is $670\,\text{km/h}$. Calculate the time during which it is losing height

6. A lift accelerates from rest uniformly at $0.9\,\text{m/s}^2$ for 1.5 s, travels at constant velocity for 7 s and then comes to rest in 3 s. Determine its velocity when travelling at constant speed and its acceleration during the final 3 s of its travel

Practice Exercise 82   Short answer questions on acceleration

1. Acceleration is defined as ....

2. Acceleration is given by $\dfrac{\cdots}{\cdots}$

3. The usual units for acceleration are ......

4. The slope of a velocity/time graph gives the ....

5. The value of free-fall acceleration for a dense object is approximately ......

6. The relationship between initial velocity u, final velocity v, acceleration a, and time t, is ....

7. A negative acceleration is called a .... or a ....

Practice Exercise 83   Multiple-choice questions on acceleration (answers on page 540)

1. If a car accelerates from rest, the acceleration is defined as the rate of change of:

   (a) energy   (b) velocity

   (c) mass   (d) displacement

2. An engine is travelling along a straight, level track at $15\,\text{m/s}$. The driver switches off the engine and applies the brakes to bring the engine to rest with uniform retardation in 5 s. The retardation of the engine is:

   (a) $3\,\text{m/s}^2$ (b) $4\,\text{m/s}^2$

   (c) $\dfrac{1}{3}\,\text{m/s}^2$ (d) $75\,\text{m/s}^2$

Six statements, (a)–(f), are given below, some of the statements being true and the remainder false.

(a)   Acceleration is the rate of change of velocity with distance

(b)   Average acceleration

$$= \frac{\text{(change of velocity)}}{\text{(time taken)}}$$

(c)   Average acceleration $= (u-v)/t$, where u is the initial velocity, v is the final velocity and t is the time

(d)   The slope of a velocity/time graph gives the acceleration

(e)   The acceleration of a dense object during free-fall is approximately $9.8\,\text{m/s}^2$ in the absence of all other forces except gravity

(f)   When the initial and final velocities are u and v, respectively, a is the acceleration and t the time, then $u = v + at$

In questions 3 and 4, select the statements required from those given.

3.   (b), (c), (d), (e) Which statement is false?

4.   (a), (c), (e), (f) Which statement is true?

A car accelerates uniformly from $5\,\text{m/s}$ to $15\,\text{m/s}$ in $20\,\text{s}$. It stays at the velocity attained at $20\,\text{s}$

for $2\,\text{min}$. Finally, the brakes are applied to give a uniform deceleration and it comes to rest in $10\,\text{s}$. Use these data in questions 5–9, selecting the correct answer from (a) to (l) given below.

(a) $-1.5\,\text{m/s}^2$ (b) $\dfrac{2}{15}\,\text{m/s}^2$ (c) 0

(d) $0.5\,\text{m/s}^2$ (e) $1.389\,\text{km/h}$ (f) $7.5\,\text{m/s}^2$

(g) $54\,\text{km/h}$ (h) $2\,\text{m/s}^2$ (i) $18\,\text{km/h}$

(j) $-\dfrac{1}{10}\,\text{m/s}^2$ (k) $1.467\,\text{km/h}$ (l) $-\dfrac{2}{3}\,\text{m/s}^2$

5.   The initial speed of the car in km/h

6.   The speed of the car after $20\,\text{s}$ in km/h

7.   The acceleration during the first $20\,\text{s}$ period

8.   The acceleration during the $2\,\text{min}$ period

9.   The acceleration during the final $10\,\text{s}$

10.   A cutting tool accelerates from $50\,\text{mm/s}$ to $150\,\text{mm/s}$ in 0.2 seconds. The average acceleration of the tool is:

(a) $500\,\text{m/s}^2$ (b) $1\,\text{m/s}^2$

(c) $20\,\text{m/s}^2$ (d) $0.5\,\text{m/s}^2$

For fully worked solutions to each of the problems in Exercises 80 to 83 in this chapter, go to the website:
www.routledge.com/cw/bird

Chapter 19

Force, mass and acceleration

Why it is important to understand: Force, mass and acceleration

When an object is pushed or pulled, a force is applied to the object. The effects of pushing or pulling an object are to cause changes in the motion and shape of the object. If a change occurs in the motion of the object then the object accelerates. Thus, acceleration results from a force being applied to an object. If a force is applied to an object and it does not move, then the object changes shape. Usually the change in shape is so small that it cannot be detected by just watching the object. However, when very sensitive measuring instruments are used, very small changes in dimensions can be detected. A force of attraction exists between all objects. If a person is taken as one object and the Earth as a second object, a force of attraction exists between the person and the Earth. This force is called the gravitational force and is the force that gives a person a certain weight when standing on the Earth's surface. It is also this force that gives freely falling objects a constant acceleration in the absence of other forces. This chapter defines force and acceleration, states Newton's three laws of motion and defines moment of inertia, all demonstrated via practical everyday situations.

At the end of this chapter, you should be able to:

- define force and state its unit
- appreciate 'gravitational force'
- state Newton's three laws of motion
- perform calculations involving force $F = ma$
- define 'centripetal acceleration'
- perform calculations involving centripetal force $= \dfrac{mv^2}{r}$
- define 'mass moment of inertia'

Science and Mathematics for Engineering. 978-0-367-20475-4. © John Bird. Published by Taylor & Francis. All rights reserved.

19.1 Introduction

When an object is pushed or pulled, a **force** is applied to the object. This force is measured in **newtons*** (**N**). The effects of pushing or pulling an object are:

(i) to cause a change in the motion of the object, and

(ii) to cause a change in the shape of the object.

If a change occurs in the motion of the object, that is, its velocity changes from u to v, then the object accelerates. Thus, it follows that acceleration results from a force being applied to an object. If a force is applied to an object and it does not move, then the object changes shape, that is, deformation of the object takes place. Usually the change in shape is so small that it cannot be detected by just watching the object. However, when very sensitive measuring instruments are used, very small changes in dimensions can be detected.

A force of attraction exists between all objects. The factors governing the size of this force F are the masses of the objects and the distances between their centres:

$$F \propto \frac{m_1 m_2}{d^2}$$

Thus, if a person is taken as one object and the Earth as a second object, a force of attraction exists between the person and the Earth. This force is called the **gravitational force**, first presented by Sir Isaac Newton, and is the force that gives a person a certain weight when standing on the Earth's surface. It is also this force that gives freely falling objects a constant acceleration in the absence of other forces.

19.2 Newton's laws of motion

To make a stationary object move or to change the direction in which the object is moving requires a force to be applied externally to the object. This concept is known as **Newton's first law of motion** and may be stated as:

---

*Sir Isaac Newton (25 December 1642–20 March 1727) was an English polymath. Newton showed that the motions of objects are governed by the same set of natural laws, by demonstrating the consistency between Kepler's laws of planetary motion and his theory of gravitation. The SI unit of force is the newton, named in his honour. To find out more about Newton go to www.routledge.com/cw/bird (For image refer to page 29)

An object remains in a state of rest, or continues in a state of uniform motion in a straight line, unless it is acted on by an externally applied force.

Since a force is necessary to produce a change of motion, an object must have some resistance to a change in its motion. The force necessary to give a stationary pram a given acceleration is far less than the force necessary to give a stationary car the same acceleration on the same surface. The resistance to a change in motion is called the **inertia** of an object and the amount of inertia depends on the mass of the object. Since a car has a much larger mass than a pram, the inertia of a car is much larger than that of a pram.

Newton's second law of motion may be stated as:

The acceleration of an object acted upon by an external force is proportional to the force and is in the same direction as the force.

Thus, force $\propto$ acceleration, or force = a constant $\times$ acceleration, this constant of proportionality being the mass of the object, i.e.

force = mass × acceleration.

The unit of force is the newton (N) and is defined in terms of mass and acceleration. One newton is the force required to give a mass of 1 kilogram an acceleration of 1 metre per second squared. Thus

$$F = ma$$

where F is the force in newtons (N), m is the mass in kilograms (kg) and a is the acceleration in metres per second squared (m/s^2), i.e. $1\,\text{N} = \dfrac{1\,\text{kg\,m}}{\text{s}^2}$

It follows that $1\,\text{m/s}^2 = 1\,\text{N/kg}$. Hence a gravitational acceleration of 9.8 m/s^2 is the same as a gravitational field of 9.8 N/kg

Newton's third law of motion may be stated as:

For every force, there is an equal and opposite reacting force.

Thus, an object on, say, a table, exerts a downward force on the table and the table exerts an equal upward force on the object, known as a **reaction force** or just a **reaction**.

Problem 1. Calculate the force needed to accelerate a boat of mass 20 tonnes uniformly from rest to a speed of 21.6 km/h in 10 minutes

The mass of the boat m is 20 t, that is 20,000 kg. The law of motion, $v = u + at$, can be used to determine the acceleration a. The initial velocity u is zero, the final velocity

$$v = 21.6 \, \text{km/h} = 21.6 \frac{\text{km}}{\text{h}} \times \frac{1 \, \text{h}}{3600 \, \text{s}} \times \frac{1000 \, \text{m}}{1 \, \text{km}}$$

$$= \frac{21.6}{3.6} = 6 \, \text{m/s},$$

and the time $t = 10 \, \text{min} = 600 \, \text{s}$

Thus, $v = u + at$, i.e. $6 = 0 + a \times 600$, from which,

$$a = \frac{6}{600} = 0.01 \, \text{m/s}^2$$

From Newton's second law, $F = ma$

i.e. **force** $= 20,000 \times 0.01 \, \text{N} = \textbf{200 N}$

Problem 2. The moving head of a machine tool requires a force of 1.2 N to bring it to rest in 0.8 s from a cutting speed of 30 m/min. Find the mass of the moving head

From Newton's second law, $F = ma$, thus $m = \dfrac{F}{a}$ where force is given as 1.2 N. The law of motion, $v = u + at$, can be used to find acceleration a, where $v = 0$, $u = 30 \, \text{m/min} = \dfrac{30}{60} \, \text{m/s} = 0.5 \, \text{m/s}$, and $t = 0.8 \, \text{s}$

Thus, $\qquad\qquad 0 = 0.5 + a \times 0.8$

from which, $\qquad a = -\dfrac{0.5}{0.8} = -0.625 \, \text{m/s}^2$

or a retardation of 0.625 m/s².

Thus the **mass** $m = \dfrac{F}{a} = \dfrac{1.2}{0.625} = \textbf{1.92 kg}$.

Problem 3. A lorry of mass 1350 kg accelerates uniformly from 9 km/h to reach a velocity of 45 km/h in 18 s. Determine (a) the acceleration of the lorry, (b) the uniform force needed to accelerate the lorry

(a) The law of motion, $v = u + at$, can be used to determine the acceleration, where final velocity

$$v = 45 \frac{\text{km}}{\text{h}} \times \frac{1 \, \text{h}}{3600 \, \text{s}} \times \frac{1000 \, \text{m}}{1 \, \text{km}} = \frac{45}{3.6} \, \text{m/s},$$

initial velocity $u = \dfrac{9}{3.6} \, \text{m/s}$ and time $t = 18 \, \text{s}$

Thus $\qquad \dfrac{45}{3.6} = \dfrac{9}{3.6} + a \times 18$

from which, $a = \dfrac{1}{18}\left(\dfrac{45}{3.6} - \dfrac{9}{3.6}\right) = \dfrac{1}{18}\left(\dfrac{36}{3.6}\right)$

$$= \frac{10}{18} = \frac{5}{9} \, \textbf{m/s}^2 \text{ or } \textbf{0.556 m/s}^2$$

(b) From Newton's second law of motion,

$$\textbf{force } F = ma = 1350 \times \frac{5}{9} = \textbf{750 N}$$

Problem 4. Find the weight of an object of mass 1.6 kg at a point on the Earth's surface where the gravitational field is 9.81 N/kg (or 9.81 m/s²)

The weight of an object is the force acting vertically downwards due to the force of gravity acting on the object. Thus:

$$\textbf{weight} = \text{force acting vertically downwards}$$

$$= \text{mass} \times \text{gravitational field}$$

$$= 1.6 \times 9.81 = \textbf{15.696 N}$$

Problem 5. A bucket of cement of mass 40 kg is tied to the end of a rope connected to a hoist. Calculate the tension in the rope when the bucket is suspended but stationary. Take the gravitational field g as 9.81 N/kg (or 9.81 m/s²)

The **tension** in the rope is the same as the force acting in the rope. The force acting vertically downwards due to the weight of the bucket must be equal to the force acting upwards in the rope, i.e. the tension. Weight of bucket of cement

$$F = mg = 40 \times 9.81 = 392.4 \, \text{N}$$

Thus, **the tension in the rope** $= \textbf{392.4 N}$

Problem 6. The bucket of cement in Problem 5 is now hoisted vertically upwards with a uniform acceleration of 0.4 m/s². Calculate the tension in the rope during the period of acceleration

With reference to Figure 19.1, the forces acting on the bucket are:

(i) a tension (or force) of T acting in the rope,

(ii) a force of mg acting vertically downwards, i.e. the weight of the bucket and cement.

The resultant force $F = T - $ mg; hence, $ma = T - $ mg

i.e. $\quad 40 \times 0.4 = T - 40 \times 9.81$

from which, **tension $T = 408.4\,$N**

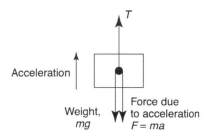

Figure 19.1

By comparing this result with that of Problem 5, it can be seen that there is an increase in the tension in the rope when an object is accelerating upwards.

Problem 7. The bucket of cement in Problem 5 is now lowered vertically downwards with a uniform acceleration of 1.4 m/s². Calculate the tension in the rope during the period of acceleration

With reference to Figure 19.2, the forces acting on the bucket are:

(i) a tension (or force) of T acting vertically upwards,

(ii) a force of mg acting vertically downwards, i.e. the weight of the bucket and cement.

The resultant force $\quad F = mg - T$

Hence, $\qquad ma = mg - T$

from which, **tension $T = m(g - a)$**

$$= 40(9.81 - 1.4)$$

$$= \mathbf{336.4\,N}$$

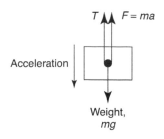

Figure 19.2

By comparing this result with that of Problem 5, it can be seen that there is a decrease in the tension in the rope when an object is accelerating downwards.

Now try the following Practice Exercise

Practice Exercise 84   Further problems on Newton's laws of motion (answers on page 540)

(Take g as 9.81 m/s², and express answers to three significant figures accuracy.)

1. A car initially at rest, accelerates uniformly to a speed of 55 km/h in 14 s. Determine the accelerating force required if the mass of the car is 800 kg

2. The brakes are applied on the car in question 1 when travelling at 55 km/h and it comes to rest uniformly in a distance of 50 m. Calculate the braking force and the time for the car to come to rest

3. The tension in a rope lifting a crate vertically upwards is 2.8 kN. Determine its acceleration if the mass of the crate is 270 kg

4. A ship is travelling at 18 km/h when it stops its engines. It drifts for a distance of 0.6 km and its speed is then 14 km/h. Determine the value of the forces opposing the motion of the ship, assuming the reduction in speed is uniform and the mass of the ship is 2000 t

5. A cage having a mass of 2 t is being lowered down a mineshaft. It moves from rest with an acceleration of 4 m/s², until it is travelling at 15 m/s. It then travels at constant speed for 700 m and finally comes to rest in 6 s

Calculate the tension in the cable supporting the cage during (a) the initial period of acceleration, (b) the period of constant speed travel, (c) the final retardation period

19.3 Centripetal acceleration

When an object moves in a circular path at constant speed, its direction of motion is continually changing and hence its velocity (which depends on both magnitude and direction) is also continually changing. Since acceleration is the (change in velocity)/(time taken) the object has an acceleration.

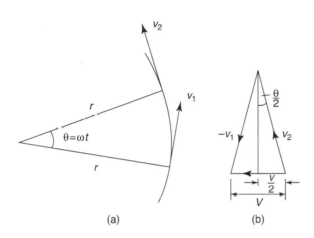

(a)   (b)

Figure 19.3

Let the object be moving with a constant angular velocity of ω and a tangential velocity of magnitude v and let the change of velocity for a small change of angle of θ ($=\omega t$) be V (see Figure 19.3(a)). Then,

$$v_2 - v_1 = V$$

The vector diagram is shown in Figure 19.3(b) and since the magnitudes of v_1 and v_2 are the same, i.e. v, the vector diagram is also an isosceles triangle.
Bisecting the angle between v_2 and v_1 gives:

$$\sin\frac{\theta}{2} = \frac{V/2}{v_2} = \frac{V}{2v}$$

i.e. $V = 2v \sin\dfrac{\theta}{2}$ (1)

Since $\theta = \omega t$, then $t = \dfrac{\theta}{\omega}$ (2)

Dividing equation (1) by (2) gives:

$$\frac{V}{t} = \frac{2v \sin\dfrac{\theta}{2}}{\dfrac{\theta}{\omega}} = \frac{v\omega \sin\dfrac{\theta}{2}}{\dfrac{\theta}{2}}$$

For small angles, $\dfrac{\sin\dfrac{\theta}{2}}{\dfrac{\theta}{2}}$ is very nearly equal to unity

Therefore, $\dfrac{V}{t} = v\omega$

or, $\dfrac{V}{t} = \dfrac{\text{change of velocity}}{\text{change of time}} = \text{acceleration}, a = v\omega$

But, $\omega = v/r$, thus $v\omega = v \times \dfrac{v}{r} = \dfrac{v^2}{r} = \text{acceleration}$.

That is, **the acceleration a is** $\dfrac{v^2}{r}$ and is towards the centre of the circle of motion (along V). It is called the **centripetal acceleration**. If the mass of the rotating object is m, then by Newton's second law, the **centripetal force** is $\dfrac{mv^2}{r}$, and its direction is towards the centre of the circle of motion.

Problem 8. A vehicle of mass 750 kg travels round a bend of radius 150 m, at 50.4 km/h. Determine the centripetal force acting on the vehicle

The centripetal force is given by $\dfrac{mv^2}{r}$ and its direction is towards the centre of the circle.

$$m = 750\,\text{kg}, v = 50.4\,\text{km/h} = \frac{50.4}{3.6}\,\text{m/s}$$

$$= 14\,\text{m/s and } r = 150\,\text{m}$$

Thus, **centripetal force** $= \dfrac{750 \times 14^2}{150} = \mathbf{980\,N}$

Problem 9. An object is suspended by a thread 250 mm long and both object and thread move in a horizontal circle with a constant angular velocity of 2.0 rad/s. If the tension in the thread is 12.5 N, determine the mass of the object

Centripetal force (i.e. tension in thread) $= \dfrac{mv^2}{r}$

$$= 12.5\,\text{N}$$

The angular velocity $\omega = 2.0$ rad/s
and radius $r = 250\,\text{mm} = 0.25\,\text{m}$

Since linear velocity $v = \omega r$, $v = 2.0 \times 0.25 = 0.5$ m/s,

and since $F = \dfrac{mv^2}{r}$, then $m = \dfrac{Fr}{v^2}$

i.e. **mass of object** $m = \dfrac{12.5 \times 0.25}{0.5^2} = \mathbf{12.5\,kg}$

Problem 10. An aircraft is turning at constant altitude, the turn following the arc of a circle of radius 1.5 km. If the maximum allowable acceleration of the aircraft is 2.5 g, determine the maximum speed of the turn in km/h. Take g as 9.8 m/s^2

The acceleration of an object turning in a circle is $\dfrac{v^2}{r}$

Thus, to determine the maximum speed of turn

$\dfrac{v^2}{r} = 2.5g$. Hence,

$$\textbf{speed of turn } v = \sqrt{2.5gr} = \sqrt{2.5 \times 9.8 \times 1500}$$

$$= \sqrt{36{,}750} = 191.7\,\text{m/s}$$

$$= 191.7 \times 3.6\,\text{km/h} = \mathbf{690\,km/h}$$

Now try the following Practice Exercises

Practice Exercise 85   Further problems on centripetal acceleration (answers on page 540)

1. Calculate the centripetal force acting on a vehicle of mass 1 t when travelling round a bend of radius 125 m at 40 km/h. If this force should not exceed 750 N, determine the reduction in speed of the vehicle to meet this requirement

2. A speed-boat negotiates an S-bend consisting of two circular arcs of radii 100 m and 150 m. If the speed of the boat is constant at 34 km/h, determine the change in acceleration when leaving one arc and entering the other

3. An object is suspended by a thread 400 mm long, and both object and thread move in a horizontal circle with a constant angular velocity of 3.0 rad/s. If the tension in the thread is 36 N, determine the mass of the object

Practice Exercise 86   Short answer questions on force, mass and acceleration

1. Force is measured in ....

2. The two effects of pushing or pulling an object are .... or ....

3. A gravitational force gives free-falling objects a .... in the absence of all other forces

4. State Newton's first law of motion

5. Describe what is meant by the inertia of an object

6. State Newton's second law of motion

7. Define the newton

8. State Newton's third law of motion

9. Explain why an object moving round a circle at a constant angular velocity has an acceleration

10. Define centripetal acceleration in symbols

11. Define centripetal force in symbols

Practice Exercise 87   Multiple-choice questions on force, mass and acceleration (answers on page 540)

1. The unit of force is the:

   (a) watt   (b) kelvin   (c) newton   (d) joule

2. If a = acceleration and F = force, then mass m is given by:

   (a) $m = a - F$ (b) $m = \dfrac{F}{a}$

   (c) $m = F - a$ (d) $m = \dfrac{a}{F}$

3. The weight of an object of mass 2 kg at a point on the Earth's surface when the gravitational field is 10 N/kg is:

   (a) 20 N   (b) 0.2 N   (c) 20 kg   (d) 5 N

4. The force required to accelerate a loaded barrow of 80 kg mass up to 0.2 m/s^2 on friction-less bearings is:

   (a) 400 N   (b) 3.2 N

   (c) 0.0025 N   (d) 16 N

5. A bucket of cement of mass 30 kg is tied to the end of a rope connected to a hoist. If the gravitational field $g = 10\,N/kg$, the tension in the rope when the bucket is suspended but stationary is:

(a) 300 N   (b) 3 N   (c) 300 kg   (d) 0.67 N

A man of mass 75 kg is standing in a lift of mass 500 kg. Use these data to determine the answers to questions 6–9. Take g as $10\,m/s^2$

6. The tension in a cable when the lift is moving at a constant speed vertically upward is:

(a) 4250 N   (b) 5750 N

(c) 4600 N   (d) 6900 N

7. The tension in the cable supporting the lift when the lift is moving at a constant speed vertically downwards is:

(a) 4250 N   (b) 5750 N

(c) 4600 N   (d) 6900 N

8. The reaction force between the man and the floor of the lift when the lift is travelling at a constant speed vertically upwards is:

(a) 750 N   (b) 900 N   (c) 600 N   (d) 475 N

9. The reaction force between the man and the floor of the lift when the lift is travelling at a constant speed vertically downwards is:

(a) 750 N   (b) 900

(c) 600 N   (d) 475 N

A ball of mass 0.5 kg is tied to a thread and rotated at a constant angular velocity of 10 rad/s in a circle of radius 1 m. Use these data to determine the answers to questions 10 and 11.

10. The centripetal acceleration is:

(a) $50\,m/s^2$ (b) $\dfrac{100}{2\pi}\,m/s^2$

(c) $\dfrac{50}{2\pi}\,m/s^2$ (d) $100\,m/s^2$

11. The tension in the thread is:

(a) 25 N   (b) $\dfrac{50}{2\pi}\,N$

(c) $\dfrac{25}{2\pi}\,N$ (d) 50 N

12. Which of the following statements is false?

(a) An externally applied force is needed to change the direction of a moving object

(b) For every force, there is an equal and opposite reaction force

(c) A body travelling at a constant velocity in a circle has no acceleration

(d) Centripetal acceleration acts towards the centre of the circle of motion

For fully worked solutions to each of the problems in Exercises 84 to 87 in this chapter, go to the website:
www.routledge.com/cw/bird

This assignment covers the material contained in chapters 15 to 19. *The marks for each question are shown in brackets at the end of each question.*

1. Express (a) a length of 125 mm in metres, (b) 25,000 mm^2 in square metres, (c) 7,500,000 mm^3 in cubic metres. (6)

2. Determine the area of a room 12 m long by 9 m wide in (a) cm^2, (b) mm^2. (4)

3. A container has a capacity of 10 litres. Calculate its volume in (a) m^3, (b) mm^3. (4)

4. The density of cast iron is 7000 kg/m^3. Calculate the mass of a block of cast iron which has a volume of 200 cm^3. (4)

5. Determine the volume, in litres, of 14 kg of petrol of density 700 kg/m^3. (4)

6. A vehicle is travelling at a constant speed of 15 metres per second for 20 kilometres. Find the time taken to cover this distance. (4)

7. A train travels from station P to station Q, a distance of 50 kilometres at an average speed of 80 kilometres per hour. It then travels from station Q to station R, a distance of 30 kilometres in 30 minutes. Finally, it travels from station R to station S at an average speed of 72 kilometres per hour in 45 minutes. Determine (a) the time taken to travel from P to Q, (b) the average speed of the train from Q to R, (c) the distance from R to S, (d) the average speed of the whole journey from P to S. (12)

8. The motion of a car is described by the speed/time graph given in Figure RT7.1. Determine the distance covered by the car when moving from 0 to Y. (6)

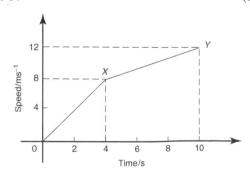

Figure RT7.1

9. For the speed/time graph of a vehicle shown in Figure RT7.2, find the acceleration for each of the three stages of the journey. (8)

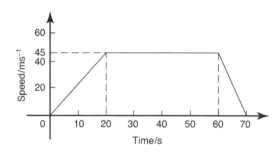

Figure RT7.2

10. A lorry travelling at 40 km/h accelerates uniformly to 76 km/h in 1 minute 40 seconds. Determine the acceleration. (6)

11. Determine the mass of the moving head of a machine tool if it requires a force of 1.5 N to bring it to rest in 0.75 s from a cutting speed of 25 m/min. (5)

12. Find the weight of an object of mass 2.5 kg at a point on the Earth's surface where the gravitational field is 9.8 N/kg. (3)

13. A van of mass 1200 kg travels round a bend of radius 120 m, at 54 km/h. Determine the centripetal force acting on the vehicle. (4)

For lecturers/instructors/teachers, fully worked solutions to each of the problems in Revision Test 7, together with a full marking scheme, are available at the website: www.routledge.com/cw/bird

Chapter 20

Forces acting at a point

Why it is important to understand: **Forces acting at a point**

In this chapter, the fundamental quantities, scalars and vectors, which form the very basis of all branches of engineering, are introduced. A force is a vector quantity and in this chapter, the resolution of forces is introduced. Resolving forces is very important in structures, where the principle is used to determine the strength of roof trusses, bridges, cranes, etc. Great lengths are gone to, to explain this very fundamental skill, where step-by-step methods are adopted to help the reader understand this very important procedure. The resolution of forces is also used in studying the motion of vehicles and other particles in dynamics and in the case of the navigation of ships, aircraft, etc., the vectors take the form of displacements, velocities and accelerations. This chapter gives a sound introduction to the manipulation and use of scalars and vectors, by both graphical and analytical methods. The importance of understanding the forces at a point is to appreciate the equilibrium of forces, whether they act on a structure or a machine or a mechanism.

At the end of this chapter, you should be able to:

- distinguish between scalar and vector quantities
- define 'centre of gravity' of an object
- define 'equilibrium' of an object
- understand the terms 'coplanar' and 'concurrent'
- determine the resultant of two coplanar forces using:
  (a) the triangle of forces method (b) the parallelogram of forces method
- calculate the resultant of two coplanar forces using:
  (a) the cosine and sine rules (b) resolution of forces
- determine the resultant of more than two coplanar forces using:
  (a) the polygon of forces method (b) calculation by resolution of forces
- determine unknown forces when three or more coplanar forces are in equilibrium

20.1 Introduction

In this chapter the fundamental quantities, scalars and vectors, which form the very basis of all branches of engineering, are introduced. A force is a vector quantity and in this chapter, the resolution of forces is introduced. Resolving forces is very important in structures, where the principle is used to determine

Science and Mathematics for Engineering. 978-0-367-20475-4, © John Bird. Published by Taylor & Francis. All rights reserved.

the strength of roof trusses, bridges, cranes, etc. The resolution of forces is also used in studying the motion of vehicles and other particles in dynamics, and in the case of the navigation of ships, aircraft, etc., the vectors take the form of displacements, velocities and accelerations. This chapter gives a sound introduction to the manipulation and use of scalars and vectors, by both graphical and analytical methods.

20.2 Scalar and vector quantities

Quantities used in engineering and science can be divided into two groups:

(a) **Scalar quantities** have a size (or magnitude) only and need no other information to specify them. Thus, 10 centimetres, 50 seconds, 7 litres and 3 kilograms are all examples of scalar quantities.

(b) **Vector quantities** have both a size or magnitude and a direction, called the line of action of the quantity. Thus, a velocity of 50 kilometres per hour due east, an acceleration of 9.81 metres per second squared vertically downwards and a force of 15 newtons at an angle of 30 degrees are all examples of vector quantities.

20.3 Centre of gravity and equilibrium

The **centre of gravity** of an object is a point where the resultant gravitational force acting on the body may be taken to act. For objects of uniform thickness lying in a horizontal plane, the centre of gravity is vertically in line with the point of balance of the object. For a thin uniform rod the point of balance and hence the centre of gravity is halfway along the rod, as shown in Figure 20.1(a).

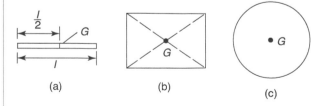

Figure 20.1

A thin flat sheet of a material of uniform thickness is called a **lamina** and the centre of gravity of a rectangular

lamina lies at the point of intersection of its diagonals, as shown in Figure 20.1(b). The centre of gravity of a circular lamina is at the centre of the circle, as shown in Figure 20.1(c).

An object is in **equilibrium** when the forces acting on the object are such that there is no tendency for the object to move. The state of equilibrium of an object can be divided into three groups.

(i) If an object is in **stable equilibrium** and it is slightly disturbed by pushing or pulling (i.e. a disturbing force is applied), the centre of gravity is raised and when the disturbing force is removed, the object returns to its original position. Thus a ball bearing in a hemispherical cup is in stable equilibrium, as shown in Figure 20.2(a).

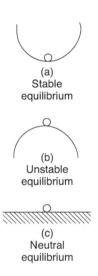

Figure 20.2

(ii) An object is in **unstable equilibrium** if, when a disturbing force is applied, the centre of gravity is lowered and the object moves away from its original position. Thus, a ball bearing balanced on top of a hemispherical cup is in unstable equilibrium, as shown in Figure 20.2(b).

(iii) When an object in **neutral equilibrium** has a disturbing force applied, the centre of gravity remains at the same height and the object does not move when the disturbing force is removed. Thus, a ball bearing on a flat horizontal surface is in neutral equilibrium, as shown in Figure 20.2(c).

20.4 Forces

When forces are all acting in the same plane, they are called **coplanar**. When forces act at the same time and at the same point, they are called **concurrent** forces.

Force is a **vector quantity** and thus has both a magnitude and a direction. A vector can be represented graphically by a line drawn to scale in the direction of the line of action of the force.

Various ways are used to distinguish between vector and scalar quantities. These include:

(i)   **bold print**,

(ii) two capital letters with an arrow above them to denote the sense of direction, for example, $\overrightarrow{AB}$ where A is the starting point and B the end point of the vector,

(iii) a line over the top of letters, for example, $\overline{AB}$ or $\overline{a}$,

(iv) letters with an arrow above, for example, $\overrightarrow{a}$, $\overrightarrow{A}$,

(v)   underlined letters, for example, $\underline{a}$,

(vi) $xi + jy$, where i and j are axes at right-angles to each other; for example, $3i + 4j$ means 3 units in the i direction and 4 units in the j direction, as shown in Figure 20.3,

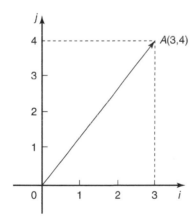

Figure 20.3

(vii) a column matrix $\begin{pmatrix} a \\ b \end{pmatrix}$; for example, the vector **0A** shown in Figure 20.3 could be represented by $\begin{pmatrix} 3 \\ 4 \end{pmatrix}$

Thus, in Figure 20.3, $\mathbf{0A} \equiv \overrightarrow{0A} \equiv \overline{0A} \equiv 3i + 4j \equiv \begin{pmatrix} 3 \\ 4 \end{pmatrix}$

The method adopted in this text is to denote vector quantities in **bold print**. Thus, **ab** in Figure 20.4 represents a force of 5 newtons acting in a direction due east.

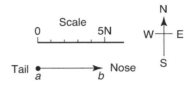

Figure 20.4

20.5 The resultant of two coplanar forces

For two forces acting at a point, there are three possibilities.

(a) For forces acting in the same direction and having the same line of action, the single force having the same effect as both of the forces, called the **resultant force** or just the **resultant**, is the arithmetic sum of the separate forces. Forces of F_1 and F_2 acting at point P, as shown in Figure 20.5(a), have exactly the same effect on point P as force F shown in Figure 20.5(b), where $F = F_1 + F_2$ and acts in the same direction as F_1 and F_2. Thus F is the resultant of F_1 and F_2.

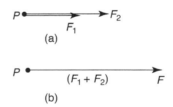

Figure 20.5

(b) For forces acting in opposite directions along the same line of action, the resultant force is the arithmetic difference between the two forces. Forces of F_1 and F_2 acting at point P as shown in Figure 20.6(a) have exactly the same effect on point P as force F shown in Figure 20.6(b), where $F = F_2 - F_1$ and acts in the direction of F_2, since F_2 is greater than F_1. Thus F is the resultant of F_1 and F_2.

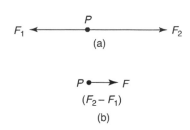

Figure 20.6

(c)   When two forces do not have the same line of action, the magnitude and direction of the resultant force may be found by a procedure called vector addition of forces. There are two graphical methods of performing **vector addition**, known as the **triangle of forces** method (see section 20.6) and the **parallelogram of forces** method (see section 20.7).

> Problem 1.   Determine the resultant force of two forces of 5 kN and 8 kN,
> (a) acting in the same direction and having the same line of action,
> (b) acting in opposite directions but having the same line of action

(a)   The vector diagram of the two forces acting in the same direction is shown in Figure 20.7(a), which assumes that the line of action is horizontal, although since it is not specified, it could be in any direction. From above, the resultant force F is given by: $F = F_1 + F_2$,
i.e. $F = (5 + 8)\,\text{kN} = \mathbf{13\,kN}$ in the direction of the original forces.

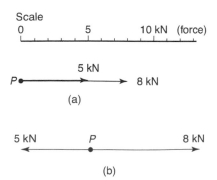

Figure 20.7

(b)   The vector diagram of the two forces acting in opposite directions is shown in Figure 20.7(b), again assuming that the line of action is in a

horizontal direction. From above, the resultant force F is given by: $F = F_2 - F_1$,
i.e. $F = (8 - 5)\,\text{kN} = \mathbf{3\,kN}$ in the direction of the 8 kN force.

20.6   Triangle of forces method

A simple procedure for the triangle of forces method of **vector addition** is as follows:

(i)   Draw a vector representing one of the forces, using an appropriate scale and in the direction of its line of action.

(ii)   From the **nose** of this vector and using the same scale, draw a vector representing the second force in the direction of its line of action.

(iii)   The resultant vector is represented in both magnitude and direction by the vector drawn from the tail of the first vector to the nose of the second vector.

> Problem 2.   Determine the magnitude and direction of the resultant of a force of 15 N acting horizontally to the right and a force of 20 N, inclined at an angle of 60° to the 15 N force. Use the triangle of forces method

Using the procedure given above and with reference to Figure 20.8:

(i)   *ab* is drawn 15 units long horizontally.

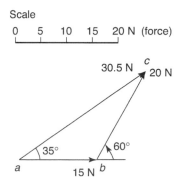

Figure 20.8

(ii)   From *b*, *bc* is drawn 20 units long, inclined at an angle of 60° to *ab*.
(Note, in angular measure, an angle of 60° from *ab* means 60° in an anticlockwise direction.)

(iii) By measurement, the resultant *ac* is 30.5 units long inclined at an angle of 35° to *ab*. That is, the resultant force is **30.5 N**, inclined at an angle of **35°** to the 15 N force.

> **Problem 3.** Find the magnitude and direction of the two forces given, using the triangle of forces method.
> First force: 1.5 kN acting at an angle of 30°
> Second force: 3.7 kN acting at an angle of −45°

From the above procedure and with reference to Figure 20.9:

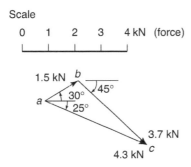

Figure 20.9

(i) *ab* is drawn at an angle of 30° and 1.5 units in length.

(ii) From *b*, *bc* is drawn at an angle of −45° and 3.7 units in length. (Note, an angle of −45° means a clockwise rotation of 45° from a line drawn horizontally to the right.)

(iii) By measurement, the resultant *ac* is 4.3 units long at an angle of −25°. That is, the resultant force is **4.3 kN** at an angle of **− 25°**

Now try the following Practice Exercise

Practice Exercise 88   Further problems on the triangle of forces method (answers on page 540)

1. Determine the magnitude and direction of the resultant of the forces 1.3 kN and 2.7 kN, having the same line of action and acting in the same direction

2. Determine the magnitude and direction of the resultant of the forces 470 N and 538 N

having the same line of action but acting in opposite directions

In questions 3–5, use the triangle of forces method to determine the magnitude and direction of the resultant of the forces given.

3. 13 N at 0° and 25 N at 30°

4. 5 N at 60° and 8 N at 90°

5. 1.3 kN at 45° and 2.8 kN at −30°

20.7   The parallelogram of forces method

A simple procedure for the parallelogram of forces method of vector addition is as follows:

(i) Draw a vector representing one of the forces, using an appropriate scale and in the direction of its line of action.

(ii) From the **tail** of this vector and using the same scale draw a vector representing the second force in the direction of its line of action.

(iii) Complete the parallelogram using the two vectors drawn in (i) and (ii) as two sides of the parallelogram.

(iv) The resultant force is represented in both magnitude and direction by the vector corresponding to the diagonal of the parallelogram drawn from the tail of the vectors in (i) and (ii).

> **Problem 4.** Use the parallelogram of forces method to find the magnitude and direction of the resultant of a force of 250 N acting at an angle of 135° and a force of 400 N acting at an angle of −120°

From the procedure given above and with reference to Figure 20.10:

(i) *ab* is drawn at an angle of 135° and 250 units in length.

(ii) *ac* is drawn at an angle of −120° and 400 units in length.

(iii) *bd* and *cd* are drawn to complete the parallelogram.

Scale

0 100 200 300 400 500 N (force)

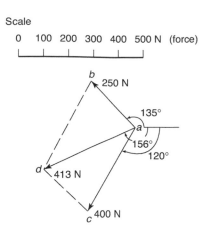

Figure 20.10

(iv) **ad** is drawn. By measurement, **ad** is 413 units long at an angle of −156°

That is, the resultant force is **413 N** at an angle of **−156°**

Now try the following Practice Exercise

Practice Exercise 89 Further problems on the parallelogram of forces method (answers on page 540)

In questions 1–5, use the parallelogram of forces method to determine the magnitude and direction of the resultant of the forces given.

1. 1.7 N at 45° and 2.4 N at –60°

2. 9 N at 126° and 14 N at 223°

3. 23.8 N at –50° and 14.4 N at 215°

4. 0.7 kN at 147° and 1.3 kN at –71°

5. 47 N at 79° and 58 N at 247°

20.8 Resultant of coplanar forces by calculation

An alternative to the graphical methods of determining the resultant of two coplanar forces is by **calculation**. This can be achieved by **trigonometry** using the **cosine rule** and the **sine rule**, as shown in Problem 5 below (see also chapter 10 on trigonometry), or by **resolution of forces** (see section 20.11).

Problem 5. Use the cosine and sine rules to determine the magnitude and direction of the resultant of a force of 8 kN acting at an angle of 50° to the horizontal and a force of 5 kN acting at an angle of −30° to the horizontal

The space diagram is shown in Figure 20.11(a). A sketch is made of the vector diagram, **0a** representing the 8 kN force in magnitude and direction and **ab** representing the 5 kN force in magnitude and direction. The resultant is given by length **0b**. By the cosine rule,

$$0b^2 = 0a^2 + ab^2 - 2(0a)(ab)\cos \angle 0ab$$

$$= 8^2 + 5^2 - 2(8)(5)\cos 100°$$

$$(\text{since } \angle 0ab = 180° - 50° - 30° = 100°)$$

$$= 64 + 25 - (-13.892) = 102.892$$

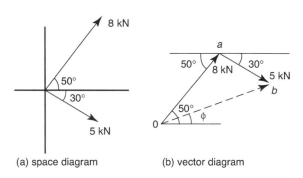

(a) space diagram   (b) vector diagram

Figure 20.11

Hence, $0b = \sqrt{102.892} = 10.14 \text{ kN}$

By the sine rule, $\dfrac{5}{\sin \angle a0b} = \dfrac{10.14}{\sin 100°}$

from which, $\sin \angle a0b = \dfrac{5 \sin 100°}{10.14} = 0.4856$

Hence $\angle a0b = \sin^{-1}(0.4856) = 29.05°$. Thus angle ϕ in Figure 20.11(b) is $50° - 29.05° = 20.95°$.

Hence the resultant of the two forces is 10.14 kN acting at an angle of 20.95° to the horizontal.

Now try the following Practice Exercise

Practice Exercise 90 Further problems on the resultant of coplanar forces by calculation (answers on page 540)

1. Forces of 7.6 kN at 32° and 11.8 kN at 143° act at a point. Use the cosine and sine rules to

calculate the magnitude and direction of their resultant

In questions 2–5, calculate the resultant of the given forces by using the cosine and sine rules.

2. 13 N at 0° and 25 N at 30°

3. 1.3 kN at 45° and 2.8 kN at –30°

4. 9 N at 126° and 14 N at 223°

5. 0.7 kN at 147° and 1.3 kN at –71°

20.9 Resultant of more than two coplanar forces

For the three coplanar forces F_1, F_2 and F_3 acting at a point as shown in Figure 20.12, the vector diagram is drawn using the nose-to-tail method of section 20.6. The procedure is:

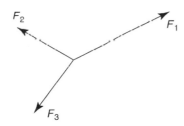

Figure 20.12

(i) Draw **0a** to scale to represent force F_1 in both magnitude and direction (see Figure 20.13).

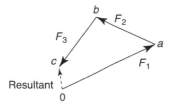

Figure 20.13

(ii) From the nose of **0a**, draw **ab** to represent force F_2

(iii) From the nose of **ab**, draw **bc** to represent force F_3

(iv) The resultant vector is given by length **0c** in Figure 20.13. The direction of resultant **0c** is from where we started, i.e. point 0, to where we finished, i.e. point c. When acting by itself, the resultant force, given by **0c**, has the same effect on the point as forces F_1, F_2 and F_3 have when acting together. The resulting vector diagram of Figure 20.13 is called the **polygon of forces**.

Problem 6. Determine graphically the magnitude and direction of the resultant of the following three coplanar forces, which may be considered as acting at a point: Force A, 12 N acting horizontally to the right; force B, 7 N inclined at 60° to force A; force C, 15 N inclined at 150° to force A

The space diagram is shown in Figure 20.14. The vector diagram shown in Figure 20.15 is produced as follows:

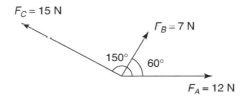

Figure 20.14

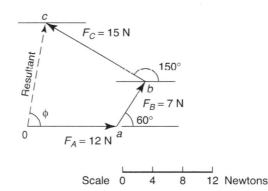

Figure 20.15

(i) **0a** represents the 12 N force in magnitude and direction.

(ii) From the nose of **0a**, **ab** is drawn inclined at 60° to **0a** and 7 units long.

(iii) From the nose of **ab**, **bc** is drawn 15 units long inclined at 150° to **0a** (i.e. 150° to the horizontal).

(iv)   **0c** represents the resultant; by measurement, the resultant is 13.8 N inclined at $\phi = 80°$ to the horizontal.

Thus the resultant of the three forces, F_A, F_B and F_C is a force of 13.8 N at 80° to the horizontal.

> **Problem 7.**   The following coplanar forces are acting at a point, the given angles being measured from the horizontal: 100 N at 30°, 200 N at 80°, 40 N at –150°, 120 N at –100° and 70 N at –60°. Determine graphically the magnitude and direction of the resultant of the five forces

The five forces are shown in the space diagram of Figure 20.16. Since the 200 N and 120 N forces have the same line of action but are in opposite sense, they can be represented by a single force of 200 – 120, i.e. 80 N acting at 80° to the horizontal. Similarly, the 100 N and 40 N forces can be represented by a force of 100 – 40, i.e. 60 N acting at 30° to the horizontal. Hence the space diagram of Figure 20.16 may be represented by the space diagram of Figure 20.17. Such a simplification of the vectors is not essential but it is easier to construct the vector diagram from a space diagram having three forces, than from one with five.

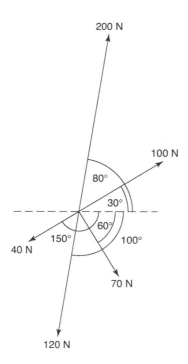

200 N

100 N

80°

30°

60°

150°

100°

40 N

70 N

120 N

Figure 20.16

The vector diagram is shown in Figure 20.18, **0a** representing the 60 N force, **ab** representing the 80 N

force and **bc** the 70 N force. The resultant **0c** is found by measurement to represent a force of 112 N and angle ϕ is 25°.

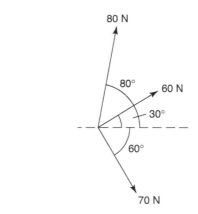

80 N

80°

60 N

30°

60°

70 N

Figure 20.17

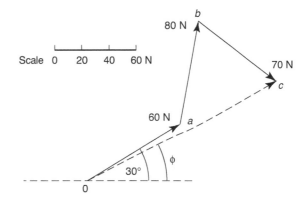

Scale 0   20   40   60 N

b

80 N

70 N

c

60 N   a

30°

ϕ

0

Figure 20.18

Thus, the five forces shown in Figure 20.16 may be represented by a single force of 112 N at 25° to the horizontal.

Now try the following Practice Exercise

> **Practice Exercise 91   Further problems on the resultant of more than two coplanar forces (answers on page 540)**

In questions 1–3, determine graphically the magnitude and direction of the resultant of the coplanar forces given which are acting at a point.

1.   Force A, 12 N acting horizontally to the right; force B, 20 N acting at 140° to force A; force C, 16 N acting 290° to force A

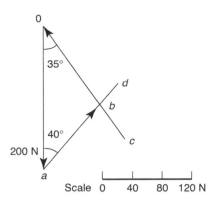

Figure 20.20

2. Force 1, 23 kN acting at 80° to the horizontal; force 2, 30 kN acting at 37° to force 1; force 3, 15 kN acting at 70° to force 2

3. Force P, 50 kN acting horizontally to the right: force Q, 20 kN at 70° to force P: force R, 40 kN at 170° to force P: force S, 80 kN at 300° to force P

4. Four horizontal wires are attached to a telephone pole and exert tensions of 30 N to the south, 20 N to the east, 50 N to the north-east and 40 N to the north-west. Determine the resultant force on the pole and its direction

20.10   Coplanar forces in equilibrium

When three or more coplanar forces are acting at a point and the vector diagram closes, there is no resultant. The forces acting at the point are in **equilibrium**.

Problem 8.   A load of 200 N is lifted by two ropes connected to the same point on the load, making angles of 40° and 35° with the vertical. Determine graphically the tensions in each rope when the system is in equilibrium

The space diagram is shown in Figure 20.19. Since the system is in equilibrium, the vector diagram must close. The vector diagram, shown in Figure 20.20, is drawn as follows:

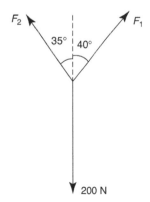

Figure 20.19

(i)   The load of 200 N is drawn vertically as shown by **0a**.

(ii)   The direction only of force F_1 is known, so from point a, **ad** is drawn at 40° to the vertical.

(iii)   The direction only of force F_2 is known, so from point 0, **0c** is drawn at 35° to the vertical.

(iv)   Lines **ad** and **0c** cross at point b; hence the vector diagram is given by triangle 0ab. By measurement, **ab** is 119 N and **0b** is 133 N.

Thus the tensions in the ropes are $F_1 = 119$ N and $F_2 = 133$ N

Problem 9.   Five coplanar forces are acting on a body and the body is in equilibrium. The forces are: 12 kN acting horizontally to the right, 18 kN acting at an angle of 75°, 7 kN acting at an angle of 165°, 16 kN acting from the nose of the 7 kN force, and 15 kN acting from the nose of the 16 kN force. Determine the directions of the 16 kN and 15 kN forces relative to the 12 kN force

With reference to Figure 20.21, **0a** is drawn 12 units long horizontally to the right. From point a, **ab** is drawn

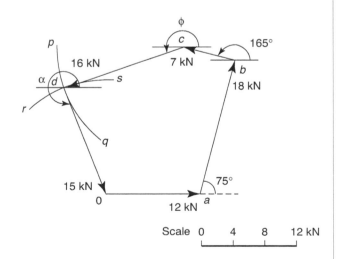

Figure 20.21

18 units long at an angle of 75°. From *b*, *bc* is drawn 7 units long at an angle of 165°. The direction of the 16 kN force is not known, thus arc *pq* is drawn with a compass, with centre at *c*, radius 16 units. Since the forces are at equilibrium, the polygon of forces must close. Using a compass with centre at 0, arc *rs* is drawn having a radius 15 units. The point where the arcs intersect is at *d*.

By measurement, angle $\phi = 198°$ and $\alpha = 291°$

Thus the 16 kN force acts at an angle of 198° (or −162°) to the 12 kN force, and the 15 kN force acts at an angle of 291° (or −69°) to the 12 kN force.

Now try the following Practice Exercise

Practice Exercise 92   Further problems on coplanar forces in equilibrium (answers on page 540)

1. A load of 12.5 N is lifted by two strings connected to the same point on the load, making angles of 22° and 31° on opposite sides of the vertical. Determine the tensions in the strings

2. A two-legged sling and hoist chain used for lifting machine parts is shown in Figure 20.22. Determine the forces in each leg of the sling if parts exerting a downward force of 15 kN are lifted

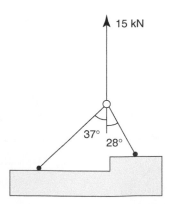

Figure 20.22

3. Four coplanar forces acting on a body are such that it is in equilibrium. The vector diagram for the forces is such that the 60 N force acts vertically upwards, the 40 N force acts at

65° to the 60 N force, the 100 N force acts from the nose of the 40 N force and the 90 N force acts from the nose of the 100 N force. Determine the direction of the 100 N and 90 N forces relative to the 60 N force

20.11   Resolution of forces

A vector quantity may be expressed in terms of its **horizontal** and **vertical components**. For example, a vector representing a force of 10 N at an angle of 60° to the horizontal is shown in Figure 20.23. If the horizontal line **0a** and the vertical line **ab** are constructed as shown, then **0a** is called the horizontal component of the 10 N force, and **ab** the vertical component of the 10 N force.

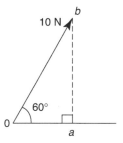

Figure 20.23

By trigonometry,

$$\cos 60° = \frac{0a}{0b}$$

hence the horizontal component,

$$0a = 10 \cos 60°$$

Also, $\sin 60° = \dfrac{ab}{0b}$

hence the vertical component, $ab = 10 \sin 60°$

This process is called **finding the horizontal and vertical components of a vector** or **the resolution of a vector**, and can be used as an alternative to graphical methods for calculating the resultant of two or more coplanar forces acting at a point.

For example, to calculate the resultant of a 10 N force acting at 60° to the horizontal and a 20 N force acting at

−30° to the horizontal (see Figure 20.24) the procedure is as follows:

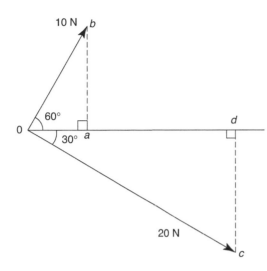

Figure 20.24

(i) Determine the horizontal and vertical components of the 10 N force, i.e.

horizontal component $0a = 10\cos 60°$

$= 5.0\,\text{N}$

and vertical component $ab = 10\sin 60°$

$= 8.66\,\text{N}$

(ii) Determine the horizontal and vertical components of the 20 N force, i.e.

horizontal component $0d = 20\cos(-30°)$

$= 17.32\,\text{N}$

and vertical component $cd = 20\sin(-30°)$

$= -10.0\,\text{N}$

(iii) Determine the total horizontal component, i.e.

$0a + 0d = 5.0 + 17.32 = 22.32\,\text{N}$

(iv) Determine the total vertical component, i.e.

$ab + cd = 8.66 + (-10.0) = -1.34\,\text{N}$

(v) Sketch the total horizontal and vertical components as shown in Figure 20.25. The resultant of the two components is given by length **0r** and, by Pythagoras's theorem,

$$\mathbf{0r} = \sqrt{22.32^2 + 1.34^2}$$

$$= 22.36\,\text{N}$$

and using trigonometry,

$$\text{angle } \phi = \tan^{-1}\frac{1.34}{22.32}$$

$$= 3.44°$$

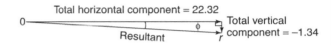

Figure 20.25

Hence the resultant of the 10 N and 20 N forces shown in Figure 20.24 is **22.36 N at an angle of − 3.44° to the horizontal**.

Problem 10. Forces of 5.0 N at 25° and 8.0 N at 112° act at a point. By resolving these forces into horizontal and vertical components, determine their resultant

The space diagram is shown in Figure 20.26.

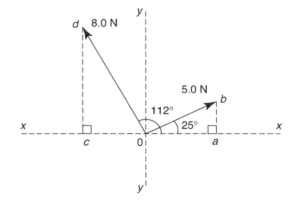

Figure 20.26

(i) The horizontal component of the 5.0 N force $0a = 5.0\cos 25° = 4.532$, and the vertical component of the 5.0 N force $ab = 5.0\sin 25° = 2.113$

(ii)   The horizontal component of the 8.0 N force $0c = 8.0\cos 112° = -2.997$ and the vertical component of the 8.0 N force $cd = 8.0\sin 112° = 7.417$

(iii)   Total horizontal component
$= 0a + 0c = 4.532 + (-2.997) = +1.535$

(iv)   Total vertical component
$= ab + cd = 2.113 + 7.417 = +9.530$

(v)   The components are shown sketched in Figure 20.27.

Figure 20.27

By Pythagoras's theorem,

$$r = \sqrt{1.535^2 + 9.530^2}$$

$$= 9.653$$

and by trigonometry, angle

$$\phi = \tan^{-1}\frac{9.530}{1.535} = 80.85°$$

Hence the resultant of the two forces shown in Figure 20.26 is a force of 9.653 N acting at 80.85° to the horizontal.

Problems 9 and 10 demonstrate the use of resolution of forces for calculating the resultant of two coplanar forces acting at a point. However, the method may be used for more than two forces acting at a point, as shown in Problem 11.

Problem 11.   Determine by resolution of forces the resultant of the following three coplanar forces acting at a point: 200 N acting at 20° to the horizontal; 400 N acting at 165° to the horizontal; 500 N acting at 250° to the horizontal

A tabular approach using a calculator may be made as shown below:

Horizontal component

Force 1	$200\cos 20°$ =	187.94
Force 2	$400\cos 165°$ =	-386.37
Force 3	$500\cos 250°$ =	-171.01
	Total horizontal component =	-369.44

Vertical component

Force 1	$200\sin 20°$ =	68.40
Force 2	$400\sin 165°$ =	103.53
Force 3	$500\sin 250°$ =	-469.85
	Total vertical component =	-297.92

The total horizontal and vertical components are shown in Figure 20.28.

$$\text{Resultant } r = \sqrt{369.44^2 + 297.92^2}$$

$$= 474.60$$

and angle

$$\phi = \tan^{-1}\frac{297.92}{369.44} = 38.88°$$

from which,

$$\alpha = 180° - 38.88° = 141.12°$$

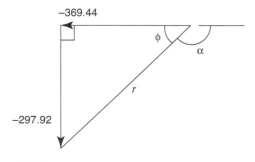

Figure 20.28

Thus the resultant of the three forces given is 474.6 N acting at an angle of −141.12° (or +218.88°) to the horizontal.

Now try the following Practice Exercise

Practice Exercise 93 Further problems on resolution of forces (answers on page 541)

1. Resolve a force of 23.0 N at an angle of 64° into its horizontal and vertical components

2. Forces of 5 N at 21° and 9 N at 126° act at a point. By resolving these forces into horizontal and vertical components, determine their resultant

In questions 3 and 4, determine the magnitude and direction of the resultant of the coplanar forces given which are acting at a point, by resolution of forces.

3. Force A, 12 N acting horizontally to the right; force B, 20 N acting at 140° to force A; force C, 16 N acting 290° to force A

4. Force 1, 23 kN acting at 80° to the horizontal; force 2, 30 kN acting at 37° to force 1; force 3, 15 kN acting at 70° to force 2

5. Determine, by resolution of forces, the resultant of the following three coplanar forces acting at a point: 10 kN acting at 32° to the horizontal; 15 kN acting at 170° to the horizontal; 20 kN acting at 240° to the horizontal

6. The following coplanar forces act at a point: force A, 15 N acting horizontally to the right: force B, 23 N at 81° to the horizontal; force C, 7 N at 210° to the horizontal; force D, 9 N at 265° to the horizontal; and force E, 28 N at 324° to the horizontal. Determine the resultant of the five forces by resolution of the forces

20.12 Summary

(a) To determine the **resultant of two coplanar forces** acting at a point, four methods are commonly used. They are:

by drawing:

(1) triangle of forces method, and

(2) parallelogram of forces method, and

by calculation:

(3) use of cosine and sine rules, and

(4) resolution of forces.

(b) To determine the **resultant of more than two coplanar forces** acting at a point, two methods are commonly used. They are:

by drawing:

(1) polygon of forces method, and

by calculation:

(2) resolution of forces.

Now try the following Practice Exercises

Practice Exercise 94 Short answer questions on forces acting at a point

1. Give one example of a scalar quantity and one example of a vector quantity

2. Explain the difference between a scalar and a vector quantity

3. What is meant by the centre of gravity of an object?

4. Where is the centre of gravity of a rectangular lamina?

5. What is meant by neutral equilibrium?

6. State the meaning of the term 'coplanar'

7. What is a concurrent force?

8. State what is meant by a triangle of forces

9. State what is meant by a parallelogram of forces

10. State what is meant by a polygon of forces

11. When a vector diagram is drawn representing coplanar forces acting at a point, and there is no resultant, the forces are in …

12. Two forces of 6 N and 9 N act horizontally to the right. The resultant is … N acting …

13. A force of 10 N acts at an angle of 50° and another force of 20 N acts at an angle of 230°. The resultant is a force … N acting at an angle of …°

14. What is meant by 'resolution of forces'?

15. A coplanar force system comprises a 20 kN force acting horizontally to the right, 30 kN at 45°, 20 kN at 180° and 25 kN at 225°. The resultant is a force of ... N acting at an angle of ...° to the horizontal

Practice Exercise 95   Multiple-choice questions on forces acting at a point (answers on page 541)

1. A physical quantity which has direction as well as magnitude is known as a:

    (a) force   (b) vector   (c) scalar   (d) weight

2. Which of the following is not a scalar quantity?

    (a) velocity   (b) potential energy
    (c) work   (d) kinetic energy

3. Which of the following is not a vector quantity?

    (a) displacement   (b) density
    (c) velocity   (d) acceleration

4. Which of the following statements is false?

    (a) Scalar quantities have size or magnitude only

    (b) Vector quantities have both magnitude and direction

    (c) Mass, length and time are all scalar quantities

    (d) Distance, velocity and acceleration are all vector quantities

5. If the centre of gravity of an object which is slightly disturbed is raised and the object returns to its original position when the disturbing force is removed, the object is said to be in:

    (a) neutral equilibrium

    (b) stable equilibrium

    (c) static equilibrium

    (d) unstable equilibrium

6. Which of the following statements is false?

7.
    (a) The centre of gravity of a lamina is at its point of balance

    (b) The centre of gravity of a circular lamina is at its centre

    (c) The centre of gravity of a rectangular lamina is at the point of intersection of its two sides

    (d) The centre of gravity of a thin uniform rod is halfway along the rod

7. The magnitude of the resultant of the vectors shown in Figure 20.29 is:

    (a) 2 N   (b) 12 N   (c) 35 N   (d) −2 N

Figure 20.29

8. The magnitude of the resultant of the vectors shown in Figure 20.30 is:

    (a) 7 N   (b) 5 N   (c) 1 N   (d) 12 N

Figure 20.30

9. Which of the following statements is false ?

    (a) There is always a resultant vector required to close a vector diagram representing a system of coplanar forces acting at a point, which are not in equilibrium

    (b) A vector quantity has both magnitude and direction

    (c) A vector diagram representing a system of coplanar forces acting at a point when in equilibrium does not close

    (d) Concurrent forces are those which act at the same time at the same point

10. Which of the following statements is false?

    (a) The resultant of coplanar forces of 1 N, 2 N and 3 N acting at a point can be 4 N

    (b) The resultant of forces of 6 N and 3 N acting in the same line of action but opposite in sense is 3 N

    (c) The resultant of forces of 6 N and 3 N acting in the same sense and having the same line of action is 9 N

    (d) The resultant of coplanar forces of 4 N at 0°, 3 N at 90° and 8 N at 180° is 15 N

11. A space diagram of a force system is shown in Figure 20.31. Which of the vector diagrams in Figure 20.32 does not represent this force system?

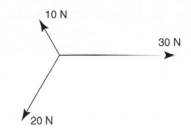

Figure 20.31

12. With reference to Figure 20.33, which of the following statements is false?

    (a) The horizontal component of F_A is 8.66 N

    (b) The vertical component of F_B is 10 N

    (c) The horizontal component of F_C is 0

    (d) The vertical component of F_D is 4 N

13. The resultant of two forces of 3 N and 4 N can never be equal to:

    (a) 2.5 N (b) 4.5 N (c) 6.5 N (d) 7.5 N

14. The magnitude of the resultant of the vectors shown in Figure 20.34 is:

    (a) 5 N (b) 13 N (c) 1 N (d) 63 N

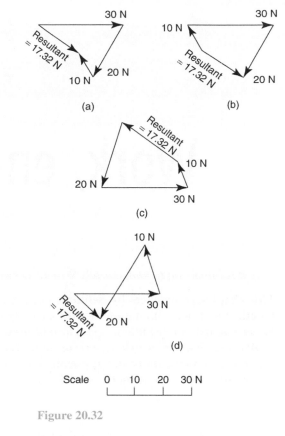

Figure 20.32

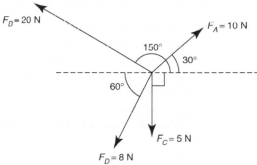

Figure 20.33

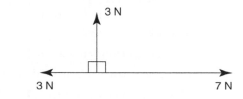

Figure 20.34

For fully worked solutions to each of the problems in Exercises 88 to 95 in this chapter, go to the website:
www.routledge.com/cw/bird

Chapter 21

Work, energy and power

Why it is important to understand: **Work, energy and power**

This chapter commences by defining work, power and energy. It also provides the mid-ordinate rule, together with an explanation on how to apply it to calculate the areas of irregular figures, such as the areas of ships' water planes. It can also be used for calculating the work done in a force-displacement or similar relationship, which may result in the form of an irregular two-dimensional shape. This chapter is fundamental to the study and application of dynamics to practical problems. This chapter is particularly important for designing motor vehicle engines.

At the end of this chapter, you should be able to:

- define work and state its unit
- perform simple calculations on work done
- appreciate that the area under a force/distance graph gives work done
- perform calculations on a force/distance graph to determine work done
- define energy and state its unit
- state several forms of energy
- state the principle of conservation of energy and give examples of conversions
- define and calculate efficiency of systems
- define power and state its unit
- understand that power = force × velocity
- perform calculations involving power, work done, energy and efficiency
- define potential energy
- perform calculations involving potential energy = mgh
- define kinetic energy
- perform calculations involving kinetic energy = $\frac{1}{2}mv^2$
- distinguish between elastic and inelastic collisions
- perform calculations involving kinetic energy in rotation = $\frac{1}{2}I\omega^2$

Science and Mathematics for Engineering. 978-0-367-20475-4, © John Bird. Published by Taylor & Francis. All rights reserved.

21.1 Introduction

This chapter commences by defining work, power and energy. It also provides the mid-ordinate rule, together with an explanation on how to apply it to calculate the areas of irregular figures, such as the areas of ship's water planes. It can also be used for calculating the work done in a force–displacement or similar relationship, which may result in the form of an irregular two-dimensional shape. This chapter is fundamental to the study and application of dynamics to practical problems.

21.2 Work

If a body moves as a result of a force being applied to it, the force is said to do work on the body. The amount of work done is the product of the applied force and the distance, i.e.

work done = force × distance moved in the

direction of the force

The unit of work is the **joule***, **J**, which is defined as the amount of work done when a force of 1 newton acts for a distance of 1 m in the direction of the force. Thus,

$$1\,J = 1\,N\,m$$

If a graph is plotted of experimental values of force (on the vertical axis) against distance moved (on the horizontal axis) a force/distance graph or work diagram is produced. **The area under the graph represents the work done**.

For example, a constant force of 20 N used to raise a load a height of 8 m may be represented on a force/distance graph as shown in Figure 21.1. The area under the graph shown shaded, represents the work done. Hence

work done = 20 N × 8 m = 160 J

Similarly, a spring extended by 20 mm by a force of 500 N may be represented by the work diagram shown in Figure 21.2, where

work done = shaded area

$$= \frac{1}{2} \times \text{base} \times \text{height}$$

$$= \frac{1}{2} \times (20 \times 10^{-3})\,\text{m} \times 500\,\text{N} = \mathbf{5\,J}$$

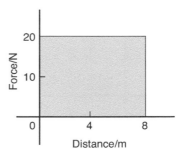

Figure 21.1

It was shown in chapter 19 that force = mass × acceleration, and that if an object is dropped from a height it has a constant acceleration of around 9.81 m/s².

*Who was Joule? – **James Prescott Joule** (24 December 1818–11 October 1889) was an English physicist and brewer. He studied the nature of heat, and discovered its relationship to mechanical work. This led to the theory of conservation of energy, which in turn led to the development of the first law of thermodynamics. The SI derived unit of energy, the joule, is named after him. To find out more about Joule go to www.routledge.com/cw/bird

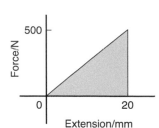

Figure 21.2

Thus if a mass of 8 kg is lifted vertically 4 m, the work done is given by:

work done = force × distance

$$= (\text{mass} \times \text{acceleration}) \times \text{distance}$$

$$= (8 \times 9.81) \times 4 = 313.92 \, \text{J}$$

The work done by a variable force may be found by determining the area enclosed by the force/distance graph using an approximate method such as the **mid-ordinate rule**.

To determine the area *ABCD* of Figure 21.3 using the mid-ordinate rule:

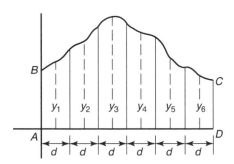

Figure 21.3

(i)   Divide base *AD* into any number of equal intervals, each of width *d* (the greater the number of intervals, the greater the precision).

(ii) Erect ordinates in the middle of each interval (shown by broken lines in Figure 21.3).

(iii) Accurately measure ordinates y_1, y_2, y_3, etc.

(iv) Area $ABCD = d(y_1 + y_2 + y_3 + y_4 + y_5 + y_6)$

In general, the mid-ordinate rule states:

Area = (width of interval) (sum of mid-ordinates)

Problem 1. Calculate the work done when a force of 40 N pushes an object a distance of 500 m in the same direction as the force

Work done = force × distance moved in the direction of the force

$$= 40 \, \text{N} \times 500 \, \text{m} = 20{,}000 \, \text{J}$$

$$(\text{since} \, 1 \, \text{J} = 1 \, \text{Nm})$$

i.e. **work done = 20 kJ**

Problem 2. Calculate the work done when a mass is lifted vertically by a crane to a height of 5 m, the force required to lift the mass being 98 N

When work is done in lifting then:

work done = (weight of the body)

× (vertical distance moved)

Weight is the downward force due to the mass of an object. Hence

work done = 98 N × 5 m = 490 J

Problem 3. A motor supplies a constant force of 1 kN which is used to move a load a distance of 5 m. The force is then changed to a constant 500 N and the load is moved a further 15 m. Draw the force/distance graph for the operation and from the graph determine the work done by the motor

The force/distance graph or work diagram is shown in Figure 21.4. Between points *A* and *B* a constant force of 1000 N moves the load 5 m; between points *C* and *D* a constant force of 500 N moves the load from 5 m to 20 m.

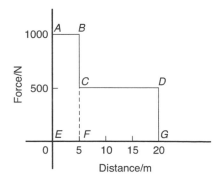

Figure 21.4

Total work done = area under the force/distance graph

$$= \text{area } ABFE + \text{area } CDGF$$

$$= (1000\,\text{N} \times 5\,\text{m}) + (500\,\text{N} \times 15\,\text{m})$$

$$= 5000\,\text{J} + 7500\,\text{J} = 12{,}500\,\text{J}$$

$$= \mathbf{12.5\,kJ}$$

Problem 4. A spring, initially in a relaxed state, is extended by 100 mm. Determine the work done by using a work diagram if the spring requires a force of 0.6 N per mm of stretch

Force required for a 100 mm extension

$$= 100\,\text{mm} \times 0.6\,\text{N/mm} = 60\,\text{N}.$$

Figure 21.5 shows the force/extension graph or work diagram representing the increase in extension in proportion to the force, as the force is increased from 0 to 60 N. The work done is the area under the graph, hence

$$\textbf{work done} = \frac{1}{2} \times \text{base} \times \text{height}$$

$$= \frac{1}{2} \times 100\,\text{mm} \times 60\,\text{N}$$

$$= \frac{1}{2} \times 100 \times 10^{-3}\,\text{m} \times 60\,\text{N} = \mathbf{3\,J}$$

(Alternatively, average force during extension $= \dfrac{(60-0)}{2} = 30\,\text{N}$ and total extension $= 100\,\text{mm} = 0.1\,\text{m}$, hence

$$\text{work done} = \text{average force} \times \text{extension}$$

$$= 30\,\text{N} \times 0.1\,\text{m} = 3\,\text{J})$$

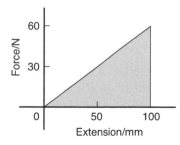

Figure 21.5

Problem 5. A spring requires a force of 10 N to cause an extension of 50 mm. Determine the work done in extending the spring (a) from zero to 30 mm, and (b) from 30 mm to 50 mm

Figure 21.6 shows the force/extension graph for the spring.

(a) Work done in extending the spring from zero to 30 mm is given by area *AB0* of Figure 21.6,

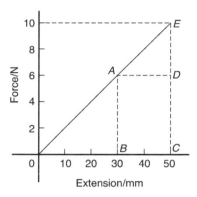

Figure 21.6

i.e. **work done** $= \dfrac{1}{2} \times \text{base} \times \text{height}$

$$= \frac{1}{2} \times 30 \times 10^{-3}\,\text{m} \times 6\,\text{N}$$

$$= 90 \times 10^{-3}\,\text{J} = \mathbf{0.09\,J}$$

(b) Work done in extending the spring from 30 mm to 50 mm is given by area *ABCE* of Figure 21.6,

i.e. **work done** = area *ABCD* + area *ADE*

$$= (20 \times 10^{-3}\,\text{m} \times 6\,\text{N})$$

$$+ \frac{1}{2}(20 \times 10^{-3}\,\text{m})(4\,\text{N})$$

$$= 0.12\,\text{J} + 0.04\,\text{J}$$

$$= \mathbf{0.16\,J}$$

Problem 6. Calculate the work done when a mass of 20 kg is lifted vertically through a distance of 5.0 m. Assume that the acceleration due to gravity is 9.81 m/s²

The force to be overcome when lifting a mass of 20 kg vertically upwards is mg, i.e. $20 \times 9.81 = 196.2\,\text{N}$ (see chapters 18 and 19).

Work done = force × distance = $196.2 \times 5.0 = \mathbf{981\,J}$

Problem 7. Water is pumped vertically upwards through a distance of 50.0 m and the work done is 294.3 kJ. Determine the number of litres of water pumped. (1 litre of water has a mass of 1 kg)

$$\text{Work done} = \text{force} \times \text{distance}$$

$$\text{i.e. } 294,300 = \text{force} \times 50.0$$

$$\text{from which, force} = \frac{294,300}{50.0} = 5886\,\text{N}$$

The force to be overcome when lifting a mass m kg vertically upwards is mg i.e. $(m \times 9.81)$ N (see chapters 18 and 19).

Thus, $5886 = m \times 9.81$

$$\text{from which, mass } m = \frac{5886}{9.81} = 600\,\text{kg}$$

Since 1 litre of water has a mass of 1 kg, **600 litres of water are pumped**.

Problem 8. The force on a cutting tool of a shaping machine varies over the length of cut as follows:

Distance (mm)	0	20	40	60	80	100
Force (kN)		60	72	65	53	44	50

Determine the work done as the tool moves through a distance of 100 mm

The force/distance graph for the given data is shown in Figure 21.7. The work done is given by the area under the graph; the area may be determined by an approximate method. Using the **mid-ordinate rule**, with each strip of width 20 mm, mid-ordinates y_1, y_2, y_3, y_4 and y_5 are erected as shown, and each is measured.

$$\text{Area under curve} = (\text{width of each strip})$$

$$\times (\text{sum of mid-ordinate values})$$

$$= (20)(69 + 69.5 + 59 + 48 + 45.5)$$

$$= (20)(291)$$

$$= 5820\,\text{kN mm} = 5820\,\text{Nm} = 5820\,\text{J}$$

Hence the work done as the tool moves through 100 mm is **5.82 kJ**

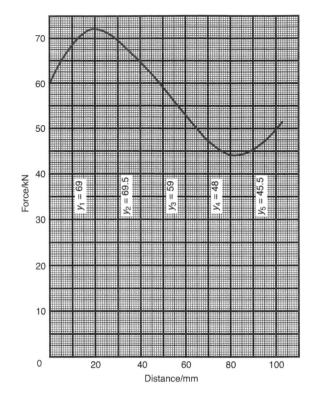

Figure 21.7

Now try the following Practice Exercise

Practice Exercise 96   Further problems on work (answers on page 541)

1. Determine the work done when a force of 50 N pushes an object 1.5 km in the same direction as the force

2. Calculate the work done when a mass of weight 200 N is lifted vertically by a crane to a height of 100 m

3. A motor supplies a constant force of 2 kN to move a load 10 m. The force is then changed to a constant 1.5 kN and the load is moved a further 20 m. Draw the force/distance graph for the complete operation, and, from the graph, determine the total work done by the motor

4. A spring, initially relaxed, is extended 80 mm. Draw a work diagram and hence determine the work done if the spring requires a force of 0.5 N/mm of stretch

5. A spring requires a force of 50 N to cause an extension of 100 mm. Determine the work done in extending the spring (a) from 0 to 100 mm, and (b) from 40 mm to 100 mm

6. The resistance to a cutting tool varies during the cutting stroke of 800 mm as follows: (i) the resistance increases uniformly from an initial 5000 N to 10,000 N as the tool moves 500 mm, and (ii) the resistance falls uniformly from 10,000 N to 6000 N as the tool moves 300 mm. Draw the work diagram and calculate the work done in one cutting stroke

21.3   Energy

Energy is the capacity, or ability, to do work. The unit of energy is the joule, the same as for work. Energy is expended when work is done. There are several forms of energy and these include:

(i) Mechanical energy

(ii) Heat or thermal energy

(iii) Electrical energy

(iv) Chemical energy

(v) Nuclear energy

(vi) Light energy

(vii) Sound energy

Energy may be converted from one form to another. **The principle of conservation of energy** states that the total amount of energy remains the same in such conversions, i.e. energy cannot be created or destroyed.

Some examples of energy conversions include:

(i) Mechanical energy is converted to electrical energy by a generator

(ii) Electrical energy is converted to mechanical energy by a motor

(iii) Heat energy is converted to mechanical energy by a steam engine

(iv) Mechanical energy is converted to heat energy by friction

(v) Heat energy is converted to electrical energy by a solar cell

(vi) Electrical energy is converted to heat energy by an electric fire

(vii) Heat energy is converted to chemical energy by living plants

(viii) Chemical energy is converted to heat energy by burning fuels

(ix) Heat energy is converted to electrical energy by a thermocouple

(x) Chemical energy is converted to electrical energy by batteries

(xi) Electrical energy is converted to light energy by a light bulb

(xii) Sound energy is converted to electrical energy by a microphone

(xiii) Electrical energy is converted to chemical energy by electrolysis

Efficiency is defined as the ratio of the useful output energy to the input energy. The symbol for efficiency is η (Greek letter eta). Hence

$$\text{efficiency } \eta = \frac{\textbf{useful output energy}}{\textbf{input energy}}$$

Efficiency has no units and is often stated as a percentage. A perfect machine would have an efficiency of 100%. However, all machines have an efficiency lower than this due to friction and other losses. Thus, if the input energy to a motor is 1000 J and the output energy is 800 J then the efficiency is

$$\frac{800}{1000} \times 100\% = \textbf{80\%}$$

Problem 9.   A machine exerts a force of 200 N in lifting a mass through a height of 6 m. If 2 kJ of energy are supplied to it, what is the efficiency of the machine?

Work done in lifting mass = force × distance moved

= weight of body × distance moved

= 200 N × 6 m = 1200 J

= useful energy output

Energy input $= 2\,kJ = 2000\,J$

Efficiency $\eta = \dfrac{\text{useful output energy}}{\text{input energy}}$

$= \dfrac{1200}{2000} = \mathbf{0.6}$ or $\mathbf{60\%}$

Problem 10. Calculate the useful output energy of an electric motor which is 70% efficient if it uses 600 J of electrical energy

Efficiency $\eta = \dfrac{\text{useful output energy}}{\text{input energy}}$

thus $\dfrac{70}{100} = \dfrac{\text{output energy}}{600\,J}$

from which, **output energy** $= \dfrac{70}{100} \times 600 = \mathbf{420\,J}$

Problem 11. 4 kJ of energy are supplied to a machine used for lifting a mass. The force required is 800 N. If the machine has an efficiency of 50%, to what height will it lift the mass?

Efficiency $\eta = \dfrac{\text{useful output energy}}{\text{input energy}}$

i.e. $\dfrac{50}{100} = \dfrac{\text{output energy}}{4000\,J}$

from which, output energy $= \dfrac{50}{100} \times 4000 = 2000\,J$

Work done $=$ force $\times$ distance moved

hence $2000\,J = 800\,N \times$ height

from which, **height** $= \dfrac{2000\,J}{800\,N} = \mathbf{2.5\,m}$

Problem 12. A hoist exerts a force of 500 N in raising a load through a height of 20 m. The efficiency of the hoist gears is 75% and the efficiency of the motor is 80%. Calculate the input energy to the hoist

The hoist system is shown diagrammatically in Figure 21.8.

Output energy $=$ work done $=$ force $\times$ distance

$= 500\,N \times 20\,m = 10,000\,J$

Figure 21.8

For the gearing,

efficiency $= \dfrac{\text{output energy}}{\text{input energy}}$

i.e. $\dfrac{75}{100} = \dfrac{10,000}{\text{input energy}}$

from which, the input energy to the gears

$= 10,000 \times \dfrac{100}{75} = 13,333\,J$

The input energy to the gears is the same as the output energy of the motor. Thus, for the motor,

efficiency $= \dfrac{\text{output energy}}{\text{input energy}}$

i.e. $\dfrac{80}{100} = \dfrac{13,333}{\text{input energy}}$

Hence,

input energy to the hoist $= 13,333 \times \dfrac{100}{80}$

$= 16,667\,J = \mathbf{16.67\,kJ}$

Now try the following Practice Exercise

Practice Exercise 97   Further problems on energy (answers on page 541)

1. A machine lifts a mass of weight 490.5 N through a height of 12 m when 7.85 kJ of energy is supplied to it. Determine the efficiency of the machine

2. Determine the output energy of an electric motor which is 60% efficient if it uses 2 kJ of electrical energy

3. A machine that is used for lifting a particular mass is supplied with 5 kJ of energy. If the machine has an efficiency of 65% and exerts a force of 812.5 N to what height will it lift the mass?

4. A load is hoisted 42 m and requires a force of 100 N. The efficiency of the hoist gear is 60% and that of the motor is 70%. Determine the input energy to the hoist

21.4   Power

Power is a measure of the rate at which work is done or at which energy is converted from one form to another.

$$\text{Power } P = \frac{\text{energy used}}{\text{time taken}} \quad \text{or} \quad P = \frac{\text{work done}}{\text{time taken}}$$

The unit of power is the **watt***, **W**, where 1 watt is equal to 1 joule per second. The watt is a small unit for many purposes and a larger unit called the kilowatt, kW, is used, where 1 kW = 1000 W.

*Who was Watt? – **James Watt** (19 January 1736–25 August 1819) was a Scottish inventor and mechanical engineer whose radically improved both the power and efficiency of steam engines. To find out more go to www.routledge.com/cw/bird

The power output of a motor, which does 120 kJ of work in 30 s, is thus given by

$$P = \frac{120\,\text{kJ}}{30\,\text{s}} = 4\,\text{kW}$$

(For electrical power, see chapter 34.)
Since work done = force × distance, then

$$\text{Power} = \frac{\text{work done}}{\text{time taken}} = \frac{\text{force} \times \text{distance}}{\text{time taken}}$$

$$= \text{force} \times \frac{\text{distance}}{\text{time taken}}$$

However, $\dfrac{\text{distance}}{\text{time taken}} = \text{velocity}$

Hence,   **power = force × velocity**

Problem 13. The output power of a motor is 8 kW. How much work does it do in 30 s?

$$\text{Power} = \frac{\text{work done}}{\text{time taken}}$$

from which,   **work done** = power × time

$$= 8000\,\text{W} \times 30\,\text{s}$$

$$= 240,000\,\text{J} = \mathbf{240\,kJ}$$

Problem 14. Calculate the power required to lift a mass through a height of 10 m in 20 s if the force required is 3924 N

Work done = force × distance moved

$$= 3924\,\text{N} \times 10\,\text{m} = 39,240\,\text{J}$$

$$\textbf{Power} = \frac{\text{work done}}{\text{time taken}} = \frac{39,240\,\text{J}}{20\,\text{s}}$$

$$= \mathbf{1962\,W} \text{ or } \mathbf{1.962\,kW}$$

Problem 15. 10 kJ of work is done by a force in moving a body uniformly through 125 m in 50 s. Determine (a) the value of the force, and (b) the power

(a) Work done = force × distance.

Hence

$$10,000\,\text{J} = \text{force} \times 125\,\text{m}$$

from which,   **force** $= \dfrac{10,000\,\text{J}}{125\,\text{m}} = \mathbf{80\,N}$

(b)   **Power** $= \dfrac{\text{work done}}{\text{time taken}} = \dfrac{10{,}000\,\text{J}}{50\,\text{s}} = \mathbf{200\,W}$

Problem 16.   A car hauls a trailer at 90 km/h when exerting a steady pull of 600 N. Calculate (a) the work done in 30 minutes and (b) the power required

(a)   Work done = force × distance moved.

The distance moved in 30 min, i.e. $\dfrac{1}{2}$h, at 90 km/h $= 45$ km. Hence,

$$\text{work done} = 600\,\text{N} \times 45{,}000\,\text{m}$$

$$= \mathbf{27{,}000\,kJ}\ \text{or}\ \mathbf{27\,MJ}$$

(b)   **Power required** $= \dfrac{\text{work done}}{\text{time taken}} = \dfrac{27 \times 10^6\,\text{J}}{30 \times 60\,\text{s}}$

$$= \mathbf{15{,}000\,W}\ \text{or}\ \mathbf{15\,kW}$$

Problem 17.   To what height will a mass of weight 981 N be raised in 40 s by a machine using a power of 2 kW?

Work done = force × distance. Hence,
work done = 981 N × height.

$$\text{Power} = \dfrac{\text{work done}}{\text{time taken}},\ \text{from which,}$$

work done = power × time taken

$$= 2000\,\text{W} \times 40\,\text{s} = 80{,}000\,\text{J}$$

Hence, 80,000 = 981 N × height, from which,

$$\mathbf{height} = \dfrac{80{,}000\,\text{J}}{981\,\text{N}} = \mathbf{81.55\,m}$$

Problem 18.   A planing machine has a cutting stroke of 2 m and the stroke takes 4 s. If the constant resistance to the cutting tool is 900 N, calculate for each cutting stroke (a) the power consumed at the tool point, and (b) the power input to the system if the efficiency of the system is 75%

(a)   Work done in each cutting stroke

$$= \text{force} \times \text{distance}$$

$$= 900\,\text{N} \times 2\,\text{m}$$

$$= 1800\,\text{J}$$

Power consumed at tool point

$$= \dfrac{\text{work done}}{\text{time taken}}$$

$$= \dfrac{1800\,\text{J}}{4\,\text{s}} = \mathbf{450\,W}$$

(b)   Efficiency $= \dfrac{\text{output energy}}{\text{input energy}} = \dfrac{\text{output power}}{\text{input power}}$

Hence, $\dfrac{75}{100} = \dfrac{450}{\text{input power}}$ from which,

$$\mathbf{input\ power} = 450 \times \dfrac{100}{75} = \mathbf{600\,W}$$

Problem 19.   An electric motor provides power to a winding machine. The input power to the motor is 2.5 kW and the overall efficiency is 60%. Calculate (a) the output power of the machine, (b) the rate at which it can raise a 300 kg load vertically upwards

(a)   Efficiency $\eta = \dfrac{\text{power output}}{\text{power input}}$

i.e. $\dfrac{60}{100} = \dfrac{\text{power output}}{2500}$ from which,

$$\mathbf{power\ output} = \dfrac{60}{100} \times 2500$$

$$= \mathbf{1500\,W}\ \text{or}\ \mathbf{1.5\,kW}$$

(b)   Power output = force × velocity, from which,

$$\text{velocity} = \dfrac{\text{power output}}{\text{force}}$$

Force acting on the 300 kg load due to gravity

$$= 300\,\text{kg} \times 9.81\,\text{m/s}^2$$

$$= 2943\,\text{N}$$

Hence, **velocity** $= \dfrac{1500}{2943} = \mathbf{0.510\,m/s}\ \text{or}\ \mathbf{510\,mm/s}$

Problem 20.   A lorry is travelling at a constant velocity of 72 km/h. The force resisting motion is 800 N. Calculate the tractive power necessary to keep the lorry moving at this speed

Power = force × velocity.

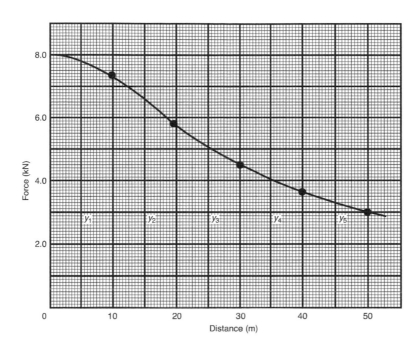

Force (kN)

Distance (m)

Figure 21.9

The force necessary to keep the lorry moving at constant speed is equal and opposite to the force resisting motion, i.e. 800 N.

$$\text{Velocity} = 72\,\text{km/h} = \frac{72 \times 1000}{60 \times 60}\,\text{m/s} = 20\,\text{m/s}.$$

Hence, $\text{power} = 800\,\text{N} \times 20\,\text{m/s} = 16{,}000\,\text{Nm/s}$

$$= 16{,}000\,\text{J/s} = 16{,}000\,\text{W or } 16\,\text{kW}$$

Thus the tractive power needed to keep the lorry moving at a constant speed of 72 km/h is 16 kW.

Problem 21.   The variation of tractive force with distance for a vehicle which is accelerating from rest is:

force (kN)	8.0	7.4	5.8	4.5	3.7	3.0
distance (m)	0	10	20	30	40	50

Determine the average power necessary if the time taken to travel the 50 m from rest is 25 s

The force/distance diagram is shown in Figure 21.9. The work done is determined from the area under the curve. Using the mid-ordinate rule with five intervals gives:

$$\text{area} = (\text{width of interval}) \times$$

$$(\text{sum of mid-ordinate})$$

$$= (10)[y_1 + y_2 + y_3 + y_4 + y_5]$$

$$= (10)[7.8 + 6.6 + 5.1 + 4.0 + 3.3]$$

$$= (10)[26.8] = 268\,\text{kN m}$$

i.e. work done $= 268\,\text{kJ}$.

$$\textbf{Average power} = \frac{\text{work done}}{\text{time taken}}$$

$$= \frac{268{,}000\,\text{J}}{25\,\text{s}}$$

$$= \textbf{10{,}720\,W or } \textbf{10.72\,kW}$$

Now try the following Practice Exercise

Practice Exercise 98   Further problems on power (answers on page 541)

1. The output power of a motor is 10 kW. How much work does it do in 1 minute?

2. Determine the power required to lift a load through a height of 20 m in 12.5 s if the force required is 2.5 kN

3. 25 kJ of work is done by a force in moving an object uniformly through 50 m in 40 s. Calculate (a) the value of the force, and (b) the power

4. A car towing another at 54 km/h exerts a steady pull of 800 N. Determine (a) the work done in $\frac{1}{4}$ h, and (b) the power required

5. To what height will a mass of weight 500 N be raised in 20 s by a motor using 4 kW of power?

6. The output power of a motor is 10 kW. Determine (a) the work done by the motor in 2 h, and (b) the energy used by the motor if it is 72% efficient

7. A car is travelling at a constant speed of 81 km/h. The frictional resistance to motion is 0.60 kN. Determine the power required to keep the car moving at this speed

8. A constant force of 2.0 kN is required to move the table of a shaping machine when a cut is being made. Determine the power required if the stroke of 1.2 m is completed in 5.0 s

9. The variation of force with distance for a vehicle that is decelerating is as follows:

Distance (m)	600	500	400	300	200	100	0
Force (kN)	24	20	16	12	8	4	0

If the vehicle covers the 600 m in 1.2 minutes, find the power needed to bring the vehicle to rest

10. A cylindrical bar of steel is turned in a lathe. The tangential cutting force on the tool is 0.5 kN and the cutting speed is 180 mm/s. Determine the power absorbed in cutting the steel

21.5 Potential and kinetic energy

Mechanical engineering is concerned principally with two kinds of energy, potential energy and kinetic energy.

Potential energy is energy due to the position of the body. The force exerted on a mass of m kg is mg N (where $g = 9.81$ m/s², the acceleration due to gravity). When the mass is lifted vertically through a height h m above some datum level, the work done is given by:

$$\text{force} \times \text{distance} = (mg)(h) \text{ J}$$

This work done is stored as potential energy in the mass.

Hence, **potential energy = mgh joules**

(the potential energy at the datum level being taken as zero).

Kinetic energy is the energy due to the motion of a body. Suppose a force F acts on an object of mass m originally at rest (i.e. $u = 0$) and accelerates it to a velocity v in a distance s:

$$\text{work done} = \text{force} \times \text{distance}$$

$$= Fs = (ma)(s) \quad \text{(if no energy is lost)}$$

where a is the acceleration.

Since $v^2 = u^2 + 2as$ (see chapter 23) and $u = 0$,

$$v^2 = 2as, \text{ from which } a = \frac{v^2}{2s}$$

hence, work done $= (ma)(s) = (m)\left(\frac{v^2}{2s}\right)(s) = \frac{1}{2}mv^2$

This energy is called the kinetic energy of the mass m, i.e.

$$\textbf{kinetic energy } = \frac{1}{2}mv^2 \textbf{ joules}$$

As stated in section 21.3, energy may be converted from one form to another. The **principle of conservation of energy** states that the total amount of energy remains the same in such conversions, i.e. energy cannot be created or destroyed.

In mechanics, the potential energy possessed by a body is frequently converted into kinetic energy, and vice versa. When a mass is falling freely, its potential energy decreases as it loses height, and its kinetic energy increases as its velocity increases. Ignoring air frictional losses, at all times:

Potential energy + kinetic energy = a constant

If friction is present, then work is done overcoming the resistance due to friction and this is dissipated as heat. Then,

Initial energy = final energy + work done

overcoming frictional resistance

Kinetic energy is not always conserved in collisions. Collisions in which kinetic energy is conserved (i.e. stays the same) are called **elastic collisions**, and those in which it is not conserved are termed **inelastic collisions**.

Problem 22. A car of mass 800 kg is climbing an incline at 10° to the horizontal. Determine the increase in potential energy of the car as it moves a distance of 50 m up the incline

With reference to Figure 21.10,

$$\sin 10° = \frac{\text{opposite}}{\text{hypotenuse}} = \frac{h}{50} \text{ (from chapter 11)}$$

from which, $h = 50 \sin 10° = 8.682$ m

Hence, increase in potential energy

$$= mgh = 800 \text{ kg} \times 9.81 \text{ m/s}^2 \times 8.682 \text{ m}$$

$$= 68140 \text{ J or } 68.14 \text{ kJ}$$

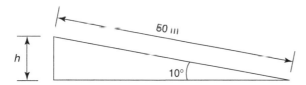

Figure 21.10

Problem 23. At the instant of striking, a hammer of mass 30 kg has a velocity of 15 m/s. Determine the kinetic energy in the hammer

Kinetic energy $= \frac{1}{2}mv^2 = \frac{1}{2}(30 \text{ kg})(15 \text{ m/s})^2$.

i.e. **kinetic energy in hammer = 3375 J or 3.375 kJ**

Problem 24. A lorry having a mass of 1.5 t is travelling along a level road at 72 km/h. When the brakes are applied, the speed decreases to 18 km/h. Determine how much the kinetic energy of the lorry is reduced

Initial velocity of lorry $v_1 = 72$ km/h

$$= 72\frac{\text{km}}{\text{h}} \times 1000\frac{\text{m}}{\text{km}} \times \frac{1 \text{ h}}{3600 \text{ s}}$$

$$= \frac{72}{3.6} = 20 \text{ m/s}$$

final velocity of lorry $v_2 = \frac{18}{3.6} = 5$ m/s and

mass of lorry $m = 1.5 \text{ t} = 1500$ kg.

Initial kinetic energy of the lorry $= \frac{1}{2}mv_1^2$

$$= \frac{1}{2}(1500)(20)^2$$

$$= 300 \text{ kJ}$$

final kinetic energy of the lorry $= \frac{1}{2}mv_2^2$

$$= \frac{1}{2}(1500)(5)^2$$

$$= 18.75 \text{ kJ}$$

Hence, **the change in kinetic energy** $= 300 - 18.75$

$$= 281.25 \text{ kJ}$$

(Part of this reduction in kinetic energy is converted into heat energy in the brakes of the lorry and is hence dissipated in overcoming frictional forces and air friction.)

Problem 25. A canister containing a meteorology balloon of mass 4 kg is fired vertically upwards from a gun with an initial velocity of 400 m/s. Neglecting the air resistance, calculate (a) its initial kinetic energy, (b) its velocity at a height of 1 km, (c) the maximum height reached

(a) **Initial kinetic energy** $= \frac{1}{2}mv^2 = \frac{1}{2}(4)(400)^2$

$$= 320 \text{ kJ}$$

(b) At a height of 1 km,
potential energy $= mgh$

$$= 4 \times 9.81 \times 1000$$

$$= 39.24 \text{ kJ}$$

By the principle of conservation of energy:
potential energy + kinetic energy at 1 km = initial kinetic energy

Hence $39,240 + \frac{1}{2}mv^2 = 320,000$

from which, $\frac{1}{2}(4)v^2 = 320,000 - 39,240$

$$= 280,760$$

Hence, $v = \sqrt{\left(\frac{2 \times 280,760}{4}\right)} = 374.7 \, \text{m/s}$

i.e. **the velocity of the canister at a height of 1 km is 374.7 m/s**

(c) At the maximum height, the velocity of the canister is zero and all the kinetic energy has been converted into potential energy. Hence,

$$\text{potential energy} = \text{initial kinetic energy}$$

$$= 320,000 \, \text{J} \quad \text{(from part (a))}$$

Then, $320,000 = mgh = (4)(9.81)(h)$

from which, height $h = \dfrac{320,000}{(4)(9.81)} = 8155 \, \text{m}$

i.e. **the maximum height reached is 8155 m or 8.155 km**

Problem 26. A pile-driver of mass 500 kg falls freely through a height of 1.5 m on to a pile of mass 200 kg. Determine the velocity with which the driver hits the pile. If, at impact, 3 kJ of energy are lost due to heat and sound, the remaining energy being possessed by the pile and driver as they are driven together into the ground a distance of 200 mm, determine (a) the common velocity immediately after impact, (b) the average resistance of the ground

---

The potential energy of the pile-driver is converted into kinetic energy.

Thus potential energy $=$ kinetic energy,

i.e. $mgh = \frac{1}{2}mv^2$

from which, velocity $v = \sqrt{2gh} = \sqrt{(2)(9.81)(1.5)}$

$$= 5.42 \, \text{m/s}$$

Hence, **the pile-driver hits the pile at a velocity of 5.42 m/s.**

(a) Before impact, kinetic energy of pile driver

$$= \frac{1}{2}mv^2 = \frac{1}{2}(500)(5.42)^2 = 7.34 \, \text{kJ}$$

Kinetic energy after impact $= 7.34 - 3 = 4.34 \, \text{kJ}$.
Thus the pile-driver and pile together have a mass of $500 + 200 = 700 \, \text{kg}$ and possess kinetic energy of 4.34 kJ.

Hence, $4.34 \times 10^3 = \frac{1}{2}mv^2 = \frac{1}{2}(700)v^2$

from which,

$$\text{velocity } v = \sqrt{\left(\frac{2 \times 4.34 \times 10^3}{700}\right)} = 3.52 \, \text{m/s}$$

Thus, **the common velocity after impact is 3.52 m/s**

(b) The kinetic energy after impact is absorbed in overcoming the resistance of the ground, in a distance of 200 mm.

$$\text{Kinetic energy} = \text{work done}$$

$$= \text{resistance} \times \text{distance}$$

i.e. $4.34 \times 10^3 = \text{resistance} \times 0.200$

from which, resistance $= \dfrac{4.34 \times 10^3}{0.200} = 21700 \, \text{N}$

Hence, **the average resistance of the ground is 21.7 kN**

Problem 27. A car of mass 600 kg reduces speed from 90 km/h to 54 km/h in 15 s. Determine the braking power required to give this change of speed

---

Change in kinetic energy of car $= \dfrac{1}{2}mv_1^2 - \dfrac{1}{2}mv_2^2$

where $m = $ mass of car $= 600 \, \text{kg}$,

$v_1 = $ initial velocity $= 90 \, \text{km/h}$

$$= \frac{90}{3.6} \, \text{m/s} = 25 \, \text{m/s},$$

and $v_2 = $ final velocity $= 54 \, \text{km/h}$

$$= \frac{54}{3.6} \, \text{m/s} = 15 \, \text{m/s}$$

Hence, change in kinetic energy $= \frac{1}{2}m(v_1^2 - v_2^2)$

$$= \frac{1}{2}(600)(25^2 - 15^2)$$

$$= 120,000 \, J$$

Braking power $= \dfrac{\text{change in energy}}{\text{time taken}}$

$$= \frac{120,000 \, J}{15 \, s} = \textbf{8000 W or 8 kW}$$

Now try the following Practice Exercises

Practice Exercise 99   Further problems on potential and kinetic energy (answers on page 541)

(Assume the acceleration due to gravity, $g = 9.81 \, m/s^2$.)

1. An object of mass 400 g is thrown vertically upwards and its maximum increase in potential energy is 32.6 J. Determine the maximum height reached, neglecting air resistance

2. A ball bearing of mass 100 g rolls down from the top of a chute of length 400 m inclined at an angle of 30° to the horizontal. Determine the decrease in potential energy of the ball bearing as it reaches the bottom of the chute

3. A vehicle of mass 800 kg is travelling at 54 km/h when its brakes are applied. Find the kinetic energy lost when the car comes to rest

4. Supplies of mass 300 kg are dropped from a helicopter flying at an altitude of 60 m. Determine the potential energy of the supplies relative to the ground at the instant of release, and its kinetic energy as it strikes the ground

5. A shell of mass 10 kg is fired vertically upwards with an initial velocity of 200 m/s. Determine its initial kinetic energy and the maximum height reached, correct to the nearest metre, neglecting air resistance

6. The potential energy of a mass is increased by 20.0 kJ when it is lifted vertically through a height of 25.0 m. It is now released and allowed to fall freely. Neglecting air resistance, find its kinetic energy and its velocity after it has fallen 10.0 m

7. A pile-driver of mass 400 kg falls freely through a height of 1.2 m on to a pile of mass 150 kg. Determine the velocity with which the driver hits the pile. If, at impact, 2.5 kJ of energy are lost due to heat and sound, the remaining energy being possessed by the pile and driver as they are driven together into the ground a distance of 150 mm, determine (a) the common velocity after impact, (b) the average resistance of the ground

Practice Exercise 100   Short answer questions on work, energy and power

1. Define work in terms of force applied and distance moved

2. Define energy, and state its unit

3. Define the joule

4. The area under a force/distance graph represents ....

5. Name five forms of energy

6. State the principle of conservation of energy

7. Give two examples of conversion of heat energy to other forms of energy

8. Give two examples of conversion of electrical energy to other forms of energy

9. Give two examples of conversion of chemical energy to other forms of energy

10. Give two examples of conversion of mechanical energy to other forms of energy

11. (a) Define efficiency in terms of energy input and energy output.
    (b) State the symbol used for efficiency

12. Define power and state its unit

13. Define potential energy

14. The change in potential energy of a body of mass m kg when lifted vertically upwards to a height h m is given by ....

15. What is kinetic energy?

16. The kinetic energy of a body of mass m kg and moving at a velocity of v m/s is given by ....

17. Distinguish between elastic and inelastic collisions

Practice Exercise 101   Multiple-choice questions on work, energy and power (answers on page 541)

1. State which of the following is incorrect:
   (a) $1\,W = 1\,J/s$
   (b) $1\,J = 1\,N/m$
   (c) $\eta = \dfrac{\text{output energy}}{\text{input energy}}$
   (d) energy = power × time

2. An object is lifted 2000 mm by a crane. If the force required is 100 N, the work done is:
   (a) $\dfrac{1}{20}\,N\,m$ (b) 200 kN m
   (c) 200 N m   (d) 20 J

3. A motor having an efficiency of 0.8 uses 800 J of electrical energy. The output energy of the motor is:
   (a) 800 J   (b) 1000 J
   (c) 640 J   (d) 6.4 J

4. 6 kJ of work is done by a force in moving an object uniformly through 120 m in 1 minute. The force applied is:
   (a) 50 N   (b) 20 N   (c) 720 N   (d) 12 N

5. For the object in question 4, the power developed is:
   (a) 6 kW   (b) 12 kW
   (c) 5/6 W   (d) 0.1 kW

6. Which of the following statements is false?
   (a) The unit of energy and work is the same
   (b) The area under a force/distance graph gives the work done
   (c) Electrical energy is converted to mechanical energy by a generator
   (d) Efficiency is the ratio of the useful output energy to the input energy

7. A machine using a power of 1 kW requires a force of 100 N to raise a mass in 10 s. The height the mass is raised in this time is:
   (a) 100 m   (b) 1 km   (c) 10 m   (d) 1 m

8. A force/extension graph for a spring is shown in Figure 21.11. Which of the following statements is false?
   The work done in extending the spring:
   (a) from 0 to 100 mm is 5 J
   (b) from 0 to 50 mm is 1.25 J
   (c) from 20 mm to 60 mm is 1.6 J
   (d) from 60 mm to 100 mm is 3.75 J

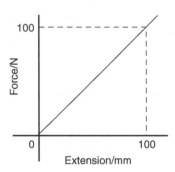

Figure 21.11

9. A vehicle of mass 1 t climbs an incline of 30° to the horizontal. Taking the acceleration due to gravity as $10\,m/s^2$, the increase in potential energy of the vehicle as it moves a distance of 200 m up the incline is:
   (a) 1 kJ   (b) 2 MJ   (c) 1 MJ   (d) 2 kJ

10. A bullet of mass 100 g is fired from a gun with an initial velocity of 360 km/h. Neglecting air resistance, the initial kinetic energy possessed by the bullet is:
    (a) 6.48 kJ   (b) 500 J
    (c) 500 kJ   (d) 6.48 MJ

11. A small motor requires 50 W of electrical power in order to produce 40 W of mechanical energy output. The efficiency of the motor is:
    (a) 10%   (b) 80%   (c) 40%   (d) 90%

12. A load is lifted 4000 mm by a crane. If the force required to lift the mass is 100 N, the work done is:

    (a) 400 J   (b) 40 N m

    (c) 25 J   (d) 400 kJ

13. A machine exerts a force of 100 N in lifting a mass through a height of 5 m. If 1 kJ of energy is supplied, the efficiency of the machine is:

    (a) 10%   (b) 20%

    (c) 100%   (d) 50%

14. At the instant of striking an object, a hammer of mass 40 kg has a velocity of 10 m/s. The kinetic energy in the hammer is:

    (a) 2 kJ   (b) 1 kJ   (c) 400 J   (d) 8 kJ

15. A machine which has an efficiency of 80% raises a load of 50 N through a vertical height of 10 m. The work input to the machine is:

    (a) 400 J   (b) 500 J   (c) 800 J   (d) 625 J

For fully worked solutions to each of the problems in Exercises 96 to 101 in this chapter, go to the website:

Chapter 22

Simply supported beams

Why it is important to understand: **Simply supported beams**

This chapter is very important for the design of beams which carry loads acting transversely to them. These structures are of importance in the design of buildings, bridges, cranes, ships, aircraft, automobiles, and so on. The chapter commences with defining the moment of a force, and then, using equilibrium considerations, demonstrates the principle of moments. This is a very important skill widely used in designing structures. Several examples are given, with increasing complexity to help the reader acquire these valuable skills. This chapter describes skills which are fundamental and extremely important in many branches of engineering. The main reason though for understanding simply supported beams is that such structures appear in the design of buildings, automobiles, aircraft, etc.

At the end of this chapter, you should be able to:

- define a 'moment' of a force and state its unit
- calculate the moment of a force from $M = F \times d$
- understand the conditions for equilibrium of a beam
- state the principle of moments
- perform calculations involving the principle of moments
- recognise typical practical applications of simply supported beams with point loadings
- perform calculations on simply supported beams having point loads

22.1 Introduction

This chapter is very important for the design of beams which carry loads, acting transversely to them. These structures are of importance in the design of buildings, bridges, cranes, ships, aircraft, automobiles and so on. The chapter commences with defining the moment of a force, and then using equilibrium considerations, demonstrates the principle of moments. This is a central skill widely used in designing structures. Several examples are given, with increasing complexity to help the reader acquire these valuable skills. This chapter describes skills which are fundamental and extremely important in many branches of engineering.

Science and Mathematics for Engineering. 978-0-367-20475-4, © John Bird. Published by Taylor & Francis. All rights reserved.

22.2 The moment of a force

When using a spanner to tighten a nut, a force tends to turn the nut in a clockwise direction. This turning effect of a force is called the **moment of a force** or more briefly, just a **moment**. The size of the moment acting on the nut depends on two factors:

(a) the size of the force acting at right angles to the shank of the spanner, and

(b) the perpendicular distance between the point of application of the force and the centre of the nut.

In general, with reference to Figure 22.1, the moment M of a force acting about a point P is: force × perpendicular distance between the line of action of the force and P, i.e.

$$M = F \times d$$

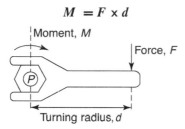

Figure 22.1

The unit of a moment is the **newton metre (N m)**. Thus, if force F in Figure 22.1 is 7 N and distance d is 3 m, then the moment M is 7 N × 3 m, i.e. 21 N m.

Problem 1. A force of 15 N is applied to a spanner at an effective length of 140 mm from the centre of a nut. Calculate (a) the moment of the force applied to the nut, (b) the magnitude of the force required to produce the same moment if the effective length is reduced to 100 mm

From above, $M = F \times d$, where M is the turning moment, F is the force applied at right angles to the spanner and d is the effective length between the force and the centre of the nut. Thus, with reference to Figure 22.2(a):

(a) Turning moment, $M = 15\,\text{N} \times 140\,\text{mm}$

$$= 2100\,\text{N\,mm}$$

$$= 2100\,\text{N\,mm} \times \frac{1\,\text{m}}{1000\,\text{mm}}$$

$$= \mathbf{2.1\,N\,m}$$

(b) Turning moment, M is 2100 N mm and the effective length d becomes 100 mm (see Figure 22.2(b)).

Applying $M = F \times d$

gives: $2100\,\text{N\,mm} = F \times 100\,\text{mm}$

from which, $\textbf{force } F = \dfrac{2100\,\text{N\,mm}}{100\,\text{mm}} = \mathbf{21\,N}$

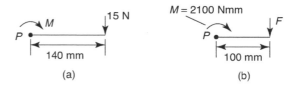

(a) (b)

Figure 22.2

Problem 2. A moment of 25 N m is required to operate a lifting jack. Determine the effective length of the handle of the jack if the force applied to it is: (a) 125 N (b) 0.4 kN

From above, moment $M = F \times d$, where F is the force applied at right angles to the handle and d is the effective length of the handle. Thus:

(a) $25\,\text{N\,m} = 125\,\text{N} \times d$, from which

$$\textbf{effective length } d = \frac{25\,\text{N\,m}}{125\,\text{N}} = \frac{1}{5}\,\text{m}$$

$$= \frac{1}{5} \times 1000\,\text{mm}$$

$$= \mathbf{200\,mm}$$

(b) Turning moment M is 25 N m and the force F becomes 0.4 kN, i.e. 400 N.

Since $M = F \times d$, then $25\,\text{N\,m} = 400\,\text{N} \times d$

Thus, $\textbf{effective length } d = \dfrac{25\,\text{N\,m}}{400\,\text{N}} = \dfrac{1}{16}\,\text{m}$

$$= \frac{1}{16} \times 1000\,\text{mm}$$

$$= \mathbf{62.5\,mm}$$

Now try the following Practice Exercise

Practice Exercise 102 Further problems on the moment of a force (answers on page 541)

1. Determine the moment of a force of 25 N applied to a spanner at an effective length of 180 mm from the centre of a nut

2. A moment of 7.5 N m is required to turn a wheel. If a force of 37.5 N applied to the rim of the wheel can just turn the wheel, calculate the effective distance from the rim to the hub of the wheel

3. Calculate the force required to produce a moment of 27 N m on a shaft, when the effective distance from the centre of the shaft to the point of application of the force is 180 mm

22.3 Equilibrium and the principle of moments

If more than one force is acting on an object and the forces do not act at a single point, then the turning effect of the forces, that is, the moment of each of the forces, must be considered.

Figure 22.3 shows a beam with its support (known as its **pivot** or **fulcrum**) at P, acting vertically upwards, and forces F_1 and F_2 acting vertically downwards at distances a and b, respectively, from the fulcrum.

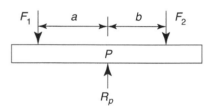

Figure 22.3

A beam is said to be in **equilibrium** when there is no tendency for it to move. There are two conditions for equilibrium:

(i) The sum of the forces acting vertically downwards must be equal to the sum of the forces acting vertically upwards, i.e. for Figure 22.3,

$$R_P = F_1 + F_2$$

(ii) The total moment of the forces acting on a beam must be zero; for the total moment to be zero:

The sum of the clockwise moments about any point must be equal to the sum of the anticlockwise, or counter-clockwise, moments about that point.

This statement is known as the **principle of moments**.

Hence, taking moments about P in Figure 22.3,

$$F_2 \times b = \text{ the clockwise moment, and}$$

$$F_1 \times a = \text{ the anticlockwise, or}$$

counter-clockwise, moment

Thus for equilibrium: $F_1 a = F_2 b$

Problem 3. A system of forces is as shown in Figure 22.4. (a) If the system is in equilibrium find the distance d. (b) If the point of application of the 5 N force is moved to point P, distance 200 mm from the support, and the 5 N force is replaced by an unknown force F, find the value of F for the system to be in equilibrium

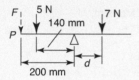

Figure 22.4

(a) From above, the clockwise moment M_1 is due to a force of 7 N acting at a distance d from the support; the support is called the **fulcrum**, i.e. $M_1 = 7\,\text{N} \times d$

The anticlockwise moment M_2 is due to a force of 5 N acting at a distance of 140 mm from the fulcrum, i.e. $M_2 = 5\,\text{N} \times 140\,\text{mm}$

Applying the principle of moments, for the system to be in equilibrium about the fulcrum:

clockwise moment = anticlockwise moment

i.e. $7\,\text{N} \times d = 5 \times 140\,\text{N mm}$

Hence, **distance** $d = \dfrac{5 \times 140\,\text{N mm}}{7\,\text{N}} = \textbf{100 mm}$

(b) When the 5 N force is replaced by force F at a distance of 200 mm from the fulcrum, the new value of the anticlockwise moment is $F \times 200$. For the system to be in equilibrium:

clockwise moment = anticlockwise moment

i.e. $(7 \times 100)\,\text{N mm} = F \times 200\,\text{mm}$

Hence, **new value of force** $F = \dfrac{700\,\text{N mm}}{200\,\text{mm}}$

$= \textbf{3.5 N}$

Problem 4. A beam is supported on its fulcrum at the point *A*, which is at mid-span, and forces act as shown in Figure 22.5. Calculate (a) force *F* for the beam to be in equilibrium, (b) the new position of the 23 N force when *F* is decreased to 21 N, if equilibrium is to be maintained

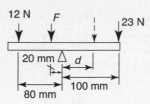

12 N   *F*   23 N

20 mm   *d*

100 mm

80 mm

Figure 22.5

(a) The clockwise moment, M_1, is due to the 23 N force acting at a distance of 100 mm from the fulcrum, i.e.

$$M_1 = 23 \times 100 = 2300 \, \text{N mm}$$

There are two forces giving the anticlockwise moment M_2. One is the force *F* acting at a distance of 20 mm from the fulcrum and the other a force of 12 N acting at a distance of 80 mm. Thus,

$$M_2 = (F \times 20) + (12 \times 80) \, \text{N mm}$$

Applying the principle of moments about the fulcrum:

clockwise moment = anticlockwise moments

i.e. $\qquad 2300 = (F \times 20) + (12 \times 80)$

Hence, $\quad F \times 20 = 2300 - 960$

i.e. $\qquad$ **force $F = \dfrac{1340}{20} = 67 \, \text{N}$**

(b) The clockwise moment is now due to a force of 23 N acting at a distance of, say, *d* from the fulcrum. Since the value of *F* is decreased to 21 N, the anticlockwise moment is $(21 \times 20) + (12 \times 80)$ N mm

Applying the principle of moments,

$$23 \times d = (21 \times 20) + (12 \times 80)$$

i.e. **distance $d = \dfrac{420 + 960}{23} = \dfrac{1380}{23} = 60 \, \text{mm}$**

Problem 5. For the centrally supported uniform beam shown in Figure 22.6, determine the values of forces F_1 and F_2 when the beam is in equilibrium

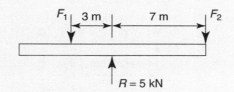

F_1 3 m   7 m F_2

$R = 5$ kN

Figure 22.6

At equilibrium: (i) $R = F_1 + F_2$ i.e. $5 = F_1 + F_2$ (1)

and $\qquad$ (ii) $F_1 \times 3 = F_2 \times 7$ $\qquad$ (2)

From equation (1), $F_2 = 5 - F_1$

Substituting for F_2 in equation (2) gives:

$$F_1 \times 3 = (5 - F_1) \times 7$$

i.e. $\qquad 3F_1 = 35 - 7F_1$

$$10F_1 = 35$$

from which, $\qquad F_1 = 3.5 \, \text{kN}$

Since $F_2 = 5 - F_1$, $\quad F_2 = 1.5 \, \text{kN}$

Thus at equilibrium, force $F_1 = 3.5 \, \text{kN}$ and force $F_2 = 1.5 \, \text{kN}$

Now try the following Practice Exercise

Practice Exercise 103   **Further problems on equilibrium and the principle of moments** (answers on page 541)

1. Determine distance *d* and the force acting at the support *A* for the force system shown in Figure 22.7, when the system is in equilibrium

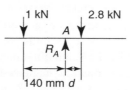

1 kN   2.8 kN

A

R_A

140 mm *d*

Figure 22.7

2. If the 1 kN force shown in Figure 22.7 is replaced by a force *F* at a distance of 250 mm to the left of R_A, find the value of *F* for the system to be in equilibrium

3. Determine the values of the forces acting at A and B for the force system shown in Figure 22.8

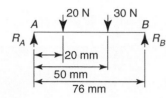

Figure 22.8

4. The forces acting on a beam are as shown in Figure 22.9. Neglecting the mass of the beam, find the value of R_A and distance d when the beam is in equilibrium

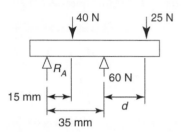

Figure 22.9

22.4 Simply supported beams having point loads

A **simply supported beam** is said to be one that rests on two knife-edge supports and is free to move horizontally. Two typical simply supported beams having loads acting at given points on the beam, called **point loading**, are shown in Figure 22.10.

A man whose mass exerts a force F vertically downwards, standing on a wooden plank which is simply supported at its ends, may, for example, be represented by the beam diagram of Figure 22.10(a) if the mass of the plank is neglected. The forces exerted by the supports on the plank, R_A and R_B, act vertically upwards, and are called **reactions**.

When the forces acting are all in one plane, the algebraic sum of the moments can be taken about **any** point.

For the beam in Figure 22.10(a) at equilibrium:

(i) $R_A + R_B = F$, and

(ii) taking moments about A, $F \times a = R_B (a + b)$.

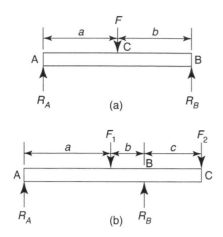

Figure 22.10

(Alternatively, taking moments about C, $R_A a = R_B b$)

For the beam in Figure 22.10(b), at equilibrium

(i) $R_A + R_B = F_1 + F_2$, and

(ii) taking moments about B, $R_A (a + b) + F_2 c = F_1 b$

Typical **practical applications** of simply supported beams with point loadings include bridges, beams in buildings, and beds of machine tools.

Problem 6. A beam is loaded as shown in Figure 22.11. Determine (a) the force acting on the beam support at B, (b) the force acting on the beam support at A, neglecting the mass of the beam

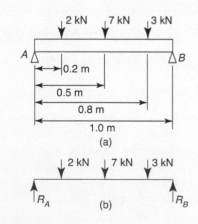

Figure 22.11

A beam supported as shown in Figure 22.11 is called a simply supported beam.

(a) Taking moments about point A and applying the principle of moments gives:

clockwise moments $=$ anticlockwise moments

$$(2 \times 0.2) + (7 \times 0.5) + (3 \times 0.8) \, \text{kN m}$$

$$= R_B \times 1.0 \, \text{m}$$

where R_B is the force supporting the beam at B, as shown in Figure 22.11(b).

Thus $(0.4 + 3.5 + 2.4) \, \text{kN m} = R_B \times 1.0 \, \text{m}$

i.e. $\qquad R_B = \dfrac{6.3 \, \text{kN m}}{1.0 \, \text{m}} = \mathbf{6.3 \, kN}$

(b) For the beam to be in equilibrium, the forces acting upwards must be equal to the forces acting downwards, thus

$R_A + R_B = (2 + 7 + 3) \, \text{kN} = 12 \, \text{kN}$

$R_B = 6.3 \, \text{kN},$

thus $\boldsymbol{R_A} = 12 - R_B = 12 - 6.3 = \mathbf{5.7 \, kN}$

Problem 7. For the beam shown in Figure 22.12 calculate (a) the force acting on support A, (b) distance d, neglecting any forces arising from the mass of the beam

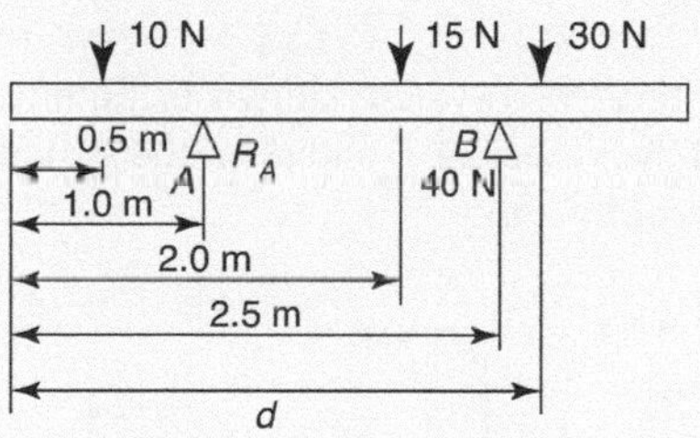

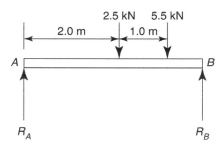

Figure 22.12

(a) From section 22.3,

(the forces acting in an upward direction)

$\qquad = $ (the forces acting in a downward direction)

Hence $\quad (R_A + 40) \, \text{N} = (10 + 15 + 30) \, \text{N}$

$$\boldsymbol{R_A} = 10 + 15 + 30 - 40$$

$$= \mathbf{15 \, N}$$

(b) Taking moments about the left-hand end of the beam and applying the principle of moments gives:

clockwise moments $=$ anticlockwise moments

$(10 \times 0.5) + (15 \times 2.0) \, \text{N m} + 30 \, \text{N} \times d$

$$= (15 \times 1.0) + (40 \times 2.5) \, \text{N m}$$

i.e. $\qquad 35 \, \text{N m} + 30 \, \text{N} \times d = 115 \, \text{N m}$

from which, $\qquad \textbf{distance } \boldsymbol{d} = \dfrac{(115 - 35) \, \text{N m}}{30 \, \text{N}}$

$$= \mathbf{2.67 \, m}$$

Problem 8. A metal bar AB is 4.0 m long and is simply supported at each end in a horizontal position. It carries loads of 2.5 kN and 5.5 kN at distances of 2.0 m and 3.0 m, respectively, from A. Neglecting the mass of the beam, determine the reactions of the supports when the beam is in equilibrium

The beam and its loads are shown in Figure 22.13. At equilibrium,

$$R_A + R_B = 2.5 + 5.5 = 8.0 \, \text{kN} \qquad (1)$$

Figure 22.13

Taking moments about A,

clockwise moments $=$ anticlockwise moment

i.e. $\qquad (2.5 \times 2.0) + (5.5 \times 3.0) = 4.0 R_B$

or $\qquad 5.0 + 16.5 = 4.0 R_B$

from which, $\qquad R_B = \dfrac{21.5}{4.0} = 5.375 \, \text{kN}$

From equation (1), $\quad R_A = 8.0 - 5.375 = 2.625 \, \text{kN}$

Thus the reactions at the supports at equilibrium are 2.625 kN at A and 5.375 kN at B.

Problem 9. A beam PQ is 5.0 m long and is simply supported at its ends in a horizontal position as shown in Figure 22.14. Its mass is equivalent to a force of 400 N acting at its centre as shown. Point loads of 12 kN and 20 kN act on the beam in the positions shown. When the beam is in equilibrium, determine (a) the reactions of the supports, R_P and R_Q, and (b) the position to which the 12 kN load must be moved for the force on the supports to be equal

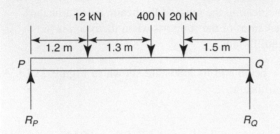

Figure 22.14

(a) At equilibrium,

$$R_P + R_Q = 12 + 0.4 + 20 = 32.4 \text{ kN} \quad (1)$$

Taking moments about P:

clockwise moments = anticlockwise moments

i.e. $(12 \times 1.2) + (0.4 \times 2.5) + (20 \times 3.5)$

$$= (R_Q \times 5.0)$$

$$14.4 + 1.0 + 70.0 = 5.0 R_Q$$

from which, $\quad R_Q = \dfrac{85.4}{5.0} = \textbf{17.08 kN}$

From equation (1),

$$R_P = 32.4 - R_Q = 32.4 - 17.08 = \textbf{15.32 kN}$$

(b) For the reactions of the supports to be equal,

$$R_P = R_Q = \frac{32.4}{2} = 16.2 \text{ kN}$$

Let the 12 kN load be at a distance d metres from P (instead of at 1.2 m from P). Taking moments about point P gives:

$$(12 \times d) + (0.4 \times 2.5) + (20 \times 3.5) = 5.0 R_Q$$

i.e. $\quad 12d + 1.0 + 70.0 = 5.0 \times 16.2$

and $\quad 12d = 81.0 - 71.0$

from which, $\quad d = \dfrac{10.0}{12} = 0.833 \text{ m}$

Hence the 12 kN load needs to be moved to a position 833 mm from P for the reactions of the supports to be equal (i.e. 367 mm to the left of its original position).

Now try the following Practice Exercises

Practice Exercise 104 Further problems on simply supported beams having point loads (answers on page 541)

1. Calculate the force R_A and distance d for the beam shown in Figure 22.15. The mass of the beam should be neglected and equilibrium conditions assumed

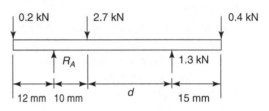

Figure 22.15

2. For the force system shown in Figure 22.16, find the values of F and d for the system to be in equilibrium

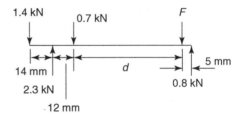

Figure 22.16

3. For the force system shown in Figure 22.17, determine distance d for the forces R_A and R_B to be equal, assuming equilibrium conditions

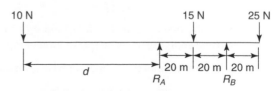

Figure 22.17

4. A simply supported beam AB is loaded as shown in Figure 22.18. Determine the load F in order that the reaction at A is zero

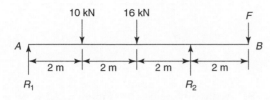

Figure 22.18

5. A uniform wooden beam, 4.8 m long, is supported at its left-hand end and also at 3.2 m from the left-hand end. The mass of the beam is equivalent to 200 N acting vertically downwards at its centre. Determine the reactions at the supports

6. For the simply supported beam PQ shown in Figure 22.19, determine (a) the reaction at each support, (b) the maximum force which can be applied at Q without losing equilibrium

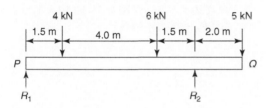

Figure 22.19

Practice Exercise 105 Short answer questions on simply supported beams

1. The moment of a force is the product of ...... and ......

2. When a beam has no tendency to move it is in ......

3. State the two conditions for equilibrium of a beam

4. State the principle of moments

5. What is meant by a simply supported beam?

6. State two practical applications of simply supported beams

Practice Exercise 106 Multiple-choice questions on simply supported beams (answers on page 541)

1. A force of 10 N is applied at right angles to the handle of a spanner, 0.5 m from the centre of a nut. The moment on the nut is:

   (a) 5 N m (b) 2 N/m (c) 0.5 m/N (d) 15 N m

2. The distance d in Figure 22.20 when the beam is in equilibrium is:

   (a) 0.5 m (b) 1.0 m (c) 4.0 m (d) 15 m

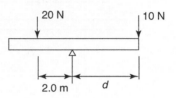

Figure 22.20

3. With reference to Figure 22.21, the clockwise moment about A is:

   (a) 70 N m (b) 10 N m

   (c) 60 N m (d) $5 \times R_B$ N m

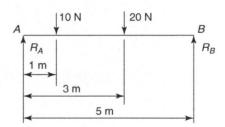

Figure 22.21

4. The force acting at B (i.e. R_B) in Figure 22.21 is:

   (a) 16 N (b) 20 N (c) 5 N (d) 14 N

5. The force acting at A (i.e. R_A) in Figure 22.21 is:

   (a) 16 N (b) 10 N (c) 15 N (d) 14 N

6. Which of the following statements is false for the beam shown in Figure 22.22 if the beam is in equilibrium?

   (a) The anticlockwise moment is 27 N

   (b) The force F is 9 N

   (c) The reaction at the support R is 18 N

   (d) The beam cannot be in equilibrium for the given conditions

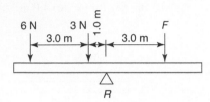

Figure 22.22

7. With reference to Figure 22.23, the reaction R_A is:

   (a) 10 N (b) 30 N (c) 20 N (d) 40 N

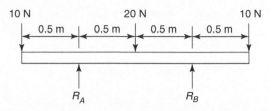

Figure 22.23

8. With reference to Figure 22.23, when moments are taken about R_A, the sum of the anticlockwise moments is:

   (a) 25 N m (b) 20 N m (c) 35 N m (d) 30 N m

9. With reference to Figure 22.23, when moments are taken about the right-hand end, the sum of the clockwise moments is:

   (a) 10 N m (b) 20 N m (c) 30 N m (d) 40 N m

10. With reference to Figure 22.23, which of the following statements is false?

    (a) $(5 + R_B) = 25$ N m   (b) $R_A = R_B$
    (c) $(10 \times 0.5) = (10 \times 1) + (10 \times 1.5) + R_A$
    (d) $R_A + R_B = 40$ N

**For fully worked solutions to each of the problems in Exercises 102 to 106 in this chapter,
go to the website:**
www.routledge.com/cw/bird

Revision Test 8: Forces acting at a point, work, energy and power and simply supported beams

This assignment covers the material contained in chapters 20 to 22. *The marks for each question are shown in brackets at the end of each question.*

1. A force of 25 N acts horizontally to the right and a force of 15 N is inclined at an angle of 30° to the 25 N force. Determine the magnitude and direction of the resultant of the two forces using (a) the triangle of forces method, (b) the parallelogram of forces method, and (c) by calculation. (14)

2. Determine graphically the magnitude and direction of the resultant of the following three coplanar forces, which may be considered as acting at a point:
   force P, 15 N acting horizontally to the right;
   force Q, 8 N inclined at 45° to force P;
   force R, 20 N inclined at 120° to force P. (8)

3. Determine by resolution of forces the resultant of the following three coplanar forces acting at a point: 120 N acting at 40° to the horizontal; 250 N acting at 145° to the horizontal; 300 N acting at 260° to the horizontal. (8)

4. A spring, initially in a relaxed state, is extended by 80 mm. Determine the work done by using a work diagram if the spring requires a force of 0.7 N per mm of stretch. (4)

5. Water is pumped vertically upwards through a distance of 40.0 m and the work done is 176.58 kJ. Determine the number of litres of water pumped. (1 litre of water has a mass of 1 kg.) (4)

6. 3 kJ of energy are supplied to a machine used for lifting a mass. The force required is 1 kN. If the machine has an efficiency of 60%, to what height will it lift the mass? (4)

7. When exerting a steady pull of 450 N, a lorry travels at 80 km/h. Calculate (a) the work done in 15 minutes, (b) the power required. (4)

8. An electric motor provides power to a winding machine. The input power to the motor is 4.0 kW and the overall efficiency is 75%. Calculate (a) the output power of the machine, (b) the rate at which it can raise a 509.7 kg load vertically upwards. (4)

9. A tank of mass 4800 kg is climbing an incline at 12° to the horizontal. Determine the increase in potential energy of the tank as it moves a distance of 40 m up the incline. (4)

10. A car of mass 500 kg reduces speed from 108 km/h to 36 km/h in 20 s. Determine the braking power required to give this change of speed. (4)

11. A moment of 18 N m is required to operate a lifting jack. Determine the effective length of the handle of the jack (in millimetres) if the force applied to it is (a) 90 N, (b) 0.36 kN. (6)

12. For the centrally supported uniform beam shown in Figure RT8.1, determine the values of forces F_1 and F_2 when the beam is in equilibrium. (8)

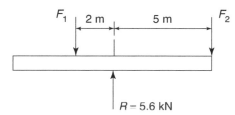

Figure RT8.1

13. For the beam shown in Figure RT8.2 calculate (a) the force acting on support Q, (b) distance d, neglecting any forces arising from the mass of the beam. (8)

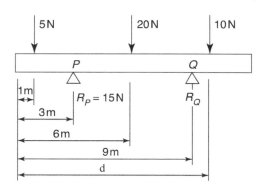

Figure RT8.2

For lecturers/instructors/teachers, fully worked solutions to each of the problems in Revision Test 8, together with a full marking scheme, are available at the website:
www.routledge.com/cw/bird

Chapter 23

Linear and angular motion

Why it is important to understand: **Linear and angular motion**

This chapter commences by defining linear and angular velocity and also linear and angular acceleration. It then derives the well-known relationships, under uniform acceleration, for displacement, velocity and acceleration, in terms of time and other parameters. The chapter then uses elementary vector analysis, similar to that used for forces in chapter 20, to determine relative velocities. This chapter deals with the basics of kinematics. A study of linear and angular motion is important for the design of moving vehicles.

At the end of this chapter, you should be able to:

- appreciate that 2π radians corresponds to $360°$
- define linear and angular velocity
- perform calculations on linear and angular velocity using $v = \omega r$ and $\omega = 2\pi n$
- define linear and angular acceleration
- perform calculations on linear and angular acceleration using $v_2 = v_1 + \alpha t$, $\omega_2 = \omega_1 + \alpha t$ and $a = r\alpha$
- select appropriate equations of motion when performing simple calculations
- appreciate the difference between scalar and vector quantities
- use vectors to determine relative velocities, by drawing and by calculation

23.1 Introduction

This chapter commences by defining linear and angular velocity and also linear and angular acceleration. It then derives the well-known relationships, under uniform acceleration, for displacement, velocity and acceleration, in terms of time and other parameters. The chapter then uses elementary vector analysis, similar to that used for forces in chapter 20, to determine relative velocities. This chapter deals with the basics of kinematics.

23.2 The radian

The unit of angular displacement is the radian, where one radian is the angle subtended at the centre of a circle by an arc equal in length to the radius, as shown in Figure 23.1.

The relationship between angle in radians θ, arc length s and radius of a circle r is:

$$s = r\boldsymbol{\theta} \tag{1}$$

Science and Mathematics for Engineering. 978-0-367-20475-4, © John Bird. Published by Taylor & Francis. All rights reserved.

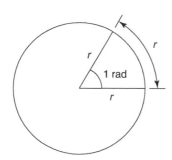

Figure 23.1

Since the arc length of a complete circle is $2\pi r$ and the angle subtended at the centre is $360°$, then from equation (1), for a complete circle,

$$2\pi r = r\theta \quad \text{or} \quad \theta = 2\pi \text{ radians}$$

Thus,   **2π radians corresponds to $360°$**   (2)

23.3   Linear and angular velocity

23.3.1   Linear velocity

Linear velocity v is defined as the rate of change of linear displacement s with respect to time t, and for motion in a straight line:

$$\text{linear velocity} = \frac{\text{change of displacement}}{\text{change of time}}$$

i.e. $$v = \frac{s}{t}$$ (3)

The unit of linear velocity is metres per second (m/s).

23.3.2   Angular velocity

The speed of revolution of a wheel or a shaft is usually measured in revolutions per minute or revolutions per second but these units do not form part of a coherent system of units. The basis used in SI units is the angle turned through (in radians) in one second.

Angular velocity is defined as the rate of change of angular displacement θ, with respect to time t, and for an object rotating about a fixed axis at a constant speed:

$$\text{angular velocity} = \frac{\text{angle turned through}}{\text{time taken}}$$

i.e. $$\omega = \frac{\theta}{t}$$ (4)

The unit of angular velocity is radians per second (rad/s). An object rotating at a constant speed of n revolutions per second subtends an angle of $2\pi n$ radians in one second, that is, its angular velocity,

$$\omega = 2\pi\text{n rad/s}$$ (5)

From equation (1), $s = r\theta$, and from equation (4), $\theta = \omega t$, hence

$$s = r\omega t \quad \text{or} \quad \frac{s}{t} = \omega r$$

However, from equation (3),

$$v = \frac{s}{t}$$

hence $$v = \omega r$$ (6)

Equation (6) gives the relationship between linear velocity v and angular velocity ω.

Problem 1.   A wheel of diameter 540 mm is rotating at $(1500/\pi)$ rev/min. Calculate the angular velocity of the wheel and the linear velocity of a point on the rim of the wheel

From equation (5), angular velocity $\omega = 2\pi n$, where n is the speed of revolution in revolutions per second, i.e.

$$n = \frac{1500}{60\pi} \text{ revolutions per second.}$$

Thus, **angular velocity** $\omega = 2\pi\left(\dfrac{1500}{60\pi}\right) = \mathbf{50\ rad/s}$

The linear velocity of a point on the rim, $v = \omega r$, where r is the radius of the wheel, i.e. $r = 0.54/2$ or 0.27 m.

Thus, **linear velocity** $v = \omega r = 50 \times 0.27 = \mathbf{13.5\ m/s}$

Problem 2.   A car is travelling at 64.8 km/h and has wheels of diameter 600 mm. (a) Find the angular velocity of the wheels in both rad/s and rev/min. (b) If the speed remains constant for 1.44 km, determine the number of revolutions made by a wheel, assuming no slipping occurs

(a) $64.8 \text{ km/h} = 64.8 \dfrac{\text{km}}{\text{h}} \times 1000 \dfrac{\text{m}}{\text{km}} \times \dfrac{1}{3600} \dfrac{\text{h}}{\text{s}}$

$$= \frac{64.8}{3.6} \text{ m/s} = 18 \text{ m/s}$$

i.e. the linear velocity v is 18 m/s.

The radius of a wheel is $(600/2)\,mm = 0.3\,m$

From equation (6), $v = \omega r$, hence $\omega = v/r$

i.e. the **angular velocity** $\omega = \dfrac{18}{0.3} = 60\ \textbf{rad/s.}$

From equation (5), angular velocity $\omega = 2\pi n$, where n is in revolutions per second. Hence $n = \omega/2\pi$ and angular speed of a wheel in revolutions per minute is $60\omega/2\pi$; but $\omega = 60\,rad/s$,

hence **angular speed** $= \dfrac{60 \times 60}{2\pi}$

$= \textbf{573 revolutions per minute (rpm).}$

(b) From equation (3), time taken to travel 1.44 km at a constant speed of 18 m/s is:

$$\frac{1440\,m}{18\,m/s} = 80s$$

Since a wheel is rotating at 573 revolutions per minute, then in 80/60 minutes it makes

$$\frac{573 \times 80}{60} = \textbf{764 revolutions.}$$

Now try the following Practice Exercise

Practice Exercise 107   Further problems on linear and angular velocity (answers on page 542)

1. A pulley driving a belt has a diameter of 360 mm and is turning at $2700/\pi$ revolutions per minute. Find the angular velocity of the pulley and the linear velocity of the belt assuming that no slip occurs

2. A bicycle is travelling at 36 km/h and the diameter of the wheels of the bicycle is 500 mm. Determine the angular velocity of the wheels of the bicycle and the linear velocity of a point on the rim of one of the wheels

23.4 Linear and angular acceleration

Linear acceleration a is defined as the rate of change of linear velocity with respect to time. For an object whose linear velocity is increasing uniformly:

$$\text{linear acceleration} = \frac{\text{change of linear velocity}}{\text{time taken}}$$

i.e. $\qquad a = \dfrac{v_2 - v_1}{t} \qquad (7)$

The unit of linear acceleration is metres per second squared (m/s^2). Rewriting equation (7) with v_2 as the subject of the formula gives:

$$v_2 = v_1 + at \qquad (8)$$

where $v_2 =$ final velocity and $v_1 =$ initial velocity.

Angular acceleration α is defined as the rate of change of angular velocity with respect to time. For an object whose angular velocity is increasing uniformly:

$$\text{angular acceleration} = \frac{\text{change of angular velocity}}{\text{time taken}}$$

i.e. $\qquad \alpha = \dfrac{\omega_2 - \omega_1}{t} \qquad (9)$

The unit of angular acceleration is radians per second squared (rad/s^2). Rewriting equation (9) with ω_2 as the subject of the formula gives:

$$\omega_2 = \omega_1 + \alpha t \qquad (10)$$

where $\omega_2 =$ final angular velocity and $\omega_1 =$ initial angular velocity.

From equation (6), $v = \omega r$. For motion in a circle having a constant radius r, $v_2 = \omega_2 r$ and $v_1 = \omega_1 r$, hence equation (7) can be rewritten as:

$$a = \frac{\omega_2 r - \omega_1 r}{t} = \frac{r(\omega_2 - \omega_1)}{t}$$

But from equation (9),

$$\frac{\omega_2 - \omega_1}{t} = \alpha$$

Hence $\qquad\qquad a = r\alpha \qquad (11)$

Problem 3. The speed of a shaft increases uniformly from 300 revolutions per minute to 800 revolutions per minute in 10 s. Find the angular acceleration, correct to 3 significant figures

From equation (9), $\alpha = \dfrac{\omega_2 - \omega_1}{t}$

Initial angular velocity

$\omega_1 = 300\,\text{rev/min} = 300/60\,\text{rev/s} = \dfrac{300 \times 2\pi}{60}\,\text{rad/s},$

final angular velocity

$\omega_2 = \dfrac{800 \times 2\pi}{60}\,\text{rad/s and time } t = 10\,\text{s}.$

Hence, **angular acceleration**

$$\alpha = \dfrac{\dfrac{800 \times 2\pi}{60} - \dfrac{300 \times 2\pi}{60}}{10}\,\text{rad/s}^2$$

$$= \dfrac{500 \times 2\pi}{60 \times 10} = \mathbf{5.24\ rad/s^2}$$

Problem 4. If the diameter of the shaft in Problem 3 is 50 mm, determine the linear acceleration of the shaft on its external surface, correct to 3 significant figures

From equation (11), $a = r\alpha$

The shaft radius is $\dfrac{50}{2}\,\text{mm} = 25\,\text{mm} = 0.025\,\text{m}$, and the angular acceleration, $\alpha = 5.24\,\text{rad/s}^2$, thus the **linear acceleration** $a = r\alpha = 0.025 \times 5.24 = \mathbf{0.131\ m/s^2}$

Now try the following Practice Exercise

Practice Exercise 108   Further problems on linear and angular acceleration (answers on page 542)

1. A flywheel rotating with an angular velocity of 200 rad/s is uniformly accelerated at a rate of 5 rad/s^2 for 15 s. Find the final angular velocity of the flywheel both in rad/s and revolutions per minute

2. A disc accelerates uniformly from 300 revolutions per minute to 600 revolutions per minute in 25 s. Determine its angular acceleration and the linear acceleration of a point on the rim of the disc, if the radius of the disc is 250 mm

23.5   Further equations of motion

From equation (3), $s = vt$, and if the linear velocity is changing uniformly from v_1 to v_2, then s = mean linear velocity × time

i.e.
$$s = \left(\dfrac{v_1 + v_2}{2}\right)t \qquad (12)$$

From equation (4), $\theta = \omega t$, and if the angular velocity is changing uniformly from ω_1 to ω_2, then θ = mean angular velocity × time

i.e.
$$\theta = \left(\dfrac{\omega_1 + \omega_2}{2}\right)t \qquad (13)$$

Two further equations of linear motion may be derived from equations (8) and (11):

$$s = v_1 t + \dfrac{1}{2}at^2 \qquad (14)$$

and
$$v_2^2 = v_1^2 + 2as \qquad (15)$$

Two further equations of angular motion may be derived from equations (10) and (13):

$$\theta = \omega_1 t + \dfrac{1}{2}\alpha t^2 \qquad (16)$$

and
$$\omega_2^2 = \omega_1^2 + 2\alpha\theta \qquad (17)$$

Table 23.1 on page 248 summarises the principal equations of linear and angular motion for uniform changes in velocities and constant accelerations, and also gives the relationships between linear and angular quantities.

Problem 5. The speed of a shaft increases uniformly from 300 rev/min to 800 rev/min in 10 s. Find the number of revolutions made by the shaft during the 10 s it is accelerating

From equation (13), angle turned through,

$$\theta = \left(\dfrac{\omega_1 + \omega_2}{2}\right)t = \left(\dfrac{\dfrac{300 \times 2\pi}{60} + \dfrac{800 \times 2\pi}{60}}{2}\right)(10)\,\text{rad}$$

However, there are 2π radians in 1 revolution, hence,

number of revolutions

$$= \left(\dfrac{\dfrac{300 \times 2\pi}{60} + \dfrac{800 \times 2\pi}{60}}{2}\right)\left(\dfrac{10}{2\pi}\right)$$

$$= \dfrac{1}{2}\left(\dfrac{1100}{60}\right)(10) = \dfrac{1100}{12}$$

$$= \mathbf{91.67\ revolutions.}$$

Table 23.1

s = arc length (m)	r = radius of circle (m)
t = time (s)	θ = angle (rad)
v = linear velocity (m/s)	ω = angular velocity (rad/s)
v_1 = initial linear velocity (m/s)	ω_1 = initial angular velocity (rad/s)
v_2 = final linear velocity (m/s)	ω_2 = final angular velocity (rad/s)
a = linear acceleration (m/s²)	α = angular acceleration (rad/s²)
n = speed of revolution (rev/s)	

Equation number	Linear motion	Angular motion
(1)		$s = r\theta$ m
(2)		2π rad $= 360°$
(3) and (4)	$v = \dfrac{s}{t}$	$\omega = \dfrac{\theta}{t}$ rad/s
(5)		$\omega = 2\pi n$ rad/s
(6)	$v = \omega r$ m/s²	
(7) and (9)	$a = \dfrac{v_2 - v_1}{t}$	$\alpha = \dfrac{\omega_2 - \omega_1}{t}$
(8) and (10)	$v_2 = (v_1 + at)$ m/s	$\omega_2 = (\omega_1 + \alpha t)$ rad/s
(11)	$a = r\alpha$ m/s²	
(12) and (13)	$s = \left(\dfrac{v_1 + v_2}{2}\right)t$	$\theta = \left(\dfrac{\omega_1 + \omega_2}{2}\right)t$
(14) and (16)	$s = v_1 t + \dfrac{1}{2}at^2$	$\theta = \omega_1 t + \dfrac{1}{2}\alpha t^2$
(15) and (17)	$v_2^2 = v_1^2 + 2as$	$\omega_2^2 = \omega_1^2 + 2\alpha\theta$

Problem 6.   The shaft of an electric motor, initially at rest, accelerates uniformly for 0.4 s at 15 rad/s². Determine the angle (in radians) turned through by the shaft in this time

From equation (16),

$$\theta = \omega_1 t + \frac{1}{2}\alpha t^2$$

Since the shaft is initially at rest,

$$\omega_1 = 0 \text{ and } \theta = \frac{1}{2}\alpha t^2$$

the angular acceleration

$$\alpha = 15 \text{ rad/s}^2 \text{ and time } t = 0.4\,\text{s}$$

Hence, **angle turned through**

$$\theta = 0 + \frac{1}{2} \times 15 \times 0.4^2 = \mathbf{1.2\ rad}$$

Problem 7.   A flywheel accelerates uniformly at 2.05 rad/s² until it is rotating at 1500 rev/min. If it completes 5 revolutions during the time it is accelerating, determine its initial angular velocity in rad/s, correct to 4 significant figures

Since the final angular velocity is 1500 rev/min,

$$\omega_2 = 1500 \, \frac{\text{rev}}{\text{min}} \times \frac{1 \text{ min}}{60 \text{ s}} \times \frac{2\pi \text{ rad}}{1 \text{ rev}} = 50\pi \text{ rad/s}$$

$$5 \text{ revolutions} = 5 \text{ rev} \times \frac{2\pi \text{ rad}}{1 \text{ rev}} = 10\pi \text{ rad.}$$

From equation (17), $\omega_2^2 = \omega_1^2 + 2\alpha\theta$

i.e. $(50\pi)^2 = \omega_1^2 + (2 \times 2.05 \times 10\pi)$

from which, $\omega_1^2 = (50\pi)^2 - (2 \times 2.05 \times 10\pi)$

$= (50\pi)^2 - 41\pi = 24{,}545$

i.e. $\omega_1 = \sqrt{24{,}545} = 156.7$ rad/s

Thus, the initial angular velocity is 156.7 rad/s, correct to 4 significant figures.

Practice Exercise 109   Further problems
on equations of motion (answers on
page 542)

1. A grinding wheel makes 300 revolutions when slowing down uniformly from 1000 rad/s to 400 rad/s. Find the time for this reduction in speed

2. Find the angular retardation for the grinding wheel in question 1

3. A disc accelerates uniformly from 300 revolutions per minute to 600 revolutions per minute in 25 s. Calculate the number of revolutions the disc makes during this accelerating period

4. A pulley is accelerated uniformly from rest at a rate of 8 rad/s². After 20 s the acceleration stops and the pulley runs at constant speed for 2 min, and then the pulley comes uniformly to rest after a further 40 s. Calculate: (a) the angular velocity after the period of acceleration, (b) the deceleration, (c) the total number of revolutions made by the pulley

23.6 Relative velocity

As stated in chapter 12, quantities used in engineering and science can be divided into two groups:

(a) **Scalar quantities** have a size or magnitude only and need no other information to specify them. Thus 20 centimetres, 5 seconds, 3 litres and 4 kilograms are all examples of scalar quantities.

(b) **Vector quantities** have both a size (or magnitude), and a direction, called the line of action of the quantity. Typical vector quantities are velocity, acceleration and force. Thus, a velocity of 30 km/h due west, and an acceleration of 7 m/s² acting vertically downwards, are both vector quantities.

A vector quantity is represented by a straight line lying along the line of action of the quantity, and having a length that is proportional to the size of the quantity, as shown in chapter 19. Thus **ab** in Figure 23.2 represents a velocity of 20 m/s, whose line of action is due west. The bold letters **ab** indicate a vector quantity and the order of the letters indicate that the line of action is from a to b.

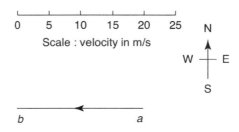

Figure 23.2

Consider two aircraft A and B flying at a constant altitude, A travelling due north at 200 m/s and B travelling 30° east of north, written N 30° E, at 300 m/s, as shown in Figure 23.3.

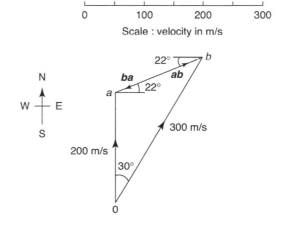

Figure 23.3

Relative to a fixed point 0, **0a** represents the velocity of A and **0b** the velocity of B. The velocity of B relative to

A, that is the velocity at which B seems to be travelling to an observer on A, is given by ab, and by measurement is 160 m/s in a direction E 22° N. The velocity of A relative to B, that is, the velocity at which A seems to be travelling to an observer on B, is given by ba and by measurement is 160 m/s in a direction W 22° S.

Problem 8. Two cars are travelling on horizontal roads in straight lines, car A at 70 km/h at N 10° E and car B at 50 km/h at W 60° N. Determine, by drawing a vector diagram to scale, the velocity of car A relative to car B

With reference to Figure 23.4(a), **0a** represents the velocity of car A relative to a fixed point 0, and **0b** represents the velocity of car B relative to a fixed point 0. The velocity of car A relative to car B is given by **ba** and by measurement is **45 km/h in a direction of E 35° N**.

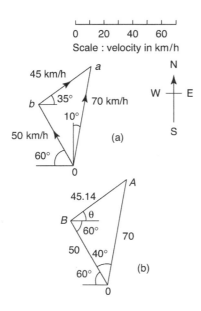

Figure 23.4

Problem 9. Verify the result obtained in Problem 8 by calculation

The triangle shown in Figure 23.4(b) is similar to the vector diagram shown in Figure 23.4(a). Angle $B0A$ is 40°. Using the cosine rule (see chapter 11):

$$BA^2 = 50^2 + 70^2 - 2 \times 50 \times 70 \times \cos 40°$$

from which, **$BA = 45.14$**

Using the sine rule:

$$\frac{50}{\sin \angle BA0} = \frac{45.14}{\sin 40°} \text{ (from chapter 11)}$$

from which,

$$\sin \angle BA0 = \frac{50 \sin 40°}{45.14} = 0.7120$$

Hence, angle $BA0 = 45.40°$; thus,
angle $AB0 = 180° - (40° + 45.40°) = 94.60°$,
and angle $\theta = 94.60° - 60° = 34.60°$
Thus, **ba** is **45.14 km/h in a direction E 34.60° N by calculation**.

Problem 10. A crane is moving in a straight line with a constant horizontal velocity of 2 m/s. At the same time it is lifting a load at a vertical velocity of 5 m/s. Calculate the velocity of the load relative to a fixed point on the earth's surface

A vector diagram depicting the motion of the crane and load is shown in Figure 23.5. **0a** represents the velocity of the crane relative to a fixed point on the Earth's surface and **ab** represents the velocity of the load relative to the crane. The velocity of the load relative to the fixed point on the Earth's surface is **0b**. By Pythagoras's theorem (see chapter 11):

$$0b^2 = 0a^2 + ab^2$$
$$= 4 + 25 = 29$$

Hence $\quad 0b = \sqrt{29} = 5.385 \text{ m/s}$

$$\text{Tan } \theta = \frac{5}{2} = 2.5$$

hence, $\quad \theta = \tan^{-1} 2.5 = 68.20°$

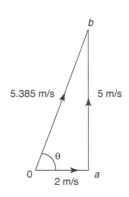

Figure 23.5

i.e. the velocity of the load relative to a fixed point on the Earth's surface is **5.385 m/s in a direction 68.20° to the motion of the crane**.

Now try the following Practice Exercises

Practice Exercise 110   Further problems on relative velocity (answers on page 542)

1. A car is moving along a straight horizontal road at 79.2 km/h and rain is falling vertically downwards at 26.4 km/h. Find the velocity of the rain relative to the driver of the car

2. Calculate the time needed to swim across a river 142 m wide when the swimmer can swim at 2 km/h in still water and the river is flowing at 1 km/h. At what angle to the bank should the swimmer swim?

3. A ship is heading in a direction N 60° E at a speed which in still water would be 20 km/h. It is carried off course by a current of 8 km/h in a direction of E 50° S. Calculate the ship's actual speed and direction

Practice Exercise 111   Short answer questions on linear and angular motion

1. State and define the unit of angular displacement

2. Write down the formula connecting an angle, arc length and the radius of a circle

3. Define linear velocity and state its unit

4. Define angular velocity and state its unit

5. Write down a formula connecting angular velocity and revolutions per second in coherent units

6. State the formula connecting linear and angular velocity

7. Define linear acceleration and state its unit

8. Define angular acceleration and state its unit

9. Write down the formula connecting linear and angular acceleration

10. Define a scalar quantity and give two examples

11. Define a vector quantity and give two examples

Practice Exercise 112   Multiple-choice questions on linear and angular motion (answers on page 542)

1. Angular displacement is measured in:
   (a) degrees   (b) radians
   (c) rev/s   (d) metres

2. An angle of $\dfrac{3\pi}{4}$ radians is equivalent to:
   (a) 270°   (b) 67.5°   (c) 135°   (d) 2.356°

3. An angle of 120° is equivalent to:
   (a) $\dfrac{2\pi}{3}$ rad   (b) $\dfrac{\pi}{3}$ rad
   (c) $\dfrac{3\pi}{4}$ rad   (d) $\dfrac{1}{3}$ rad

4. An angle of 2 rad at the centre of a circle subtends an arc length of 40 mm at the circumference of the circle. The radius of the circle is:
   (a) 40π mm   (b) 80 mm
   (c) 20 mm   (d) $(40/\pi)$ mm

5. A point on a wheel has a constant angular velocity of 3 rad/s. The angle turned through in 15 seconds is:
   (a) 45 rad   (b) 10π rad
   (c) 5 rad   (d) 90π rad

6. An angular velocity of 60 revolutions per minute is the same as:
   (a) $(1/2\pi)$ rad/s   (b) 120π rad/s
   (c) $(30/\pi)$ rad/s   (d) 2π rad/s

7. A wheel of radius 15 mm has an angular velocity of 10 rad/s. A point on the rim of the wheel has a linear velocity of:
   (a) 300π mm/s   (b) 2/3 mm/s
   (c) 150 mm/s   (d) 1.5 mm/s

8. The shaft of an electric motor is rotating at 20 rad/s and its speed is increased uniformly to 40 rad/s in 5 s. The angular acceleration of the shaft is:
   (a) 4000 rad/s^2 (b) 4 rad/s^2
   (c) 160 rad/s^2 (d) 12 rad/s^2

9. A point on a flywheel of radius 0.5 m has a uniform linear acceleration of 2 m/s^2. Its angular acceleration is:
   (a) 2.5 rad/s^2 (b) 0.25 rad/s^2
   (c) 1 rad/s^2 (d) 4 rad/s^2

   Questions 10–13 refer to the following data. A car accelerates uniformly from 10 m/s to 20 m/s over a distance of 150 m. The wheels of the car each have a radius of 250 mm.

10. The time the car is accelerating is:
    (a) 0.2 s (b) 15 s (c) 10 s (d) 5 s

11. The initial angular velocity of each of the wheels is:
    (a) 20 rad/s (b) 40 rad/s
    (c) 2.5 rad/s (d) 0.04 rad/s

12. The angular acceleration of each of the wheels is:
    (a) 1 rad/s^2 (b) 0.25 rad/s^2
    (c) 400 rad/s^2 (d) 4 rad/s^2

13. The linear acceleration of a point on each of the wheels is:
    (a) 1 m/s^2 (b) 4 m/s^2
    (c) 3 m/s^2 (d) 100 m/s^2

**For fully worked solutions to each of the problems in Exercises 107 to 112 in this chapter,
go to the website:**
www.routledge.com/cw/bird

Chapter 24

Friction

Why it is important to understand: **Friction**

When a block is placed on a flat surface and sufficient force is applied to the block, the force being parallel to the surface, the block slides across the surface. When the force is removed, motion of the block stops: thus there is a force which resists sliding. In this chapter, both dynamic and static frictions are explained, together with the factors that affect the size and direction of frictional forces. A low coefficient of friction is desirable in bearings, pistons moving within cylinders and on ski runs; however, for a force being transmitted by belt drives and braking systems, a high value of coefficient is necessary. Advantages and disadvantages of frictional forces are discussed. Knowledge of friction is of importance for the static and dynamic behaviour of stationary and moving bodies.

At the end of this chapter, you should be able to:

- understand dynamic or sliding friction
- appreciate factors which affect the size and direction of frictional forces
- define coefficient of friction, μ
- perform calculations involving $F = \mu N$
- state practical applications of friction
- state advantages and disadvantages of frictional forces

24.1 Introduction to friction

When an object, such as a block of wood, is placed on a floor and sufficient force is applied to the block, the force being parallel to the floor, the block slides across the floor. When the force is removed, motion of the block stops; thus there is a force which resists sliding. This force is called **dynamic** or **sliding friction**. A force may

be applied to the block, which is insufficient to move it. In this case, the force resisting motion is called the **static friction** or **stiction**. Thus there are two categories into which a frictional force may be split:

(i) dynamic or sliding friction force which occurs when motion is taking place, and

(ii) static friction force which occurs before motion takes place.

Science and Mathematics for Engineering. 978-0-367-20475-4, © John Bird. Published by Taylor & Francis. All rights reserved.

There are three factors that affect the size and direction of frictional forces.

(i) The size of the frictional force depends on the type of surface (a block of wood slides more easily on a polished metal surface than on a rough concrete surface).

(ii) The size of the frictional force depends on the size of the force acting at right angles to the surfaces in contact, called the **normal force**; thus, if the weight of a block of wood is doubled, the frictional force is doubled when it is sliding on the same surface.

(iii) The direction of the frictional force is always opposite to the direction of motion. Thus the frictional force opposes motion, as shown in Figure 24.1.

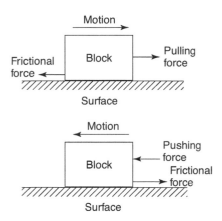

Figure 24.1

24.2 Coefficient of friction

The **coefficient of friction** μ is a measure of the amount of friction existing between two surfaces. A low value of coefficient of friction indicates that the force required for sliding to occur is less than the force required when the coefficient of friction is high. The value of the coefficient of friction is given by:

$$\mu = \frac{\text{frictional force } (F)}{\text{normal force } (N)}$$

Transposing gives: frictional force $= \mu \times$ normal force i.e.

$$F = \mu N$$

The direction of the forces given in this equation is as shown in Figure 24.2.

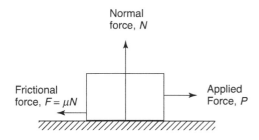

Figure 24.2

The coefficient of friction is the ratio of a force to a force, and hence has no units. Typical values for the coefficient of friction when sliding is occurring, i.e. the dynamic coefficient of friction, are:

For polished oiled metal surfaces	less than 0.1
For glass on glass	0.4
For rubber on tarmac	close to 1.0

The coefficient of friction μ for dynamic friction is, in general, a little less than that for static friction. However, for dynamic friction, μ increases with speed; additionally, it is dependent on the area of the surface in contact.

Problem 1. A block of steel requires a force of 10.4 N applied parallel to a steel plate to keep it moving with constant velocity across the plate. If the normal force between the block and the plate is 40 N, determine the dynamic coefficient of friction

As the block is moving at constant velocity, the force applied must be that required to overcome frictional forces, i.e. frictional force $F = 10.4$ N; the normal force is 40 N, and since $F = \mu N$,

$$\mu = \frac{F}{N} = \frac{10.4}{40} = 0.26$$

i.e. **the dynamic coefficient of friction is 0.26**

Problem 2. The surface between the steel block and plate of Problem 1 is now lubricated and the dynamic coefficient of friction falls to 0.12. Find the new value of force required to push the block at a constant speed

The normal force depends on the weight of the block and remains unaltered at 40 N. The new value of the dynamic coefficient of friction is 0.12 and since the frictional force $F = \mu N$,

$$F = 0.12 \times 40 = 4.8\,\text{N}$$

The block is sliding at constant speed, thus the force required to overcome the frictional force is also 4.8 N, i.e. **the required applied force is 4.8 N**

Problem 3. The material of a brake is being tested and it is found that the dynamic coefficient of friction between the material and steel is 0.91. Calculate the normal force when the frictional force is 0.728 kN

The dynamic coefficient of friction $\mu = 0.91$ and the frictional force

$$F = 0.728\,\text{kN} = 728\,\text{N}.$$

Since $F = \mu N$, then normal force

$$N = \frac{F}{\mu} = \frac{728}{0.91} = 800\,\text{N}$$

i.e. **the normal force is 800 N**

Now try the following Practice Exercise

Practice Exercise 113   Further problems on the coefficient of friction (answers on page 542)

1. The coefficient of friction of a brake pad and a steel disc is 0.82. Determine the normal force between the pad and the disc if the frictional force required is 1025 N

2. A force of 0.12 kN is needed to push a bale of cloth along a chute at a constant speed. If the normal force between the bale and the chute is 500 N, determine the dynamic coefficient of friction

3. The normal force between a belt and its driver wheel is 750 N. If the static coefficient of friction is 0.9 and the dynamic coefficient of friction is 0.87, calculate (a) the maximum force which can be transmitted, and (b) maximum force which can be transmitted when the belt is running at a constant speed

24.3   Applications of friction

In some applications, a low coefficient of friction is desirable, for example, in bearings, pistons moving within cylinders, on ski runs and so on. However, for such applications as force being transmitted by belt drives and braking systems, a high value of coefficient is necessary.

Problem 4. State three advantages, and three disadvantages of frictional forces

Instances where frictional forces are an advantage include:

(i) Almost all fastening devices, which rely on frictional forces to keep them in place once secured, examples being screws, nails, nuts, clips and clamps.

(ii) Satisfactory operation of brakes and clutches, which rely on frictional forces being present.

(iii) In the absence of frictional forces, most accelerations along a horizontal surface are impossible; for example, a person's shoes just slip when walking is attempted and the tyres of a car just rotate with no forward motion of the car being experienced.

Disadvantages of frictional forces include:

(i) Energy is wasted in the bearings associated with shafts, axles and gears due to heat being generated.

(ii) Wear is caused by friction, for example, in shoes, brake lining materials and bearings.

(iii) Energy is wasted when motion through air occurs (it is much easier to cycle with the wind rather than against it).

Problem 5. Discuss briefly two design implications that arise due to frictional forces and how lubrication may or may not help

(i) Bearings are made of an alloy called white metal, which has a relatively low melting point. When the rotating shaft rubs on the white metal bearing, heat is generated by friction, often in one spot and the white metal may melt in this area, rendering the bearing useless. Adequate lubrication (oil or grease) separates the shaft from the white metal, keeps the coefficient of friction small and prevents damage to the bearing. For very large bearings, oil is pumped under pressure into the bearing and the oil is used to rcmove the heat generated, often passing through oil coolers before being re-circulated. Designers should ensure that the heat generated by friction can be dissipated.

(ii)   Wheels driving belts, to transmit force from one place to another, are used in many workshops. The coefficient of friction between the wheel and the belt must be high, and it may be increased by dressing the belt with a tar-like substance. Since frictional force is proportional to the normal force, a slipping belt is made more efficient by tightening it, thus increasing the normal and hence the frictional force. Designers should incorporate some belt tension mechanism into the design of such a system.

> **Problem 6.**   Explain what is meant by the terms (a) the limiting or static coefficient of friction, and (b) the sliding or dynamic coefficient of friction

(a)   When an object is placed on a surface and a force is applied to it in a direction parallel to the surface, if no movement takes place, then the applied force is balanced exactly by the frictional force. As the size of the applied force is increased, a value is reached such that the object is just on the point of moving. The limiting or static coefficient of friction is given by the ratio of this applied force to the normal force, where the normal force is the force acting at right angles to the surfaces in contact.

(b)   Once the applied force is sufficient to overcome the stiction or static friction, its value can be reduced slightly and the object moves across the surface. A particular value of the applied force is then sufficient to keep the object moving at a constant velocity. The sliding or dynamic coefficient of friction is the ratio of the applied force, to maintain constant velocity, to the normal force.

Now try the following Practice Exercises

Practice Exercise 114   Short answer questions on friction

1.   The .... of frictional force depends on the .... of surfaces in contact

2.   The .... of frictional force depends on the size of the .... to the surfaces in contact

3.   The .... of frictional force is always .... to the direction of motion

4.   The coefficient of friction between surfaces should be a .... value for materials concerned with bearings

5.   The coefficient of friction should have a .... value for materials concerned with braking systems

6.   The coefficient of dynamic or sliding friction is given by $\dfrac{\cdots\cdots}{\cdots\cdots}$

7.   The coefficient of static or limiting friction is given by $\dfrac{\cdots\cdots}{\cdots\cdots}$ when .... is just about to take place

8.   Lubricating surfaces in contact result in a .... of the coefficient of friction

9.   Briefly discuss the factors affecting the size and direction of frictional forces

10.   Name three practical applications where a low value of coefficient of friction is desirable and state briefly how this is achieved in each case

11.   Name three practical applications where a high value of coefficient of friction is required when transmitting forces and discuss how this is achieved

12.   For an object on a surface, two different values of coefficient of friction are possible. Give the names of these two coefficients of friction and state how their values may be obtained

Practice Exercise 115   Multiple-choice questions on friction (answers on page 542)

1.   A block of metal requires a frictional force F to keep it moving with constant velocity across a surface. If the coefficient of friction is μ, then the normal force N is given by:

(a) $\dfrac{\mu}{F}$ (b) μF (c) $\dfrac{F}{\mu}$ (d) F

2. The unit of the linear coefficient of friction is:

(a) newtons   (b) radians
(c) dimensionless   (d) newton/metre

Questions 3–7 refer to the statements given below. Select the statement required from each group given.

(a) The coefficient of friction depends on the type of surfaces in contact

(b) The coefficient of friction depends on the force acting at right angles to the surfaces in contact

(c) The coefficient of friction depends on the area of the surfaces in contact

(d) Frictional force acts in the opposite direction to the direction of motion

(e) Frictional force acts in the direction of motion

(f) A low value of coefficient of friction is required between the belt and the wheel in a belt drive system

(g) A low value of coefficient of friction is required for the materials of a bearing

(h) The dynamic coefficient of friction is given by (normal force)/(frictional force) at constant speed

(i) The coefficient of static friction is given by (applied force) $\div$ (frictional force) as sliding is just about to start

(j) Lubrication results in a reduction in the coefficient of friction

3. Which statement is false from (a), (b), (f) and (i)?

4. Which statement is false from (b), (e), (g) and (j)?

5. Which statement is true from (c), (f), (h) and (i)?

6. Which statement is false from (b), (c), (e) and (j)?

7. Which statement is false from (a), (d), (g) and (h)?

8. The normal force between two surfaces is 100 N and the dynamic coefficient of friction is 0.4. The force required to maintain a constant speed of sliding is:

(a) 100.4 N   (b) 40 N
(c) 99.6 N   (d) 250 N

9. The normal force between two surfaces is 50 N and the force required to maintain a constant speed of sliding is 25 N. The dynamic coefficient of friction is:

(a) 25   (b) 2
(c) 75   (d) 0.5

10. The maximum force, which can be applied to an object without sliding occurring, is 60 N, and the static coefficient of friction is 0.3. The normal force between the two surfaces is:

(a) 200 N   (b) 18 N
(c) 60.3 N   (d) 59.7 N

For fully worked solutions to each of the problems in Exercises 113 to 115 in this chapter, go to the website:
www.routledge.com/cw/bird

Chapter 25

Simple machines

Why it is important to understand: Simple machines

This chapter commences by defining load, effort, mechanical advantage, velocity ratio and efficiency, where efficiency is then defined in terms of mechanical advantage and velocity ratio. These terms, together with other terms defined in earlier chapters, are applied to simple and quite complex pulley systems, screw-jacks, gear trains and levers. This chapter is fundamental to the study of the behaviour of machines and is thus important in studying the fundamental concepts of machine motion and behaviour.

At the end of this chapter, you should be able to:

- define a simple machine
- define force ratio, movement ratio, efficiency and limiting efficiency
- understand and perform calculations with pulley systems
- understand and perform calculations with a simple screw-jack
- understand and perform calculations with gear trains
- understand and perform calculations with levers

25.1 Machines

A machine is a device that can change the magnitude or line of action, or both magnitude and line of action of a force. A simple machine usually amplifies an input force, called the **effort**, to give a larger output force, called the **load**. Some typical examples of simple machines include pulley systems, screw-jacks, gear trains and levers. This chapter is fundamental to the study of the behaviour of machines.

25.2 Force ratio, movement ratio and efficiency

The **force ratio** or **mechanical advantage** is defined as the ratio of load to effort, i.e.

$$\text{Force ratio} = \frac{\text{load}}{\text{effort}} = \text{mechanical advantage} \quad (1)$$

Science and Mathematics for Engineering. 978-0-367-20475-4, © John Bird. Published by Taylor & Francis. All rights reserved.

Since both load and effort are measured in newtons, force ratio is a ratio of the same units and thus is a dimension-less quantity.

The **movement ratio** or **velocity ratio** is defined as the ratio of the distance moved by the effort to the distance moved by the load, i.e.

$$\textbf{Movement ratio} = \frac{\textbf{distance moved by the effort}}{\textbf{distance moved by the load}}$$

$$= \textbf{velocity ratio} \qquad (2)$$

Since the numerator and denominator are both measured in metres, movement ratio is a ratio of the same units and thus is a dimension-less quantity.

The **efficiency of a simple machine** is defined as the ratio of the force ratio to the movement ratio, i.e.

$$\text{Efficiency} = \frac{\text{force ratio}}{\text{movement ratio}} = \frac{\text{mechanical advantage}}{\text{velocity ratio}}$$

Since the numerator and denominator are both dimension-less quantities, efficiency is a dimension-less quantity. It is usually expressed as a percentage, thus:

$$\textbf{Efficiency} = \frac{\textbf{force ratio}}{\textbf{movement ratio}} \times \textbf{100\%} \qquad (3)$$

Due to the effects of friction and inertia associated with the movement of any object, some of the input energy to a machine is converted into heat and losses occur. Since losses occur, the energy output of a machine is less than the energy input, thus the mechanical efficiency of any machine cannot reach 100%.

For simple machines, the relationship between effort and load is of the form: $F_e = aF_l + b$, where F_e is the effort, F_l is the load and a and b are constants.

From equation (1),

$$\text{force ratio} = \frac{\text{load}}{\text{effort}} = \frac{F_l}{F_e} = \frac{F_l}{aF_l + b}$$

Dividing both numerator and denominator by F_l gives:

$$\frac{F_l}{aF_l + b} = \frac{1}{a + \dfrac{b}{F_l}}$$

When the load is large, F_l is large and $\dfrac{b}{F_l}$ is small compared with a. The force ratio then becomes

approximately equal to $\dfrac{1}{a}$ and is called the **limiting force ratio**, i.e.

$$\textbf{limiting ratio} = \frac{1}{a}$$

The limiting efficiency of a simple machine is defined as the ratio of the limiting force ratio to the movement ratio, i.e.

$$\textbf{Limiting efficiency} = \frac{1}{a \times \textbf{movement ratio}} \times \textbf{100\%}$$

where a is the constant for the **law of the machine**:

$$F_e = aF_l + b$$

Due to friction and inertia, the limiting efficiency of simple machines is usually well below 100%.

Problem 1. A simple machine raises a load of 160 kg through a distance of 1.6 m. The effort applied to the machine is 200 N and moves through a distance of 16 m. Taking g as 9.8 m/s^2, determine the force ratio, movement ratio and efficiency of the machine

From equation (1),

$$\textbf{force ratio} = \frac{\text{load}}{\text{effort}} = \frac{160\,\text{kg}}{200\,\text{N}}$$

$$= \frac{160 \times 9.81\,\text{N}}{200\,\text{N}} = \textbf{7.84}$$

From equation (2),

$$\textbf{movement ratio} = \frac{\text{distance moved by the effort}}{\text{distance moved by the load}}$$

$$= \frac{16\,\text{m}}{1.6\,\text{m}} = \textbf{10}$$

From equation (3),

$$\textbf{efficiency} = \frac{\text{force ratio}}{\text{movement ratio}} \times 100\%$$

$$= \frac{7.84}{10} \times 100 = \textbf{78.4\%}$$

Problem 2. For the simple machine of Problem 1, determine: (a) the distance moved by the effort to move the load through a distance of 0.9 m, (b) the effort which would be required to raise a load of 200 kg, assuming the same efficiency, (c) the efficiency if, due to lubrication, the effort to raise the 160 kg load is reduced to 180 N

(a) Since the movement ratio is 10, then from equation (2), **distance moved by the effort**

$$= 10 \times \text{distance moved by the load}$$

$$= 10 \times 0.9 = \mathbf{9\,m}$$

(b) Since the force ratio is 7.84, then from equation (1),

$$\mathbf{effort} = \frac{\text{load}}{7.84} = \frac{200 \times 9.8}{7.84} = \mathbf{250\,N}$$

(c) The new force ratio is given by

$$\frac{\text{load}}{\text{effort}} = \frac{160 \times 9.8}{180} = 8.711$$

Hence, **the new efficiency after lubrication**

$$= \frac{8.711}{10} \times 100 = \mathbf{87.11\%}$$

Problem 3. In a test on a simple machine, the effort/load graph was a straight line of the form $F_e = aF_l + b$. Two values lying on the graph were at $F_e = 10\,$N, $F_l = 30\,$N, and at $F_e = 74\,$N, $F_l = 350\,$N. The movement ratio of the machine was 17. Determine: (a) the limiting force ratio, (b) the limiting efficiency of the machine

(a) The equation $F_e = aF_l + b$ is of the form $y = mx + c$, where m is the gradient of the graph. The slope of the line passing through points (x_1, y_1) and (x_2, y_2) of the graph $y = mx + c$ is given by:

$$m = \frac{y_2 - y_1}{x_2 - x_1} \text{ (from chapter 10, page 100).}$$

Thus for $F_e = aF_l + b$, the slope a is given by:

$$a = \frac{74 - 10}{350 - 30} = \frac{64}{320} = 0.2$$

The **limiting force ratio** is $\dfrac{1}{a}$, that is $\dfrac{1}{0.2} = \mathbf{5}$

(b) The **limiting efficiency**

$$= \frac{1}{a \times \text{movement ratio}} \times 100$$

$$= \frac{1}{0.2 \times 17} \times 100$$

$$= \mathbf{29.4\%}$$

Now try the following Practice Exercise

Practice Exercise 116   Further problems on force ratio, movement ratio and efficiency (answers on page 491)

1. A simple machine raises a load of 825 N through a distance of 0.3 m. The effort is 250 N and moves through a distance of 3.3 m. Determine: (a) the force ratio, (b) the movement ratio, (c) the efficiency of the machine at this load

2. The efficiency of a simple machine is 50%. If a load of 1.2 kN is raised by an effort of 300 N, determine the movement ratio

3. An effort of 10 N applied to a simple machine moves a load of 40 N through a distance of 100 mm, the efficiency at this load being 80%. Calculate: (a) the movement ratio, (b) the distance moved by the effort

4. The effort required to raise a load using a simple machine, for various values of load is as shown:

Load F_l (N)	2050	4120	7410	8240	10,300
Effort F_e (N)	252	340	465	505	580

If the movement ratio for the machine is 30, determine (a) the law of the machine, (b) the limiting force ratio, (c) the limiting efficiency

5. For the data given in question 4, determine the values of force ratio and efficiency for each value of the load. Hence plot graphs of effort, force ratio and efficiency to a base of load. From the graphs, determine the effort required to raise a load of 6 kN and the efficiency at this load

25.3 Pulleys

A **pulley system** is a simple machine. A single-pulley system, shown in Figure 25.1(a), changes the line of action of the effort, but does not change the magnitude of the force. A two-pulley system, shown in Figure 25.1(b), changes both the line of action and the magnitude of the force.

Theoretically, each of the ropes marked (i) and (ii) share the load equally, thus the theoretical effort is only half of the load, i.e. the theoretical force ratio is 2. In practice the actual force ratio is less than 2 due to losses. A three-pulley system is shown in Figure 25.1(c). Each of the ropes marked (i), (ii) and (iii) carry one-third of the load, thus the theoretical force ratio is 3. In general, for a multiple pulley system having a total of n pulleys, the theoretical force ratio is n. Since the theoretical efficiency of a pulley system (neglecting losses) is 100 and since from equation (3),

$$\text{efficiency} = \frac{\text{force ratio}}{\text{movement ratio}} \times 100\%$$

it follows that when the force ratio is n,

$$100 = \frac{n}{\text{movement ratio}} \times 100\%$$

that is, the movement ratio is also n.

Problem 4. A load of 80 kg is lifted by a three-pulley system similar to that shown in Figure 25.1(c) and the applied effort is 392 N. Calculate (a) the force ratio, (b) the movement ratio, (c) the efficiency of the system. Take g to be 9.8 m/s^2

(a) From equation (1), the force ratio is given by $\dfrac{\text{load}}{\text{effort}}$. The load is 80 kg, i.e. (80×9.8) N, hence,

$$\textbf{force ratio} = \frac{80 \times 9.8}{392} = \textbf{2}$$

(b) From above, for a system having n pulleys, the movement ratio is n. Thus, for a three-pulley system, the **movement ratio is 3**

(c) From equation (3),

$$\textbf{efficiency} = \frac{\text{force ratio}}{\text{movement ratio}} \times 100\%$$

$$= \frac{2}{3} \times 100 = \textbf{66.67\%}$$

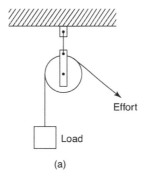

Effort

Load

(a)

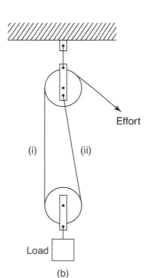

Effort

(i)   (ii)

Load

(b)

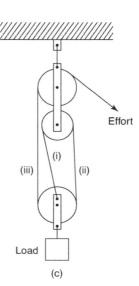

Effort

(i)

(iii)   (ii)

Load

(c)

Figure 25.1

Problem 5.   A pulley system consists of two blocks, each containing three pulleys and connected as shown in Figure 25.2. An effort of 400 N is required to raise a load of 1500 N. Determine (a) the force ratio, (b) the movement ratio, (c) the efficiency of the pulley system

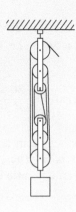

Figure 25.2

(a)   From equation (1),

$$\textbf{force ratio} = \frac{\text{load}}{\text{effort}} = \frac{1500}{400} = \textbf{3.75}$$

(b)   An n-pulley system has a movement ratio of n, hence this 6-pulley system has a **movement ratio of 6**

(c)   From equation (3),

$$\textbf{efficiency} = \frac{\text{force ratio}}{\text{movement ratio}} \times 100\%$$

$$= \frac{3.75}{6} \times 100 = \textbf{62.5\%}$$

Now try the following Practice Exercise

Practice Exercise 117   Further problems on pulleys (answers on page 542)

1.   A pulley system consists of four pulleys in an upper block and three pulleys in a lower block. Make a sketch of this arrangement showing how a movement ratio of 7 may be obtained. If the force ratio is 4.2, what is the efficiency of the pulley?

2.   A three-pulley lifting system is used to raise a load of 4.5 kN. Determine the effort required to raise this load when losses are neglected. If the actual effort required is 1.6 kN, determine the efficiency of the pulley system at this load

25.4   The screw-jack

A **simple screw-jack** is shown in Figure 25.3 and it is a simple machine since it changes both the magnitude and the line of action of a force.

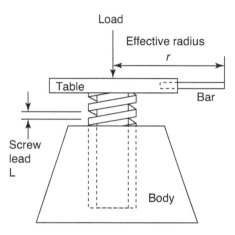

Figure 25.3

The screw of the table of the jack is located in a fixed nut in the body of the jack. As the table is rotated by means of a bar, it raises or lowers a load placed on the table. For a single-start thread, as shown, for one complete revolution of the table, the effort moves through a distance $2\pi r$, and the load moves through a distance equal to the lead of the screw, say L.

$$\textbf{Movement ratio} = \frac{2\pi r}{L} \qquad (4)$$

Problem 6.   A screw-jack is being used to support the axle of a car, the load on it being 2.4 kN. The screw jack has an effort of effective radius 200 mm and a single-start square thread, having a lead of 5 mm. Determine the efficiency of the jack if an effort of 60 N is required to raise the car axle

From equation (3),

$$\text{efficiency} = \frac{\text{force ratio}}{\text{movement ratio}} \times 100\%$$

where force ratio $= \dfrac{\text{load}}{\text{effort}} = \dfrac{2400\,\text{N}}{60\,\text{N}} = 40$

From equation (4),

$$\text{movement ratio} = \frac{2\pi r}{L} = \frac{2\pi(200)\,\text{mm}}{5\,\text{mm}} = 251.3$$

Hence,

$$\textbf{efficiency} = \frac{\text{force ratio}}{\text{movement ratio}} \times 100\%$$

$$= \frac{40}{251.3} \times 100 = \textbf{15.9\%}$$

Now try the following Practice Exercise

Practice Exercise 118   Further problems on the screw-jack (answers on page 542)

1. Sketch a simple screw-jack. The single-start screw of such a jack has a lead of 6 mm and the effective length of the operating bar from the centre of the screw is 300 mm. Calculate the load which can be raised by an effort of 150 N if the efficiency at this load is 20%

2. A load of 1.7 kN is lifted by a screw-jack having a single-start screw of lead 5 mm. The effort is applied at the end of an arm of effective length 320 mm from the centre of the screw. Calculate the effort required if the efficiency at this load is 25%

25.5   Gear trains

A **simple gear train** is used to transmit rotary motion and can change both the magnitude and the line of action of a force, hence it is a simple machine. The gear train shown in Figure 25.4 consists of **spur gears** and has an effort applied to one gear, called the driver, and a load applied to the other gear, called the **follower**.

In such a system, the teeth on the wheels are so spaced that they exactly fill the circumference with a

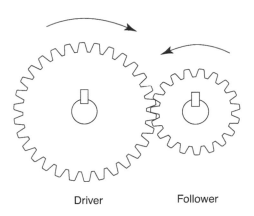

Driver   Follower

Figure 25.4

whole number of identical teeth, and the teeth on the driver and follower mesh without interference. Under these conditions, the number of teeth on the driver and follower are in direct proportion to the circumference of these wheels, i.e.

$$\frac{\textbf{number of teeth on driver}}{\textbf{number of teeth on follower}}$$
$$= \frac{\textbf{circumference of driver}}{\textbf{circumference of follower}} \qquad (5)$$

If there are, say, 40 teeth on the driver and 20 teeth on the follower, then the follower makes two revolutions for each revolution of the driver. In general:

$$\frac{\text{number of revolutions made by driver}}{\text{number of revolutions made by the follower}}$$
$$= \frac{\text{number of teeth on follower}}{\text{number of teeth on driver}} \qquad (6)$$

It follows from equation (6) that the speeds of the wheels in a gear train are inversely proportional to the number of teeth. The ratio of the speed of the driver wheel to that of the follower is the movement ratio, i.e.

$$\textbf{Movement ratio} = \frac{\textbf{speed of driver}}{\textbf{speed of follower}}$$
$$= \frac{\textbf{teeth on follower}}{\textbf{teeth on driver}} \qquad (7)$$

When the same direction of rotation is required on both the driver and the follower an **idler wheel** is used as shown in Figure 25.5.

Let the driver, idler and follower be A, B and C, respectively, and let N be the speed of rotation and T be the number of teeth. Then from equation (7),

$$\frac{N_B}{N_A} = \frac{T_A}{T_B} \quad \text{or} \quad N_A = N_B \frac{T_B}{T_A}$$

$$\text{and} \quad \frac{N_C}{N_B} = \frac{T_B}{T_C} \quad \text{or} \quad N_C = N_B \frac{T_B}{T_C}$$

Thus, $\dfrac{\text{speed of } A}{\text{speed of } C} = \dfrac{N_A}{N_C} = \dfrac{N_B \dfrac{T_B}{T_A}}{N_B \dfrac{T_B}{T_C}} = \dfrac{T_B}{T_A} \times \dfrac{T_C}{T_B} = \dfrac{T_C}{T_A}$

This shows that the movement ratio is independent of the idler, only the direction of the follower being altered.

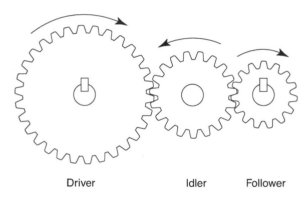

Driver   Idler   Follower

Figure 25.5

A **compound gear train** is shown in Figure 25.6, in which gear wheels B and C are fixed to the same shaft and hence $N_B = N_C$

From equation (7),

$$\frac{N_A}{N_B} = \frac{T_B}{T_A} \quad \text{i.e.} \quad N_B = N_A \times \frac{T_A}{T_B}$$

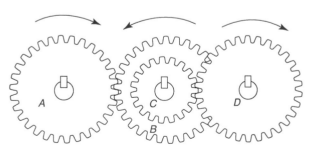

Figure 25.6

Also,

$$\frac{N_D}{N_C} = \frac{T_C}{T_D} \quad \text{i.e.} \quad N_D = N_C \times \frac{T_C}{T_D}$$

But $N_B = N_C$ and

$$N_D = N_B \times \frac{T_C}{T_D}$$

therefore, $\qquad N_D = N_A \times \dfrac{T_A}{T_B} \times \dfrac{T_C}{T_D} \qquad$ (8)

For compound gear trains having, say, P gear wheels,

$$N_P = N_A \times \frac{T_A}{T_B} \times \frac{T_C}{T_D} \times \frac{T_E}{T_F} \cdots\cdots \times \frac{T_O}{T_P}$$

from which,

$$\textbf{movement ratio} = \frac{N_A}{N_P} = \frac{T_B}{T_A} \times \frac{T_D}{T_C} \cdots\cdots \times \frac{T_P}{T_O}$$

Problem 7. A driver gear on a shaft of a motor has 35 teeth and meshes with a follower having 98 teeth. If the speed of the motor is 1400 revolutions per minute, find the speed of rotation of the follower

From equation (7),

$$\frac{\text{speed of driver}}{\text{speed of follower}} = \frac{\text{teeth on follower}}{\text{teeth on driver}}$$

i.e. $\qquad \dfrac{1400}{\text{speed of follower}} = \dfrac{98}{35}$

Hence, **speed of follower** $= \dfrac{1400 \times 35}{98} = \textbf{500 rev/min.}$

Problem 8. A compound gear train similar to that shown in Figure 25.6 consists of a driver gear A, having 40 teeth, engaging with gear B, having 160 teeth. Attached to the same shaft as B, gear C has 48 teeth and meshes with gear D on the output shaft, having 96 teeth. Determine (a) the movement ratio of this gear system and (b) the efficiency when the force ratio is 6

(a) From equation (8),

$$\text{the speed of } D = \text{speed of } A \times \frac{T_A}{T_B} \times \frac{T_C}{T_D}$$

From equation (7),

$$\text{movement ratio} = \frac{\text{speed of } A}{\text{speed of } D} = \frac{T_B}{T_A} \times \frac{T_D}{T_C}$$

$$= \frac{160}{40} \times \frac{96}{48} = 8$$

(b) The efficiency of any simple machine

$$= \frac{\text{force ratio}}{\text{movement ratio}} \times 100\%$$

Thus, **efficiency** $= \dfrac{6}{8} \times 100 = \textbf{75\%}$

Now try the following Practice Exercise

Practice Exercise 119 Further problems on gear trains (answers on page 542)

1. The driver gear of a gear system has 28 teeth and meshes with a follower gear having 168 teeth. Determine the movement ratio and the speed of the follower when the driver gear rotates at 60 revolutions per second

2. A compound gear train has a 30-tooth driver gear A, meshing with a 90-tooth follower gear B. Mounted on the same shaft as B and attached to it is a gear C with 60 teeth, meshing with a gear D on the output shaft having 120 teeth. Calculate the movement and force ratios if the overall efficiency of the gears is 72%

3. A compound gear train is as shown in Figure 25.6. The movement ratio is 6 and the numbers of teeth on gears A, C and D are 25, 100 and 60, respectively. Determine the number of teeth on gear B and the force ratio when the efficiency is 60%

25.6 Levers

A **lever** can alter both the magnitude and the line of action of a force and is thus classed as a simple machine. There are three types or orders of levers, as shown in Figure 25.7.

A lever of the first order has the **fulcrum** placed between the effort and the load, as shown in Figure 25.7(a).

A lever of the second order has the load placed between the effort and the fulcrum, as shown in Figure 25.7(b).

A lever of the third order has the effort applied between the load and the fulcrum, as shown in Figure 25.7(c).

Problems on levers can largely be solved by applying the principle of moments (see chapter 22). Thus for the lever shown in Figure 25.7(a), when the lever is in equilibrium,

$$\text{anticlockwise moment} = \text{clockwise moment}$$

i.e.
$$a \times F_l = b \times F_e$$

Thus, **force ratio** $= \dfrac{F_l}{F_e} = \dfrac{b}{a}$

$$= \frac{\textbf{distance of effort from fulcrum}}{\textbf{distance of load from fulcrum}}$$

Problem 9. The load on a first-order lever, similar to that shown in Figure 25.7(a), is 1.2 kN. Determine the effort, the force ratio and the movement ratio when the distance between the fulcrum and the load is 0.5 m and the distance between the fulcrum and effort is 1.5 m. Assume the lever is 100% efficient

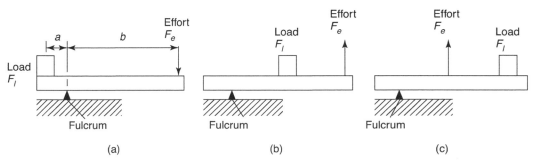

Figure 25.7

(a)   (b)   (c)

Applying the principle of moments, for equilibrium:

anticlockwise moment = clockwise moment

i.e. $1200\,N \times 0.5\,m = \text{effort} \times 1.5\,m$

Hence, $\text{effort} = \dfrac{1200 \times 0.5}{1.5} = \mathbf{400\,N}$

$\text{force ratio} = \dfrac{F_l}{F_e} = \dfrac{1200}{400} = \mathbf{3}$

Alternatively, $\text{force ratio} = \dfrac{b}{a} = \dfrac{1.5}{0.5} = \mathbf{3}$

This result shows that to lift a load of, say, 300 N, an effort of 100 N is required.

Since, from equation (3),

$$\text{efficiency} = \dfrac{\text{force ratio}}{\text{movement ratio}} \times 100\%$$

then, $\text{movement ratio} = \dfrac{\text{force ratio}}{\text{efficiency}} \times 100\%$

$$= \dfrac{3}{100} \times 100 = \mathbf{3}$$

This result shows that to raise the load by, say, 100 mm, the effort has to move 300 mm.

> **Problem 10.** A second-order lever *AB* is in a horizontal position. The fulcrum is at point *C*. An effort of 60 N applied at *B* just moves a load at point *D*, when *BD* is 0.5 m and *BC* is 1.25 m. Calculate the load and the force ratio of the lever

A second-order lever system is shown in Figure 25.7(b). From chapter 22, taking moments about the fulcrum as the load is just moving, gives:

anticlockwise moment = clockwise moment

i.e. $60\,N \times 1.25\,m = \text{load} \times 0.75\,m$

Thus, $\text{load} = \dfrac{60 \times 1.25}{0.75} = \mathbf{100\,N}$

From equation (1),

$\text{force ratio} = \dfrac{\text{load}}{\text{effort}} = \dfrac{100}{60} = \mathbf{1.67}$

Alternatively,

$\text{force ratio} = \dfrac{\text{distance of effort from fulcrum}}{\text{distance of load from fulcrum}}$

$$= \dfrac{1.25}{0.75} = \mathbf{1.67}$$

Now try the following Practice Exercises

Practice Exercise 120 Further problems on levers (answers on page 542)

1. In a second-order lever system, the force ratio is 2.5. If the load is at a distance of 0.5 m from the fulcrum, find the distance that the effort acts from the fulcrum if losses are negligible

2. A lever *AB* is 2 m long and the fulcrum is at a point 0.5 m from *B*. Find the effort to be applied at *A* to raise a load of 0.75 kN at *B* when losses are negligible

3. The load on a third-order lever system is at a distance of 750 mm from the fulcrum and the effort required to just move the load is 1 kN when applied at a distance of 250 mm from the fulcrum. Determine the value of the load and the force ratio if losses are negligible

Practice Exercise 121 Short answer questions on simple machines

1. State what is meant by a simple machine

2. Define force ratio

3. Define movement ratio

4. Define the efficiency of a simple machine in terms of the force and movement ratios

5. State briefly why the efficiency of a simple machine cannot reach 100%

6. With reference to the law of a simple machine, state briefly what is meant by the term 'limiting force ratio'

7. Define limiting efficiency

8. Explain why a four-pulley system has a force ratio of 4, when losses are ignored

9. Give the movement ratio for a screw-jack in terms of the effective radius of the effort and the screw lead

10. Explain the action of an idler gear

11. Define the movement ratio for a two-gear system in terms of the teeth on the wheels

12. Show that the action of an idler wheel does not affect the movement ratio of a gear system

13. State the relationship between the speed of the first gear and the speed of the last gear in a compound train of four gears, in terms of the teeth on the wheels

14. Define the force ratio of a first-order lever system in terms of the distances of the load and effort from the fulcrum

15. Use sketches to show what is meant by: (a) a first-order, (b) a second-order, (c) a third-order lever system. Give one practical use for each type of lever

Practice Exercise 122 Multiple-choice questions on simple machines (answers on page 492)

A simple machine requires an effort of 250 N moving through 10 m to raise a load of 1000 N through 2 m. Use this data to find the correct answers to questions 1–3, selecting the answers from:

(a) 0.25 (b) 4 (c) 80% (d) 20%

(e) 100 (f) 5 (g) 100% (h) 0.2

(i) 25%

1. Find the force ratio

2. Find the movement ratio

3. Find the efficiency

The law of a machine is of the form $F_e = aF_l + b$. An effort of 12 N is required to raise a load of 40 N and an effort of 6 N is required to raise a load of 16 N. The movement ratio of the machine is 5. Use this data to find the correct answers to questions 4 to 6, selecting these answers from:

(a) 80% (b) 4 (c) 2.8 (d) 0.25

(e) $\dfrac{1}{2.8}$ (f) 25% (g) 100% (h) 2

(i) 25%

4. Determine the constant a

5. Find the limiting force ratio

6. Find the limiting efficiency

7. Which of the following statements is false?

(a) A single-pulley system changes the line of action of the force but does not change the magnitude of the force, when losses are neglected

(b) In a two-pulley system, the force ratio is $\dfrac{1}{2}$ when losses are neglected

(c) In a two-pulley system, the movement ratio is 2

(d) The efficiency of a two-pulley system is 100% when losses are neglected

8. Which of the following statements concerning a screw-jack is false?

(a) A screw-jack changes both the line of action and the magnitude of the force

(b) For a single-start thread, the distance moved in 5 revolutions of the table is $5l$, where l is the lead of the screw

(c) The distance moved by the effort is $2\pi r$, where r is the effective radius of the effort

(d) The movement ratio is given by $\dfrac{2\pi r}{5l}$

9. In a simple gear train, a follower has 50 teeth and the driver has 30 teeth. The movement ratio is:
(a) 0.6 (b) 20 (c) 1.67 (d) 80

10. Which of the following statements is true?

(a) An idler wheel between a driver and a follower is used to make the direction of the follower opposite to that of the driver

(b) An idler wheel is used to change the movement ratio

(c) An idler wheel is used to change the force ratio

(d) An idler wheel is used to make the direction of the follower the same as that of the driver

11. Which of the following statements is false?

(a) In a first-order lever, the fulcrum is between the load and the effort

(b) In a second-order lever, the load is between the effort and the fulcrum

(c) In a third-order lever, the effort is applied between the load and the fulcrum

(d) The force ratio for a first-order lever system is given by:

$$\frac{\text{distance of load from fulcrum}}{\text{distance of effort from fulcrum}}$$

12. In a second-order lever system, the load is 200 mm from the fulcrum and the effort is 500 mm from the fulcrum. If losses are neglected, an effort of 100 N will raise a load of:

(a) 100 N   (b) 250 N   (c) 400 N   (d) 40 N

**For fully worked solutions to each of the problems in Exercises 116 to 122 in this chapter,
go to the website:**
www.routledge.com/cw/bird

Revision Test 9: Linear and angular motion, friction and simple machines

This assignment covers the material contained in chapters 23 to 25. *The marks for each question are shown in brackets at the end of each question.*

1. A train is travelling at 90 km/h and has wheels of diameter 1600 mm.

    (a) Find the angular velocity of the wheels in both rad/s and rev/min.

    (b) If the speed remains constant for 2 km, determine the number of revolutions made by a wheel, assuming no slipping occurs. (9)

2. The speed of a shaft increases uniformly from 200 revolutions per minute to 700 revolutions per minute in 12 s. Find the angular acceleration, correct to 3 significant figures. (5)

3. The shaft of an electric motor, initially at rest, accelerates uniformly for 0.3 s at 20 rad/s². Determine the angle (in radians) turned through by the shaft in this time. (4)

4. The material of a brake is being tested and it is found that the dynamic coefficient of friction between the material and steel is 0.90. Calculate the normal force when the frictional force is 0.630 kN. (5)

5. A simple machine raises a load of 120 kg through a distance of 1.2 m. The effort applied to the machine is 150 N and moves through a distance of 12 m.

Taking g as 10 m/s², determine the force ratio, movement ratio and efficiency of the machine. (6)

6. A load of 30 kg is lifted by a three-pulley system and the applied effort is 140 N. Calculate, taking g to be 9.8 m/s², (a) the force ratio, (b) the movement ratio, (c) the efficiency of the system. (5)

7. A screw-jack is being used to support the axle of a lorry, the load on it being 5.6 kN. The screw-jack has an effort of effective radius 318 mm and a single-start square thread, having a lead of 5 mm. Determine the efficiency of the jack if an effort of 70 N is required to raise the car axle. (6)

8. A driver gear on a shaft of a motor has 32 teeth and meshes with a follower having 96 teeth. If the speed of the motor is 1410 revolutions per minute, find the speed of rotation of the follower. (4)

9. The load on a first-order lever is 1.5 kN. Determine the effort, the force ratio and the movement ratio when the distance between the fulcrum and the load is 0.4 m and the distance between the fulcrum and effort is 1.6 m. Assume the lever is 100% efficient. (6)

For lecturers/instructors/teachers, fully worked solutions to each of the problems in Revision Test 9, together with a full marking scheme, are available at the website:

www.routledge.com/cw/bird

Chapter 26

The effects of forces
on materials

Why it is important to understand: **The effects of forces on materials**

A good knowledge of some of the constants used in the study of the properties of materials is vital in most branches of engineering, especially in mechanical, manufacturing, aeronautical and civil and structural engineering. For example, most steels look the same, but steels used for the pressure hull of a submarine are about five times stronger than those used in the construction of a small building, and it is very important for the professional and chartered engineer to know what steel to use for what construction; this is because the cost of the high-tensile steel used to construct a submarine pressure hull is considerably higher than the cost of the mild steel, or similar material, used to construct a small building. The engineer must not only take into consideration the ability of the chosen material of construction to do the job, but also its cost. Similar arguments lie in manufacturing engineering, where the engineer must be able to estimate the ability of his/her machines to bend, cut or shape the artefact s/he is trying to produce, and at a competitive price! This chapter provides explanations of the different terms that are used in determining the properties of various materials. The importance of knowing about the effects of forces on materials is to aid the design and construction of structures in an efficient and trustworthy manner.

At the end of this chapter, you should be able to:

- define force and state its unit
- recognise a tensile force and state relevant practical examples
- recognise a compressive force and state relevant practical examples
- recognise a shear force and state relevant practical examples
- define stress and state its unit
- calculate stress σ from $\sigma = \dfrac{F}{A}$
- define strain
- calculate strain e from $\varepsilon = \dfrac{x}{L}$
- define elasticity, plasticity, limit of proportionality and elastic limit

Science and Mathematics for Engineering. 978-0-367-20475-4. © John Bird. Published by Taylor & Francis. All rights reserved.

- state Hooke's law
- define Young's modulus of elasticity E and stiffness
- appreciate typical values for E
- calculate E from $E = \dfrac{\sigma}{\varepsilon}$
- perform calculations using Hooke's law
- plot a load/extension graph from given data
- define ductility, brittleness and malleability, with examples of each

26.1 Introduction

A good knowledge of some of the constants used in the study of the properties of materials is vital in most branches of engineering, especially in mechanical, manufacturing, aeronautical and civil and structural engineering. For example, most steels look the same, but steels used for the pressure hull of a submarine are about 5 times stronger than those used in the construction of a small building, and it is very important for the professional and chartered engineer to know what steel to use for what construction; this is because the cost of the high-tensile steel used to construct a submarine pressure hull is considerably higher than the cost of the mild steel, or similar material, used to construct a small building. The engineer must not only take into consideration the ability of the chosen material of construction to do the job, but also its cost. Similar arguments lie in manufacturing engineering, where the engineer must be able to estimate the ability of his/her machines to bend, cut or shape the artefact s/he is trying to produce, and at a competitive price! This chapter provides explanations of the different terms that are used in determining the properties of various materials.

26.2 Forces

A **force** exerted on a body can cause a change in either the shape or the motion of the body. The unit of force is the **newton***, **N**.

No solid body is perfectly rigid and when forces are applied to it, changes in dimensions occur. Such changes are not always perceptible to the human eye since they are so small. For example, the span of a bridge will sag under the weight of a vehicle and a spanner will bend slightly when tightening a nut. It is important

for engineers and designers to appreciate the effects of forces on materials, together with their mechanical properties.

The three main types of mechanical force that can act on a body are:

(i) tensile,

(ii) compressive, and

(iii) shear.

26.3 Tensile force

Tension is a force that tends to stretch a material, as shown in Figure 26.1. For example,

(i) the rope or cable of a crane carrying a load is in tension,

Force   Force

Figure 26.1

(ii) rubber bands, when stretched, are in tension,

(iii) when a nut is tightened, a bolt is under tension.

A tensile force, i.e. one producing tension, increases the length of the material on which it acts.

---

*Sir Isaac Newton (25 December 1642–20 March 1727) was an English polymath. Newton showed that the motions of objects are governed by the same set of natural laws, by demonstrating the consistency between Kepler's laws of planetary motion and his theory of gravitation. The SI unit of force is the newton, named in his honour. To find out more about Newton go to www.routledge.com/cw/bird (For image refer to page 29)

26.4 Compressive force

Compression is a force that tends to squeeze or crush a material, as shown in Figure 26.2. For example,

Figure 26.2

(i)   a pillar supporting a bridge is in compression,

(ii) the sole of a shoe is in compression,

(iii) the jib of a crane is in compression.

A compressive force, i.e. one producing compression, will decrease the length of the material on which it acts.

26.5 Shear force

Shear is a force that tends to slide one face of the material over an adjacent face. For example,

(i)   a rivet holding two plates together is in shear if a tensile force is applied between the plates – as shown in Figure 26.3,

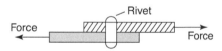

Figure 26.3

(ii) a guillotine cutting sheet metal, or garden shears, each provide a shear force,

(iii) a horizontal beam is subject to shear force,

(iv) transmission joints on cars are subject to shear forces.

A shear force can cause a material to bend, slide or twist.

Problem 1.   Figure 26.4(a) represents a crane and Figure 26.4(b) a transmission joint. State the types of forces acting, labelled A to F

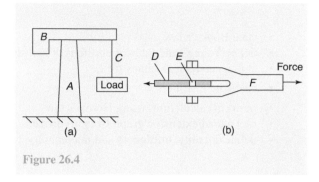

Figure 26.4

(a)   For the crane, A, a supporting member, is in **compression**, B, a horizontal beam, is in **shear**, and C, a rope, is in **tension**.

(b)   For the transmission joint, parts D and F are in **tension**, and E, the rivet or bolt, is in **shear**.

26.6 Stress

Forces acting on a material cause a change in dimensions and the material is said to be in a state of **stress**. Stress is the ratio of the applied force F to cross-sectional area A of the material. The symbol used for tensile and compressive stress is σ (Greek letter sigma). The unit of stress is the **pascal***, **Pa**, where 1 Pa = 1 N/m^2. Hence

$$\sigma = \frac{F}{A} \text{ Pa}$$

where F is the force in newtons and A is the cross-sectional area in square metres. For tensile and compressive forces, the cross-sectional area is that which is at right angles to the direction of the force.

For a shear force the shear stress is equal to $\frac{F}{A}$, where the cross-sectional area A is that which is parallel to the direction of the force. The symbol used for shear stress is τ (Greek letter tau).

Problem 2.   A rectangular bar having a cross-sectional area of 75 mm^2 has a tensile force of 15 kN applied to it. Determine the stress in the bar

Cross-sectional area $A = 75\,\text{mm}^2 = 75 \times 10^{-6}\,\text{m}^2$ and force $F = 15\,\text{kN} = 15 \times 10^3\,\text{N}$.

Stress in bar $\sigma = \dfrac{F}{A} = \dfrac{15 \times 10^3\,\text{N}}{75 \times 10^{-6}\,\text{m}^2}$

$$= 0.2 \times 10^9\,\text{Pa} = \textbf{200 MPa}.$$

Problem 3. A wire of circular cross-section, has a tensile force of 60.0 N applied to it and this force produces a stress of 3.06 MPa in the wire. Determine the diameter of the wire

Force $F = 60.0\,\text{N}$ and

stress $\sigma = 3.06\,\text{MPa} = 3.06 \times 10^6\,\text{Pa}$.

Since $\sigma = \dfrac{F}{A}$ then area

$$A = \frac{F}{\sigma} = \frac{60.0\,\text{N}}{3.06 \times 10^6\,\text{Pa}} = 19.61 \times 10^{-6}\,\text{m}^2$$

$$= 19.61\,\text{mm}^2$$

Cross-sectional area $A = \dfrac{\pi d^2}{4}$; hence $19.61 = \dfrac{\pi d^2}{4}$

from which,

$$d^2 = \frac{4 \times 19.61}{\pi} \quad \text{and} \quad d = \sqrt{\left(\frac{4 \times 19.61}{\pi}\right)} = 5.0$$

i.e.   **diameter of wire = 5.0 mm**.

Now try the following Practice Exercise

Practice Exercise 123   Further problems on stress (answers on page 543)

1. A rectangular bar having a cross-sectional area of 80 mm^2 has a tensile force of 20 kN applied to it. Determine the stress in the bar

2. A circular section cable has a tensile force of 1 kN applied to it and the force produces a stress of 7.8 MPa in the cable. Calculate the diameter of the cable

3. A square-sectioned support of side 12 mm is loaded with a compressive force of 10 kN. Determine the compressive stress in the support

4. A bolt having a diameter of 5 mm is loaded so that the shear stress in it is 120 MPa. Determine the value of the shear force on the bolt

5. A split pin requires a force of 400 N to shear it. The maximum shear stress before shear occurs is 120 MPa. Determine the minimum diameter of the pin

6. A tube of outside diameter 60 mm and inside diameter 40 mm is subjected to a tensile load of 60 kN. Determine the stress in the tube

*Who was Pascal? – **Blaise Pascal** (19 June 1623–19 August 1662) was a French polymath who made important contributions to the study of fluids and clarified the concepts of pressure and vacuum. He corresponded with Pierre de Fermat on probability theory, strongly influencing the development of modern economics and social science. The unit of pressure, the *pascal*, is named in his honour. To find out more about Pascal go to www.routledge.com/cw/bird

26.7 Strain

The fractional change in a dimension of a material produced by a force is called the **strain**. For a tensile or compressive force, strain is the ratio of the change of length to the original length. The symbol used for strain is ε (Greek letter epsilon). For a material of length L metres which changes in length by an amount x metres when subjected to stress,

$$\varepsilon = \frac{x}{L}$$

Strain is dimension-less and is often expressed as a percentage, i.e.

$$\text{percentage strain} = \frac{x}{L} \times 100$$

For a shear force, strain is denoted by the symbol γ (Greek letter gamma) and, with reference to Figure 26.5, is given by:

$$\gamma = \frac{x}{l}$$

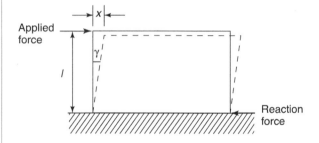

Figure 26.5

Problem 4. A bar 1.60 m long contracts axially by 0.1 mm when a compressive load is applied to it. Determine the strain and the percentage strain

$$\textbf{Strain } \varepsilon = \frac{\text{contraction}}{\text{original length}} = \frac{0.1 \text{ mm}}{1.60 \times 10^3 \text{ mm}}$$

$$= \frac{0.1}{1600} = \textbf{0.0000625}$$

Percentage strain $= 0.0000625 \times 100 = \textbf{0.00625\%}$

Problem 5. A wire of length 2.50 m has a percentage strain of 0.012% when loaded with a tensile force. Determine the extension of the wire

Original length of wire $= 2.50 \text{ m} = 2500 \text{ mm}$ and

$$\text{strain} = \frac{0.012}{100} = 0.00012$$

$$\text{Strain } \varepsilon = \frac{\text{extension } x}{\text{original length } L}$$

hence,

$$\textbf{extension } x = \varepsilon L = (0.00012)(2500) = \textbf{0.30 mm}$$

Problem 6. (a) A rectangular section metal bar has a width of 10 mm and can support a maximum compressive stress of 20 MPa; determine the minimum breadth of the bar when loaded with a force of 3 kN. (b) If the bar in (a) is 2 m long and decreases in length by 0.25 mm when the force is applied, determine the strain and the percentage strain

(a) Since stress $\sigma = \dfrac{\text{force } F}{\text{area } A}$ then,

$$\text{area } A = \frac{F}{\sigma} = \frac{3000 \text{ N}}{20 \times 10^6 \text{ Pa}}$$

$$= 150 \times 10^{-6} \text{ m}^2 = 150 \text{ mm}^2$$

Cross-sectional area $=$ width $\times$ breadth, hence

$$\textbf{breadth} = \frac{\text{area}}{\text{width}} = \frac{150}{10} = \textbf{15 mm}.$$

(b) **Strain** $\varepsilon = \dfrac{\text{contraction}}{\text{original length}}$

$$= \frac{0.25}{2000} = \textbf{0.000125}$$

Percentage strain $= 0.000125 \times 100 = \textbf{0.0125\%}$

Problem 7. A rectangular block of plastic material 500 mm long by 20 mm wide by 300 mm high has its lower face glued to a bench and a force of 200 N is applied to the upper face and in line with it. The upper face moves 15 mm relative to the lower face. Determine (a) the shear stress, and (b) the shear strain in the upper face, assuming the deformation is uniform

(a) Shear stress

$$\tau = \frac{\text{force}}{\text{area parallel to the force}}$$

Area of any face parallel to the force

$$= 500\,\text{mm} \times 20\,\text{mm}$$

$$= (0.5 \times 0.02)\,\text{m}^2 = 0.01\,\text{m}^2$$

Hence, **shear stress**

$$\tau = \frac{200\,\text{N}}{0.01\,\text{m}^2} = \textbf{20,000 Pa} \quad \text{or} \quad \textbf{20 kPa.}$$

(b) Shear strain

$$\gamma = \frac{x}{l} \quad \text{(see side view in Figure 26.6)}$$

$$= \frac{15}{300} = \textbf{0.05 (or 5\%)}$$

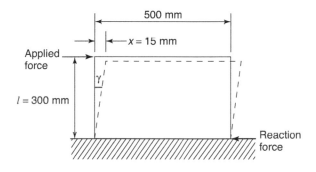

Figure 26.6

Now try the following Practice Exercise

Practice Exercise 124   Further problems on strain (answers on page 543)

1. A wire of length 4.5 m has a percentage strain of 0.050% when loaded with a tensile force. Determine the extension in the wire

2. A metal bar 2.5 m long extends by 0.05 mm when a tensile load is applied to it. Determine (a) the strain and (b) the percentage strain

3. An 80 cm long bar contracts axially by 0.2 mm when a compressive load is applied to it. Determine the strain and the percentage strain

4. A pipe has an outside diameter of 20 mm, an inside diameter of 10 mm and length 0.30 m and it supports a compressive load of 50 kN. The pipe shortens by 0.6 mm when the load is applied. Determine (a) the compressive stress and (b) the compressive strain in the pipe when supporting this load

5. A rectangular block of plastic material 400 mm long by 15 mm wide by 300 mm high has its lower face fixed to a bench and a force of 150 N is applied to the upper face and in line with it. The upper face moves 12 mm relative to the lower face. Determine (a) the shear stress, and (b) the shear strain in the upper face, assuming the deformation is uniform

26.8   Elasticity, limit of proportionality and elastic limit

Elasticity is the ability of a material to return to its original shape and size on the removal of external forces.

Plasticity is the property of a material of being permanently deformed by a force without breaking. Thus if a material does not return to the original shape, it is said to be plastic.

Within certain load limits, mild steel, copper, polythene and rubber are examples of elastic materials; lead and plasticine are examples of plastic materials.

If a tensile force applied to a uniform bar of mild steel is gradually increased and the corresponding extension of the bar is measured, then provided the applied force is not too large, a graph depicting these results is likely to be as shown in Figure 26.7. Since the graph is a straight line, **extension is directly proportional to the applied force**.

The point on the graph where extension is no longer proportional to the applied force is known as the **limit of proportionality**. Just beyond this point the material can behave in a non-linear elastic manner, until the **elastic limit** is reached. If the applied force is large, it is found that the material becomes plastic and no longer

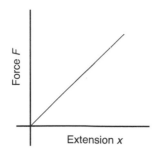

Figure 26.7

returns to its original length when the force is removed. The material is then said to have passed its elastic limit and the resulting graph of force/extension is no longer a straight line. Stress $\sigma = \dfrac{F}{A}$, from section 26.6, and since, for a particular bar, area A can be considered as a constant, then $F \propto \sigma$.

Strain $\varepsilon = \dfrac{x}{L}$, from section 26.7, and since for a particular bar L is constant, then $x \propto \varepsilon$. Hence, for stress applied to a material below the limit of proportionality a graph of stress/strain will be as shown in Figure 26.8, and is a similar shape to the force/extension graph of Figure 26.7.

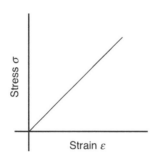

Figure 26.8

26.9 Hooke's law

Hooke's law* states:

Within the limit of proportionality, the extension of a material is proportional to the applied force.

It follows, from section 26.8, that:

Within the limit of proportionality of a material, the strain produced is directly proportional to the stress producing it.

26.9.1 Young's modulus of elasticity

Within the limit of proportionality, stress $\propto$ strain, hence

$$\text{stress} = (\text{a constant}) \times \text{strain}$$

This constant of proportionality is called **Young's modulus of elasticity*** and is given the symbol E. The value of E may be determined from the gradient of the straight line portion of the stress/strain graph. The dimensions of E are pascals (the same as for stress, since strain is dimension-less).

$$E = \frac{\sigma}{\varepsilon} \text{ Pa}$$

Some **typical values** for Young's modulus of elasticity E include: aluminium alloy 70 GPa (i.e. 70×10^9 Pa), brass 90 GPa, copper 96 GPa, titanium alloy 110 GPa, diamond 1200 GPa, mild steel 210 GPa, lead 18 GPa, tungsten 410 GPa, cast iron 110 GPa, zinc 85 GPa, glass fibre 72 GPa, carbon fibre 300 GPa.

26.9.2 Stiffness

A material having a large value of Young's modulus is said to have a high value of material stiffness, where stiffness is defined as:

$$\text{Stiffness} = \frac{\text{force } F}{\text{extension } x}$$

For example, mild steel is a much stiffer material than lead.

***Who was Hooke?** – **Robert Hooke** (28 July 1635–3 March 1703) was an English natural philosopher, architect and polymath who, amongst other things, discovered the law of elasticity. To find out more about Hooke go to www.routledge.com/cw/bird (For image refer to page 101)

***Who was Young?** – **Thomas Young** (13 June 1773–10 May 1829) was an English polymath. He is perhaps best known for his work on Egyptian hieroglyphics and the Rosetta Stone, but Young also made notable scientific contributions to the fields of vision, light, solid mechanics, energy, physiology, language and musical harmony. *Young's modulus* relates the stress in a body to its associated strain. To find out more about Young go to www.routledge.com/cw/bird (For image refer to page 102)

Since $E = \dfrac{\sigma}{\varepsilon}$, $\sigma = \dfrac{F}{A}$

and $\varepsilon = \dfrac{x}{L}$, then

$$E = \frac{\dfrac{F}{A}}{\dfrac{x}{L}} = \frac{FL}{Ax} = \left(\frac{F}{x}\right)\left(\frac{L}{A}\right)$$

i.e. $E = (\textbf{stiffness}) \times \left(\dfrac{L}{A}\right)$

Stiffness $\left(=\dfrac{F}{x}\right)$ is also the gradient of the force/extension graph, hence

$$E = (\textbf{gradient of force/extension graph})\left(\frac{L}{A}\right)$$

Since L and A for a particular specimen are constant, the greater Young's modulus the greater the material stiffness.

Problem 8. A wire is stretched 2 mm by a force of 250 N. Determine the force that would stretch the wire 5 mm, assuming that the limit of proportionality is not exceeded

Hooke's law states that extension x is proportional to force F, provided that the limit of proportionality is not exceeded, i.e. $x \propto F$ or $x = kF$, where k is a constant. When $x = 2$ mm, $F = 250$ N, thus $2 = k(250)$, from which, constant $k = \dfrac{2}{250} = \dfrac{1}{125}$

When $x = 5$ mm, then $5 = kF$, i.e. $5 = \left(\dfrac{1}{125}\right)F$ from which, force $F = 5(125) = 625$ N.
Thus to stretch the wire 5 mm, a force of 625 N is required.

Problem 9. A copper rod of diameter 20 mm and length 2.0 m has a tensile force of 5 kN applied to it. Determine (a) the stress in the rod, (b) by how much the rod extends when the load is applied. Take the modulus of elasticity for copper as 96 GPa

(a) Force $F = 5$ kN $= 5000$ N and cross-sectional area
$$A = \frac{\pi d^2}{4} = \frac{\pi (0.020)^2}{4} = 0.000314 \text{ m}^2.$$

$$\textbf{Stress } \sigma = \frac{F}{A} = \frac{5000 \text{ N}}{0.000314 \text{ m}^2} = 15.92 \times 10^6 \text{ Pa}$$

$$= \textbf{15.92 MPa}.$$

(b) Since $E = \dfrac{\sigma}{\varepsilon}$ then

$$\text{strain } \varepsilon = \frac{\sigma}{E} = \frac{15.92 \times 10^6 \text{ Pa}}{96 \times 10^9 \text{ Pa}} = 0.000166$$

Strain $\varepsilon = \dfrac{x}{L}$, hence extension $x = \varepsilon L$

$$= (0.000166)(2.0) = 0.000332 \text{ m}$$

i.e. **extension of rod is 0.332 mm**

Problem 10. A bar of thickness 15 mm and having a rectangular cross-section carries a load of 120 kN. Determine the minimum width of the bar to limit the maximum stress to 200 MPa. The bar, which is 1.0 m long, extends by 2.5 mm when carrying a load of 120 kN. Determine the modulus of elasticity of the material of the bar

Force $F = 120$ kN $= 120{,}000$ N and cross-sectional area $A = (15x)10^{-6}$ m², where x is the width of the rectangular bar in millimetres.
Stress $\sigma = \dfrac{F}{A}$, from which,

$$A = \frac{F}{\sigma} = \frac{120{,}000 \text{ N}}{200 \times 10^6 \text{ Pa}} = 6 \times 10^{-4} \text{ m}^2$$

$$= 6 \times 10^{-4} \times 10^6 \text{ mm}^2$$

$$= 6 \times 10^2 \text{ mm}^2 = 600 \text{ mm}^2.$$

Hence, $600 = 15x$, from which,

$$\textbf{width of bar } x = \frac{600}{15} = \textbf{40 mm}.$$

Extension of bar $= 2.5$ mm $= 0.0025$ m

Strain $\varepsilon = \dfrac{x}{L} = \dfrac{0.0025}{1.0} = 0.0025$

$$\textbf{Modulus of elasticity } E = \frac{\text{stress}}{\text{strain}} = \frac{200 \times 10^6}{0.0025}$$

$$= 80 \times 10^9 = \textbf{80 GPa}$$

Problem 11.   In an experiment to determine the modulus of elasticity of a sample of mild steel, a wire is loaded and the corresponding extension noted. The results of the experiment are as shown.

Load (N)	0	40	110	160	200	250	290	340
Extension (mm)	0	1.2	3.3	4.8	6.0	7.5	10.0	16.2

Draw the load/extension graph. The mean diameter of the wire is 1.3 mm and its length is 8.0 m. Determine the modulus of elasticity E of the sample, and the stress at the limit of proportionality

A graph of load/extension is shown in Figure 26.9.

$$E = \frac{\sigma}{\varepsilon} = \frac{\frac{F}{A}}{\frac{x}{L}} = \left(\frac{F}{x}\right)\left(\frac{L}{A}\right)$$

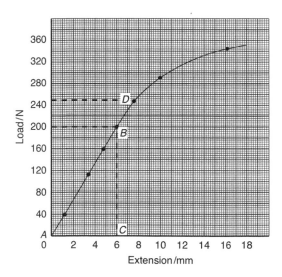

Figure 26.9

$\dfrac{F}{x}$ is the gradient of the straight line part of the load/extension graph.

$$\text{Gradient } \frac{F}{x} = \frac{BC}{AC} = \frac{200\,\text{N}}{6 \times 10^{-3}\,\text{m}} = 33.33 \times 10^3\,\text{N/m}.$$

$$\text{Modulus of elasticity} = (\text{gradient of graph})\left(\frac{L}{A}\right)$$

Length of specimen $L = 8.0\,\text{m}$ and

cross-sectional area $A = \dfrac{\pi d^2}{4} = \dfrac{\pi (0.0013)^2}{4}$

$$= 1.327 \times 10^{-6}\,\text{m}^2.$$

Hence **modulus of elasticity**

$$E = (33.33 \times 10^3)\left(\frac{8.0}{1.327 \times 10^{-6}}\right) = \textbf{201 GPa}.$$

The limit of proportionality is at point D in Figure 26.9 where the graph no longer follows a straight line. This point corresponds to a load of 250 N as shown.

Stress at the limit of proportionality

$$= \frac{\text{force}}{\text{area}} = \frac{250}{1.327 \times 10^{-6}}$$

$$= 188.4 \times 10^6\,\text{Pa} = \textbf{188.4 MPa}$$

Now try the following Practice Exercise

Practice Exercise 125   Further problems on Hooke's law (answers on page 543)

1. A wire is stretched 1.5 mm by a force of 300 N. Determine the force that would stretch the wire 4 mm, assuming the elastic limit of the wire is not exceeded

2. A rubber band extends 50 mm when a force of 300 N is applied to it. Assuming the band is within the elastic limit, determine the extension produced by a force of 60 N

3. A force of 25 kN applied to a piece of steel produces an extension of 2 mm. Assuming the elastic limit is not exceeded, determine (a) the force required to produce an extension of 3.5 mm and (b) the extension when the applied force is 15 kN

4. A test to determine the load/extension graph for a specimen of copper gave the following results:

Load (kN)	8.5	15.0	23.5	30.0
Extension (mm)	0.04	0.07	0.11	0.14

Plot the load/extension graph, and from the graph determine (a) the load at an extension of 0.09 mm and (b) the extension corresponding to a load of 12.0 kN

5. A circular section bar is 2.5 m long and has a diameter of 60 mm. When subjected to a compressive load of 30 kN it shortens by 0.20 mm. Determine Young's modulus of elasticity for the material of the bar

6. A bar of thickness 20 mm and having a rectangular cross-section carries a load of 82.5 kN. Determine (a) the minimum width of the bar to limit the maximum stress to 150 MPa and (b) the modulus of elasticity of the material of the bar if the 150 mm long bar extends by 0.8 mm when carrying a load of 200 kN

7. A metal rod of cross-sectional area 100 mm^2 carries a maximum tensile load of 20 kN. The modulus of elasticity for the material of the rod is 200 GPa. Determine the percentage strain when the rod is carrying its maximum load

26.10 Ductility, brittleness and malleability

Ductility is the ability of a material to be plastically deformed by elongation, without fracture. This is a property that enables a material to be drawn out into wires. For ductile materials such as mild steel, copper and gold, large extensions can result before fracture occurs with increasing tensile force. Ductile materials usually have a percentage elongation value of about 15% or more.

Brittleness is the property of a material manifested by fracture without appreciable prior plastic deformation. Brittleness is a lack of ductility, and brittle materials such as cast iron, glass, concrete, brick and ceramics, have virtually no plastic stage, the elastic stage being followed by immediate fracture. Little or no 'waist' occurs before fracture in a brittle material undergoing a tensile test.

Malleability is the property of a material whereby it can be shaped when cold by hammering or rolling. A malleable material is capable of undergoing plastic deformation without fracture. Examples of malleable materials include lead, gold, putty and mild steel.

Problem 12. Sketch typical load/extension curves for (a) an elastic non-metallic material, (b) a brittle material and (c) a ductile material. Give a typical example of each type of material

(a) A typical load/extension curve for an elastic non-metallic material is shown in Figure 26.10(a), and an example of such a material is polythene.

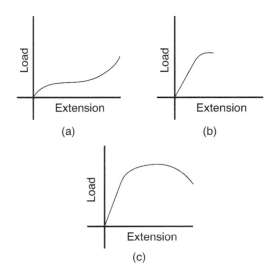

Figure 26.10

(b) A typical load/extension curve for a brittle material is shown in Figure 26.10(b), and an example of such a material is cast iron.

(c) A typical load/extension curve for a ductile material is shown in Figure 26.10(c), and an example of such a material is mild steel.

Now try the following Practice Exercises

Practice Exercise 126   Short answer questions on the effects of forces on materials

1. Name three types of mechanical force that can act on a body

2. What is a tensile force? Name two practical examples of such a force

3. What is a compressive force? Name two practical examples of such a force

4. Define a shear force and name two practical examples of such a force

5. Define elasticity and state two examples of elastic materials

6. Define plasticity and state two examples of plastic materials

7. Define the limit of proportionality

8. State Hooke's law

9. What is the difference between a ductile and a brittle material?

10. Define stress. What is the symbol used for
    (a) a tensile stress   (b) a shear stress?

11. Strain is the ratio $\dfrac{\ldots\ldots}{\ldots\ldots}$

12. The ratio $\dfrac{\text{stress}}{\text{strain}}$ is called ....

13. State the units of (a) stress, (b) strain, (c) Young's modulus of elasticity

14. Stiffness is the ratio $\dfrac{\ldots\ldots}{\ldots\ldots}$

15. Sketch on the same axes a typical load/extension graph for a ductile and a brittle material

16. Define (a) ductility, (b) brittleness, (c) malleability

Practice Exercise 127   Multiple-choice questions on the effects of forces on materials (answers on page 543)

1. The unit of strain is:
   (a) pascals   (b) metres
   (c) dimension-less   (d) newtons

2. The unit of stiffness is:
   (a) newtons   (b) pascals
   (c) newtons per metre (d) dimension-less

3. The unit of Young's modulus of elasticity is:
   (a) pascals   (b) metres
   (c) dimension-less   (d) newtons

4. A wire is stretched 3 mm by a force of 150 N. Assuming the elastic limit is not exceeded, the force that will stretch the wire 5 mm is:
   (a) 150 N   (b) 250 N
   (c) 90 N   (d) 450 N

5. For the wire in question 4, the extension when the applied force is 450 N is:
   (a) 1 mm   (b) 3 mm
   (c) 9 mm   (d) 12 mm

6. Due to the forces acting, a horizontal beam is in:
   (a) tension   (b) compression   (c) shear

7. Due to forces acting, a pillar supporting a bridge is in:
   (a) tension   (b) compression   (c) shear

8. Which of the following statements is false?
   (a) Elasticity is the ability of a material to return to its original dimensions after deformation by a load
   (b) Plasticity is the ability of a material to retain any deformation produced in it by a load
   (c) Ductility is the ability to be permanently stretched without fracturing
   (d) Brittleness is the lack of ductility and a brittle material has a long plastic stage

9. A circular rod of cross-sectional area 100 mm² has a tensile force of 100 kN applied to it. The stress in the rod is:
   (a) 1 MPa   (b) 1 GPa
   (c) 1 kPa   (d) 100 MPa

10. A metal bar 5.0 m long extends by 0.05 mm when a tensile load is applied to it. The percentage strain is:
    (a) 0.1   (b) 0.01   (c) 0.001   (d) 0.0001

An aluminium rod of length 1.0 m and cross-sectional area 500 mm² is used to support a load of 5 kN which causes the rod to contract by 100 μm. For questions 11–13, select the correct answer from the following list:

(a) 100 MPa   (b) 0.001   (c) 10 kPa

(d) 100 GPa   (e) 0.01   (f) 10 MPa

(g) 10 GPa   (h) 0.0001   (i) 10 Pa

11. The stress in the rod

12. The strain in the rod

13. Young's modulus of elasticity

**For fully worked solutions to each of the problems in Exercises 123 to 127 in this chapter,
go to the website:**
www.routledge.com/cw/bird

Chapter 27

Linear momentum and impulse

Why it is important to understand: **Linear momentum and impulse**

This chapter is of considerable importance in the study of the motion and collision of vehicles, ships and so on. The chapter commences with defining momentum and impulse, together with Newton's laws of motion. It then applies these laws to solving practical problems in the fields of ballistics, pile drivers, etc. A study of linear and angular momentum is of importance when designing the motion and the crash worthiness of cars, buses, etc.

At the end of this chapter, you should be able to:

- define momentum and state its unit
- state Newton's first law of motion
- calculate momentum given mass and velocity
- state Newton's second law of motion
- define impulse and appreciate when impulsive forces occur
- state Newton's third law of motion
- calculate impulse and impulsive force
- use the equation of motion $v^2 = u^2 + 2as$ in calculations

Science and Mathematics for Engineering. 978-0-367-20475-4, © John Bird. Published by Taylor & Francis. All rights reserved.

27.1 Introduction

This chapter is of considerable importance in the study of the motion and collision of vehicles, ships and so on. The chapter commences with defining momentum, impulse, together with Newton's laws of motion*. It then applies these laws to solving practical problems in the fields of ballistics, pile drivers, etc.

27.2 Linear momentum

The **momentum** of a body is defined as the product of its mass and its velocity, i.e.

$$\textbf{momentum} = \textbf{\textit{mu}}$$

where m = mass (in kg) and u = velocity (in m/s). The unit of momentum is kg m/s.

Since velocity is a vector quantity, **momentum is a vector quantity**, i.e. it has both magnitude and direction. **Newton's first law of motion** states:

A body continues in a state of rest or in a state of uniform motion in a straight line unless acted on by some external force.

Hence the momentum of a body remains the same provided no external forces act on it.

The principle of conservation of momentum for a closed system (i.e. one on which no external forces act) may be stated as:

The total linear momentum of a system is a constant.

The total momentum of a system before collision in a given direction is equal to the total momentum of the system after collision in the same direction. In Figure 27.1, masses m_1 and m_2 are travelling in the same direction with velocity $u_1 > u_2$. A collision will

occur, and applying the principle of conservation of momentum:

total momentum before impact = total momentum after impact

i.e.
$$m_1 u_1 + m_2 u_2 = m_1 v_1 + m_2 v_2$$

where v_1 and v_2 are the velocities of m_1 and m_2 after impact.

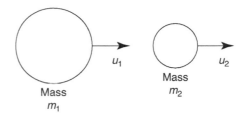

Figure 27.1

Problem 1. Determine the momentum of a pile driver of mass 400 kg when it is moving downwards with a speed of 12 m/s

Momentum = mass × velocity

$$= 400 \, \text{kg} \times 12 \, \text{m/s}$$

$$= \textbf{4800 kg m/s downwards}.$$

Problem 2. A cricket ball of mass 150 g has a momentum of 4.5 kg m/s. Determine the velocity of the ball in km/h

Momentum = mass × velocity, hence

$$\text{velocity} = \frac{\text{momentum}}{\text{mass}} = \frac{4.5 \, \text{kg m/s}}{150 \times 10^{-3} \, \text{kg}} = 30 \, \text{m/s}.$$

$$30 \, \text{m/s} = 30 \, \frac{\text{m}}{\text{s}} \times 3600 \, \frac{\text{s}}{\text{h}} \times \frac{1 \, \text{km}}{1000 \, \text{m}}$$

$$= 30 \times 3.6 \, \text{km/h}$$

$$= \textbf{108 km/h}$$

$$= \textbf{velocity of cricket ball}.$$

Problem 3. Determine the momentum of a railway wagon of mass 50 t moving at a velocity of 72 km/h

*Sir Isaac Newton (25 December 1642–20 March 1727) was an English polymath. Newton showed that the motions of objects are governed by the same set of natural laws, by demonstrating the consistency between Kepler's laws of planetary motion and his theory of gravitation. The SI unit of force is the newton, named in his honour. To find out more about Newton go to www.routledge.com/cw/bird (For image refer to page 29)

Momentum = mass × velocity.

Mass = 50 t = 50,000 kg (since 1 t = 1000 kg) and

$$\text{velocity} = 72\,\text{km/h} = 72\,\frac{\text{km}}{\text{h}} \times \frac{1\,\text{h}}{3600\,\text{s}} \times \frac{1000\,\text{m}}{1\,\text{km}}$$

$$= \frac{72}{3.6}\,\text{m/s} = 20\,\text{m/s}.$$

Hence, **momentum** = 50,000 kg × 20 m/s

$$= 1,000,000\,\text{kg m/s} = \mathbf{10^6\ kg\ m/s}.$$

Problem 4. A wagon of mass 10 t is moving at a speed of 6 m/s and collides with another wagon of mass 15 t, which is stationary. After impact, the wagons are coupled together. Determine the common velocity of the wagons after impact

Mass $m_1 = 10\,\text{t} = 10,000\,\text{kg}$, $m_2 = 15,000\,\text{kg}$ and velocity $u_1 = 6\,\text{m/s}$, $u_2 = 0$

Total momentum before impact

$$= m_1 u_1 + m_2 u_2 = (10,000 \times 6) + (15,000 \times 0)$$

$$= 60,000\,\text{kg m/s}.$$

Let the common velocity of the wagons after impact be v m/s.

Since total momentum before impact = total momentum after impact:

$$60,000 = m_1 v + m_2 v$$

$$= v(m_1 + m_2)$$

$$= v(25,000)$$

Hence, $$v = \frac{60,000}{25,000} = 2.4\,\text{m/s}$$

i.e. **the common velocity after impact is 2.4 m/s in the direction in which the 10 t wagon is initially travelling**.

Problem 5. A body has a mass of 30 g and is moving with a velocity of 20 m/s. It collides with a second body which has a mass of 20 g and which is moving with a velocity of 15 m/s. Assuming that the bodies both have the same velocity after impact, determine this common velocity, (a) when the initial velocities have the same line of action and the same sense and (b) when the initial velocities have the same line of action but are opposite in sense

Mass $m_1 = 30\,\text{g} = 0.030\,\text{kg}$, $m_2 = 20\,\text{g} = 0.020\,\text{kg}$, velocity $u_1 = 20$ m/s and $u_2 = 15$ m/s.

(a) When the velocities have the same line of action and the same sense, both u_1 and u_2 are considered as positive values.

Total momentum before impact

$$= m_1 u_1 + m_2 u_2 = (0.030 \times 20) + (0.020 \times 15)$$

$$= 0.60 + 0.30 = 0.90\,\text{kg m/s}.$$

Let the common velocity after impact be v m/s.

Total momentum before impact

$$= \text{total momentum after impact}$$

i.e. $$0.90 = m_1 v + m_2 v = v(m_1 + m_2)$$

$$0.90 = v(0.030 + 0.020)$$

from which,

common velocity $v = \dfrac{0.90}{0.050} = \mathbf{18\,m/s\ in\ the}$

direction in which the bodies are initially travelling.

(b) When the velocities have the same line of action but are opposite in sense, one is considered as positive and the other negative because velocities are vector quantities. Taking the direction of mass m_1 as positive gives: velocity $u_1 = +20$ m/s and $u_2 = -15$ m/s.

Total momentum before impact

$$= m_1 u_1 + m_2 u_2 = (0.030 \times 20) + (0.020 \times -15)$$

$$= 0.60 - 0.30 = +0.30\,\text{kg m/s}.$$

and since it is positive this indicates a momentum in the same direction as that of mass m_1. If the common velocity after impact is v m/s then

$$0.30 = v(m_1 + m_2) = v(0.050)$$

from which, **common velocity** $v = \dfrac{0.30}{0.050} = \mathbf{6\,m/s}$

in the direction that the 30 g mass is initially travelling.

Now try the following Practice Exercise

Practice Exercise 128 Further problems on linear momentum (answers on page 543)

(Where necessary, take g as 9.81 m/s^2.)

1. Determine the momentum in a mass of 50 kg having a velocity of 5 m/s

2. A milling machine and its component have a combined mass of 400 kg. Determine the momentum of the table and component when the feed rate is 360 mm/min

3. The momentum of a body is 160 kg m/s when the velocity is 2.5 m/s. Determine the mass of the body

4. Calculate the momentum of a car of mass 750 kg moving at a constant velocity of 108 km/h

5. A football of mass 200 g has a momentum of 5 kg m/s. What is the velocity of the ball in km/h?

6. A wagon of mass 8 t is moving at a speed of 5 m/s and collides with another wagon of mass 12 t, which is stationary. After impact, the wagons are coupled together. Determine the common velocity of the wagons after impact

7. A car of mass 800 kg was stationary when hit head-on by a lorry of mass 2000 kg travelling at 15 m/s. Assuming no brakes are applied and the car and lorry move as one, determine the speed of the wreckage immediately after collision

8. A body has a mass of 25 g and is moving with a velocity of 30 m/s. It collides with a second body which has a mass of 15 g and which is moving with a velocity of 20 m/s. Assuming that the bodies both have the same speed after impact, determine their common velocity (a) when the speeds have the same line of action and the same sense and (b) when the speeds have the same line of action but are opposite in sense

27.3 Impulse and impulsive forces

Newton's second law of motion states:

The rate of change of momentum is directly proportional to the applied force producing the change, and takes place in the direction of this force.

In the SI system, the units are such that:

the applied force = rate of change of momentum

$$= \frac{\text{change of momentum}}{\text{time taken}} \qquad (1)$$

When a force is suddenly applied to a body due to either a collision with another body or being hit by an object such as a hammer, the time taken in equation (1) is very small and difficult to measure. In such cases, the total effect of the force is measured by the change of momentum it produces.

Forces that act for very short periods of time are called **impulsive forces**. The product of the impulsive force and the time during which it acts is called the **impulse** of the force and is equal to the change of momentum produced by the impulsive force, i.e.

impulse = applied force × time

= change in linear momentum

Examples where impulsive forces occur include when a gun recoils and when a free-falling mass hits the ground. Solving problems associated with such occurrences often requires the use of the equation of motion: $v^2 = u^2 + 2as$, from chapter 23.

When a pile is being hammered into the ground, the ground resists the movement of the pile and this resistance is called a **resistive force**.

Newton's third law of motion may be stated as:

For every action there is an equal and opposite reaction.

The force applied to the pile is the resistive force; the pile exerts an equal and opposite force on the ground.

In practice, when impulsive forces occur, energy is not entirely conserved and some energy is changed into heat, noise and so on.

Problem 6. The average force exerted on the work-piece of a press-tool operation is 150 kN, and the tool is in contact with the work-piece for 50 ms. Determine the change in momentum

From above, change of linear momentum

$$= \text{applied force} \times \text{time} (= \text{impulse}).$$

Hence, **change in momentum of work-piece**

$$= 150 \times 10^3 \, \text{N} \times 50 \times 10^{-3} \, \text{s}$$

$$= \mathbf{7500 \, kg \, m/s} \quad (\text{since } 1 \, \text{N} = 1 \, \text{kg m/s}^2)$$

Problem 7. A force of 15 N acts on a body of mass 4 kg for 0.2 s. Determine the change in velocity

Impulse = applied force × time

$$= \text{change in linear momentum}$$

i.e. $\quad 15 \, \text{N} \times 0.2 \, \text{s} = \text{mass} \times \text{change in velocity}$

$$= 4 \, \text{kg} \times \text{change in velocity}$$

from which, **change in velocity**

$$= \frac{15 \, \text{N} \times 0.2 \, \text{s}}{4 \, \text{kg}}$$

$$= \mathbf{0.75 \, m/s} \quad (\text{since } 1 \, \text{N} = 1 \, \text{kg m/s}^2)$$

Problem 8. A mass of 8 kg is dropped vertically on to a fixed horizontal plane and has an impact velocity of 10 m/s. The mass rebounds with a velocity of 6 m/s. If the mass–plane contact time is 40 ms, calculate (a) the impulse and (b) the average value of the impulsive force on the plane

(a) Impulse = change in momentum = $m(u_1 - v_1)$ where u_1 = impact velocity = 10 m/s and v_1 = rebound velocity = −6 m/s. (v_1 is negative since it acts in the opposite direction to u_1 and velocity is a vector quantity.)

Thus, **impulse** = $m(u_1 - v_1)$

$$= 8 \, \text{kg}(10 - -6) \, \text{m/s}$$

$$= 8 \times 16 = \mathbf{128 \, kg \, m/s}$$

(b) **Impulsive force** $= \dfrac{\text{impulse}}{\text{time}}$

$$= \frac{128 \, \text{kg m/s}}{40 \times 10^{-3} \, \text{s}}$$

$$= \mathbf{3200 \, N} \quad \text{or} \quad \mathbf{3.2 \, kN}$$

Problem 9. The hammer of a pile driver of mass 1 t falls a distance of 1.5 m onto a pile. The blow takes place in 25 ms and the hammer does not rebound. Determine the average applied force exerted on the pile by the hammer

Initial velocity $u = 0$, acceleration due to gravity $g = 9.81 \, \text{m/s}^2$ and distance $s = 1.5 \, \text{m}$.

Using the equation of motion: $v^2 = u^2 + 2gs$

gives: $\qquad\qquad v^2 = 0^2 + 2(9.81)(1.5)$

from which, impact velocity, $\quad v = \sqrt{(2)(9.81)(1.5)}$

$$= 5.425 \, \text{m/s}$$

Neglecting the small distance moved by the pile and hammer after impact,

momentum lost by hammer

$$= \text{the change of momentum}$$

$$= mv = 1000 \, \text{kg} \times 5.425 \, \text{m/s}$$

Rate of change of momentum $= \dfrac{\text{change of momentum}}{\text{change of time}}$

$$= \frac{1000 \times 5.425}{25 \times 10^{-3}}$$

$$= 217,000 \, \text{N}$$

Since the impulsive force is the rate of change of momentum, **the average force exerted on the pile is 217 kN**

Problem 10. A mass of 40 g having a velocity of 15 m/s collides with a rigid surface and rebounds with a velocity of 5 m/s. The duration of the impact is 0.20 ms. Determine (a) the impulse and (b) the impulsive force at the surface

Mass $m = 40 \, \text{g} = 0.040 \, \text{kg}$, initial velocity $u = 15 \, \text{m/s}$ and final velocity $v = -5 \, \text{m/s}$ (negative since the rebound is in the opposite direction to velocity u).

(a) Momentum before impact $= mu = 0.040 \times 15$

$$= 0.6 \, \text{kg m/s}$$

Momentum after impact $= mv = 0.040 \times -5$

$$= -0.2 \, \text{kg m/s}$$

Impulse = change of momentum

$$= 0.6 - (-0.2)$$

$$= \mathbf{0.8 \, kg \, m/s}$$

(b) **Impulsive force** $= \dfrac{\text{change of momentum}}{\text{change of time}}$

$= \dfrac{0.8\,\text{kg m/s}}{0.20 \times 10^{-3}\,\text{s}}$

$= \mathbf{4000\,N}$ or $\mathbf{4\,kN}$

Now try the following Practice Exercises

Practice Exercise 129 Further problems on impulse and impulsive forces (answers on page 543)

(Where necessary, take g as $9.81\,\text{m/s}^2$)

1. The sliding member of a machine tool has a mass of 200 kg. Determine the change in momentum when the sliding speed is increased from 10 mm/s to 50 mm/s

2. A force of 48 N acts on a body of mass 8 kg for 0.25 s. Determine the change in velocity

3. The speed of a car of mass 800 kg is increased from 54 km/h to 63 km/h in 2 s. Determine the average force in the direction of motion necessary to produce the change in speed

4. A 10 kg mass is dropped vertically on to a fixed horizontal plane and has an impact velocity of 15 m/s. The mass rebounds with a velocity of 5 m/s. If the contact time of mass and plane is 0.025 s, calculate (a) the impulse and (b) the average value of the impulsive force on the plane

5. The hammer of a pile driver of mass 1.2 t falls 1.4 m on to a pile. The blow takes place in 20 ms and the hammer does not rebound. Determine the average applied force exerted on the pile by the hammer

6. A tennis ball of mass 60 g is struck from rest with a racket. The contact time of ball on racket is 10 ms and the ball leaves the racket with a velocity of 25 m/s. Calculate (a) the impulse and (b) the average force exerted by a racket on the ball

7. In a press-tool operation, the tool is in contact with the work piece for 40 ms. If the average force exerted on the work piece is 90 kN, determine the change in momentum

Practice Exercise 130 Short answer questions on linear momentum and impulse

1. Define momentum

2. State Newton's first law of motion

3. State the principle of the conservation of momentum

4. State Newton's second law of motion

5. Define impulse

6. What is meant by an impulsive force?

7. State Newton's third law of motion

Practice Exercise 131 Multiple-choice questions on linear momentum and impulse (answers on page 543)

1. A mass of 100 g has a momentum of 100 kg m/s. The velocity of the mass is:

    (a) 10 m/s   (b) 10^2 m/s

    (c) 10^{-3} m/s   (d) 10^3 m/s

2. A rifle bullet has a mass of 50 g. The momentum when the muzzle velocity is 108 km/h is:

    (a) 54 kg m/s   (b) 1.5 kg m/s

    (c) 15,000 kg m/s   (d) 21.6 kg m/s

    A body P of mass 10 kg has a velocity of 5 m/s and the same line of action as a body Q of mass 2 kg and having a velocity of 25 m/s. The bodies collide, and their velocities are the same after impact. In questions 3–6, select the correct answer from the following:

    (a) 25/3 m/s (b) 360 kg m/s (c) 0

    (d) 30 m/s   (e) 160 kg m/s (f) 100 kg m/s

    (g) 20 m/s

3. Determine the total momentum of the system before impact when P and Q have the same sense

4. Determine the total momentum of the system before impact when P and Q have the opposite sense

5. Determine the velocity of P and Q after impact if their sense is the same before impactc

6. Determine the velocity of P and Q after impact if their sense is opposite before impact

7. A force of 100 N acts on a body of mass 10 kg for 0.1 s. The change in velocity of the body is:

    (a) 1 m/s   (b) 100 m/s

    (c) 0.1 m/s   (d) 0.01 m/s

A vertical pile of mass 200 kg is driven 100 mm into the ground by the blow of a 1 t hammer which falls through 1.25 m. In questions 8–12, take g as 10 m/s² and select the correct answer from the following:

(a) 25 m/s   (b) 25/6 m/s   (c) 5 kg m/s

(d) 0   (e) 625/6 kN   (f) 5000 kg m/s

(g) 5 m/s   (h) 12 kN

8. Calculate the velocity of the hammer immediately before impact

9. Calculate the momentum of the hammer just before impact

10. Calculate the momentum of the hammer and pile immediately after impact assuming they have the same velocity

11. Calculate the velocity of the hammer and pile immediately after impact assuming they have the same velocity

12. Calculate the resistive force of the ground, assuming it to be uniform

For fully worked solutions to each of the problems in Exercises 128 to 131 in this chapter, go to the website:
www.routledge.com/cw/bird

This assignment covers the material contained in chapters 26 and 27. *The marks for each question are shown in brackets at the end of each question.*

1. A metal bar having a cross-sectional area of 80 mm^2 has a tensile force of 20 kN applied to it. Determine the stress in the bar. (4)

2. (a) A rectangular metal bar has a width of 16 mm and can support a maximum compressive stress of 15 MPa; determine the minimum breadth of the bar when loaded with a force of 6 kN.

   (b) If the bar in (a) is 1.5 m long and decreases in length by 0.18 mm when the force is applied, determine the strain and the percentage strain. (7)

3. A wire is stretched 2.50 mm by a force of 400 N. Determine the force that would stretch the wire 3.50 mm, assuming that the elastic limit is not exceeded. (5)

4. A copper tube has an internal diameter of 140 mm and an outside diameter of 180 mm and is used to support a load of 4 kN. The tube is 600 mm long before the load is applied. Determine, in micrometres, by how much the tube contracts when loaded, taking the modulus of elasticity for copper as 96 GPa. (8)

5. Determine the momentum of a lorry of mass 10 tonnes moving at a velocity of 81 km/h. (4)

6. A ball of mass 50 g is moving with a velocity of 4 m/s when it strikes a stationary ball of mass 25 g. The velocity of the 50 g ball after impact is 2.5 m/s in the same direction as before impact. Determine the velocity of the 25 g ball after impact. (8)

7. A force of 24 N acts on a body of mass 6 kg for 150 ms. Determine the change in velocity. (4)

8. The hammer of a pile-driver of mass 800 kg falls a distance of 1.0 m on to a pile. The blow takes place in 20 ms and the hammer does not rebound. Determine (a) the velocity of impact, (b) the momentum lost by the hammer, (c) the average applied force exerted on the pile by the hammer. (10)

For lecturers/instructors/teachers, fully worked solutions to each of the problems in Revision Test 10, together with a full marking scheme, are available at the website:

www.routledge.com/cw/bird

Chapter 28

Torque

Why it is important to understand: **Torque**

This chapter commences by defining a couple and a torque; it then shows how the energy and work done can be calculated from these terms. This chapter then derives the expression which relates torque to the product of mass moment of inertia and the angular acceleration. The expression for kinetic energy due to rotation is also derived. These expressions are then used for calculating the power transmitted from one shaft to another, via a belt. This work is very important for calculating the power transmitted in rotating shafts and other similar artefacts in many branches of engineering. For example, torque is important when designing propeller shafts for ships and automobiles, and also for helicopters, etc

At the end of this chapter, you should be able to:

- define a couple
- define a torque and state its unit
- calculate torque given force and radius
- calculate work done, given torque and angle turned through
- calculate power, given torque and angle turned through
- appreciate kinetic energy $= \dfrac{I\omega^2}{2}$ where I is the moment of inertia
- appreciate that torque $T = I\alpha$ where a is the angular acceleration
- calculate torque given I and α
- calculate kinetic energy given I and ω
- understand power transmission by means of belt and pulley
- perform calculations involving torque, power and efficiency of belt drives

28.1 Introduction

This chapter commences by defining a couple and a torque. It next shows how the energy and work done can be calculated from these terms. It then derives the expression which relates torque to the product of mass moment of inertia and the angular acceleration. The expression for kinetic energy due to rotation is also derived. These expressions are then used for calculating the power transmitted from one shaft to another, via a belt. This work is very important for calculating the power transmitted in rotating shafts and other similar artefacts in many branches of engineering.

Science and Mathematics for Engineering. 978-0-367-20475-4, © John Bird. Published by Taylor & Francis. All rights reserved.

28.2 Couple and torque

When two equal forces act on a body as shown in Figure 28.1, they cause the body to rotate, and the system of forces is called a **couple**. The turning moment of a couple is called a **torque** T.

In Figure 28.1, torque = magnitude of either force × perpendicular distance between the forces,

i.e. $$T = Fd$$

Figure 28.1

The unit of torque is the **newton metre, N m**.

When a force F newtons is applied at a radius r metres from the axis of, say, a nut to be turned by a spanner, as shown in Figure 28.2, the torque T applied to the nut is given by:

$$T = Fr \text{ N m}$$

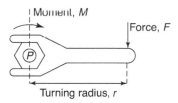

Figure 28.2

Problem 1. Determine the torque when a pulley wheel of diameter 300 mm has a force of 80 N applied at the rim

Torque $T = Fr$, where force $F = 80$ N and radius

$$r = \frac{300}{2} = 150 \text{ mm} = 0.15 \text{ m}$$

Hence, **torque** $T = (80)(0.15) = \mathbf{12\,N\,m}$

Problem 2. Determine the force applied tangentially to a bar of a screw jack at a radius of 800 mm, if the torque required is 600 N m

Torque T = force × radius, from which

$$\mathbf{force} = \frac{\text{torque}}{\text{radius}} = \frac{600 \text{ N m}}{800 \times 10^{-3} \text{ m}} = \mathbf{750\,N}$$

Problem 3. The circular hand-wheel of a valve of diameter 500 mm has a couple applied to it composed of two forces, each of 250 N. Calculate the torque produced by the couple

Torque produced by couple, $T = Fd$, where force $F = 250$ N and distance between the forces $d = 500$ mm $= 0.5$ m

Hence, **torque** $T = (250)(0.5) = \mathbf{125\,N\,m}$.

Now try the following Practice Exercise

Practice Exercise 132   Further problems on torque (answers on page 543)

1. Determine the torque developed when a force of 200 N is applied tangentially to a spanner at a distance of 350 mm from the centre of the nut

2. During a machining test on a lathe, the tangential force on the tool is 150 N. If the torque on the lathe spindle is 12 N m, determine the diameter of the work-piece

28.3 Work done and power transmitted by a constant torque

Figure 28.3(a) shows a pulley wheel of radius r metres attached to a shaft and a force F newtons applied to the rim at point P.

Figure 28.3(b) shows the pulley wheel having turned through an angle θ radians as a result of the force F being applied. The force moves through a distance s, where arc length $s = r\theta$.

Work done = force × distance moved by the force

$$= F \times r\theta = Fr\theta \text{ N m} = Fr\theta \text{ J}$$

However, Fr is the torque T, hence,

$$\mathbf{work\ done} = \mathbf{\mathit{T}\theta\ joules}$$

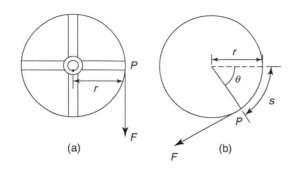

Figure 28.3

$$\text{Average power} = \frac{\text{work done}}{\text{time taken}} = \frac{T\theta}{\text{time taken}}$$

for a constant torque T.

However, (angle θ)/(time taken) = angular velocity ω rad/s.
Hence,

$$\textbf{power } P = T\omega \textbf{ watts} \tag{1}$$

Angular velocity $\omega = 2\pi n$ rad/s where n is the speed in rev/s.
Hence,

$$\textbf{power } P = 2\pi n T \textbf{ watts} \tag{2}$$

Sometimes power is in units of horsepower (hp), where 1 horsepower = 745.7 watts.

i.e. **1 hp = 745.7 watts**

Problem 4. A constant force of 150 N is applied tangentially to a wheel of diameter 140 mm. Determine the work done, in joules, in 12 revolutions of the wheel

Torque $T = Fr$, where $F = 150$ N and radius

$$r = \frac{140}{2} = 70 \text{ mm} = 0.070 \text{ m}$$

Hence, torque $T = (150)(0.070) = 10.5$ N m
Work done $= T\theta$ joules, where torque, $T = 10.5$ N m and angular displacement, $\theta = 12$ revolutions $= 12 \times 2\pi$ rad $= 24\pi$ rad.
Hence, **work done** $= T\theta = (10.5)(24\pi) = \textbf{792 J}$

Problem 5. Calculate the torque developed by a motor whose spindle is rotating at 1000 rev/min and developing a power of 2.50 kW

Power $P = 2\pi n T$ (from above), from which, torque

$$T = \frac{P}{2\pi n} \text{ N m}$$

where power $P = 2.50$ kW $= 2500$ W and speed, $n = 1000/60$ rev/s.
Thus,

$$\textbf{torque } T = \frac{P}{2\pi n} = \frac{2500}{2\pi \left(\dfrac{1000}{60} \right)}$$

$$= \frac{2500 \times 60}{2\pi \times 1000} = \textbf{23.87 N m}$$

Problem 6. An electric motor develops a power of 5 hp and a torque of 12.5 N m. Determine the speed of rotation of the motor in rev/min

Power $P = 2\pi n T$, from which,

$$\text{speed } n = \frac{P}{2\pi T} \text{ rev/s}$$

where   power $P = 5$ hp $= 5 \times 745.7 = 3728.5$ W
and   torque $T = 12.5$ N m.

Hence,   speed $n = \dfrac{3728.5}{2\pi(12.5)} = 47.47$ rev/s.

The speed of rotation of the motor $= 47.47 \times 60$
$$= \textbf{2848 rev/min}.$$

Problem 7. In a turning-tool test, the tangential cutting force is 50 N. If the mean diameter of the work-piece is 40 mm, calculate (a) the work done per revolution of the spindle, (b) the power required when the spindle speed is 300 rev/min

(a) Work done $= T\theta$, where $T = Fr$.

Force $F = 50$ N, radius $r = \dfrac{40}{2} = 20$ mm $= 0.02$ m and angular displacement $\theta = 1$ rev $= 2\pi$ rad.

Hence, **work done per revolution of spindle**
$$= Fr\theta = (50)(0.02)(2\pi) = \textbf{6.28 J}$$

(b) Power $P = 2\pi nT$, where torque $T = Fr =$
(50)(0.02) = 1 N m and speed, $n = \dfrac{300}{60} = 5$ rev/s.

Hence, **power required $P = 2\pi(5)(1) = 31.42$ W**

Problem 8. A motor connected to a shaft develops a torque of 5 kN m. Determine the number of revolutions made by the shaft if the work done is 9 MJ

Work done $= T\theta$, from which, angular displacement

$$\theta = \frac{\text{work done}}{\text{torque}}$$

Work done $= 9\,\text{MJ} = 9 \times 10^6\,\text{J}$

and   torque $= 5\,\text{kN m} = 5000\,\text{N m}$.

Hence, angular displacement

$$\theta = \frac{9 \times 10^6}{5000} = 1800\,\text{rad}.$$

2π rad = 1 rev,

hence, **the number of revolutions made by the shaft**

$$= \frac{1800}{2\pi} = 286.5\,\text{rev}$$

Now try the following Practice Exercise

Practice Exercise 133 Further problems on work done and power transmitted by a constant torque (answers on page 543)

1. A constant force of 4 kN is applied tangentially to the rim of a pulley wheel of diameter 1.8 m attached to a shaft. Determine the work done, in joules, in 15 revolutions of the pulley wheel

2. A motor connected to a shaft develops a torque of 3.5 kN m. Determine the number of revolutions made by the shaft if the work done is 11.52 MJ

3. A wheel is turning with an angular velocity of 18 rad/s and develops a power of 810 W at this speed. Determine the torque developed by the wheel

4. Calculate the torque provided at the shaft of an electric motor that develops an output power of 3.2 hp at 1800 rev/min

5. Determine the angular velocity of a shaft when the power available is 2.75 kW and the torque is 200 N m

6. The drive shaft of a ship supplies a torque of 400 kN m to its propeller at 400 rev/min. Determine the power delivered by the shaft

7. A motor is running at 1460 rev/min and produces a torque of 180 N m. Determine the average power developed by the motor

8. A wheel is rotating at 1720 rev/min and develops a power of 600 W at this speed. Calculate (a) the torque, (b) the work done, in joules, in a quarter of an hour

28.4 Kinetic energy and moment of inertia

The tangential velocity v of a particle of mass m moving at an angular velocity ω rad/s at a radius r metres (see Figure 28.4) is given by:

$$v = \omega r \text{ m/s}$$

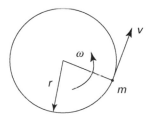

Figure 28.4

The kinetic energy of a particle of mass m is given by:

$$\textbf{Kinetic energy} = \frac{1}{2}mv^2 \text{ (from chapter 21)}$$
$$= \frac{1}{2}m(\omega r)^2$$
$$= \frac{1}{2}m^2\omega^2 r^2 \text{ joules}$$

The total kinetic energy of a system of masses rotating at different radii about a fixed axis but with the same angular velocity, as shown in Figure 28.5, is given by:

$$\text{Total kinetic energy} = \frac{1}{2}m_1\omega^2 r_1^2 + \frac{1}{2}m_2\omega^2 r_2^2 + \frac{1}{2}m_3\omega^2 r_3^2$$

$$= (m_1 r_1^2 + m_2 r_2^2 + m_3 r_3^2)\frac{\omega^2}{2}$$

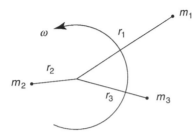

Figure 28.5

In general, this may be written as:

$$\textbf{Total kinetic energy} = \left(\sum mr^2\right)\frac{\omega^2}{2} = I\frac{\omega^2}{2}$$

where $I \left(= \sum mr^2\right)$ is called the **moment of inertia** of the system about the axis of rotation and has units of kg m². The moment of inertia of a system is a measure of the amount of work done to give the system an angular velocity of ω rad/s, or the amount of work that can be done by a system turning at ω rad/s.

From section 28.2, work done $= T\theta$, and if this work is available to increase the kinetic energy of a rotating body of moment of inertia I, then:

$$T\theta = I\left(\frac{\omega_2^2 - \omega_1^2}{2}\right)$$

where ω_1 and ω_2 are the initial and final angular velocities,

i.e. $$T\theta = I\left(\frac{\omega_2 + \omega_1}{2}\right)(\omega_2 - \omega_1)$$

However,

$$\left(\frac{\omega_2 + \omega_1}{2}\right)$$

is the mean angular velocity, i.e. $\dfrac{\theta}{t}$, where t is the time, and $(\omega_2 - \omega_1)$ is the change in angular velocity, i.e. αt, where α is the angular acceleration.

Hence,

$$T\theta = I\left(\frac{\theta}{t}\right)(\alpha t)$$

from which,

torque $T = I\alpha$

where I is the moment of inertia in kg m², α is the angular acceleration in rad/s² and T is the torque in N m.

Problem 9. A shaft system has a moment of inertia of 37.5 kg m². Determine the torque required to give it an angular acceleration of 5.0 rad/s²

Torque $T = I\alpha$, where moment of inertia $I = 37.5$ kg m² and angular acceleration $\alpha = 5.0$ rad/s².

Hence, **torque $T = I\alpha = (37.5)(5.0) = 187.5\,\text{N\,m}$**

Problem 10. A shaft has a moment of inertia of 31.4 kg m². What angular acceleration of the shaft would be produced by an accelerating torque of 495 N m?

Torque $T = I\alpha$, from which, angular acceleration $\alpha = \dfrac{T}{I}$, where torque $T = 495$ N m and moment of inertia $I = 31.4$ kg m².
Hence,

$$\textbf{angular acceleration } \alpha = \frac{495}{31.4} = \textbf{15.76 rad/s}^2$$

Problem 11. A body of mass 100 g is fastened to a wheel and rotates in a circular path of 500 mm in diameter. Determine the increase in kinetic energy of the body when the speed of the wheel increases from 450 rev/min to 750 rev/min

From above, kinetic energy $= I\dfrac{\omega^2}{2}$

Thus, increase in kinetic energy $= I\left(\dfrac{\omega_2^2 - \omega_1^2}{2}\right)$

where moment of inertia $I = mr^2$,
mass $m = 100$ g $= 0.1$ kg and
radius

$$r = \frac{500}{2} = 250\,\text{mm} = 0.25\,\text{m}$$

Initial angular velocity

$$\omega_1 = 450\,\text{rev/min} = \frac{450 \times 2\pi}{60}\,\text{rad/s}$$

$$= 47.12\,\text{rad/s},$$

and final angular velocity

$$\omega_2 = 750 \text{ rev/min} = \frac{750 \times 2\pi}{60} \text{ rad/s}$$

$$= 78.54 \text{ rad/s}.$$

Thus, **increase in kinetic energy**

$$= I\left(\frac{\omega_2^2 - \omega_1^2}{2}\right) = (mr^2)\left(\frac{\omega_2^2 - \omega_1^2}{2}\right)$$

$$= (0.1)(0.25^2)\left(\frac{78.54^2 - 47.12^2}{2}\right) = \mathbf{12.34\,J}$$

Problem 12. A system consists of three small masses rotating at the same speed about the same fixed axis. The masses and their radii of rotation are: 15 g at 250 mm, 20 g at 180 mm and 30 g at 200 mm. Determine (a) the moment of inertia of the system about the given axis and (b) the kinetic energy in the system if the speed of rotation is 1200 rev/min

(a) Moment of inertia of the system $I = \sum mr^2$

i.e. $I = [(15 \times 10^{-3} \text{ kg})(0.25 \text{ m})^2]$

$\qquad + [(20 \times 10^{-3} \text{ kg})(0.18 \text{ m})^2]$

$\qquad + [(30 \times 10^{-3} \text{ kg})(0.20 \text{ m})^2]$

$\qquad = (9.375 \times 10^{-4}) + (6.48 \times 10^{-4})$

$\qquad\qquad\qquad\qquad\qquad + (12 \times 10^{-4})$

$\qquad = 27.855 \times 10^{-4} \text{ kg m}^2$

$\qquad = \mathbf{2.7855 \times 10^{-3} \text{ kg m}^2}$

(b) Kinetic energy $= I\dfrac{\omega^2}{2}$, where moment of inertia $I = 2.7855 \times 10^{-3} \text{kg m}^2$ and angular velocity

$$\omega = 2\pi n = 2\pi\left(\frac{1200}{60}\right) \text{ rad/s} = 40\pi \text{ rad/s}$$

Hence,

kinetic energy in the system

$$= (2.7855 \times 10^{-3})\frac{(40\pi)^2}{2} = \mathbf{21.99\,J}$$

Now try the following Practice Exercise

Practice Exercise 134   Further problems on kinetic energy and moment of inertia (answers on page 543)

1. A shaft system has a moment of inertia of 51.4 kg m². Determine the torque required to give it an angular acceleration of 5.3 rad/s²

2. A shaft has an angular acceleration of 20 rad/s² and produces an accelerating torque of 600 N m. Determine the moment of inertia of the shaft

3. A uniform torque of 3.2 kN m is applied to a shaft while it turns through 25 revolutions. Assuming no frictional or other resistances, calculate the increase in kinetic energy of the shaft (i.e. the work done). If the shaft is initially at rest and its moment of inertia is 24.5 kg m², determine its rotational speed, in rev/min, at the end of the 25 revolutions

4. An accelerating torque of 30 N m is applied to a motor, while it turns through 10 revolutions. Determine the increase in kinetic energy. If the moment of inertia of the rotor is 15 kg m² and its speed at the beginning of the 10 revolutions is 1200 rev/min, determine its speed at the end

5. A shaft with its associated rotating parts has a moment of inertia of 48 kg m². Determine the uniform torque required to accelerate the shaft from rest to a speed of 1500 rev/min while it turns through 15 revolutions

6. A small body, of mass 82 g, is fastened to a wheel and rotates in a circular path of 456 mm diameter. Calculate the increase in kinetic energy of the body when the speed of the wheel increases from 450 rev/min to 950 rev/min

7. A system consists of three small masses rotating at the same speed about the same fixed axis. The masses and their radii of rotation are: 16 g at 256 mm, 23 g at 192 mm and 31 g at 176 mm. Determine (a) the moment of inertia of the system about the given axis, and (b) the kinetic energy in the system if the speed of rotation is 1250 rev/min

28.5 Power transmission and efficiency

A common and simple method of transmitting power from one shaft to another is by means of a **belt** passing over pulley wheels which are keyed to the shafts, as shown in Figure 28.6. Typical applications include an electric motor driving a lathe or a drill, and an engine driving a pump or generator.

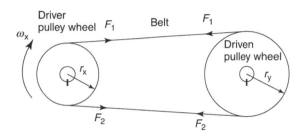

Figure 28.6

For a belt to transmit power between two pulleys there must be a difference in tensions in the belt on either side of the driving and driven pulleys. For the direction of rotation shown in Figure 28.6, $F_2 > F_1$

The torque T available at the driving wheel to do work is given by:

$$T = (F_2 - F_1)r_x \, \text{N m}$$

and the available power P is given by:

$$P = T\omega = (F_2 - F_1)r_x\omega_x \, \text{watts}$$

From section 28.4, the linear velocity of a point on the driver wheel $v_x = r_x\omega_x$

Similarly, the linear velocity of a point on the driven wheel $v_y = r_y\omega_y$. Assuming no slipping,

$$v_x = v_y \quad \text{i.e.} \quad r_x\omega_x = r_y\omega_y$$

Hence $\qquad r_x(2\pi n_x) = r_y(2\pi n_y)$

from which, $\qquad \dfrac{r_x}{r_y} = \dfrac{n_y}{n_x}$

$$\text{Percentage efficiency} = \frac{\text{useful work output}}{\text{energy output}} \times 100$$

or $\qquad$ **efficiency** $= \dfrac{\textbf{power output}}{\textbf{power input}} \times \textbf{100\%}$

Problem 13. An electric motor has an efficiency of 75% when running at 1450 rev/min. Determine the output torque when the power input is 3.0 kW

$$\text{Efficiency} = \frac{\text{power output}}{\text{power input}} \times 100\% \quad \text{hence}$$

$$75 = \frac{\text{power output}}{3000} \times 100 \quad \text{from which,}$$

$$\text{power output} = \frac{75}{100} \times 3000 = 2250 \, \text{W}.$$

From section 28.3, power output $P = 2\pi nT$ from which, torque,

$$T = \frac{P}{2\pi n}, \quad \text{where } n = (1450/60) \, \text{rev/s}.$$

Hence, **output torque** $= \dfrac{2250}{2\pi\left(\dfrac{1450}{60}\right)} = \textbf{14.82 N m}$

Problem 14. A 15 kW motor is driving a shaft at 1150 rev/min by means of pulley wheels and a belt. The tensions in the belt on each side of the driver pulley wheel are 400 N and 50 N. The diameters of the driver and driven pulley wheels are 500 mm and 750 mm respectively. Determine (a) the efficiency of the motor, (b) the speed of the driven pulley wheel

(a) From above,
power output from motor $= (F_2 - F_1)r_x\omega_x$
Force $F_2 = 400$ N and $F_1 = 50$ N, hence
$(F_2 - F_1) = 350$ N,

$$\text{radius } r_x = \frac{500}{2} = 250 \, \text{mm} = 0.25 \, \text{m}$$

and angular velocity

$$\omega_x = \frac{1150 \times 2\pi}{60} \, \text{rad/s}$$

Hence power output from motor $= (F_2 - F_1)r_x\omega_x$

$$= (350)(0.25)\left(\frac{1150 \times 2\pi}{60}\right) = 10.54 \, \text{kW}$$

Power input = 15 kW

Hence, **efficiency of the motor** $= \dfrac{\text{power output}}{\text{power input}}$

$$= \dfrac{10.54}{15} \times 100$$

$$= \mathbf{70.27\%}$$

(b) From above, $\dfrac{r_x}{r_y} = \dfrac{n_y}{n_x}$ from which,

speed of driven pulley wheel

$$n_y = \dfrac{n_x r_x}{r_y} = \dfrac{1150 \times 0.25}{\dfrac{0.750}{2}} = \mathbf{767 \ rev/min}.$$

Problem 15. A crane lifts a load of mass 5 tonne to a height of 25 m. If the overall efficiency of the crane is 65% and the input power to the hauling motor is 100 kW, determine how long the lifting operation takes

---

The increase in potential energy is the work done and is given by mgh (see chapter 21), where mass $m = 5\,t = 5000$ kg, $g = 9.81$ m/s^2 and height $h = 25$ m.

$$\text{Hence, work done} = mgh = (5000)(9.81)(25)$$

$$= 1.226 \ \text{MJ}$$

$$\text{Input power} = 100 \ \text{kW} = 100{,}000 \ \text{W}$$

$$\text{Efficiency} = \dfrac{\text{output power}}{\text{input power}} \times 100$$

$$\text{hence,} \quad 65 = \dfrac{\text{output power}}{100{,}000} \times 100$$

from which,

$$\text{output power} = \dfrac{65}{100} \times 100{,}000 = 65{,}000 \ \text{W}$$

$$= \dfrac{\text{work done}}{\text{time taken}}$$

Thus, time taken for lifting operation

$$= \dfrac{\text{work done}}{\text{output power}} = \dfrac{1.226 \times 10^6 \ \text{J}}{65000 \ \text{W}} = \mathbf{18.86 \ s}$$

Problem 16. The tool of a shaping machine has a mean cutting speed of 250 mm/s and the average cutting force on the tool in a certain shaping operation is 1.2 kN. If the power input to the motor driving the machine is 0.75 kW, determine the overall efficiency of the machine

---

Velocity $v = 250$ mm/s $= 0.25$ m/s and force $F = 1.2$ kN $= 1200$ N.

From chapter 21, power output required at the cutting tool (i.e. power output), $P = \text{force} \times \text{velocity}$

$$= 1200 \ \text{N} \times 0.25 \ \text{m/s} = 300 \ \text{W}$$

Power input $= 0.75$ kW $= 750$ W
Hence,

efficiency of the machine $= \dfrac{\text{output power}}{\text{input power}} \times 100$

$$= \dfrac{300}{750} \times 100 = \mathbf{40\%}$$

Problem 17. Calculate the input power of the motor driving a train at a constant speed of 72 km/h on a level track, if the efficiency of the motor is 80% and the resistance due to friction is 20 kN

Force resisting motion $= 20$ kN $= 20{,}000$ N and

$$\text{velocity} = 72 \ \text{km/h} = \dfrac{72}{3.6} = 20 \ \text{m/s}.$$

Output power from motor $=$ resistive force $\times$ velocity of train (from chapter 21) $= 20{,}000 \times 20 = 400$ kW

$$\text{Efficiency} = \dfrac{\text{power output}}{\text{power input}} \times 100$$

$$\text{hence,} \quad 80 = \dfrac{400}{\text{power input}} \times 100$$

from which, **power input** $= 400 \times \dfrac{100}{80} = \mathbf{500 \ kW}$

Now try the following Practice Exercises

Practice Exercise 135 Further problems on power transmission and efficiency (answers on page 543)

1. A motor has an efficiency of 72% when running at 2600 rev/min. If the output torque is 16 N m at this speed, determine the power supplied to the motor

2. The difference in tensions between the two sides of a belt round a driver pulley of radius 240 mm is 200 N. If the driver pulley wheel is on the shaft of an electric motor running at 700 rev/min and the power input to the motor is 5 kW, determine the efficiency of

the motor. Determine also the diameter of the driven pulley wheel if its speed is to be 1200 rev/min

3. A winch is driven by a 4 kW electric motor and is lifting a load of 400 kg to a height of 5.0 m. If the lifting operation takes 8.6 s, calculate the overall efficiency of the winch and motor

4. A belt and pulley system transmits a power of 5 kW from a driver to a driven shaft. The driver pulley wheel has a diameter of 200 mm and rotates at 600 rev/min. The diameter of the driven wheel is 400 mm. Determine the speed of the driven pulley and the tension in the slack side of the belt when the tension in the tight side of the belt is 1.2 kN

5. The average force on the cutting tool of a lathe is 750 N and the cutting speed is 400 mm/s. Determine the power input to the motor driving the lathe if the overall efficiency is 55%

6. A ship's anchor has a mass of 5 t. Determine the work done in raising the anchor from a depth of 100 m. If the hauling gear is driven by a motor whose output is 80 kW and the efficiency of the haulage is 75%, determine how long the lifting operation takes

Practice Exercise 136   Short answer questions on torque

1. In engineering, what is meant by a couple?

2. Define torque

3. State the unit of torque

4. State the relationship between work, torque T and angular displacement θ

5. State the relationship between power P, torque T and angular velocity ω

6. Complete the following: 1 horsepower = ... watts

7. Define moment of inertia and state the symbol used

8. State the unit of moment of inertia

9. State the relationship between torque, moment of inertia and angular acceleration

10. State one method of power transmission commonly used

11. Define efficiency

Practice Exercise 137   Multiple-choice questions on torque (answers on page 544)

1. The unit of torque is:

   (a) N   (b) Pa   (c) N/m   (d) N m

2. The unit of work is:

   (a) N   (b) J   (c) W   (d) N/m

3. The unit of power is:

   (a) N   (b) J   (c) W   (d) N/m

4. The unit of the moment of inertia is:

   (a) kg m^2 (b) kg   (c) kg/m^2 (d) N m

5. A force of 100 N is applied to the rim of a pulley wheel of diameter 200 mm. The torque is:

   (a) 2 N m   (b) 20 kN m

   (c) 10 N m   (d) 20 N m

6. The work done on a shaft to turn it through 5π radians is 25π J. The torque applied to the shaft is:

   (a) 0.2 N m   (b) $125\pi^2$ N m

   (c) 30π N m   (d) 5 N m

7. A 5 kW electric motor is turning at 50 rad/s. The torque developed at this speed is:

   (a) 100 N m   (b) 250 N m

   (c) 0.01 N m   (d) 0.1 N m

8. The force applied tangentially to a bar of a screw-jack at a radius of 500 mm, if the torque required is 1 kN m, is:

   (a) 2 N   (b) 2 kN   (c) 500 N   (d) 0.5 N

9. A 10 kW motor developing a torque of $(200/\pi)$ N m is running at a speed of:

   (a) $(\pi/20)$ rev/s   (b) 50π rev/s

   (c) 25 rev/s   (d) $(20/\pi)$ rev/s

10. A shaft and its associated rotating parts has a moment of inertia of $50 \, \text{kg} \, \text{m}^2$. The angular acceleration of the shaft to produce an accelerating torque of $5 \, \text{kN} \, \text{m}$ is:

    (a) $10 \, \text{rad/s}^2$ (b) $250 \, \text{rad/s}^2$

    (c) $0.01 \, \text{rad/s}^2$ (d) $100 \, \text{rad/s}^2$

11. A motor has an efficiency of 25% when running at 3000 rev/min. If the output torque is $10 \, \text{N} \, \text{m}$, the power input is:

    (a) $4\pi \, \text{kW}$ (b) $0.25\pi \, \text{kW}$

    (c) $15\pi \, \text{kW}$ (d) $75\pi \, \text{kW}$

12. In a belt–pulley wheel system, the effective tension in the belt is $500 \, \text{N}$ and the diameter of the driver wheel is $200 \, \text{mm}$. If the power output from the driving motor is $5 \, \text{kW}$, the driver pulley wheel turns at:

    (a) $50 \, \text{rad/s}$ (b) $2500 \, \text{rad/s}$

    (c) $100 \, \text{rad/s}$ (d) $0.1 \, \text{rad/s}$

For fully worked solutions to each of the problems in Exercises 132 to 137 in this chapter, go to the website:

Chapter 29

Pressure in fluids

Why it is important to understand: **Pressure in fluids**

This chapter describes fluid pressure and also defines Archimedes' Principle, which is used to determine the buoyancy of boats, yachts, ships, etc. The chapter also describe gauges used in fluid mechanics, such as barometers, manometers, the Bourdon pressure gauge, and vacuum gauges. These gauges are used to determine the properties and behaviour of fluids when they are met in practice. Knowledge of hydrostatics is of importance in boats, ships and other floating bodies and structures.

At the end of this chapter, you should be able to:

- define pressure and state its unit
- understand pressure in fluids
- distinguish between atmospheric, absolute and gauge pressures
- state and apply Archimedes' principle
- describe the construction and principle of operation of different types of barometer
- describe the construction and principle of operation of different types of manometer
- describe the construction and principle of operation of the Bourdon pressure gauge
- describe the construction and principle of operation of different types of vacuum gauge

29.1 Pressure

The pressure acting on a surface is defined as the perpendicular force per unit area of surface. The unit of pressure is the **pascal***, **Pa** where 1 pascal is equal to 1 newton per square metre. Thus pressure,

$$p = \frac{F}{A} \text{ pascals}$$

where F is the force in newtons acting at right-angles to a surface of area A square metres.

When a force of 20 N acts uniformly over, and perpendicular to, an area of 4 m^2, then the pressure on the area p is given by:

$$p = \frac{20\,\text{N}}{4\,\text{m}^2} = 5\,\text{Pa}$$

It should be noted that for **irregular shaped flat surfaces**, such as the water planes of ships, their areas

Science and Mathematics for Engineering. 978-0-367-20475-4, © John Bird. Published by Taylor & Francis. All rights reserved.

can be calculated using the mid-ordinate rule, described in chapter 21.

> **Problem 1.** A table loaded with books has a force of 250 N acting in each of its legs. If the contact area between each leg and the floor is 50 mm², find the pressure each leg exerts on the floor

From above, pressure $p = \dfrac{\text{force}}{\text{area}}$

Hence,

$$p = \frac{250\,\text{N}}{50\,\text{mm}^2} = \frac{250\,\text{N}}{50 \times 10^{-6}\text{m}^2}$$

$$= 5 \times 10^6\,\text{N/m}^2 = \textbf{5 MPa}$$

That is, **the pressure exerted by each leg on the floor is 5 MPa.**

> **Problem 2.** Calculate the force exerted by the atmosphere on a pool of water that is 30 m long by 10 m wide, when the atmospheric pressure is 100 kPa

From above, pressure $= \dfrac{\text{force}}{\text{area}}$

hence, force = pressure × area.
The area of the pool is 30 m × 10 m = 300 m².
Thus, force on pool F = pressure × area
$$= 100\,\text{kPa} \times 300\,\text{m}^2$$
and since 1 Pa = 1 N/m²,

$$F = (100 \times 10^3)\,\frac{\text{N}}{\text{m}^2} \times 300\,\text{m}^2 = 3 \times 10^7\,\text{N}$$

$$= 30 \times 10^6\,\text{N} = \textbf{30 MN}$$

That is, **the force on the pool of water is 30 MN.**

---

*Who was Pascal? – **Blaise Pascal** (19 June 1623–19 August 1662) was a French polymath who made important contributions to the study of fluids and clarified the concepts of pressure and vacuum. He corresponded with Pierre de Fermat on probability theory, strongly influencing the development of modern economics and social science. The unit of pressure, the pascal, is named in his honour. To find out more about *Pascal* go to www.routledge.com/cw/bird (For image refer to page 273)

> **Problem 3.** A circular piston exerts a pressure of 80 kPa on a fluid, when the force applied to the piston is 0.2 kN. Find the diameter of the piston

From above, pressure $= \dfrac{\text{force}}{\text{area}}$

hence, area $= \dfrac{\text{force}}{\text{pressure}}$

Force in newtons = 0.2 kN = 0.2×10^3 N = 200 N, and pressure in pascals = 80 kPa = 80,000 Pa = 80,000 N/m².
Hence,

$$\text{area} = \frac{\text{force}}{\text{pressure}} = \frac{200\,\text{N}}{80,000\,\text{N/m}^2} = 0.0025\,\text{m}^2$$

Since the piston is circular, its area is given by $\pi d^2/4$, where d is the diameter of the piston. Hence,

$$\text{area} = \frac{\pi d^2}{4} = 0.0025$$

$$\text{from which,}\quad d^2 = 0.0025 \times \frac{4}{\pi} = 0.003183$$

i.e. $d = \sqrt{0.003183} = 0.0564\,\text{m} = 56.4\,\text{mm}$

Hence, **the diameter of the piston is 56.4 mm.**

Now try the following Practice Exercise

Practice Exercise 138   Further problems on pressure (answers on page 544)

1. A force of 280 N is applied to a piston of a hydraulic system of cross-sectional area 0.010 m². Determine the pressure produced by the piston in the hydraulic fluid

2. Find the force on the piston of question 1 to produce a pressure of 450 kPa

3. If the area of the piston in question 1 is halved and the force applied is 280 N, determine the new pressure in the hydraulic fluid

29.2 Fluid pressure

A fluid is either a liquid or a gas and there are four basic factors governing the pressure within fluids.

(a) The pressure at a given depth in a fluid is equal in all directions; see Figure 29.1(a).

(b) The pressure at a given depth in a fluid is independent of the shape of the container in which the fluid is held. In Figure 29.1(b), the pressure at X is the same as the pressure at Y.

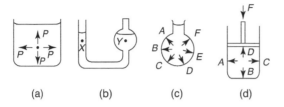

(a)   (b)   (c)   (d)

Figure 29.1

(c) Pressure acts at right angles to the surface containing the fluid. In Figure 29.1(c), the pressures at points A to F all act at right angles to the container.

(d) When a pressure is applied to a fluid, this pressure is transmitted equally in all directions. In Figure 29.1(d), if the mass of the fluid is neglected, the pressures at points A to D are all the same.

The pressure p at any point in a fluid depends on three factors:

(a) the density of the fluid ρ in kg/m^3,

(b) the gravitational acceleration g taken as approximately 9.8 m/s^2 (or the gravitational field force in N/kg), and

(c) the height of fluid vertically above the point h metres.

The relationship connecting these quantities is:

$$p = \rho gh \text{ pascals}$$

When the container shown in Figure 29.2 is filled with water of density 1000 kg/m^3, the pressure due to the water at a depth of 0.03 m below the surface is given by:

$$p = \rho gh = (1000 \times 9.8 \times 0.03) \text{ Pa} = 294 \text{ Pa}$$

In the case of the **Mariana Trench**, which is situated in the Pacific ocean, near Guam, the hydrostatic pressure is about 115.2 MPa or 1152 bar, where 1 bar = 10^5 Pa and the density of sea water is 1020 kg/m^3.

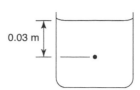

Figure 29.2

Problem 4. A tank contains water to a depth of 600 mm. Calculate the water pressure (a) at a depth of 350 mm and (b) at the base of the tank. Take the density of water as 1000 kg/m^3 and the gravitational acceleration as 9.8 m/s^2

From above, pressure p at any point in a fluid is given by $p = \rho gh$ pascals, where ρ is the density in kg/m^3, g is the gravitational acceleration in m/s^2 and h is the height of fluid vertically above the point in metres.

(a) At a depth of 350 mm = 0.35 m,

$$p = \rho gh = 1000 \times 9.8 \times 0.35$$
$$= \textbf{3430 Pa} = \textbf{3.43 kPa}$$

(b) At the base of the tank, the vertical height of the water is 600 mm = 0.6 m. Hence,

$$p = 1000 \times 9.8 \times 0.6 = \textbf{5880 Pa} = \textbf{5.88 kPa}$$

Problem 5. A storage tank contains petrol to a height of 4.7 m. If the pressure at the base of the tank is 32.3 kPa, determine the density of the petrol. Take the gravitational field force as 9.8 m/s^2

From above, pressure $p = \rho gh$ pascals, where ρ is the density in kg/m^3, g is the gravitational acceleration in m/s^2 and h is the vertical height of the petrol in metres.

Transposing gives: $\rho = \dfrac{p}{gh}$

Pressure p is 32.2 kPa = 32,200 Pa

hence,   density $\rho = \dfrac{32{,}200}{9.8 \times 4.7} = 699 \text{ kg/m}^3$.

That is, **the density of the petrol is 699 kg/m^3**.

Problem 6. A vertical tube is partly filled with mercury of density 13,600 kg/m^3. Find the height, in millimetres, of the column of mercury, when the pressure at the base of the tube is 101 kPa. Take the gravitational field force as 9.8 m/s^2

From above, pressure $p = \rho g h$, hence vertical height h is given by:

$$h = \frac{p}{\rho g}$$

Pressure $p = 101\,\text{kPa} = 101,000\,\text{Pa}$, thus,

$$h = \frac{101,000}{13,600 \times 9.8} = 0.758\,\text{m}$$

That is, **the height of the column of mercury is 758 mm**.

Now try the following Practice Exercise

Practice Exercise 139   Further problems on fluid pressure (answers on page 544)

Take the gravitational acceleration as $9.8\,\text{m/s}^2$.

1. Determine the pressure acting at the base of a dam, when the surface of the water is 35 m above base level. Take the density of water as $1000\,\text{kg/m}^3$

2. An uncorked bottle is full of sea water of density $1030\,\text{kg/m}^3$. Calculate, correct to 3 significant figures, the pressures on the side wall of the bottle at depths of (a) 30 mm and (b) 70 mm below the top of the bottle

3. A U-tube manometer is used to determine the pressure at a depth of 500 mm below the free surface of a fluid. If the pressure at this depth is 6.86 kPa, calculate the density of the liquid used in the manometer

29.3 Atmospheric pressure

The air above the Earth's surface is a fluid, having a density ρ which varies from approximately $1.225\,\text{kg/m}^3$ at sea level to zero in outer space. Since $p = \rho g h$, where height h is several thousands of metres, the air exerts a pressure on all points on the Earth's surface. This pressure, called **atmospheric pressure**, has a value of approximately 100 kilopascals (or 1 bar). Two terms are commonly used when measuring pressures:

(a) **absolute pressure**, meaning the pressure above that of an absolute vacuum (i.e. zero pressure), and

(b) **gauge pressure**, meaning the pressure above that normally present due to the atmosphere.

Thus, **absolute pressure = atmospheric pressure + gauge pressure**.

Thus, a gauge pressure of 50 kPa is equivalent to an absolute pressure of $(100 + 50)\,\text{kPa}$, i.e. 150 kPa, since the atmospheric pressure is approximately 100 kPa.

Problem 7. Calculate the absolute pressure at a point on a submarine, at a depth of 30 m below the surface of the sea, when the atmospheric pressure is 101 kPa. Take the density of sea water as $1030\,\text{kg/m}^3$ and the gravitational acceleration as $9.8\,\text{m/s}^2$

From section 29.2, the pressure due to the sea, that is, the gauge pressure (p_g) is given by:

$$p_g = \rho g h \text{ pascals i.e.}$$

$$p_g = 1030 \times 9.8 \times 30 = 302,820\,\text{Pa} = 302.82\,\text{kPa}$$

From above, absolute pressure

$$= \text{atmospheric pressure} + \text{gauge pressure}$$

$$= (101 + 302.82)\,\text{kPa} = 403.82\,\text{kPa}$$

That is, **the absolute pressure at a depth of 30 m is 403.82 kPa**.

Now try the following Practice Exercise

Practice Exercise 140   Further problems on atmospheric pressure (answers on page 544)

Take the gravitational acceleration as $9.8\,\text{m/s}^2$, the density of water as $1000\,\text{kg/m}^3$, and the density of mercury as $13,600\,\text{kg/m}^3$.

1. The height of a column of mercury in a barometer is 750 mm. Determine the atmospheric pressure, correct to 3 significant figures

2. A U-tube manometer containing mercury gives a height reading of 250 mm of mercury when connected to a gas cylinder. If the barometer reading at the same time is 756 mm of mercury, calculate the absolute pressure of the gas in the cylinder, correct to 3 significant figures

3. A water manometer connected to a condenser shows that the pressure in the condenser is 350 mm below atmospheric pressure. If the barometer is reading 760 mm of mercury, determine the absolute pressure in the condenser, correct to 3 significant figures

4. A Bourdon pressure gauge shows a pressure of 1.151 MPa. If the absolute pressure is 1.25 MPa, find the atmospheric pressure in millimetres of mercury

i.e. $V\rho g$, where V is the volume of the body and ρ is the density of water, i.e.

$$0.835 \, \text{N} = V \times 1000 \, \text{kg/m}^3 \times 9.81 \, \text{m/s}^2$$

$$= V \times 9.81 \, \text{kN/m}^3$$

$$\text{Hence, } V = \frac{0.835}{9.81 \times 10^3} \, \text{m}^3$$

$$= 8.512 \times 10^{-5} \text{m}^3$$

$$= \mathbf{8.512 \times 10^4 \, mm^3}$$

29.4 Archimedes' principle

Archimedes' principle* states that:

If a solid body floats, or is submerged, in a liquid, the liquid exerts an upthrust on the body equal to the gravitational force on the liquid displaced by the body.

In other words, if a solid body is immersed in a liquid, the apparent loss of weight is equal to the weight of liquid displaced.

If V is the volume of the body below the surface of the liquid, then the apparent loss of weight W is given by:

$$W = V\omega = V\rho g$$

where ω is the specific weight (i.e. weight per unit volume) and ρ is the density.

If a body floats on the surface of a liquid all of its weight appears to have been lost. The weight of liquid displaced is equal to the weight of the floating body.

Problem 8. A body weighs 2.760 N in air and 1.925 N when completely immersed in water of density 1000 kg/m³. Calculate (a) the volume of the body, (b) the density of the body and (c) the relative density of the body. Take the gravitational acceleration as 9.81 m/s²

(a) The apparent loss of weight is 2.760 N − 1.925 N = 0.835 N. This is the weight of water displaced,

*Who was Archimedes? – **Archimedes of Syracuse** (c.287 BC–c.212 BC) was a Greek polymath, credited with advances in hydrostatics, statics, groundbreaking work on levers, and designing innovative machines, including the Archimedes screw pump that is still in use today. He is probably most famous for shouting 'Eureka' though, when he discovered a method for determining the volume of an object with an irregular shape whilst in the bathtub. To find out more about Archimedes go to www.routledge.com/cw/bird

(b)   The density of the body

$$= \frac{\text{mass}}{\text{volume}} = \frac{\text{weight}}{g \times V}$$

$$= \frac{2.760\,\text{N}}{9.81\,\text{m/s}^2 \times 8.512 \times 10^{-5}\,\text{m}^3}$$

$$= \frac{\dfrac{2.760}{9.81}\,\text{kg} \times 10^5}{8.512\,\text{m}^3} = \mathbf{3305\,kg/m^3}$$

$$= \mathbf{3.305\,tonne/m^3}$$

(c)   Relative density $= \dfrac{\text{density}}{\text{density of water}}$

Hence, **the relative density of the body**

$$= \frac{3305\,\text{kg/m}^3}{1000\,\text{kg/m}^3} = \mathbf{3.305}$$

> **Problem 9.** A rectangular watertight box is 560 mm long, 420 mm wide and 210 mm deep. It weighs 223 N. (a) If it floats with its sides and ends vertical in water of density 1030 kg/m³, what depth of the box will be submerged? (b) If the box is held completely submerged in water of density 1030 kg/m³, by a vertical chain attached to the underside of the box, what is the force in the chain?

(a)   The apparent weight of a floating body is zero. That is, the weight of the body is equal to the weight of liquid displaced. This is given by: $V\rho g$ where V is the volume of liquid displaced, and ρ is the density of the liquid.

Here,

$$223\,\text{N} = V \times 1030\,\text{kg/m}^3 \times 9.81\,\text{m/s}^2$$

$$= V \times 10.104\,\text{kN/m}^3$$

Hence,

$$V = \frac{223\,\text{N}}{10.104\,\text{kN/m}^3} = 22.07 \times 10^{-3}\,\text{m}^3$$

This volume is also given by Lbd, where $L =$ length of box, $b =$ breadth of box and $d =$ depth of box submerged, i.e.

$$22.07 \times 10^{-3}\,\text{m}^3 = L \times b \times d$$

$$= 0.56\,\text{m} \times 0.42\,\text{m} \times d$$

Hence, **depth submerged**

$$d = \frac{22.07 \times 10^{-3}}{0.56 \times 0.42} = 0.09384\,\text{m} = \mathbf{93.84\,mm}.$$

(b)   The volume of water displaced is the total volume of the box. The upthrust or buoyancy of the water, i.e. the 'apparent loss of weight', is greater than the weight of the box. The force in the chain accounts for the difference.

Volume of water displaced

$$V = 0.56\,\text{m} \times 0.42\,\text{m} \times 0.21\,\text{m}$$

$$= 4.9392 \times 10^{-2}\,\text{m}^3$$

Weight of water displaced

$$= V\rho g = 4.9392 \times 10^{-2}\,\text{m}^3$$

$$\times 1030\,\text{kg/m}^3 \times 9.81\,\text{m/s}^2$$

$$= 499.1\,\text{N}$$

The **force in the chain**

$$= \text{weight of water displaced} - \text{weight of box}$$

$$= 499.1\,\text{N} - 223\,\text{N} = \mathbf{276.1\,N}$$

Now try the following Practice Exercise

Practice Exercise 141   Further problems on Archimedes' principle (answers on page 544)

Take the gravitational acceleration as 9.8 m/s², the density of water as 1000 kg/m³ and the density of mercury as 13600 kg/m³.

1.   A body of volume 0.124 m³ is completely immersed in water of density 1000 kg/m³. What is the apparent loss of weight of the body?

2.   A body of weight 27.4 N and volume 1240 cm³ is completely immersed in water of specific weight 9.81 kN/m³. What is its apparent weight?

3.   A body weighs 512.6 N in air and 256.8 N when completely immersed in oil of density 810 kg/m³. What is the volume of the body?

4. A body weighs 243 N in air and 125 N when completely immersed in water. What will it weigh when completely immersed in oil of relative density 0.8?

5. A watertight rectangular box, 1.2 m long and 0.75 m wide, floats with its sides and ends vertical in water of density 1000 kg/m^3. If the depth of the box in the water is 280 mm, what is its weight?

6. A body weighs 18 N in air and 13.7 N when completely immersed in water of density 1000 kg/m^3. What is the density and relative density of the body?

7. A watertight rectangular box is 660 mm long and 320 mm wide. Its weight is 336 N. If it floats with its sides and ends vertical in water of density 1020 kg/m^3, what will be its depth in the water?

8. A watertight drum has a volume of 0.165 m^3 and a weight of 115 N. It is completely submerged in water of density 1030 kg/m^3, held in position by a single vertical chain attached to the underside of the drum. What is the force in the chain?

29.5 Measurement of pressure

As stated earlier, pressure is the force exerted by a fluid per unit area. A fluid (i.e. liquid, vapour or gas) has a negligible resistance to a shear force, so that the force it exerts always acts at right angles to its containing surface.

The SI unit of pressure is the **pascal**, Pa, which is unit force per unit area, i.e. **1 Pa = 1 N/m^2**.

The pascal is a very small unit and a commonly used larger unit is the bar, where **1 bar = 10^5 Pa**.

Atmospheric pressure is due to the mass of the air above the Earth's surface, being attracted by Earth's gravity. Atmospheric pressure changes continuously. A standard value of atmospheric pressure, called 'standard atmospheric pressure', is often used, having a value of 101,325 Pa or 1.01325 bars or 1013.25 millibars. This latter unit, the millibar, is usually used in the measurement of meteorological pressures. (Note that when atmospheric pressure varies from 101,325 Pa it is no longer standard.)

Pressure indicating instruments are made in a wide variety of forms because of their many different applications. Apart from the obvious criteria such as pressure range, accuracy and response, many measurements also require special attention to material, sealing and temperature effects. The fluid whose pressure is being measured may be corrosive or may be at high temperatures. Pressure indicating devices used in science and industry include:

(i) barometers (see section 29.6),

(ii) manometers (see section 29.8),

(iii) Bourdon pressure gauges (see section 29.9), and

(iv) McLeod and Pirani gauges (see section 29.10).

29.6 Barometers

29.6.1 Introduction

A barometer is an instrument for measuring atmospheric pressure. It is affected by seasonal changes of temperature. Barometers are therefore also used for the measurement of altitude and also as one of the aids in weather forecasting. The value of atmospheric pressure will thus vary with climatic conditions, although not usually by more than about 10% of standard atmospheric pressure.

29.6.2 Construction and principle of operation

A simple barometer consists of a glass tube, just less than 1 m in length, sealed at one end, filled with mercury and then inverted into a trough containing more mercury. Care must be taken to ensure that no air enters the tube during this latter process. Such a barometer is shown in Figure 29.3(a) and it is seen that the level of the mercury column falls, leaving an empty space, called a vacuum. Atmospheric pressure acts on the surface of the mercury in the trough as shown and this pressure is equal to the pressure at the base of the column of mercury in the inverted tube, i.e. the pressure of the atmosphere is supporting the column of mercury. If the atmospheric pressure falls the barometer height h decreases. Similarly, if the atmospheric pressure rises then h increases. Thus atmospheric pressure can be measured in terms of the height of the mercury column. It may be shown that for mercury the height h is 760 mm at standard atmospheric pressure, i.e. a vertical column of mercury 760 mm high exerts a pressure equal to the standard value of atmospheric pressure.

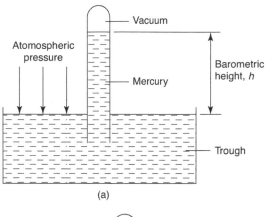

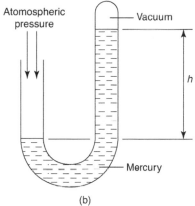

Figure 29.3

There are thus several ways in which atmospheric pressure can be expressed:

Standard atmospheric pressure

$$= 101,325 \text{ Pa or } 101.325 \text{ kPa}$$

$$= 101,325 \text{ N/m}^2 \text{ or } 101.325 \text{ kN/m}^2$$

$$= 1.01325 \text{ bars or } 1013.25 \text{ mbars}$$

$$= 760 \text{ mm of mercury}$$

Another arrangement of a typical barometer is shown in Figure 29.3(b) where a U-tube is used instead of an inverted tube and trough, the principle being similar.

If, instead of mercury, water was used as the liquid in a barometer, then the barometric height h at standard atmospheric pressure would be 13.6 times more than for mercury, i.e. about 10.4 m high, which is not very

practicable. This is because the relative density of mercury is 13.6.

29.6.3 Types of barometer

The **Fortin*** **barometer** is an example of a mercury barometer that enables barometric heights to be measured to a high degree of accuracy (in the order of one-tenth of a millimetre or less). Its construction is merely a more sophisticated arrangement of the inverted tube and trough shown in Figure 29.3(a), with the addition of a vernier scale to measure the barometric height with great accuracy. A disadvantage of this type of barometer is that it is not portable.

A Fortin barometer is shown in Figure 29.4. Mercury is contained in a leather bag at the base of the mercury reservoir, and height H of the mercury in the reservoir can be adjusted using the screw at the base of the barometer to depress or release the leather bag. To measure the atmospheric pressure the screw is adjusted until the pointer at H is just touching the surface of the mercury

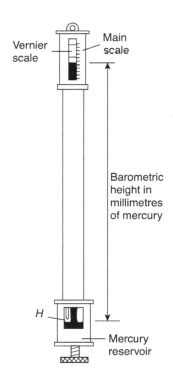

Figure 29.4

*Who was Fortin? – **Jean Nicolas Fortin** (1750–1831) was a French maker of scientific instruments. Fortin is chiefly remembered for his design of barometer, now called a Fortin barometer, which he introduced in about 1800. To find out more about Fortin go to www.routledge.com/cw/bird

and the height of the mercury column is then read using the main and vernier scales. The measurement of atmospheric pressure using a Fortin barometer is much more accurate than using a simple barometer.

A portable type often used is the **aneroid barometer**. Such a barometer consists basically of a circular, hollow, sealed vessel S usually made from thin flexible metal. The air pressure in the vessel is reduced to nearly zero before sealing, so that a change in atmospheric pressure will cause the shape of the vessel to expand or contract. These small changes can be magnified by means of a lever and be made to move a pointer over a calibrated scale. Figure 29.5 shows a typical arrangement of an aneroid barometer. The scale is usually circular and calibrated in millimetres of mercury. These instruments require frequent calibration.

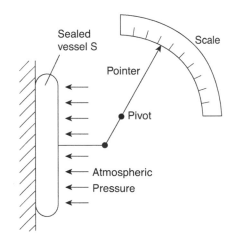

Figure 29.5

29.7 Absolute and gauge pressure

A barometer measures the true or absolute pressure of the atmosphere. The term absolute pressure means the pressure above that of an absolute vacuum (which is zero pressure), as stated earlier. In Figure 29.6 a pressure scale is shown with the line AB representing absolute zero pressure (i.e. a vacuum) and the line CD representing atmospheric pressure. With most practical pressure-measuring instruments the part of the instrument that is subjected to the pressure being measured is also subjected to atmospheric pressure. Thus practical instruments actually determine the difference between the pressure being measured and atmospheric pressure. The pressure that the instrument

is measuring is then termed the gauge pressure. In Figure 29.6, the line EF represents an absolute pressure which has a value greater than atmospheric pressure, i.e. the 'gauge' pressure is positive.

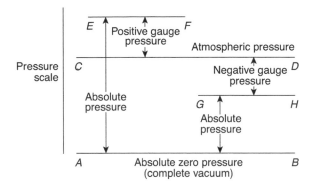

Figure 29.6

Thus, **absolute pressure = gauge pressure + atmospheric pressure**.

Hence a gauge pressure of, say, 60 kPa recorded on an indicating instrument when the atmospheric pressure is 101 kPa is equivalent to an absolute pressure of 60 kPa + 101 kPa, or 161 kPa.

Pressure-measuring indicating instruments are referred to generally as **pressure gauges** (which acts as a reminder that they measure 'gauge' pressure).

It is possible, of course, for the pressure indicated on a pressure gauge to be below atmospheric pressure, i.e. the gauge pressure is negative. Such a gauge pressure is often referred to as a vacuum, even though it does not necessarily represent a complete vacuum at absolute zero pressure. Such a pressure is shown by the line GH in Figure 29.6. An indicating instrument used for measuring such pressures is called a **vacuum gauge**.

A vacuum gauge indication of, say, 0.4 bar means that the pressure is 0.4 bar less than atmospheric pressure. If atmospheric pressure is 1 bar, then the absolute pressure is 1 − 0.4 or 0.6 bar.

29.8 The manometer

A manometer is a device for measuring or comparing fluid pressures, and is the simplest method of indicating such pressures.

29.8.1 U-tube manometer

A U-tube manometer consists of a glass tube bent into a U shape and containing a liquid such as mercury. A U-tube manometer is shown in Figure 29.7(a). If limb A is connected to a container of gas whose pressure is above atmospheric, then the pressure of the gas will cause the levels of mercury to move as shown in Figure 29.7(b), such that the difference in height is h_1. The measuring scale can be calibrated to give the gauge pressure of the gas as h_1 mm of mercury.

If limb A is connected to a container of gas whose pressure is below atmospheric then the levels of mercury will move as shown in Figure 29.7(c), such that their pressure difference is h_2 mm of mercury.

It is also possible merely to compare two pressures, say P_A and P_B, using a U-tube manometer. Figure 29.7(d) shows such an arrangement with $(P_B - P_A)$ equivalent to h mm of mercury. One application of this differential pressure-measuring device is in determining the velocity of fluid flow in pipes.

For the measurement of lower pressures, water or paraffin may be used instead of mercury in the U-tube to give larger values of h and thus greater sensitivity.

29.8.2 Inclined manometers

For the measurement of very low pressures, greater sensitivity is achieved by using an inclined manometer, a typical arrangement of which is shown in Figure 29.8. With the inclined manometer the liquid used is water and the scale attached to the inclined tube is calibrated in terms of the vertical height h. Thus when a vessel containing gas under pressure is connected to the reservoir, movement of the liquid levels of the manometer occurs. Since small-bore tubing is used the movement of the liquid in the reservoir is very small compared with the movement in the inclined tube and is thus neglected. Hence the scale on the manometer is usually in the range 0.2 mbar to 2 mbar.

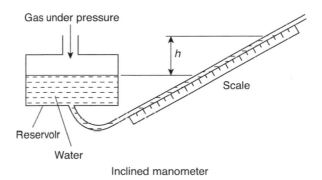

Figure 29.8

The pressure of a gas that a manometer is capable of measuring is naturally limited by the length of tube used. Most manometer tubes are less than 2 m in length and this restricts measurement to a maximum pressure of about 2.5 bar (or 250 kPa) when mercury is used.

29.9 The Bourdon pressure gauge

Pressures many times greater than atmospheric can be measured by the Bourdon pressure gauge, which is the most extensively used of all pressure-indicating instruments. It is a robust instrument. Its main component is a piece of metal tube (called the Bourdon tube), usually made of phosphor bronze or alloy steel, of oval or elliptical cross-section, sealed at one end and bent into an arc. In some forms the tube is bent into a spiral for greater sensitivity. A typical arrangement is shown in Figure 29.9(a). One end E of the Bourdon tube is fixed and the fluid whose pressure is to be measured is

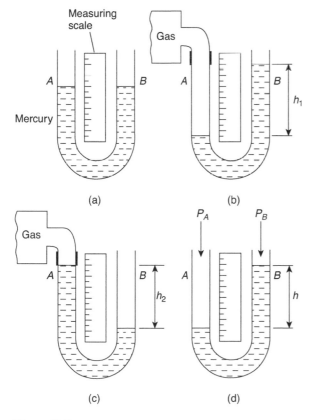

Figure 29.7

connected to this end. The pressure acts at right angles to the metal tube wall as shown in the cross-section of the tube in Figure 29.9(b). Because of its elliptical shape it is clear that the sum of the pressure components, i.e. the total force acting on the sides A and C, exceeds the sum of the pressure components acting on ends B and D. The result is that sides A and C tend to move outwards and B and D inwards tending to form a circular cross-section. As the pressure in the tube is increased the tube tends to uncurl, or if the pressure is reduced the tube curls up further. The movement of the free end of the tube is, for practical purposes, proportional to the pressure applied to the tube, this pressure, of course, being the gauge pressure (i.e. the difference between atmospheric pressure acting on the outside of the tube and the applied pressure acting on the inside of the tube). By using a link, a pivot and a toothed segment as shown in Figure 29.9(a), the movement can be converted into the rotation of a pointer over a graduated calibrated scale.

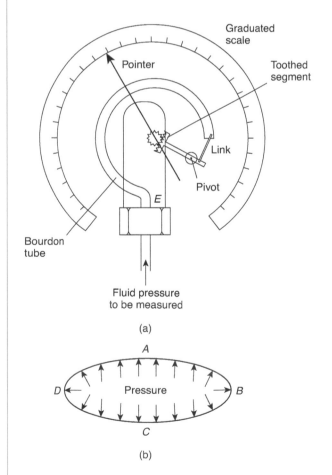

Figure 29.9

The Bourdon tube pressure gauge is capable of measuring high pressures up to 10^4 bar (i.e. 7600 m of mercury) with the addition of special safety features.

A pressure gauge must be calibrated, and this is done either by a manometer, for low pressures, or by a piece of equipment called a **'dead weight tester'**. This tester consists of a piston operating in an oil-filled cylinder of known bore, and carrying accurately known weights as shown in Figure 29.10. The gauge under test is attached to the tester and a screwed piston or ram applies the required pressure, until the weights are just lifted. While the gauge is being read, the weights are turned to reduce friction effects.

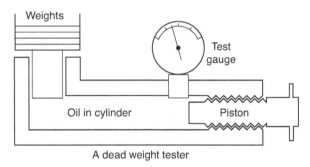

A dead weight tester

Figure 29.10

29.10 Vacuum gauges

Vacuum gauges are instruments for giving a visual indication, by means of a pointer, of the amount by which the pressure of a fluid applied to the gauge is less than the pressure of the surrounding atmosphere. Two examples of vacuum gauges are the McLeod gauge and the Pirani gauge.

29.10.1 McLeod gauges

The **McLeod* gauge** is normally regarded as a standard and is used to calibrate other forms of vacuum gauges. The basic principle of this gauge is that it takes a known volume of gas at a pressure so low that it cannot be measured, then compresses the gas in a known ratio until the pressure becomes large enough to be measured by an ordinary manometer. This device is used to measure low

*Who was McLeod? – **Herbert McLeod** (February 1841– October 1923) was a British chemist, noted for the invention of the McLeod gauge and for the invention of a sunshine recorder. To find out more about McLeod go to www.routledge.com/cw/bird

pressures, often in the range 10^{-6} to 1.0 mm of mercury. A disadvantage of the McLeod gauge is that it does not give a continuous reading of pressure and is not suitable for registering rapid variations in pressure.

29.10.2 Pirani gauges

The **Pirani*** **gauge** measures the resistance and thus the temperature of a wire through which current is flowing. The thermal conductivity decreases with the pressure in the range 10^{-1} to 10^{-4} mm of mercury so that the increase in resistance can be used to measure pressure in this region. The Pirani gauge is calibrated by comparison with a McLeod gauge.

Now try the following Practice Exercises

Practice Exercise 142 Short answer questions on pressure in fluids

1. Define pressure

2. State the unit of pressure

3. Define a fluid

4. State the four basic factors governing the pressure in fluids

5. Write down a formula for determining the pressure at any point in a fluid in symbols, defining each of the symbols and giving their units

6. What is meant by atmospheric pressure?

7. State the approximate value of atmospheric pressure

8. State what is meant by gauge pressure

9. State what is meant by absolute pressure

10. State the relationship between absolute, gauge and atmospheric pressures

11. State Archimedes' principle

---

*Who was Pirani? – **Marcello Stefano Pirani** (1 July, 1880–11 January, 1968) was a German physicist known for his invention of the Pirani vacuum gauge, a vacuum gauge based on the principle of heat loss measurement. Throughout his career, he worked on advancing lighting technology and pioneered work on the physics of gas discharge. To find out more about Pirani go to www.routledge.com/cw/bird

---

12. Name four pressure measuring devices

13. Standard atmospheric pressure is 101325 Pa. State this pressure in millibars

14. Briefly describe how a barometer operates

15. State the advantage of a Fortin barometer over a simple barometer

16. What is the main disadvantage of a Fortin barometer?

17. Briefly describe an aneroid barometer

18. What is a vacuum gauge?

19. Briefly describe the principle of operation of a U-tube manometer

20. When would an inclined manometer be used in preference to a U-tube manometer?

21. Briefly describe the principle of operation of a Bourdon pressure gauge

22. What is a 'dead weight tester'?

23. What is a Pirani gauge?

24. What is a McLeod gauge used for?

Practice Exercise 143 Multiple-choice questions on pressure in fluids (answers on page 544)

1. A force of 50 N acts uniformly over and at right angles to a surface. When the area of the surface is 5 m^2, the pressure on the area is:

   (a) 250 Pa   (b) 10 Pa

   (c) 45 Pa   (d) 55 Pa

2. Which of the following statements is false? The pressure at a given depth in a fluid

   (a) is equal in all directions

   (b) is independent of the shape of the container

   (c) acts at right angles to the surface containing the fluid

   (d) depends on the area of the surface

3. A container holds water of density $1000\,kg/m^3$. Taking the gravitational acceleration as $10\,m/s^2$, the pressure at a depth of $100\,mm$ is:

   (a) 1 kPa (b) 1 MPa

   (c) 100 Pa (d) 1 Pa

4. If the water in question 3 is now replaced by a fluid having a density of $2000\,kg/m^3$, the pressure at a depth of $100\,mm$ is:

   (a) 2 kPa (b) 500 kPa

   (c) 200 Pa (d) 0.5 Pa

5. The gauge pressure of fluid in a pipe is $70\,kPa$ and the atmospheric pressure is $100\,kPa$. The absolute pressure of the fluid in the pipe is:

   (a) 7 MPa (b) 30 kPa

   (c) 170 kPa (d) 10/7 kPa

6. A U-tube manometer contains mercury of density $13,600\,kg/m^3$. When the difference in the height of the mercury levels is $100\,mm$ and taking the gravitational acceleration as $10\,m/s^2$, the gauge pressure is:

   (a) 13.6 Pa (b) 13.6 MPa

   (c) 13 710 Pa (d) 13.6 kPa

7. The mercury in the U-tube of question 6 is to be replaced by water of density $1000\,kg/m^3$. The height of the tube to contain the water for the same gauge pressure is:

   (a) (1/13.6) of the original height

   (b) 13.6 times the original height

   (c) 13.6 m more than the original height

   (d) 13.6 m less than the original height

8. Which of the following devices does not measure pressure?
   (a) barometer (b) McLeod gauge

   (c) thermocouple (d) manometer

9. A pressure of $10\,kPa$ is equivalent to:
   (a) 10 millibars (b) 1 bar

   (c) 0.1 bar (d) 0.1 millibars

10. A pressure of $1000\,mbars$ is equivalent to:
    (a) $0.1\,kN/m^2$ (b) $10\,kPa$

    (c) $1000\,Pa$ (d) $100\,kN/m^2$

11. Which of the following statements is false?

    (a) Barometers may be used for the measurement of altitude

    (b) Standard atmospheric pressure is the pressure due to the mass of the air above the ground

    (c) The maximum pressure that a mercury manometer, using a 1 m length of glass tubing, is capable of measuring is in the order of 130 kPa

    (d) An inclined manometer is designed to measure higher values of pressure than the U-tube manometer

In questions 12 and 13 assume that atmospheric pressure is 1 bar.

12. A Bourdon pressure gauge indicates a pressure of 3 bars. The absolute pressure of the system being measured is:
    (a) 1 bar (b) 2 bars

    (c) 3 bars (d) 4 bars

13. In question 12, the gauge pressure is:
    (a) 1 bar (b) 2 bars

    (c) 3 bars (d) 4 bars

In questions 14–18 select the most suitable pressure-indicating device from the following list:

    (a) mercury-filled U-tube manometer

    (b) Bourdon gauge

    (c) McLeod gauge

    (d) aneroid barometer

    (e) Pirani gauge

    (f) Fortin barometer

    (g) water-filled inclined barometer

14. A robust device to measure high pressures in the range 0–30 MPa

15. Calibration of a Pirani gauge

16. Measurement of gas pressures comparable with atmospheric pressure

17. To measure pressures of the order of 1 MPa

18. Measurement of atmospheric pressure to a high degree of accuracy

19. Figure 29.7(b), on page 309, shows a U-tube manometer connected to a gas under pressure. If atmospheric pressure is 76 cm of mercury and h_1 is measured in centimetres then the gauge pressure (in cm of mercury) of the gas is:

    (a) h_1 (b) $h_1 + 76$

    (c) $h_1 - 76$ (d) $76 - h_1$

20. In question 19 the absolute pressure of the gas (in cm of mercury) is:

    (a) h_1 (b) $h_1 + 76$ (c) $h_1 - 76$ (d) $76 - h_1$

21. Which of the following statements is true?

    (a) Atmospheric pressure of $101.325 \, \text{kN/m}^2$ is equivalent to 101.325 millibars

    (b) An aneroid barometer is used as a standard for calibration purposes

    (c) In engineering, 'pressure' is the force per unit area exerted by fluids

    (d) Water is normally used in a barometer to measure atmospheric pressure

For fully worked solutions to each of the problems in Exercises 138 to 143 in this chapter, go to the website:

Chapter 30

Heat energy and transfer

Why it is important to understand: **Heat energy and transfer**

This chapter defines sensible and latent heat and provides the appropriate formulae to calculate the amount of energy required to convert a solid to a gas and vice-versa, and also for other combinations of solids, liquids and gases. This information is often required by engineers if they are required to design an artefact, say, to convert ice to steam via the state of liquid water. An example of a household requirement of when this type of calculation is required is that of the simple domestic kettle. When the designer is required to design a domestic electric kettle, it is important that the design is such that the powering arrangement is just enough to boil the required amount of water in a reasonable time. If the powering were too low, you may have great difficulty in boiling the water when the kettle is full. Similar calculations are required for large water containers, which are required to boil large quantities of water for other uses, including for kitchens in schools, to make tea/coffee, etc, and for large hotels, which have many uses for hot water. The chapter also describes the three main methods of heat transfer, namely conduction, convection and radiation, together with their uses. All of this is fundamental in the design of heat engines, compressors, refrigerators, etc.

At the end of this chapter, you should be able to:

- distinguish between heat and temperature
- appreciate that temperature is measured on the Celsius or the thermodynamic scale
- convert temperatures from Celsius into Kelvin and vice versa
- recognise several temperature measuring devices
- define specific heat capacity, c and recognise typical values
- calculate the quantity of heat energy Q using $Q = mc(t_2 - t_1)$
- understand change of state from solid to liquid to gas, and vice versa
- distinguish between sensible and latent heat
- define specific latent heat of fusion
- define specific latent heat of vaporisation
- recognise typical values of latent heats of fusion and vaporisation
- calculate quantity of heat Q using $Q = mL$
- describe the principle of operation of a simple refrigerator
- understand conduction, convection and radiation
- describe the principle of operation of a vacuum flask
- define thermal efficiency, calorific value and combustion
- understand heat exchangers through practical calculations

Science and Mathematics for Engineering. 978-0-367-20475-4. © John Bird. Published by Taylor & Francis. All rights reserved.

30.1 Introduction

This chapter defines sensible and latent heat and provides the appropriate formulae to calculate the amount of energy required to convert a solid to a gas and vice-versa, and also for other combinations of solids, liquids and gases. This information is often required by engineers if they are required to design an artefact, say, to convert ice to steam via the state of liquid water. An example of a household requirement of when this type of calculation is required is that of the simple domestic kettle. When the designer is required to design a domestic electric kettle, it is important that the design is such that the powering arrangement is (just) enough to boil the required amount of water in a reasonable time. If the powering were too low, you may have great difficulty in boiling the water when the kettle is full. Similar calculations are required for large water containers, which are required to boil large quantities of water for other uses, including for kitchens in schools, to make tea/coffee, etc., and for large hotels, which have many uses for hot water. The chapter also describes the three main methods of heat transfer, namely conduction, convection and radiation, together with their uses.

30.2 Heat and temperature

Heat is a form of energy and is measured in joules.
Temperature is the degree of hotness or coldness of a substance. Heat and temperature are thus not the same thing. For example, twice the heat energy is needed to boil a full container of water than half a container – that is, different amounts of heat energy are needed to cause an equal rise in the temperature of different amounts of the same substance.

Temperature is measured either (i) on the **Celsius*** (°C) scale (formerly Centigrade), where the temperature at which ice melts, i.e. the freezing point of water, is taken as 0°C and the point at which water boils under normal atmospheric pressure is taken as 100°C, or (ii) on the **thermodynamic scale**, in which the unit of temperature is the **kelvin*** (K). The Kelvin scale uses the same temperature interval as the Celsius scale but as its zero takes the 'absolute zero of temperature' which is at about – 273°C. Hence,

Kelvin temperature = degree Celsius + 273

i.e. $$K = (°C) + 273$$

Thus, for example, $0°C = 273$ K, $25°C = 298$ K and $100°C = 373$ K.

Problem 1.   Convert the following temperatures into the Kelvin scale: (a) 37°C, (b) −28°C

From above,

· Kelvin temperature = degrees Celsius + 273.

(a) 37°C corresponds to a Kelvin temperature of $37 + 273$, i.e. **310 K**

(b) −28°C corresponds to a Kelvin temperature of $-28 + 273$, i.e. **245 K**

Problem 2.   Convert the following temperatures into the Celsius scale:

(a) 365 K   (b) 213 K

*Who was Celsius? – go to www.routledgr.com/cw/bird (for image of Celsius refer to page 51)
*William Thomson, 1st Baron Kelvin (26 June 1824–17 December 1907), was an Irish mathematical physicist and engineer. Lord Kelvin is widely known for determining the correct value of absolute zero as approximately – 273.15 Celsius. Absolute temperatures are stated in units of kelvin in his honour. To find out more go to: www.routledge.com/cw/bird

From above, K = degrees Celsius + 273.
Hence, degrees Celsius = Kelvin temperature −273

(a) 365 K corresponds to 365 − 273, i.e. **92°C**

(b) 213 K corresponds to 213 − 273, i.e. **−60°C**

Now try the following Practice Exercise

Practice Exercise 144   Further problems on
temperature scales (answers on page 544)

1. Convert the following temperatures into the
    Kelvin scale:

    (a) 51°C   (b) −78°C   (c) 183°C

2. Convert the following temperatures into the
    Celsius scale:

    (a) 307 K   (b) 237 K   (c) 415 K

30.3 The measurement of temperature

A **thermometer** is an instrument that measures
temperature. Any substance that possesses one or more
properties that vary with temperature can be used to
measure temperature. These properties include changes
in length, area or volume, in electrical resistance or
in colour. Examples of temperature-measuring devices
include:

(i) **liquid-in-glass thermometers**, which use the
     expansion of a liquid with increase in temperature
     as their principle of operation,

(ii) **thermocouples**, which use the electromotive
      force set up when the junction of two dissimilar
      metals is heated,

(iii) **resistance thermometers**, which use the change
       in electrical resistance caused by temperature
       change, and

(iv) **pyrometers**, which are devices for measuring
      very high temperatures, using the principle that
      all substances emit radiant energy when hot, the
      rate of emission depending on their temperature.

Each of these temperature-measuring devices, together
with others, are described in chapter 33, page 349.

30.4 Specific heat capacity

The **specific heat capacity** of a substance is the quantity
of heat energy required to raise the temperature of 1 kg
of the substance by 1°C. The symbol used for specific
heat capacity is c and the units are J/(kgC) or J/(kg K).
(Note that these units may also be written as $J\,kg^{-1}°C^{-1}$
or $J\,kg^{-1}K^{-1}$.)
Some typical values of specific heat capacity for the
range of temperature 0°C to 100°C include:

Water	4190 J/(kgC)	Ice	2100 J/(kgC)
Aluminium	950 J/(kgC)	Copper	390 J/(kgC)
Iron	500 J/(kgC)	Lead	130 J/(kgC)

Hence to raise the temperature of 1 kg of iron by 1°C
requires 500 J of energy, to raise the temperature of 5 kg
of iron by 1°C requires (500 × 5) J of energy, and to
raise the temperature of 5 kg of iron by 40°C requires
(500 × 5 × 40) J of energy, i.e. 100 kJ.
In general, the quantity of heat energy Q required to
raise a mass m kg of a substance with a specific heat
capacity of c J/(kgC), from temperature t_1°C to t_2°C is
given by:

$$Q = mc(t_2 - t_1) \text{ joules}$$

Problem 3.   Calculate the quantity of heat
required to raise the temperature of 5 kg of water
from 0°C to 100°C. Assume the specific heat
capacity of water is 4200 J/(kgC)

Quantity of heat energy
$$Q = mc(t_2 - t_1)$$

$$= 5\,kg × 4200\,J/(kgC) × (100 - 0)°C$$

$$= 5 × 4200 × 100$$

$$= \textbf{2,100,000 J} \text{ or } \textbf{2100 kJ} \text{ or } \textbf{2.1 MJ}$$

Problem 4.   A block of cast iron having a mass of
10 kg cools from a temperature of 150°C to 50°C.
How much energy is lost by the cast iron? Assume
the specific heat capacity of iron is 500 J/(kgC)

Quantity of heat energy
$$Q = mc(t_2 - t_1)$$

$$= 10\,kg × 500\,J/(kgC) × (50 - 150)°C$$

$$= 10 × 500 × (-100)$$

$$= \textbf{−500,000 J} \text{ or } \textbf{−500 kJ} \text{ or } \textbf{−0.5 MJ}$$

(Note that the minus sign indicates that heat is given out or lost.)

Problem 5. Some lead having a specific heat capacity of 130 J/(kgC) is heated from 27°C to its melting point at 327°C. If the quantity of heat required is 780 kJ, determine the mass of the lead

Quantity of heat $Q = mc(t_2 - t_1)$, hence,

$$780 \times 10^3 \text{ J} = m \times 130 \text{ J/(kgC)}$$
$$\times (327 - 27)°C$$

i.e. $$780,000 = m \times 130 \times 300$$

from which, $$\textbf{mass } \textbf{\textit{m}} = \frac{780,000}{130 \times 300} \text{ kg} = \textbf{20 kg}.$$

Problem 6. 273 kJ of heat energy are required to raise the temperature of 10 kg of copper from 15°C to 85°C. Determine the specific heat capacity of copper

Quantity of heat $Q = mc(t_2 - t_1)$, hence:

$$273 \times 10^3 \text{ J} = 10 \text{ kg} \times c \times (85 - 15)°C$$

where c is the specific heat capacity,

i.e. $$273,000 = 10 \times c \times 70$$

from which, **specific heat capacity**

$$c = \frac{273,000}{10 \times 70} = \textbf{390 J/(kgC)}.$$

Problem 7. 5.7 MJ of heat energy are supplied to 30 kg of aluminium that is initially at a temperature of 20°C. If the specific heat capacity of aluminium is 950 J(kgC), determine its final temperature

Quantity of heat $Q = mc(t_2 - t_1)$, hence,

$$5.7 \times 10^6 \text{ J} = 30 \text{ kg} \times 950 \text{ J/(kgC)}$$
$$\times (t_2 - 20)°C$$

from which, $$(t_2 - 20) = \frac{5.7 \times 10^6}{30 \times 950} = 200$$

Hence, the **final temperature**

$$t_2 = 200 + 20 = \textbf{220°C}$$

Problem 8. A copper container of mass 500 g contains 1 litre of water at 293 K. Calculate the quantity of heat required to raise the temperature of the water and container to boiling point, assuming there are no heat losses. Assume that the specific heat capacity of copper is 390 J/(kg K), the specific heat capacity of water is 4.2 kJ(kg K) and 1 litre of water has a mass of 1 kg

Heat is required to raise the temperature of the water, and also to raise the temperature of the copper container.

For the water: $m = 1 \text{ kg}, \quad t_1 = 293 \text{ K},$

$$t_2 = 373 \text{ K (i.e. boiling point)}$$

and $$c = 4.2 \text{ kJ/(kg K)}.$$

Quantity of heat required for the water is given by:

$$Q_W = mc(t_2 - t_1)$$
$$= (1 \text{ kg}) \left(4.2 \frac{\text{kJ}}{\text{kg K}} \right) (373 - 293) \text{ K}$$
$$= 4.2 \times 80 \text{ kJ}$$

i.e. $\quad \textbf{\textit{Q}}_W = \textbf{336 kJ}$

For the copper container:

$$m = 500 \text{ g} = 0.5 \text{ kg}, t_1 = 293 \text{ K},$$
$$t_2 = 373 \text{ K and}$$
$$c = 390 \text{ J/(kg K)} = 0.39 \text{ kJ(kg K)}.$$

Quantity of heat required for the copper container is given by:

$$Q_C = mc(t_2 - t_1)$$
$$= (0.5 \text{ kg})(0.39 \text{ kJ/(kg K)})(80 \text{ K})$$

i.e. $\textbf{\textit{Q}}_C = \textbf{15.6 kJ}$

Total quantity of heat required,

$$Q = Q_W + Q_C = 336 + 15.6 = \textbf{351.6 kJ}$$

Now try the following Practice Exercise

Practice Exercise 145 Further problems on specific heat capacity (answers on page 544)

1. Determine the quantity of heat energy (in megajoules) required to raise the temperature of 10 kg of water from 0°C to 50°C. Assume the specific heat capacity of water is 4200 J/(kgC)

2. Some copper, having a mass of 20 kg, cools from a temperature of 120°C to 70°C. If the specific heat capacity of copper is 390 J/(kgC), how much heat energy is lost by the copper?

3. A block of aluminium having a specific heat capacity of 950 J/(kgC) is heated from 60°C to its melting point at 660°C. If the quantity of heat required is 2.85 MJ, determine the mass of the aluminium block

4. 20.8 kJ of heat energy is required to raise the temperature of 2 kg of lead from 16°C to 96°C. Determine the specific heat capacity of lead

5. 250 kJ of heat energy is supplied to 10 kg of iron which is initially at a temperature of 15°C. If the specific heat capacity of iron is 500 J/(kgC) determine its final temperature

30.5 Change of state

A material may exist in any one of three states – solid, liquid or gas. If heat is supplied at a constant rate to some ice initially at, say, −30°C, its temperature rises as shown in Figure 30.1. Initially the temperature increases from −30°C to 0°C as shown by the line AB. It then remains constant at 0°C for the time BC required for the ice to melt into water.

When melting commences, the energy gained by continual heating is offset by the energy required for the change of state, and the temperature remains constant even though heating is continued. When the ice is completely melted to water, continual heating raises the temperature to 100°C, as shown by CD in Figure 30.1. The water then begins to boil and the temperature again remains constant at 100°C, shown as DE, until all the water has vaporised. Continual heating raises the temperature of the steam as shown by EF, in the region where the steam is termed superheated.

Changes of state from solid to liquid or liquid to gas occur without change of temperature, and such changes are reversible processes. When heat energy flows to or from a substance and causes a change of temperature,

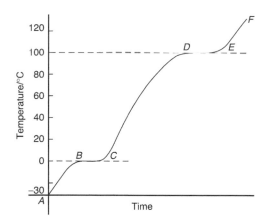

Figure 30.1

such as between A and B, between C and D and between E and F in Figure 30.1, it is called **sensible heat** (since it can be 'sensed' by a thermometer).

Heat energy which flows to or from a substance while the temperature remains constant, such as between B and C and between D and E in Figure 30.1, is called **latent heat** (latent means concealed or hidden).

Problem 9. Steam initially at a temperature of 130°C is cooled to a temperature of 20°C below the freezing point of water, the loss of heat energy being at a constant rate. Make a sketch, and briefly explain, the expected temperature/time graph representing this change

A temperature/time graph representing the change is shown in Figure 30.2. Initially steam cools until it reaches the boiling point of water at 100°C. Temperature then remains constant, i.e. between A and B, even though it is still giving off heat (i.e. latent heat). When all the

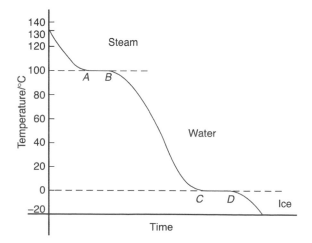

Figure 30.2

steam at 100°C has changed to water at 100°C it starts to cool again until it reaches the freezing point of water at 0°C.

From C to D the temperature again remains constant (i.e. latent heat), until all the water is converted to ice. The temperature of the ice then decreases as shown.

Now try the following Practice Exercise

Practice Exercise 146 **A further problem on change of state (answer on page 544)**

1. Some ice, initially at −40°C, has heat supplied to it at a constant rate until it becomes superheated steam at 150°C. Sketch a typical temperature/time graph expected and use it to explain the difference between sensible and latent heat

30.6 Latent heats of fusion and vaporisation

The **specific latent heat of fusion** is the heat required to change 1 kg of a substance from the solid state to the liquid state (or vice versa) at constant temperature.

The **specific latent heat of vaporisation** is the heat required to change 1 kg of a substance from a liquid to a gaseous state (or vice versa) at constant temperature. The units of the specific latent heats of fusion and vaporisation are J/kg, or more often kJ/kg, and some typical values are shown in Table 30.1.

The quantity of heat Q supplied or given out during a change of state is given by:

$$Q = mL$$

where m is the mass in kilograms and L is the specific latent heat.

Thus, for example, the heat required to convert 10 kg of ice at 0°C to water at 0°C is given by 10 kg × 335 kJ/kg = 3350 kJ or 3.35 MJ.

Besides changing temperature, the effects of supplying heat to a material can involve changes in dimensions, as well as in colour, state and electrical resistance. Most substances expand when heated and contract when cooled, and there are many practical applications and design implications of thermal movement (see chapter 31).

Table 30.1

	Latent heat of fusion (kJ/kg)	Melting point (°C)
Mercury	11.8	−39
Lead	22	327
Silver	100	957
Ice	335	0
Aluminium	387	660

	Latent heat of vaporisation (kJ/kg)	Boiling point (°C)
Oxygen	214	−183
Mercury	286	357
Ethyl alcohol	857	79
Water	2257	100

Problem 10. How much heat is needed to melt completely 12 kg of ice at 0°C? Assume the latent heat of fusion of ice is 335 kJ/kg

Quantity of heat required

$$Q = mL = 12\,kg \times 335\,kJ/kg$$

$$= \mathbf{4020\,kJ}\ or\ \mathbf{4.02\,MJ}$$

Problem 11. Calculate the heat required to convert 5 kg of water at 100°C to superheated steam at 100°C. Assume the latent heat of vaporisation of water is 2260 kJ/kg

Quantity of heat required

$$Q = mL = 5\,kg \times 2260\,kJ/kg$$

$$= \mathbf{11,300\,kJ}\ or\ \mathbf{11.3\,MJ}$$

Problem 12. Determine the heat energy needed to convert 5 kg of ice initially at −20°C completely to water at 0°C. Assume the specific heat capacity of ice is 2100 J/(kgC) and the specific latent heat of fusion of ice is 335 kJ/kg

Quantity of heat energy needed

$$Q = sensible\ heat + latent\ heat.$$

The quantity of heat needed to raise the temperature of ice from −20°C to 0°C, i.e. sensible heat

$$Q_1 = mc(t_2 - t_1)$$
$$= 5 \text{ kg} \times 2100 \text{ J/(kgC)} \times (0 - -20)^\circ\text{C}$$
$$= (5 \times 2100 \times 20) \text{ J} = \textbf{210 kJ}$$

The quantity of heat needed to melt 5 kg of ice at 0°C, i.e. the latent heat

$$Q_2 = mL = 5 \text{ kg} \times 335 \text{ kJ/kg} = \textbf{1675 kJ}$$

Total heat energy needed

$$Q = Q_1 + Q_2 = 210 + 1675 = \textbf{1885 kJ}$$

Problem 13. Calculate the heat energy required to convert completely 10 kg of water at 50°C into steam at 100°C, given that the specific heat capacity of water is 4200 J/(kgC) and the specific latent heat of vaporisation of water is 2260 kJ/kg

Quantity of heat required = sensible heat + latent heat.

Sensible heat $Q_1 = mc(t_2 - t_1)$
$$= 10 \text{ kg} \times 4200 \text{ J/(kgC)}$$
$$\times (100 - 50)^\circ\text{C} = \textbf{2100 kJ}$$

Latent heat $Q_2 = mL = 10 \text{ kg} \times 2260 \text{ kJ/kg}$
$$= \textbf{22,600 kJ}$$

Total heat energy required,

$$Q = Q_1 + Q_2 = (2100 + 22,600) \text{ kJ}$$
$$= \textbf{24,700 kJ or 24.70 MJ}$$

Problem 14. Determine the amount of heat energy needed to change 400 g of ice, initially at −20°C, into steam at 120°C. Assume the following: latent heat of fusion of ice = 335 kJ/kg, latent heat of vaporisation of water = 2260 kJ/kg, specific heat capacity of ice = 2.14 kJ/(kgC), specific heat capacity of water = 4.2 kJ/(kgC) and specific heat capacity of steam = 2.01 kJ/(kgC)

The energy needed is determined in five stages:

(i) Heat energy needed to change the temperature of ice from −20°C to 0°C is given by:

$$Q_1 = mc(t_2 - t_1)$$
$$= 0.4 \text{ kg} \times 2.14 \text{ kJ/(kgC)} \times (0 - -20)^\circ\text{C}$$
$$= \textbf{17.12 kJ}$$

(ii) Latent heat needed to change ice at 0°C into water at 0°C is given by:

$$Q_2 = mL_f = 0.4 \text{ kg} \times 335 \text{ kJ/kg}$$
$$= \textbf{134 kJ}$$

(iii) Heat energy needed to change the temperature of water from 0°C (i.e. melting point) to 100°C (i.e. boiling point) is given by:

$$Q_3 = mc(t_2 - t_1)$$
$$= 0.4 \text{ kg} \times 4.2 \text{ kJ/(kgC)} \times 100^\circ\text{C}$$
$$= \textbf{168 kJ}$$

(iv) Latent heat needed to change water at 100°C into steam at 100°C is given by:

$$Q_4 = mL_v = 0.4 \text{ kg} \times 2260 \text{ kJ/kg}$$
$$= \textbf{904 kJ}$$

(v) Heat energy needed to change steam at 100°C into steam at 120°C is given by:

$$Q_5 = mc(t_1 - t_2)$$
$$= 0.4 \text{ kg} \times 2.01 \text{ kJ/(kgC)} \times 20^\circ\text{C}$$
$$= \textbf{16.08 kJ}$$

Total heat energy needed
$$Q = Q_1 + Q_2 + Q_3 + Q_4 + Q_5$$
$$= 17.12 + 134 + 168 + 904 + 16.08$$
$$= \textbf{1239.2 kJ}$$

Now try the following Practice Exercise

Practice Exercise 147   Further problems on the latent heats of fusion and vaporisation (answers on page 544)

1. How much heat is needed to melt completely 25 kg of ice at 0°C. Assume the specific latent heat of fusion of ice is 335 kJ/kg

2. Determine the heat energy required to change 8 kg of water at 100°C to superheated steam at 100°C. Assume the specific latent heat of vaporisation of water is 2260 kJ/kg

3. Calculate the heat energy required to convert 10 kg of ice initially at −30°C completely into water at 0°C. Assume the specific heat capacity of ice is 2.1 kJ/(kgC) and the specific latent heat of fusion of ice is 335 kJ/kg

4. Determine the heat energy needed to convert completely 5 kg of water at 60°C to steam at 100°C, given that the specific heat capacity of water is 4.2 kJ/(kgC) and the specific latent heat of vaporisation of water is 2260 kJ/kg

30.7   A simple refrigerator

The boiling point of most liquids may be lowered if the pressure is lowered. In a simple refrigerator a working fluid, such as ammonia or freon, has the pressure acting on it reduced. The resulting lowering of the boiling point causes the liquid to vaporise. In vaporising, the liquid takes in the necessary latent heat from its surroundings, i.e. the freezer, which thus becomes cooled. The vapour is immediately removed by a pump to a condenser that is outside of the cabinet, where it is compressed and changed back into a liquid, giving out latent heat. The cycle is repeated when the liquid is pumped back to the freezer to be vaporised.

30.8   Conduction, convection and radiation

Heat may be **transferred** from a hot body to a cooler body by one or more of three methods, these being: (a) by **conduction**, (b) by **convection**, or (c) by **radiation**.

30.8.1   Conduction

Conduction is the transfer of heat energy from one part of a body to another (or from one body to another) without the particles of the body moving.

Conduction is associated with solids. For example, if one end of a metal bar is heated, the other end will become hot by conduction. Metals and metallic alloys are good conductors of heat, whereas air, wood, plastic, cork, glass and gases are examples of poor conductors (i.e. they are heat insulators).

Practical applications of conduction include:

(i) A domestic saucepan or dish conducts heat from the source to the contents. Also, since wood and plastic are poor conductors of heat they are used for saucepan handles.

(ii) The metal of a radiator of a central heating system conducts heat from the hot water inside to the air outside.

30.8.2   Convection

Convection is the transfer of heat energy through a substance by the actual movement of the substance itself. Convection occurs in liquids and gases, but not in solids. When heated, a liquid or gas becomes less dense. It then rises and is replaced by a colder liquid or gas and the process repeats. For example, electric kettles and central heating radiators always heat up at the top first.

Examples of convection are:

(i) Natural circulation hot water heating systems depend on the hot water rising by convection to the top of the house and then falling back to the bottom of the house as it cools, releasing the heat energy to warm the house as it does so.

(ii) Convection currents cause air to move and therefore affect climate.

(iii) When a radiator heats the air around it, the hot air rises by convection and cold air moves in to take its place.

(iv) A cooling system in a car radiator relies on convection.

(v) Large electrical transformers dissipate waste heat to an oil tank. The heated oil rises by convection to the top, then sinks through cooling fins, losing heat as it does so.

(vi) In a refrigerator, the cooling unit is situated near the top. The air surrounding the cold pipes become heavier as it contracts and sinks towards the bottom. Warmer, less dense air is pushed upwards and in turn is cooled. A cold convection current is thus created.

30.8.3 Radiation

Radiation is the transfer of heat energy from a hot body to a cooler one by electromagnetic waves.

Heat radiation is similar in character to light waves – it travels at the same speed and can pass through a vacuum – except that the frequency of the waves is different. Waves are emitted by a hot body, are transmitted through space (even a vacuum) and are not detected until they fall onto another body. Radiation is reflected from shining, polished surfaces but absorbed by dull, black surfaces.

Practical applications of radiation include:

(i)   Heat from the sun reaching earth.

(ii)   Heat felt from a flame.

(iii)   Cooker grills.

(iv)   Industrial furnaces.

(v)   Infra-red space heaters.

30.9   Vacuum flask

A cross-section of a typical vacuum flask is shown in Figure 30.3 and it is seen to be a double-walled bottle with a vacuum space between the walls, the whole supported in a protective outer case.

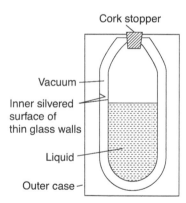

Cork stopper

Vacuum

Inner silvered surface of thin glass walls

Liquid

Outer case

Figure 30.3

Very little heat can be transferred by conduction because of the vacuum space and the cork stopper (cork is a bad conductor of heat). Also, because of the vacuum space, no convection is possible. Radiation is minimised by silvering the two glass surfaces (radiation is reflected off shining surfaces).

Thus a vacuum flask is an example of prevention of all three types of heat transfer, and is therefore able to keep hot liquids hot and cold liquids cold.

30.10   Use of insulation in conserving fuel

Fuel used for heating a building is becoming increasingly expensive. By the careful use of insulation, heat can be retained in a building for longer periods and the cost of heating thus minimised.

(i)   Since convection causes hot air to rise it is important to insulate the roof space, which is probably the greatest source of heat loss in the home. This can be achieved by laying fibre-glass insulation between the wooden joists in the roof space.

(ii)   Glass is a poor conductor of heat. However, large losses can occur through thin panes of glass and such losses can be reduced by using double-glazing. Two sheets of glass, separated by air, are used. Air is a very good insulator but the air space must not be too large otherwise convection currents can occur which would carry heat across the space.

(iii)   Hot water tanks should be lagged to prevent conduction and convection of heat to the surrounding air.

(iv)   Brick, concrete, plaster and wood are all poor conductors of heat. A house is made from two walls with an air gap between them. Air is a poor conductor and trapped air minimises losses through the wall. Heat losses through the walls can be prevented almost completely by using cavity wall insulation, i.e. plastic-foam.

Besides changing temperature, the effects of supplying heat to a material can involve changes in dimensions, as well as in colour, state and electrical resistance.

Most substances expand when heated and contract when cooled, and there are many practical applications and design implications of thermal movement as explained in chapter 31.

30.11   Thermal efficiency

Petrol engines, marine engines, steam engines and any other power plants obtaining energy directly from a source of heat, inevitably waste some of the heat energy into its surroundings – by conduction, convection or radiation. Such losses are unavoidable but may be reduced by suitable lagging with a material of low thermal conductivity. Efficiency can be improved by making use of the heat energy in waste or exhaust gases – the feed water in some steam plants, for example, may be pre-heated by the hot flue gases. In general,

(Heat energy supplied from a fuel)

= (Useful heat energy) + (unavoidable losses)

Thermal efficiency, $\eta = \dfrac{\textbf{useful heat energy}}{\textbf{heat energy supplied}} \times \textbf{100\%}$

30.12 Calorific value and combustion

Combustion is the chemical combination of oxygen with the combustible elements of a fuel. Combustion of a fuel causes the release of heat energy and the amount of energy released when a unit mass of fuel is efficiently burned is known as the **calorific value** of the fuel.

The calorific value of a typical fuel oil is 45 MJ/kg, which means that when 1 kg of this fuel is efficiently burned it will release 45×10^6 joules of heat energy.

30.13 Heat Exchangers

A **heat exchanger** is a device designed to efficiently transfer or 'exchange' heat from one matter to another. When a fluid is used to transfer heat, the fluid could be a liquid, such as water or oil, or could be moving air. A heat exchanger usually consists of one or more tubes within a shell; one fluid flows through the tubes, the other through the shell and over the outer surface of the tubes. If the fluids are at different temperatures, heat will flow from the hotter to the colder and an exchange of heat will occur – hence the name 'heat exchanger'.

The most well-known type of heat exchanger is a **car radiator**. Here the heat source is hot engine coolant and this heat is transferred to the air flowing through the radiator. **Coolers** and **heaters** are also examples of heat exchangers.

With **steam plant**, a boiler is a heat exchanger where the hotter fluid is the flue gas. The economiser and superheater within steam plant are both heat exchangers in which the water or steam passes through tubes and the hot gases pass over the outer surface of the tubes to heat the media within them.

Heavy industry applications of heat exchangers are found in mining, coke industries, pulp and paper works, iron and steel-making industries, and with metallurgy.

Problem 15. A press tool of mass 5 kg and specific heat capacity 0.5 kJ/(kgC) is to be heated from 15C to 850C in a hardening process. To initiate the necessary temperature rise, the furnace uses 0.40 kg of gas with a calorific value of 18 MJ/kg. Calculate the thermal efficiency of the furnace

Heat energy required, $Q = mc(t_2 - t_1)$
$$= 5 \text{ kg} \times 0.5 \times 10^3 \text{ J/(kgC)} \times (850 - 15)°C$$
i.e. $\qquad Q = 2087500 \text{ J} = 2.0875 \text{ MJ}$

Heat energy supplied = mass × calorific value
$$= 0.40 \text{ kg} \times 18 \times 10^6 \text{ J/kg}$$
$$= 7200000 \text{ J} = 7.2 \text{ MJ}$$

Efficiency of furnace $= \dfrac{\text{heat energy required}}{\text{heat energy supplied}}$

$$= \dfrac{2.0875 \times 10^6}{7.2 \times 10^6} = 0.290$$

Hence, **furnace efficiency**, $\eta = 0.290 \times 100 = \textbf{29.0\%}$
29.0% efficiency shows that 71.0% of the heat energy is wasted; this is not very efficient, but not untypical.

Problem 16. An electric furnace has an efficiency of 60% and is rated at 8 kW. 15 kg of copper at 25C is placed in the furnace until it melts at 1085C. If the specific heat capacity of copper is assumed to be 0.39 kJ/(kgC) and the latent heat of fusion is 180 kJ/(kgC), calculate the time taken for the furnace to melt the copper

Sensible heat energy required, $Q_S = mc(t_2 - t_1) =$
$15 \text{ kg} \times 0.39 \times 10^3 \text{ J/(kgC)} \times (1085 - 25)°C$
i.e. $\qquad Q_S = 6201000 \text{ J}$

Latent heat energy required, $Q_L = mL$
$$= 15 \text{ kg} \times 180 \times 10^3 \text{ J/(kgC)}$$
i.e. $\qquad Q_L = 2700000 \text{ J}$

Hence, total heat energy required, $Q_T = Q_S + Q_L$
$$= 6201000 + 2700000$$
i.e. $\qquad Q_T = 8901000 \text{ J} = \textbf{8.901 MJ}$

However, the furnace is only 60% efficient, so the true total heat energy required
$$= \dfrac{100}{60} \times 8901000 = 14835000 \text{ J} = \textbf{14.835 MJ}$$

Now, Power $= \dfrac{\text{energy}}{\text{time}}$ from which,

time taken $= \dfrac{\text{energy}}{\text{power}} = \dfrac{14835000 \text{ J}}{8 \times 10^3 \text{ W}} = 1854.375 \text{s}$

$$1854.375 \text{ s} = \frac{1854.375 \text{ s}}{60 \text{ s/min}} = 30.90625 \text{ minutes}$$

$$0.90625 \text{ minutes} = 0.90625 \text{ min} \times 60 \text{ s/min} = 54.375 \text{ s}$$

Hence, **the time taken to melt the copper = 30 minutes 54 seconds**

Problem 17. An ice-making machine takes in water at a temperature of 16°C and produces ice cubes at −6°C. Water is taken in at the rate of 1 kg every 6 minutes and the input power to the machine itself is 3 kW. Assuming the specific heat capacity of water is 4200 J/(kgC), the specific heat capacity of ice is 2100 J/(kgC) and the specific latent heat of fusion of the ice is 335 kJ/kg, determine the efficiency of the ice-making machine

---

Heat energy extracted per kg to cool the water from 16C down to 0C,

$$Q_1 = mc(t_2 - t_1) = 1 \text{ kg} \times 4200 \text{ J/(kgC)} \times (16 - 0)°C$$

i.e. $\qquad Q_1 = 67200 \text{ J/kg}$

Latent heat energy extracted per kg to convert the water to ice,

$$Q_2 = mL = 1 \text{kg} \times 335 \times 10^3 \text{J/(kgC)}$$

i.e. $\qquad Q_2 = 335000 \text{ J/kg}$

Heat energy extracted per kg to cool the ice from 0°C down to −6°C,

$$Q_3 = mc(t_2 - t_1) = 1 \text{ kg} \times 2100 \text{J/(kgC)} \times (0 - 6)°C$$

i.e. $\qquad Q_3 = 12600 \text{ J/kg}$

Hence, **total heat energy extracted per kilogram,** $Q_T = 67200 + 335000 + 12600 = $ **414800J/kg**

Mass flow rate, $\dot{m}$ is given by $\dfrac{\text{mass}}{\text{time}}$ in kg/s

i.e. $\quad \dot{m} = \dfrac{1 \text{ kg}}{6 \times 60 \text{ s}} = \dfrac{1}{360}$ **kg/s**

Power output, P_{OUT} = heat energy extracted per second per kg = $\dot{m} \times Q_T$

i.e. $\quad P_{OUT} = \dfrac{1}{360} \text{ kg/s} \times 414800 \text{ J/kg} = $ **1152.22 W**

Thermal efficiency, $\eta_{TH} = \dfrac{P_{OUT}}{P_{IN}} \times 100\%$

$$= \dfrac{1152.22}{3000} \times 100\%$$

since the input power, $P_{IN} = 3 \text{ kW} = 3000 \text{ W}$

i.e. **the thermal efficiency of the ice making machine,** $\boldsymbol{\eta_{TH} = 38.41\%}$

Problem 18. A heat exchanger produces dry steam at 100C from feed water at 16C at a rate of 2 kg/s. The heat exchanger receives heat energy at a rate of 6.4 MW from the fuel used. The specific heat capacity of water is 4190 J/(kgC) and its latent heat of vaporisation is 2257 kJ/kg. Determine (a) the heat energy received per kilogram of steam produced, (b) the output power of the heat exchanger, and (c) its thermal efficiency

---

(a) Total heat energy received per kilogram of steam,

$$Q_T = \text{sensible heat received} + \text{latent heat received}$$

$$= mc_W (t_2 - t_1) + mL \quad \text{where m} = 1 \text{ kg}$$

$$= [1 \times 4190 \times (100 - 16)] + [1 \times 2257 \times 10^3]$$

$$= 351960 + 2257000$$

i.e. $\qquad Q_T = 2.60896 \times 10^6 \text{ J/kg} = $ **2.61 MJ/kg**

(b) Mass flow rate, $\dot{m} = 2$ kg/s

Output power of the heat exchanger

$$= \dot{m} \times Q_T = 2 \text{ kg/s} \times 2.61 \times 10^6 \text{ J/kg}$$

$$= 5.22 \times 10^6 \text{ J/s} = \textbf{5.22 MW}$$

(c) **Thermal efficiency of the heat exchanger**
$$= \dfrac{\text{output power}}{\text{input power}} \times 100\%$$

$$= \dfrac{5.22 \times 10^6}{6.4 \times 10^6} \times 100\% = 81.56\%$$

Now try the following Practice Exercise

Practice Exercise 148 Further problems on heat exchangers (answers on page 544)

1. An aluminium alloy has a specific heat capacity of 900 J/(kgC). The alloy melts at 650C and the latent heat of fusion is 400 kJ/kg. (a) Calculate the amount of heat energy required to melt an ingot of the alloy which has a mass of 6.25 kg, starting from a room temperature of 18C. (b) Find the time taken to melt the ingot if the heating is performed by an electric furnace rated at 12 kW having an overall efficiency of 75%

2. A press tool of mass 8 kg and specific heat capacity 0.5 kJ/(kgC) is to be heated from 15C to 900C in a hardening process. To initiate the necessary temperature rise, the furnace uses 0.50 kg of gas with a calorific value of 20 MJ/kg. Calculate the thermal efficiency of the furnace

3. An electric furnace has an efficiency of 70% and is rated at 10 kW. 20 kg of copper at 20°C is placed in the furnace until it melts at 1100°C. If the specific heat capacity of copper is assumed to be 450 J/(kgC) and the latent heat of fusion is 180 kJ/(kgC), calculate the time taken for the furnace to melt the copper

4. A heat exchanger is rated at 12 MW and is supplied with water at 15C. The exchanger heats the water to 100C at which point it is turned into dry steam. The heat exchanger supplies dry steam at 2.5 kg/s. If the specific heat capacity of water is 4.19 kJ/(kgC) and the specific heat of vaporisation is 2.26 MJ/kg, determine (a) the total heat supplied per kilogram, and (b) the thermal efficiency

5. An ice-making machine takes in water at a temperature of 15C and produces ice cubes at − 5C. Water is taken in at the rate of 1 kg every 5 minutes and the input power to the machine itself is 3 kW. Assuming the specific heat capacity of water is 4.2 kJ/(kgC), the specific heat capacity of ice is 2.1 kJ/(kgC) and the specific latent heat of fusion of the ice is 335 kJ/kg, determine the efficiency of the ice-making machine

6. A heat exchanger produces dry steam at 100°C from feed water at 17C at a rate of 3 kg/s. The heat exchanger receives heat energy at a rate of 10 MW from the fuel used. The specific heat capacity of water is 4190 J/(kgC) and its latent heat of vaporisation is 2257 kJ/kg. Determine (a) the heat energy received per kilogram of steam produced, and (b) the output power of the heat exchanger and its thermal efficiency

Practice Exercise 149   Short answer questions on heat energy

1. Differentiate between temperature and heat

2. Name two scales on which temperature is measured

3. Name any four temperature-measuring devices

4. Define specific heat capacity and name its unit

5. Differentiate between sensible and latent heat

6. The quantity of heat Q required to raise a mass m kg from temperature t_1°C to t_2°C, the specific heat capacity being c, is given by $Q = $ ......

7. What is meant by the specific latent heat of fusion?

8. Define the specific latent heat of vaporisation

9. Explain briefly the principle of operation of a simple refrigerator

10. State three methods of heat transfer

11. Define conduction and state two practical examples of heat transfer by this method

12. Define convection and give three examples of heat transfer by this method

13. What is meant by radiation? Give three uses

14. How can insulation conserve fuel in a typical house?

15. Define thermal efficiency

16. Define combustion

17. What is meant by calorific value?

18. What is a heat exchanger? State three practical examples

Practice Exercise 150   Multiple-choice questions on heat energy (answers on page 544)

1. Heat energy is measured in:
   (a) kelvin   (b) watts
   (c) kilograms   (d) joules

2. A change of temperature of 20°C is equivalent to a change in thermodynamic temperature of:
   (a) 293 K   (b) 20 K
   (c) 80 K   (d) 120 K

3. A temperature of 20°C is equivalent to:

   (a) 293 K   (b) 20 K

   (c) 80 K   (d) 120 K

4. The unit of specific heat capacity is:

   (a) joules per kilogram

   (b) joules

   (c) joules per kilogram kelvin

   (d) cubic metres

5. The quantity of heat required to raise the temperature of 500 g of iron by 2°C, given that the specific heat capacity is 500 J/(kgC), is:

   (a) 500 kJ   (b) 0.5 kJ

   (c) 2 J   (d) 250 kJ

6. The heat energy required to change 1 kg of a substance from a liquid to a gaseous state at the same temperature is called:

   (a) specific heat capacity

   (b) specific latent heat of vaporisation

   (c) sensible heat

   (d) specific latent heat of fusion

7. The temperature of pure melting ice is:

   (a) 373 K   (b) 273 K

   (c) 100 K   (d) 0 K

8. 1.95 kJ of heat is required to raise the temperature of 500 g of lead from 15°C to its final temperature. Taking the specific heat capacity of lead to be 130 J/(kgC), the final temperature is:

   (a) 45°C   (b) 37.5°C

   (c) 30°C   (d) 22.5°C

9. Which of the following temperatures is absolute zero?

   (a) 0°C   (b) −173°C

   (c) −23°C   (d) −373°C

10. When two wires of different metals are twisted together and heat applied to the junction, an electromotive force (e.m.f.) is produced. This effect is used in a thermocouple to measure:

    (a) e.m.f.   (b) temperature

    (c) expansion   (d) heat

11. Which of the following statements is false?

    (a) −30°C is equivalent to 243 K

    (b) Convection only occurs in liquids and gases

    (c) Conduction and convection cannot occur in a vacuum

    (d) Radiation is absorbed by a silver surface

12. The transfer of heat through a substance by the actual movement of the particles of the substance is called:

    (a) conduction   (b) radiation

    (c) convection   (d) specific heat capacity

13. Which of the following statements is true?

    (a) Heat is the degree of hotness or coldness of a body

    (b) Heat energy that flows to or from a substance while the temperature remains constant is called sensible heat

    (c) The unit of specific latent heat of fusion is J/(kg K)

    (d) A cooker-grill is a practical application of radiation

For fully worked solutions to each of the problems in Exercises 144 to 150 in this chapter, go to the website:

This assignment covers the material contained in chapters 28 to 30. *The marks for each question are shown in brackets at the end of each question.*

When required take the density of water to be $1000 \, kg/m^3$ and gravitational acceleration as $9.81 \, m/s^2$.

1. Determine the force applied tangentially to a bar of a screw-jack at a radius of 60 cm, if the torque required is 750 N m. (4)

2. Calculate the torque developed by a motor whose spindle is rotating at 900 rev/min and developing a power of 4.20 kW. (5)

3. A motor connected to a shaft develops a torque of 8 kN m. Determine the number of revolutions made by the shaft if the work done is 7.2 MJ. (6)

4. Determine the angular acceleration of a shaft that has a moment of inertia of 32 kg m^2 produced by an accelerating torque of 600 N m. (5)

5. An electric motor has an efficiency of 72% when running at 1400 rev/min. Determine the output torque when the power input is 2.50 kW. (6)

6. A circular piston exerts a pressure of 150 kPa on a fluid when the force applied to the piston is 0.5 kN. Calculate the diameter of the piston, correct to the nearest millimetre. (7)

7. A tank contains water to a depth of 500 mm. Determine the water pressure

   (a) at a depth of 300 mm,

   (b) at the base of the tank. (6)

8. When the atmospheric pressure is 101 kPa, calculate the absolute pressure, to the nearest kilopascal, at a point on a submarine which is 50 m below the sea water surface. Assume that the density of seawater is $1030 \, kg/m^3$. (5)

9. A body weighs 2.85 N in air and 2.35 N when completely immersed in water. Determine (a) the volume of the body, (b) the density of the body, (c) the relative density of the body. (9)

10. A submarine dives to a depth of 700 m. What is the gauge pressure on its surface if the density of seawater is $1020 \, kg/m^3$? (5)

11. A block of aluminium having a mass of 20 kg cools from a temperature of 250°C to 80°C. How much energy is lost by the aluminium? Assume the specific heat capacity of aluminium is 950 J/(kgC). (5)

12. Calculate the heat energy required to convert completely 12 kg of water at 30°C to superheated steam at 100°C. Assume that the specific heat capacity of water is 4200 J/(kgC) and the specific latent heat of vaporisation of water is 2260 kJ/(kgC). (7)

13. An electric furnace has an efficiency of 62% and is rated at 7.5 kW. 20 kg of copper at 23C is placed in the furnace until it melts at 1079C. If the specific heat capacity of copper is assumed to be 0.39 kJ/(kgC) and the latent heat of fusion is 180 kJ/(kgC), calculate the time taken for the furnace to melt the copper. (10)

For lecturers/instructors/teachers, fully worked solutions to each of the problems in Revision Test 11, together with a full marking scheme, are available at the website:

www.routledge.com/cw/bird

Chapter 31

Thermal expansion

Why it is important to understand: **Thermal expansion**

Thermal expansion and contraction are very important features in engineering science. For example, if the metal railway lines of a railway track are heated or cooled due to weather conditions or time of day, their lengths can increase or decrease accordingly. If the metal lines are heated due to the weather effects, then the railway lines will attempt to expand, and depending on their construction, they can buckle, rendering the track useless for transporting trains. In countries with large temperature variations, this effect can be much worse, and the engineer may have to choose a superior metal to withstand these changes. The effect of metals expanding and contracting due to the rise and fall of temperatures, accordingly, can also be put to good use. A classic example of this is the simple, humble domestic thermostat, which when the water gets too hot, will cause the metal thermostat to expand and switch off the electric heater; conversely, when the water becomes too cool, the metal thermostat shrinks, causing the electric heater to switch on again. All sorts of materials, besides metals, are affected by thermal expansion and contraction. The chapter also defines the coefficients of linear, superficial and cubic expansion. Thermal expansion is of importance in the design and construction of structures, switches and other bodies.

At the end of this chapter, you should be able to:

- appreciate that expansion and contraction occur with change of temperature
- describe practical applications where expansion and contraction must be allowed for
- understand the expansion and contraction of water
- define the coefficient of linear expansion α
- recognise typical values for the coefficient of linear expansion
- calculate the new length L_2, after expansion or contraction, using $L_2 = L_1[1 + \alpha(t_2 - t_1)]$
- define the coefficient of superficial expansion β
- calculate the new surface area A_2, after expansion or contraction, using $A_2 = A_1[1 + \beta(t_2 - t_1)]$
- appreciate that $\beta \approx 2\alpha$
- define the coefficient of cubic expansion γ
- recognise typical values for the coefficient of cubic expansion
- appreciate that $\gamma \approx 3\alpha$
- calculate the new volume V_2, after expansion or contraction, using $V_2 = V_1[1 + \gamma(t_2 - t_1)]$

Science for Engineering. 978-1-138-82688-5, © John Bird. Published by Taylor & Francis. All rights reserved.

31.1 Introduction

When heat is applied to most materials, **expansion** occurs in all directions. Conversely, if heat energy is removed from a material (i.e. the material is cooled) **contraction** occurs in all directions. The effects of expansion and contraction each depend on the **change of temperature** of the material.

Thermal expansion and contraction are very important features in engineering science. For example, if the metal railway lines of a railway track are heated or cooled due to weather conditions or time of day, their lengths can increase or decrease accordingly. If the metal lines are heated due to the weather effects, then the railway lines will attempt to expand, and depending on their construction, they can buckle, rendering the track useless for transporting trains. In countries with large temperature variations, this effect can be much worse, and the engineer may have to choose a superior metal to withstand these changes. Accordingly, the effect of metals expanding and contracting due to the rise and fall of temperatures can also be put to good use. A classic example of this is the simple humble domestic thermostat, which when the water gets too hot, will cause the metal thermostat to expand and switch off an electric heater; conversely, when the water becomes too cool, the metal thermostat shrinks, causing the electric heater to switch on again. All sorts of materials, besides metals, are affected by thermal expansion and contraction. This chapter also defines the coefficients of linear, superficial and cubic expansion.

31.2 Practical applications of thermal expansion

Some practical applications where expansion and contraction of solid materials must be allowed for include:

(i)   Overhead electrical transmission lines are hung so that they are slack in summer, otherwise their contraction in winter may snap the conductors or bring down pylons.

(ii)   Gaps need to be left in lengths of railway lines to prevent buckling in hot weather (except where these are continuously welded).

(iii) Ends of large bridges are often supported on rollers to allow them to expand and contract freely.

(iv)   Fitting a metal collar to a shaft or a steel tyre to a wheel is often achieved by first heating the collar or tyre so that they expand, fitting them in position, and then cooling them so that the contraction holds them firmly in place; this is known as a 'shrink-fit'. By a similar method hot rivets are used for joining metal sheets.

(v)   The amount of expansion varies with different materials. Figure 31.1(a) shows a bimetallic strip at room temperature (i.e. two different strips of metal riveted together). When heated, brass expands more than steel, and since the two metals are riveted together the bimetallic strip is forced into an arc as shown in Figure 31.1(b). Such a movement can be arranged to make or break an electric circuit, and bimetallic strips are used, in particular, in thermostats (which are temperature-operated switches) used to control central heating systems, cookers, refrigerators, toasters, irons, hot-water and alarm systems.

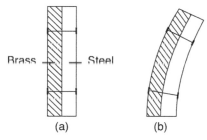

Figure 31.1

(vi)   Motor engines use the rapid expansion of heated gases to force a piston to move.

(vii) Designers must predict, and allow for, the expansion of steel pipes in a steam-raising plant so as to avoid damage and consequent danger to health.

31.3 Expansion and contraction of water

Water is a liquid that at low temperature displays an unusual effect. If cooled, contraction occurs until, at about 4°C, the volume is at a minimum. As the temperature is further decreased from 4°C to 0°C expansion occurs, i.e. the volume increases. (For cold, deep fresh water, the temperature at the bottom is more likely to be about 4°C, somewhat warmer than less-deep water.) When ice is formed, considerable expansion occurs, and it is this expansion that often causes frozen water pipes to burst.

A practical application of the expansion of a liquid is with thermometers, where the expansion of a liquid, such as mercury or alcohol, is used to measure temperature.

31.4 Coefficient of linear expansion

The amount by which unit length of a material expands when the temperature is raised one degree is called the **coefficient of linear expansion** of the material and is represented by α (Greek alpha).

The units of the coefficient of linear expansion are m/(mK), although it is usually quoted as just /K or K^{-1}. For example, copper has a coefficient of linear expansion value of $17 \times 10^{-6} \, K^{-1}$, which means that a 1 m long bar of copper expands by 0.000017 m if its temperature is increased by 1 K (or 1°C). If a 6 m long bar of copper is subjected to a temperature rise of 25 K then the bar will expand by $(6 \times 0.000017 \times 25)$ m, i.e. 0.00255 m or 2.55 mm. (Since the Kelvin scale uses the same temperature interval as the Celsius scale, a **change** of temperature of, say, 50°C, is the same as a change of temperature of 50 K.)

If a material, initially of length L_1 and at a temperature of t_1 and having a coefficient of linear expansion α, has its temperature increased to t_2, then the new length L_2 of the material is given by:

New length = original length + expansion

i.e.
$$L_2 = L_1 + L_1\alpha(t_2 - t_1)$$

i.e.
$$L_2 = L_1[1 + \alpha(t_2 - t_1)] \qquad (1)$$

Some typical values for the coefficient of linear expansion include:

Aluminium	$23 \times 10^{-6} \, K^{-1}$
Brass	$18 \times 10^{-6} \, K^{-1}$
Concrete	$12 \times 10^{-6} \, K^{-1}$
Copper	$17 \times 10^{-6} \, K^{-1}$
Gold	$14 \times 10^{-6} \, K^{-1}$
Invar (nickel-steel alloy)	$0.9 \times 10^{-6} \, K^{-1}$
Iron	$11–12 \times 10^{-6} \, K^{-1}$
Nylon	$100 \times 10^{-6} \, K^{-1}$
Steel	$15–16 \times 10^{-6} \, K^{-1}$
Tungsten	$4.5 \times 10^{-6} \, K^{-1}$
Zinc	$31 \times 10^{-6} \, K^{-1}$

Problem 1. The length of an iron steam pipe is 20.0 m at a temperature of 18°C. Determine the length of the pipe under working conditions when the temperature is 300°C. Assume the coefficient of linear expansion of iron is $12 \times 10^{-6} \, K^{-1}$

Length $L_1 = 20.0$ m, temperature $t_1 = 18°C$, $t_2 = 300°C$ and $\alpha = 12 \times 10^{-6} \, K^{-1}$.

Length of pipe at 300°C is given by:

$$L_2 = L_1[1 + \alpha(t_2 - t_1)]$$
$$= 20.0[1 + (12 \times 10^{-6})(300 - 18)]$$
$$= 20.0[1 + 0.003384] = 20.0[1.003384]$$
$$= \mathbf{20.06768 \, m}$$

i.e. an increase in length of 0.06768 m or **67.68 mm**.

In practice, allowances are made for such expansions. U-shaped expansion joints are connected into pipelines carrying hot fluids to allow some 'give' to take up the expansion.

Problem 2. An electrical overhead transmission line has a length of 80.0 m between its supports at 15°C. Its length increases by 92 mm at 65°C. Determine the coefficient of linear expansion of the material of the line

Length $L_1 = 80.0$ m, $L_2 = 80.0$ m + 92 mm = 80.092 m, temperature $t_1 = 15°C$ and temperature $t_2 = 65°C$.

$$\text{Length } L_2 = L_1[1 + \alpha(t_2 - t_1)]$$

i.e.
$$80.092 = 80.0[1 + \alpha(65 - 15)]$$
$$80.092 = 80.0 + (80.0)(\alpha)(50)$$

i.e.
$$80.092 - 80.0 = (80.0)(\alpha)(50)$$

Hence, the coefficient of linear expansion

$$\alpha = \frac{0.092}{(80.0)(50)} = 0.000023$$

i.e.
$$\mathbf{\alpha = 23 \times 10^{-6} \, K^{-1}}$$

(which is aluminium – see above).

Problem 3. A measuring tape made of copper measures 5.0 m at a temperature of 288 K. Calculate the percentage error in measurement when the temperature has increased to 313 K. Take the coefficient of linear expansion of copper as $17 \times 10^{-6} \, K^{-1}$

Length $L_1 = 5.0$ m, temperature $t_1 = 288$ K, $t_2 = 313$ K and $\alpha = 17 \times 10^{-6} \, K^{-1}$.

Length at 313 K is given by:

$$\text{Length } L_2 = L_1[1 + \alpha(t_2 - t_1)]$$

$$= 5.0[1 + (17 \times 10^{-6})(313 - 288)]$$

$$= 5.0[1 + (17 \times 10^{-6})(25)]$$

$$= 5.0[1 + 0.000425]$$

$$= 5.0[1.000425] = 5.002125 \text{ m}$$

i.e. the length of the tape has increased by 0.002125 m.
Percentage error in measurement at 313 K

$$-\frac{\text{increase in length}}{\text{original length}} \times 100\%$$

$$= \frac{0.002125}{5.0} \times 100 = \mathbf{0.0425\%}$$

Problem 4. The copper tubes in a boiler are 4.20 m long at a temperature of 20°C. Determine the length of the tubes (a) when surrounded only by feed water at 10°C, (b) when the boiler is operating and the mean temperature of the tubes is 320°C. Assume the coefficient of linear expansion of copper to be $17 \times 10^{-6} \, K^{-1}$

(a) Initial length $L_1 = 4.20$ m, initial temperature $t_1 = 20°C$, final temperature $t_2 = 10°C$ and $\alpha = 17 \times 10^{-6} \, K^{-1}$.

Final length at 10°C is given by:

$$L_2 = L_1[1 + \alpha(t_2 - t_1)]$$

$$= 4.20[1 + (17 \times 10^{-6})(10 - 20)]$$

$$= 4.20[1 - 0.00017] = \mathbf{4.1993 \text{ m}}$$

i.e. **the tube contracts by 0.7 mm when the temperature decreases from 20°C to 10°C.**

(b) Length $L_1 = 4.20$ m, $t_1 = 20°C$, $t_2 = 320°C$ and $\alpha = 17 \times 10^{-6} \, K^{-1}$.

Final length at 320°C is given by:

$$L_2 = L_1[1 + \alpha(t_2 - t_1)]$$

$$= 4.20[1 + (17 \times 10^{-6})(320 - 20)]$$

$$= 4.20[1 + 0.0051] = \mathbf{4.2214 \text{ m}}$$

i.e. **the tube extends by 21.4 mm when the temperature rises from 20°C to 320°C.**

Now try the following Practice Exercise

Practice Exercise 151   Further problems on the coefficient of linear expansion (answers on page 545)

1. A length of lead piping is 50.0 m long at a temperature of 16°C. When hot water flows through it the temperature of the pipe rises to 80°C. Determine the length of the hot pipe if the coefficient of linear expansion of lead is $29 \times 10^{-6} \, K^{-1}$

2. A rod of metal is measured at 285 K and is 3.521 m long. At 373 K the rod is 3.523 m long. Determine the value of the coefficient of linear expansion for the metal

3. A copper overhead transmission line has a length of 40.0 m between its supports at 20°C. Determine the increase in length at 50°C if the coefficient of linear expansion of copper is $17 \times 10^{-6} \, K^{-1}$

4. A brass measuring tape measures 2.10 m at a temperature of 15°C. Determine (a) the increase in length when the temperature has increased to 40°C, (b) the percentage error in measurement at 40°C. Assume the coefficient of linear expansion of brass to be $18 \times 10^{-6} \, K^{-1}$

5. A pendulum of a 'grandfather' clock is 2.0 m long and made of steel. Determine the change in length of the pendulum if the temperature rises by 15 K. Assume the coefficient of linear expansion of steel to be $15 \times 10^{-6} \, K^{-1}$

6. A temperature control system is operated by the expansion of a zinc rod which is 200 mm long at 15°C. If the system is set so that

the source of heat supply is cut off when the rod has expanded by 0.20 mm, determine the temperature to which the system is limited. Assume the coefficient of linear expansion of zinc to be $31 \times 10^{-6} \, K^{-1}$

7. A length of steel railway line is 31.0 m long when the temperature is 288 K. Determine the increase in length of the line when the temperature is raised to 303 K. Assume the coefficient of linear expansion of steel to be $15 \times 10^{-6} \, K^{-1}$

8. A brass shaft is 15.02 mm in diameter and has to be inserted in a hole of diameter 15.0 mm Determine by how much the shaft must be cooled to make this possible, without using force. Take the coefficient of linear expansion of brass as $18 \times 10^{-6} \, K^{-1}$

31.5 Coefficient of superficial expansion

The amount by which unit area of a material increases when the temperature is raised by one degree is called the **coefficient of superficial (i.e. area) expansion** and is represented by β (Greek beta).

If a material having an initial surface area A_1 at temperature t_1 and having a coefficient of superficial expansion β, has its temperature increased to t_2, then the new surface area A_2 of the material is given by:

New surface area = original surface area

+ increase in area

i.e. $\qquad A_2 = A_1 + A_1 \beta (t_2 - t_1)$

i.e. $\qquad \boldsymbol{A_2 = A_1[1 + \beta(t_2 - t_1)]} \qquad (2)$

It is shown in Problem 5 below that the coefficient of superficial expansion is twice the coefficient of linear expansion, i.e. $\beta = 2\alpha$, to a very close approximation.

Problem 5. Show that for a rectangular area of material having dimensions L by b the coefficient of superficial expansion $\beta \approx 2\alpha$, where α is the coefficient of linear expansion

Initial area $A_1 = Lb$. For a temperature rise of 1 K, side L will expand to $(L + L\alpha)$ and side b will expand to $(b + b\alpha)$. Hence the new area of the rectangle A_2 is

given by:

$$A_2 = (L + L\alpha)(b + b\alpha) = L(1 + \alpha)b(1 + \alpha)$$
$$= Lb(1 + \alpha)^2 = Lb(1 + 2\alpha + \alpha^2)$$
$$\approx Lb(1 + 2\alpha)$$

since α^2 is very small (see typical values in section 31.4)
Hence, $A_2 \approx A_1(1 + 2\alpha)$
For a temperature rise of $(t_2 - t_1) \, K$

$$A_2 \approx A_1[1 + 2\alpha(t_2 - t_1)]$$

Thus, from equation (2), $\boldsymbol{\beta \approx 2\alpha}$

31.6 Coefficient of cubic expansion

The amount by which unit volume of a material increases for a one degree rise of temperature is called the **coefficient of cubic (or volumetric) expansion** and is represented by γ (Greek gamma).
If a material having an initial volume V_1 at temperature t_1 and having a coefficient of cubic expansion γ, has its temperature raised to t_2, then the new volume V_2 of the material is given by:

New volume = initial volume + increase in volume

i.e. $\qquad V_2 = V_1 + V_1 \gamma (t_2 - t_1)$

i.e. $\qquad \boldsymbol{V_2 = V_1[1 + \gamma(t_2 - t_1)]} \qquad (3)$

It is shown in Problem 6 below that the coefficient of cubic expansion is three times the coefficient of linear expansion, i.e. $\gamma = 3\alpha$, to a very close approximation. A liquid has no definite shape and only its cubic or volumetric expansion need be considered. Thus with expansions in liquids, equation (3) is used.

Problem 6. Show that for a rectangular block of material having dimensions L, b and h, the coefficient of cubic expansion $\gamma \approx 3\alpha$, where α is the coefficient of linear expansion

Initial volume, $V_1 = Lbh$. For a temperature rise of 1 K, side L expands to $(L + L\alpha)$, side b expands to $(b + b\alpha)$ and side h expands to $(h + h\alpha)$.

Hence the new volume of the block V_2 is given by:

$$V_2 = (L + L\alpha)(b + b\alpha)(h + h\alpha)$$

$$= L(1 + \alpha)b(1 + \alpha)h(1 + \alpha)$$

$$= Lbh(1 + \alpha)^3 = Lbh(1 + 3\alpha + 3\alpha^2 + \alpha^3)$$

$$\approx Lbh(1 + 3\alpha) \quad \text{since terms in } \alpha^2 \text{ and } \alpha^3 \text{ are}$$

very small

Hence, $V_2 \approx V_1(1 + 3\alpha)$

For a temperature rise of $(t_2 - t_1)\,\text{K}$,

$$V_2 \approx V_1[1 + 3\alpha(t_2 - t_1)]$$

Thus, from equation (3), $\boldsymbol{\gamma \approx 3\alpha}$

Some **typical values** for the coefficient of cubic expansion measured at $20°C$ (i.e. 293 K) include:

Ethyl alcohol	$1.1 \times 10^{-3}\,\text{K}^{-1}$
Mercury	$1.82 \times 10^{-4}\,\text{K}^{-1}$
Paraffin oil	$9 \times 10^{-2}\,\text{K}^{-1}$
Water	$2.1 \times 10^{-4}\,\text{K}^{-1}$

The coefficient of cubic expansion γ is only constant over a limited range of temperature.

Problem 7. A brass sphere has a diameter of 50 mm at a temperature of 289 K. If the temperature of the sphere is raised to 789 K, determine the increase in (a) the diameter, (b) the surface area, (c) the volume of the sphere. Assume the coefficient of linear expansion for brass is $18 \times 10^{-6}\,\text{K}^{-1}$

(a) Initial diameter $L_1 = 50\,\text{mm}$, initial temperature $t_1 = 289\,\text{K}$, final temperature $t_2 = 789\,\text{K}$ and $\alpha = 18 \times 10^{-6}\,\text{K}^{-1}$.

New diameter at 789 K is given by:

$$L_2 = L_1[1 + \alpha(t_2 - t_1)] \quad \text{from equation (1)}$$

i.e. $L_2 = 50[1 + (18 \times 10^{-6})(789 - 289)]$

$$= 50[1 + 0.009]$$

$$= 50.45\,\text{mm}$$

Hence, the increase in the diameter is 0.45 mm

(b) Initial surface area of sphere

$$A_1 = 4\pi r^2 = 4\pi \left(\frac{50}{2}\right)^2 = 2500\,\pi\,\text{mm}^2$$

New surface area at 789 K is given by:

$$A_2 = A_1[1 + \beta(t_2 - t_1)] \quad \text{from equation (2)}$$

i.e. $A_2 = A_1[1 + 2\alpha(t_2 - t_1)]$ since $\beta = 2\alpha$, to a very close approximation

Thus $A_2 = 2500\pi[1 + 2(18 \times 10^{-6})(500)]$

$$= 2500\pi[1 + 0.018]$$

$$= 2500\pi + 2500\pi(0.018)$$

Hence, the increase in surface area

$$= 2500\pi(0.018) = \mathbf{141.4\,mm^2}$$

(c) Initial volume of sphere,

$$V_1 = \frac{4}{3}\pi r^3 = \frac{4}{3}\pi \left(\frac{50}{2}\right)^3\,\text{mm}^3$$

New volume at 789 K is given by:

$$V_2 = V_1[1 + \gamma(t_2 - t_1)] \quad \text{from equation (3)}$$

i.e. $V_2 = V_1[1 + 3\alpha(t_2 - t_1)]$ since $\gamma = 3\alpha$, to a very close approximation

Thus $V_2 = \frac{4}{3}\pi(25)^3[1 + 3(18 \times 10^{-6})(500)]$

$$= \frac{4}{3}\pi(25)^3[1 + 0.027]$$

$$= \frac{4}{3}\pi(25)^3 + \frac{4}{3}\pi(25)^3(0.027)$$

Hence, the increase in volume

$$= \frac{4}{3}\pi(25)^3(0.027) = \mathbf{1767\,mm^3}$$

Problem 8. Mercury contained in a thermometer has a volume of 476 mm^3 at 15°C. Determine the temperature at which the volume of mercury is 478 mm^3, assuming the coefficient of cubic expansion for mercury to be $1.8 \times 10^{-4}\,\text{K}^{-1}$

Initial volume $V_1 = 476\,\text{mm}^3$, final volume $V_2 = 478\,\text{mm}^3$, initial temperature $t_1 = 15°C$ and $\gamma = 1.8 \times 10^{-4}\,\text{K}^{-1}$.

Final volume $V_2 = V_1[1 + \gamma(t_2 - t_1)]$

from equation (3)

i.e. $\qquad V_2 = V_1 + V_1\gamma(t_2 - t_1)$ from which,

$$(t_2 - t_1) = \frac{V_2 - V_1}{V_1\gamma}$$

$$= \frac{478 - 476}{(476)(1.8 \times 10^{-4})}$$

$$= 23.34°C$$

Hence, $t_2 = 23.34 + t_1 = 23.34 + 15 = 38.34°C$

Hence, the temperature at which the volume of mercury is 478 mm³ is 38.34°C

Problem 9.   A rectangular glass block has a length of 100 mm, width 50 mm and depth 20 mm at 293 K. When heated to 353 K its length increases by 0.054 mm. What is the coefficient of linear expansion of the glass? Find also (a) the increase in surface area (b) the change in volume resulting from the change of length

Final length, $L_2 = L_1[1 + \alpha(t_2 - t_1)]$ from equation (1); hence, increase in length is given by:

$$L_2 - L_1 = L_1\alpha(t_2 - t_1)$$

Hence, $0.054 = (100)(\alpha)(353 - 293)$

from which, the **coefficient of linear expansion** is given by:

$$\alpha = \frac{0.054}{(100)(60)} = \mathbf{9 \times 10^{-6} K^{-1}}$$

(a)   Initial surface area of glass

$$A_1 = (2 \times 100 \times 50) + (2 \times 50 \times 20) + (2 \times 100 \times 20)$$

$$= 10,000 + 2000 + 4000 = 16,000 \text{ mm}^2$$

Final surface area of glass

$$A_2 = A_1[1 + \beta(t_2 - t_1)] = A_1[1 + 2\alpha(t_2 - t_1)]$$

since $\beta = 2\alpha$ to a very close approximation

Hence, **increase in surface area**

$$= A_1(2\alpha)(t_2 - t_1)$$

$$= (16,000)(2 \times 9 \times 10^{-6})(60)$$

$$= \mathbf{17.28 \text{ mm}^2}$$

(b)   Initial volume of glass

$$V_1 = 100 \times 50 \times 20 = 100,000 \text{ mm}^3$$

Final volume of glass $V_2 = V_1[1 + \gamma(t_2 - t_1)]$

$$= V_1[1 + 3\alpha(t_2 - t_1)]$$

since $\gamma = 3\alpha$ to a very close approximation

Hence, **increase in volume of glass**

$$= V_1(3\alpha)(t_2 - t_1)$$

$$= (100,000)(3 \times 9 \times 10^{-6})(60)$$

$$= \mathbf{162 \text{ mm}^3}$$

Now try the following Practice Exercises

Practice Exercise 152   Further questions on the coefficients of superficial and cubic expansion (answers on page 545)

1.   A silver plate has an area of 800 mm² at 15°C. Determine the increase in the area of the plate when the temperature is raised to 100°C. Assume the coefficient of linear expansion of silver to be $19 \times 10^{-6} K^{-1}$

2.   At 283 K a thermometer contains 440 mm³ of alcohol. Determine the temperature at which the volume is 480 mm³ assuming that the coefficient of cubic expansion of the alcohol is $12 \times 10^{-4} K^{-1}$

3.   A zinc sphere has a radius of 30.0 mm at a temperature of 20°C. If the temperature of the sphere is raised to 420°C, determine the increase in: (a) the radius, (b) the surface area, (c) the volume of the sphere. Assume the coefficient of linear expansion for zinc to be $31 \times 10^{-6} K^{-1}$

4.   A block of cast iron has dimensions of 50 mm by 30 mm by 10 mm at 15°C. Determine the increase in volume when the temperature of the block is raised to 75°C. Assume the coefficient of linear expansion of cast iron to be $11 \times 10^{-6} K^{-1}$

5.   Two litres of water, initially at 20°C, is heated to 40°C. Determine the volume of water at

40°C if the coefficient of volumetric expansion of water within this range is $30 \times 10^{-5}\,\text{K}^{-1}$

6. Determine the increase in volume, in litres, of $3\,\text{m}^3$ of water when heated from 293 K to boiling point if the coefficient of cubic expansion is $2.1 \times 10^{-4}\,\text{K}^{-1}$ (1 litre $\approx 10^{-3}\,\text{m}^3$)

7. Determine the reduction in volume when the temperature of 0.5 litre of ethyl alcohol is reduced from 40°C to −15°C. Take the coefficient of cubic expansion for ethyl alcohol as $1.1 \times 10^{-3}\,\text{K}^{-1}$

Practice Exercise 153   Short answer questions on thermal expansion

1. When heat is applied to most solids and liquids .... occurs

2. When solids and liquids are cooled they usually ....

3. State three practical applications where the expansion of metals must be allowed for

4. State a practical disadvantage where the expansion of metals occurs

5. State one practical advantage of the expansion of liquids

6. What is meant by the 'coefficient of expansion'

7. State the symbol and the unit used for the coefficient of linear expansion

8. Define the 'coefficient of superficial expansion' and state its symbol

9. Describe how water displays an unexpected effect between 0°C and 4°C

10. Define the 'coefficient of cubic expansion' and state its symbol

Practice Exercise 154   Multiple-choice questions on thermal expansion (answers on page 545)

1. When the temperature of a rod of copper is increased, its length:

   (a) stays the same (b) increases (c) decreases

2. The amount by which unit length of a material increases when the temperature is raised one degree is called the coefficient of:

   (a) cubic expansion (b) superficial expansion (c) linear expansion

3. The symbol used for volumetric expansion is:

   (a) γ (b) β (c) L (d) α

4. A material of length L_1 at temperature θ_1 K is subjected to a temperature rise of θ K. The coefficient of linear expansion of the material is $\alpha\,\text{K}^{-1}$. The material expands by:

   (a) $L_2(1 + \alpha\theta)$ (b) $L_1\alpha(\theta - \theta_1)$

   (c) $L_1[1 + \alpha(\theta - \theta_1)]$ (d) $L_1\alpha\theta$

5. Some iron has a coefficient of linear expansion of $12 \times 10^{-6}\,\text{K}^{-1}$. A 100 mm length of iron piping is heated through 20 K. The pipe extends by:

   (a) 0.24 mm   (b) 0.024 mm

   (c) 2.4 mm   (d) 0.0024 mm

6. If the coefficient of linear expansion is A, the coefficient of superficial expansion is B and the coefficient of cubic expansion is C, which of the following is false?

   (a) $C = 3A$ (b) $A = B/2$

   (c) $B = \dfrac{3}{2}C$ (d) $A = C/3$

7. The length of a 100 mm bar of metal increases by 0.3 mm when subjected to a temperature rise of 100 K. The coefficient of linear expansion of the metal is:

   (a) $3 \times 10^{-3}\,\text{K}^{-1}$ (b) $3 \times 10^{-4}\,\text{K}^{-1}$

   (c) $3 \times 10^{-5}\,\text{K}^{-1}$ (d) $3 \times 10^{-6}\,\text{K}^{-1}$

8. A liquid has a volume V_1 at temperature θ_1. The temperature is increased to θ_2. If γ is the coefficient of cubic expansion, the increase in volume is given by:

   (a) $V_1\gamma(\theta_2 - \theta_1)$ (b) $V_1\gamma\theta_2$

   (c) $V_1 + V_1\gamma\theta_2$ (d) $V_1[1 + \gamma(\theta_2 - \theta_1)]$

9. Which of the following statements is false?

   (a) Gaps need to be left in lengths of railway lines to prevent buckling in hot weather

   (b) Bimetallic strips are used in thermostats, a thermostat being a temperature-operated switch

   (c) As the temperature of water is decreased from 4°C to 0°C contraction occurs

   (d) A change of temperature of 15°C is equivalent to a change of temperature of 15 K

10. The volume of a rectangular block of iron at a temperature t_1 is V_1. The temperature is raised to t_2 and the volume increases to V_2. If the coefficient of linear expansion of iron is α, then volume V_1 is given by:

   (a) $V_2[1 + \alpha(t_2 - t_1)]$ (b) $\dfrac{V_2}{1 + 3\alpha\,(t_2 - t_1)}$

   (c) $3V_2\alpha(t_2 - t_1)$ (d) $\dfrac{1 + \alpha\,(t_2 - t_1)}{V_2}$

**For fully worked solutions to each of the problems in Exercises 151 to 154 in this chapter,
go to the website:**

Chapter 32

Ideal gas laws

Why it is important to understand: **Ideal gas laws**

The relationships that exist between pressure, volume and temperature in a gas are given in a set of laws called the gas laws, the most fundamental being those of Boyle, Charles, and the pressure or Gay-Lussac's law, together with Dalton's law of partial pressures and the characteristic gas equation. These laws are used for all sorts of practical applications, including for designing pressure vessels, in the form of circular cylinders and spheres, which are used for storing and transporting gases. Another example of this is the pressure in car tyres, which can increase due to a temperature increase, and can decrease due to a temperature decrease. Other examples are large and medium size gas storage cylinders and domestic spray cans, which can explode if they are heated. In the case of domestic spray cans, these can explode dangerously in a domestic situation if they are left on a window sill where the sunshine acting on them causes them to heat up or, if they are thrown on to a fire. In these cases, the consequence can be disastrous, so don't throw your 'full' spray can on to a fire; you may very sadly and deeply regret it! Another example of a gas storage vessel is that used by your 'local' gas companies, which supply natural gas (methane) to domestic properties, businesses, etc.

At the end of this chapter, you should be able to:

- state and perform calculations involving Boyle's law
- understand the term isothermal
- state and perform calculations involving Charles' law
- understand the term isobaric
- state and perform calculations involving the pressure or Gay-Lussac law
- state and perform calculations on Dalton's law of partial pressures
- state and perform calculations on the characteristic gas equation
- understand the term STP

Science and Mathematics for Engineering. 978-0-367-20475-4, © John Bird. Published by Taylor & Francis. All rights reserved.

32.1 Boyle's law

Boyle's law* states:

The volume V of a fixed mass of gas is inversely proportional to its absolute pressure p at constant temperature

i.e. $p \propto \dfrac{1}{V}$ or $p = \dfrac{k}{V}$ or $pV = k$ at constant temperature, where $p =$ absolute pressure in pascals (Pa), $V =$ volume in m^3 and $k =$ a constant.

Changes that occur at constant temperature are called **isothermal** changes. When a fixed mass of gas at constant temperature changes from pressure p_1 and volume V_1 to pressure p_2 and volume V_2 then:

$$p_1 V_1 = p_2 V_2$$

*Who was Boyle? – **Robert Boyle** (25 January 1627–31 December 1691) was a natural philosopher, chemist, physicist, and inventor. Regarded today as the first modern chemist, he is best known for **Boyle's law**, which describes the inversely proportional relationship between the absolute pressure and volume of a gas, providing the temperature is kept constant within a closed system. To find out more about Boyle go to www.routledge.com/cw/bird

Problem 1. A gas occupies a volume of $0.10\,m^3$ at a pressure of $1.8\,MPa$. Determine (a) the pressure if the volume is changed to $0.06\,m^3$ at constant temperature and (b) the volume if the pressure is changed to $2.4\,MPa$ at constant temperature

(a) Since the change occurs at constant temperature (i.e. an isothermal change), Boyle's law applies, i.e. $p_1 V_1 = p_2 V_2$ where $p_1 = 1.8\,MPa$, $V_1 = 0.10\,m^3$ and $V_2 = 0.06\,m^3$

Hence, $(1.8)(0.10) = p_2(0.06)$ from which,

$$\textbf{pressure } p_2 = \frac{1.8 \times 0.10}{0.06} = \textbf{3 MPa}$$

(b) $p_1 V_1 = p_2 V_2$ where $p_1 = 1.8\,MPa$, $V_1 = 0.10\,m^3$ and $p_2 = 2.4\,MPa$.

Hence, $(1.8)(0.10) = (2.4)V_2$ from which,

$$\textbf{volume } V_2 = \frac{1.8 \times 0.10}{2.4}$$
$$= \textbf{0.075 } \mathbf{m^3}$$

Problem 2. In an isothermal process, a mass of gas has its volume reduced from $3200\,mm^3$ to $2000\,mm^3$. If the initial pressure of the gas is $110\,kPa$, determine the final pressure

Since the process is isothermal, it takes place at constant temperature and hence Boyle's law applies, i.e. $p_1 V_1 = p_2 V_2$, where $p_1 = 110\,kPa$, $V_1 = 3200\,mm^3$ and $V_2 = 2000\,mm^3$.

Hence, $(110)(3200) = p_2(2000)$ from which,

$$\textbf{final pressure } p_2 = \frac{110 \times 3200}{2000}$$
$$= \textbf{176 kPa}$$

Problem 3. Some gas occupies a volume of $1.5\,m^3$ in a cylinder at a pressure of $250\,kPa$. A piston, sliding in the cylinder, compresses the gas isothermally until the volume is $0.5\,m^3$. If the area of the piston is $300\,cm^2$, calculate the force on the piston when the gas is compressed

An isothermal process means constant temperature and thus Boyle's law applies, i.e. $p_1 V_1 = p_2 V_2$, where $V_1 = 1.5\,m^3$, $V_2 = 0.5\,m^3$ and $p_1 = 250\,kPa$.

Hence, $(250)(1.5) = p_2 (0.5)$ from which,

$$\text{pressure } p_2 = \frac{250 \times 1.5}{0.5} = 750 \text{ kPa}$$

$$\text{Pressure} = \frac{\text{force}}{\text{area}}, \text{ from which,}$$

force = pressure × area.

Hence, **force on the piston**

$$= (750 \times 10^3 \text{Pa}) (300 \times 10^{-4} \text{m}^2) = \textbf{22.5 kN}$$

Problem 4. The gas in a syringe has a pressure of 600 mm of mercury (Hg) and a volume of quantity 20 mL. If the syringe is compressed to a volume of 3 mL, what will be the pressure of the gas, assuming that its temperature does not change?

Since

$$P_1 V_1 = P_2 V_2$$

then

$$(600 \text{mm Hg})(20 \text{mL}) = (P_2)(3 \text{mL})$$

from which,

$$\text{new pressure, } \textbf{P}_2 = \frac{600 \text{ mm Hg} \times 20 \text{mL}}{3 \text{mL}}$$

$$= \textbf{4000 mm of mercury}$$

Now try the following Practice Exercise

Practice Exercise 155   Further problems on Boyle's law (answers on page 545)

1. The pressure of a mass of gas is increased from 150 kPa to 750 kPa at constant temperature. Determine the final volume of the gas, if its initial volume is 1.5 m^3

2. In an isothermal process, a mass of gas has its volume reduced from 50 cm^3 to 32 cm^3. If the initial pressure of the gas is 80 kPa, determine its final pressure

3. The piston of an air compressor compresses air to $\frac{1}{4}$ of its original volume during its stroke. Determine the final pressure of the air if the original pressure is 100 kPa, assuming an isothermal change

4. A quantity of gas in a cylinder occupies a volume of 2 m^3 at a pressure of 300 kPa. A piston slides in the cylinder and compresses the gas, according to Boyle's law, until the volume is 0.5 m^3. If the area of the piston is 0.02 m^2, calculate the force on the piston when the gas is compressed

5. The gas in a simple pump has a pressure of 400 mm of mercury (Hg) and a volume of 10 mL. If the pump is compressed to a volume of 2 mL, calculate the pressure of the gas, assuming that its temperature does not change?

32.2 Charles' law

Charles' law* states:

For a given mass of gas at constant pressure, the volume V is directly proportional to its thermodynamic temperature T

i.e. $V \propto T$ or $V = kT$ or $\dfrac{V}{T} = k$ at constant pressure,

where T = thermodynamic temperature in kelvin (K). A process that takes place at constant pressure is called an **isobaric** process.

The relationship between the Celsius scale of temperature and the thermodynamic or absolute scale is given by:

$$\text{kelvin} = \text{degrees Celsius} + 273$$

i.e. $\textbf{K} = \textbf{°C} + \textbf{273}$ or $\textbf{°C} = \textbf{K} - \textbf{273}$

(as stated in chapter 30).
If a given mass of gas at a constant pressure occupies a volume V_1 at a temperature T_1 and a volume V_2 at temperature T_2, then

$$\frac{V_1}{T_1} = \frac{V_2}{T_2}$$

Problem 5. A gas occupies a volume of 1.2 litres at 20°C. Determine the volume it occupies at 130°C if the pressure is kept constant

Since the change occurs at constant pressure (i.e. an isobaric process), Charles' law applies,

i.e.
$$\frac{V_1}{T_1} = \frac{V_2}{T_2}$$

where $V_1 = 1.2$ litre, $T_1 = 20°C = (20 + 273)K = 293$ K and $T_2 = (130 + 273)K = 403$ K.

Hence,
$$\frac{1.2}{293} = \frac{V_2}{403}$$

from which, **volume at 130°C, $V_2 = \dfrac{(1.2)(403)}{293}$**

$$= \textbf{1.65 litres}$$

*Who was Charles? – **Jacques Alexandre César Charles** (12 November 1746–7 April 1823) was a French inventor, scientist, mathematician, and balloonist. Charles' Law describes how gases expand when heated. To find out more about Charles go to www.routledge.com/cw/bird

Problem 6. Gas at a temperature of 150°C has its volume reduced by one-third in an isobaric process. Calculate the final temperature of the gas

Since the process is isobaric it takes place at constant pressure and hence Charles' law applies,

i.e.
$$\frac{V_1}{T_1} = \frac{V_2}{T_2}$$

where $T_1 = (150 + 273) K = 423$ K and $V_2 = \dfrac{2}{3} V_1$

Hence
$$\frac{V_1}{423} = \frac{\frac{2}{3} V_1}{T_2}$$

from which, **final temperature**

$$T_2 = \frac{2}{3}(423) = \textbf{282 K} \text{ or } (282 - 273)°C \text{ i.e. } \textbf{9°C}$$

Problem 7. A balloon is under a constant internal pressure. If its volume is 15 litres at a temperature of 35°C, what will be its volume at a temperature of 10°C

Since

$$\frac{V_1}{T_1} = \frac{V_2}{T_2}$$

Where

$$T_1 = (35 + 273)K = 308K$$

and

$$T_2 = (10 + 273)K = 283K$$

Hence,

$$\frac{15 \text{litres}}{308 \text{ K}} = \frac{V_2}{283 \text{ K}}$$

and

new volume, $V_2 = \dfrac{15 \text{ litres} \times 283K}{308 \text{ K}} = \textbf{13.8 litres}$

Now try the following Practice Exercise

Practice Exercise 156   Further problems on Charles' law (answers on page 545)

1. Some gas initially at 16°C is heated to 96°C at constant pressure. If the initial volume of the gas is 0.8 m³, determine the final volume of the gas

2. A gas is contained in a vessel of volume 0.02 m³ at a pressure of 300 kPa and a temperature of 15°C. The gas is passed into a vessel of volume 0.015 m³. Determine to what temperature the gas must be cooled for the pressure to remain the same

3. In an isobaric process gas at a temperature of 120°C has its volume reduced by a sixth. Determine the final temperature of the gas

4. The volume of a balloon is 30 litres at a temperature of 27°C. If the balloon is under a constant internal pressure, calculate its volume at a temperature of 12°C

32.3 The pressure or Gay-Lussac's law

The **pressure** or **Gay-Lussac's*** **law** states:

The pressure p of a fixed mass of gas is directly proportional to its thermodynamic temperature T at constant volume

i.e. $p \propto T$ or $p = kT$ or $\dfrac{p}{T} = k$

When a fixed mass of gas at constant volume changes from pressure p_1 and temperature T_1, to pressure p_2 and temperature T_2 then:

$$\frac{p_1}{T_1} = \frac{p_2}{T_2}$$

Problem 8. Gas initially at a temperature of 17°C and pressure 150 kPa is heated at constant volume until its temperature is 124°C. Determine the final pressure of the gas, assuming no loss of gas

Since the gas is at constant volume, the pressure law applies,

i.e. $\dfrac{p_1}{T_1} = \dfrac{p_2}{T_2}$

where $T_1 = (17 + 273)\,\text{K} = 290\,\text{K}$, $T_2 = (124 + 273)\,\text{K} = 397\,\text{K}$ and $p_1 = 150\,\text{kPa}$

Hence, $\dfrac{150}{290} = \dfrac{p_2}{397}$ from which, **final pressure**

$$p_2 = \frac{(150)(397)}{290} = \mathbf{205.3\,kPa}$$

Problem 9. A rigid pressure vessel is subjected to a gas pressure of 10 atmospheres at a temperature of 20°C. The pressure vessel can withstand a maximum pressure of 30 atmospheres. What gas temperature increase will this vessel withstand? (Note that 1 atmosphere of pressure means 1.01325 bar or 1.01325 × 10⁵ Pa or 14.5 psi)

Since

$$\frac{P_1}{T_1} = \frac{P_2}{T_2}$$

*Who was Gay-Lussac? – **Joseph Louis Gay-Lussac** (6 December 1778–9 May 1850) was a French chemist and physicist. He is known mostly for two laws related to gases, and for his work on alcohol-water mixtures, which led to the Gay-Lussac scale used to measure alcoholic beverages in many countries. To find out more about Gay-Lussac go to www.routledge.com/cw/bird

where

$$T_1 = (20 + 273)K = 293K,$$

then

$$\frac{10 \, \text{atmospheres}}{293K} = \frac{30 \, \text{atmospheres}}{T_2}$$

from which, new temperature,

$$T_2 = \frac{30 \, \text{atmospheres} \times 293 \, K}{10 \, \text{atmospheres}}$$

$$= \mathbf{879 \, K} \, \text{or} \, (879 - 273)°\text{C} = \mathbf{606°C}$$

Hence, **temperature rise** = (879–293)K = **586** K
or **temperature rise** = (606–20)°**C** = **586°C**
Note that a temperature **change** of 586 K is equal to a temperature **change** of 586°C

___

Now try the following Practice Exercise

Practice Exercise 157   A further problem on the pressure law (answer on page 545)

1. Gas, initially at a temperature of 27°C and pressure 100 kPa, is heated at constant volume until its temperature is 150°C. Assuming no loss of gas, determine the final pressure of the gas

2. A pressure vessel is subjected to a gas pressure of 8 atmospheres at a temperature of 15°C. The vessel can withstand a maximum pressure of 28 atmospheres. Calculate the gas temperature increase the vessel can withstand

32.4   Dalton's law of partial pressure

Dalton's law* of partial pressure states:

The total pressure of a mixture of gases occupying a given volume is equal to the sum of the pressures of each gas, considered separately, at constant temperature

The pressure of each constituent gas when occupying a fixed volume alone is known as the **partial pressure** of that gas.

An **ideal gas** is one that completely obeys the gas laws given in sections 32.2–32.5. In practice no gas is an ideal gas, although air is very close to being one. For calculation purposes the difference between an ideal and an actual gas is very small.

> **Problem 10.**   A gas R in a container exerts a pressure of 200 kPa at a temperature of 18°C. Gas Q is added to the container and the pressure increases to 320 kPa at the same temperature. Determine the pressure that gas Q alone exerts at the same temperature

Initial pressure $p_R = 200$ kPa, and the pressure of gases R and Q together, $p = p_R + p_Q = 320$ kPa.

*Who was Dalton? – **John Dalton** (6 September 1766– 27 July 1844) was an English chemist, meteorologist and physicist. He is best known for his work in the development of modern atomic theory in chemistry, and his law of partial pressures is now known as Dalton's law. To find out more about Dalton go to www.routledge.com/cw/bird

By Dalton's law of partial pressure, **the pressure of gas Q alone is** $p_Q = p - p_R = 320 - 200 = \textbf{120 kPa}$.

Now try the following Practice Exercise

Practice Exercise 158   A further problem on Dalton's law of partial pressure (answer on page 545)

1. A gas A in a container exerts a pressure of 120 kPa at a temperature of 20°C. Gas B is added to the container and the pressure increases to 300 kPa at the same temperature. Determine the pressure that gas B alone exerts at the same temperature

32.5   Characteristic gas equation

Frequently, when a gas is undergoing some change, the pressure, temperature and volume all vary simultaneously. Provided there is no change in the mass of a gas, the above gas laws can be combined, giving

$$\frac{p_1 V_1}{T_1} = \frac{p_2 V_2}{T_2} = k \quad \text{where } k \text{ is a constant.}$$

For an ideal gas, constant $k = mR$, where m is the mass of the gas in kg, and R is the **characteristic gas constant**,

i.e.
$$\frac{pV}{T} = mR$$

or
$$pV = mRT$$

This is called the **characteristic gas equation**. In this equation, p = absolute pressure in pascals, V = volume in m³, m = mass in kg, R = characteristic gas constant in J/(kg K), and T = thermodynamic temperature in kelvin.

Some typical values of the characteristic gas constant R include: air, 287 J/(kg K), hydrogen 4160 J/(kg K), oxygen 260 J/(kg K) and carbon dioxide 184 J/(kg K).
Standard temperature and pressure (**STP**) refers to a temperature of 0°C, i.e. 273 K, and normal atmospheric pressure of 101.325 kPa.

32.6   Worked problems on the characteristic gas equation

Problem 11.   A gas occupies a volume of 2.0 m³ when at a pressure of 100 kPa and a temperature of 120°C. Determine the volume of the gas at 15°C if the pressure is increased to 250 kPa

Using the combined gas law:

$$\frac{p_1 V_1}{T_1} = \frac{p_2 V_2}{T_2}$$

where $V_1 = 2.0$ m³, $p_1 = 100$ kPa, $p_2 = 250$ kPa, $T_1 = (120 + 273)$ K $= 393$ K and $T_2 = (15 + 273)$ K $= 288$ K, gives:

$$\frac{(100)(2.0)}{393} = \frac{(250) V_2}{288}$$

from which, **volume at 15°C**

$$V_2 = \frac{(100)(2.0)(288)}{(393)(250)} = \textbf{0.586 m}^3$$

Problem 12.   20,000 mm³ of air initially at a pressure of 600 kPa and temperature 180°C is expanded to a volume of 70,000 mm³ at a pressure of 120 kPa. Determine the final temperature of the air, assuming no losses during the process

Using the combined gas law:

$$\frac{p_1 V_1}{T_1} = \frac{p_2 V_2}{T_2}$$

where $V_1 = 20,000$ mm³, $V_2 = 70,000$ mm³, $p_1 = 600$ kPa, $p_2 = 120$ kPa, and $T_1 = (180 + 273)$ K $= 453$ K, gives:

$$\frac{(600)(20,000)}{453} = \frac{(120)(70,000)}{T_2}$$

from which, **final temperature**

$$T_2 = \frac{(120)(70,000)(453)}{(600)(20,000)} = \textbf{317 K or 44°C.}$$

Problem 13.   Some air at a temperature of 40°C and pressure 4 bar occupies a volume of 0.05 m³. Determine the mass of the air assuming the characteristic gas constant for air to be 287 J/(kg K)

From above, $pV = mRT$, where $p = 4$ bar $= 4 \times 10^5$ Pa (since 1 bar $= 10^5$ Pa – see chapter 29), $V = 0.05\,\text{m}^3$, $T = (40 + 273)K = 313\,\text{K}$ and $R = 287\,\text{J/(kg K)}$.

Hence $(4 \times 10^5)(0.05) = m(287)(313)$

from which, **mass of air**

$$m = \frac{(4 \times 10^5)(0.05)}{(287)(313)} = \textbf{0.223 kg or 223 g}$$

Problem 14. A cylinder of helium has a volume of $600\,\text{cm}^3$. The cylinder contains 200 g of helium at a temperature of 25°C. Determine the pressure of the helium if the characteristic gas constant for helium is 2080 J/(kg K)

From the characteristic gas equation, $pV = mRT$, $V = 600\,\text{cm}^3 = 600 \times 10^{-6}\text{m}^3$, $m = 200\,\text{g} = 0.2\,\text{kg}$, $T = (25 + 273)\,\text{K} = 298\,\text{K}$ and $R = 2080\,\text{J/(kg K)}$.

Hence $(p)(600 \times 10^{-6}) = (0.2)(2080)(298)$

from which, **pressure** $p = \dfrac{(0.2)(2080)(298)}{(600 \times 10^{-6})}$

$$= 206{,}613{,}333\,\text{Pa}$$

$$= \textbf{206.6 MPa}$$

Problem 15. A spherical vessel has a diameter of 1.2 m and contains oxygen at a pressure of 2 bar and a temperature of −20°C. Determine the mass of oxygen in the vessel. Take the characteristic gas constant for oxygen to be 0.260 kJ/(kg K)

From the characteristic gas equation, $pV = mRT$

$V = $ volume of spherical vessel

$$= \frac{4}{3}\pi r^3 = \frac{4}{3}\pi \left(\frac{1.2}{2}\right)^3 = 0.905\,\text{m}^3$$

$p = 2$ bar $= 2 \times 10^5$ Pa,

$T = (-20 + 273)\,\text{K} = 253\,\text{K}$

and $R = 0.260\,\text{kJ/(kg K)} = 260\,\text{J/(kg K)}$

Hence, $(2 \times 10^5)(0.905) = m(260)(253)$

from which, **mass of oxygen**

$$m = \frac{(2 \times 10^5)(0.905)}{(260)(253)} = \textbf{2.75 kg}$$

Problem 16. Determine the characteristic gas constant of a gas which has a specific volume of $0.5\,\text{m}^3/\text{kg}$ at a temperature of 20°C and pressure 150 kPa

From the characteristic gas equation, $pV = mRT$

from which, $R = \dfrac{pV}{mT}$

where $p = 150 \times 10^3$ Pa,

$T = (20 + 273)\,\text{K} = 293\,\text{K}$

and specific volume $V/m = 0.5\,\text{m}^3/\text{kg}$.

Hence the **characteristic gas constant**

$$R = \left(\frac{p}{T}\right)\left(\frac{V}{m}\right) = \left(\frac{150 \times 10^3}{293}\right)(0.5)$$

$$= \textbf{256 J/(kg K)}$$

Now try the following Practice Exercise

Practice Exercise 159   Further problems on the characteristic gas equation (answers on page 545)

1. A gas occupies a volume of $1.20\,\text{m}^3$ when at a pressure of 120 kPa and a temperature of 90°C. Determine the volume of the gas at 20°C if the pressure is increased to 320 kPa

2. A given mass of air occupies a volume of $0.5\,\text{m}^3$ at a pressure of 500 kPa and a temperature of 20°C. Find the volume of the air at STP

3. A balloon is under an internal pressure of 110 kPa with a volume of 16 litres at a temperature of 22°C. If the balloon's internal pressure decreases to 50 kPa, what will be its volume if the temperature also decreases to 12°C

4. A spherical vessel has a diameter of 2.0 m and contains hydrogen at a pressure of 300 kPa and a temperature of −30°C. Determine the mass of hydrogen in the vessel. Assume the characteristic gas constant R for hydrogen is 4160 J/(kg K)

5. A cylinder 200 mm in diameter and 1.5 m long contains oxygen at a pressure of 2 MPa and a temperature of 20°C. Determine the

mass of oxygen in the cylinder. Assume the characteristic gas constant for oxygen is 260 J/(kg K)

6. A gas is pumped into an empty cylinder of volume 0.1 m³ until the pressure is 5 MPa. The temperature of the gas is 40°C. If the cylinder mass increases by 5.32 kg when the gas has been added, determine the value of the characteristic gas constant

7. The mass of a gas is 1.2 kg and it occupies a volume of 13.45 m³ at STP. Determine its characteristic gas constant

8. 30 cm³ of air initially at a pressure of 500 kPa and temperature 150°C is expanded to a volume of 100 cm³ at a pressure of 200 kPa. Determine the final temperature of the air, assuming no losses during the process

9. A quantity of gas in a cylinder occupies a volume of 0.05 m³ at a pressure of 400 kPa and a temperature of 27°C. It is compressed according to Boyle's law until its pressure is 1 MPa, and then expanded according to Charles' law until its volume is 0.03 m³. Determine the final temperature of the gas

10. Some air at a temperature of 35°C and pressure 2 bar occupies a volume of 0.08 m³. Determine the mass of the air assuming the characteristic gas constant for air to be 287 J/(kg K). (1 bar = 10^5 Pa)

11. Determine the characteristic gas constant R of a gas that has a specific volume of 0.267 m³/kg at a temperature of 17°C and pressure 200 kPa

32.7 Further worked problems on the characteristic gas equation

Problem 17. A vessel has a volume of 0.80 m³ and contains a mixture of helium and hydrogen at a pressure of 450 kPa and a temperature of 17°C. If the mass of helium present is 0.40 kg determine (a) the partial pressure of each gas and (b) the mass of hydrogen present. Assume the characteristic gas constant for helium to be 2080 J/(kg K) and for hydrogen 4160 J/(kg K)

(a) $V = 0.80\,\text{m}^3, p = 450\,\text{kPa}$,

$T = (17 + 273)\,\text{K} = 290\,\text{K}, m_{\text{He}} = 0.40\,\text{kg}$,

$R_{\text{He}} = 2080\,\text{J/(kg K)}$.

If p_{He} is the partial pressure of the helium, then using the characteristic gas equation,

$p_{\text{He}}V = m_{\text{He}}R_{\text{He}}T$ gives:

$$(p_{\text{He}})(0.80) = (0.40)(2080)(290)$$

from which, **the partial pressure of the helium**

$$p_{\text{He}} = \frac{(0.40)(2080)(290)}{(0.80)}$$

$$= \mathbf{301.6\,kPa}$$

By Dalton's law of partial pressure the total pressure p is given by the sum of the partial pressures, i.e. $p = p_{\text{H}} + p_{\text{He}}$, from which, **the partial pressure of the hydrogen**

$$p_{\text{H}} = p - p_{\text{He}} = 450 - 301.6$$

$$= \mathbf{148.4\,kPa}$$

(b) From the characteristic gas equation,

$$p_{\text{H}}V = m_{\text{H}}R_{\text{H}}T$$

Hence, $(148.4 \times 10^3)(0.8) = m_{\text{H}}(4160)(290)$

from which, **mass of hydrogen**

$$m_{\text{H}} = \frac{(148.4 \times 10^3)(0.8)}{(4160)(290)}$$

$$= \mathbf{0.098\,kg}\ \text{or}\ \mathbf{98\,g}$$

Problem 18. A compressed air cylinder has a volume of 1.2 m³ and contains air at a pressure of 1 MPa and a temperature of 25°C. Air is released from the cylinder until the pressure falls to 300 kPa and the temperature is 15°C. Determine (a) the mass of air released from the container and (b) the volume it would occupy at STP. Assume the characteristic gas constant for air to be 287 J/(kg K)

$V_1 = 1.2\,\text{m}^3\ (= V_2), p_1 = 1\,\text{MPa} = 10^6\,\text{Pa}$,

$T_1 = (25 + 273)\,\text{K} = 298\,\text{K}$,

$T_2 = (15 + 273)\,\text{K} = 288\,\text{K}$,

$p_2 = 300\,\text{kPa} = 300 \times 10^3\,\text{Pa}$

and $R = 287\,\text{J/(kg K)}$.

(a)   Using the characteristic gas equation,
$p_1 V_1 = m_1 R T_1$, to find the initial mass of air in the cylinder gives:

$$(10^6)(1.2) = m_1(287)(298)$$

from which, mass $m_1 = \dfrac{(10^6)(1.2)}{(287)(298)}$

$$= 14.03 \, \text{kg}$$

Similarly, using $p_2 V_2 = m_2 R T_2$ to find the final mass of air in the cylinder gives:

$$(300 \times 10^3)(1.2) = m_2(287)(288)$$

from which, mass $m_2 = \dfrac{(300 \times 10^3)(1.2)}{(287)(288)}$

$$= 4.36 \, \text{kg}$$

Mass of air released from cylinder

$$= m_1 - m_2 = 14.03 - 4.36 = \mathbf{9.67 \, kg}$$

(b)   At STP, $T = 273$ K and $p = 101.325$ kPa. Using the characteristic gas equation

$$pV = mRT$$

volume, $V = \dfrac{mRT}{p} = \dfrac{(9.67)(287)(273)}{101,325}$

$$= \mathbf{7.48 \, m^3}$$

Problem 19.   A vessel X contains gas at a pressure of 750 kPa at a temperature of 27°C. It is connected via a valve to vessel Y that is filled with a similar gas at a pressure of 1.2 MPa and a temperature of 27°C. The volume of vessel X is 2.0 m³ and that of vessel Y is 3.0 m³. Determine the final pressure at 27°C when the valve is opened and the gases are allowed to mix. Assume R for the gas to be 300 J/(kg K)

For vessel X:

$$p_X = 750 \times 10^3 \, \text{Pa}, \; T_X = (27 + 273) \, \text{K} = 300 \, \text{K},$$

$$V_X = 2.0 \, \text{m}^3 \text{ and } R = 300 \, \text{J/(kg K)}$$

From the characteristic gas equation,

$$p_X V_X = m_X R T_X$$

Hence $(750 \times 10^3)(2.0) = m_X(300)(300)$

from which, mass of gas in vessel X,

$$m_X = \dfrac{(750 \times 10^3)(2.0)}{(300)(300)} = 16.67 \, \text{kg}$$

For vessel Y:

$$p_Y = 1.2 \times 10^6 \, \text{Pa}, \; T_Y = (27 + 273) \, \text{K} = 300 \, \text{K},$$

$$V_Y = 3.0 \, \text{m}^3 \text{ and } R = 300 \, \text{J/(kg K)}$$

From the characteristic gas equation,

$$p_Y V_Y = m_Y R T_Y$$

Hence $(1.2 \times 10^6)(3.0) = m_Y(300)(300)$

from which, mass of gas in vessel Y,

$$m_Y = \dfrac{(1.2 \times 10^6)(3.0)}{(300)(300)} = 40 \, \text{kg}$$

When the valve is opened, mass of mixture

$$m = m_X + m_Y$$

$$= 16.67 + 40 = 56.67 \, \text{kg}.$$

Total volume $V = V_X + V_Y = 2.0 + 3.0 = 5.0 \, \text{m}^3$,
$R = 300$ J/(kg K), $T = 300$ K.
From the characteristic gas equation,

$$pV = mRT$$

$$p(5.0) = (56.67)(300)(300)$$

from which, **final pressure**

$$p = \dfrac{(56.67)(300)(300)}{5.0} = \mathbf{1.02 \, MPa}$$

Now try the following Practice Exercises

Practice Exercise 160   Further questions on ideal gas laws (answers on page 545)

1.   A vessel P contains gas at a pressure of 800 kPa at a temperature of 25°C. It is connected via a valve to vessel Q that is filled with similar gas at a pressure of 1.5 MPa and a temperature of 25°C. The volume of vessel P is 1.5 m³, and that of vessel R is 2.5 m³. Determine the final pressure at 25°C when the valve is opened and the gases are allowed to mix. Assume R for the gas to be 297 J/(kg K)

2. A vessel contains 4 kg of air at a pressure of 600 kPa and a temperature of 40°C. The vessel is connected to another by a short pipe and the air exhausts into it. The final pressure in both vessels is 250 kPa and the temperature in both is 15°C. If the pressure in the second vessel before the air entered was zero, determine the volume of each vessel. Assume R for air is 287 J/(kg K)

3. A vessel has a volume of 0.75 m^3 and contains a mixture of air and carbon dioxide at a pressure of 200 kPa and a temperature of 27°C. If the mass of air present is 0.5 kg determine (a) the partial pressure of each gas and (b) the mass of carbon dioxide. Assume the characteristic gas constant for air to be 287 J/(kg K) and for carbon dioxide 184 J/(kg K)

4. A mass of gas occupies a volume of 0.02 m^3 when its pressure is 150 kPa and its temperature is 17°C. If the gas is compressed until its pressure is 500 kPa and its temperature is 57°C, determine (a) the volume it will occupy and (b) its mass, if the characteristic gas constant for the gas is 205 J/(kg K)

5. A compressed air cylinder has a volume of 0.6 m^3 and contains air at a pressure of 1.2 MPa absolute and a temperature of 37°C. After use the pressure is 800 kPa absolute and the temperature is 17°C. Calculate (a) the mass of air removed from the cylinder, and (b) the volume the mass of air removed would occupy at STP conditions. Take R for air as 287 J/(kg K) and atmospheric pressure as 100 kPa

Practice Exercise 161   Short answer questions on ideal gas laws

1. State Boyle's law

2. State Charles' law

3. State the pressure law

4. State Dalton's law of partial pressures

5. State the relationship between the Celsius and the thermodynamic scale of temperature

6. What is (a) an isothermal change and (b) an isobaric change?

7. Define an ideal gas

8. State the characteristic gas equation

9. What is meant by STP?

Practice Exercise 162   Multiple-choice questions on ideal gas laws (answers on page 545)

1. Which of the following statements is false?

    (a) At constant temperature, Charles' law applies

    (b) The pressure of a given mass of gas decreases as the volume is increased at constant temperature

    (c) Isobaric changes are those which occur at constant pressure

    (d) Boyle's law applies at constant temperature

2. A gas occupies a volume of 4 m^3 at a pressure of 400 kPa. At constant temperature, the pressure is increased to 500 kPa. The new volume occupied by the gas is:
    (a) 5 m^3 (b) 0.3 m^3
    (c) 0.2 m^3 (d) 3.2 m^3

3. A gas at a temperature of 27°C occupies a volume of 5 m^3. The volume of the same mass of gas at the same pressure but at a temperature of 57°C is:
    (a) 10.56 m^3 (b) 5.50 m^3
    (c) 4.55 m^3 (d) 2.37 m^3

4. Which of the following statements is false?

    (a) An ideal gas is one that completely obeys the gas laws

    (b) Isothermal changes are those that occur at constant volume

    (c) The volume of a gas increases when the temperature increases at constant pressure

(d) Changes that occur at constant pressure are called isobaric changes.

A gas has a volume of $0.4\,m^3$ when its pressure is $250\,kPa$ and its temperature is 400 K. Use these data in questions 5 and 6.

5. The temperature when the pressure is increased to $400\,kPa$ and the volume is increased to $0.8\,m^3$ is:
(a) 400 K   (b) 80 K
(c) 1280 K   (d) 320 K

6. The pressure when the temperature is raised to 600 K and the volume is reduced to $0.2\,m^3$ is:
(a) 187.5 kPa   (b) 250 kPa
(c) 333.3 kPa   (d) 750 kPa

7. A gas has a volume of $3\,m^3$ at a temperature of 546 K and a pressure of 101.325 kPa. The volume it occupies at STP is:
(a) $3\,m^3$ (b) $1.5\,m^3$
(c) $6\,m^3$

8. Which of the following statements is false?
(a) A characteristic gas constant has units of J/(kg K)
(b) STP conditions are 273 K and 101.325 kPa
(c) All gases are ideal gases
(d) An ideal gas is one that obeys the gas laws

A mass of 5 kg of air is pumped into a container of volume $2.87\,m^3$. The characteristic gas constant for air is 287 J/(kg K). Use these data in questions 9 and 10.

9. The pressure when the temperature is $27°C$ is:
(a) 1.6 kPa   (b) 6 kPa
(c) 150 kPa   (d) 15 kPa

10. The temperature when the pressure is 200 kPa is:
(a) $400°C$ (b) $127°C$
(c) 127 K   (d) 283 K

For fully worked solutions to each of the problems in Exercises 155 to 162 in this chapter, go to the website:
www.routledge.com/cw/bird

Chapter 33

The measurement of temperature

Why it is important to understand: **The measurement of temperature**

A change in temperature of a substance can often result in a change in one or more of its physical properties. Thus, although temperature cannot be measured directly, its effects can be measured. Some properties of substances used to determine changes in temperature include changes in dimensions, electrical resistance, state, type and volume of radiation and colour. Temperature measuring devices available are many and varied. Those described in this chapter are those most often used in science and industry. The measurement of temperature is important in medicines and very many branches of science and engineering.

At the end of this chapter, you should be able to:

- describe the construction, principle of operation and practical applications of the following temperature measuring devices:

    (a) liquid-in-glass thermometer (including advantages of mercury, and sources of error)

    (b) thermocouple (including advantages and sources of error)

    (c) resistance thermometer (including limitations and advantages of platinum coil)

    (d) thermistor

    (e) pyrometer (total radiation and optical types, including advantages and disadvantages)

- describe the principle of operation of:

    (a) temperature indicating paints and crayons

    (b) bimetallic thermometer

    (c) mercury-in-steel thermometer

    (d) gas thermometer

- select the appropriate temperature measuring device for a particular application

Science for Engineering. 978-1-138-82688-5, © John Bird. Published by Taylor & Francis. All rights reserved.

33.1 Introduction

A change in temperature of a substance can often result in a change in one or more of its physical properties. Thus, although temperature cannot be measured directly, its effects can be measured. Some properties of substances used to determine changes in temperature include changes in dimensions, electrical resistance, state, type and volume of radiation and colour.

Temperature-measuring devices available are many and varied. Those described in sections 33.2–33.10 are those most often used in science and industry.

33.2 Liquid-in-glass thermometers

A **liquid-in-glass thermometer** uses the expansion of a liquid with increase in temperature as its principle of operation.

33.2.1 Construction

A typical liquid-in-glass thermometer is shown in Figure 33.1 and consists of a sealed stem of uniform small-bore tubing, called a capillary tube, made of glass, with a cylindrical glass bulb formed at one end. The bulb and part of the stem are filled with a liquid such as mercury or alcohol and the remaining part of the tube is evacuated. A temperature scale is formed by etching graduations on the stem. A safety reservoir is usually provided, into which the liquid can expand without bursting the glass if the temperature is raised beyond the upper limit of the scale.

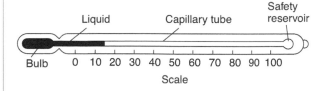

Figure 33.1

33.2.2 Principle of operation

The operation of a liquid-in-glass thermometer depends on the liquid expanding with increase in temperature and contracting with decrease in temperature. The position of the end of the column of liquid in the tube is a measure of the temperature of the liquid in the bulb –

shown as 15°C in Figure 33.1, which is about room temperature. Two fixed points are needed to calibrate the thermometer, with the interval between these points being divided into 'degrees'. In the first thermometer, made by Celsius, the fixed points chosen were the temperature of melting ice (0°C) and that of boiling water at standard atmospheric pressure (100°C), in each case the blank stem being marked at the liquid level. The distance between these two points, called the fundamental interval, was divided into 100 equal parts, each equivalent to 1°C, thus forming the scale.

The **clinical thermometer**, with a limited scale around body temperature, the **maximum and/or minimum thermometer**, recording the maximum day temperature and minimum night temperature, and the **Beckman thermometer***, which is used only in accurate measurement of temperature change, and has no fixed points, are particular types of liquid-in-glass thermometer which all operate on the same principle.

*Who was Beckman? – **Ernst Otto Beckmann** (4 July, 1853–12 July, 1923) was a German chemist who is remembered for his invention of the Beckmann differential thermometer. To find out more about Beckman go to www.routledge.com/cw/bird

33.2.3   Advantages

The liquid-in-glass thermometer is simple in construction, relatively inexpensive, easy to use and portable, and is the most widely used method of temperature measurement having industrial, chemical, clinical and meteorological applications.

33.2.4   Disadvantages

Liquid-in-glass thermometers tend to be fragile and hence easily broken, can only be used where the liquid column is visible, cannot be used for surface temperature measurements, cannot be read from a distance and are unsuitable for high temperature measurements.

33.2.5   Advantages of mercury

The use of mercury in a thermometer has many advantages, for mercury:

(i)   is clearly visible,

(ii) has a fairly uniform rate of expansion,

(iii) is readily obtainable in the pure state,

(iv) does not 'wet' the glass,

(v)   is a good conductor of heat.

Mercury has a freezing point of $-39°C$ and cannot be used in a thermometer below this temperature. Its boiling point is $357°C$, but before this temperature is reached some distillation of the mercury occurs if the space above the mercury is a vacuum. To prevent this, and to extend the upper temperature limits to over $500°C$, an inert gas such as nitrogen under pressure is used to fill the remainder of the capillary tube. Alcohol, often dyed red to be seen in the capillary tube, is considerably cheaper than mercury and has a freezing point of $-113°C$, which is considerably lower than for mercury. However, it has a low boiling point at about $79°C$.

33.2.6   Errors

Typical errors in liquid-in-glass thermometers may occur due to:

(i)   the slow cooling rate of glass,

(ii) incorrect positioning of the thermometer,

(iii) a delay in the thermometer becoming steady (i.e. slow response time),

(iv) non-uniformity of the bore of the capillary tube, which means that equal intervals marked on the stem do not correspond to equal temperature intervals.

33.3   Thermocouples

Thermocouples use the electromotive force (e.m.f.) set up when the junction of two dissimilar metals is heated.

33.3.1   Principle of operation

At the junction between two different metals, say, copper and constantan, there exists a difference in electrical potential, which varies with the temperature of the junction. This is known as the 'thermo-electric effect'. If the circuit is completed with a second junction at a different temperature, a current will flow round the circuit. This principle is used in the thermocouple. Two different metal conductors having their ends twisted together are shown in Figure 33.2. If the two junctions are at different temperatures, a current I flows round the circuit.

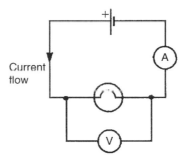

Figure 33.2

The deflection on the galvanometer G depends on the difference in temperature between junctions X and Y and is caused by the difference between voltages V_x and V_y. The higher temperature junction is usually called the 'hot junction' and the lower temperature junction the 'cold junction'. If the cold junction is kept at a constant known temperature, the galvanometer can be calibrated to indicate the temperature of the hot junction directly. The cold junction is then known as the reference junction.

In many instrumentation situations, the measuring instrument needs to be located far from the point at which the measurements are to be made. Extension leads are then used, usually made of the same material as the thermocouple but of smaller gauge. The reference junction is then effectively moved to their ends. The thermocouple is used by positioning the hot junction where the temperature is required. The meter will indicate the temperature of the hot junction only if the reference junction is at 0°C for:

$$\text{(temperature of hot junction)}$$
$$= \text{(temperature of the cold junction)}$$
$$+ \text{(temperature difference)}$$

In a laboratory the reference junction is often placed in melting ice, but in industry it is often positioned in a thermostatically controlled oven or buried underground where the temperature is constant.

33.3.2 Construction

Thermocouple junctions are made by twisting together the ends of two wires of dissimilar metals before welding them. The construction of a typical copper–constantan thermocouple for industrial use is shown in Figure 33.3. Apart from the actual junction the two conductors used must be insulated electrically from each other with appropriate insulation, shown in Figure 33.3 as twin-holed tubing. The wires and insulation are usually inserted into a sheath for protection from environments in which they might be damaged or corroded.

33.3.3 Applications

A copper–constantan thermocouple can measure temperature from −250°C up to about 400°C, and is used typically with boiler flue gases, food processing and with sub-zero temperature measurement. An iron–constantan thermocouple can measure temperature from −200°C to about 850°C, and is used typically in paper and pulp mills, re-heat and annealing furnaces and in chemical reactors. A chromel–alumel thermocouple can measure temperatures from −200°C to about 1100°C and is used

typically with blast furnace gases, brick kilns and in glass manufacture.

For the measurement of temperatures above 1100°C radiation pyrometers are normally used. However, thermocouples are available made of platinum–platinum/rhodium, capable of measuring temperatures up to 1400°C, or tungsten–molybdenum which can measure up to 2600°C.

33.3.4 Advantages

A thermocouple:

(i)   has a very simple, relatively inexpensive construction,

(ii) can be made very small and compact,

(iii) is robust,

(iv) is easily replaced if damaged,

(v)   has a small response time,

(vi) can be used at a distance from the actual measuring instrument and is thus ideal for use with automatic and remote-control systems.

33.3.5 Sources of error

Sources of error in the thermocouple, which are difficult to overcome, include:

(i)   voltage drops in leads and junctions,

(ii) possible variations in the temperature of the cold junction,

(iii) stray thermoelectric effects, which are caused by the addition of further metals into the 'ideal' two-metal thermocouple circuit.

Additional leads are frequently necessary for extension leads or voltmeter terminal connections.

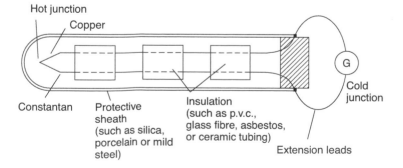

Figure 33.3

A thermocouple may be used with a battery- or mains-operated electronic thermometer instead of a millivoltmeter. These devices amplify the small e.m.f.s from the thermocouple before feeding them to a multi-range voltmeter calibrated directly with temperature scales. These devices have great accuracy and are almost unaffected by voltage drops in the leads and junctions.

Problem 1. A chromel–alumel thermocouple generates an e.m.f. of 5 mV. Determine the temperature of the hot junction if the cold junction is at a temperature of 15°C and the sensitivity of the thermocouple is 0.04 mV/°C

Temperature difference for 5 mV

$$= \frac{5\,\text{mV}}{0.04\,\text{mV/}^\circ\text{C}} = 125^\circ\text{C}$$

Temperature at hot junction

$$= \text{temperature of cold junction} + \text{temperature difference}$$
$$= 15^\circ\text{C} + 125^\circ\text{C} = \mathbf{140^\circ C}$$

Now try the following Practice Exercise

Practice Exercise 163   Further problem on the thermocouple (answer on page 545)

1. A platinum–platinum/rhodium thermocouple generates an e.m.f. of 7.5 mV. If the cold junction is at a temperature of 20°C, determine the temperature of the hot junction. Assume the sensitivity of the thermocouple to be 6 µV/°C

33.4   Resistance thermometers

Resistance thermometers use the change in electrical resistance caused by temperature change.

33.4.1   Construction

Resistance thermometers are made in a variety of sizes, shapes and forms depending on the application for which they are designed. A typical resistance thermometer is shown diagrammatically in Figure 33.4. The most common metal used for the coil in such thermometers is platinum even though its sensitivity is not as high as that of other metals such as copper and nickel. However, platinum is a very stable metal and provides repro-ducible results in a resistance thermometer. A platinum resistance thermometer is often used as a calibrating device. Since platinum is expensive, connecting leads of another metal, usually copper, are used with the thermometer to connect it to a measuring circuit.

The platinum and the connecting leads are shown joined at *A* and *B* in Figure 33.4, although sometimes this junction may be made outside of the sheath. However, these leads often come into close contact with the heat source which can introduce errors into the measurements. These may be eliminated by including a pair of identical leads, called dummy leads, which experience the same temperature change as the extension leads.

33.4.2   Principle of operation

With most metals a rise in temperature causes an increase in electrical resistance, and since resistance can be measured accurately, this property can be used to measure temperature. If the resistance of a length of wire at 0°C is R_0, and its resistance at θ°C is R_θ, then

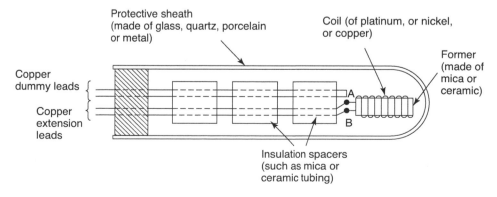

Figure 33.4

$R_\theta = R_0(1 + \alpha\theta)$, where α is the temperature coefficient of resistance of the material (see chapter 35).

Rearranging gives:

$$\text{temperature } \theta = \frac{R_\theta - R_0}{\alpha R_0}$$

Values of R_0 and α may be determined experimentally or obtained from existing data. Thus, if R_θ can be measured, temperature θ can be calculated. This is the principle of operation of a resistance thermometer. Although a sensitive ohmmeter can be used to measure R_θ, for more accurate determinations a **Wheatstone* bridge circuit is used** as shown in Figure 33.5. This circuit compares an unknown resistance R_θ with others of known values, R_1 and R_2 being fixed values and R_3 being variable. Galvanometer G is a sensitive centre-zero microammeter. R_3 is varied until zero deflection is obtained on the galvanometer, i.e. no current flows through G and the bridge is said to be 'balanced'.

At balance: $R_2R_\theta = R_1R_3$

from which, $$R_\theta = \frac{R_1 R_3}{R_2}$$

and if R_1 and R_2 are of equal value, then $R_\theta = R_3$

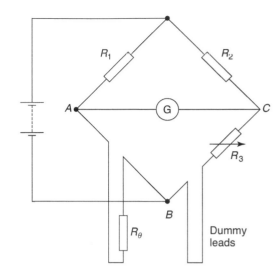

Figure 33.5

*Who was Wheatstone? – **Sir Charles Wheatstone** (6 February 1802–19 October 1875), was an English scientist best known for his contributions in the development of the Wheatstone bridge, which is used to measure an unknown electrical resistance. To find out more about Wheatstone go to www.routledge.com/cw/bird

A resistance thermometer may be connected between points A and B in Figure 33.5 and its resistance R_θ at any temperature θ accurately measured. Dummy leads included in arm BC help to eliminate errors caused by the extension leads which are normally necessary in such a thermometer.

33.4.3   Limitations

Resistance thermometers using a nickel coil are used mainly in the range $-100°C$ to $300°C$, whereas platinum resistance thermometers are capable of measuring with greater accuracy temperatures in the range $-200°C$ to about $800°C$. The upper range may be extended to about $1500°C$ if high melting point materials are used for the sheath and coil construction.

33.4.4   Advantages and disadvantages of a platinum coil

Platinum is commonly used in resistance thermometers since it is chemically inert, i.e. un-reactive, resists corrosion and oxidation and has a high melting point of $1769°C$. A disadvantage of platinum is its slow response to temperature variation.

33.4.5   Applications

Platinum resistance thermometers may be used as calibrating devices or in applications such as heat-treating and annealing processes, and can be adapted easily for use with automatic recording or control systems. Resistance thermometers tend to be fragile and easily damaged especially when subjected to excessive vibration or shock.

Problem 2. A platinum resistance thermometer has a resistance of 25 Ω at 0°C. When measuring the temperature of an annealing process a resistance value of 60 Ω is recorded. To what temperature does this correspond? Take the temperature coefficient of resistance of platinum as 0.0038/°C

$R_\theta = R_0(1 + \alpha\theta)$, where $R_0 = 25\ \Omega$, $R_\theta = 60\ \Omega$ and $\alpha = 0.0038/°C$.
Rearranging gives:

$$\textbf{temperature } \theta = \frac{R_\theta - R_0}{\alpha R_0} = \frac{60 - 25}{(0.0038)(25)}$$
$$= \textbf{368.4}\ \overset{\circ}{\textbf{C}}$$

Now try the following Practice Exercise

Practice Exercise 164   Further problem on the resistance thermometer (answer on page 545)

1. A platinum resistance thermometer has a resistance of 100 Ω at 0°C. When measuring the temperature of a heat process a resistance value of 177 Ω is measured using a Wheatstone bridge. Given that the temperature coefficient of resistance of platinum is 0.0038/°C, determine the temperature of the heat process, correct to the nearest degree

33.5   Thermistors

A thermistor is a semi-conducting material – such as mixtures of oxides of copper, manganese, cobalt, etc. – in the form of a fused bead connected to two leads. As its temperature is increased its resistance rapidly decreases. Typical resistance/temperature curves for a thermistor and common metals are shown in Figure 33.6. The resistance of a typical thermistor can vary from 400 Ω at 0°C to 100 Ω at 140°C.

33.5.1   Advantages

The main advantages of a thermistor are its high sensitivity and small size. It provides an inexpensive method of measuring and detecting small changes in temperature.

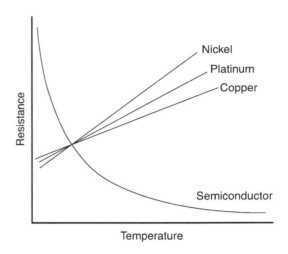

Figure 33.6

33.6   Pyrometers

A **pyrometer** is a device for measuring very high temperatures and uses the principle that all substances emit radiant energy when hot, the rate of emission depending on their temperature. The measurement of thermal radiation is therefore a convenient method of determining the temperature of hot sources and is particularly useful in industrial processes. There are two main types of pyrometer, namely the total radiation pyrometer and the optical pyrometer.

Pyrometers are very convenient instruments since they can be used at a safe and comfortable distance from the hot source. Thus applications of pyrometers are found in measuring the temperature of molten metals, the interiors of furnaces or the interiors of volcanoes. Total radiation pyrometers can also be used in conjunction with devices which record and control temperature continuously.

33.6.1   Total radiation pyrometers

A typical arrangement of a total radiation pyrometer is shown in Figure 33.7. Radiant energy from a hot source, such as a furnace, is focused on to the hot junction of a thermocouple after reflection from a concave mirror. The temperature rise recorded by the thermocouple depends on the amount of radiant energy received, which in turn depends on the temperature of the hot source. The galvanometer G shown connected to the thermocouple records the current which results from the e.m.f. developed and may be calibrated to give a direct reading of the temperature of the hot source. The

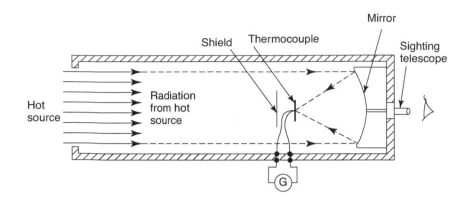

Figure 33.7

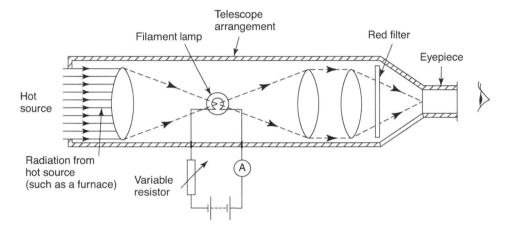

Figure 33.8

thermocouple is protected from direct radiation by a shield as shown, and the hot source may be viewed through the sighting telescope. For greater sensitivity, a thermopile may be used, a **thermopile** being a number of thermocouples connected in series. Total radiation pyrometers are used to measure temperature in the range $700°C$ to $2000°C$.

33.6.2 Optical pyrometers

When the temperature of an object is raised sufficiently, two visual effects occur; the object appears brighter and there is a change in colour of the light emitted. These effects are used in the optical pyrometer where a comparison or matching is made between the brightness of the glowing hot source and the light from a filament of known temperature.

The most frequently used optical pyrometer is the disappearing filament pyrometer, and a typical arrangement is shown in Figure 33.8. A filament lamp is built into a telescope arrangement which receives radiation from a hot source, an image of which is seen through an eyepiece. A red filter is incorporated as a protection to the eye.

The current flowing through the lamp is controlled by a variable resistor. As the current is increased, the temperature of the filament increases and its colour changes. When viewed through the eyepiece, the filament of the lamp appears superimposed on the image of the radiant energy from the hot source. The current is varied until the filament glows as brightly as the background. It will then merge into the background and seem to disappear. The current required to achieve this is a measure of the temperature of the hot source, and the ammeter can be calibrated to read the temperature directly. Optical pyrometers may be used to measure temperatures up to, and even in excess of, $3000°C$.

33.6.3   Advantages of pyrometers

(i)   There is no practical limit to the temperature that a pyrometer can measure.

(ii)   A pyrometer need not be brought directly into the hot zone and so is free from the effects of heat and chemical attack that can often cause other measuring devices to deteriorate in use.

(iii)   Very fast rates of change of temperature can be followed by a pyrometer.

(iv)   The temperature of moving bodies can be measured.

(v)   The lens system makes the pyrometer virtually independent of its distance from the source.

33.6.4   Disadvantages of pyrometers

(i)   A pyrometer is often more expensive than other temperature-measuring devices.

(ii)   A direct view of the heat process is necessary.

(iii)   Manual adjustment is necessary.

(iv)   A reasonable amount of skill and care is required in calibrating and using a pyrometer. For each new measuring situation the pyrometer must be re-calibrated.

(v)   The temperature of the surroundings may affect the reading of the pyrometer and such errors are difficult to eliminate.

33.7   Temperature-indicating paints and crayons

Temperature-indicating paints contain substances which change their colour when heated to certain temperatures. This change is usually due to chemical decomposition, such as loss of water, in which the change in colour of the paint after having reached the particular temperature will be a permanent one. However, in some types the original colour returns after cooling. Temperature-indicating paints are used where the temperature of inaccessible parts of apparatus and machines is required to be known. They are particularly useful in heat-treatment processes where the temperature of the component needs to be known before a quenching operation. There are several such paints available, and most have only a small temperature range so that different paints have to be used for different temperatures. The usual range of temperatures covered by these paints is from about 30°C to 700°C.

Temperature-sensitive crayons consist of fusible solids compressed into the form of a stick. The melting point of such crayons is used to determine when a given temperature has been reached. The crayons are simple to use but indicate a single temperature only, i.e. its melting point temperature. There are over 100 different crayons available, each covering a particular range of temperatures. Crayons are available for temperatures within the range of 50°C to 1400°C. Such crayons are used in metallurgical applications such as preheating before welding, hardening, annealing or tempering, or in monitoring the temperature of critical parts of machines or for checking mould temperatures in the rubber and plastics industries.

33.8   Bimetallic thermometers

Bimetallic thermometers depend on the expansion of metal strips which operate an indicating pointer. Two thin metal strips of differing thermal expansion are welded or riveted together and the curvature of the bimetallic strip changes with temperature change. For greater sensitivity the strips may be coiled into a flat spiral or helix, one end being fixed and the other being made to rotate a pointer over a scale. Bimetallic thermometers are useful for alarm and over-temperature applications where extreme accuracy is not essential. If the whole is placed in a sheath, protection from corrosive environments is achieved but with a reduction in response characteristics. The normal upper limit of temperature measurement by this kind of thermometer is about 200°C, although with special metals the range can be extended to about 400°C.

33.9   Mercury-in-steel thermometer

The **mercury-in-steel thermometer** is an extension of the principle of the mercury-in-glass thermometer. Mercury in a steel bulb expands via a small bore capillary tube into a pressure-indicating device, say a Bourdon gauge, the position of the pointer indicating the amount of expansion and thus the temperature. The advantages of this instrument are that it is robust and, by increasing the length of the capillary tube, the gauge can be placed some distance from the bulb and can thus be used to monitor temperatures in positions which are inaccessible to the liquid-in-glass thermometer. Such

358 Section II

thermometers may be used to measure temperatures up to 600°C.

33.10 Gas thermometers

The gas thermometer consists of a flexible U-tube of mercury connected by a capillary tube to a vessel containing gas. The change in the volume of a fixed mass of gas at constant pressure, or the change in pressure of a fixed mass of gas at constant volume, may be used to measure temperature. This thermometer is cumbersome and rarely used to measure temperature directly, but it is often used as a standard with which to calibrate other types of thermometer. With pure hydrogen the range of the instrument extends from −240°C to 1500°C and measurements can be made with extreme accuracy.

33.11 Choice of measuring device

Problem 3. State which device would be most suitable:

(a) to measure metal in a furnace, in the range 50°C to 1600°C

(b) to measure the air in an office in the range 0°C to 40°C

(c) to measure boiler flue gas in the range 15°C to 300°C

(d) to measure a metal surface, where a visual indication is required when it reaches 425°C

(e) to measure materials in a high-temperature furnace in the range 2000°C to 2800°C

(f) to calibrate a thermocouple in the range −100°C to 500°C

(g) to measure brick in a kiln up to 900°C

(h) as an inexpensive method for food processing applications in the range −25°C to −75°C

(a) Radiation pyrometer

(b) Mercury-in-glass thermometer

(c) Copper–constantan thermocouple

(d) Temperature-sensitive crayon

(e) Optical pyrometer

(f) Platinum resistance thermometer or gas thermometer

(g) Chromel–alumel thermocouple

(h) Alcohol-in-glass thermometer

Now try the following Practice Exercises

Practice Exercise 165 Short answer questions on the measurement of temperature

For each of the temperature measuring devices listed in 1–10, state very briefly its principle of operation and the range of temperatures that it is capable of measuring.

1. Mercury-in-glass thermometer

2. Alcohol-in-glass thermometer

3. Thermocouple

4. Platinum resistance thermometer

5. Total radiation pyrometer

6. Optical pyrometer

7. Temperature-sensitive crayons

8. Bimetallic thermometer

9. Mercury-in-steel thermometer

10. Gas thermometer

Practice Exercise 166 Multiple-choice questions on the measurement of temperature (answers on page 546)

1. The most suitable device for measuring very small temperature changes is a

(a) thermopile (b) thermocouple
(c) thermistor

2. When two wires of different metals are twisted together and heat applied to the junction, an e.m.f. is produced. This effect is used in a thermocouple to measure:

(a) e.m.f.   (b) temperature
(c) expansion   (d) heat

3. A cold junction of a thermocouple is at room temperature of 15°C. A voltmeter connected to the thermocouple circuit indicates 10 mV. If the voltmeter is calibrated as 20°C/mV, the temperature of the hot source is:

   (a) 185°C   (b) 200°C   (c) 35°C   (d) 215°C

4. The e.m.f. generated by a copper–constantan thermometer is 15 mV. If the cold junction is at a temperature of 20°C, the temperature of the hot junction when the sensitivity of the thermocouple is 0.03 mV/°C is:

   (a) 480°C   (b) 520°C   (c) 20.45°C   (d) 500°C

   In questions 5–12, select the most appropriate temperature measuring device from this list.

   (a) copper–constantan thermocouple

   (b) thermistor

   (c) mercury-in-glass thermometer

   (d) total radiation pyrometer

   (e) platinum resistance thermometer

   (f) gas thermometer

   (g) temperature-sensitive crayon

   (h) alcohol-in-glass thermometer

   (i) bimetallic thermometer

   (j) mercury-in-steel thermometer

   (k) optical pyrometer

5. Over-temperature alarm at about 180°C

6. Food processing plant in the range −250°C to +250°C

7. Automatic recording system for a heat treating process in the range 90°C to 250°C

8. Surface of molten metals in the range 1000°C to 1800°C

9. To calibrate accurately a mercury-in-glass thermometer

10. Furnace up to 3000°C

11. Inexpensive method of measuring very small changes in temperature

12. Metal surface where a visual indication is required when the temperature reaches 520°C

For fully worked solutions to each of the problems in Exercises 163 to 166 in this chapter, go to the website:
www.routledge.com/cw/bird

This assignment covers the material contained in chapters 31 to 33. *The marks for each question are shown in brackets at the end of each question.*

1. A copper overhead transmission line has a length of 60 m between its supports at 15°C. Calculate its length at 40°C, if the coefficient of linear expansion of copper is $17 \times 10^{-6}K^{-1}$. (6)

2. A gold sphere has a diameter of 40 mm at a temperature of 285 K. If the temperature of the sphere is raised to 785 K, determine the increase in (a) the diameter, (b) the surface area, (c) the volume of the sphere. Assume the coefficient of linear expansion for gold is $14 \times 10^{-6}K^{-1}$. (12)

3. Some gas occupies a volume of 2.0 m³ in a cylinder at a pressure of 200 kPa. A piston, sliding in the cylinder, compresses the gas isothermally until the volume is 0.80 m³. If the area of the piston is 240 cm², calculate the force on the piston when the gas is compressed. (5)

4. Gas at a temperature of 180°C has its volume reduced by a quarter in an isobaric process. Determine the final temperature of the gas. (4)

5. Some air at a pressure of 3 bar and at a temperature of 60°C occupies a volume of 0.08 m³. Calculate the mass of the air, correct to the nearest gram, assuming the characteristic gas constant for air is 287 J/(kg K). (5)

6. A compressed air cylinder has a volume of 1.0 m³ and contains air at a temperature of 24°C and a pressure of 1.2 MPa. Air is released from the cylinder until the pressure falls to 400 kPa and the temperature is 18°C. Calculate (a) the mass of air released from the container, (b) the volume it would occupy at STP. Assume the characteristic gas constant for air to be 287 J/(kg K). (10)

7. A platinum resistance thermometer has a resistance of 24 Ω at 0°C. When measuring the temperature of an annealing process a resistance value of 68 Ω is recorded. To what temperature does this correspond? Take the temperature coefficient of resistance of platinum as 0.0038/°C. (4)

8. State which device would be most suitable to measure the following:

    (a) materials in a high-temperature furnace in the range 1800°C to 3000°C,

    (b) the air in a factory in the range 0°C to 35°C,

    (c) an inexpensive method for food processing applications in the range −20°C to −80°C,

    (d) boiler flue gas in the range 15°C to 250°C. (4)

For lecturers/instructors/teachers, fully worked solutions to each of the problems in Revision Test 12, together with a full marking scheme, are available at the website:

www.routledge.com/cw/bird

Section III

Electrical applications

Chapter 34

An introduction to electric circuits

Why it is important to understand: **An introduction to electric circuits**

Electric circuits are a part of the basic fabric of modern technology. A circuit consists of electrical elements connected together, and we can use symbols to draw circuits. Engineers use electrical circuits to solve problems that are important in modern society such as in the generation, transmission and consumption of electrical power and energy. The outstanding characteristics of electricity compared with other power sources are its mobility and flexibility. The elements in an electric circuit include sources of energy, resistors, capacitors, inductors and so on. Analysis of electric circuits means determining the unknown quantities such as voltage, current and power associated with one or more elements in the circuit. Basic d.c. electric circuit analysis and laws are explained in this chapter and knowledge of these is essential in the solution of engineering problems.

At the end of this chapter, you should be able to:

- recognise common electrical circuit diagram symbols
- understand that electric current is the rate of movement of charge and is measured in amperes
- appreciate that the unit of charge is the coulomb
- calculate charge or quantity of electricity Q from $Q = I \times t$
- understand that a potential difference between two points in a circuit is required for current to flow
- appreciate that the unit of p.d. is the volt
- understand that resistance opposes current flow and is measured in ohms
- appreciate what an ammeter, voltmeter, ohmmeter, multimeter, oscilloscope, wattmeter, BM80, 420 MIT megger, tachometer, and a stroboscope measure
- state Ohm's law as $V = I \times R$ or $I = \dfrac{V}{R}$ or $R = \dfrac{V}{I}$

Science and Mathematics for Engineering. 978-0-367-20475-4, © John Bird. Published by Taylor & Francis. All rights reserved.

- use Ohm's law in calculations, including multiples and sub-multiples of units
- describe a conductor and an insulator, giving examples of each
- appreciate that electrical power P is given by $P = V \times I = I^2 \times R = \dfrac{V^2}{R}$ watts
- calculate electrical power
- define electrical energy and state its unit
- calculate electrical energy
- state the three main effects of an electric current, giving practical examples of each
- explain the importance of fuses in electrical circuits

34.1 Introduction

This chapter provides an introduction to electric circuits, initially introducing basic electrical diagram symbols, defining electric current, charge, potential difference and resistance and listing typical measuring instruments. The most important law in electrical engineering, Ohm's law, is explained with calculations, as are multiples and sub-multiples of units. Electrical power and energy calculations are demonstrated and the main effects of an electric current are listed. In addition, conductors, insulators and the use of fuses are explained.

34.2 Standard symbols for electrical components

Symbols are used for components in electrical circuit diagrams and some of the more common ones are shown in Figure 34.1.

34.3 Electric current and quantity of electricity

All **atoms** consist of **protons**, **neutrons** and **electrons**. The protons, which have positive electrical charges, and the neutrons, which have no electrical charge, are contained within the **nucleus**. Removed from the nucleus are minute negatively charged particles called electrons. Atoms of different materials differ from one another by having different numbers of protons, neutrons and electrons. An equal number of protons and electrons exist within an atom, which is

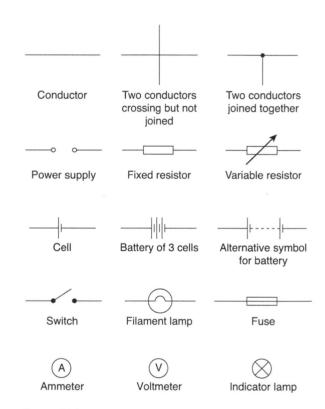

Figure 34.1

said to be electrically balanced, as the positive and negative charges cancel each other out. When there are more than two electrons in an atom the electrons are arranged into **shells** at various distances from the nucleus.

All atoms are bound together by powerful forces of attraction existing between the nucleus and its electrons. Electrons in the outer shell of an atom, however, are attracted to their nucleus less powerfully than are electrons whose shells are nearer the nucleus.

It is possible for an atom to lose an electron; the atom, which is now called an **ion**, is not now electrically balanced, but is positively charged and is thus able to attract an electron to itself from another atom. Electrons that move from one atom to another are called free electrons and such random motion can continue indefinitely. However, if an electric pressure or **voltage** is applied across any material there is a tendency for electrons to move in a particular direction. This movement of free electrons, known as **drift**, constitutes an electric current flow.

Thus current is the rate of movement of charge

Conductors are materials that contain electrons that are loosely connected to the nucleus and can easily move through the material from one atom to another. **Insulators** are materials whose electrons are held firmly to their nucleus.

The unit used to measure the **quantity of electrical charge Q** is called the **coulomb* C** (where 1 coulomb $= 6.24 \times 10^{18}$ electrons).

If the drift of electrons in a conductor takes place at the rate of one coulomb per second the resulting current is said to be a current of one **ampere***.

Thus 1 ampere = 1 coulomb per second or 1 A = 1 C/s

Hence 1 coulomb = 1 ampere second or 1 C = 1 As

Generally, if I is the current in amperes and t the time in seconds during which the current flows, then $I \times t$ represents the quantity of electrical charge in coulombs, i.e. quantity of electrical charge transferred,

$$Q = I \times t \text{ coulombs}$$

Problem 1. What current must flow if 0.24 coulombs is to be transferred in 15 ms?

Since the quantity of electricity $Q = I \times t$, then

$$\textbf{current } I = \frac{Q}{t} = \frac{0.24}{15 \times 10^{-3}} = \frac{0.24 \times 10^3}{15}$$

$$= \frac{240}{15} = \textbf{16 A}$$

*Who was Coulomb? – **Charles-Augustin de Coulomb** (14 June 1736–23 August 1806) was best known for developing Coulomb's law, the definition of the electrostatic force of attraction and repulsion. To find out more go to www.routledge.com/cw/bird

*Who was Ampere? – **André-Marie Ampère** (1775–1836) is generally regarded as one of the founders of classical electromagnetism. To find out more go to www.routledge.com/cw/bird

Problem 2. If a current of 10 A flows for four minutes, find the quantity of electricity transferred

Quantity of electricity $Q = I \times t$ coulombs

$I = 10\,\text{A}$ and $t = 4 \times 60 = 240\,\text{s}$

Hence, **charge** $Q = I \times t = 10 \times 240 = \mathbf{2400\ C}$

Now try the following Practice Exercise

Practice Exercise 167 Further problems on charge (answers on page 546)

1. In what time would a current of 10 A transfer a charge of 50 C?

2. A current of 6 A flows for 10 minutes. What charge is transferred?

3. How long must a current of 100 mA flow so as to transfer a charge of 80 C?

34.4 Potential difference and resistance

For a continuous current to flow between two points in a circuit a **potential difference (p.d.)** or **voltage V** is required between them; a complete conducting path is necessary to and from the source of electrical energy. The unit of p.d. is the **volt, V**, (named in honour of the Italian physicist **Alessandro Volta***).
Figure 34.2 shows a cell connected across a filament lamp. Current flow, by convention, is considered as flowing from the positive terminal of the cell, around the circuit to the negative terminal.
The flow of electric current is subject to friction. This friction, or opposition, is called **resistance R** and is the property of a conductor that limits current. The unit of resistance is the **ohm***; 1 ohm is defined as the resistance which will have a current of 1 ampere flowing through it when 1 volt is connected across it,

i.e. **resistance $R = \dfrac{\text{potential difference}}{\text{current}}$**

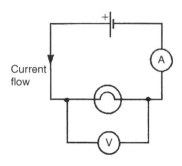

Figure 34.2

34.5 Basic electrical measuring instruments

An **ammeter** is an instrument used to measure current and must be connected **in series** with the circuit. Figure 34.2 shows an ammeter connected in series with the lamp to measure the current flowing through it. Since all the current in the circuit passes through the ammeter it must have a very **low resistance**.

*Who was Volta? – **Alessandro Giuseppe Antonio Anastasio Volta** (18 February 1745–5 March 1827) was the Italian physicist who invented the battery. To find out more go to www.routledge.com/cw/bird

A **voltmeter** is an instrument used to measure p.d. and must be connected **in parallel** with the part of the circuit whose p.d. is required. In Figure 34.2, a voltmeter is connected in parallel with the lamp to measure the p.d. across it. To avoid a significant current flowing through it a voltmeter must have a very **high resistance**.

An **ohmmeter** is an instrument for measuring resistance. A **multimeter**, or universal instrument, may be used to measure voltage, current and resistance. An 'Avometer' is a typical older example; a '**fluke**' is a much used multimeter.

An **oscilloscope** may be used to observe waveforms and to measure voltages and currents. The display of an oscilloscope involves a spot of light moving across a screen. The amount by which the spot is deflected from its initial position depends on the p.d. applied to the terminals of the oscilloscope and the range selected.

The displacement is calibrated in 'volts per cm'. For example, if the spot is deflected 3 cm and the volts/cm switch is on 10 V/cm then the magnitude of the p.d. is 3 cm × 10 V/cm, i.e. 30 V.

A **wattmeter** is an instrument for the measurement of power in an electrical circuit.

A **BM80** or a **420 MIT megger** or a **bridge megger** may be used to measure both continuity and insulation resistance. **Continuity testing** is the measurement of the resistance of a cable to discover if the cable is continuous, i.e. that it has no breaks or high resistance joints. **Insulation resistance testing** is the measurement of resistance of the insulation between cables, individual cables to earth or metal plugs and sockets, and so on. An insulation resistance in excess of 1 MΩ is normally acceptable.

A **tachometer** is an instrument that indicates the speed, usually in revolutions per minute, at which an engine shaft is rotating.

A **stroboscope** is a device for viewing a rotating object at regularly recurring intervals, by means of either (a) a rotating or vibrating shutter, or (b) a suitably designed lamp which flashes periodically. If the period between successive views is exactly the same as the time of one revolution of the revolving object, and the duration of the view very short, the object will appear to be stationary. (See chapter 43 for more detail about electrical measuring instruments and measurements.)

34.6   Ohm's law

Ohm's law* states that the current I flowing in a circuit is directly proportional to the applied voltage V and inversely proportional to the resistance R, provided the temperature remains constant. Thus,

$$I = \frac{V}{R} \quad \text{or} \quad V = IR \quad \text{or} \quad R = \frac{V}{I}$$

*Who was Ohm? – **Georg Simon Ohm** (16 March 1789– 6 July 1854) was a Bavarian physicist and mathematician who wrote a complete theory of electricity, in which he stated his law for electromotive force. To find out more go to www.routledge.com/cw/bird

> Problem 3.   The current flowing through a resistor is 0.8 A when a p.d. of 20 V is applied. Determine the value of the resistance

From Ohm's law,

$$\text{resistance } R = \frac{V}{I} = \frac{20}{0.8} = \frac{200}{8} = \textbf{25 } \boldsymbol{\Omega}$$

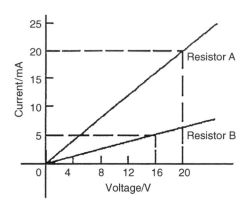

Figure 34.3

34.7 Multiples and sub-multiples

Currents, voltages and resistances can often be very large or very small. Thus **multiples and sub-multiples** of units are often used, as stated in chapters 3 and 15. The most common ones, with an example of each, are listed in Table 34.1.

Problem 4. Determine the p.d. which must be applied to a 2 kΩ resistor in order that a current of 10 mA may flow

Resistance $R = 2\,k\Omega = 2 \times 10^3 = 2000\,\Omega$, and

current $I = 10\,mA = 10 \times 10^{-3}\,A \quad$ or $\quad \dfrac{10}{10^3}\,A \quad$ or

$\dfrac{10}{1000}\,A = 0.01\,A$

From Ohm's law, **potential difference**

$$V = IR = (0.01)(2000) = \textbf{20 V}$$

Problem 5. A coil has a current of 50 mA flowing through it when the applied voltage is 12 V. What is the resistance of the coil?

Resistance

$$R = \frac{V}{I} = \frac{12}{50 \times 10^{-3}} = \frac{12 \times 10^3}{50} = \frac{12,000}{50} = \textbf{240 } \boldsymbol{\Omega}$$

Problem 6. The current/voltage relationship for two resistors A and B is as shown in Figure 34.3. Determine the value of the resistance of each resistor

For resistor A, $\quad \boldsymbol{R} = \dfrac{V}{I} = \dfrac{20\,V}{20\,mA} = \dfrac{20}{0.02} = \dfrac{2000}{2}$

$$= \textbf{1000 } \boldsymbol{\Omega} \quad \text{or} \quad \textbf{1 k}\boldsymbol{\Omega}$$

For resistor B, $\quad \boldsymbol{R} = \dfrac{V}{I} = \dfrac{16\,V}{5\,mA} = \dfrac{16}{0.005} = \dfrac{16,000}{5}$

$$= \textbf{3200 } \boldsymbol{\Omega} \quad \text{or} \quad \textbf{3.2 k}\boldsymbol{\Omega}$$

Now try the following Practice Exercise

Practice Exercise 168   Further problems on Ohm's law (answers on page 546)

1. The current flowing through a heating element is 5 A when a p.d. of 35 V is applied across it. Find the resistance of the element

2. An electric light bulb of resistance 960 Ω is connected to a 240 V supply. Determine the current flowing in the bulb

3. Graphs of current against voltage for two resistors P and Q are shown in Figure 34.4. Determine the value of each resistor

4. Determine the p.d. which must be applied to a 5 kΩ resistor such that a current of 6 mA may flow

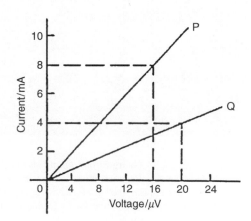

Figure 34.4

Table 34.1

Prefix	Name	Meaning	Example
M	mega	multiply by 1,000,000 (i.e. $\times 10^6$)	$2\,M\Omega = 2,000,000$ ohms
k	kilo	multiply by 1000 (i.e. $\times 10^3$)	$10\,kV = 10,000$ volts
m	milli	Divide by 1000 (i.e. $\times 10^{-3}$)	$25\,mA = \dfrac{25}{1000}A = 0.025$ amperes
μ	micro	Divide by 1,000,000 (i.e. $\times 10^{-6}$)	$50\,\mu V = \dfrac{50}{1,000,000}V = 0.00005$ volts

34.8 Conductors and insulators

A **conductor** is a material having a low resistance which allows electric current to flow in it. All metals are conductors and some examples include copper, aluminium, brass, platinum, silver, gold and carbon.

An **insulator** is a material having a high resistance which does not allow electric current to flow in it. Some examples of insulators include plastic, rubber, glass, porcelain, air, paper, cork, mica, ceramics and certain oils.

34.9 Electrical power and energy

34.9.1 Electrical power

Power P in an electrical circuit is given by the product of potential difference V and current I. The unit of power is the **watt***, **W**.

Hence, $P = V \times I$ **watts**   (1)

From Ohm's law, $V = IR$.
Substituting for V in equation (1) gives:

$$P = (IR) \times I$$

i.e. $P = I^2R$ **watts**

Also, from Ohm's law, $I = \dfrac{V}{R}$

*Who was Watt? – **James Watt** (19 January 1736–25 August 1819) was a Scottish inventor and mechanical engineer whose radically improved both the power and efficiency of steam engines. To find out more go to www.routledge.com/cw/bird (For image refer to page 225)

Substituting for I in equation (1) gives:

$$P = V \times \frac{V}{R}$$

i.e. $$P = \frac{V^2}{R} \text{ watts}$$

There are thus three possible formulae which may be used for calculating power.

Problem 7. A 100 W electric light bulb is connected to a 250 V supply. Determine (a) the current flowing in the bulb and (b) the resistance of the bulb

Power $P = V \times I$, from which, current $I = \dfrac{P}{V}$

(a) **Current** $I = \dfrac{100}{250} = \dfrac{10}{25} = \dfrac{2}{5} = \textbf{0.4 A}$

(b) **Resistance** $R = \dfrac{V}{I} = \dfrac{250}{0.4} = \dfrac{2500}{4} = \textbf{625}\,\boldsymbol{\Omega}$

Problem 8. Calculate the power dissipated when a current of 4 mA flows through a resistance of 5 kΩ

Power $P = I^2R = (4 \times 10^{-3})^2(5 \times 10^3)$

$$= 16 \times 10^{-6} \times 5 \times 10^3 = 80 \times 10^{-3}$$

$$= \textbf{0.08 W} \quad \text{or} \quad \textbf{80 mW}$$

Alternatively, since

$$I = 4 \times 10^{-3} \quad \text{and} \quad R = 5 \times 10^3$$

then from Ohm's law, voltage

$$V = IR = 4 \times 10^{-3} \times 5 \times 10^3 = 20\,V$$

Hence **power** $P = V \times I = 20 \times 4 \times 10^{-3} = \textbf{80 mW}$

Problem 9.   An electric kettle has a resistance of $30\,\Omega$. What current will flow when it is connected to a $240\,V$ supply? Find also the power rating of the kettle

Current $I = \dfrac{V}{R} = \dfrac{240}{30} = \textbf{8 A}$

Power $P = VI = 240 \times 8 = 1920\,W = \textbf{1.92 kW}$

$$= \textbf{power rating of kettle}.$$

Problem 10.   A current of 5 A flows in the winding of an electric motor, the resistance of the winding being $100\,\Omega$. Determine (a) the p.d. across the winding and (b) the power dissipated by the coil

(a)   Potential difference across winding

$$V = IR = 5 \times 100 = \textbf{500 V}$$

(b)   Power dissipated by coil

$$P = I^2 R = 5^2 \times 100 = \textbf{2500 W} \quad \text{or} \quad \textbf{2.5 kW}$$

(Alternatively, $P = V \times I = 500 \times 5$

$$= \textbf{2500 W or 2.5 kW})$$

34.9.2   Electrical energy

Electrical energy = power × time

If the power is measured in watts and the time in seconds then the unit of energy is watt-seconds or **joules***. If the power is measured in kilowatts and the time in hours then the unit of energy is **kilowatt-hours**, often called the '**unit of electricity**'. The 'electricity meter' in the home records the number of kilowatt-hours used and is thus an energy meter.

*Who was Joule? – **James Prescott Joule** (24 December 1818–11 October 1889) was an English physicist and brewer. He studied the nature of heat, and discovered its relationship to mechanical work. To find out more go to www.routledge.com/cw/bird (For image refer to page 219)

Problem 11.   A 12 V battery is connected across a load having a resistance of $40\,\Omega$. Determine the current flowing in the load, the power consumed and the energy dissipated in 2 minutes

Current $I = \dfrac{V}{R} = \dfrac{12}{40} = \textbf{0.3 A}$

Power consumed $P = VI = (12)(0.3) = \textbf{3.6 W}$

Energy dissipated = power × time

$$= (3.6\,W)(2 \times 60\,s) = \textbf{432 J}$$

(since $1\,J = 1\,W\,s$).

Problem 12.   A source of electromotive force (e.m.f.) of 15 V supplies a current of 2 A for 6 minutes. How much energy is provided in this time?

Energy = power × time, and power = voltage × current

Hence, **energy** = $VIt = 15 \times 2 \times (6 \times 60)$

$$= 10{,}800\,W\,s \text{ or } J = \textbf{10.8 kJ}$$

Problem 13.   Electrical equipment in an office takes a current of 13 A from a 240 V supply. Estimate the cost per week of electricity if the equipment is used for 30 hours each week and 1 kWh of energy costs 12.5p

Power = VI watts = $240 \times 13 = 3120\,W = 3.12\,kW$

Energy used per week = power × time

$$= (3.12\,kW) \times (30\,h)$$

$$= 93.6\,kWh$$

Cost at 12.5p per kWh = $93.6 \times 12.5 = 1170p$

Hence, **weekly cost of electricity = £11.70**

Problem 14.   An electric heater consumes 3.6 MJ when connected to a 250 V supply for 40 minutes. Find the power rating of the heater and the current taken from the supply

Power = $\dfrac{\text{energy}}{\text{time}} = \dfrac{3.6 \times 10^6\,J}{40 \times 60\,s}$ (or W) $= 1500\,W$

i.e. **power rating of heater = 1.5 kW**

Power $P = VI$, thus $I = \dfrac{P}{V} = \dfrac{1500}{250} = \textbf{6 A}$

Hence the current taken from the supply is **6 A**

Problem 15. A business uses two 3 kW fires for an average of 20 hours each per week, and six 150 W lights for 30 hours each per week. If the cost of electricity is 14p per unit, determine the weekly cost of electricity to the business

Energy = power × time

Energy used by one 3 kW fire in 20 hours

$= 3\,kW \times 20\,h = 60\,kWh$

Hence weekly energy used by two 3 kW fires

$= 2 \times 60 = 120\,kWh$

Energy used by one 150 W light for 30 hours

$= 150\,W \times 30\,h = 4500\,Wh = 4.5\,kWh$

Hence weekly energy used by six 150 W lamps

$= 6 \times 4.5 = 27\,kWh$

Total energy used per week $= 120 + 27 = 147\,kWh$

1 unit of electricity = 1 kWh of energy

Thus, **weekly cost of energy** at 14p per kWh

$= 14 \times 147 = 2058p = \mathbf{£20.58}$

Now try the following Practice Exercise

Practice Exercise 169   Further problems on power and energy (answers on page 546)

1. The hot resistance of a 250 V filament lamp is 625 Ω. Determine the current taken by the lamp and its power rating

2. Determine the resistance of an electric fire which takes a current of 12 A from a 240 V supply. Find also the power rating of the fire and the energy used in 20 h

3. Determine the power dissipated when a current of 10 mA flows through an appliance having a resistance of 8 kΩ

4. 85.5 J of energy are converted into heat in 9 s. What power is dissipated?

5. A current of 4 A flows through a conductor and 10 W is dissipated. What p.d. exists across the ends of the conductor?

6. Find the power dissipated when:

   (a) a current of 5 mA flows through a resistance of 20 kΩ

   (b) a voltage of 400 V is applied across a 120 kΩ resistor

   (c) a voltage applied to a resistor is 10 kV and the current flow is 4 mA

7. A d.c. electric motor consumes 72 MJ when connected to 400 V supply for 2 h 30 min. Find the power rating of the motor and the current taken from the supply

8. A p.d. of 500 V is applied across the winding of an electric motor and the resistance of the winding is 50 Ω. Determine the power dissipated by the coil

9. In a household during a particular week three 2 kW fires are used on average 25 h each and eight 100 W light bulbs are used on average 35 h each. Determine the cost of electricity for the week if 1 unit of electricity costs 15p

10. Calculate the power dissipated by the element of an electric fire of resistance 30 Ω when a current of 10 A flows in it. If the fire is on for 30 hours in a week determine the energy used. Determine also the weekly cost of energy if electricity costs 13.5p per unit

34.10   Main effects of electric current

The three main effects of an electric current are:

(a) magnetic effect, (b) chemical effect,

(c) heating effect.

Some practical applications of the effects of an electric current include:

- **Magnetic effect**: bells, relays, motors, generators, transformers, telephones, car-ignition and lifting magnets (see chapter 39).

- **Chemical effect**: primary and secondary cells and electroplating (see chapter 36).

- **Heating effect**: cookers, water heaters, electric fires, irons, furnaces, kettles and soldering irons.

34.11 Fuses

If there is a fault in a piece of equipment then excessive current may flow. This will cause overheating and possibly a fire; **fuses** protect against this happening. Current from the supply to the equipment flows through the fuse. The fuse is a piece of wire which can carry a stated current; if the current rises above this value it will melt. If the fuse melts (blows) then there is an open circuit and no current can then flow – thus protecting the equipment by isolating it from the power supply.

The fuse must be able to carry slightly more than the normal operating current of the equipment to allow for tolerances and small current surges. With some equipment there is a very large surge of current for a short time at switch on. If a fuse is fitted to withstand this large current there would be no protection against faults which cause the current to rise slightly above the normal value. Therefore special anti-surge fuses are fitted. These can stand 10 times the rated current for 10 milliseconds. If the surge lasts longer than this the fuse will blow.

A circuit diagram symbol for a fuse is shown in Figure 34.1 on page 364.

Problem 16. If 5 A, 10 A and 13 A fuses are available, state which is most appropriate for the following appliances which are both connected to a 240 V supply:

(a) an electric toaster having a power rating of 1 kW,

(b) an electric fire having a power rating of 3 kW

Power $P = VI$, from which, current $I = \dfrac{P}{V}$

(a) For the toaster, current

$$I = \frac{P}{V} = \frac{1000}{240} = \frac{100}{24} = 4.17\,\text{A}$$

Hence a **5 A fuse** is most appropriate.

(b) For the fire, current

$$I = \frac{P}{V} = \frac{3000}{240} = \frac{300}{24} = 12.5\,\text{A}$$

Hence a **13 A fuse** is most appropriate.

Now try the following Practice Exercises

Practice Exercise 170 Further problem on fuses (answer on page 546)

1. A television set having a power rating of 120 W and an electric lawnmower of power rating 1 kW are both connected to a 250 V supply. If 3 A, 5 A and 13 A fuses are available state which is the most appropriate for each appliance

Practice Exercise 171 Short answer questions on the introduction to electric circuits

1. Draw the preferred symbols for the following components used when drawing electrical circuit diagrams:

    (a) fixed resistor,   (b) cell,

    (c) filament lamp,   (d) fuse,

    (e) voltmeter

2. State the unit of (a) current, (b) potential difference, (c) resistance

3. State an instrument used to measure, (a) current, (b) potential difference, (c) resistance

4. What is a multimeter?

5. State an instrument used to measure: (a) engine rotational speed, (b) continuity and insulation testing, (c) electrical power

6. State Ohm's law

7. State the meaning of the following abbreviations of prefixes used with electrical units:

    (a) k   (b) μ   (c) m   (d) M

8. What is a conductor? Give four examples

9. What is an insulator? Give four examples

10. Complete the following statement: 'An ammeter has a .... resistance and must be connected .... with the load'

11. Complete the following statement: 'A voltmeter has a .... resistance and must be connected .... with the load'

12. State the unit of electrical power. State three formulae used to calculate power

13. State two units used for electrical energy

14. State the three main effects of an electric current and give two examples of each

15. What is the function of a fuse in an electrical circuit?

Practice Exercise 172   Multiple-choice problems on the introduction to electric circuits (answers on page 546)

1. The ohm is the unit of:
   (a) charge   (b) electric potential
   (c) current   (d) resistance

2. The current which flows when 0.1 coulomb is transferred in 10 ms is:
   (a) 10 A   (b) 1 A   (c) 10 mA   (d) 100 mA

3. The p.d. applied to a 1 kΩ resistance in order that a current of 100 μA may flow is:
   (a) 1 V   (b) 100 V   (c) 0.1 V   (d) 10 V

4. Which of the following formulae for electrical power is incorrect?
   (a) VI (b) $\dfrac{V}{I}$ (c) I^2R (d) $\dfrac{V^2}{R}$

5. The power dissipated by a resistor of 4 Ω when a current of 5 A passes through it is:
   (a) 6.25 W   (b) 20 W   (c) 80 W   (d) 100 W

6. Which of the following statements is true?
   (a) Electric current is measured in volts
   (b) 200 kΩ resistance is equivalent to 2 MΩ
   (c) An ammeter has a low resistance and must be connected in parallel with a circuit
   (d) An electrical insulator has a high resistance

7. A current of 3 A flows for 50 h through a 6 Ω resistor. The energy consumed by the resistor is:
   (a) 0.9 kWh   (b) 2.7 kWh
   (c) 9 kWh   (d) 27 kWh

8. What must be known in order to calculate the energy used by an electrical appliance?
   (a)   voltage and current
   (b)   current and time of operation
   (c)   power and time of operation
   (d)   current and resistance

9. Voltage drop is the:
   (a)   difference in potential between two points
   (b)   maximum potential
   (c)   voltage produced by a source
   (d)   voltage at the end of a circuit

10. The energy used by a 3 kW heater in 1 minute is:
   (a) 180,000 J   (b) 3000 J   (c) 180 J   (d) 50 J

11. Electromotive force is provided by:
   (a)   resistances
   (b)   a conducting path
   (c)   an electrical supply source
   (d)   an electric current

12. A 240 V, 60 W lamp has a working resistance of:
   (a) 1400 ohm   (b) 60 ohm
   (c) 960 ohm   (d) 325 ohm

13. When an atom loses an electron, the atom:
   (a)   experiences no effect at all
   (b)   becomes positively charged
   (c)   disintegrates
   (d)   becomes negatively charged

14. The largest number of 60 W electric light bulbs which can be operated from a 240 V supply fitted with 5 A fuse is:

    (a) 20   (b) 48   (c) 12   (d) 4

15. The unit of current is the:

    (a) joule   (b) coulomb   (c) ampere   (d) volt

16. State which of the following is incorrect.

    (a) $1\,W = 1\,J\,s^{-1}$

    (b) $1\,J = 1\,N/m$

    (c) $\eta = \dfrac{\text{output energy}}{\text{input energy}}$

    (d)   energy $=$ power $\times$ time

17. The coulomb is a unit of:

    (a) voltage   (b) power

    (c) energy   (d) quantity of electricity

18. If in the circuit shown in Figure 34.5, the reading on the voltmeter is 4 V and the reading on the ammeter is 20 mA, the resistance of resistor R is:

    (a) 0.005 Ω   (b) 5 Ω
    (c) 80 Ω   (d) 200 Ω

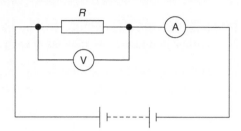

Figure 34.5

For fully worked solutions to each of the problems in Exercises 167 to 172 in this chapter, go to the website:
www.routledge.com/cw/bird

Chapter 35

Resistance variation

Why it is important to understand: Resistance variation

An electron travelling through the wires and loads of an electric circuit encounters resistance. Resistance is the hindrance to the flow of charge. The flow of charge through wires is often compared to the flow of water through pipes. The resistance to the flow of charge in an electric circuit is analogous to the frictional effects between water and the pipe surfaces as well as the resistance offered by obstacles that are present in its path. It is this resistance that hinders the water flow and reduces both its flow rate and its drift speed. Like the resistance to water flow, the total amount of resistance to charge flow within a wire of an electric circuit is affected by some clearly identifiable variables. Factors which affect resistance are length, cross-sectional area and type of material. The value of a resistor also changes with changing temperature, but this is not, as we might expect, mainly due to a change in the dimensions of the component as it expands or contracts. It is due mainly to a change in the resistivity of the material caused by the changing activity of the atoms that make up the resistor. Resistance variation due to length, cross-sectional area, type of material and temperature variation are explained in this chapter, with calculations to aid understanding. In addition, the resistor colour coding/ohmic values are explained.

At the end of this chapter, you should be able to:

- State four methods of resistor construction
- appreciate that electrical resistance depends on four factors
- appreciate that resistance $R = \dfrac{\rho 1}{a}$, where ρ is the resistivity
- recognise typical values of resistivity and its unit
- perform calculations using $R = \dfrac{\rho 1}{a}$
- define the temperature coefficient of resistance, α
- recognise typical values for α
- perform calculations using $R_\theta = R_0(1 + \alpha\theta)$
- determine the resistance and tolerance of a fixed resistor from its colour code
- determine the resistance and tolerance of a fixed resistor from its letter and digit code

Science and Mathematics for Engineering. 978-0-367-20475-4, © John Bird. Published by Taylor & Francis. All rights reserved.

35.1   Resistor construction

There is a wide range of resistor types. Four of the most common methods of construction are:

(i)   **Surface Mount Technology (SMT)**
Many modern circuits use SMT resistors. Their manufacture involves depositing a film of resistive material such as tin oxide on a tiny ceramic chip. The edges of the resistor are then accurately ground or cut with a laser to give a precise resistance across the ends of the device. Tolerances may be as low as ±0.02% and SMT resistors normally have very low power dissipation. Their main advantage is that very high component density can be achieved.

(ii) **Wire wound resistors**
A length of wire such as nichrome or manganin, whose resistive value per unit length is known, is cut to the desired value and wound around a ceramic former prior to being lacquered for protection. This type of resistor has a large physical size, which is a disadvantage; however, they can be made with a high degree of accuracy, and can have a **high power rating**.
Wire wound resistors are used in **power circuits** and **motor starters**.

(iii) **Metal film resistors**
Metal film resistors are made from small rods of ceramic coated with metal, such as a nickel alloy. The value of resistance is controlled firstly by the thickness of the coating layer (the thicker the layer, the lower the value of resistance), and secondly by cutting a fine spiral groove along the rod using a laser or diamond cutter to cut the metal coating into a long spiral strip, which forms the resistor.
Metal film resistors are low tolerance, precise resistors (±1% or less) and are used in **electronic circuits**.

(iv) **Carbon film resistors**
Carbon film resistors have a similar construction to metal film resistors but generally with wider

tolerance, typically ±5%. They are inexpensive, in common use, and are used in **electronic circuits**.

Some typical resistors are shown in Figure 35.1

35.2   Resistance and resistivity

The resistance of an electrical conductor depends on four factors, these being: (a) the length of the conductor, (b) the cross-sectional area of the conductor, (c) the type of material and (d) the temperature of the material.

Resistance R is directly proportional to length l of a conductor, i.e. $R \propto l$. Thus, for example, if the length of a piece of wire is doubled, then the resistance is doubled.

Resistance R is inversely proportional to cross-sectional area a of a conductor, i.e. $R \propto \dfrac{1}{a}$. Thus, for example, if the cross-sectional area of a piece of wire is doubled then the resistance is halved.

Since $R \propto l$ and $R \propto \dfrac{1}{a}$ then $R \propto \dfrac{l}{a}$. By inserting a constant of proportionality into this relationship the type of material used may be taken into account. The constant of proportionality is known as the **resistivity** of the material and is given the symbol ρ (Greek rho).

Thus, **resistance $R = \dfrac{\rho l}{a}$ ohms**

ρ is measured in ohm metres ($\Omega\,m$).

The value of the resistivity is that resistance of a unit cube of the material measured between opposite faces of the cube.

Resistivity varies with temperature and some typical values of resistivities measured at about room temperature are given below:

Copper $1.7 \times 10^{-8}\,\Omega\,m$ (or $0.017\,\mu\Omega\,m$)

Aluminium $2.6 \times 10^{-8}\,\Omega\,m$ (or $0.026\,\mu\Omega\,m$)

Carbon (graphite) $10 \times 10^{-8}\,\Omega\,m$ ($0.10\,\mu\Omega\,m$)

Glass $1 \times 10^{10}\,\Omega\,m$ (or $10^{4}\,\mu\Omega\,m$)

Mica $1 \times 10^{13}\,\Omega\,m$ (or $10^{7}\,\mu\Omega\,m$)

Note that good conductors of electricity have a low value of resistivity and good insulators have a high value of resistivity.

Problem 1.   The resistance of a 5 m length of wire is 600 Ω. Determine (a) the resistance of an 8 m length of the same wire and (b) the length of the same wire when the resistance is 420 Ω

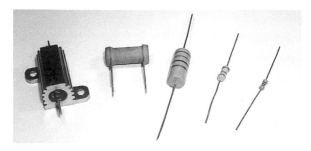

Figure 35.1

(a) Resistance R is directly proportional to length l i.e. $R \propto l$. Hence, $600\,\Omega \propto 5$ m or $600 = (k)(5)$, where k is the coefficient of proportionality.

Hence, $k = \dfrac{600}{5} = 120$

When the length l is 8 m, then resistance

$$R = kl = (120)(8) = \mathbf{960\,\Omega}$$

(b) When the resistance is $420\,\Omega$, $420 = kl$

from which, length $l = \dfrac{420}{k} = \dfrac{420}{120} = \mathbf{3.5\,m}$

Problem 2. A piece of wire of cross-sectional area $2\,mm^2$ has a resistance of $300\,\Omega$. Find (a) the resistance of a wire of the same length and material if the cross-sectional area is $5\,mm^2$, (b) the cross-sectional area of a wire of the same length and material of resistance $750\,\Omega$

Resistance R is inversely proportional to cross-sectional area a, i.e. $R \propto \dfrac{1}{a}$

Hence, $300\,\Omega \propto \dfrac{1}{2\,mm^2}$ or $300 = (k)\left(\dfrac{1}{2}\right)$

from which, the coefficient of proportionality

$k = 300 \times 2 = 600$

(a) When the cross-sectional area $a = 5\,mm^2$, then

$$\textbf{resistance } R = (k)\left(\dfrac{1}{5}\right) = (600)\left(\dfrac{1}{5}\right) = \mathbf{120\,\Omega}$$

(Note that resistance has decreased as the cross-sectional is increased.)

(b) When the resistance is $750\,\Omega$ then $750 = (k)\left(\dfrac{1}{a}\right)$

from which, **cross-sectional area**

$a = \dfrac{k}{750} = \dfrac{600}{750} = \mathbf{0.8\,mm^2}$

Problem 3. A wire of length 8 m and cross-sectional area $3\,mm^2$ has a resistance of $0.16\,\Omega$. If the wire is drawn out until its cross-sectional area is $1\,mm^2$, determine the resistance of the wire

Resistance R is directly proportional to length l, and inversely proportional to the cross-sectional area a,

i.e. $R \propto \dfrac{l}{a}$ or $R = k\left(\dfrac{l}{a}\right)$, where k is the coefficient of proportionality.

Since $R = 0.16$, $l = 8$ and $a = 3$, then $0.16 = (k)\left(\dfrac{8}{3}\right)$

from which, $k = 0.16 \times \dfrac{3}{8} = 0.06$

If the cross-sectional area is reduced to $\dfrac{1}{3}$ of its original area then the length must be tripled to 3×8, i.e. 24 m.

New resistance $R = k\left(\dfrac{l}{A}\right) = 0.06\left(\dfrac{24}{1}\right) = \mathbf{1.44\,\Omega}$

Problem 4. Calculate the resistance of a 2 km length of aluminium overhead power cable if the cross-sectional area of the cable is $100\,mm^2$. Take the resistivity of aluminium to be $0.03 \times 10^{-6}\,\Omega\,m$

Length $l = 2\,km = 2000\,m$,
area $a = 100\,mm^2 = 100 \times 10^{-6}\,m^2$ and
resistivity $\rho = 0.03 \times 10^{-6}\,\Omega\,m$.

Resistance $R = \dfrac{\rho l}{a} = \dfrac{(0.03 \times 10^{-6}\,\Omega\,m)(2000\,m)}{100 \times 10^{-6}\,m^2}$

$= \dfrac{0.03 \times 2000}{100}\,\Omega = \mathbf{0.6\,\Omega}$

Problem 5. Calculate the cross-sectional area, in mm^2, of a piece of copper wire, 40 m in length and having a resistance of $0.25\,\Omega$. Take the resistivity of copper as $0.02 \times 10^{-6}\,\Omega\,m$

Resistance $R = \dfrac{\rho l}{a}$ hence

cross-sectional area

$a = \dfrac{\rho l}{R} = \dfrac{(0.02 \times 10^{-6}\,\Omega\,m)(40\,m)}{0.25\,\Omega}$

$= 3.2 \times 10^{-6}\,m^2$

$= (3.2 \times 10^{-6}) \times 10^6\,mm^2$

$= \mathbf{3.2\,mm^2}$

Problem 6. The resistance of 1.5 km of wire of cross-sectional area $0.17\,mm^2$ is $150\,\Omega$. Determine the resistivity of the wire

Resistance $R = \dfrac{\rho l}{a}$ hence

resistivity $\rho = \dfrac{Ra}{l} = \dfrac{(150\,\Omega)(0.17 \times 10^{-6}\,m^2)}{(1500\,m)}$

$= \mathbf{0.017 \times 10^{-6}\,\Omega\,m \text{ or } 0.017\mu\Omega\,m}$

Problem 7. Determine the resistance of 1200 m of copper cable having a diameter of 12 mm if the resistivity of copper is $1.7 \times 10^{-8}\ \Omega$ m

Cross-sectional area of cable

$$a = \pi r^2 = \pi \left(\frac{12}{2}\right)^2 = 36\pi\ \text{mm}^2 = 36\pi \times 10^{-6}\ \text{m}^2.$$

Resistance $R = \dfrac{\rho l}{a} = \dfrac{(1.7 \times 10^{-8}\ \Omega\,\text{m})(1200\,\text{m})}{36\pi \times 10^{-6}\text{m}^2}$

$$= \frac{1.7 \times 1200 \times 10^6}{10^8 \times 36\pi}\ \Omega$$

$$= \frac{1.7 \times 12}{36\pi}\ \Omega = \mathbf{0.180\ \Omega}$$

Now try the following Practice Exercise

Practice Exercise 173   Further problems on resistance and resistivity (answers on page 546)

1. The resistance of a 2 m length of cable is 2.5 Ω. Determine (a) the resistance of a 7 m length of the same cable and (b) the length of the same wire when the resistance is 6.25 Ω

2. Some wire of cross-sectional area 1 mm^2 has a resistance of 20 Ω. Determine (a) the resistance of a wire of the same length and material if the cross-sectional area is 4 mm^2 and (b) the cross-sectional area of a wire of the same length and material if the resistance is 32 Ω

3. Some wire of length 5 m and cross-sectional area 2 mm^2 has a resistance of 0.08 Ω. If the wire is drawn out until its cross-sectional area is 1 mm^2, determine the resistance of the wire

4. Find the resistance of 800 m of copper cable of cross-sectional area 20 mm^2. Take the resistivity of copper as 0.02 $\mu\Omega$ m

5. Calculate the cross-sectional area, in mm^2, of a piece of aluminium wire 100 m long and having a resistance of 2 Ω. Take the resistivity of aluminium as $0.03 \times 10^{-6}\ \Omega$ m

6. The resistance of 500 m of wire of cross-sectional area 2.6 mm^2 is 5 Ω. Determine the resistivity of the wire in $\mu\Omega$ m

7. Find the resistance of 1 km of copper cable having a diameter of 10 mm if the resistivity of copper is $0.017 \times 10^{-6}\ \Omega$ m

35.3 Temperature coefficient of resistance

In general, as the temperature of a material increases, most conductors increase in resistance, insulators decrease in resistance, whilst the resistance of some special alloys remain almost constant.

The **temperature coefficient of resistance** of a material is the increase in the resistance of a 1 Ω resistor of that material when it is subjected to a rise of temperature of 1°C. The symbol used for the temperature coefficient of resistance is α (Greek alpha). Thus, if some copper wire of resistance 1 Ω is heated through 1°C and its resistance is then measured as 1.0043 Ω then $\alpha = 0.0043\ \Omega/\Omega°$C for copper. The units are usually expressed only as 'per °C', i.e. $\alpha = 0.0043/°$C for copper.

If the 1 Ω resistor of copper is heated through 100°C then the resistance at 100°C would be $1 + 100 \times 0.0043 = 1.43\ \Omega$.

Some typical values of temperature coefficient of resistance measured at 0°C are given below:

Copper	0.0043/°C
Nickel	0.0062/°C
Constantan	0
Aluminium	0.0038/°C
Carbon	−0.00048/°C
Eureka	0.00001/°C

(Note that the negative sign for carbon indicates that its resistance falls with increase of temperature.)

If the resistance of a material at 0°C is known, the resistance at any other temperature can be determined from:

$$R_\theta = R_0(1 + \alpha_0\theta)$$

where R_0 = resistance at 0°C,

R_θ = resistance at temperature θ°C

and α_0 = temperature coefficient of resistance at 0°C.

Problem 8. A coil of copper wire has a resistance of 100 Ω when its temperature is 0°C. Determine its resistance at 70°C if the temperature coefficient of resistance of copper at 0°C is 0.0043/°C

Resistance $R_\theta = R_0(1 + \alpha_0\theta)$

Hence, **resistance at 70°C**

$R_{70} = 100[1 + (0.0043)(70)]$

$= 100[1 + 0.301] = 100(1.301) = $ **130.1 Ω**

Problem 9. An aluminium cable has a resistance of 27 Ω at a temperature of 35°C. Determine its resistance at 0°C. Take the temperature coefficient of resistance at 0°C to be 0.0038/°C

Resistance at θ°C, $R_\theta = R_0(1 + \alpha_0\theta)$.

Hence, resistance at 0°C

$$R_0 = \frac{R_\theta}{(1 + \alpha\theta)} = \frac{27}{[1 + (0.0038)(35)]}$$

$$= \frac{27}{1 + 0.133} = \frac{27}{1.133} = \textbf{23.83 Ω}$$

Problem 10. A carbon resistor has a resistance of 1 kΩ at 0°C. Determine its resistance at 80°C. Assume that the temperature coefficient of resistance for carbon at 0°C is −0.0005/°C

Resistance at temperature θ°C

$R_\theta = R_0(1 + \alpha_0\theta)$

i.e. $R_\theta = 1000[1 + (-0.0005)(80)]$

$= 1000[1 - 0.040] = 1000(0.96) = $ **960 Ω**

If the resistance of a material at room temperature (approximately 20°C) R_{20} and the temperature coefficient of resistance at 20°C α_{20} are known then the resistance R_θ at temperature θ°C is given by:

$$R_\theta = R_{20}[1 + \alpha_{20}(\theta - 20)]$$

Problem 11. A coil of copper wire has a resistance of 10 Ω at 20°C. If the temperature coefficient of resistance of copper at 20°C is 0.004/°C determine the resistance of the coil when the temperature rises to 100°C

Resistance at θ°C $R_\theta = R_{20}[1 + \alpha_{20}(\theta - 20)]$.

Hence, resistance at 100°C

$R_{100} = 10[1 + (0.004)(100 - 20)]$

$= 10[1 + (0.004)(80)]$

$= 10[1 + 0.32]$

$= 10(1.32) = $ **13.2 Ω**

Problem 12. The resistance of a coil of aluminium wire at 18°C is 200 Ω. The temperature of the wire is increased and the resistance rises to 240 Ω. If the temperature coefficient of resistance of aluminium is 0.0039/°C at 18°C determine the temperature to which the coil has risen

Let the temperature rise to θ°C.

Resistance at θ°C $R_\theta = R_{18}[1 + \alpha_{18}(\theta - 18)]$

i.e. $240 = 200[1 + (0.0039)(\theta - 18)]$

$240 = 200 + (200)(0.0039)(\theta - 18)$

$240 - 200 = 0.78(\theta - 18)$

$40 = 0.78(\theta - 18)$

$\dfrac{40}{0.78} = \theta - 18$

$51.28 = \theta - 18$

from which, $\theta = 51.28 + 18 = 69.28$°C

Hence the temperature of the coil increases to 69.28°C

If the resistance at 0°C is not known, but is known at some other temperature θ_1, then the resistance at any temperature can be found as follows:

$$R_1 = R_0(1 + \alpha_0\theta_1) \quad \text{and} \quad R_2 = R_0(1 + \alpha_0\theta_2)$$

Dividing one equation by the other gives:

$$\frac{R_1}{R_2} = \frac{1 + \alpha_0\theta_1}{1 + \alpha_0\theta_2}$$

where $R_2 = $ resistance at temperature θ_2

Problem 13. Some copper wire has a resistance of 200 Ω at 20°C. A current is passed through the wire and the temperature rises to 90°C. Determine the resistance of the wire at 90°C, correct to the nearest ohm, assuming that the temperature coefficient of resistance is 0.004/°C at 0°C

$R_{20} = 200$ Ω, $\alpha_0 = 0.004$/°C and

$$\frac{R_{20}}{R_{90}} = \frac{[1 + \alpha_0(20)]}{[1 + \alpha_0(90)]}$$

Hence, $R_{90} = \dfrac{R_{20}\left[1 + 90\alpha_0\right]}{\left[1 + 20\alpha_0\right]} = \dfrac{200[1 + 90(0.004)]}{[1 + 20(0.004)]}$

$= \dfrac{200[1 + 0.36]}{[1 + 0.08]}$

$= \dfrac{200(1.36)}{(1.08)} = \mathbf{251.85\ \Omega}$

i.e. the resistance of the wire at 90°C is 252 Ω, correct to the nearest ohm.

Now try the following Practice Exercises

Practice Exercise 174   Further problems on the temperature coefficient of resistance (answers on page 546)

1. A coil of aluminium wire has a resistance of 50 Ω when its temperature is 0°C. Determine its resistance at 100°C if the temperature coefficient of resistance of aluminium at 0°C is 0.0038/°C

2. A copper cable has a resistance of 30 Ω at a temperature of 50°C. Determine its resistance at 0°C. Take the temperature coefficient of resistance of copper at 0°C as 0.0043/°C

3. The temperature coefficient of resistance for carbon at 0°C is −0.00048/°C. What is the significance of the minus sign? A carbon resistor has a resistance of 500 Ω at 0°C. Determine its resistance at 50°C

4. A coil of copper wire has a resistance of 20 Ω at 18°C. If the temperature coefficient of resistance of copper at 18°C is 0.004/°C, determine the resistance of the coil when the temperature rises to 98°C

5. The resistance of a coil of nickel wire at 20°C is 100 Ω. The temperature of the wire is increased and the resistance rises to 130 Ω. If the temperature coefficient of resistance of nickel is 0.006/°C at 20°C, determine the temperature to which the coil has risen

6. Some aluminium wire has a resistance of 50 Ω at 20°C. The wire is heated to a temperature of 100°C. Determine the resistance of the wire at 100°C, assuming that the temperature coefficient of resistance at 0°C is 0.004/°C

7. A copper cable is 1.2 km long and has a cross-sectional area of 5 mm². Find its resistance at 80°C if at 20°C the resistivity of copper is 0.02×10^{-6} Ω m and its temperature coefficient of resistance is 0.004/°C

35.4   Resistor colour coding and ohmic values

(a) Colour code for fixed resistors

The colour code for fixed resistors is given in Table 35.1

(i)   For a **four-band fixed resistor** (i.e. resistance values with two significant figures): yellow-violet-orange-red indicates 47 kΩ with a tolerance of ±2%

(Note that the first band is the one nearest the end of the resistor.)

(ii) For a **five-band fixed resistor** (i.e. resistance values with three significant figures): red-yellow-white-orange-brown indicates 249 kΩ with a tolerance of ±1%

(Note that the fifth band is 1.5 to 2 times wider than the other bands.)

Problem 14.   Determine the value and tolerance of a resistor having a colour coding of: orange-orange-silver-brown.

The first two bands, i.e. orange-orange, give 33 from Table 35.1.
The third band, silver, indicates a multiplier of 10^2 from Table 35.1, which means that the value of the resistor is $33 \times 10^{-2} = 0.33$ Ω.
The fourth band, i.e. brown, indicates a tolerance of ±1% from Table 35.1. Hence a colour coding of orange-orange-silver-brown represents a resistor of value **0.33 Ω with a tolerance of ±1%**

Problem 15.   Determine the value and tolerance of a resistor having a colour coding of: brown-black-brown.

The first two bands, i.e. brown-black, give 10 from Table 35.1.
The third band, brown, indicates a multiplier of 10 from Table 35.1, which means that the value of the resistor is $10 \times 10 = 100$ Ω

Table 35.1

Colour	Significant figures	Multiplier	Tolerance
Silver	–	10^{-2}	$\pm10\%$
Gold	–	10^{-1}	$\pm5\%$
Black	0	1	–
Brown	1	10	$\pm1\%$
Red	2	10^2	$\pm2\%$
Orange	3	10^3	–
Yellow	4	10^4	–
Green	5	10^5	$\pm0.5\%$
Blue	6	10^6	$\pm0.25\%$
Violet	7	10^7	$\pm0.1\%$
Grey	8	10^8	–
White	9	10^9	–
None	–	–	$\pm20\%$

There is no fourth band colour in this case; hence, from Table 35.1, the tolerance is $\pm20\%$. Hence a colour coding of brown-black-brown represents a resistor of value **100 Ω with a tolerance of $\pm20\%$**

Problem 16. Between what two values should a resistor with colour coding brown-black-brown-silver lie?

From Table 35.1, brown-black-brown-silver indicates 10×10, i.e. 100 Ω, with a tolerance of $\pm10\%$
This means that the value could lie between

$$(100 - 10\% \text{ of } 100)\,\Omega$$
and
$$(100 + 10\% \text{ of } 100)\,\Omega$$

i.e. brown-black-brown-silver indicates any value **between 90 Ω and 110 Ω**

Problem 17. Determine the colour coding for a 47 kΩ resistor having a tolerance of $\pm5\%$

From Table 35.1, $47\,\text{k}\Omega = 47 \times 10^3$ has a colour coding of yellow-violet-orange. With a tolerance of $\pm5\%$, the fourth band will be gold.

Hence 47 kΩ $\pm 5\%$ has a colour coding of:
yellow-violet-orange-gold.

Problem 18. Determine the value and tolerance of a resistor having a colour coding of: orange-green-red-yellow-brown.

Orange-green-red-yellow-brown is a five-band fixed resistor and from Table 35.1, indicates: $352 \times 10^4\,\Omega$ with a tolerance of $\pm1\%$

$$352 \times 10^4\,\Omega = 3.52 \times 10^6\,\Omega, \text{ i.e. } 3.52\,\text{M}\Omega$$

Therefore, orange-green-red-yellow-brown indicates **3.52 MΩ $\pm 1\%$**

(b) Letter and digit code for resistors

Another way of indicating the value of resistors is the letter and digit code shown in Table 35.2.

Table 35.2

Resistance value	Marked as
0.47 Ω	R47
1 Ω	1R0
4.7 Ω	4R7
47 Ω	47R
100 Ω	100R
1 kΩ	1K0
10 kΩ	10 K
10 MΩ	10 M

Tolerance is indicated as follows: $F = \pm1\%$, $G = \pm2\%$, $J = \pm5\%$, $K = \pm10\%$ and $M = \pm20\%$.
Thus, for example,

$$R33M = 0.33\,\Omega \pm 20\%$$
$$4R7K = 4.7\,\Omega \pm 10\%$$
$$390RJ = 390\,\Omega \pm 5\%$$

Problem 19. Determine the value of a resistor marked as 6K8F

From Table 35.2, 6K8F is equivalent to: **6.8 kΩ $\pm 1\%$**

Problem 20. Determine the value of a resistor marked as 4M7M

From Table 35.2, 4M7M is equivalent to: **4.7 MΩ ± 20%**

Problem 21. Determine the letter and digit code for a resistor having a value of 68 kΩ ± 10%

From Table 35.2, 68 kΩ ± 10% has a letter and digit code of: **68 KK**

Now try the following Practice Exercise

Practice Exercise 175   Resistor colour coding and ohmic values (answers on page 546)

1. Determine the value and tolerance of a resistor having a colour coding of: blue-grey-orange-red

2. Determine the value and tolerance of a resistor having a colour coding of: yellow-violet-gold

3. Determine the value and tolerance of a resistor having a colour coding of: blue-white-black-black-gold

4. Determine the colour coding for a 51 kΩ four-band resistor having a tolerance of ±2%

5. Determine the colour coding for a 1 MΩ four-band resistor having a tolerance of ±10%

6. Determine the range of values expected for a resistor with colour coding: red-black-green-silver

7. Determine the range of values expected for a resistor with colour coding: yellow-black-orange-brown

8. Determine the value of a resistor marked as (a) R22G (b) 4K7F

9. Determine the letter and digit code for a resistor having a value of 100 kΩ ± 5%

10. Determine the letter and digit code for a resistor having a value of 6.8 MΩ ± 20%

Practice Exercise 176   Short answer questions on resistance variation

1. Name three types of resistor construction and state one practical application of each

2. Name four factors which can effect the resistance of a conductor

3. If the length of a piece of wire of constant cross-sectional area is halved, the resistance of the wire is ......

4. If the cross-sectional area of a certain length of cable is trebled, the resistance of the cable is ......

5. What is resistivity? State its unit and the symbol used

6. Complete the following: Good conductors of electricity have a ...... value of resistivity and good insulators have a ...... value of resistivity

7. What is meant by the 'temperature coefficient of resistance'? State its units and the symbols used

8. If the resistance of a metal at $0°C$ is R_0, R_θ is the resistance at $\theta°C$ and α_0 is the temperature coefficient of resistance at $0°C$ then: $R_\theta = ...$

9. Explain briefly the colour coding on resistors

10. Explain briefly the letter and digit code for resistors

Practice Exercise 177   Multiple-choice questions on resistance variation (answers on page 546)

1. The unit of resistivity is:

   (a) ohms (b) ohm millimetre
   (c) ohm metre (d) ohm/metre

2. The length of a certain conductor of resistance $100 \, \Omega$ is doubled and its cross-sectional area is halved. Its new resistance is:

   (a) $100 \, \Omega$ (b) $200 \, \Omega$
   (c) $50 \, \Omega$ (d) $400 \, \Omega$

3. The resistance of a 2 km length of cable of cross-sectional area $2\,mm^2$ and resistivity of $2 \times 10^{-8}\,\Omega\,m$ is:

   (a) $0.02\,\Omega$ (b) $20\,\Omega$
   (c) $0.02\,m\Omega$ (d) $200\,\Omega$

4. A piece of graphite has a cross-sectional area of $10\,mm^2$. If its resistance is $0.1\,\Omega$ and its resistivity $10 \times 10^{-8}\,\Omega\,m$, its length is:

   (a) $10\,km$ (b) $10\,cm$
   (c) $10\,mm$ (d) $10\,m$

5. The symbol for the unit of temperature coefficient of resistance is:

   (a) $\Omega/°C$ (b) Ω
   (c) $°C$ (d) $\Omega/\Omega°C$

6. A coil of wire has a resistance of $10\,\Omega$ at $0°C$. If the temperature coefficient of resistance for the wire is $0.004/°C$, its resistance at $100°C$ is:

   (a) $0.4\,\Omega$ (b) $1.4\,\Omega$
   (c) $14\,\Omega$ (d) $10\,\Omega$

7. A nickel coil has a resistance of $13\,\Omega$ at $50°C$. If the temperature coefficient of resistance at $0°C$ is $0.006/°C$, the resistance at $0°C$ is:

   (a) $16.9\,\Omega$ (b) $10\,\Omega$
   (c) $43.3\,\Omega$ (d) $0.1\,\Omega$

8. A colour coding of red-violet-black on a resistor indicates a value of:

   (a) $27\,\Omega \pm 20\%$ (b) $270\,\Omega$
   (c) $270\,\Omega \pm 20\%$ (d) $27\,\Omega \pm 10\%$

9. A resistor marked as 4K7G indicates a value of:

   (a) $47\,\Omega \pm 20\%$ (b) $4.7\,k\Omega \pm 20\%$
   (c) $0.47\,\Omega \pm 10\%$ (d) $4.7\,k\Omega \pm 2\%$

For fully worked solutions to each of the problems in Exercises 173 to 177 in this chapter, go to the website:

www.routledge.com/cw/bird

Chapter 36

Batteries and alternative sources of energy

Why it is important to understand: Batteries and alternative sources of energy

Batteries store electricity in a chemical form, inside a closed-energy system. They can be re-charged and re-used as a power source in small appliances, machinery and remote locations. Batteries can store d.c. electrical energy produced by renewable sources such as solar, wind and hydro power in chemical form. Because renewable energy-charging sources are often intermittent in their nature, batteries provide energy storage in order to provide a relatively constant supply of power to electrical loads regardless of whether the sun is shining or the wind is blowing. In an off-grid photovoltaic (PV) system, for example, battery storage provides a way to power common household appliances regardless of the time of day or the current weather conditions. In a grid-tie with battery backup PV system batteries provide uninterrupted power in case of utility power failure. Energy causes movement; every time something moves, energy is being used. Energy moves cars, makes machines run, heats ovens and lights our homes. One form of energy can be changed into another form. When petrol is burned in a vehicle engine, the energy stored in petrol is changed into heat energy. When we stand in the sun, light energy is changed into heat. When a torch or flashlight is turned on, chemical energy stored in the battery is changed into light and heat. To find energy, look for motion, heat, light, sound, chemical reactions or electricity. The sun is the source of all energy. The sun's energy is stored in coal, petroleum, natural gas, food, water and wind. While there are two types of energy, renewable and non-renewable, most of the energy we use comes from burning non-renewable fuels – coal, petroleum or oil, or natural gas. These supply the majority of our energy needs because we have designed ways to transform their energy on a large scale to meet consumer needs. Regardless of the energy source, the energy contained in them is changed into a more useful form of electricity. This chapter explores the increasingly important area of battery use and briefly looks at some alternative sources of energy.

At the end of this chapter you should be able to:

- list practical applications of batteries
- understand electrolysis and its applications, including electroplating
- appreciate the purpose and construction of a simple cell

Science and Mathematics for Engineering. 978-0-367-20475-4, © John Bird. Published by Taylor & Francis. All rights reserved.

- explain polarisation and local action
- explain corrosion and its effects
- define the terms e.m.f., E, and internal resistance, r, of a cell
- perform calculations using $V = E - Ir$
- determine the total e.m.f. and total internal resistance for cells connected in series and in parallel
- distinguish between primary and secondary cells
- explain the construction and practical applications of the Leclanché, mercury, lead–acid and alkaline cells
- list the advantages and disadvantages of alkaline cells over lead–acid cells
- understand the importance of lithium-ion batteries – applications, advantages and disadvantages
- understand the term 'cell capacity' and state its unit
- understand the importance of safe battery disposal
- appreciate advantages of fuel cells and their likely future applications
- understand the implications of alternative energy sources and state six examples
- appreciate the importance and uses of solar energy, its advantages and disadvantages and list some practical applications
- appreciate that glass batteries may be successful technology of the future

36.1   Introduction to batteries

A battery is a device that **converts chemical energy to electricity**. If an appliance is placed between its terminals the current generated will power the device. Batteries are an indispensable item for many electronic devices and are essential for devices that require power when no mains power is available. For example, without the battery, there would be no mobile phones or laptop computers.

The battery is now over 200 years old and batteries are found almost everywhere in consumer and industrial products. Some **practical examples** where batteries are used include:

in laptops, in cameras, in mobile phones, in cars, in watches and clocks, for security equipment, in electronic meters, for smoke alarms, for meters used to read gas, water and electricity consumption at home, to power a camera for an endoscope looking internally at the body, and for transponders used for toll collection on highways throughout the world

Batteries tend to be split into two categories – **primary**, which are not designed to be electrically re-charged, i.e. are disposable (see Section 36.6), and **secondary batteries**, which are designed to be re-charged, such as those used in mobile phones (see Section 36.7).

In more recent years it has been necessary to design batteries with reduced size, but with increased lifespan and capacity.

If an application requires small size and high power then the 1.5 V battery is used. If longer lifetime is required then the 3 to 3.6 V battery is used. In the 1970s the 1.5 V **manganese battery** was gradually replaced by the **alkaline battery**. **Silver oxide batteries** were gradually introduced in the 1960s and are still the preferred technology for watch batteries today.

Lithium-ion batteries were introduced in the 1970s because of the need for longer lifetime applications. Indeed, some such batteries have been known to last well over ten years before replacement, a characteristic that means that these batteries are still very much in demand today for digital cameras, and sometimes for watches and computer clocks. Lithium batteries are capable of delivering high currents but tend to be expensive.

For more on lithium-ion batteries see Section 36, page 392.

More types of batteries and their uses are listed in Table 36.2 on page 393.

36.2   Some chemical effects of electricity

A material must contain **charged particles** to be able to conduct electric current. In **solids**, the current is carried by **electrons**. Copper, lead, aluminium, iron and carbon are some examples of solid conductors. In **liquids and gases**, the current is carried by the part of a molecule which has acquired an electric charge, called **ions**. These can possess a positive or negative charge, and examples include hydrogen ion H^+, copper ion Cu^{++} and hydroxyl ion OH^-. Distilled water contains no ions and is a poor conductor of electricity, whereas salt water contains ions and is a fairly good conductor of electricity. **Electrolysis** is the decomposition of a liquid compound by the passage of electric current through it. Practical applications of electrolysis include the electroplating of metals (see below), the refining of copper and the extraction of aluminium from its ore.

An **electrolyte** is a compound which will undergo electrolysis. Examples include salt water, copper sulphate and sulphuric acid.

The **electrodes** are the two conductors carrying current to the electrolyte. The positive-connected electrode is called the **anode** and the negative-connected electrode the **cathode**.

When two copper wires connected to a battery are placed in a beaker containing a salt water solution, current will flow through the solution. Air bubbles appear around the wires as the water is changed into hydrogen and oxygen by electrolysis.

Electroplating uses the principle of electrolysis to apply a thin coat of one metal to another metal. Some practical applications include the tin-plating of steel, silver-plating of nickel alloys and chromium-plating of steel. If two copper electrodes connected to a battery are placed in a beaker containing copper sulphate as the electrolyte it is found that the cathode (i.e. the electrode connected to the negative terminal of the battery) gains copper whilst the anode loses copper.

36.3 The simple cell

The purpose of an **electric cell** is to convert chemical energy into electrical energy.

A **simple cell** comprises two dissimilar conductors (electrodes) in an electrolyte. Such a cell is shown in Figure 36.1, comprising copper and zinc electrodes. An electric current is found to flow between the electrodes. Other possible electrode pairs exist, including zinc–lead and zinc–iron. The electrode potential (i.e. the p.d. measured between the electrodes) varies for each pair of metals. By knowing the e.m.f. of each metal with respect to some standard electrode, the e.m.f. of any pair of metals may be determined. The standard used is the hydrogen electrode. The **electrochemical series** is a way of listing elements in order of electrical potential,

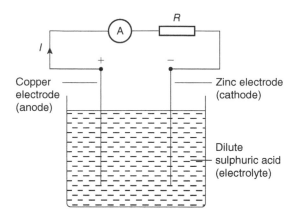

Figure 36.1

and Table 36.1 shows a number of elements in such a series.

In a simple cell two faults exist – those due to **polarization** and **local action**.

Polarization

If the simple cell shown in Figure 36.1 is left connected for some time, the current I decreases fairly rapidly. This is because of the formation of a film of hydrogen bubbles on the copper anode. This effect is known as the polarization of the cell. The hydrogen prevents full contact between the copper electrode and the electrolyte and this increases the internal resistance of the cell. The effect can be overcome by using a chemical depolarizing agent or depolarizer, such as potassium dichromate, which removes the hydrogen bubbles as they form. This allows the cell to deliver a steady current.

Table 36.1 Part of the electro-chemical series

Potassium
Sodium
Aluminium
Zinc
Iron
Lead
Hydrogen
Copper
Silver
Carbon

Local action

When commercial zinc is placed in dilute sulphuric acid, hydrogen gas is liberated from it and the zinc dissolves. The reason for this is that impurities, such as traces of iron, are present in the zinc which set up small primary cells with the zinc. These small cells are short-circuited by the electrolyte, with the result that localized currents flow, causing corrosion. This action is known as local action of the cell. This may be prevented by rubbing a small amount of mercury on the zinc surface, which forms a protective layer on the surface of the electrode. When two metals are used in a simple cell the electrochemical series may be used to predict the behaviour of the cell:

(i) The metal that is higher in the series acts as the negative electrode, and vice versa. For example, the zinc electrode in the cell shown in Figure 36.1 is negative and the copper electrode is positive.

(ii) The greater the separation in the series between the two metals the greater is the e.m.f. produced by the cell.

The electrochemical series is representative of the order of reactivity of the metals and their compounds:

(i) The higher metals in the series react more readily with oxygen and vice versa.

(ii) When two metal electrodes are used in a simple cell the one that is higher in the series tends to dissolve in the electrolyte.

36.4 Corrosion

Corrosion is the gradual destruction of a metal in a damp atmosphere by means of simple cell action. In addition to the presence of moisture and air required for rusting, an electrolyte, an anode and a cathode are required for corrosion. Thus, if metals widely spaced in the electrochemical series are used in contact with each other in the presence of an electrolyte, corrosion will occur. For example, if a brass valve is fitted to a heating system made of steel, corrosion will occur.

The **effects of corrosion** include the weakening of structures, the reduction of the life of components and materials, the wastage of materials and the expense of replacement.

Corrosion may be **prevented** by coating with paint, grease, plastic coatings and enamels, or by plating with tin or chromium. Also, iron may be galvanized, i.e. plated with zinc, the layer of zinc helping to prevent the iron from corroding.

36.5 E.m.f. and internal resistance of a cell

The **electromotive force (e.m.f.)**, E, of a cell is the p.d. between its terminals when it is not connected to a load (i.e. the cell is on 'no load').

The e.m.f. of a cell is measured by using a **high resistance voltmeter** connected in parallel with the cell. The voltmeter must have a high resistance otherwise it will pass current and the cell will not be on 'no-load'. For example, if the resistance of a cell is $1\,\Omega$ and that of a voltmeter $1\,M\Omega$ then the equivalent resistance of the circuit is $1\,M\Omega + 1\,\Omega$, i.e. approximately $1\,M\Omega$, hence no current flows and the cell is not loaded.

The voltage available at the terminals of a cell falls when a load is connected. This is caused by the **internal resistance** of the cell which is the opposition of the material of the cell to the flow of current. The internal resistance acts in series with other resistances in the circuit. Figure 36.2 shows a cell of e.m.f. E volts and internal resistance, r, and XY represents the terminals of the cell.

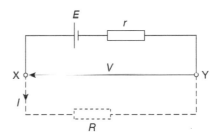

Figure 36.2

When a load (shown as resistance R) is not connected, no current flows and the terminal p.d., $V = E$. When R is connected a current I flows which causes a voltage drop in the cell, given by Ir. The p.d. available at the cell terminals is less than the e.m.f. of the cell and is given by:

$$V = E - Ir$$

Thus if a battery of e.m.f. 12 volts and internal resistance $0.01\,\Omega$ delivers a current of $100\,A$, the terminal p.d.,

$$V = 12 - (100)(0.01)$$
$$= 12 - 1 = 11\,V$$

When different values of potential difference V across a cell or power supply are measured for different values of current I, a graph may be plotted as shown in Figure 36.3. Since the e.m.f. E of the cell or power

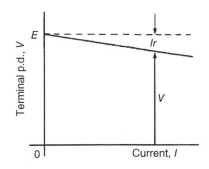

Figure 36.3

supply is the p.d. across its terminals on no load (i.e. when $I = 0$), then E is as shown by the broken line. Since $V = E - Ir$ then the internal resistance may be calculated from

$$r = \frac{E - V}{I}$$

When a current is flowing in the direction shown in Figure 36.2 the cell is said to be **discharging** ($E > V$). When a current flows in the opposite direction to that shown in Figure 36.2 the cell is said to be **charging** ($V > E$).

A **battery** is a combination of more than one cell. The cells in a battery may be connected in series or in parallel.

(i) **For cells connected in series:**

Total e.m.f. = sum of cells' e.m.f.s

Total internal resistance = sum of cells' internal resistances

(ii) **For cells connected in parallel:**

If each cell has the same e.m.f. and internal resistance:

Total e.m.f. = e.m.f. of one cell

Total internal resistance of n cells

$$= \frac{1}{n} \times \text{internal resistance of one cell}$$

Problem 1.   Eight cells, each with an internal resistance of $0.2\,\Omega$ and an e.m.f. of $2.2\,V$ are connected (a) in series, (b) in parallel. Determine the e.m.f. and the internal resistance of the batteries so formed

(a) When connected in series, total e.m.f.

= sum of cells' e.m.f.

$= 2.2 \times 8 = \mathbf{17.6\,V}$

Total internal resistance

= sum of cells' internal resistance

$= 0.2 \times 8 = \mathbf{1.6\,\Omega}$

(b) When connected in parallel, total e.m.f

= e.m.f. of one cell

$= \mathbf{2.2\,V}$

Total internal resistance of 8 cells

$= \frac{1}{8} \times \text{internal resistance of one cell}$

$= \frac{1}{8} \times 0.2 = \mathbf{0.025\,\Omega}$

Problem 2.   A cell has an internal resistance of $0.02\,\Omega$ and an e.m.f. of $2.0\,V$. Calculate its terminal p.d. if it delivers (a) $5\,A$, (b) $50\,A$

(a) Terminal p.d. $V = E - Ir$ where E = e.m.f. of cell, I = current flowing and r = internal resistance of cell

$E = 2.0\,V, I = 5\,A$ and $r = 0.02\,\Omega$

Hence **terminal p.d.**

$V = 2.0 - (5)(0.02) = 2.0 - 0.1 = \mathbf{1.9\,V}$

(b) When the current is $50\,A$, terminal p.d.,

$V = E - Ir = 2.0 - 50(0.02)$

i.e. $V = 2.0 - 1.0 = \mathbf{1.0\,V}$

Thus the terminal p.d. decreases as the current drawn increases.

Problem 3.   The p.d. at the terminals of a battery is $25\,V$ when no load is connected and $24\,V$ when a load taking $10\,A$ is connected. Determine the internal resistance of the battery

When no load is connected the e.m.f. of the battery, E, is equal to the terminal p.d., V, i.e. $E = 25\,V$.

When current $I = 10\,A$ and terminal p.d.

$V = 24\,V$, then $V = E - Ir$

i.e. $24 = 25 - (10)r$

Hence, rearranging gives

$10r = 25 - 24 = 1$

and the internal resistance,

$$r = \frac{1}{10} = \mathbf{0.1\,\Omega}$$

Problem 4. Ten 1.5 V cells, each having an internal resistance of 0.2 Ω, are connected in series to a load of 58 Ω. Determine (a) the current flowing in the circuit and (b) the p.d. at the battery terminals

(a) For ten cells, battery e.m.f., $E = 10 \times 1.5 = 15$ V, and the total internal resistance, $r = 10 \times 0.2 = 2$ Ω. When connected to a 58 Ω load the circuit is as shown in Figure 36.4

$$\text{Current } I = \frac{\text{e.m.f.}}{\text{total resistance}}$$

$$= \frac{15}{58 + 2}$$

$$= \frac{15}{60} = \mathbf{0.25\,A}$$

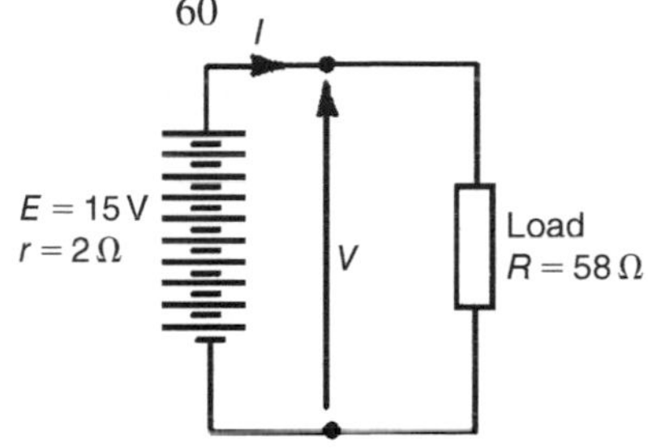

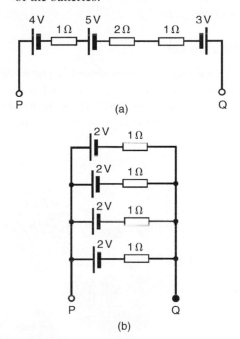

Figure 36.4

(b) P.d. at battery terminals, $V = E - Ir$

i.e. $V = 15 - (0.25)(2) = \mathbf{14.5\,V}$

Now try the following Practice Exercise

Practice Exercise 178   E.m.f. and internal resistance of a cell (Answers on page 546)

1. Twelve cells, each with an internal resistance of 0.24 Ω and an e.m.f. of 1.5 V are connected (a) in series, (b) in parallel. Determine the e.m.f. and internal resistance of the batteries so formed

2. A cell has an internal resistance of 0.03 Ω and an e.m.f. of 2.2 V. Calculate its terminal p.d. if it delivers:

   (a) 1 A   (b) 20 A   (c) 50 A

3. The p.d. at the terminals of a battery is 16 V when no load is connected and 14 V when a load taking 8 A is connected. Determine the internal resistance of the battery

4. A battery of e.m.f. 20 V and internal resistance 0.2 Ω supplies a load taking 10 A. Determine the p.d. at the battery terminals and the resistance of the load

5. Ten 2.2 V cells, each having an internal resistance of 0.1 Ω are connected in series to a load of 21 Ω. Determine (a) the current flowing in the circuit and (b) the p.d. at the battery terminals

6. For the circuits shown in Figure 36.5 the resistors represent the internal resistance of the batteries. Find, in each case:

   (i) the total e.m.f. across PQ

   (ii) the total equivalent internal resistances of the batteries.

Figure 36.5

7. The voltage at the terminals of a battery is 52 V when no load is connected and 48.8 V when a load taking 80 A is connected. Find the internal resistance of the battery. What would be the terminal voltage when a load taking 20 A is connected?

36.6   Primary cells

Primary cells cannot be recharged, that is, the conversion of chemical energy to electrical energy is irreversible and the cell cannot be used once the chemicals are exhausted. Examples of primary cells include the Leclanché cell and the mercury cell.

Leclanché cell

A typical dry **Leclanché*** is shown in Figure 36.6. Such a cell has an e.m.f. of about 1.5 V when new, but this falls rapidly if in continuous use due to polarization. The hydrogen film on the carbon electrode forms faster than can be dissipated by the depolarizer. The Leclanché cell is suitable only for intermittent use, applications including torches, transistor radios, bells, indicator circuits, gas lighters, controlling switch-gear and so on. The cell is the most commonly used of primary cells, is cheap, requires little maintenance and has a shelf life of about two years.

Mercury cell

A typical mercury cell is shown in Figure 36.7. Such a cell has an e.m.f. of about 1.3 V which remains constant for a relatively long time. Its main advantages over the Leclanché cell is its smaller size and its long shelf

*Who was Leclanché? – **Georges Leclanché** (1839–14 September 1882) was the French electrical engineer who invented the Leclanché cell, the forerunner of the modern battery. To find out more go to www.routledge.com/cw/bird

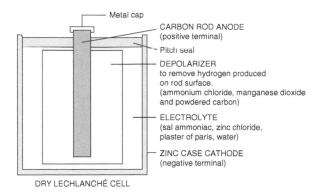

DRY LECHLANCHÉ CELL

Figure 36.6

life. Typical practical applications include hearing aids, medical electronics, cameras and for guided missiles.

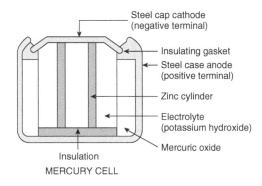

MERCURY CELL

Figure 36.7

36.7 Secondary cells

Secondary cells can be recharged after use, that is, the conversion of chemical energy to electrical energy is reversible and the cell may be used many times. Examples of secondary cells include the lead–acid cell and the nickel cadmium and nickel–metal cells. Practical applications of such cells include car batteries, telephone circuits and for traction purposes – such as milk delivery vans and fork-lift trucks.

Lead–acid cell

A typical lead–acid cell is constructed of:

(i)   A container made of glass, ebonite or plastic.

(ii) **Lead plates**

    (a)   the negative plate (cathode) consists of spongy lead

    (b)   the positive plate (anode) is formed by pressing lead peroxide into the lead grid.

The plates are interleaved as shown in the plan view of Figure 36.8 to increase their effective cross-sectional area and to minimize internal resistance.

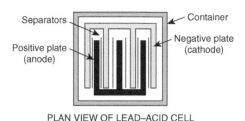

PLAN VIEW OF LEAD–ACID CELL

Figure 36.8

(iii)   **Separators** made of glass, celluloid or wood.

(iv)   An **electrolyte** which is a mixture of sulphuric acid and distilled water.

The relative density (or specific gravity) of a lead–acid cell, which may be measured using a hydrometer, varies between about 1.26 when the cell is fully charged to about 1.19 when discharged. The terminal p.d. of a lead–acid cell is about 2 V.

When a cell supplies current to a load it is said to be **discharging**. During discharge:

(i)   the lead peroxide (positive plate) and the spongy lead (negative plate) are converted into lead sulphate, and

(ii)   the oxygen in the lead peroxide combines with hydrogen in the electrolyte to form water. The electrolyte is therefore weakened and the relative density falls.

The terminal p.d. of a lead–acid cell when fully discharged is about 1.8 V. A cell is **charged** by connecting a d.c. supply to its terminals, the positive terminal of the cell being connected to the positive terminal of the supply. The charging current flows in the reverse direction to the discharge current and the chemical action is reversed. During charging:

(i)   the lead sulphate on the positive and negative plates is converted back to lead peroxide and lead, respectively, and

(ii)   the water content of the electrolyte decreases as the oxygen released from the electrolyte combines with the lead of the positive plate. The relative density of the electrolyte thus increases.

The colour of the positive plate when fully charged is dark brown and when discharged is light brown. The colour of the negative plate when fully charged is grey and when discharged is light grey.

To help maintain lead-acid cells, always store them in a charged condition, never let the open cell voltage drop much below 2.10 V, apply a topping charge every six months or when recommended, avoid repeated deep discharges, charge more often or use a larger battery, prevent sulphation and grid corrosion by choosing the correct charge and float voltages, and avoid operating them at elevated ambient temperatures.

Nickel cadmium and nickel–metal cells

In both types the positive plate is made of nickel hydroxide enclosed in finely perforated steel tubes, the resistance being reduced by the addition of pure nickel or graphite. The tubes are assembled into nickel–steel plates.

In the nickel–metal cell (sometimes called the **Edison* cell** or **nife cell**), the negative plate is made of iron oxide, with the resistance being reduced by a little mercuric oxide, the whole being enclosed in perforated steel tubes and assembled in steel plates.

*Who was Edison? – **Thomas Alva Edison** (11 February 1847–18 October 1931) was an American inventor and businessman. Edison is the fourth most prolific inventor in history, holding well over 1,000 US patents in his name, as well as many patents elsewhere. To find out more go to www.routledge.com/cw/bird

In the nickel cadmium cell the negative plate is made of cadmium. The electrolyte in each type of cell is a solution of potassium hydroxide which does not undergo any chemical change and thus the quantity can be reduced to a minimum. The plates are separated by insulating rods and assembled in steel containers which are then enclosed in a non-metallic crate to insulate the cells from one another. The average discharge p.d. of an alkaline cell is about 1.2 V.

Advantages of a nickel cadmium cell or a nickel–metal cell over a lead–acid cell include:

(i)   more robust construction

(ii) capable of withstanding heavy charging and discharging currents without damage

(iii) has a longer life

(iv) for a given capacity is lighter in weight

(v)   can be left indefinitely in any state of charge or discharge without damage

(vi) is not self-discharging

Disadvantages of nickel cadmium and nickel–metal cells over a lead–acid cell include:

(i)   is relatively more expensive

(ii) requires more cells for a given e.m.f.

(iii) has a higher internal resistance

(iv) must be kept sealed

(v)   has a lower efficiency

Nickel cells may be used in extremes of temperature, in conditions where vibration is experienced or where duties require long idle periods or heavy discharge currents. Practical examples include traction and marine work, lighting in railway carriages, military portable radios and for starting diesel and petrol engines. See also Table 36.2, page 393.

36.8 Lithium-ion batteries

Lithium-ion batteries are incredibly popular and may be found in laptops, hand-held PCs, mobile phones and iPods.

A **lithium-ion battery** (sometimes **Li-ion battery** or **LIB**) is a member of a family of rechargeable battery types in which lithium ions move from the negative electrode to the positive electrode during discharge and back when charging. Li-ion batteries use an intercalated lithium compound as one electrode material, compared to the metallic lithium used in a non-rechargeable lithium battery. The electrolyte, which allows for ionic movement, and the two electrodes are the constituent components of a lithium-ion battery cell.

Lithium-ion batteries are common in **consumer electronics**. They are one of the most popular types of rechargeable batteries for portable electronics, with a high energy density, small memory effect and only a slow loss of charge when not in use. Beyond consumer electronics, LIBs are also growing in popularity for **military**, **battery electric vehicle** and **aerospace applications**. Also, lithium-ion batteries are becoming a common replacement for the lead-acid batteries that have been used historically for **golf carts** and **utility vehicles**. Instead of heavy lead plates and acid electrolyte, the trend is to use lightweight lithium-ion battery packs that can provide the same voltage as lead-acid batteries, so no modification to the vehicle's drive system is required.

Chemistry, performance, cost and safety characteristics vary across LIB types. Handheld electronics mostly use LIBs based on lithium cobalt oxide ($LiCoO_2$), which offers high energy density, but presents safety risks, especially when damaged. Lithium iron phosphate ($LiFePO_4$), lithium manganese oxide (LMnO or LMO) and lithium nickel manganese cobalt oxide ($LiNiMnCoO_2$ or NMC) offer lower energy density, but longer lives and inherent safety. Such batteries are widely used for **electric tools**, **medical equipment** and other roles. NMC in particular is a leading contender for **automotive applications**. Lithium nickel cobalt aluminium oxide ($LiNiCoAlO_2$ or NCA) and lithium titanate ($Li_4Ti_5O_{12}$ or LTO) are specialty designs aimed at particular niche roles. The new lithium sulphur batteries promise the highest performance to weight ratio.

Lithium-ion batteries can be dangerous under some conditions and can pose a safety hazard since they contain, unlike other rechargeable batteries, a flammable electrolyte and are also kept pressurised. Because of this the testing standards for these batteries are more stringent than those for acid-electrolyte batteries, requiring both a broader range of test conditions and additional battery-specific tests. This is in response to reported accidents and failures, and there have been battery-related recalls by some companies.

For many years, nickel-cadmium had been the only suitable battery for **portable equipment** from **wireless communications** to **mobile computing**. Nickel-metal-hydride and lithium-ion emerged in the early 1990s and today, lithium-ion is the fastest growing and most

Table 36.2

Type of battery	Common uses	Hazardous component	Disposal recycling options
Wet cell (i.e. a primary cell that has a liquid electrolyte)			
Lead–acid batteries	Electrical energy supply for vehicles including cars, trucks, boats, tractors and motorcycles. Small sealed lead–acid batteries are used for emergency lighting and uninterruptible power supplies	Sulphuric acid and lead	Recycle – most petrol stations and garages accept old car batteries, and council waste facilities have collection points for lead–acid batteries
Dry cell: Non-chargeable – single use (for example, AA, AAA, C, D, lantern and miniature watch sizes)			
Zinc carbon	Torches, clocks, shavers, radios, toys and smoke alarms	Zinc	Not classed as hazardous waste – can be disposed with household waste
Zinc chloride	Torches, clocks, shavers, radios, toys and smoke alarms	Zinc	Not classed as hazardous waste – can be disposed with household waste
Alkaline manganese	Personal stereos and radio/cassette players	Manganese	Not classed as hazardous waste – can be disposed with household waste
Primary button cells (i.e. a small flat battery shaped like a 'button' used in small electronic devices)			
Mercuric oxide	Hearing aids, pacemakers and cameras	Mercury	Recycle at council waste facility, if available
Zinc–air	Hearing aids, pagers and cameras	Zinc	Recycle at council waste facility, if available
Silver oxide	Calculators, watches and cameras	Silver	Recycle at council waste facility, if available
Lithium	Computers, watches and cameras	Lithium (explosive and flammable)	Recycle at council waste facility, if available
Dry cell rechargeable – secondary batteries			
Nickel cadmium (NiCd)	Mobile phones, cordless power tools, laptop computers, shavers, motorised toys, personal stereos	Cadmium	Recycle at council waste facility, if available
Nickel–metal hydride (NiMH)	Alternative to NiCd batteries, but longer life	Nickel	Recycle at council waste facility, if available
Lithium-ion (Li-ion)	Alternative to NiCd and NiMH batteries, but greater energy storage capacity	Lithium	Recycle at council waste facility, if available

promising battery chemistry. Pioneer work with the lithium battery began in 1912 under G.N. Lewis but it was not until the early 1970s when the first non-rechargeable lithium batteries became commercially available. Lithium is the lightest of all metals, has the greatest electrochemical potential and provides the largest energy density for weight.

Attempts to develop rechargeable lithium batteries initially failed due to safety problems. Because of the inherent instability of lithium metal, especially during charging, research shifted to a non-metallic lithium battery using lithium ions. Although slightly lower in energy density than lithium metal, lithium-ion is safe, provided certain precautions are met when charging and discharging. In 1991, the Sony Corporation commercialized the first lithium-ion battery and other manufacturers followed suit.

The energy density of lithium-ion is typically twice that of the standard nickel-cadmium. There is potential for higher energy densities. The load characteristics are reasonably good and behave similarly to nickel-cadmium in terms of discharge. The high cell voltage of 3.6 volts allows battery pack designs with only one cell. Most of today's mobile phones run on a single cell. A nickel-based pack would require three 1.2-volt cells connected in series.

Lithium-ion is a low maintenance battery, an advantage that most other chemistries cannot claim. There is no memory and no scheduled cycling is required to prolong the battery's life. In addition, the self-discharge is less than half compared to nickel-cadmium, making lithium-ion well suited for modern fuel gauge applications. Lithium-ion cells cause little harm when disposed.

Despite its overall advantages, lithium-ion has its drawbacks. It is fragile and requires a protection circuit to maintain safe operation. Built into each pack, the protection circuit limits the peak voltage of each cell during charge and prevents the cell voltage from dropping too low on discharge. In addition, the cell temperature is monitored to prevent temperature extremes. The maximum charge and discharge current on most packs are limited to between 1C and 2C. With these precautions in place, the possibility of metallic lithium plating occurring due to overcharge is virtually eliminated.

Ageing is a concern with most lithium-ion batteries and many manufacturers remain silent about this issue. Some capacity deterioration is noticeable after one year, whether the battery is in use or not. The battery frequently fails after two or three years. It should be noted that other chemistries also have age-related degenerative effects. This is especially true for nickel-metal-hydride if exposed to high ambient temperatures. At the same time, lithium-ion packs are known to have served for five years in some applications.

Manufacturers are constantly improving lithium-ion. New and enhanced chemical combinations are introduced every six months or so. With such rapid progress, it is difficult to assess how well the revised battery will age.

Storage in a cool place slows the ageing process of lithium-ion (and other chemistries). Manufacturers recommend storage temperatures of 15°C (59°F). In addition, the battery should be partially charged during storage; the manufacturer recommends a 40% charge.

The most economical lithium-ion battery in terms of cost-to-energy ratio is the cylindrical 18650 (size is 18mm x 65.2mm). This cell is used for mobile computing and other applications that do not demand ultra-thin geometry. If a slim pack is required, the prismatic lithium-ion cell is the best choice. These cells come at a higher cost in terms of stored energy.

Summary of advantages of lithium-ion batteries

(i) High energy density – potential for yet higher capacities.

(ii) Does not need prolonged priming when new; one regular charge is all that's needed.

(iii) Relatively low self-discharge – is less than half that of nickel-based batteries.

(iv) Low maintenance – no periodic discharge is needed; there is no memory.

(v) Speciality cells can provide very high current to applications such as power tools.

Summary of limitations of lithium-ion batteries

(i) Requires protection circuit to maintain voltage and current within safe limits.

(ii) Subject to ageing, even if not in use – storage in a cool place at 40% charge reduces the ageing effect.

(iii) Expensive to manufacture – about 40% higher in cost than nickel-cadmium.

(iv) Not fully mature – metals and chemicals are changing on a continuing basis.

To extend the battery life of lithium-ion batteries

(i) **Keep batteries at room temperature**, meaning between 20°C and 25°C; heat is by far the greatest factor in reducing lithium-ion battery life.

(ii) **Obtain a high-capacity lithium-ion battery, rather than carrying a spare**; batteries deteriorate over time, whether they're being used or not so a spare battery won't last much longer than the one in use.

(iii) **Allow partial discharges and avoid full ones**; unlike NiCad batteries, lithium-ion batteries do not have a charge memory, which means that deep-discharge cycles are not required – in fact, it's better for the battery to use partial-discharge cycles.

(iv) **Avoid completely discharging lithium-ion batteries**; if a lithium-ion battery is discharged below 2.5 volts per cell, a safety circuit built into the battery opens and the battery appears to be dead and the original charger will be of no use – only battery analyzers with the boost function have a chance of recharging the battery.

(v) **For extended storage, discharge a lithium-ion battery to about 40% and store it in a cool place**.

Lithium-ion batteries are a huge improvement over previous types of batteries and getting 500 charge/discharge cycles from a lithium-ion battery is becoming commonplace.

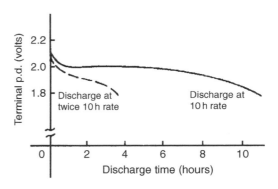

Figure 36.9

36.9   Cell capacity

The **capacity** of a cell is measured in ampere-hours (Ah). A fully charged 50 Ah battery rated for 10 h discharge can be discharged at a steady current of 5 A for 10 h, but if the load current is increased to 10 A then the battery is discharged in 3–4 h, since the higher the discharge current, the lower is the effective capacity of the battery. Typical discharge characteristics for a lead–acid cell are shown in Figure 36.9.

36.10   Safe disposal of batteries

Battery disposal has become a topical subject in the UK because of greater awareness of the dangers and implications of depositing up to 300 million batteries per annum – a waste stream of over 20 000 tonnes – into landfill sites.

Certain batteries contain substances which can be a hazard to humans, wildlife and the environment, as well as posing a fire risk. Other batteries can be recycled for their metal content.

Waste batteries are a concentrated source of toxic heavy metals such as mercury, lead and cadmium. If batteries containing heavy metals are disposed of incorrectly, the metals can leach out and pollute the soil and groundwater, endangering humans and wildlife. Long-term exposure to cadmium, a known human carcinogen (i.e. a substance producing cancerous growth), can cause liver and lung disease. Mercury can cause damage to the human brain, spinal system, kidneys and liver. Sulphuric acid in lead–acid batteries can cause severe skin burns or irritation upon contact. It is increasingly important to correctly dispose of all types of batteries.

Table 36.2 lists types of batteries, their common uses, their hazardous components and disposal recycling options.

Battery disposal has become more regulated since the Landfill Regulations 2002 and Hazardous Waste Regulations 2005. From the Waste Electrical and Electronic Equipment (WEEE) Regulations 2006, commencing July 2007 all producers (manufacturers and importers) of electrical and electronic equipment have been responsible for the cost of collection, treatment and recycling of obligated WEEE generated in the UK.

36.11   Fuel cells

A **fuel cell** is an electrochemical energy conversion device, similar to a battery, but differing from the latter in that it is designed for continuous replenishment of the reactants consumed, i.e. it produces electricity from an external source of fuel and oxygen, as opposed to the limited energy storage capacity of a battery. Also, the electrodes within a battery react and change as a battery is charged or discharged, whereas a fuel cell's electrodes are catalytic (i.e. not permanently changed) and relatively stable.

Typical reactants used in a fuel cell are hydrogen on the anode side and oxygen on the cathode side (i.e. a **hydrogen cell**). Usually, reactants flow in and reaction products flow out. Virtually continuous long-term operation is feasible as long as these flows are maintained.

Fuel cells are very attractive in modern applications for their high efficiency and ideally emission-free use, in contrast to currently more modern fuels such as methane or natural gas that generate carbon dioxide. The only by-product of a fuel cell operating on pure hydrogen is water vapour.

Currently, fuel cells are a very expensive alternative to internal combustion engines. However, continued research and development is likely to make fuel cell vehicles available at market prices within a few years.

Fuel cells are very useful as power sources in remote locations, such as spacecraft, remote weather stations, and in certain military applications. A fuel cell running on hydrogen can be compact, lightweight and has no moving parts.

36.12 Alternative and renewable energy sources

Alternative energy refers to energy sources which could replace coal, traditional gas and oil, all of which increase the atmospheric carbon when burned as fuel.

Renewable energy implies that it is derived from a source which is automatically replenished or one that is effectively infinite so that it is not depleted as it is used. Coal, gas and oil are not renewable because, although the fields may last for generations, their time span is finite and will eventually run out.

There are many means of harnessing energy which have less damaging impacts on our environment and include the following:

1. **Solar energy** is one of the most resourceful sources of energy for the future. The reason for this is that the total energy received each year from the sun is around 35 000 times the total energy used by man. However, about one-third of this energy is either absorbed by the outer atmosphere or reflected back into space. Solar energy could be used to run cars, power plants and space ships. **Solar panels** on roofs capture heat in water storage systems. **Photovoltaic cells**, when suitably positioned, convert sunlight to electricity.

   For more on solar energy, see Section 36.13 following.

2. **Wind power** is another alternative energy source that can be used without producing by-products that are harmful to nature. The fins of a windmill rotate in a vertical plane which is kept vertical to the wind by means of a tail fin and as wind flow crosses the blades of the windmill it is forced to rotate and can be used to generate electricity (see chapter 11). Like solar power, harnessing the wind is highly dependent upon weather and location. The average wind velocity of Earth is around 9 m/s, and the power that could be produced when a windmill is facing a wind of 10 m.p.h. (i.e. around 4.5 m/s) is around 50 watts.

3. **Hydroelectricity** is achieved by the damming of rivers and utilizing the potential energy in the water. As the water stored behind a dam is released at high pressure, its kinetic energy is transferred onto turbine blades and used to generate electricity. The system has enormous initial costs but has relatively low maintenance costs and provides power quite cheaply.

4. **Tidal power** utilizes the natural motion of the tides to fill reservoirs which are then slowly discharged through electricity-producing turbines.

5. **Geothermal energy** is obtained from the internal heat of the planet and can be used to generate steam to run a steam turbine which, in turn, generates electricity. The radius of the Earth is about 4000 miles with an internal core temperature of around 4000°C at the centre. Drilling three miles from the surface of the Earth, a temperature of 100°C is encountered; this is sufficient to boil water to run a steam-powered electric power plant. Although drilling three miles down is possible, it is not easy. Fortunately, however, volcanic features called **geothermal hotspots** are found all around the world. These are areas which transmit excess internal heat from the interior of the Earth to the outer crust, which can be used to generate electricity.

   Advantages of geothermal power include being environmentally friendly, global warming effects are mitigated, has no fuel costs, gives predictable 24/7 power, has a high load factor and no pollution.

   Disadvantages of geothermal power include having a long gestation time leading to possible cost overruns, slow technology improvement, problems with financing, innumerable regulations, and having limited locations.

6. **Biomass** is fuel that is developed from organic materials, a renewable and sustainable source of energy used to create electricity or other forms of power. Some examples of materials that make up biomass fuels are scrap lumber, forest debris, certain crops, manure and some types of waste residues. With a constant supply of waste – from construction and demolition activities, to wood not used in papermaking, to municipal solid waste – green energy production can continue indefinitely. There are several methods to convert biomass into electricity and these are discussed in chapter 44.

36.13 Solar energy

Solar energy – power from the sun – is a vast and inexhaustible resource. Once a system is in place to convert it into useful energy, the fuel is free and will never be subject to the variations of energy markets. Furthermore, it represents a clean alternative to the fossil fuels that currently pollute air and water, threaten public health and contribute to global warming. Given the abundance and the appeal of solar energy, this resource will play a prominent role in our energy future.

In the broadest sense, solar energy supports all life on Earth and is the basis for almost every form of energy used. The sun makes plants grow, which can be burned as 'biomass' fuel or, if left to rot in swamps and compressed underground for millions of years, in the form of coal and oil. Heat from the sun causes temperature differences between areas, producing wind that can power turbines. Water evaporates because of the sun, falls on high elevations and rushes down to the sea, spinning hydroelectric turbines as it passes. However, solar energy usually refers to ways the sun's energy can be used to directly generate heat, lighting and electricity.

The amount of energy from the sun that falls on the Earth's surface is enormous. All the energy stored in Earth's reserves of coal, oil and natural gas is matched by the energy from just 20 days of sunshine.

Originally developed for energy requirements for an orbiting earth satellite, solar power has expanded in recent years for use in domestic and industrial needs. Solar power is produced by collecting sunlight and converting it into electricity. This is achieved by using solar panels, which are large flat panels made up of many individual solar cells.

Some advantages of solar power

(a) The major advantage of solar power is that no pollution is created in the process of generating electricity. Environmentally it is the most clean and green energy. It is renewable (unlike gas, oil and coal) and sustainable, helping to protect the environment.

(b) Solar energy does not require any fuel.

(c) Solar energy does not pollute air by releasing carbon dioxide, nitrogen oxide, sulphur dioxide or mercury into the atmosphere like many traditional forms of electrical generation do.

(d) Solar energy does not contribute to global warming, acid rain or smog. It actively contributes to the decrease of harmful green house gas emissions.

(e) There is no on-going cost for the power solar energy generates, as solar radiation is free everywhere; once installed, there are no recurring costs.

(f) Solar energy can be flexibly applied to a variety of stationary or portable applications. Unlike most forms of electrical generation, the panels can be made small enough to fit pocket-size electronic devices, such as a calculator, or sufficiently large to charge an automobile battery or supply electricity to entire buildings.

(g) Solar energy offers much more self-reliance than depending upon a power utility for all electricity.

(h) Solar energy is quite economical in the long run. After the initial investment has been recovered, the energy from the sun is practically free. Solar energy systems are virtually maintenance free and will last for decades.

(i) Solar energy is unaffected by the supply and demand of fuel and is therefore not subjected to the ever-increasing price of fossil fuel.

(j) By not using any fuel, solar energy does not contribute to the cost and problems of the recovery and transportation of fuel or the storage of radioactive waste.

(k) Solar energy is generated where it is needed, hence large scale transmission cost is minimized.

(l) Solar energy can be utilized to offset utility-supplied energy consumption. It does not only reduce electricity bills, but will also continue to supply homes/businesses with electricity in the event of a power outage.

(m) A solar energy system can operate entirely independently, not requiring a connection to a power or gas grid at all. Systems can therefore be installed in remote locations, making it more practical and cost-effective than the supply of utility electricity to a new site.

(n) Solar energy projects operate silently, have no moving parts, do not release offensive smells and do not require the addition of any additional fuel.

(o) Solar energy projects support local job and wealth creation, fuelling local economies.

Some disadvantages of solar power

(a) The initial cost is the main disadvantage of installing a solar energy system, largely because of the high cost of the semi-conducting materials used in building solar panels.

(b) The cost of solar energy is also high compared to non-renewable utility-supplied electricity. As energy shortages are becoming more common, solar energy is becoming more price-competitive.

(c) Solar panels require quite a large area for installation to achieve a good level of efficiency.

(d) The efficiency of the system also relies on the location of the sun, although this problem can be overcome with the installation of certain components.

(e) The production of solar energy is influenced by the presence of clouds or pollution in the air. Similarly, no solar energy will be produced during the night although a battery backup system and/or net metering will solve this problem.

Some applications of solar energy

(i) **Concentrating Solar Power (CSP):** Concentrating solar power (CSP) plants are utility-scale generators that produce electricity using mirrors or lenses to efficiently concentrate the sun's energy. The four principal CSP technologies are parabolic troughs, dish-Stirling engine systems, central receivers and concentrating photovoltaic systems (CPV).

(ii) **Solar Thermal Electric Power Plants:** Solar thermal energy involves harnessing solar power for practical applications from solar heating to electrical power generation. Solar thermal collectors, such as solar hot water panels, are commonly used to generate **solar hot water for domestic and light industrial applications**. This energy system is also used in architecture and building design to **control heating and ventilation** in both active solar and passive solar designs.

(iii) **Photovoltaics:** Photovoltaic or PV technology employs solar cells or solar photovoltaic arrays to convert energy from the sun into electricity. Solar cells produce direct current electricity from the sun's rays, which can be used to power equipment or to recharge batteries. Many **pocket calculators** incorporate a single solar cell, but for larger applications, cells are generally grouped together to form PV modules that are in turn arranged in solar arrays. Solar arrays can be used to power **orbiting satellites** and other **spacecraft**, and in remote areas as a source of power for **roadside emergency telephones**, **remote sensing**, **school crossing warning signs** and **cathodic protection of pipelines**.

(iv) **Solar Heating Systems:** Solar hot water systems use sunlight to heat water. The systems are composed of solar thermal collectors and a storage tank, and they may be active, passive or batch systems.

(v) **Passive Solar Energy:** Building designers are concerned to maintain the building environment at a comfortable temperature through the sun's daily and annual cycles. This can be achieved by: (a) **direct gain** in which the positioning of windows, skylights and shutters to control the amount of direct solar radiation reaching the interior and warming the air and surfaces within a building; (b) **indirect gain** in which solar radiation is captured by a part of the building envelope and then transmitted indirectly to the building through conduction and convection; and (c) **isolated gain** which involves passively capturing solar heat and then moving it passively into or out of the building via a liquid or air directly or using a thermal store. Sunspaces, greenhouses and solar closets are alternative ways of capturing isolated heat gain from which warmed air can be taken.

(vi) **Solar Lighting:** Also known as day-lighting, solar lighting is the use of natural light to provide illumination to offset energy use in electric lighting systems and reduce the cooling load on high voltage a.c. systems. Day-lighting features include building orientation, window orientation, exterior shading, saw-tooth roofs, clerestory windows, light shelves, skylights and light tubes.

Architectural trends increasingly recognise day-lighting as a cornerstone of sustainable design.

(vii) **Solar Cars:** A solar car is an electric vehicle powered by energy obtained from solar panels on the surface of the car which convert the sun's energy directly into electrical energy. Solar cars are not currently a practical form of transportation. Although they can operate for limited distances without sun, the solar cells are generally very fragile. Development teams have focused their efforts on optimising the efficiency of the vehicle, but many have only enough room for one or two people.

(viii) **Solar Power Satellite:** A solar power satellite (SPS) is a proposed satellite built in high Earth orbit that uses microwave power transmission to beam solar power to a very large antenna on Earth where it can be used in place of conventional power sources. The advantage of placing the solar collectors in space is the unobstructed view of the sun, unaffected by the day/night cycle, weather or seasons. However, the costs of construction are very high, and SPSs will not be able to compete with conventional sources unless low launch costs can be achieved or unless a space-based manufacturing industry develops and they can be built in orbit from off-earth materials.

(ix) **Solar Updraft Tower:** A solar updraft tower is a proposed type of renewable-energy power plant. Air is heated in a very large circular greenhouse-like structure, and the resulting convection causes the air to rise and escape through a tall tower. The moving air drives turbines, which produce electricity. There are no solar updraft towers in operation at present. A research prototype operated in Spain in the 1980s, and EnviroMission is proposing to construct a full-scale power station using this technology in Australia.

(x) **Renewable Solar Power Systems with Regenerative Fuel Cell Systems:** NASA in the USA has long recognized the unique advantages of regenerative fuel cell (RFC) systems to provide energy storage for solar power systems in space. RFC systems are uniquely qualified to provide the necessary energy storage for solar surface power systems on the moon or Mars during long periods of darkness, i.e. during the 14-day lunar night or the 12-hour Martian night. The nature of the RFC and its inherent design flexibility enables it to effectively meet the requirements of space missions. In the course of implementing the NASA RFC Programme, researchers recognized that there are numerous applications in government, industry, transportation and the military for RFC systems as well.

36.14 Glass batteries

Due to the advancements in technology, the electric car industry is witnessing exponential growth. The growth in this industry has led to higher adoption of **glass batteries**. A glass battery is a type of solid-state battery which essentially uses a glass electrolyte and sodium or lithium metal electrodes. Solid-state batteries are much more effective than conventional ones and are particularly long-lasting and lightweight as they use electrolyte and electrode materials. As the battery electrolyte, the glass battery technology generally uses a form of glass which is fixed with reactive alkali metals, such as sodium or lithium. Glass electrolyte is made of a specific mixture of sodium or lithium along with oxygen, chlorine and barium. When charged quickly, the battery avoids the formation of needle-like dendrites (a crystal with a branching treelike structure) on the anode. The battery can also be made using low-cost sodium instead of lithium. Glass batteries have a much shorter charging time than Li-ion batteries and have many benefits. For instance, these kinds of batteries experience safer, faster-charging, as well as retaining long-lasting charge.

Glass batteries are a low-cost battery, they are non-combustible and have long battery life with high volumetric energy density along with fast rates of charge and discharge. As compared to lithium-ion battery, sodium and lithium glass batteries have three times the energy storage capacity.

Increasing demand for batteries that have high energy density and longer cycles as well as fast charging capacity is anticipated to be the major factor expected to drive the growth of the glass battery market in the future. Moreover, growing demand for long-life and safe-design batteries is also expected to add to the growth of the glass battery market in the near future. Also, in coming years, glass batteries will be **used to store energy produced by renewable sources**. This energy could then be used to provide power for homes or to run electric vehicles. This will further act as a driving factor for the growth of the glass battery market. As rechargeable batteries are the most appropriate way of storing electric power, the glass battery technology can be also used to store intermittent solar and wind power on the electric grid.

Quick shut down when discharged is anticipated to be a major factor expected to maybe hamper the growth of the glass battery market in the near future. Moreover, overheating and corrosion decreases battery life, and this will be another restraining factor for the growth of glass battery market in the near future.

Now try the following Practice Exercise

Practice Exercise 179 Short answer questions on batteries and alternative sources of energy

1. Define a battery.

2. State five practical applications of batteries.

3. State advantages of lithium-ion batteries over alkaline batteries.

4. What is electrolysis?

5. What is an electrolyte?

6. Conduction in electrolytes is due to . . . . . .

7. A positive-connected electrode is called the . . . . . . and the negative-connected electrode the . . . . . .

8. State two practical applications of electrolysis.

9. The purpose of an electric cell is to convert . . . . . . to . . . . . .

10. Make a labelled sketch of a simple cell.

11. What is the electrochemical series?

12. With reference to a simple cell, explain briefly what is meant by
(a) polarization, (b) local action.

13. What is corrosion? Name two effects of corrosion and state how they may be prevented.

14. What is meant by the e.m.f. of a cell? How may the e.m.f. of a cell be measured?

15. Define internal resistance.

16. If a cell has an e.m.f. of E volts, an internal resistance of r ohms and supplies a current I amperes to a load, the terminal p.d. V volts is given by: $V = $ . . . . . .

17. Name the two main types of cells.

18. Explain briefly the difference between primary and secondary cells.

19. Name two types of primary cells.

20. Name two types of secondary cells.

21. State three typical applications of primary cells.

22. State three typical applications of secondary cells.

23. State four advantages of lithium-ion batteries.

24. State three disadvantages of lithium-ion batteries.

25. State four ways of extending the battery life of lithium-ion batteries.

26. In what unit is the capacity of a cell measured?

27. Why is safe disposal of batteries important?

28. Name any six types of battery and state three common applications for each.

29. What is a 'fuel cell'? How does it differ from a battery?

30. State the advantages of fuel cells.

31. State three practical applications of fuel cells.

32. What is meant by (a) alternative energy, (b) renewable energy.

33. State five alternative energy sources and briefly describe each.

34. State five advantages of solar power.

35. State three disadvantages of solar power.

36. State and briefly describe seven practical applications of solar power.

37. What is a glass battery?

38. State four potential advantages of a glass battery compared with a conventional battery.

39. State what may hamper growth of glass batteries in the near future.

Practice Exercise 180 Multi-choice questions on batteries and alternative sources of energy (Answers on page 546)

1. A battery consists of:

    (a) a cell   (b) a circuit

    (c) a generator   (d) a number of cells

2. The terminal p.d. of a cell of e.m.f. 2 V and internal resistance 0.1 Ω when supplying a current of 5 A will be:

    (a) 1.5 V   (b) 2 V

    (c) 1.9 V   (d) 2.5 V

3. Five cells, each with an e.m.f. of 2 V and internal resistance 0.5 Ω are connected in series. The resulting battery will have:

    (a) an e.m.f. of 2 V and an internal resistance of 0.5 Ω

    (b) an e.m.f. of 10 V and an internal resistance of 2.5 Ω

    (c) an e.m.f. of 2 V and an internal resistance of 0.1 Ω

    (d) an e.m.f. of 10 V and an internal resistance of 0.1 Ω

4. If the five cells of question 3 are connected in parallel, the resulting battery will have:

    (a) an e.m.f. of 2 V and an internal resistance of 0.5 Ω

    (b) an e.m.f. of 10 V and an internal resistance of 2.5 Ω

    (c) an e.m.f. of 2 V and an internal resistance of 0.1 Ω

    (d) an e.m.f. of 10 V and an internal resistance of 0.1 Ω

5. Which of the following statements is false?

    (a) A Leclanché cell is suitable for use in torches

    (b) A nickel cadmium cell is an example of a primary cell

    (c) When a cell is being charged its terminal p.d. exceeds the cell e.m.f.

    (d) A secondary cell may be recharged after use

6. Which of the following statements is false? When two metal electrodes are used in a simple cell, the one that is higher in the electrochemical series:

    (a) tends to dissolve in the electrolyte

    (b) is always the negative electrode

    (c) reacts most readily with oxygen

    (d) acts as an anode

7. Five 2 V cells, each having an internal resistance of 0.2 Ω, are connected in series to a load of resistance 14 Ω. The current flowing in the circuit is:

    (a) 10 A   (b) 1.4 A

    (c) 1.5 A   (d) $\frac{2}{3}$ A

8. For the circuit of question 7, the p.d. at the battery terminals is:

    (a) 10 V   (b) $9\frac{1}{3}$ V

    (c) 0 V   (d) $10\frac{2}{3}$ V

9. Which of the following statements is true?

    (a) The capacity of a cell is measured in volts

    (b) A primary cell converts electrical energy into chemical energy

    (c) Galvanizing iron helps to prevent corrosion

    (d) A positive electrode is termed the cathode

10. The greater the internal resistance of a cell:

    (a) the greater the terminal p.d.

    (b) the less the e.m.f.

    (c) the greater the e.m.f.

    (d) the less the terminal p.d.

11. The negative pole of a dry cell is made of:

    (a) carbon

    (b) copper

    (c) zinc

    (d) mercury

12. The energy of a secondary cell is usually renewed:

    (a) by passing a current through it

    (b) it cannot be renewed at all

    (c) by renewing its chemicals

    (d) by heating it

13. Which of the following statements is true?

    (a) A zinc carbon battery is rechargeable and is not classified as hazardous

    (b) A nickel cadmium battery is not rechargeable and is classified as hazardous

    (c) A lithium battery is used in watches and is not rechargeable

    (d) An alkaline manganese battery is used in torches and is classified as hazardous

**For fully worked solutions to each of the problems in Exercises 166 to 168 in this chapter,
go to the website:**
www.routledge.com/cw/bird

Chapter 37

Series and parallel networks

Why it is important to understand: Series and parallel networks

There are two ways in which components may be connected together in an electric circuit. One way is in series' where components are connected end-to-end'; another way is in parallel' where components are connected across each other'. When a circuit is more complicated than two or three elements, it is very likely to be a network of individual series and parallel circuits. A firm understanding of the basic principles associated with series and parallel circuits is a sufficient background to begin an investigation of any single-source d.c. network having a combination of series and parallel elements or branches. Confidence in the analysis of series-parallel networks comes only through exposure, practice, and experience. At first glance, these circuits may seem complicated, but with methodical analysis the functionality of the circuit can become obvious. This chapter explains, with examples, series, parallel and series/parallel networks The relationships between voltages, currents and resistances for these networks are considered through calculations.

At the end of this chapter, you should be able to:

- calculate unknown voltages, current and resistances in a series circuit
- understand voltage division in a series circuit
- calculate unknown voltages, currents and resistances in a parallel network
- calculate unknown voltages, currents and resistances in series-parallel networks
- understand current division in a two-branch parallel network
- describe the advantages and disadvantages of series and parallel connection of lamps

37.1 Introduction

It is important to be able to analyse an electrical circuit, i.e. to calculate the currents and potential differences (p.d.s) within a circuit. There are two ways of connecting circuits. One way is 'in series', where the components are connected end to end, and the other way is 'in parallel', where the components are connected across each other. Often a circuit is a mixture of both series and parallel connections. This chapter shows how currents and voltages are calculated in series, parallel and

Science and Mathematics for Engineering. 978-0-367-20475-4, © John Bird. Published by Taylor & Francis. All rights reserved.

series-parallel circuits, including using voltage and current division. Advantages and disadvantages of series and parallel connection of lamps is also described.

37.2 Series circuits

Figure 37.1 shows three resistors R_1, R_2 and R_3 connected end to end, i.e. in series, with a battery source of V volts. Since the circuit is closed a current I will flow and the p.d. across each resistor may be determined from the voltmeter readings V_1, V_2 and V_3

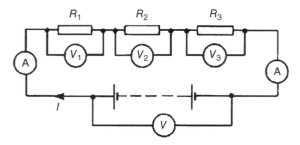

Figure 37.1

In a series circuit

(a) the current I is the same in all parts of the circuit and hence the same reading is found on each of the ammeters shown, and

(b) the sum of the voltages V_1, V_2 and V_3 is equal to the total applied voltage V

i.e. $V = V_1 + V_2 + V_3$

From Ohm's law: $V_1 = IR_1$, $V_2 = IR_2$, $V_3 = IR_3$ and $V = IR$ where R is the total circuit resistance. Since $V = V_1 + V_2 + V_3$ then $IR = IR_1 + IR_2 + IR_3$. Dividing throughout by I gives

$$R = R_1 + R_2 + R_3$$

Thus for a series circuit, the total resistance is obtained by adding together the values of the separate resistances.

Problem 1. For the circuit shown in Figure 37.2, determine (a) the battery voltage V, (b) the total resistance of the circuit, and (c) the values of resistors R_1, R_2 and R_3, given that the p.d.s across R_1, R_2 and R_3 are 5 V, 2 V and 6 V, respectively

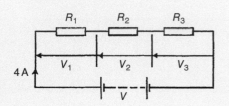

Figure 37.2

(a) Battery voltage

$$V = V_1 + V_2 + V_3 = 5 + 2 + 6 = \mathbf{13\,V}$$

(b) Total circuit resistance

$$R = \frac{V}{I} = \frac{13}{4} = \mathbf{3.25\,\Omega}$$

(c) Resistance $R_1 = \dfrac{V_1}{I} = \dfrac{5}{4} = \mathbf{1.25\,\Omega}$,

resistance $R_2 = \dfrac{V_2}{I} = \dfrac{2}{4} = \mathbf{0.5\,\Omega}$ and

resistance $R_3 = \dfrac{V_3}{I} = \dfrac{6}{4} = \mathbf{1.5\,\Omega}$

(Check: $R_1 + R_2 + R_3 = 1.25 + 0.5 + 1.5$
$= 3.25\,\Omega = R$)

Problem 2. For the circuit shown in Figure 37.3, determine the p.d. across resistor R_3. If the total resistance of the circuit is 100 Ω, determine the current flowing through resistor R_1. Find also the value of resistor R_2

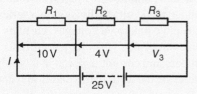

Figure 37.3

P.d. across R_3, $V_3 = 25 - 10 - 4 = \mathbf{11\,V}$

$$\text{Current } I = \frac{V}{R} = \frac{25}{100} = \mathbf{0.25\,A}$$

which is the current flowing in each resistor.

$$\text{Resistance } R_2 = \frac{V_2}{I} = \frac{4}{0.25} = \mathbf{16\,\Omega}$$

Problem 3. A 12 V battery is connected in a circuit having three series-connected resistors having resistances of 4 Ω, 9 Ω and 11 Ω. Determine the current flowing through, and the p.d. across the 9 Ω resistor. Find also the power dissipated in the 11 Ω resistor

The circuit diagram is shown in Figure 37.4.

Total resistance $R = 4 + 9 + 11 = 24\ \Omega$.

$$\text{Current } I = \frac{V}{R} = \frac{12}{24} = \textbf{0.5 A},$$

which is the current in the 9 Ω resistor.

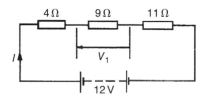

Figure 37.4

P.d. across the 9 Ω resistor,

$$V_1 = I \times 9 = 0.5 \times 9 = \textbf{4.5 V}$$

Power dissipated in the 11 Ω resistor,

$$P - I^2R - (0.5)^2(11) = (0.25)(11) = \textbf{2.75 W}$$

37.3 Potential divider

The voltage distribution for the circuit shown in Figure 37.5(a) is given by:

$$V_1 = \left(\frac{R_1}{R_1 + R_2}\right) V \text{ and } V_2 = \left(\frac{R_2}{R_1 + R_2}\right) V$$

The circuit shown in Figure 37.5(b) is often referred to as a **potential divider** circuit. Such a circuit can consist of a number of similar elements in series connected across a voltage source, voltages being taken from connections between the elements. Frequently the divider consists of two resistors as shown in Figure 37.5(b), where

$$V_{\textbf{OUT}} = \left(\frac{R_2}{R_1 + R_2}\right) V_{\textbf{IN}}$$

A potential divider is the simplest way of producing a source of lower e.m.f. from a source of higher e.m.f. and is the basic operating mechanism of the **potentiometer**,

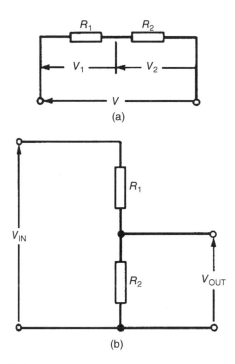

Figure 37.5

a measuring device for accurately measuring potential differences (see page 491).

Problem 4. Determine the value of voltage V shown in Figure 37.6

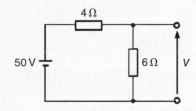

Figure 37.6

Figure 37.6 may be redrawn as shown in Figure 37.7, and voltage

$$V = \left(\frac{6}{6 + 4}\right)(50) = \textbf{30 V}$$

Problem 5. Two resistors are connected in series across a 24 V supply and a current of 3 A flows in the circuit. If one of the resistors has a resistance of 2 Ω determine (a) the value of the other resistor, and (b) the p.d. across the 2 Ω resistor. If the circuit is connected for 50 hours, how much energy is used?

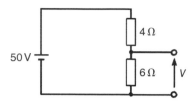

Figure 37.7

The circuit diagram is shown in Figure 37.8.

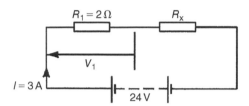

Figure 37.8

(a) Total circuit resistance

$$R = \frac{V}{I} = \frac{24}{3} = 8\,\Omega$$

Value of unknown resistance

$$R_x = 8 - 2 = \mathbf{6\,\Omega}$$

(b) P.d. across $2\,\Omega$ resistor

$$V_1 = IR_1 = 3 \times 2 = \mathbf{6\,V}$$

Alternatively, from above,

$$V_1 = \left(\frac{R_1}{R_1 + R_X}\right) V$$

$$= \left(\frac{2}{2 + 6}\right)(24) = \mathbf{6\,V}$$

Energy used = power × time

$$= (V \times I) \times t$$

$$= (24 \times 3\,\text{W})(50\,\text{h})$$

$$= 3600\,\text{Wh} = \mathbf{3.6\,kWh}$$

Now try the following Practice Exercise

Practice Exercise 181   Further problems on series circuits (answers on page 496)

1. The p.d.s measured across three resistors connected in series are 5 V, 7 V and 10 V, and the supply current is 2 A. Determine (a) the supply voltage, (b) the total circuit resistance and (c) the values of the three resistors

2. For the circuit shown in Figure 37.9, determine the value of V_1. If the total circuit resistance is $36\,\Omega$ determine the supply current and the value of resistors R_1, R_2 and R_3

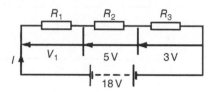

Figure 37.9

3. When the switch in the circuit in Figure 37.10 is closed the reading on voltmeter 1 is 30 V and that on voltmeter 2 is 10 V. Determine the reading on the ammeter and the value of resistor R_x

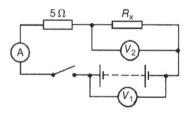

Figure 37.10

4. Calculate the value of voltage V in Figure 37.11

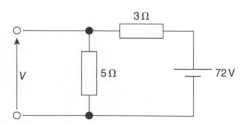

Figure 37.11

5. Two resistors are connected in series across an 18 V supply and a current of 5 A flows. If one of the resistors has a value of 2.4 Ω determine (a) the value of the other resistor and (b) the p.d. across the 2.4 Ω resistor

6. An arc lamp takes 9.6 A at 55 V. It is operated from a 120 V supply. Find the value of the stabilising resistor to be connected in series

7. An oven takes 15 A at 240 V. It is required to reduce the current to 12 A. Find (a) the resistor which must be connected in series and (b) the voltage across the resistor

37.4 Parallel networks

Figure 37.12 shows three resistors R_1, R_2 and R_3 connected across each other, i.e. in parallel, across a battery source of V volts.

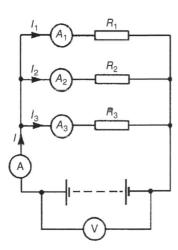

Figure 37.12

In a parallel circuit:

(a) the sum of the currents I_1, I_2 and I_3 is equal to the total circuit current I

 i.e. $I = I_1 + I_2 + I_3$ and

(b) the source p.d., V volts, is the same across each of the resistors.

From Ohm's law: $I_1 = \dfrac{V}{R_1}, I_2 = \dfrac{V}{R_2}, I_3 = \dfrac{V}{R_3}$ and $I = \dfrac{V}{R}$

where R is the total circuit resistance.

Since $I = I_1 + I_2 + I_3$ then $\dfrac{V}{R} = \dfrac{V}{R_1} + \dfrac{V}{R_2} + \dfrac{V}{R_3}$

Dividing throughout by V gives:

$$\frac{1}{R} = \frac{1}{R_1} + \frac{1}{R_2} + \frac{1}{R_3}$$

This equation must be used when finding the total resistance R of a parallel circuit.
For the special case of **two resistors in parallel**

$$\frac{1}{R} = \frac{1}{R_1} + \frac{1}{R_2} = \frac{R_2 + R_1}{R_1 R_2}$$

Hence, $$R = \frac{R_1 R_2}{R_1 + R_2} \qquad \left(\text{i.e. } \frac{\text{product}}{\text{sum}}\right)$$

Problem 6. For the circuit shown in Figure 37.13, determine (a) the reading on the ammeter, and (b) the value of resistor R_2

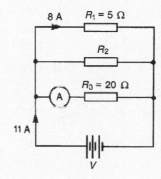

Figure 37.13

P.d. across R_1 is the same as the supply voltage V.

Hence supply voltage $V = 8 \times 5 = 40\,\text{V}$

(a) **Reading on ammeter**

$$I = \frac{V}{R_3} = \frac{40}{20} = 2\,\text{A}$$

(b) Current flowing through R_2

$$= 11 - 8 - 2 = 1\,\text{A}$$

Hence, $R_2 = \dfrac{V}{I_2} = \dfrac{40}{1} = \mathbf{40\,\Omega}$

Problem 7. Two resistors, of resistance 3 Ω and 6 Ω, are connected in parallel across a battery having a voltage of 12 V. Determine (a) the total circuit resistance and (b) the current flowing in the 3 Ω resistor

The circuit diagram is shown in Figure 37.14.

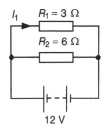

Figure 37.14

(a) The total circuit resistance R is given by

$$\frac{1}{R} = \frac{1}{R_1} + \frac{1}{R_2}$$

$$= \frac{1}{3} + \frac{1}{6} = \frac{2+1}{6} = \frac{3}{6}$$

Since $\frac{1}{R} = \frac{3}{6}$ then **total circuit resistance,**

$R = 2\,\Omega$

(Alternatively, $R = \dfrac{R_1 R_2}{R_1 + R_2}$

$$= \frac{3 \times 6}{3 + 6} = \frac{18}{9} = 2\,\Omega)$$

(b) Current in the 3 Ω resistance

$$I_1 = \frac{V}{R_1} = \frac{12}{3} = 4\,A$$

Problem 8. For the circuit shown in Figure 37.15, find (a) the value of the supply voltage V and (b) the value of current I

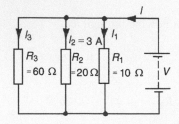

Figure 37.15

(a) P.d. across 20 Ω resistor $= I_2 R_2 = 3 \times 20 = 60\,V$. Hence, **supply voltage $V = 60\,V$** since the circuit is connected in parallel.

(b) Current $\quad I_1 = \dfrac{V}{R_1} = \dfrac{60}{10} = 6\,A,$

$I_2 = 3\,A$ and

$$I_3 = \frac{V}{R_3} = \frac{60}{60} = 1\,A.$$

Current $\quad I = I_1 + I_2 + I_3$

hence **current $I = 6 + 3 + 1 = 10\,A$**

Alternatively,

$$\frac{1}{R} = \frac{1}{60} + \frac{1}{20} + \frac{1}{10} = \frac{1+3+6}{60} = \frac{10}{60}$$

Hence, total resistance

$$R = \frac{60}{10} = 6\,\Omega$$

and **current**

$$I = \frac{V}{R} = \frac{60}{6} = 10\,A$$

Problem 9. Given four 1 Ω resistors, state how they must be connected to give an overall resistance of (a) $\frac{1}{4}$ Ω, (b) 1 Ω, (c) $1\frac{1}{3}$ Ω, (d) $2\frac{1}{2}$ Ω, all four resistors being connected in each case

(a) **All four in parallel** (see Figure 37.16),

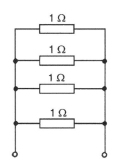

Figure 37.16

Since $\quad \dfrac{1}{R} = \dfrac{1}{1} + \dfrac{1}{1} + \dfrac{1}{1} + \dfrac{1}{1} = \dfrac{4}{1}$

i.e. $\quad R = \dfrac{1}{4}\,\Omega.$

(b) **Two in series, in parallel with another two in series** (see Figure 37.17), since 1 Ω and 1 Ω in series gives 2 Ω, and 2 Ω in parallel with 2 Ω gives

$$\frac{2 \times 2}{2 + 2} = \frac{4}{4} = 1\,\Omega$$

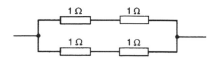

Figure 37.17

(c) **Three in parallel, in series with one** (see Figure 37.18), since for the three in parallel,

$$\frac{1}{R} = \frac{1}{1} + \frac{1}{1} + \frac{1}{1} = \frac{3}{1}$$

i.e. $R = \frac{1}{3}\,\Omega$ and $\frac{1}{3}\,\Omega$ in series with $1\,\Omega$ gives $1\frac{1}{3}\,\Omega$

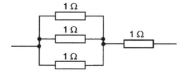

Figure 37.18

(d) **Two in parallel, in series with two in series** (see Figure 37.19), since for the two in parallel

$$R = \frac{1 \times 1}{1 + 1} = \frac{1}{2}\,\Omega$$

and $\frac{1}{2}\,\Omega$, $1\,\Omega$ and $1\,\Omega$ in series gives $2\frac{1}{2}\,\Omega$

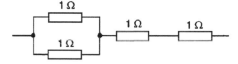

Figure 37.19

Problem 10. Find the equivalent resistance for the circuit shown in Figure 37.20

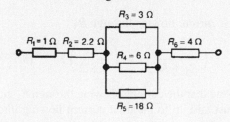

Figure 37.20

R_3, R_4 and R_5 are connected in parallel and their equivalent resistance R is given by:

$$\frac{1}{R} = \frac{1}{3} + \frac{1}{6} + \frac{1}{18} = \frac{6 + 3 + 1}{18} = \frac{10}{18}$$

hence $R = \frac{18}{10} = 1.8\,\Omega$

The circuit is now equivalent to four resistors in series and the **equivalent circuit resistance**

$$= 1 + 2.2 + 1.8 + 4 = \mathbf{9\,\Omega}$$

Problem 11. Resistances of $10\,\Omega$, $20\,\Omega$ and $30\,\Omega$ are connected (a) in series and (b) in parallel to a $240\,V$ supply. Calculate the supply current in each case

(a) The series circuit is shown in Figure 37.21.

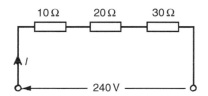

Figure 37.21

The equivalent resistance

$$R_T = 10\,\Omega + 20\,\Omega + 30\,\Omega = 60\,\Omega.$$

Supply current

$$I = \frac{V}{R_T} = \frac{240}{60} = \mathbf{4\,A}$$

(b) The parallel circuit is shown in Figure 37.22. The equivalent resistance R_T of $10\,\Omega$, $20\,\Omega$ and $30\,\Omega$ resistances connected in parallel is given by:

$$\frac{1}{R_T} = \frac{1}{10} + \frac{1}{20} + \frac{1}{30} = \frac{6 + 3 + 2}{60} = \frac{11}{60}$$

hence, $R_T = \frac{60}{11}\,\Omega$

Supply current

$$I = \frac{V}{R_T} = \frac{240}{\frac{60}{11}} = \frac{240 \times 11}{60} = \mathbf{44\,A}$$

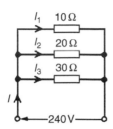

Figure 37.22

(Check:

$$I_1 = \frac{V}{R_1} = \frac{240}{10} = 24 \, \text{A},$$

$$I_2 = \frac{V}{R_2} = \frac{240}{20} = 12 \, \text{A and}$$

$$I_3 = \frac{V}{R_3} = \frac{240}{30} = 8 \, \text{A}$$

For a parallel circuit, $I = I_1 + I_2 + I_3$

$$= 24 + 12 + 8 = \textbf{44 A, as above.})$$

37.5   Current division

For the circuit shown in Figure 37.23, the total circuit resistance R_T is given by:

$$R_T = \frac{R_1 R_2}{R_1 + R_2}$$

and $V = IR_T = I\left(\dfrac{R_1 R_2}{R_1 + R_2}\right)$

$\text{Current}\, I_1 = \dfrac{V}{R_1} = \dfrac{I}{R_1}\left(\dfrac{R_1 R_2}{R_1 + R_2}\right) = \left(\dfrac{R_2}{\boldsymbol{R_1 + R_2}}\right)(I)$

Similarly,

$\text{Current}\, I_2 = \dfrac{V}{R_2} = \dfrac{I}{R_2}\left(\dfrac{R_1 R_2}{R_1 + R_2}\right) = \left(\dfrac{R_1}{\boldsymbol{R_1 + R_2}}\right)(I)$

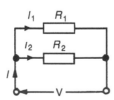

Figure 37.23

Summarising, with reference to Figure 37.23:

$$I_1 = \left(\frac{R_2}{\boldsymbol{R_1 + R_2}}\right)(I) \quad \text{and} \quad I_2 = \left(\frac{R_1}{\boldsymbol{R_1 + R_2}}\right)(I)$$

It is important to note that current division can only be applied to **two** parallel resistors. If there are more than two parallel resistors, then current division cannot be determined using the above formulae.

Problem 12.   For the series-parallel arrangement shown in Figure 37.24, find (a) the supply current, (b) the current flowing through each resistor and (c) the p.d. across each resistor

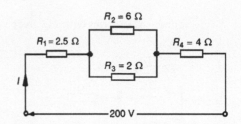

Figure 37.24

(a)   The equivalent resistance R_x of R_2 and R_3 in parallel is:

$$R_x = \frac{6 \times 2}{6 + 2} = 1.5 \, \Omega$$

The equivalent resistance R_T of R_1, R_x and R_4 in series is:

$$R_T = 2.5 + 1.5 + 4 = 8 \, \Omega$$

Supply current $I = \dfrac{V}{R_T} = \dfrac{200}{8} = \textbf{25 A}$

(b)   The current flowing through R_1 and R_4 is 25 A

The current flowing through R_2

$$= \left(\frac{R_3}{R_2 + R_3}\right) I = \left(\frac{2}{6 + 2}\right) 25 = \textbf{6.25 A}$$

The current flowing through R_3

$$= \left(\frac{R_2}{R_2 + R_3}\right) I = \left(\frac{6}{6 + 2}\right) 25 = \textbf{18.75 A}$$

(Note that the currents flowing through R_2 and R_3 must add up to the total current flowing into the parallel arrangement, i.e. 25 A.)

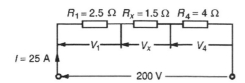

Figure 37.25

(c)   The equivalent circuit of Figure 37.24 is shown in Figure 37.25.

p.d. across R_1, i.e. $V_1 = IR_1 = (25)(2.5) = $ **62.5 V**

p.d. across R_x, i.e. $V_x = IR_x = (25)(1.5) = $ **37.5 V**

p.d. across R_4, i.e. $V_4 = IR_4 = (25)(4) = $ **100 V**

Hence, the p.d. across $R_2 = $ p.d. across R_3
$= $ **37.5 V**

Problem 13.   For the circuit shown in Figure 37.26 calculate (a) the value of resistor R_x such that the total power dissipated in the circuit is 2.5 kW, (b) the current flowing in each of the four resistors

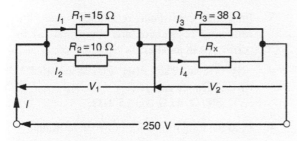

Figure 37.26

(a)   Power dissipated $P = VI$ watts, hence
$2500 = (250)(I)$

i.e. $I = \dfrac{2500}{250} = 10\,\text{A}$

From Ohm's law,

$$R_\text{T} = \frac{V}{I} = \frac{250}{10} = 25\,\Omega,$$

where R_T is the equivalent circuit resistance.

The equivalent resistance of R_1 and R_2 in parallel is

$$\frac{15 \times 10}{15 + 10} = \frac{150}{25} = 6\,\Omega$$

The equivalent resistance of resistors R_3 and R_x in parallel is equal to $25\,\Omega - 6\,\Omega = 19\,\Omega$.

There are three methods whereby R_x can be determined.

Method 1
The voltage $V_1 = IR$, where R is $6\,\Omega$, from above,

i.e. $V_1 = (10)(6) = 60\,\text{V}$

Hence, $V_2 = 250\,\text{V} - 60\,\text{V} = 190\,\text{V}$

$= $ p.d. across R_3

$= $ p.d. across R_x

$$I_3 = \frac{V_2}{R_3} = \frac{190}{38} = 5\,\text{A}.$$

Thus $I_4 = 5\,\text{A}$ also, since $I = 10\,\text{A}$

Thus, $R_x = \dfrac{V_2}{I_4} = \dfrac{190}{5} = $ **38 Ω**

Method 2
Since the equivalent resistance of R_3 and R_x in parallel is $19\,\Omega$,

$$19 = \frac{38R_X}{38 + R_X} \quad \left(\text{i.e. } \frac{\text{product}}{\text{sum}}\right)$$

Hence, $19(38 + R_x) = 38R_x$

$722 + 19R_x = 38R_x$

$722 = 38R_x - 19R_x = 19R_x$

Thus, $R_x = \dfrac{722}{19} = $ **38 Ω**

Method 3
When two resistors having the same value are connected in parallel the equivalent resistance is always half the value of one of the resistors. Thus, in this case, since $R_\text{T} = 19\,\Omega$ and $R_3 = 38\,\Omega$, then $R_x = 38\,\Omega$ could have been deduced on sight.

(b)   Current $I_1 = \left(\dfrac{R_2}{R_1 + R_2}\right)I = \left(\dfrac{10}{15 + 10}\right)(10)$

$$= \left(\frac{2}{5}\right)(10) = \textbf{4 A}$$

Current $I_2 = \left(\dfrac{R_1}{R_1 + R_2}\right)I = \left(\dfrac{15}{15 + 10}\right)(10)$

$$= \left(\frac{3}{5}\right)(10) = \textbf{6 A}$$

From part (a), method 1, $I_3 = I_4 = 5\,\text{A}$

Problem 14. For the arrangement shown in Figure 37.27, find the current I_x

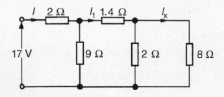

Figure 37.27

Commencing at the right-hand side of the arrangement shown in Figure 37.27, the circuit is gradually reduced in stages as shown in Figure 37.28(a)–(d).

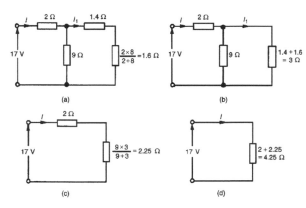

Figure 37.28

From Figure 37.28(d),

$$I = \frac{17}{4.25} = 4\,\text{A}$$

From Figure 37.28(b),

$$I_1 = \left(\frac{9}{9+3}\right)(I) = \left(\frac{9}{12}\right)(4) = 3\,\text{A}$$

From Figure 37.27,

$$I_x = \left(\frac{2}{2+8}\right)(I_1) = \left(\frac{2}{10}\right)(3) = \mathbf{0.6\,A}$$

Now try the following Practice Exercise

Practice Exercise 182   Further problems on parallel networks (answers on page 496)

1. Resistances of $4\,\Omega$ and $12\,\Omega$ are connected in parallel across a 9 V battery. Determine (a) the equivalent circuit resistance, (b) the supply current and (c) the current in each resistor

2. For the circuit shown in Figure 37.29 determine (a) the reading on the ammeter, and (b) the value of resistor R

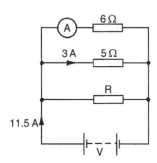

Figure 37.29

3. Find the equivalent resistance when the following resistances are connected (a) in series, (b) in parallel:

   (i) $3\,\Omega$ and $2\,\Omega$,   (ii) $20\,\text{k}\,\Omega$ and $40\,\text{k}\Omega$,
   (iii) $4\,\Omega$, $8\,\Omega$ and $16\,\Omega$,
   (iv) $800\,\Omega$, $4\,\text{k}\Omega$ and $1500\,\Omega$

4. Find the total resistance between terminals A and B of the circuit shown in Figure 37.30(a)

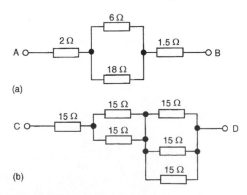

Figure 37.30

5. Find the equivalent resistance between terminals C and D of the circuit shown in Figure 37.30(b)

6. Resistors of $20\,\Omega$, $20\,\Omega$ and $30\,\Omega$ are connected in parallel. What resistance must be added in series with the combination to obtain a total resistance of $10\,\Omega$. If the complete circuit expends a power of $0.36\,\text{kW}$, find the total current flowing

7. (a) Calculate the current flowing in the $30\,\Omega$ resistor shown in Figure 37.31. (b) What additional value of resistance would have to be placed in parallel with the $20\,\Omega$ and $30\,\Omega$ resistors to change the supply current to $8\,\text{A}$, the supply voltage remaining constant?

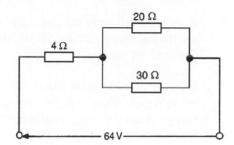

Figure 37.31

8. For the circuit shown in Figure 37.32, find (a) V_1, (b) V_2, without calculating the current flowing

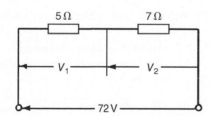

Figure 37.32

9. Determine the currents and voltages indicated in the circuit shown in Figure 37.33

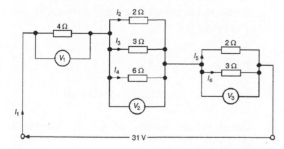

Figure 37.33

10. Find the current I in Figure 37.34

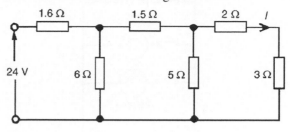

Figure 37.34

37.6   Wiring lamps in series and in parallel

37.6.1   Series connection

Figure 37.35 shows three lamps, each rated at $240\,\text{V}$, connected in series across a $240\,\text{V}$ supply.

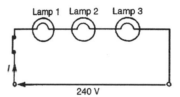

Figure 37.35

(i)   Each lamp has only $\dfrac{240}{3}$ V, i.e. 80 V, across it and thus each lamp glows dimly.

(ii) If another lamp of similar rating is added in series with the other three lamps then each lamp now has $\dfrac{240}{4}$ V, i.e. 60 V, across it and each now glows even more dimly.

(iii) If a lamp is removed from the circuit or if a lamp develops a fault (i.e. an open circuit) or if the switch is opened, then the circuit is broken, no current flows, and the remaining lamps will not light up.

(iv) Less cable is required for a series connection than for a parallel one.

The series connection of lamps is usually limited to decorative lighting such as for Christmas tree lights.

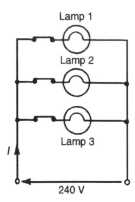

Lamp 1

Lamp 2

Lamp 3

240 V

Figure 37.36

37.6.2 Parallel connection

Figure 37.36 shows three similar lamps, each rated at 240 V, connected in parallel across a 240 V supply.

(i) Each lamp has 240 V across it and thus each will glow brilliantly at their rated voltage.

(ii) If any lamp is removed from the circuit or develops a fault (open circuit) or a switch is opened, the remaining lamps are unaffected.

(iii) The addition of further similar lamps in parallel does not affect the brightness of the other lamps.

(iv) More cable is required for parallel connection than for a series one.

The parallel connection of lamps is the most widely used in electrical installations.

Problem 15.   If three identical lamps are connected in parallel and the combined resistance is 150 Ω, find the resistance of one lamp

Let the resistance of one lamp be R, then,

$$\frac{1}{150} = \frac{1}{R} + \frac{1}{R} + \frac{1}{R} = \frac{3}{R}$$

from which,

$$R = 3 \times 150 = \textbf{450}\ \boldsymbol{\Omega}$$

Problem 16.   Three identical lamps A, B and C are connected in series across a 150 V supply. State (a) the voltage across each lamp and (b) the effect of lamp C failing

(a) Since each lamp is identical and they are connected in series there is $\dfrac{150}{3}$ V, i.e. **50 V across each**.

(b) If lamp C fails, i.e. open circuits, no current will flow and **lamps A and B will not operate**.

Now try the following Practice Exercises

Practice Exercise 183   Further problems on wiring lamps in series and in parallel (answers on page 547)

1. If four identical lamps are connected in parallel and the combined resistance is 100 Ω, find the resistance of one lamp

2. Three identical filament lamps are connected (a) in series, (b) in parallel across a 210 V supply. State for each connection the p.d. across each lamp

Practice Exercise 184   Short answer questions on series and parallel networks

1. Name three characteristics of a series circuit

2. Show that for three resistors R_1, R_2 and R_3 connected in series the equivalent resistance R is given by:

$$R = R_1 + R_2 + R_3$$

3. Name three characteristics of a parallel network

4. Show that for three resistors R_1, R_2 and R_3 connected in parallel the equivalent resistance R is given by:

$$\frac{1}{R} = \frac{1}{R_1} + \frac{1}{R_2} + \frac{1}{R_3}$$

5. Explain the potential divider circuit

6. Compare the merits of wiring lamps in (a) series, (b) parallel

Practice Exercise 185 Multiple-choice
questions on series and parallel networks
(answers on page 547)

1. If two $4\,\Omega$ resistors are connected in series the
    effective resistance of the circuit is:

    (a) $8\,\Omega$ (b) $4\,\Omega$ (c) $2\,\Omega$ (d) $1\,\Omega$

2. If two $4\,\Omega$ resistors are connected in parallel
    the effective resistance of the circuit is:

    (a) $8\,\Omega$ (b) $4\,\Omega$ (c) $2\,\Omega$ (d) $1\,\Omega$

3. With the switch in Figure 37.37 closed, the
    ammeter reading will indicate:

    (a) $1\dfrac{2}{3}\,A$ (b) $75\,A$

    (c) $\dfrac{1}{3}\,A$ (d) $3\,A$

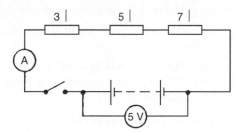

Figure 37.37

4. The effect of connecting an additional parallel
    load to an electrical supply source is to increase
    the:

    (a) resistance of the load
    (b) voltage of the source
    (c) current taken from the source
    (d) p.d. across the load

5. The equivalent resistance when a resistor of
    $\dfrac{1}{3}\,\Omega$ is connected in parallel with a $\dfrac{1}{4}\,\Omega$
    resistance is:

    (a) $\dfrac{1}{7}\,\Omega$ (b) $7\,\Omega$

    (c) $\dfrac{1}{12}\,\Omega$ (d) $\dfrac{3}{4}\,\Omega$

6. A $6\,\Omega$ resistor is connected in parallel with the
    three resistors of Figure 37.38. With the switch
    closed the ammeter reading will indicate:

    (a) $\dfrac{3}{4}\,A$ (b) $4\,A$

    (c) $\dfrac{1}{4}\,A$ (d) $1\dfrac{1}{3}\,A$

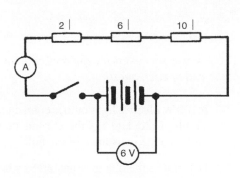

Figure 37.38

7. A $10\,\Omega$ resistor is connected in parallel
    with a $15\,\Omega$ resistor and the combination in
    series with a $12\,\Omega$ resistor. The equivalent
    resistance of the circuit is:

    (a) $37\,\Omega$ (b) $18\,\Omega$
    (c) $27\,\Omega$ (d) $4\,\Omega$

8. When three $3\,\Omega$ resistors are connected in
    parallel, the total resistance is:

    (a) $3\,\Omega$ (b) $9\,\Omega$
    (c) $1\,\Omega$ (d) $0.333\,\Omega$

9. The total resistance of two resistors R_1 and
    R_2 when connected in parallel is given by:

    (a) $R_1 + R_2$ (b) $\dfrac{1}{R_1} + \dfrac{1}{R_2}$

    (c) $\dfrac{R_1 + R_2}{R_1 R_2}$ (d) $\dfrac{R_1 R_2}{R_1 + R_2}$

10. If in the circuit shown in Figure 37.39,
    the reading on the voltmeter is $5\,V$ and
    the reading on the ammeter is $25\,mA$, the
    resistance of resistor R is:

    (a) $0.005\,\Omega$ (b) $3\,\Omega$
    (c) $125\,\Omega$ (d) $200\,\Omega$

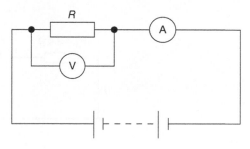

Figure 37.39

**For fully worked solutions to each of the problems in Exercises 181 to 185 in this chapter,
go to the website:** www.routledge.com/cw/bird

Revision Test 13: Electric circuits, resistance variation, batteries and series and parallel networks

This assignment covers the material contained in chapters 34 to 37. *The marks for each question are shown in brackets at the end of each question.*

1. A 100 W electric light bulb is connected to a 200 V supply. Calculate (a) the current flowing in the bulb, (b) the resistance of the bulb. (4)

2. Determine the charge transferred when a current of 5 mA flows for 10 minutes. (4)

3. A current of 12 A flows in the element of an electric fire of resistance 10 Ω. Determine the power dissipated by the element. If the fire is on for 5 hours every day, calculate for a one week period (a) the energy used, (b) cost of using the fire if electricity cost 11p per unit. (6)

4. Calculate the resistance of 1200 m of copper cable of cross-sectional area 15 mm². Take the resistivity of copper as 0.02 μΩ m. (4)

5. At a temperature of 40°C, an aluminium cable has a resistance of 25 Ω. If the temperature coefficient of resistance at 0°C is 0.0038/°C, calculate its resistance at 0°C. (5)

6. The resistance of a coil of copper wire at 20°C is 150 Ω. The temperature of the wire is increased and the resistance rises to 200 Ω. If the temperature coefficient of resistance of copper is 0.004/°C at 20°C determine the temperature to which the coil has risen, correct to the nearest degree. (7)

7. (a) Determine the values of the resistors with the following colour coding:

    (i) red-red-orange-silver

    (ii) orange-orange-black-blue-green.

    (b) What is the value of a resistor marked as 47 KK? (6)

8. Four cells, each with an internal resistance of 0.40 Ω and an e.m.f. of 2.5 V are connected in series to a load of 38.4 Ω. Determine the current flowing in the circuit and the p.d. at the battery terminals. (5)

9. (a) State six typical applications of primary cells.

    (b) State six typical applications of secondary cells.

    (c) State the advantages of a fuel cell over a conventional battery and state three practical applications. (12)

10. State for lithium-ion batteries

    (a) three typical practical applications

    (b) four advantages compared with other batteries

    (c) three limitations. (10)

11. Name six alternative, renewable energy sources, and give a brief description of each. (18)

12. For solar power, briefly state (a) seven advantages (b) five disadvantages. (12)

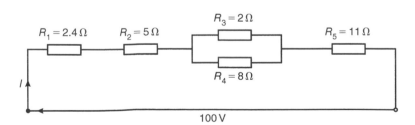

Figure RT13.1

13. Resistances of $5\,\Omega$, $7\,\Omega$ and $8\,\Omega$ are connected in series. If a $10\,V$ supply voltage is connected across the arrangement determine the current flowing through and the p.d. across the $7\,\Omega$ resistor. Calculate also the power dissipated in the $8\,\Omega$ resistor. (6)

14. For the series–parallel network shown in Figure RT13.1, find (a) the supply current, (b) the current flowing through each resistor, (c) the p.d. across each resistor, (d) the total power dissipated in the circuit, (e) the cost of energy if the circuit is connected for 80 hours. Assume electrical energy costs 12 p per unit. (16)

15. For the arrangement shown in Figure RT13.2, determine the current I_x. (8)

16. Four identical filament lamps are connected (a) in series, (b) in parallel across a $240\,V$ supply. State for each connection the p.d. across each lamp. (2)

17. State three advantages and two disadvantages of glass batteries.

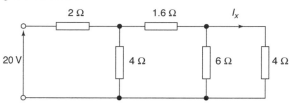

Figure RT13.2

For lecturers/instructors/teachers, fully worked solutions to each of the problems in Revision Test 13, together with a full marking scheme, are available at the website: www.routledge.com/cw/bird

Chapter 38

Kirchhoff's laws

Why it is important to understand: **Kirchhoff's laws**

In the previous chapter it was seen that a single equivalent resistance can be found when two or more resistors are connected together in series, parallel or combinations of both, and that these circuits obey Ohm's Law. However, sometimes in more complex circuits we cannot simply use Ohm's Law alone to find the voltages or currents circulating within the circuit. For these types of calculations we need certain rules which allow us to obtain the circuit equations and for this we can use Kirchhoff's laws. In this chapter, Kirchhoff's laws are explained in detail using a number of numerical worked examples. An electrical/electronic engineer often needs to be able to analyse an electrical network to determine currents flowing in each branch and the voltage across each branch.

At the end of this chapter, you should be able to:

- state Kirchhoff's laws
- use Kirchhoff's laws to determine unknown currents and voltages in d.c. circuits

38.1 Introduction

Complex d.c. circuits cannot always be solved by Ohm's law and the formulae for series and parallel resistors alone. **Kirchhoff*** (a German physicist) developed two laws which further help the determination of unknown currents and voltages in d.c. series/parallel networks. This chapter states Kirchhoff's laws and demonstrates how unknown currents and voltages may be evaluated in series/parallel circuits.

38.2 Kirchhoff's current and voltage laws

38.2.1 Current law

At any junction in an electric circuit the total current flowing towards that junction is equal to the total current flowing away from the junction, i.e. $\sum I = 0$

Science and Mathematics for Engineering. 978-0-367-20475-4, © John Bird. Published by Taylor & Francis. All rights reserved.

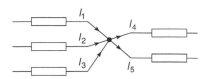

Figure 38.1

Thus referring to Figure 38.1:

$$I_1 + I_2 + I_3 = I_4 + I_5$$

or $$I_1 + I_2 + I_3 - I_4 - I_5 = 0$$

38.2.2 Voltage law

In any closed loop in a network, the algebraic sum of the voltage drops (i.e. products of current and resistance) taken around the loop is equal to the resultant e.m.f. acting in that loop.

Thus referring to Figure 38.2:

$$E_1 - E_2 = IR_1 + IR_2 + IR_3$$

*Who was Kirchhoff? – **Gustav Robert Kirchhoff** (12 March 1824–17 October 1887) was a German physicist. Concepts in circuit theory and thermal emission are named 'Kirchhoff's laws' after him, as well as a law of thermochemistry. To find out more go to www.routledge.com/cw/bird*

(Note that if current flows away from the positive terminal of a source, that source is considered by convention to be positive. Thus moving anticlockwise around the loop of Figure 19.2, E_1 is positive and E_2 is negative.)

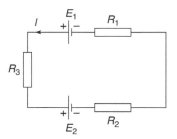

Figure 38.2

38.3 Worked problems on Kirchhoff's laws

Problem 1. Determine the value of the unknown currents marked in Figure 38.3

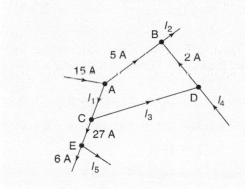

Figure 38.3

Applying Kirchhoff's current law to each junction in turn gives:

For junction A: $\qquad 15 = 5 + I_1$

Hence $\qquad \boldsymbol{I_1 = 10\,A}$

For junction B: $\qquad 5 + 2 = I_2$

Hence $\qquad \boldsymbol{I_2 = 7\,A}$

For junction C: $\qquad I_1 = 27 + I_3$

i.e. $\qquad 10 = 27 + I_3$

Hence $\qquad \boldsymbol{I_3 = 10 - 27 = -17A}$

(i.e. in the opposite direction to that shown in Figure 38.3)

For junction D: $\qquad I_3 + I_4 = 2$

i.e. $\qquad\qquad -17 + I_4 = 2$

Hence $\qquad\qquad I_4 = 17 + 2 = \textbf{19 A}$

For junction E: $\qquad 27 = 6 + I_5$

Hence $\qquad\qquad I_5 = 27 - 6 = \textbf{21 A}$

Problem 2. Determine the value of e.m.f. E in Figure 38.4

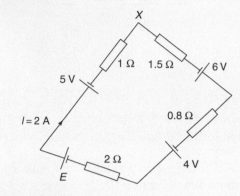

Figure 38.4

Applying Kirchhoff's voltage law and moving clockwise around the loop of Figure 38.4 starting at point X gives:

$$6 + 4 + E - 5 = I(1.5) + I(0.8) + I(2) + I(1)$$

$$5 + E = I(5.3) = 2(5.3) \text{ since current, } I \text{ is 2 A}$$

Hence $\qquad 5 + E = 10.6$

and $\qquad$ **e.m.f.** $E = 10.6 - 5 = \textbf{5.6 V}$

Problem 3. Use Kirchhoff's laws to determine the current flowing in the 4 Ω resistance of the network shown in Figure 38.5

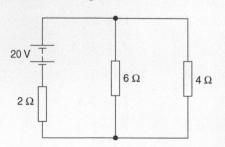

Figure 38.5

Step 1

Label current I_1 flowing from the positive terminal of the 20 V source and current I_2 flowing through the 6 V resistance as shown in Figure 38.6. By **Kirchhoff's current law**, the current flowing in the 4 V resistance must be $(I_1 - I_2)$

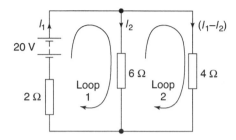

Figure 38.6

Step 2

Label loops 1 and 2 as shown in Figure 38.6 (both loops have been shown clockwise, although they do not need to be in the same direction). **Kirchhoffs voltage law** is now applied to each loop in turn:

For loop 1: $\quad 20 = 2I_1 + 6I_2 \qquad\qquad$ (1)

For loop 2: $\quad 0 = 4(I_1 - I_2) - 6I_2 \qquad$ (2)

Note the zero on the left-hand side of equation (2) since there is no voltage source in loop 2. Note also the minus sign in front of $6I_2$. This is because loop 2 is moving through the 6 V resistance in the opposite direction to current I_2

Equation (2) simplifies to:

$$0 = 4I_1 - 10I_2 \qquad\qquad (3)$$

Step 3

Solve the simultaneous equations (1) and (3) for currents I_1 and I_2 (see chapter 8):

$$20 = 2I_1 + 6I_2 \qquad\qquad (1)$$

$$0 = 4I_1 - 10I_2 \qquad\qquad (3)$$

$2 \times$ equation (1) gives:

$$40 = 4I_1 + 12I_2 \qquad\qquad (4)$$

equation (4) – equation (3) gives:

$$40 = 0 + (12I_2 - -10I_2)$$

i.e. $\qquad\qquad 40 = 22I_2$

Hence, **current $I_2 = \dfrac{40}{22} = 1.818$ A**

Substituting $I_2 = 1.818$ into equation (1) gives:

$$20 = 2I_1 + 6(1.818)$$

$$20 = 2I_1 + 10.908$$

and $\qquad 20 - 10.908 = 2I_1$

from which, $\qquad I_1 = \dfrac{20 - 10.908}{2}$

$$= \dfrac{9.092}{2} = \textbf{4.546 A}$$

Hence, **the current flowing in the 4 V resistance** is

$$I_1 - I_2 - (4.546 - 1.818) = \textbf{2.728 A}$$

The currents and their directions are as shown in Figure 38.7.

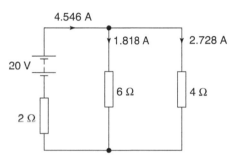

Figure 38.7

Problem 4. Use Kirchhoff's laws to determine the current flowing in each branch of the network shown in Figure 38.8

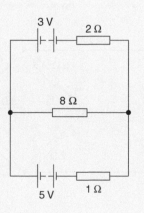

Figure 38.8

Step 1

The currents I_1 and I_2 are labelled as shown in Figure 38.9 and by Kirchhoff's current law the current in the 8 V resistance is $(I_1 + I_2)$.

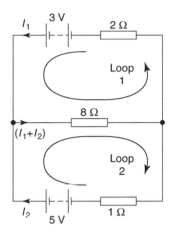

Figure 38.9

Step 2

Loops 1 and 2 are labelled as shown in Figure 38.9. Kirchhoff's voltage law is now applied to each loop in turn.

For loop 1: $\quad 3 = 2I_1 + 8(I_1 + I_2)$ $\qquad$ (1)

For loop 2: $\quad 5 = 8(I_1 + I_2) + (1)(I_2)$ $\qquad$ (2)

Equation (1) simplifies to:

$$3 = 10I_1 + 8I_2 \qquad (3)$$

Equation (2) simplifies to:

$$5 = 8I_1 + 9I_2 \qquad (4)$$

$4 \times$ equation (3) gives:

$$12 = 40I_1 + 32I_2 \qquad (5)$$

$5 \times$ equation (4) gives:

$$25 = 40I_1 + 45I_2 \qquad (6)$$

equation (6) − equation (5) gives:

$$13 = 13I_2$$

and $\qquad\qquad I_2 = \textbf{1 A}$

Substituting $I_2 = 1$ in equation (3) gives:

$$3 = 10I_1 + 8(1)$$

$$3 - 8 = 10I_1$$

and

$$I_1 = \frac{-5}{10} = -0.5A$$

(i.e. I_1 is flowing in the opposite direction to that shown in Figure 38.9.)

The **current in the 8 V resistance** is

$$(I_1 + I_2) = (-0.5 + 1) = 0.5 \text{ A}$$

___

Now try the following Practice Exercises

Practice Exercise 186   Further problems on Kirchhoff's laws (answers on page 547)

1. Find currents I_3, I_4 and I_6 in Figure 38.10

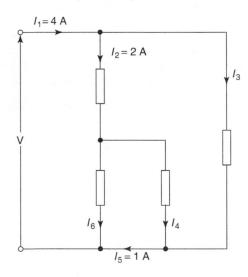

Figure 38.10

2. For the networks shown in Figure 38.11, find the values of the currents marked

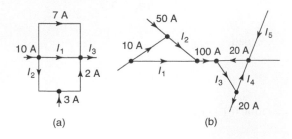

Figure 38.11

3. Use Kirchhoff's laws to find the current flowing in the 6 Ω resistor of Figure 38.12 and the power dissipated in the 4 Ω resistor

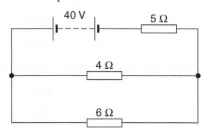

Figure 38.12

4. Find the current flowing in the 3 Ω resistor for the network shown in Figure 38.13(a). Find also the p.d. across the 10 Ω and 2 Ω resistors

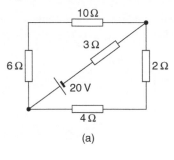

(a)

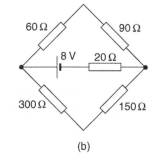

(b)

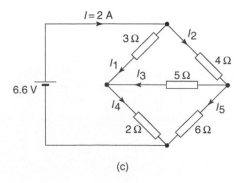

(c)

Figure 38.13

5. For the network shown in Figure 38.13(b), find: (a) the current in the battery, (b) the current in the 300 Ω resistor, (c) the current in the 90 Ω resistor and (d) the power dissipated in the 150 Ω resistor

6. For the bridge network shown in Figure 38.13(c), find the currents I_1 to I_5

Practice Exercise 187   Short answer questions on Kirchhoff's laws

1. State Kirchhoff's current law

2. State Kirchhoff's voltage law

Practice Exercise 188   Multiple-choice questions on Kirchhoff's laws (answers on page 547)

1. The current flowing in the branches of a d.c. circuit may be determined using:

    (a) Kirchhoff's laws
    (b) Lenz's law
    (c) Faraday's laws
    (d) Fleming's left-hand rule

2. Which of the following statements is true? For the junction in the network shown in Figure 38.14:

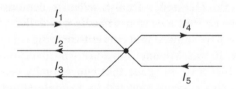

Figure 38.14

   (a) $I_5 - I_4 = I_3 - I_2 + I_1$
   (b) $I_1 + I_2 + I_3 = I_4 + I_5$
   (c) $I_2 + I_3 + I_5 = I_1 + I_4$
   (d) $I_1 - I_2 - I_3 - I_4 + I_5 = 0$

3. Which of the following statements is true? For the circuit shown in Figure 38.15:

   (a) $E_1 + E_2 + E_3 = Ir_1 + Ir_2 + Ir_3$
   (b) $E_2 + E_3 - E_1 - I(r_1 + r_2 + r_3) = 0$
   (c) $I(r_1 + r_2 + r_3) = E_1 - E_2 - E_3$
   (d) $E_2 + E_3 - E_1 = Ir_1 + Ir_2 + Ir_3$

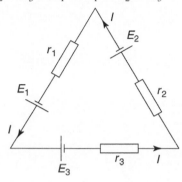

Figure 38.15

4. The current I flowing in resistor R in the circuit shown in Figure 38.16 is:

   (a) $I_2 I_1$ (b) $I_1 - I_2$
   (c) $I_1 + I_2$ (d) I_1

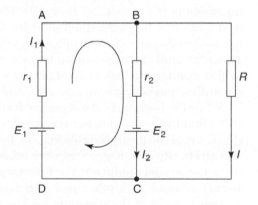

Figure 38.16

5. Applying Kirchhoff's voltage law clockwise around loop *ABCD* in the circuit shown in Figure 38.16 gives:

   (a) $E_1 - E_2 = I_1r_1 + I_2r_2$
   (b) $E_2 = E_1 + I_1r_1 + I_2r_2$
   (c) $E_1 + E_2 = I_1r_1 + I_2r_2$
   (d) $E_1 + I_1r_1 = E_2 + I_2r_2$

For fully worked solutions to each of the problems in Exercises 186 to 188 in this chapter, go to the website: www.routledge.com/cw/bird

Chapter 39

Magnetism and electromagnetism

Why it is important to understand: **Magnetism and electromagnetism**

Many common devices rely on magnetism. Familiar examples include computer disk drives, tape recorders, VCRs, transformers, motors, generators, and so on. Practically all transformers and electric machinery uses magnetic material for shaping and directing the magnetic fields which act as a medium for transferring and connecting energy. It is therefore important to be able to analyse and describe magnetic field quantities for understanding these devices. Magnetic materials are significant in determining the properties of a piece of electromagnetic equipment or an electric machine, and affect its size and efficiency. To understand their operation, knowledge of magnetism and magnetic circuit principles is required. In this chapter, we look at fundamentals of magnetism, relationships between electrical and magnetic quantities, magnetic circuit concepts, and methods of analysis.

While the basic facts about magnetism have been known since ancient times, it was not until the early 1800s that the connection between electricity and magnetism was made and the foundations of modern electromagnetic theory established. In 1819, Hans Christian Oersted, a Danish scientist, demonstrated that electricity and magnetism were related when he showed that a compass needle was deflected by a current-carrying conductor. The following year, Andre Ampere showed that current-carrying conductors attract or repel each other just like magnets. However, it was Michael Faraday who developed our present concept of the magnetic field as a collection of flux lines in space that conceptually represent both the intensity and the direction of the field. It was this concept that led to an understanding of magnetism and the development of important practical devices such as the transformer and the electric generator. In electrical machines, the magnetic circuits may be formed by ferromagnetic materials (as in transformers), or by ferromagnetic materials in conjunction with air (as in rotating machines). In most electrical machines, the magnetic field is produced by passing an electric current through coils wound on ferromagnetic material. In this chapter important concepts of electromagnetism are explained and simple calculations performed.

Science and Mathematics for Engineering. 978-0-367-20475-4, © John Bird. Published by Taylor & Francis. All rights reserved.

At the end of this chapter, you should be able to:

- describe the magnetic field around a permanent magnet
- state the laws of magnetic attraction and repulsion for two magnets in close proximity
- apply the screw rule to determine direction of magnetic field
- recognise that the magnetic field around a solenoid is similar to a magnet
- apply the screw rule or grip rule to a solenoid to determine magnetic field direction
- understand that magnetic fields are produced by electric currents
- recognise and describe practical applications of an electromagnet, i.e. electric bell, relay, lifting magnet, telephone receiver
- define magnetic flux, Φ, and magnetic flux density, B, and state their units
- perform simple calculations involving $B = \dfrac{\Phi}{A}$
- appreciate factors upon which the force F on a current-carrying conductor depends
- perform calculations using $F = BIl$ and $F = Bil \sin \theta$
- recognise that a loudspeaker is a practical application of force F
- use Fleming's left-hand rule to pre-determine direction of force in a current carrying conductor
- describe the principle of operation of a simple d.c. motor
- appreciate that force F on a charge in a magnetic field is given by $F = QvB$
- perform calculations using $F = QvB$

39.1 Introduction to magnetism and magnetic circuits

The study of magnetism began in the thirteenth century with many eminent scientists and physicists such as William Gilbert*, Hans Christian Oersted*, Michael Faraday*, James Maxwell*, André Ampère* and Wilhelm Weber* all having some input on the subject since. The association between electricity and magnetism is a fairly recent finding in comparison with the very first understanding of basic magnetism.

Today, magnets have **many varied practical applications**. For example, they are used in motors and generators, telephones, relays, loudspeakers, computer hard drives and floppy disks, anti-lock brakes, cameras, fishing reels, electronic ignition systems, keyboards, TV and radio components and in transmission equipment.

The full theory of magnetism is one of the most complex of subjects; this chapter provides an introduction to the topic.

39.2 Magnetic fields

A **permanent magnet** is a piece of ferromagnetic material (such as iron, nickel or cobalt) which has properties of attracting other pieces of these materials. The area around a magnet is called the **magnetic field** and it is in this area that the effects of the **magnetic force** produced by the magnet can be detected. The magnetic field of a bar magnet can be represented pictorially by the 'lines of force' (or lines of 'magnetic flux' as they are called) as shown in Figure 39.1. Such a field pattern can be produced by placing iron filings in the vicinity of the magnet.

The field direction at any point is taken as that in which the north-seeking pole of a compass needle points when suspended in the field. External to the magnet the direction of the field is north to south.

The laws of magnetic attraction and repulsion can be demonstrated by using two bar magnets. In Figure 39.2(a), with **unlike poles** adjacent, **attraction**

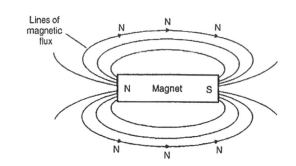

Figure 39.1

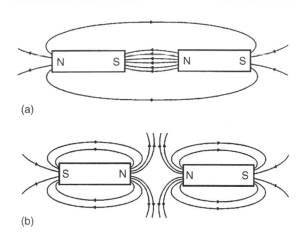

(a)

(b)

Figure 39.2

occurs. In Figure 39.2(b), with **like poles** adjacent, **repulsion** occurs.

*Who was Gilbert? – **William Gilbert** (24 May 1544–30 November 1603) was an English physician, physicist and natural philosopher who is credited as one of the originators of the term *electricity*. A unit of magnetomotive force was named the *gilbert* in his honour. To find out more about Gilbert go to www.routledge.com/cw/bird

*Who was Ørsted? – **Hans Christian Ørsted** (14 August 1777–9 March 1851) was the Danish physicist and chemist who discovered that electric currents create magnetic fields. The unit, the *oersted*, is named after him. To find out more about Ørsted go to www.routledge.com/cw/bird

*Who was Faraday? – **Michael Faraday** (22 September 1791–25 August 1867) was an English scientist whose main discoveries include electromagnetic induction, diamagnetism and electrolysis. The SI unit of capacitance, the farad, is named in his honour. To find out more about Faraday go to www.routledge.com/cw/bird

Magnetic fields are produced by electric currents as well as by permanent magnets. The field forms a circular pattern with the current-carrying conductor at the centre. The effect is portrayed in Figure 39.3 where the convention adopted is:

(a) Current flowing away from viewer   (b) Current flowing towards viewer

Figure 39.3

James Clerk Maxwell.

*Who was Maxwell? – **James Clerk Maxwell** (13 June 1831–5 November 1879) was a Scottish theoretical physicist who formulated classical electromagnetic theory. His equations demonstrate that electricity, magnetism and light are all manifestations of the same thing – the electromagnetic field. To find out more about Maxwell go to www.routledge.com/cw/bird

*Who was Ampere? – **André-Marie Ampère** (20 January 1775–10 June 1836) is generally regarded as one of the main founders of classical electromagnetism. The SI unit of measurement of electric current, the ampere, is named after him. To find out more about Ampère go to www.routledge.com/cw/bird (For image refer to page 343)

(i) Current flowing **away** from the viewer, i.e. into the paper, is indicated by $\oplus$. This may be thought of as the feathered end of the shaft of an arrow. See Figure 39.3(a).

(ii) Current flowing **towards** the viewer, i.e. out of the paper, is indicated by •. This may be thought of as the point of an arrow. See Figure 39.3(b).

The direction of the magnetic lines of flux is best remembered by the **screw rule** which states that:

If a normal right-hand thread screw is screwed along the conductor in the direction of the current, the direction of rotation of the screw is in the direction of the magnetic field.

For example, with current flowing away from the viewer (Figure 39.3(a)) a right-hand thread screw driven into the paper has to be rotated clockwise. Hence the direction of the magnetic field is clockwise.

A magnetic field set up by a long coil, or **solenoid**, is shown in Figure 39.4 and is seen to be similar to that of a bar magnet. If the solenoid is wound on an iron bar an even stronger magnetic field is produced, the iron becoming magnetised and behaving like a permanent

*Who was Weber? – **Wilhelm Eduard Weber** (24 October 1804–23 June 1891) was a German physicist who, together with Gauss, invented the first electromagnetic telegraph. The SI unit of magnetic flux, the weber (Wb), is named after him. To find out more about Weber go to www.routledge.com/cw/bird

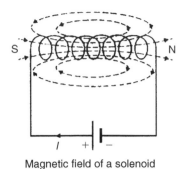

Magnetic field of a solenoid

Figure 39.4

magnet. The **direction** of the magnetic field produced by the current I in the solenoid may be found by either of two methods, i.e. the screw rule or the grip rule.

(a) **The screw rule** states that if a normal right-hand thread screw is placed along the axis of the solenoid and is screwed in the direction of the current it moves in the direction of the magnetic field **inside** the solenoid (i.e. points in the direction of the north pole).

(b) **The grip rule** states that if the coil is gripped with the **right** hand, with the fingers pointing in the direction of the current, then the thumb, outstretched parallel to the axis of the solenoid, points in the direction of the magnetic field **inside** the solenoid (i.e. points in the direction of the north pole).

Problem 1. Figure 39.5 shows a coil of wire wound on an iron core connected to a battery. Sketch the magnetic field pattern associated with the current carrying coil and determine the polarity of the field

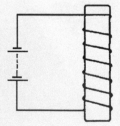

Figure 39.5

The magnetic field associated with the solenoid in Figure 39.5 is similar to the field associated with a bar magnet and is as shown in Figure 39.6. The polarity of the field is determined either by the screw rule or by the grip rule. Thus the north pole is at the bottom and the south pole at the top.

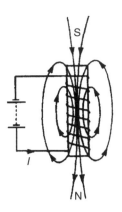

Figure 39.6

39.3 Electromagnets

The solenoid is very important in electromagnetic theory since the magnetic field inside the solenoid is practically uniform for a particular current, and is also versatile, inasmuch that a variation of the current can alter the strength of the magnetic field. An electromagnet, based on the solenoid, provides the basis of many items of electrical equipment, examples of which include electric bells, relays, lifting magnets and telephone receivers.

39.3.1 Electric bells

There are various types of electric bell, including the single-stroke bell, the trembler bell, the buzzer and a continuously ringing bell, but all depend on the attraction exerted by an electromagnet on a soft iron armature. A typical single-stroke bell circuit is shown in Figure 39.7. When the push button is operated, a current passes through the coil. Since the iron-cored coil is energised the soft iron armature is attracted to the electromagnet. The armature also carries a striker which hits the gong. When the circuit is broken the coil becomes demagnetised and the spring steel strip pulls the armature back to its original position. The striker will only operate when the push is operated.

39.3.2 Relays

A relay is similar to an electric bell except that contacts are opened or closed by its operation instead of a gong being struck. A typical simple relay is shown in Figure 39.8, which consists of a coil wound on a soft iron core.

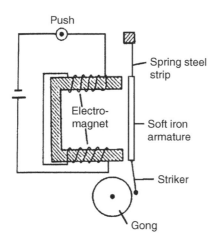

Figure 39.7

When the coil is energised the hinged soft iron armature is attracted to the electromagnet and pushes against two fixed contacts so that they are connected together, thus closing some other electrical circuit.

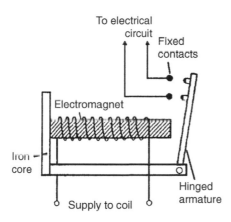

Figure 39.8

39.3.3 Lifting magnets

Lifting magnets, incorporating large electromagnets, are used in iron and steel works for lifting scrap metal. A typical robust lifting magnet, capable of exerting large attractive forces, is shown in the elevation and plan view of Figure 39.9 where a coil C is wound round a central core P of the iron casting. Over the face of the electromagnet is placed a protective non-magnetic sheet of material R. The load Q which must be of magnetic material is lifted when the coils are energised, the magnetic flux paths M being shown by the broken lines.

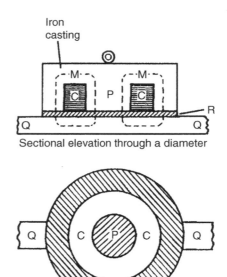

Sectional elevation through a diameter

Plan view

Figure 39.9

39.3.4 Telephone receivers

Whereas a transmitter or microphone changes sound waves into corresponding electrical signals, a telephone receiver converts the electrical waves back into sound waves. A typical telephone receiver is shown in Figure 39.10 and consists of a permanent magnet with coils wound on its poles. A thin, flexible diaphragm of magnetic material is held in position near to the magnetic poles but not touching them. Variation in current from the transmitter varies the magnetic field and the diaphragm consequently vibrates. The vibration produces sound variations corresponding to those transmitted.

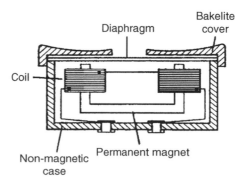

Figure 39.10

39.4   Magnetic flux and flux density

Magnetic flux is the amount of magnetic field (or the number of lines of force) produced by a magnetic source. The symbol for magnetic flux is Φ (Greek letter phi). The unit of magnetic flux is the **weber***, **Wb**.

Magnetic flux density is the amount of flux passing through a defined area that is perpendicular to the direction of the flux:

$$\textbf{Magnetic flux density} = \frac{\textbf{magnetic flux}}{\textbf{area}}$$

The symbol for magnetic flux density is B. The unit of magnetic flux density is the **tesla***, **T,** where $1\ \text{T} = 1\ \text{Wb/m}^2$. Hence

$$B = \frac{\Phi}{A}\textbf{tesla}\quad\text{where } A(\text{m}^2)\text{ is the area}$$

> **Problem 2.** A magnetic pole face has a rectangular section having dimensions 200 mm by 100 mm. If the total flux emerging from the pole is 150 μWb, calculate the flux density

Flux $\Phi = 150\,\mu\text{Wb} = 150 \times 10^{-6}\,\text{Wb}$, and cross-sectional area $A = 200 \times 100 = 20{,}000\,\text{mm}^2 = 20{,}000 \times 10^{-6}\,\text{m}^2$.

Flux density $B = \dfrac{\Phi}{A} = \dfrac{150 \times 10^{-6}}{20{,}000 \times 10^{-6}}$

$$= \textbf{0.0075 T}\ \text{or}\ \ \textbf{7.5 mT}$$

> **Problem 3.**   The maximum working flux density of a lifting electromagnet is 1.8 T and the effective area of a pole face is circular in cross-section. If the total magnetic flux produced is 353 mWb, determine the radius of the pole face

Flux density $B = 1.8$ T and

flux $\Phi = 353\,\text{mWb} = 353 \times 10^{-3}$ Wb.

Since $B = \dfrac{\Phi}{A}$, cross-sectional area

$$A = \frac{\Phi}{B} = \frac{353 \times 10^{-3}}{1.8}\ \text{m}^2 = 0.1961\ \text{m}^2$$

The pole face is circular, hence area $= \pi r^2$, where r is the radius.

*Who was Tesla? – **Nikola Tesla** (10 July 1856–7 January 1943) started working in the telephony and electrical fields before emigrating to the United States in 1884 to work for Thomas Edison. He soon struck out on his own though, setting up laboratories and companies to develop a range of electrical devices. In 1960, in honour of Tesla, the term 'tesla' was coined for the SI unit of magnetic field strength. To find out more about Tesla go to www.routledge.com/cw/bird

Hence, $\pi r^2 = 0.1961$, from which,

$$r^2 = \frac{0.1961}{\pi} \text{ and radius}$$

$$r = \sqrt{\frac{0.1961}{\pi}} = 0.250\,\text{m}$$

i.e. **the radius of the pole face is 250 mm**.

Now try the following Practice Exercise

Practice Exercise 189   Further problems on magnetic circuits (answers on page 547)

1. What is the flux density in a magnetic field of cross-sectional area 20 cm^2 having a flux of 3 mWb?

2. Determine the total flux emerging from a magnetic pole face having dimensions 5 cm by 6 cm, if the flux density is 0.9 T

3. The maximum working flux density of a lifting electromagnet is 1.9 T and the effective area of a pole face is circular in cross-section. If the total magnetic flux produced is 611 mWb determine the radius of the pole face

4. An electromagnet of square cross section produces a flux density of 0.45 T. If the magnetic flux is 720 μWb find the dimensions of the electromagnet cross-section

39.5 Force on a current-carrying conductor

If a current-carrying conductor is placed in a magnetic field produced by permanent magnets, then the fields due to the current-carrying conductor and the permanent magnets interact and cause a force to be exerted on the conductor. The force on the current-carrying conductor in a magnetic field depends upon:

(a) the flux density of the field, B teslas,

(b) the strength of the current, I amperes,

(c) the length of the conductor perpendicular to the magnetic field, l metres, and

(d) the directions of the field and the current.

When the magnetic field, the current and the conductor are mutually at right-angles then:

Force $F = BIl$ newtons

When the conductor and the field are at an angle $\theta°$ to each other then:

Force $F = BIl \sin \theta$ newtons

Since when the magnetic field, current and conductor are mutually at right-angles, $F = BIl$, the magnetic flux density B may be defined by $B = \dfrac{F}{Il}$, i.e. the flux density is 1 T if the force exerted on 1 m of a conductor when the conductor carries a current of 1 A is 1 N.

39.5.1 Loudspeaker

A simple application of the above force is the moving-coil loudspeaker. The loudspeaker is used to convert electrical signals into sound waves.

Figure 39.11 shows a typical loudspeaker having a magnetic circuit comprising a permanent magnet and soft iron pole pieces so that a strong magnetic field is available in the short cylindrical air-gap. A moving coil, called the voice or speech coil, is suspended from the end of a paper or plastic cone so that it lies in the gap. When an electric current flows through the coil it produces a force which tends to move the cone backwards and forwards according to the direction of the current. The cone acts as a piston, transferring this force to the air, and producing the required sound waves.

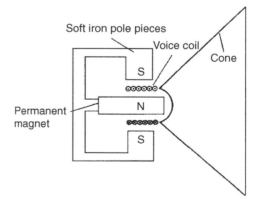

Figure 39.11

Problem 4. A conductor carries a current of 20 A and is at right-angles to a magnetic field having a flux density of 0.9 T. If the length of the conductor in the field is 30 cm, calculate the force acting on the conductor. Determine also the value of the force if the conductor is inclined at an angle of 30° to the direction of the field

$B = 0.9$ T, I $= 20$ A and $l = 30$ cm $= 0.30$ m.

Force $F = BIl = (0.9)(20)(0.30)$ newtons when the conductor is at right-angles to the field, as shown in Figure 39.12(a),

i.e. $F = 5.4$ N

When the conductor is inclined at 30° to the field, as shown in Figure 39.12(b), then force,

$$F = BIl \sin \theta$$

$$= (0.9)(20)(0.30) \sin 30° = 2.7$$ N

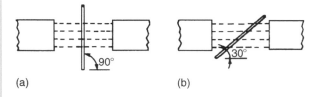

(a) (b)

Figure 39.12

If the current-carrying conductor shown in Figure 39.13(a) is placed in the magnetic field shown in Figure 39.13(a), then the two fields interact and cause a force to be exerted on the conductor as shown in Figure 39.13(b). The field is strengthened above the conductor and weakened below, thus tending to move the conductor downwards. This is the basic principle of operation of the electric motor (see section 39.6).

The direction of the force exerted on a conductor can be pre-determined by using **Fleming's*** **left-hand rule** (often called the **motor rule**) which states:

Let the thumb, first finger and second finger of the left hand be extended such that they are all at right-angles to each other (as shown in Figure 39.14). If the first finger points in the direction of the magnetic field, the second finger points in the direction of the current, then the thumb will point in the direction of the motion of the conductor.

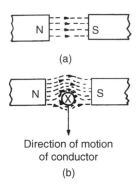

(a)

Direction of motion
of conductor

(b)

Figure 39.13

Summarising:

First finger – Field

seCond finger – Current

thuMb – Motion

*Who was Fleming? – **Sir John Ambrose Fleming** (29 November 1849–18 April 1945) was the English electrical engineer and physicist best known for inventing the vacuum tube, an invention that many credit as starting the modern electronics movement. To find out more about Fleming go to www.routledge.com/cw/bird

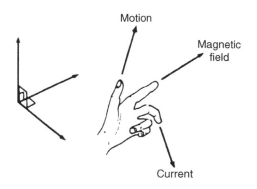

Figure 39.14

Problem 5. Determine the current required in a 400 mm length of conductor of an electric motor, when the conductor is situated at right-angles to a magnetic field of flux density 1.2 T, if a force of 1.92 N is to be exerted on the conductor. If the conductor is vertical, the current flowing downwards and the direction of the magnetic field is from left to right, what is the direction of the force?

Force $= 1.92$ N, $l = 400$ mm $= 0.40$ m and $B = 1.2$ T.

Since $F = BIl$, then $I = \dfrac{F}{Il}$ hence

$$\text{current } I = \frac{1.92}{(1.2)(0.4)} = \textbf{4 A}$$

If the current flows downwards, the direction of its magnetic field due to the current alone will be clockwise when viewed from above. The lines of flux will reinforce (i.e. strengthen) the main magnetic field at the back of the conductor and will be in opposition in the front (i.e. weaken the field). **Hence the force on the conductor will be from back to front (i.e. toward the viewer).** This direction may also have been deduced using Fleming's left-hand rule.

Problem 6. A conductor 350 mm long carries a current of 10 A and is at right-angles to a magnetic field lying between two circular pole faces each of radius 60 mm. If the total flux between the pole faces is 0.5 mWb, calculate the magnitude of the force exerted on the conductor

$l = 350$ mm $= 0.35$ m, $I = 10$ A, area of pole face $A = \pi r^2 = \pi (0.06)^2$ m^2 and $\Phi = 0.5$ mWb $= 0.5 \times 10^{-3}$ Wb.

Force $F = BIl$, and $B = \dfrac{\Phi}{A}$, hence force

$$F = \frac{\Phi}{A} Il = \frac{(0.5 \times 10^{-3})}{\pi (0.06)^2} (10)(0.35) \text{ newtons}$$

i.e.   **force $= 0.155$ N**

Problem 7. With reference to Figure 39.15 determine (a) the direction of the force on the conductor in Figure 39.15(a), (b) the direction of the force on the conductor in Figure 39.15(b), (c) the direction of the current in Figure 39.15(c), (d) the polarity of the magnetic system in Figure 39.15(d)

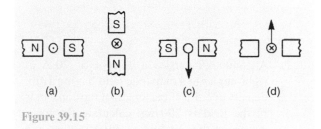

Figure 39.15

(a) The direction of the main magnetic field is from north to south, i.e. left to right. The current is flowing towards the viewer, and using the screw rule, the direction of the field is anticlockwise. Hence, either by Fleming's left-hand rule, or by sketching the interacting magnetic field as shown in Figure 39.16(a), the direction of the force on the conductor is seen to be upward.

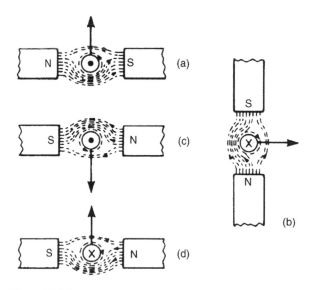

Figure 39.16

(b) Using a similar method to part (a) it is seen that the force on the conductor is to the right – see Figure 39.16(b).

(c) Using Fleming's left-hand rule, or by sketching as in Figure 39.16(c), it is seen that the current is toward the viewer, i.e. out of the paper.

(d) Similar to part (c), the polarity of the magnetic system is as shown in Figure 39.16(d).

Now try the following Practice Exercise

Practice Exercise 190   Further problems on the force on a current-carrying conductor (answers on page 547)

1. A conductor carries a current of 70 A at right-angles to a magnetic field having a flux density of 1.5 T. If the length of the conductor in the field is 200 mm calculate the force acting on the conductor. What is the force when the conductor and field are at an angle of 45°?

2. Calculate the current required in a 240 mm length of conductor of a d.c. motor when the conductor is situated at right-angles to the magnetic field of flux density 1.25 T, if a force of 1.20 N is to be exerted on the conductor

3. A conductor 30 cm long is situated at right-angles to a magnetic field. Calculate the strength of the magnetic field if a current of 15 A in the conductor produces a force on it of 3.6 N

4. A conductor 300 mm long carries a current of 13 A and is at right-angles to a magnetic field between two circular pole faces, each of diameter 80 mm. If the total flux between the pole faces is 0.75 mWb calculate the force exerted on the conductor

5. (a) A 400 mm length of conductor carrying a current of 25 A is situated at right-angles to a magnetic field between two poles of an electric motor. The poles have a circular cross-section. If the force exerted on the conductor is 80 N and the total flux between the pole faces is 1.27 mWb, determine the diameter of a pole face. (b) If the conductor in part (a) is vertical, the current flowing downwards and the direction of the magnetic field is from left to right, what is the direction of the 80 N force?

39.6   Principle of operation of a simple d.c. motor

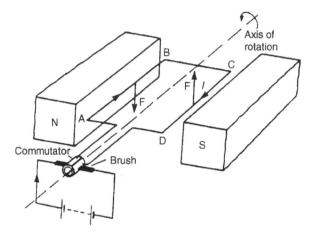

Figure 39.17

A rectangular coil which is free to rotate about a fixed axis is shown placed inside a magnetic field produced by permanent magnets in Figure 39.17. A direct current is fed into the coil via carbon brushes bearing on a commutator, which consists of a metal ring split into two halves separated by insulation. When current flows in the coil a magnetic field is set up around the coil which interacts with the magnetic field produced by the magnets. This causes a force F to be exerted on the current-carrying conductor which, by Fleming's left-hand rule, is downwards between points A and B and upward between C and D for the current direction shown. This causes a torque and the coil rotates anticlockwise. When the coil has turned through 90° from the position shown in Figure 39.17 the brushes connected to the positive and negative terminals of the supply make contact with different halves of the commutator ring, thus reversing the direction of the current flow in the conductor. If the current is not reversed and the coil rotates past this position the forces acting on it change direction and it rotates in the opposite direction thus never making more than half a revolution. The current direction is reversed every time the coil swings through the vertical position and thus the coil rotates anticlockwise for as long as the current flows. This is the principle of operation of a d.c. motor, which is thus a device that takes in electrical energy and converts it into mechanical energy.

39.7 Force on a charge

When a charge of Q coulombs is moving at a velocity of v m/s in a magnetic field of flux density B teslas, the charge moving perpendicular to the field, then the magnitude of the force F exerted on the charge is given by:

$$F = QvB \text{ newtons}$$

Problem 8. An electron in a television tube has a charge of 1.6×10^{-19} coulombs and travels at 3×10^7 m/s perpendicular to a field of flux density $18.5\ \mu$T. Determine the force exerted on the electron in the field

From above, force $F = QvB$ newtons, where

$$Q = \text{charge in coulombs} = 1.6 \times 10^{-19} \text{C},$$
$$v = \text{velocity of charge} = 3 \times 10^7 \text{ m/s}$$
and $$B = \text{flux density} = 18.5 \times 10^{-6} \text{ T}$$

Hence, **force on electron**

$$F = 1.6 \times 10^{-19} \times 3 \times 10^7 \times 18.5 \times 10^{-6}$$
$$= 1.6 \times 3 \times 18.5 \times 10^{-18}$$
$$= 88.8 \times 10^{-18} = \textbf{8.88} \times \textbf{10}^{-17} \textbf{N}$$

Now try the following Practice Exercises

Practice Exercise 191 Further problems on the force on a charge (answers on page 547)

1. Calculate the force exerted on a charge of 2×10^{-18} C travelling at 2×10^6 m/s perpendicular to a field of density 2×10^{-7} T

2. Determine the speed of a 10^{-19} C charge travelling perpendicular to a field of flux density 10^{-7} T, if the force on the charge is 10^{-20} N

Practice Exercise 192 Short answer questions on electromagnetism

1. State six practical applications of magnets
2. What is a permanent magnet?
3. Sketch the pattern of the magnetic field associated with a bar magnet. Mark the direction of the field
4. The direction of the magnetic field around a current-carrying conductor may be remembered using the ..... rule
5. Sketch the magnetic field pattern associated with a solenoid connected to a battery and wound on an iron bar. Show the direction of the field
6. Name three applications of electromagnetism
7. State what happens when a current-carrying conductor is placed in a magnetic field between two magnets
8. Define magnetic flux
9. The symbol for magnetic flux is ..... and the unit of flux is the .....
10. Define magnetic flux density
11. The symbol for magnetic flux density is ..... and the unit of flux density is .....
12. The force on a current-carrying conductor in a magnetic field depends on four factors. Name them
13. The direction of the force on a conductor in a magnetic field may be predetermined using Fleming's ...... rule
14. State two applications of the force on a current-carrying conductor
15. Explain, with the aid of a sketch, the action of a simplified d.c. motor

Practice Exercise 193 Multiple-choice questions on electromagnetism (answers on page 547)

1. The unit of magnetic flux density is the:
   (a) weber (b) weber per metre
   (c) ampere per metre (d) tesla

2. An electric bell depends for its action on:

   (a) a permanent magnet

   (b) reversal of current

   (c) a hammer and a gong

   (d) an electromagnet

3. A relay can be used to:

   (a) decrease the current in a circuit

   (b) control a circuit more readily

   (c) increase the current in a circuit

   (d) control a circuit from a distance

4. There is a force of attraction between two current-carrying conductors when the current in them is:

   (a) in opposite directions

   (b) in the same direction

   (c) of different magnitude

   (d) of the same magnitude

5. The magnetic field due to a current-carrying conductor takes the form of:

   (a) rectangles

   (b) concentric circles

   (c) wavy lines

   (d) straight lines radiating outwards

6. The total flux in the core of an electrical machine is 20 mWb and its flux density is 1 T. The cross-sectional area of the core is:

   (a) $0.05\,m^2$ (b) $0.02\,m^2$

   (c) $20\,m^2$ (d) $50\,m^2$

7. A conductor carries a current of 10 A at right-angles to a magnetic field having a flux density of 500 mT. If the length of the conductor in the field is 20 cm, the force on the conductor is:

   (a) 100 kN (b) 1 kN

   (c) 100 N (d) 1 N

8. If a conductor is horizontal, the current flowing from left to right and the direction of the surrounding magnetic field is from above to below, the force exerted on the conductor is:

   (a) from left to right

   (b) from below to above

   (c) away from the viewer

   (d) towards the viewer

9. For the current-carrying conductor lying in the magnetic field shown in Figure 39.18(a), the direction of the force on the conductor is:

   (a) to the left   (b) upwards

   (c) to the right   (d) downwards

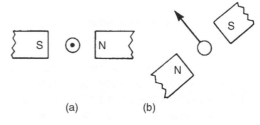

(a)   (b)

Figure 39.18

10. For the current-carrying conductor lying in the magnetic field shown in Figure 39.18(b), the direction of the current in the conductor is:

    (a) towards the viewer

    (b) away from the viewer

11. Figure 39.19 shows a rectangular coil of wire placed in a magnetic field and free to rotate about axis *AB*. If the current flows into the coil at *C*, the coil will:

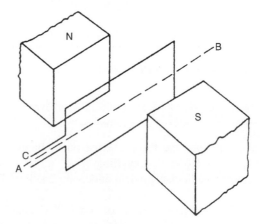

Figure 39.19

(a) commence to rotate anticlockwise

(b) commence to rotate clockwise

(c) remain in the vertical position

(d) experience a force towards the north pole

12. The force on an electron travelling at 10^7 m/s in a magnetic field of density $10 \, \mu T$ is 1.6×10^{-17} N. The electron has a charge of:

(a) 1.6×10^{-28}C   (b) 1.6×10^{-15}C

(c) 1.6×10^{-19}C   (d) 1.6×10^{-25}C

For fully worked solutions to each of the problems in Exercises 189 to 193 in this chapter, go to the website:
www.routledge.com/cw/bird

Chapter 40

Electromagnetic induction

Why it is important to understand: **Electromagnetic induction**

Electromagnetic induction is the production of a potential difference (voltage) across a conductor when it is exposed to a varying magnetic field. Michael Faraday is generally credited with the discovery of induction in the 1830s. Faraday's law of induction is a basic law of electromagnetism that predicts how a magnetic field will interact with an electric circuit to produce an electromotive force (e.m.f.). It is the fundamental operating principle of transformers, inductors, and many types of electrical motors, generators and solenoids. A.c. generators use Faraday's law to produce rotation and thus convert electrical and magnetic energy into rotational kinetic energy. This idea can be used to run all kinds of motors. Probably one of the greatest inventions of all time is the transformer. Alternating current from the primary coil moves quickly back and forth across the secondary coil. The moving magnetic field caused by the changing field (flux) induces a current in the secondary coil. This chapter explains electromagnetic induction, Faraday's laws, Lenz's law and Fleming's rule and develops various calculations to help understanding of the concepts.

The transformer is one of the simplest of electrical devices. Its basic design, materials, and principles have changed little over the last one hundred years, yet transformer designs and materials continue to be improved. Transformers are essential in high voltage power transmission providing an economical means of transmitting power over large distances. A major application of transformers is to increase voltage before transmitting electrical energy over long distances through cables. Cables have resistance and so dissipate electrical energy. By transforming electrical power to a high-voltage, and therefore low-current form, for transmission and back again afterward, transformers enable economical transmission of power over long distances. Consequently, transformers have shaped the electricity supply industry, permitting generation to be located remotely from points of demand. All but a tiny fraction of the world's electrical power has passed through a series of transformers by the time it reaches the consumer. Transformers are also used extensively in electronic products to step down the supply voltage to a level suitable for the low voltage circuits they contain. The transformer also electrically isolates the end user from contact with the supply voltage. Signal and audio transformers are used to couple stages of amplifiers and to match devices such as microphones and record players to the input of amplifiers. Audio transformers allowed telephone circuits to carry on a two-way conversation over a single pair of wires. This chapter explains the principle of operation of a transformer.

Science and Mathematics for Engineering. 978-0-367-20475-4, © John Bird. Published by Taylor & Francis. All rights reserved.

At the end of this chapter, you should be able to:

- understand how an e.m.f. may be induced in a conductor
- state Faraday's laws of electromagnetic induction
- state Lenz's law
- use Fleming's right-hand rule for relative directions
- appreciate that the induced e.m.f., $E = Blv$ or $E = Blv \sin \theta$
- calculate induced e.m.f. given B, l, v and θ and determine relative directions
- define inductance L and state its unit
- define mutual inductance
- calculate mutual inductance using $E_2 = -M \dfrac{dI_1}{dt}$
- understand the principle of operation of a transformer
- understand the term 'rating' of a transformer
- use $\dfrac{V_1}{V_2} = \dfrac{N_1}{N_2} = \dfrac{I_2}{I_1}$ in calculations on transformers

40.1 Introduction to electromagnetic induction

When a conductor is moved across a magnetic field so as to cut through the lines of force (or flux), an electromotive force (e.m.f.) is produced in the conductor. If the conductor forms part of a closed circuit then the e.m.f. produced causes an electric current to flow round the circuit. Hence an e.m.f. (and thus current) is 'induced' in the conductor as a result of its movement across the magnetic field. This effect is known as '**electromagnetic induction**'.

Figure 40.1(a) shows a coil of wire connected to a centre-zero galvanometer, which is a sensitive ammeter with the zero-current position in the centre of the scale.

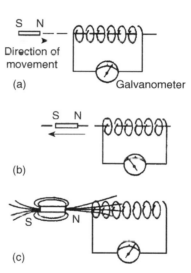

Figure 40.1

(a) When the magnet is moved at constant speed towards the coil (Figure 40.1(a)), a deflection is noted on the galvanometer showing that a current has been produced in the coil.

(b) When the magnet is moved at the same speed as in (a) but away from the coil the same deflection is noted but is in the opposite direction (see Figure 40.1(b)).

(c) When the magnet is held stationary, even within the coil, no deflection is recorded.

(d) When the coil is moved at the same speed as in (a) and the magnet held stationary the same galvanometer deflection is noted.

(e) When the relative speed is, say, doubled, the galvanometer deflection is doubled.

(f) When a stronger magnet is used, a greater galvanometer deflection is noted.

(g) When the number of turns of wire of the coil is increased, a greater galvanometer deflection is noted.

Figure 40.1(c) shows the magnetic field associated with the magnet. As the magnet is moved towards the coil, the magnetic flux of the magnet moves across, or cuts, the coil. **It is the relative movement of the magnetic flux and the coil that causes an e.m.f. and thus current, to be induced in the coil.** This effect is known as **electromagnetic induction**. The laws of electromagnetic induction stated in section 40.2 evolved from experiments such as those described above.

40.2 Laws of electromagnetic induction

Faraday's* laws of electromagnetic induction state:

(i) *An induced e.m.f. is set up whenever the magnetic field linking that circuit changes.*

(ii) *The magnitude of the induced e.m.f. in any circuit is proportional to the rate of change of the magnetic flux linking the circuit.*

Lenz's* law states:

The direction of an induced e.m.f. is always such that it tends to set up a current opposing the motion or the change of flux responsible for inducing that e.m.f.

An alternative method to Lenz's law of determining relative directions is given by **Fleming's* Right-hand rule** (often called the geneRator rule) which states:

Let the thumb, first finger and second finger of the right hand be extended such that they are all at right angles to each other (as shown in Figure 40.2). If the first finger points in the direction of the magnetic field and the thumb points in the direction of motion of the conductor relative to the magnetic field, then

the second finger will point in the direction of the induced e.m.f.

Summarising:

First finger – Field

thuMb – Motion

sEcond finger – E.m.f.

In a generator, conductors forming an electric circuit are made to move through a magnetic field. By Faraday's law an e.m.f. is induced in the conductors and thus a source of e.m.f. is created. A generator converts mechanical energy into electrical energy. (The action of a simple a.c. generator is described in chapter 41.) The induced e.m.f. E set up between the ends of the conductor shown in Figure 40.3 is given by:

$$E = Blv \text{ volts}$$

*Who was Lenz? – **Heinrich Friedrich Emil Lenz** (12 February 1804–10 February 1865) was a Russian physicist remembered for formulating Lenz's law in electrodynamics. To find out more about Lenz go to www.routledge.com/cw/bird

*Who was Fleming? – **Sir John Ambrose Fleming** (29 November 1849–18 April 1945) was the English electrical engineer and physicist best known for inventing the vacuum tube, an invention that many credit as starting the modern electronics movement. To find out more about Fleming go to www.routledge.com/cw/bird (For image refer to page 432)

*Who was Faraday? – **Michael Faraday** (22 September 1791–25 August 1867) was an English scientist whose main discoveries include electromagnetic induction, diamagnetism and electrolysis. The SI unit of capacitance, the farad, is named in his honour. To find out more about Faraday go to www.routledge.com/cw/bird (For image refer to page 426)

Motion

Magnetic field

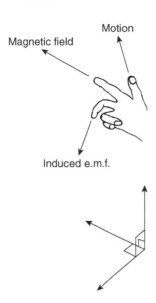

Induced e.m.f.

Figure 40.2

where B the flux density, is measured in teslas, l the length of conductor in the magnetic field, is measured in metres, and v the conductor velocity, is measured in metres per second.

Magnetic flux density B

Conductor

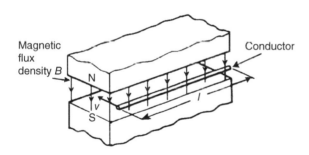

Figure 40.3

If the conductor moves at an angle $\theta°$ to the magnetic field (instead of at 90° as assumed above) then

$$E = Blv \sin \theta \text{ volts}$$

Problem 1. A conductor 300 mm long moves at a uniform speed of 4 m/s at right-angles to a uniform magnetic field of flux density 1.25 T. Determine the current flowing in the conductor when (a) its ends are open-circuited, (b) its ends are connected to a load of 20 Ω resistance

When a conductor moves in a magnetic field it will have an e.m.f. induced in it but this e.m.f. can only produce a current if there is a closed circuit.

Induced e.m.f. $E = Blv = (1.25)\left(\dfrac{300}{1000}\right)(4) = 1.5\text{ V}$

(a) If the ends of the conductor are open-circuited **no current will flow** even though 1.5 V has been induced.

(b) From Ohm's law, current

$$I = \frac{E}{R} = \frac{1.5}{20} = \textbf{0.075 A} \text{ or } \textbf{75 mA}$$

Problem 2. At what velocity must a conductor 75 mm long cut a magnetic field of flux density 0.6 T if an e.m.f. of 9 V is to be induced in it? Assume the conductor, the field and the direction of motion are mutually perpendicular

Induced e.m.f. $E = Blv$, hence velocity $v = \dfrac{E}{Bl}$

Hence velocity

$$v = \frac{9}{(0.6)(75 \times 10^{-3})}$$

$$= \frac{9 \times 10^3}{0.6 \times 75} = \textbf{200 m/s}$$

Problem 3. A conductor moves with a velocity of 15 m/s at an angle of (a) 90°, (b) 60° and (c) 30° to a magnetic field produced between two square-faced poles of side length 2 cm. If the flux leaving a pole face is 5 μWb, find the magnitude of the induced e.m.f. in each case

$v = 15$ m/s, length of conductor in magnetic field $l = 2$ cm $= 0.02$ m, $A = 2 \times 2$ cm$^2 = 4 \times 10^{-4}$ m^2 and $\Phi = 5 \times 10^{-6}$ Wb.

(a) $E_{90} = Blv \sin 90° = \left(\dfrac{\Phi}{A}\right)lv \sin 90°$

$$= \left(\frac{5 \times 10^{-6}}{4 \times 10^{-4}}\right)(0.02)(15)(1) = \textbf{3.75 mV}$$

(b) $E_{60} = Blv \sin 60° = E_{90} \sin 60°$

$$= 3.75 \sin 60° = \textbf{3.25 mV}$$

(c) $E_{30} = Blv \sin 30° = E_{90} \sin 30°$

   $= 3.75 \sin 30° = \mathbf{1.875\ mV}$

Problem 4.   The wing span of a metal aeroplane is 36 m. If the aeroplane is flying at 400 km/h, determine the e.m.f. induced between its wing tips. Assume the vertical component of the Earth's magnetic field is 40 μT

Induced e.m.f. across wing tips, $E = Blv$,

   $B = 40\,\mu T = 40 \times 10^{-6}\,T, \quad l = 36\,m$

and $v = 400\dfrac{km}{h} \times 1000\dfrac{m}{km} \times \dfrac{1\,h}{60 \times 60\,s}$

   $= \dfrac{(400)(1000)}{3600} = \dfrac{4000}{36}\,m/s.$

Hence,

   $E = Blv = (40 \times 10^{-6})(36)\left(\dfrac{4000}{36}\right) = \mathbf{0.16\ V}$

Problem 5.   The diagram shown in Figure 40.4 represents the generation of e.m.f.s. Determine (a) the direction in which the conductor has to be moved in Figure 40.4(a), (b) the direction of the induced e.m.f. in Figure 40.4(b), (c) the polarity of the magnetic system in Figure 40.4(c)

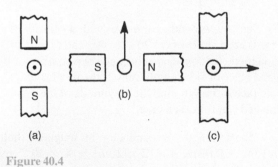

Figure 40.4

The direction of the e.m.f., and thus the current due to the e.m.f. may be obtained by either Lenz's law or Fleming's right-hand rule (i.e. GeneRator rule).

(a)   *Using Lenz's law:* The field due to the magnet and the field due to the current-carrying conductor are shown in Figure 40.5(a) and are seen to reinforce to the left of the conductor. Hence the force on the conductor is to the right. However, Lenz's law states that the direction of the induced e.m.f. is always such as to oppose the effect producing it.

Thus the conductor will have to be moved to the left.

(b)   *Using Fleming's right-hand rule:*

   First finger – Field, i.e. north to south, or right to left

   thuMb – Motion, i.e. upwards

   sEcond finger – E.m.f.

i.e. **towards the viewer or out of the paper**, as shown in Figure 40.5(b).

(c)   The polarity of the magnetic system of Figure 40.4(c) is shown in Figure 40.5(c) and is obtained using Fleming's right-hand rule.

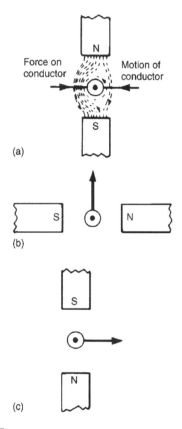

Figure 40.5

Now try the following Practice Exercise

Practice Exercise 194   Further problems on induced e.m.f. (answers on page 548)

1.   A conductor of length 15 cm is moved at 750 mm/s at right-angles to a uniform flux density of 1.2 T. Determine the e.m.f. induced in the conductor

2. Find the speed that a conductor of length 120 mm must be moved at right-angles to a magnetic field of flux density 0.6 T to induce in it an e.m.f. of 1.8 V

3. A 25 cm long conductor moves at a uniform speed of 8 m/s through a uniform magnetic field of flux density 1.2 T. Determine the current flowing in the conductor when (a) its ends are open-circuited, (b) its ends are connected to a load of 15 ohms resistance

4. A straight conductor 500 mm long is moved with constant velocity at right-angles both to its length and to a uniform magnetic field. Given that the e.m.f. induced in the conductor is 2.5 V and the velocity is 5 m/s, calculate the flux density of the magnetic field. If the conductor forms part of a closed circuit of total resistance 5 ohms, calculate the force on the conductor

5. A car is travelling at 80 km/h. Assuming the back axle of the car is 1.76 m in length and the vertical component of the Earth's magnetic field is 40 μT, find the e.m.f. generated in the axle due to motion

6. A conductor moves with a velocity of 20 m/s at an angle of (a) 90°, (b) 45°, (c) 30° to a magnetic field produced between two square-faced poles of side length 2.5 cm. If the flux on the pole face is 60 mWb, find the magnitude of the induced e.m.f. in each case

7. A conductor 400 mm long is moved at 70° to a 0.85 T magnetic fields. If it has a velocity of 115 km/h, calculate (a) the induced voltage, (b) the force acting on the conductor if connected to an 8 Ω resistor

40.3 Self inductance

Inductance is the name given to the property of a circuit whereby there is an e.m.f. induced into the circuit by the change of flux linkages produced by a current change, as stated in chapter 42, page 471.

When the e.m.f. is induced in the same circuit as that in which the current is changing, the property is called **self inductance L**.

The unit of inductance is the **henry**[*], **H**.

40.4 Mutual inductance

When an e.m.f. is induced in a circuit by a change of flux due to current changing in an adjacent circuit, the property is called **mutual inductance M**.
Mutually induced e.m.f. in the second coil

$$E_2 = -M\frac{dI_1}{dt} \text{ volts}$$

where M is the **mutual inductance** between two coils, in henries, and $\frac{dI_1}{dt}$ is the rate of change of current in the first coil.

Problem 6. Calculate the mutual inductance between two coils when a current changing at 200 A/s in one coil induces an e.m.f. of 1.5 V in the other

Induced e.m.f. $\left| E_2 \right| = M\frac{dI_1}{dt}$, i.e. $1.5 = M(200)$

Thus, **mutual inductance**

$$M = \frac{1.5}{200} = 0.0075 \text{ H} \quad \text{or} \quad 7.5 \text{ mH}$$

Problem 7. The mutual inductance between two coils is 18 mH. Calculate the steady rate of change of current in one coil to induce an e.m.f. of 0.72 V in the other

Induced e.m.f. $\left| E_2 \right| = M\frac{dI_1}{dt}$

Hence rate of change of current

$$\frac{dI_1}{dt} = \frac{\left| E_2 \right|}{M} = \frac{0.72}{0.018} = 40 \text{ A/s}$$

Problem 8. Two coils have a mutual inductance of 0.2 H. If the current in one coil is changed from 10 A to 4 A in 10 ms, calculate (a) the average induced e.m.f. in the second coil, (b) the change of flux linked with the second coil if it is wound with 500 turns

[*]Who was Henry? – Joseph Henry (17 December 1797–13 May 1878) was an American scientist who discovered the electromagnetic phenomenon of self-inductance. To find out more about Henry go to www.routledge.com/cw/bird (For image refer to page 30)

(a) **Induced e.m.f.** $E_2 = -M\dfrac{dI_1}{dt}$

$$= -(0.2)\left(\frac{10-4}{10 \times 10^{-3}}\right)$$

$$= \mathbf{-120\,V}$$

(b) From section 41.8, induced e.m.f. $|E_2| = N\dfrac{d\Phi}{dt}$

hence $d\Phi = \dfrac{|E_2|\,dt}{N}$

Thus the **change of flux**

$$d\Phi = \frac{(120)(10 \times 10^{-3})}{500} = \mathbf{2.4\,mWb}$$

Now try the following Practice Exercise

Practice Exercise 195   Further problems on
mutual inductance (answers on page 548)

1. The mutual inductance between two coils is 150 mH. Find the magnitude of the e.m.f. induced in one coil when the current in the other is increasing at a rate of 30 A/s

2. Determine the mutual inductance between two coils when a current changing at 50 A/s in one coil induces an e.m.f. of 80 mV in the other

3. Two coils have a mutual inductance of 0.75 H. Calculate the magnitude of the e.m.f. induced in one coil when a current of 2.5 A in the other coil is reversed in 15 ms

4. The mutual inductance between two coils is 240 mH. If the current in one coil changes from 15 A to 6 A in 12 ms, calculate (a) the average e.m.f. induced in the other coil, (b) the change of flux linked with the other coil if it is wound with 400 turns

---

40.5   The transformer

A transformer is a device which uses the phenomenon of mutual induction to change the values of alternating voltages and currents. In fact, one of the main advantages of a.c. transmission and distribution is the ease with which an alternating voltage can be increased or decreased by transformers.

Losses in transformers are generally low and thus efficiency is high. Being static they have a long life and are very stable.

Transformers range in size from the miniature units used in electronic applications to the large power transformers used in power stations. The principle of operation is the same for each.

A transformer is represented in Figure 40.6(a) as consisting of two electrical circuits linked by a common ferromagnetic core. One coil is termed the **primary winding** which is connected to the supply of electricity, and the other the **secondary winding**, which may be connected to a load. A circuit diagram symbol for a transformer is shown in Figure 40.6(b).

40.5.1   Transformer principle of operation

When the secondary is an open-circuit and an alternating voltage V_1 is applied to the primary winding, a small current – called the no-load current I_0 – flows, which sets up a magnetic flux in the core. This alternating flux links with both primary and secondary coils and induces in them e.m.f.s of E_1 and E_2 respectively by mutual induction.

The induced e.m.f. E in a coil of N turns is given by

$E = -N\dfrac{d\Phi}{dt}$ volts, where $\dfrac{d\Phi}{dt}$ is the rate of change of

flux (see chapter 42, page 447). In an ideal transformer, the rate of change of flux is the same for both primary and secondary coils and thus $\dfrac{E_1}{N_1} = \dfrac{E_2}{N_2}$, i.e. **the induced e.m.f. per turn is constant**.

Assuming no losses, $E_1 = V_1$ and $E_2 = V_2$

Hence $\qquad \dfrac{V_1}{N_1} = \dfrac{V_2}{N_2}$ or $\dfrac{V_1}{V_2} = \dfrac{N_1}{N_2}$ (1)

$\dfrac{V_1}{V_2}$ is called the voltage ratio and $\dfrac{N_1}{N_2}$ the turns ratio, or the '**transformation ratio**' of the transformer. If N_2 is less than N_1 then V_2 is less than V_1 and the device is termed a **step-down transformer**. If N_2 is greater than N_1 then V_2 is greater than V_1 and the device is termed a **step-up transformer**.

When a load is connected across the secondary winding, a current I_2 flows. In an ideal transformer losses are neglected and a transformer is considered to be 100% efficient.

Hence input power = output power, or $V_1I_1 = V_2I_2$, i.e. in an ideal transformer the **primary and secondary ampere-turns are equal**.

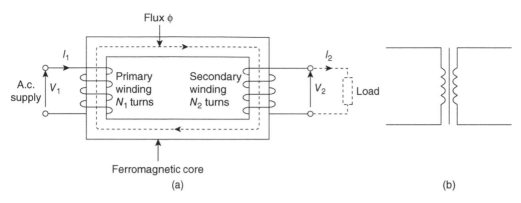

Figure 40.6

Thus $\dfrac{V_1}{V_2} = \dfrac{I_2}{I_1}$ (2)

Combining equations (1) and (2) gives:

$$\frac{V_1}{V_2} = \frac{N_1}{N_2} = \frac{I_2}{I_1}$$ (3)

The **rating** of a transformer is stated in terms of the volt-amperes that it can transform without overheating. With reference to Figure 40.6(a), the transformer rating is either $V_1 I_1$ or $V_2 I_2$, where I_2 is the full-load secondary current.

Problem 9. A transformer has 500 primary turns and 3000 secondary turns. If the primary voltage is 240 V, determine the secondary voltage, assuming an ideal transformer

For an ideal transformer, voltage ratio = turns ratio

i.e. $\dfrac{V_1}{V_2} = \dfrac{N_1}{N_2}$ hence $\dfrac{240}{V_2} = \dfrac{500}{3000}$

Thus, **secondary voltage**

$$V_2 = \frac{(240)(3000)}{500} = \textbf{1440 V} \text{ or } \textbf{1.44 kV}$$

Problem 10. An ideal transformer with a turns ratio of 2:7 is fed from a 240 V supply. Determine its output voltage

A turns ratio of 2:7 means that the transformer has 2 turns on the primary coil for every 7 turns on the secondary coil (i.e. it is a step-up transformer); thus $\dfrac{N_1}{N_2} = \dfrac{2}{7}$

For an ideal transformer,

$$\frac{N_1}{N_2} = \frac{V_1}{V_2}   \text{ hence }   \frac{2}{7} = \frac{240}{V_2}$$

Thus the secondary voltage

$$V_2 = \frac{(240)(7)}{2} = \textbf{840 V}$$

Problem 11. An ideal transformer has a turns ratio of 8:1 and the primary current is 3 A when it is supplied at 240 V. Calculate the secondary voltage and current

A turns ratio of 8:1 means $\dfrac{N_1}{N_2} = \dfrac{8}{1}$ (i.e. it is a step-down transformer).

$$\frac{N_1}{N_2} = \frac{V_1}{V_2}$$

hence, secondary voltage

$$V_2 = V_1 \left(\frac{N_2}{N_1}\right) = 240 \left(\frac{1}{8}\right) = \textbf{30 volts}$$

Also, $\dfrac{N_1}{N_2} = \dfrac{I_2}{I_1}$ hence, secondary current

$$I_2 = I_1 \left(\frac{N_1}{N_2}\right) = 3 \left(\frac{8}{1}\right) = \textbf{24 A}$$

Problem 12. An ideal transformer, connected to a 240 V mains, supplies a 12 V, 150 W lamp. Calculate the transformer turns ratio and the current taken from the supply

$V_1 = 240$ V, $V_2 = 12$ V, and since power $P = VI$, then

$$I_2 = \frac{P}{V_2} = \frac{150}{12} = 12.5\,\text{A}$$

Turns ratio $= \dfrac{N_1}{N_2} = \dfrac{V_1}{V_2} = \dfrac{240}{12} = \mathbf{20}$

$\dfrac{V_1}{V_2} = \dfrac{I_2}{I_1}$ from which, $I_1 = I_2 \left(\dfrac{V_2}{V_1}\right)$

$= 12.5 \left(\dfrac{12}{240}\right)$

Hence, **the current taken from the supply**

$$I_1 = \dfrac{12.5}{20} = \mathbf{0.625\ A}$$

Problem 13. A 12 Ω resistor is connected across the secondary winding of an ideal transformer whose secondary voltage is 120 V. Determine the primary voltage if the supply current is 4 A

Secondary current

$$I_2 = \dfrac{V_2}{R_2} = \dfrac{120}{12} = 10\,A$$

$$\dfrac{V_1}{V_2} = \dfrac{I_2}{I_1}$$

from which, the primary voltage

$$V_1 = V_2 \left(\dfrac{I_2}{I_1}\right) = 120 \left(\dfrac{10}{4}\right) = \mathbf{300\ volts}$$

Now try the following Practice Exercises

Practice Exercise 196   Further problems on the transformer principle of operation (answers on page 548)

1. A transformer has 600 primary turns connected to a 1.5 kV supply. Determine the number of secondary turns for a 240 V output voltage, assuming no losses

2. An ideal transformer with a turns ratio of 2:9 is fed from a 220 V supply. Determine its output voltage

3. An ideal transformer has a turns ratio of 12:1 and is supplied at 192 V. Calculate the secondary voltage

4. A transformer primary winding connected across a 415 V supply has 750 turns. Determine how many turns must be wound on the secondary side if an output of 1.66 kV is required

5. An ideal transformer has a turns ratio of 15:1 and is supplied at 180 V when the primary current is 4 A. Calculate the secondary voltage and current

6. A step-down transformer having a turns ratio of 20:1 has a primary voltage of 4 kV and a load of 10 kW. Neglecting losses, calculate the value of the secondary current

7. A transformer has a primary to secondary turns ratio of 1:15. Calculate the primary voltage necessary to supply a 240 V load. If the load current is 3 A determine the primary current. Neglect any losses

8. A 20 Ω resistance is connected across the secondary winding of a single-phase power transformer whose secondary voltage is 150 V. Calculate the primary voltage and the turns ratio if the supply current is 5 A, neglecting losses

Practice Exercise 197   Short answer questions on electromagnetic induction

1. What is electromagnetic induction?

2. State Faraday's laws of electromagnetic induction

3. State Lenz's law

4. Explain briefly the principle of the generator

5. The direction of an induced e.m.f. in a generator may be determined using Fleming's ...... rule

6. The e.m.f. E induced in a moving conductor may be calculated using the formula $E = Blv$. Name the quantities represented and their units

7. What is self-inductance? State its symbol and unit

8. What is mutual inductance? State its symbol and unit

9. The mutual inductance between two coils is M. The e.m.f. E_2 induced in one coil by the current changing at $\dfrac{dI_1}{dt}$ in the other is given by: $E_2 = $ ...... volts

10. What is a transformer?

11. Explain briefly how a voltage is induced in the secondary winding of a transformer

12. Draw the circuit diagram symbol for a transformer

13. State the relationship between turns and voltage ratios for a transformer

14. How is a transformer rated?

15. Briefly describe the principle of operation of a transformer

Practice Exercise 198   Multiple-choice questions on electromagnetic induction (Answers on page 548)

1. A current changing at a rate of 5 A/s in a coil of inductance 5 H induces an e.m.f. of:

   (a) 25 V in the same direction as the applied voltage

   (b) 1 V in the same direction as the applied voltage

   (c) 25 V in the opposite direction to the applied voltage

   (d) 1 V in the opposite direction to the applied voltage

2. A bar magnet is moved at a steady speed of 1.0 m/s towards a coil of wire which is connected to a centre-zero galvanometer. The magnet is now withdrawn along the same path at 0.5 m/s. The deflection of the galvanometer is in the:

   (a) same direction as previously, with the magnitude of the deflection doubled

   (b) opposite direction as previously, with the magnitude of the deflection halved

   (c) same direction as previously, with the magnitude of the deflection halved

   (d) opposite direction as previously, with the magnitude of the deflection doubled

3. An e.m.f. of 1 V is induced in a conductor moving at 10 cm/s in a magnetic field of 0.5 T. The effective length of the conductor in the magnetic field is:

   (a) 20 cm   (b) 5 m   (c) 20 m   (d) 50 m

4. Which of the following is false?

   (a) Fleming's left-hand rule or Lenz's law may be used to determine the direction of an induced e.m.f.

   (b) An induced e.m.f. is set up whenever the magnetic field linking that circuit changes

   (c) The direction of an induced e.m.f. is always such as to oppose the effect producing it

   (d) The induced e.m.f. in any circuit is proportional to the rate of change of the magnetic flux linking the circuit

5. A strong permanent magnet is plunged into a coil and left in the coil. What is the effect produced on the coil after a short time?

   (a) The coil winding becomes hot

   (b) The insulation of the coil burns out

   (c) A high voltage is induced

   (d) There is no effect

6. Self-inductance occurs when:

   (a) the current is changing

   (b) the circuit is changing

   (c) the flux is changing

   (d) the resistance is changing

7. Faraday's laws of electromagnetic induction are related to:

   (a) the e.m.f. of a chemical cell

   (b) the e.m.f. of a generator

   (c) the current flowing in a conductor

   (d) the strength of a magnetic field

8. The mutual inductance between two coils, when a current changing at 20 A/s in one coil induces an e.m.f. of 10 mV in the other, is:

   (a) 0.5 H   (b) 200 mH   (c) 0.5 mH   (d) 2 H

9. A transformer has 800 primary turns and 100 secondary turns. To obtain 40 V from the secondary winding, the voltage applied to the primary winding must be:

   (a) 5 V   (b) 20 V   (c) 2.5 V   (d) 320 V

10. A step-up transformer has a turns ratio of 10. If the output current is 5 A, the input current is:

    (a) 50 A   (b) 5 A   (c) 2.5 A   (d) 0.5 A

11. A 440 V/110 V transformer has 1000 turns on the primary winding. The number of turns on the secondary winding is:

    (a) 550   (b) 250   (c) 4000   (d) 25

12. A 1 kV/250 V transformer has 500 turns on the secondary winding. The number of turns on the primary winding is:

    (a) 2000   (b) 125   (c) 1000   (d) 250

13. The power input to a mains transformer is 200 W. If the primary current is 2.5 A, the secondary voltage is 2 V and assuming no losses in the transformer, the turns ratio is:

    (a) 80:1 step-up   (b) 40:1 step-up

    (c) 80:1 step-down   (d) 40:1 step-down

14. An ideal transformer has a turns ratio of 1:5 and is supplied at 200 V when the primary current is 3 A. Which of the following statements is false?

    (a)   The turns ratio indicates a step-up transformer

    (b)   The secondary voltage is 40 V

    (c)   The secondary current is 15 A

    (d)   The transformer rating is 0.6 kVA

    (e)   The secondary voltage is 1 kV

    (f)   The secondary current is 0.6 A

For fully worked solutions to each of the problems in Exercises 194 to 198 in this chapter, go to the website:

www.routledge.com/cw/bird

This assignment covers the material contained in chapters 38 to 40. *The marks for each question are shown in brackets at the end of each question.*

1. Determine the value of currents I_1 to I_5 shown in Figure RT14.1. (5)

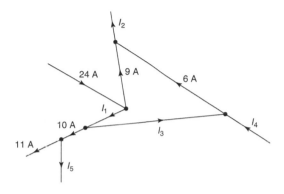

Figure RT14.1

2. Find the current flowing in the 5 Ω resistor of the circuit shown in Figure RT14.2 using Kirchhoff's laws. Find also the current flowing in each of the other two branches of the circuit. (11)

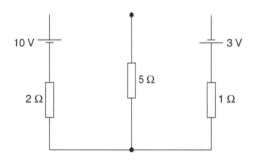

Figure RT14.2

3. The maximum working flux density of a lifting electromagnet is 1.7 T and the effective area of a pole face is circular in cross-section. If the total magnetic flux produced is 214 mWb, determine the radius of the pole face. (6)

4. A conductor, 25 cm long is situated at right angles to a magnetic field. Determine the strength of the magnetic field if a current of 12 A in the conductor produces a force on it of 4.5 N. (5)

5. An electron in a television tube has a charge of 1.5×10^{-19}C and travels at 3×10^7 m/s perpendicular to a field of flux density 20 μT. Calculate the force exerted on the electron in the field. (5)

6. A lorry is travelling at 100 km/h. Assuming the vertical component of the Earth's magnetic field is 40 μT and the back axle of the lorry is 1.98 m, find the e.m.f. generated in the axle due to motion. (6)

7. Two coils, P and Q, have a mutual inductance of 100 mH. If a current of 3 A in coil P is reversed in 20 ms, determine (a) the average e.m.f. induced in coil Q, (b) the flux change linked with coil Q if it is wound with 200 turns. (6)

8. An ideal transformer connected to a 250 V mains, supplies a 25 V, 200 W lamp. Calculate the transformer turns ratio and the current taken from the supply. (6)

For lecturers/instructors/teachers, fully worked solutions to each of the problems in Revision Test 14, together with a full marking scheme, are available at the website:

www.routledge.com/cw/bird

Chapter 41

Alternating voltages and currents

Why it is important to understand: **Alternating voltages and currents**

With alternating current (a.c.), the flow of electric charge periodically reverses direction, whereas with direct current (d.c.), the flow of electric charge is only in one direction. In a power station, electricity can be made most easily by using a gas or steam turbine or water impeller to drive a generator consisting of a spinning magnet inside a set of coils. The resultant voltage is always 'alternating' by virtue of the magnet's rotation. Now, alternating voltage can be carried around the country via cables far more effectively than direct current because a.c. can be passed through a transformer and a high voltage can be reduced to a low voltage, suitable for use in homes. The electricity arriving at your home is alternating voltage. Electric light bulbs and toasters can operate perfectly from 230 volts a.c. Other equipment such as televisions have an internal power supply which converts the 230 volts a.c. to a low d.c. voltage for the electronic circuits. How is this done? There are several ways but the simplest is to use a transformer to reduce the voltage to, say 12 volts a.c. This lower voltage can be fed through a 'rectifier' which combines the negative and positive alternating cycles so that only positive cycles emerge. A.c. is the form in which electric power is delivered to businesses and residences. The usual waveform of an a.c. power circuit is a sine wave. In certain applications, different waveforms are used, such as triangular or square waves. Audio and radio signals carried on electrical wires are also examples of alternating current. The frequency of the electrical system varies by country; most electric power is generated at either 50 or 60 hertz. Some countries have a mixture of 50 Hz and 60 Hz supplies, notably Japan. A low frequency eases the design of electric motors, particularly for hoisting, crushing and rolling applications, and commutator-type traction motors for applications such as railways. However, low frequency also causes noticeable flicker in arc lamps and incandescent light bulbs. The use of lower frequencies also provides the advantage of lower impedance losses, which are proportional to frequency. 16.7 Hz frequency is still used in some European rail systems, such as in Austria, Germany, Norway, Sweden and Switzerland. Off-shore, military, textile industry, marine, computer mainframe, aircraft, and spacecraft applications sometimes use 400 Hz, for benefits of reduced weight of apparatus or higher motor speeds. This chapter introduces alternating current and voltages, with its terminology and values.

Science and Mathematics for Engineering. 978-0-367-20475-4, © John Bird. Published by Taylor & Francis. All rights reserved.

At the end of this chapter, you should be able to:

- appreciate why a.c. is used in preference to d.c.
- describe the principle of operation of an a.c. generator
- distinguish between unidirectional and alternating waveforms
- define cycle, period or periodic time T and frequency f of a waveform
- perform calculations involving $T = \dfrac{1}{f}$
- define instantaneous, peak, mean and r.m.s. values, and form and peak factors for a sine wave
- calculate mean and r.m.s. values and form and peak factors for given waveforms

41.1 Introduction

Electricity is produced by generators at power stations and then distributed by a vast network of transmission lines (called the National Grid system) to industry and for domestic use. It is easier and cheaper to generate **alternating current (a.c.)** than direct current (d.c.) and a.c. is more conveniently distributed than d.c. since its voltage can be readily altered using transformers. Whenever d.c. is needed in preference to a.c., devices called **rectifiers** are used for conversion.

41.2 The a.c. generator

Let a single turn coil be free to rotate at constant angular velocity symmetrically between the poles of a magnet system as shown in Figure 41.1.

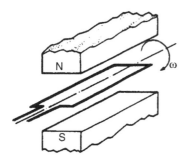

Figure 41.1

An e.m.f. is generated in the coil (from Faraday's laws) which varies in magnitude and reverses its direction at regular intervals. The reason for this is shown in Figure 41.2. In positions (a), (e) and (i) the conductors of the loop are effectively moving along the magnetic field, no flux is cut and hence no e.m.f. is induced. In position (c) maximum flux is cut and hence maximum e.m.f. is induced. In position (g), maximum flux is cut and hence maximum e.m.f. is again induced. However, using Fleming's right-hand rule, the induced e.m.f. is in the opposite direction to that in position (c) and is thus shown as $-E$. In positions (b), (d), (f) and (h) some flux is cut and hence some e.m.f. is induced. If all such positions of the coil are considered, in one revolution of the coil, one cycle of alternating e.m.f. is produced as shown. This is the principle of operation of the **a.c. generator** (i.e. the **alternator**).

41.3 Waveforms

If values of quantities which vary with time t are plotted to a base of time, the resulting graph is called a **waveform**. Some typical waveforms are shown in Figure 41.3. Waveforms (a) and (b) are **unidirectional waveforms**, for, although they vary considerably with time, they flow in one direction only (i.e. they do not cross the time axis and become negative). Waveforms (c) to (g) are called **alternating waveforms** since their quantities are continually changing in direction (i.e. alternately positive and negative).

A waveform of the type shown in Figure 41.3(g) is called a **sine wave**. It is the shape of the waveform of e.m.f. produced by an alternator and thus the mains electricity supply is of 'sinusoidal' form. One complete series of values is called a **cycle** (i.e. from O to P in Figure 41.3(g)). The time taken for an alternating quantity to complete one cycle is called the **period** or the **periodic time T** of the waveform.

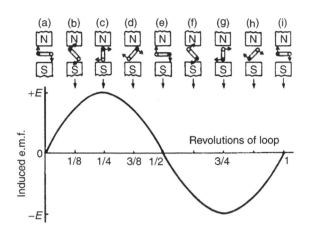

Figure 41.2

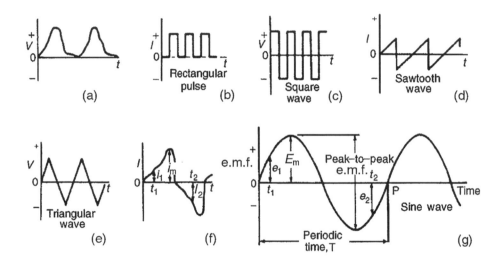

Figure 41.3

The number of cycles completed in one second is called the **frequency** f of the supply and is measured in **hertz***, **Hz**. The standard frequency of the electricity supply in the UK is 50 Hz.

$$T = \frac{1}{f} \quad \text{or} \quad f = \frac{1}{T}$$

Problem 1. Determine the periodic time for frequencies of (a) 50 Hz and (b) 20 kHz

*Who was Hertz? – **Heinrich Rudolf Hertz** (22 February 1857–1 January 1894) was the first person to conclusively prove the existence of electromagnetic waves. The scientific unit of frequency was named the hertz in his honour. To find out more about Hertz go to www.routledge.com/cw/bird (For image refer to page 30)

(a) Periodic time $T = \dfrac{1}{f} = \dfrac{1}{50} = \mathbf{0.02\,s}$ or **20 ms**

(b) Periodic time $T = \dfrac{1}{f} = \dfrac{1}{20,000} = \mathbf{0.00005\,s}$ or

$$\mathbf{50\,\mu s}$$

Problem 2. Determine the frequencies for periodic times of (a) 4 ms and (b) 8 μs

(a) Frequency $f = \dfrac{1}{T} = \dfrac{1}{4 \times 10^{-3}} = \dfrac{1000}{4}$

$$= \mathbf{250\,Hz}$$

(b) Frequency $f = \dfrac{1}{T} = \dfrac{1}{8 \times 10^{-6}} = \dfrac{1,000,000}{8}$

$$= \textbf{125,000 Hz} \quad \text{or} \quad \textbf{125 kHz}$$

$$\text{or} \qquad \textbf{0.125 MHz}$$

Problem 3. An alternating current completes 5 cycles in 8 ms. What is its frequency?

Time for one cycle $= \dfrac{8}{5}$ ms $= 1.6$ ms $=$ periodic time T.

Frequency $f = \dfrac{1}{T} = \dfrac{1}{1.6 \times 10^{-3}} = \dfrac{1000}{1.6} = \dfrac{10,000}{16}$

$$= \textbf{625 Hz}$$

Now try the following Practice Exercise

Practice Exercise 199   Further problems on frequency and periodic time (answers on page 548)

1. Determine the periodic time for the following frequencies:

   (a) 2.5 Hz,   (b) 100 Hz,   (c) 40 kHz

2. Calculate the frequency for the following periodic times:

   (a) 5 ms,   (b) 50 µs,   (c) 0.2 s

3. An alternating current completes 4 cycles in 5 ms. What is its frequency?

41.4   a.c. values

Instantaneous values are the values of the alternating quantities at any instant of time. They are represented by small letters, i, v, e, etc. (see Figures 41.3(f) and (g)). The largest value reached in a half cycle is called the **peak value** or the **maximum value** or the **crest value** or the **amplitude** of the waveform. Such values are represented by V_m, I_m, E_m, etc. (see Figures 41.3(f) and (g)). A **peak-to-peak** value of e.m.f. is shown in Figure 41.3(g) and is the difference between the maximum and minimum values in a cycle.

The **average** or **mean value** of a symmetrical alternating quantity (such as a sine wave), is the average value

measured over a half cycle (since over a complete cycle the average value is zero).

$$\textbf{Average or mean value} = \frac{\textbf{area under the curve}}{\textbf{length of base}}$$

The area under the curve is found by approximate methods such as the trapezoidal rule, the mid-ordinate rule or Simpson's rule. Average values are represented by V_{AV}, I_{AV}, E_{AV}, etc.

For a sine wave:

$$\textbf{average value} = \textbf{0.637} \times \textbf{maximum value}$$

$$\left(\text{i.e. } \frac{2}{\pi} \times \textbf{maximum value} \right)$$

The **effective value** of an alternating current is that current which will produce the same heating effect as an equivalent direct current. The effective value is called the **root mean square (r.m.s.) value** and whenever an alternating quantity is given, it is assumed to be the r.m.s. value. For example, the domestic mains supply in the UK is 230 V+10% and is assumed to mean '230 V r.m.s.'. The symbols used for r.m.s. values are I, V, E, etc. For a non-sinusoidal waveform as shown in Figure 41.4 the r.m.s. value is given by:

$$I = \sqrt{\left(\frac{i_1^2 + i_2^2 + \ldots + i_n^2}{n} \right)}$$

where n is the number of intervals used.

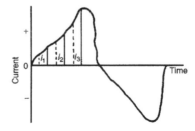

Figure 41.4

For a sine wave:

$$\textbf{r.m.s. value} = \textbf{0.707} \times \textbf{maximum value}$$

$$\left(\text{i.e.} \frac{1}{\sqrt{2}} \times \textbf{maximum value} \right)$$

$$\textbf{Form factor} = \frac{\textbf{r.m.s. value}}{\textbf{average value}}$$

For a sine wave, form factor = 1.11

$$\textbf{Peak factor} = \frac{\textbf{maximum value}}{\textbf{r.m.s. value}}$$

For a sine wave, peak factor = 1.41
The values of form and peak factors give an indication of the shape of waveforms.

> **Problem 4.** For the periodic waveforms shown in Figure 41.5 determine for each: (i) the frequency, (ii) the average value over half a cycle, (iii) the r.m.s. value, (iv) the form factor, (v) the peak factor

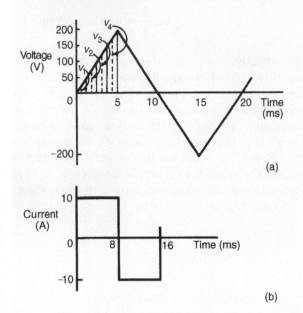

Figure 41.5

Average value of waveform

$$= \frac{\text{area under curve}}{\text{length of base}}$$

$$= \frac{1 \text{ volt second}}{10 \times 10^{-3} \text{ second}}$$

$$= \frac{1000}{10} = \textbf{100 V}$$

(iii) In Figure 41.5(a), the first $\frac{1}{4}$ cycle is divided into 4 intervals.

Thus r.m.s. value

$$= \sqrt{\left(\frac{v_1^2 + v_2^2 + v_3^2 + v_4^2}{4}\right)}$$

$$= \sqrt{\left(\frac{25^2 + 75^2 + 125^2 + 175^2}{4}\right)}$$

$$= \textbf{114.6 V}$$

(Note that the greater the number of intervals chosen, the greater the accuracy of the result. For example, if twice the number of ordinates as that chosen above are used, the r.m.s. value is found to be 115.6 V)

(iv) Form factor $= \dfrac{\text{r.m.s. value}}{\text{average value}} = \dfrac{114.6}{100}$

$$= \textbf{1.15}$$

(v) Peak factor $= \dfrac{\text{maximum value}}{\text{r.m.s. value}} = \dfrac{200}{114.6}$

$$= \textbf{1.75}$$

(a) **Triangular waveform** (Figure 41.5(a))

(i) Time for one complete cycle
= 20 ms = periodic time T

Hence frequency $f = \dfrac{1}{T} = \dfrac{1}{20 \times 10^{-3}}$

$$= \frac{1000}{20} = \textbf{50 Hz}$$

(ii) Area under the triangular waveform for a half cycle $= \dfrac{1}{2} \times$ base $\times$ height

$$= \frac{1}{2} \times (10 \times 10^{-3}) \times 200$$

$$= 1 \text{ volt second}$$

(b) **Rectangular waveform** (Figure 41.5(b))

(i) Time for one complete cycle = 16 ms = periodic time T

Hence frequency $f = \dfrac{1}{T} = \dfrac{1}{16 \times 10^{-3}}$

$$= \frac{1000}{16} = \textbf{62.5 Hz}$$

(ii) Average value over half a cycle

$$= \frac{\text{area under curve}}{\text{length of base}} = \frac{10 \times (8 \times 10^{-3})}{8 \times 10^{-3}}$$

$$= \textbf{10 A}$$

(iii)   The r.m.s. value $= \sqrt{\left(\dfrac{i_1^2 + i_2^2 + i_3^2 + i_4^2}{4}\right)}$

$= \mathbf{10\,A}$ however many intervals are chosen, since the waveform is rectangular.

(iv)   Form factor $= \dfrac{\text{r.m.s. value}}{\text{average value}} = \dfrac{10}{10} = \mathbf{1}$

(v)   Peak factor $= \dfrac{\text{maximum value}}{\text{r.m.s. value}} = \dfrac{10}{10} = \mathbf{1}$

Problem 5.   The following table gives the corresponding values of current and time for a half cycle of alternating current:

time t (ms)	0	0.5	1.0	1.5	2.0	2.5
current i (A)	0	7	14	23	40	56
time t (ms)	3.0	3.5	4.0	4.5	5.0	
current i (A)	68	76	60	5	0	

Assuming the negative half cycle is identical in shape to the positive half cycle, plot the waveform and find (a) the frequency of the supply, (b) the instantaneous values of current after 1.25 ms and 3.8 ms, (c) the peak or maximum value, (d) the mean or average value, (e) the r.m.s. value of the waveform

The half cycle of alternating current is shown plotted in Figure 41.6.

(a)   Time for a half cycle $= 5$ ms; hence the time for one cycle, i.e. the periodic time $T = 10$ ms or 0.01 s.

Frequency $f = \dfrac{1}{T} = \dfrac{1}{0.01} = \mathbf{100\,Hz}$.

(b)   Instantaneous value of current after 1.25 ms is **19 A**, from Figure 41.6. Instantaneous value of current after 3.8 ms is **70 A**, from Figure 41.6.

(c)   Peak or maximum value $= \mathbf{76\,A}$

(d)   Mean or average value $= \dfrac{\text{area under curve}}{\text{length of base}}$
Using the mid-ordinate rule with 10 intervals, each of width 0.5 ms gives: area under curve

$= (0.5 \times 10^{-3})[3 + 10 + 19 + 30 + 49 + 63$

$+ 73 + 72 + 30 + 2]$ (see Figure 41.6)

$= (0.5 \times 10^{-3})(351)$

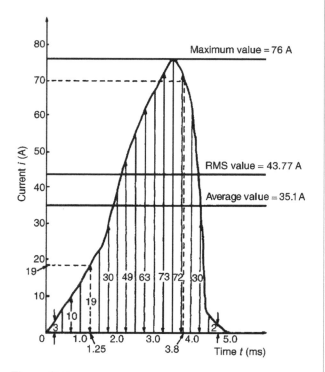

Figure 41.6

Hence mean or average value

$= \dfrac{(0.5 \times 10^{-3})(351)}{5 \times 10^{-3}}$

$= \mathbf{35.1\,A}$

(e)   r.m.s. value $= \sqrt{\left(\dfrac{\begin{array}{c}3^2 + 10^2 + 19^2 + 30^2 \\ + 49^2 + 63^2 + 73^2 + 72^2 \\ + 30^2 + 2^2\end{array}}{10}\right)}$

$= \sqrt{\left(\dfrac{19157}{10}\right)} = \mathbf{43.8\,A}$

Problem 6.   Calculate the r.m.s. value of a sinusoidal current of maximum value 20 A

For a sine wave,

r.m.s. value $= 0.707 \times$ maximum value

$= 0.707 \times 20 = \mathbf{14.14\,A}$

Problem 7.   Determine the peak and mean values for a 240 V mains supply

For a sine wave, r.m.s. value of voltage $V = 0.707 \times V_m$

A 240 V mains supply means that 240 V is the r.m.s. value,

hence $V_m = \dfrac{V}{0.707} = \dfrac{240}{0.707} = \mathbf{339.5\,V} = \mathbf{peak\ value}$

Mean value $V_{AV} = 0.637 V_m = 0.637 \times 339.5$

$$= \mathbf{216.3\,V}$$

Problem 8. A supply voltage has a mean value of 150 V. Determine its maximum value and its r.m.s. value

For a sine wave, mean value $= 0.637 \times$ maximum value.

Hence **maximum value** $= \dfrac{\text{mean value}}{0.637} = \dfrac{150}{0.637}$

$$= \mathbf{235.5\,V}$$

r.m.s. value $= 0.707 \times$ maximum value

$$= 0.707 \times 235.5$$

$$= \mathbf{166.5\,V}$$

Now try the following Practice Exercises

Practice Exercise 200 Further problems on a.c. values of waveforms (answers on page 548)

1. An alternating current varies with time over half a cycle as follows:

Current (A)	0	0.7	2.0	4.2	8.4	8.2
time (ms)	0	1	2	3	4	5

Current (A)	2.5	1.0	0.4	0.2	0
time (ms)	6	7	8	9	10

The negative half cycle is similar. Plot the curve and determine: (a) the frequency, (b) the instantaneous values at 3.4 ms and 5.8 ms, (c) its mean value, (d) its r.m.s. value

2. For the waveforms shown in Figure 41.7 determine for each (i) the frequency, (ii) the average value, over half a cycle, (iii) the r.m.s. value, (iv) the form factor, (v) the peak factor

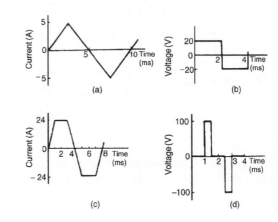

Figure 41.7

3. An alternating voltage is triangular in shape, rising at a constant rate to a maximum of 300 V in 8 ms and then falling to zero at a constant rate in 4 ms. The negative half cycle is identical in shape to the positive half cycle. Calculate (a) the mean voltage over half a cycle, (b) the r.m.s. voltage

4. Calculate the r.m.s. value of a sinusoidal curve of maximum value 300 V

5. Find the peak and mean values for a 200 V mains supply

6. A sinusoidal voltage has a maximum value of 120 V. Calculate its r.m.s. and average values

7. A sinusoidal current has a mean value of 15.0 A. Determine its maximum and r.m.s. values

41.5 Electrical safety – insulation and fuses

Insulation is used to prevent 'leakage', and when determining what type of insulation should be used, the maximum voltage present must be taken into account. For this reason, **peak values are always considered when choosing insulation materials**.

Fuses are the weak link in a circuit and are used to break the circuit if excessive current is drawn. Excessive current could lead to a fire. Fuses rely on the heating effect of the current, and for this reason **r.m.s. values must always be used when calculating the appropriate fuse size**.

41.6 Semiconductor diodes

The most important semiconductors used in the electronics industry are **silicon** and **germanium**. Adding extremely small amounts of impurities to pure semiconductors in a controlled manner is called **doping**. Antimony, arsenic and phosphorus are called n-type impurities and form an **n-type material** when any of these impurities are added to silicon or germanium. **n** type semiconductor material contains an excess of **n**egative charge carriers, Indium, aluminium and boron are called p-type impurities and form a **p-type material** when any of these impurities are added to a semiconductor. **p** type material contains an excess of **p**ositive charge carriers.

A **p-n** junction is a piece of semiconductor material in which part of the material is p-type and part is n-type. When a junction is formed between n-type and p-type semiconductor materials, the resulting device is called a **diode**.

The word '**diode**' is made up from **di** (which means two) and elect**rode**, and it is therefore a device that has two electrodes or terminals. These terminals are called **anode** and **cathode** respectively, and the circuit diagram symbol is shown in Figure 41.8. (Note that sometimes the circle is omitted in the symbol).

The diode is a **one-way device**, i.e. it will allow current to flow in one direction only when its anode is **positive** with respect to its cathode; the direction in which current flows is called the **forward** direction. When the anode is **negative** with respect to the cathode, no current should flow; this is called the **reverse** direction.

The diode can be regarded as a **switch**, with current able to flow only when the switch is on. Automatic switching in circuits is performed by **diodes**, and their main application is found in rectifying circuits. A **diode** is the simplest possible semiconductor device.

41.7 Rectification

The process of obtaining unidirectional currents and voltages from alternating currents and voltages is called **rectification**. Automatic switching in circuits is achieved using diodes.

Half-wave rectification

Using a single diode, D, as shown in Figure 41.8, **half-wave rectification** is obtained. When P is sufficiently positive with respect to Q, diode D is switched on and current i flows. When P is negative with respect to Q, diode D is switched off. Transformer T isolates the equipment from direct connection with the mains supply and enables the mains voltage to be changed.

Thus, an alternating, sinusoidal waveform applied to the transformer primary is rectified into a unidirectional waveform. Unfortunately, the output waveform shown in Figure 41.8 is not constant (i.e. steady), and as such, would be unsuitable as a d.c. power supply for electronic equipment. It would, however, be satisfactory as a battery charger. In Section 41.9, methods of smoothing the output waveform are discussed.

Full-wave rectification using a centre-tapped transformer

Two diodes may be used as shown in Figure 41.10 to obtain **full-wave rectification** where a centre-tapped transformer T is used. When P is sufficiently positive with respect to Q, diode D_1 conducts and current flows (shown by the broken line in Figure 41.10). When S is positive with respect to Q, diode D_2 conducts and current flows (shown by the continuous line in Figure 41.10).

The current flowing in the load R is in the same direction for both half-cycles of the input. The output waveform is thus as shown in Figure 41.10. The output is unidirectional, but is not constant; however, it is better than the output waveform produced with a half-wave rectifier. Section 41.8 explains how the waveform may be improved so as to be of more use.

A **disadvantage** of this type of rectifier is that centre-tapped transformers are expensive.

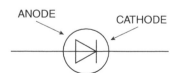

Figure 41.8

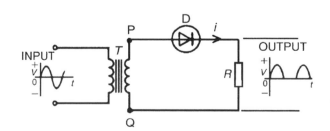

Figure 41.9

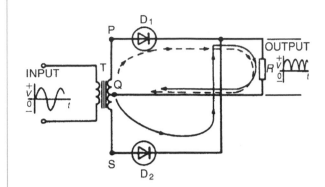

Figure 41.10

Full-wave bridge rectification

Four diodes may be used in a **bridge rectifier** circuit, as shown in Figure 41.11, to obtain **full-wave rectification**. (Note, the term 'bridge' means a network of four elements connected to form a square, the input being applied to two opposite corners and the output being taken from the remaining two corners.) As for the rectifier shown in Figure 41.10, the current flowing in load R is in the same direction for both half-cycles of the input giving the output waveform shown.

Following the broken line in Figure 41.11:
When P is positive with respect to Q, current flows from the transformer to point E, through diode D_4 to point F, then through load R to point H, through D_2 to point G, and back to the transformer.

Following the full line in Figure 41.11:
When Q is positive with respect to P, current flows from the transformer to point G, through diode D_3 to point F, then through load R to point H, through D_1 to point E, and back to the transformer. The output waveform is not steady and needs improving; a method of smoothing is explained in the next section.

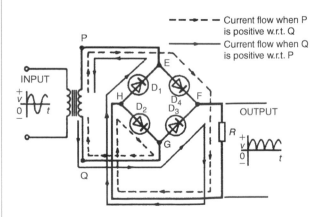

---- ▶ Current flow when P is positive w.r.t. Q
——— Current flow when Q is positive w.r.t. P

Figure 41.11

41.8 Smoothing of the rectified output waveform

The pulsating outputs obtained from the half- and full-wave rectifier circuits are not suitable for the operation of equipment that requires a steady d.c. output, such as would be obtained from batteries. For example, for applications such as audio equipment, a supply with a large variation is unacceptable since it produces 'hum' in the output. **Smoothing** is the process of removing the worst of the output waveform variations.

To smooth out the pulsations a large capacitor, C, is connected across the output of the rectifier, as shown in Figure 41.12; the effect of this is to maintain the output voltage at a level which is very near to the peak of the output waveform. The improved waveforms for half-wave and full-wave rectifiers are shown in more detail in Figure 41.13.

During each pulse of output voltage, the capacitor C charges to the same potential as the peak of the waveform, as shown as point X in Figure 41.13. As the waveform dies away, the capacitor discharges across the load, as shown by XY. The output voltage is then restored to the peak value the next time the rectifier conducts, as shown by YZ. This process continues as shown in Figure 41.13.

Capacitor C is called a **reservoir capacitor** since it stores and releases charge between the peaks of the rectified waveform.

The variation in potential between points X and Y is called **ripple**, as shown in Figure 41.13; the object is to reduce ripple to a minimum. Ripple may be reduced even further by the addition of inductance and another capacitor in a '**filter**' circuit arrangement, as shown in Figure 41.14.

The output voltage from the rectifier is applied to capacitor C_1 and the voltage across points AA is shown in Figure 41.14, similar to the waveforms of Figure 41.13. The load current flows through the inductance L; when current is changing, e.m.f.s are induced, as explained in chapter 40. By Lenz's law, the induced voltages will oppose those causing the current changes.

As the ripple voltage increases and the load current increases, the induced e.m.f. in the inductor will oppose the increase. As the ripple voltage falls and the load current falls, the induced e.m.f. will try to maintain the current flow.

The voltage across points BB in Figure 41.14 and the current in the inductance are almost ripple-free. A further capacitor, C_2, completes the process.

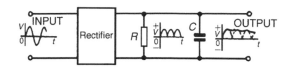

Figure 41.12

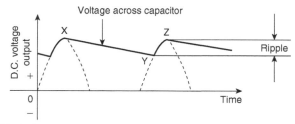

(a) Half-wave rectifier

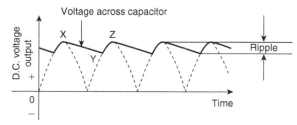

(b) Full-wave rectifier

Figure 41.13

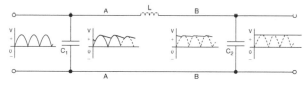

Figure 41.14

Now try the following Practice Exercises

Practice Exercise 201   Short answer questions on alternating voltages and currents

1. Briefly explain the principle of operation of the simple alternator

2. What is meant by (a) waveform, (b) cycle?

3. What is the difference between an alternating and a unidirectional waveform?

4. The time to complete one cycle of a waveform is called the ......

5. What is frequency? Name its unit

6. The mains supply voltage has a special shape of waveform called a ......

7. Define peak value

8. What is meant by the r.m.s. value?

9. What is the mean value of a sinusoidal alternating e.m.f. which has a maximum value of 100 V?

10. The effective value of a sinusoidal waveform is ..... × maximum value

11. Complete the statement:
Form factor = ... ÷ ......, and for a sine wave, form factor = ...

12. Complete the statement:
Peak factor = ... ÷ ......, and for a sine wave, peak factor = ...

13. How is switching obtained when converting a.c. to d.c.?

14. Draw an appropriate circuit diagram suitable for half-wave rectifications and explain its operation.

15. Explain, with a diagram, how full-wave rectification is obtained using a centre-tapped transformer.

16. Explain, with a diagram, how full-wave rectification is obtained using a bridge rectifier circuit.

17. Explain a simple method of smoothing the output of a rectifier.

Practice Exercise 202   Multiple-choice questions on alternating voltages and currents (answers on page 548)

1. The number of complete cycles of an alternating current occurring in one second is known as:

(a) the maximum value of the alternating current

(b) the frequency of the alternating current

(c) the peak value of the alternating current

(d) the r.m.s. or effective value

2. The value of an alternating current at any given instant is:

(a) a maximum value

(b) a peak value

(c) an instantaneous value

(d) an r.m.s. value

3. An alternating current completes 100 cycles in 0.1 s. Its frequency is:

(a) 20 Hz   (b) 100 Hz

(c) 0.002 Hz   (d) 1 kHz

4. In Figure 41.8, at the instant shown, the generated e.m.f. will be:

(a) zero   (b) an r.m.s. value

(c) an average value   (d) a maximum value

Figure 41.15

5. The supply of electrical energy for a consumer is usually by a.c. because:

(a) transmission and distribution are more easily effected

(b) it is most suitable for variable speed motors

(c) the volt drop in cables is minimal

(d) cable power losses are negligible

6. Which of the following statements is false?

(a) It is cheaper to use a.c. than d.c.

(b) Distribution of a.c. is more convenient than with d.c. since voltages may be readily altered using transformers

(c) An alternator is an a.c. generator

(d) A rectifier changes d.c. to a.c.

7. An alternating voltage of maximum value 100 V is applied to a lamp. Which of the following direct voltages, if applied to the lamp, would cause the lamp to light with the same brilliance?

(a) 100 V   (b) 63.7 V

(c) 70.7 V   (d) 141.4 V

8. The value normally stated when referring to alternating currents and voltages is the:

(a) instantaneous value   (b) r.m.s. value

(c) average value   (d) peak value

9. State which of the following is false. For a sine wave:

(a) the peak factor is 1.414

(b) the r.m.s. value is $0.707 \times$ peak value

(c) the average value is $0.637 \times$ r.m.s. value

(d) the form factor is 1.11

10. An a.c. supply is 70.7 V, 50 Hz. Which of the following statements is false?

(a) The periodic time is 20 ms

(b) The peak value of the voltage is 70.7 V

(c) The r.m.s. value of the voltage is 70.7 V

(d) The peak value of the voltage is 100 V

For fully worked solutions to each of the problems in Exercises 199 to 202 in this chapter, go to the website:

www.routledge.com/cw/bird

Chapter 42

Capacitors and inductors

Why it is important to understand: Capacitors and inductors

The capacitor is a widely used electrical component and it has several features that make it useful and important. A capacitor can store energy, so capacitors are often found in power supplies. Capacitors are used for timing, controlling the charging and discharging, for smoothing in a power supply, for coupling between stages of an audio system and a loudspeaker, for filtering in the tone control of an audio system, for tuning in a radio system, and for storing energy in a camera flash circuit. Capacitors find uses in virtually every form of electronic circuits from analogue circuits, including amplifiers and power supplies, through to oscillators, integrators and many more. Capacitors are also used in logic circuits, primarily for providing decoupling to prevent spikes and ripple on the supply lines which could cause spurious triggering of the circuits. So capacitors are very important components in electrical and electronic circuits, and this chapter introduces the terminology and calculations to aid understanding. Inductors are used extensively in analogue circuits and signal processing with applications ranging from the use of large inductors in power supplies, which in conjunction with filter capacitors remove mains hum or other fluctuations from the direct current output, to the small inductance of the ferrite bead installed around a cable to prevent radio frequency interference from being transmitted down the wire. Inductors are used as the energy storage device in many switched-mode power supplies to produce d.c. current. An inductor connected to a capacitor forms a tuned circuit, which acts as a resonator for oscillating current. Tuned circuits are widely used in radio frequency equipment such as radio transmitters and receivers, as narrow band-pass filters to select a single frequency from a composite signal, and in electronic oscillators to generate sinusoidal signals. Two (or more) inductors in proximity that have coupled magnetic flux (mutual inductance) form a transformer, which is a fundamental component of every electric utility power grid.

At the end of this chapter, you should be able to:

- describe an electrostatic field
- define electric field strength E and state its unit
- define capacitance and state its unit
- describe a capacitor and draw the circuit diagram symbol
- perform simple calculations involving $C = \dfrac{Q}{V}$ and $Q = It$

Science and Mathematics for Engineering. 978-0-367-20475-4. © John Bird. Published by Taylor & Francis. All rights reserved.

- define electric flux density D and state its unit
- define permittivity, distinguishing between ε_0, ε_r and ε
- perform simple calculations involving $D = \dfrac{Q}{A}$, $E = \dfrac{V}{d}$ and $\dfrac{D}{E} = \varepsilon_0\,\varepsilon_r$
- understand that for a parallel plate capacitor, $C = \dfrac{\varepsilon_0 \varepsilon_r A(n-1)}{d}$
- perform calculations involving capacitors connected in parallel and in series
- define dielectric strength and state its unit
- state that the energy stored in a capacitor is given by $W = \dfrac{1}{2}CV^2$ joules
- describe practical types of capacitor
- understand the precautions needed when discharging capacitors
- appreciate the property of inductance
- define inductance L and state its unit
- appreciate that e.m.f. $E = -N\dfrac{d\Phi}{dt} = -L\dfrac{dI}{dt}$
- calculate induced e.m.f. given N, t, L, change of flux or change of current
- appreciate factors which affect the inductance of an inductor
- draw the circuit diagram symbols for inductors
- calculate the energy stored in an inductor using $W = \dfrac{1}{2}LI^2$ joules
- calculate inductance L of a coil, given $L = \dfrac{N\Phi}{I}$

42.1   Capacitors and capacitance

A **capacitor** is a device capable of storing electrical energy. Next to the resistor, the capacitor is the most commonly encountered component in electrical circuits. Capacitors are used extensively in electrical and electronic circuits. For example, capacitors are used to smooth rectified a.c. outputs, they are used in telecommunication equipment – such as radio receivers – for tuning to the required frequency, they are used in time delay circuits, in electrical filters, in oscillator circuits, and in magnetic resonance imaging (MRI) in medical body scanners, to name but a few practical applications.

Some typical small capacitors are shown in Figure 42.1. Figure 42.2 shows a capacitor consisting of a pair of parallel metal plates X and Y separated by an insulator, which could be air. Since the plates are electrical conductors each will contain a large number of mobile electrons. Because the plates are connected to a d.c. supply the electrons on plate X, which have a small

negative charge, will be attracted to the positive pole of the supply and will be repelled from the negative pole of the supply on to plate Y. X will become positively charged due to its shortage of electrons whereas Y will have a negative charge due to its surplus of electrons.

Figure 42.1

The difference in charge between the plates results in a p.d. existing between them, the flow of electrons dying away and ceasing when the p.d. between the plates

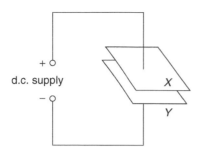

Figure 42.2

equals the supply voltage. The plates are then said to be **charged** and there exists an **electric field** between them. Figure 42.3 shows a side view of the plates with the field represented by 'lines of electrical flux'. If the plates are disconnected from the supply and connected together through a resistor the surplus of electrons on the negative plate will flow through the resistor to the positive plate. This is called **discharging.** The current flow decreases to zero as the charges on the plates reduce. The current flowing in the resistor causes it to liberate heat showing that **energy is stored in the electric field**.

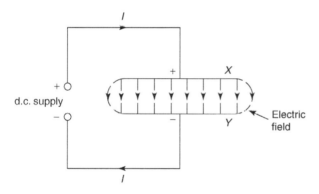

Figure 42.3

The symbols for a fixed capacitor and a variable capacitor used in electrical circuit diagrams are shown in Figure 42.4.

Fixed capacitor   Variable capacitor

Figure 42.4

*Who was Coulomb? – **Charles-Augustin de Coulomb** (14 June 1736–23 August 1806) was best known for developing Coulomb's law, the definition of the electrostatic force of attraction and repulsion To find out more about Coulomb go to www.routledge.com/cw/bird (For image refer to page 365)

42.1.1 Summary of important formulae and definitions

From chapter 34, charge Q is given by:

$$Q = I \times t \text{ coulombs}^*$$

where I is the current in amperes and t the time in seconds.

A **dielectric** is an insulating medium separating charged surfaces.

Electric field strength, electric force, or voltage gradient,

$$E = \frac{\text{p.d. across dielectric}}{\text{thickness of dielectric}}$$

i.e. $$E = \frac{V}{d} \text{ volts/m}$$

Electric flux density

$$D = \frac{Q}{A} \text{ C/m}^2$$

Charge Q on a capacitor is proportional to the applied voltage V, i.e. $Q \propto V$.

$$Q = CV$$

where the constant of proportionality C is the capacitance.

Capacitance $$C = \frac{Q}{V}$$

The unit farad is named after **Michael Faraday***. The unit of capacitance is the **farad, F** (or more usually $\mu F = 10^{-6} \text{ F}$ or $pF = 10^{-12} \text{ F}$), which is defined as the capacitance of a capacitor when a potential difference (p.d.) of one volt appears across the plates when charged with one coulomb.

Every system of electrical conductors possesses capacitance. For example, there is capacitance between the

*Who was Faraday? – **Michael Faraday** (22 September 1791–25 August 1867) was an English scientist whose main discoveries include electromagnetic induction, diamagnetism and electrolysis. The SI unit of capacitance, the farad, is named in his honour. To find out more about Faraday go to www.routledge.com/cw/bird (For image refer to page 426)

conductors of overhead transmission lines and also between the wires of a telephone cable. In these examples the capacitance is undesirable but has to be accepted, minimised or compensated for. There are other situations, such as in capacitors, where capacitance is a desirable property.

The ratio of electric flux density D to electric field strength E is called **absolute permittivity ε** of a dielectric. Thus $\dfrac{D}{E} = \varepsilon$

Permittivity of free space is a constant, given by

$$\varepsilon_0 = 8.85 \times 10^{-12} \text{F/m}.$$

Relative permittivity

$$\varepsilon_r = \frac{\text{flux density of the field in a dielectric}}{\text{flux density of the field in a vacuum}}$$

(ε_r has no units). Examples of the values of ε_r include: air = 1.00, polythene = 2.3, mica = 3–7, glass = 5–10, ceramics = 6–1000.

Absolute permittivity $\varepsilon = \varepsilon_0 \varepsilon_r$, thus

$$\frac{D}{E} = \varepsilon_0 \varepsilon_r$$

Problem 1. (a) Determine the p.d. across a 4 μF capacitor when charged with 5 mC. (b) Find the charge on a 50 pF capacitor when the voltage applied to it is 2kV

(a) $\qquad C = 4\mu F = 4 \times 10^{-6} F \quad$ and
$\qquad Q = 5 \text{ mC} = 5 \times 10^{-3} \text{ C}$

Since $C = \dfrac{Q}{V}$

then $V = \dfrac{Q}{C} = \dfrac{5 \times 10^{-3}}{4 \times 10^{-6}} = \dfrac{5 \times 10^{6}}{4 \times 10^{3}} = \dfrac{5000}{4}$

Hence, **p.d.** $V = \mathbf{1250\,V}$ or $\mathbf{1.25\,kV}$

(b) $\qquad C = 50 \text{ pF} = 50 \times 10^{-12} \text{ F} \quad$ and
$\qquad V = 2 \text{ kV} = 2000 \text{ V}$

$\qquad Q = CV = 50 \times 10^{-12} \times 2000$

$$= \frac{5 \times 2}{10^{8}} = 0.1 \times 10^{-6}$$

Hence, **charge** $Q = \mathbf{0.1\ \mu C}$

Problem 2. A direct current of 4 A flows into a previously uncharged 20 μF capacitor for 3 ms. Determine the p.d. between the plates

$I = 4$ A, $C = 20\,\mu F = 20 \times 10^{-6}$ F and
$t = 3$ ms $= 3 \times 10^{-3}$s

$Q = I \times t = 4 \times 3 \times 10^{-3}$C

$V = \dfrac{Q}{C} = \dfrac{4 \times 3 \times 10^{-3}}{20 \times 10^{-6}} = \dfrac{12 \times 10^{6}}{20 \times 10^{3}} = 0.6 \times 10^{3}$

$$= 600 \text{ V}$$

Hence, the p.d. between the plates is 600 V

Problem 3. A 5 μF capacitor is charged so that the p.d. between its plates is 800 V. Calculate how long the capacitor can provide an average discharge current of 2 mA

$C = 5\,\mu F = 5 \times 10^{-6}$ F, $V = 800$ V and

$I = 2 \text{mA} = 2 \times 10^{-3}$A

$Q = CV = 5 \times 10^{-6} \times 800 = 4 \times 10^{-3}$C

Also, $Q = I \times t$

Thus, $t = \dfrac{Q}{I} = \dfrac{4 \times 10^{-3}}{2 \times 10^{-3}} = 2$ s

Hence, the capacitor can provide an average discharge current of 2 mA for 2 s.

Problem 4. Two parallel rectangular plates measuring 20 cm by 40 cm carry an electric charge of 0.2 μC. Calculate the electric flux density. If the plates are spaced 5 mm apart and the voltage between them is 0.25 kV determine the electric field strength

Area $A = 20$ cm $\times 40$ cm $= 800$ cm$^2 = 800 \times 10^{-4}$ m^2.
Charge $Q = 0.2$ μC $= 0.2 \times 10^{-6}$C.

Electric flux density

$$D = \frac{Q}{A} = \frac{0.2 \times 10^{-6}}{800 \times 10^{-4}} = \frac{0.2 \times 10^{4}}{800 \times 10^{6}}$$

$$= \frac{2000}{800} \times 10^{-6} = \mathbf{2.5\ \mu C/m^2}$$

Voltage $V = 0.25$ kV $= 250$ V and plate spacing
$d = 5$ mm $= 5 \times 10^{-3}$m.

Electric field strength

$$E = \frac{V}{d} = \frac{250}{5 \times 10^{-3}} = \mathbf{50\,kV/m}$$

Now try the following Practice Exercise

Practice Exercise 203　Further problems on capacitors and capacitance (answers on page 548)

(Where appropriate take ε_0 as 8.85×10^{-12}F/m.)

1. Find the charge on a 10 µF capacitor when the applied voltage is 250V

2. Determine the voltage across a 1000 pF capacitor to charge it with 2 µC

3. The charge on the plates of a capacitor is 6 mC when the potential between them is 2.4 kV. Determine the capacitance of the capacitor

4. For how long must a charging current of 2 A be fed to a 5 µF capacitor to raise the p.d. between its plates by 500 V?

5. A direct current of 10 A flows into a previously uncharged 5 µF capacitor for 1 ms. Determine the p.d. between the plates

6. A capacitor uses a dielectric 0.04 mm thick and operates at 30 V. What is the electric field strength across the dielectric at this voltage?

7. A charge of 1.5 µC is carried on two parallel rectangular plates each measuring 60 mm by 80 mm. Calculate the electric flux density. If the plates are spaced 10 mm apart and the voltage between them is 0.5 kV determine the electric field strength

42.2　The parallel-plate capacitor

For a parallel-plate capacitor, experiments show that capacitance C is proportional to the area A of a plate, inversely proportional to the plate spacing d (i.e. the dielectric thickness) and depends on the nature of the dielectric:

$$\textbf{Capacitance } C = \frac{\varepsilon_0 \varepsilon_r A(n-1)}{d} \textbf{ farads}$$

where $\varepsilon_0 = 8.85 \times 10^{-12}$F/m (constant),
　　　ε_r = relative permittivity,
　　　A = area of one of the plates in m^2,
　　　d = thickness of dielectric in m,
and　　n = number of plates.

Problem 5. (a) A ceramic capacitor has an effective plate area of 4 cm^2 separated by 0.1 mm of ceramic of relative permittivity 100. Calculate the capacitance of the capacitor in picofarads. (b) If the capacitor in part (a) is given a charge of 1.2 µC what will be the p.d. between the plates?

(a)　Area $A = 4$ cm$^2 = 4 \times 10^{-4}$ m^2,
$d = 0.1$ mm $= 0.1 \times 10^{-3}$ m, $\varepsilon_0 = 8.85 \times 10^{-12}$ F/m
and $\varepsilon_r = 100$

$$\textbf{Capacitance } C = \frac{\varepsilon_0 \varepsilon_r A}{d} \text{ farads}$$

$$= \frac{8.85 \times 10^{-12} \times 100 \times 4 \times 10^{-4}}{0.1 \times 10^{-3}} \text{ F}$$

$$= \frac{8.85 \times 4}{10^{10}} \text{F} = \frac{8.85 \times 4 \times 10^{12}}{10^{10}} \text{ pF}$$

$$= \textbf{3540 pF}$$

(b)　$Q = CV$ thus $V = \dfrac{Q}{C} = \dfrac{1.2 \times 10^{-6}}{3540 \times 10^{-12}} V = \textbf{339 V}$

Problem 6. A waxed paper capacitor has two parallel plates, each of effective area 800 cm^2. If the capacitance of the capacitor is 4425 pF determine the effective thickness of the paper if its relative permittivity is 2.5

$A = 800$ cm$^2 = 800 \times 10^{-4}$ m$^2 = 0.08$ m^2,
$C = 4425$ pF $= 4425 \times 10^{-12}$ F, $\varepsilon_0 = 8.85 \times 10^{-12}$ F/m
and $\varepsilon_r = 2.5$

$$\text{Since } C = \frac{\varepsilon_0 \varepsilon_r A}{d} \text{ then}$$

$$d = \frac{\varepsilon_0 \varepsilon_r A}{C}$$

$$= \frac{8.85 \times 10^{-12} \times 2.5 \times 0.08}{4425 \times 10^{-12}}$$

$$= 0.0004 \text{ m}$$

Hence, the thickness of the paper is 0.4 mm

Problem 7. A parallel plate capacitor has nineteen interleaved plates each 75 mm by 75 mm separated by mica sheets 0.2 mm thick. Assuming the relative permittivity of the mica is 5, calculate the capacitance of the capacitor

$n = 19$ thus $n - 1 = 18$,
$A = 75 \times 75 = 562$ mm$^2 = 5625 \times 10^{-6}$m^2,
$\varepsilon_r = 5, \varepsilon_0 = 8.85 \times 10^{-12}$F/m and
$d = 0.2$ mm $= 0.2 \times 10^{-3}$m.

Capacitance $C = \dfrac{\varepsilon_0 \varepsilon_r A(n-1)}{d}$

$= \dfrac{8.85 \times 10^{-12} \times 5 \times 5625 \times 10^{-6} \times 18}{0.2 \times 10^{-3}}$ F

$= \mathbf{0.0224 \,\mu F}$ or $\mathbf{22.4 \, nF}$

Now try the following Practice Exercise

Practice Exercise 204   Further problems on parallel plate capacitors (answers on page 548)

(Where appropriate take ε_0 as 8.85×10^{-12} F/m.)

1. A capacitor consists of two parallel plates each of area 0.01 m², spaced 0.1 mm in air. Calculate the capacitance in picofarads

2. A waxed paper capacitor has two parallel plates, each of effective area 0.2 m². If the capacitance is 4000 pF determine the effective thickness of the paper if its relative permittivity is 2

3. How many plates has a parallel-plate capacitor having a capacitance of 5 nF, if each plate is 40 mm by 40 mm and each dielectric is 0.102 mm thick with a relative permittivity of 6

4. A parallel-plate capacitor is made from 25 plates, each 70 mm by 120 mm interleaved with mica of relative permittivity 5. If the capacitance of the capacitor is 3000 pF determine the thickness of the mica sheet

5. A capacitor is constructed with parallel plates and has a value of 50 pF. What would be the capacitance of the capacitor if the plate area is doubled and the plate spacing is halved?

42.3 Capacitors connected in parallel and series

For n **parallel-connected capacitors**, the equivalent capacitance C is given by:

$$C = C_1 + C_2 + C_3 ..... + C_n$$

(similar to **resistors** connected in **series**).
Also, total charge Q is given by:

$$Q = Q_1 + Q_2 + Q_3$$

For n **series-connected capacitors** the equivalent capacitance C is given by:

$$\frac{1}{C} = \frac{1}{C_1} + \frac{1}{C_2} + \frac{1}{C_3} + ... + \frac{1}{C_n}$$

(similar to **resistors** connected in **parallel**).
When connected in series the charge on each capacitor is the same.

Problem 8.   Calculate the equivalent capacitance of two capacitors of 6 μF and 4 μF connected (a) in parallel and (b) in series

(a)   In parallel, equivalent capacitance

$$C = C_1 + C_2 = 6\,\mu F + 4\,\mu F = \mathbf{10\,\mu F}$$

(b)   For the special case of **two capacitors in series:**

$$\frac{1}{C} = \frac{1}{C_1} + \frac{1}{C_2} = \frac{C_2 + C_1}{C_1 C_2}$$

Hence $C = \dfrac{C_1 C_2}{C_1 + C_2}$ $\left(\text{i.e. } \dfrac{\text{product}}{\text{sum}} \right)$

Thus, $C = \dfrac{6 \times 4}{6 + 4} = \dfrac{24}{10} = \mathbf{2.4 \;\mu F}$

Problem 9.   What capacitance must be connected in series with a 30 μF capacitor for the equivalent capacitance to be 12 μF?

Let $C = 12\,\mu F$ (the equivalent capacitance), $C_1 = 30\,\mu F$ and C_2 be the unknown capacitance.
For two capacitors in series

$$\frac{1}{C} = \frac{1}{C_1} + \frac{1}{C_2}$$

Hence $\dfrac{1}{C_2} = \dfrac{1}{C} - \dfrac{1}{C_1} = \dfrac{C_1 - C}{C C_1}$

and $C_2 = \dfrac{C C_1}{C_1 - C} = \dfrac{12 \times 30}{30 - 12} = \dfrac{360}{18} = \mathbf{20\,\mu F}$

Problem 10. Capacitances of 3 μF, 6 μF and 12 μF are connected in series across a 350 V supply. Calculate (a) the equivalent circuit capacitance, (b) the charge on each capacitor and (c) the p.d. across each capacitor

The circuit diagram is shown in Figure 42.5.

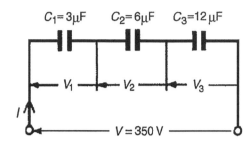

Figure 42.5

(a) The equivalent circuit capacitance C for three capacitors in series is given by:

$$\frac{1}{C} = \frac{1}{C_1} + \frac{1}{C_2} + \frac{1}{C_3}$$

i.e. $\quad \frac{1}{C} = \frac{1}{3} + \frac{1}{6} + \frac{1}{12} = \frac{4+2+1}{12} = \frac{7}{12}$

Hence, the equivalent circuit capacitance

$$C = \frac{12}{7} = 1\frac{5}{7}\ \mu\text{F} \quad \text{or} \quad \textbf{1.714 μF}$$

(b) Total charge $Q_T = CV$,

hence

$$Q_T = \frac{12}{7} \times 10^{-6} \times 350 = 600\mu\text{ C} \quad \text{or} \quad 0.6\ \text{mC}$$

Since the capacitors are connected in series, 0.6 mC is the charge on each of them.

(c) The voltage across the 3 μF capacitor

$$V_1 = \frac{Q}{C_1} = \frac{0.6 \times 10^{-3}}{3 \times 10^{-6}} = \textbf{200 V}$$

The voltage across the 6 μF capacitor

$$V_2 = \frac{Q}{C_2} = \frac{0.6 \times 10^{-3}}{6 \times 10^{-6}} = \textbf{100 V}$$

The voltage across the 12 μF capacitor

$$V_3 = \frac{Q}{C_3} = \frac{0.6 \times 10^{-3}}{12 \times 10^{-6}} = \textbf{50 V}$$

(Check: In a series circuit $V = V_1 + V_2 + V_3$

$$V_1 + V_2 + V_3 = 200 + 100 + 50$$

$$= 350\,\text{V} = \text{ supply voltage.})$$

In practice, capacitors are rarely connected in series unless they are of the same capacitance. The reason for this can be seen from the above problem where the lowest-valued capacitor (i.e. 3 μF) has the highest p.d. across it (i.e. 200 V) which means that if all the capacitors have an identical construction they must all be rated at the highest voltage.

Now try the following Practice Exercise

Practice Exercise 205 Further problems on capacitors in parallel and series (answers on page 549)

1. Capacitors of 2 μF and 6 μF are connected (a) in parallel and (b) in series. Determine the equivalent capacitance in each case

2. Find the capacitance to be connected in series with a 10 μF capacitor for the equivalent capacitance to be 6 μF

3. What value of capacitance would be obtained if capacitors of 0.15 μF and 0.10 μF are connected (a) in series and (b) in parallel?

4. Two 6 μF capacitors are connected in series with one having a capacitance of 12 μF. Find the total equivalent circuit capacitance. What capacitance must be added in series to obtain a capacitance of 1.2 μF?

5. For the arrangement shown in Figure 42.6 find (a) the equivalent circuit capacitance and (b) the voltage across a 4.5 μF capacitor

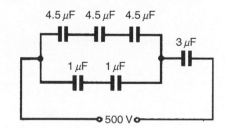

Figure 42.6

6. In Figure 42.7 capacitors P, Q and R are identical and the total equivalent capacitance of the circuit is $3\,\mu F$. Determine the values of P, Q and R

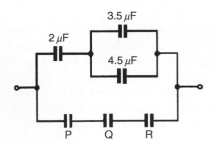

Figure 42.7

42.4 Dielectric strength

The maximum amount of field strength that a dielectric can withstand is called the dielectric strength of the material.

$$\textbf{Dielectric strength}\quad E_m = \frac{V_m}{d}$$

Problem 11. A capacitor is to be constructed so that its capacitance is $0.2\,\mu F$ and to take a p.d. of $1.25\,kV$ across its terminals. The dielectric is to be mica which has a dielectric strength of $50\,MV/m$. Find (a) the thickness of the mica needed, and (b) the area of a plate assuming a two-plate construction (assume ε_r for mica to be 6)

(a) Dielectric strength $E = \dfrac{V}{d}$

i.e. $d = \dfrac{V}{E} = \dfrac{1.25 \times 10^3}{50 \times 10^6}\,\text{m} = \textbf{0.025\,mm}$

(b) Capacitance $C = \dfrac{\varepsilon_0 \varepsilon_r A}{d}$ hence,

$\text{area } A = \dfrac{Cd}{\varepsilon_0 \varepsilon_r} = \dfrac{0.2 \times 10^{-6} \times 0.025 \times 10^{-3}}{8.85 \times 10^{-12} \times 6}\,\text{m}^2$

$= 0.09416\,\text{m}^2$

$= \textbf{941.6\,cm}^2$

42.5 Energy stored in capacitors

The energy W stored by a capacitor is given by:

$$W = \frac{1}{2}CV^2 \textbf{ joules}$$

Problem 12. (a) Determine the energy stored in a $3\,\mu F$ capacitor when charged to $400\,V$. (b) Find also the average power developed if this energy is dissipated in a time of $10\,\mu s$

(a) **Energy stored**

$$W = \frac{1}{2}CV^2 \text{ joules}$$

$$= \frac{1}{2} \times 3 \times 10^{-6} \times 400^2$$

$$= \frac{3}{2} \times 16 \times 10^{-2} = \textbf{0.24\,J}$$

(b) **Power** $= \dfrac{\text{energy}}{\text{time}} = \dfrac{0.24}{10 \times 10^{-6}}\text{W} = \textbf{24\,kW}$

Problem 13. A $12\,\mu F$ capacitor is required to store $4\,J$ of energy. Find the p.d. to which the capacitor must be charged

Energy stored $W = \dfrac{1}{2}CV^2$ hence, $V^2 = \dfrac{2W}{C}$

and p.d. $V = \sqrt{\dfrac{2W}{C}}$

$$= \sqrt{\dfrac{2 \times 4}{12 \times 10^{-6}}} = \sqrt{\dfrac{2 \times 10^6}{3}} = \textbf{816.5\,V}$$

Now try the following Practice Exercise

Practice Exercise 206 **Further problems on energy stored in capacitors (answers on page 549)**

(Assume $\varepsilon_0 = 8.85 \times 10^{-12}$ F/m.)

1. When a capacitor is connected across a $200\,V$ supply the charge is $4\,\mu C$. Find (a) the capacitance and (b) the energy stored

2. Find the energy stored in a 10 μF capacitor when charged to 2 kV

3. A 3300 pF capacitor is required to store 0.5 mJ of energy. Find the p.d. to which the capacitor must be charged

4. A Bakelite capacitor is to be constructed to have a capacitance of 0.04 μF and to have a steady working potential of 1 kV maximum. Allowing a safe value of field stress of 25 MV/m find (a) the thickness of Bakelite required, (b) the area of plate required if the relative permittivity of Bakelite is 5, (c) the maximum energy stored by the capacitor and (d) the average power developed if this energy is dissipated in a time of 20 μs

42.6 Practical types of capacitor

Practical types of capacitor are characterised by the material used for their dielectric. The main types include: variable air, mica, paper, ceramic, plastic, titanium oxide and electrolytic.

1. **Variable air capacitors**. These usually consist of two sets of metal plates (such as aluminium), one fixed, the other variable. The set of moving plates rotate on a spindle as shown by the end view of Figure 42.8. As the moving plates are rotated through half a revolution, the meshing, and therefore the capacitance, varies from a minimum to a maximum value. Variable air capacitors are used in radio and electronic circuits where very low losses are required, or where a variable capacitance is needed. The maximum value of such capacitors is between 500 pF and 1000 pF.

2. **Mica capacitors**. A typical practical variable air capacitor is shown in Figure 42.9. A typical older type construction is shown in Figure 42.10. Usually

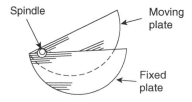

Figure 42.8

Figure 42.9

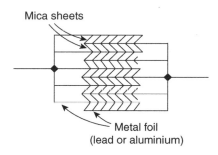

Figure 42.10

the whole capacitor is impregnated with wax and placed in a Bakelite case. Mica is easily obtained in thin sheets and is a good insulator. However, mica is expensive and is not used in capacitors above about 0.2 μF. A modified form of mica capacitor is the silvered mica type. The mica is coated on both sides with a thin layer of silver which forms the plates. Capacitance is stable and less likely to change with age. Such capacitors have a constant capacitance with change of temperature, a high working voltage rating and a long service life and are used in high frequency circuits with fixed values of capacitance up to about 1000 pF.

3. **Paper capacitors**. A typical paper capacitor is shown in Figure 42.11 where the length of the roll corresponds to the capacitance required. The whole is usually impregnated with oil or wax to exclude moisture, and then placed in a plastic or aluminium container for protection. Paper capacitors are made in various working voltages up to about 150 kV

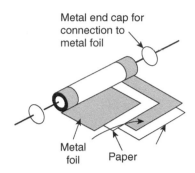

Figure 42.11

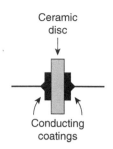

Figure 42.14

and are used where loss is not very important. The maximum value of this type of capacitor is between 500 pF and 10 μF. Disadvantages of paper capacitors include variation in capacitance with temperature change and a shorter service life than most other types of capacitor.

4. **Ceramic capacitors**. These are made in various forms, each type of construction depending on the value of capacitance required. For high values, a tube of ceramic material is used as shown in the cross-section of Figure 42.12. For smaller values the cup construction is used as shown in Figure 42.13, and for still smaller values the disc construction shown in Figure 42.14 is used. Certain ceramic materials have a very high permittivity and this enables capacitors of high capacitance to be made which are of small physical size with a high working voltage rating. Ceramic capacitors are available in the range 1 pF to 0.1 μF and may be used in high

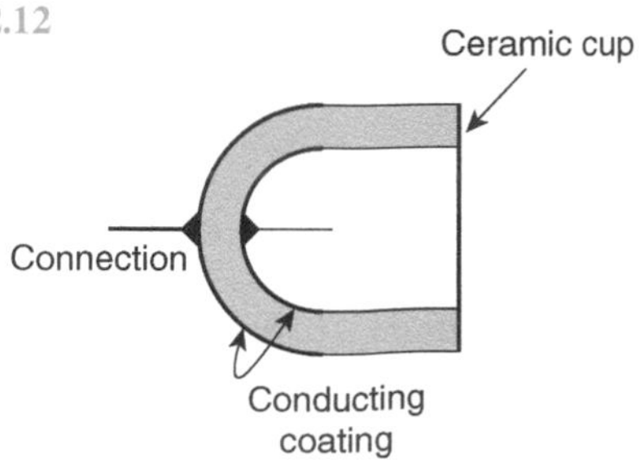

frequency electronic circuits subject to a wide range of temperatures.

5. **Plastic capacitors**. Some plastic materials such as polystyrene and Teflon can be used as dielectrics. Construction is similar to the paper capacitor but using a plastic film instead of paper. Plastic capacitors operate well under conditions of high temperature, provide a precise value of capacitance, a very long service life and high reliability.

6. **Titanium oxide capacitors** have a very high capacitance with a small physical size when used at a low temperature.

7. **Electrolytic capacitors**. Construction is similar to the paper capacitor with aluminium foil used for the plates and with a thick absorbent material, such as paper, impregnated with an electrolyte (ammonium borate), separating the plates. The finished capacitor is usually assembled in an aluminium container and hermetically sealed. Its operation depends on the formation of a thin aluminium oxide layer on the positive plate by electrolytic action when a suitable direct potential is maintained between the plates. This oxide layer is very thin and forms the dielectric. (The absorbent paper between the plates is a conductor and does not act as a dielectric.) Such capacitors **must always be used on d.c. and must be connected with the correct polarity**; if this is not done the capacitor will be destroyed since the oxide layer will be destroyed. Electrolytic capacitors are manufactured with working voltages from 6 V to 600 V, although accuracy is generally not very high. These capacitors **possess a much larger capacitance than other types of capacitors of similar dimensions** due to the oxide film being only a few microns thick. The fact that they can be used only on d.c. supplies limit their usefulness.

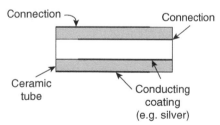

Figure 42.12

Figure 42.13

42.7 Supercapacitors

Electrical double-layer capacitors (EDLC) are, together with pseudocapacitors, part of a new type of electro-chemical capacitor called **supercapacitors**, also known as **ultracapacitors**.

Supercapacitors have the **highest available capacitance** values per unit volume and the greatest energy density of all capacitors. Supercapacitor support up to 12,000 F/1.2 V, with specific capacitance values up to 10,000 times that of electrolytic capacitors. Supercapacitors bridge the gap between capacitors and batteries. However, batteries still have about 10 times the capacity of supercapacitors.

Supercapacitors are **polarised** and must operate with the correct polarity.

Applications of supercapacitors for power and energy requirements, include long, small currents for static memory (SRAM) in electronic equipment, power electronics that require very short, high currents as in the KERS system in Formula 1 cars, and recovery of braking energy in vehicles.

Advantages of supercapacitors include:

(i) Long life, with little degradation over hundreds of thousands of charge cycles

(ii) Low cost per cycle

(iii) Good reversibility

(iv) Very high rates of charge and discharge

(v) Extremely low internal resistance (ESR) and consequent high cycle efficiency (95% or more) and extremely low heating levels

(vi) High output power

(vii) High specific power

(viii) Improved safety, no corrosive electrolyte and low toxicity of materials

(ix) Simple charge methods – no full-charge detection is needed; no danger of overcharging

(x) When used in conjunction with rechargeable batteries, in some applications the EDLC can supply energy for a short time, reducing battery cycling duty and extending life

Disadvantages of supercapacitors include:

(i) The amount of energy stored per unit weight is generally lower than that of an electrochemical battery

(ii) Has the highest dielectric absorption of any type of capacitor

(iii) High self-discharge – the rate is considerably higher than that of an electrochemical battery

(iv) Low maximum voltage – series connections are needed to obtain higher voltages and voltage balancing may be required

(v) Unlike practical batteries, the voltage across any capacitor, including EDLCs, drops significantly as it discharges. Effective storage and recovery of energy requires complex electronic control and switching equipment, with consequent energy loss

(vi) Very low internal resistance allows extremely rapid discharge when shorted, resulting in a spark hazard similar to any other capacitor of similar voltage and capacitance (generally much higher than electrochemical cells)

Summary

Supercapacitors are some of the best devices available for delivering a quick surge of power. Because an ultracapacitor stores energy in an electric field, rather than in a chemical reaction, it can survive hundreds of thousands more charge and discharge cycles than a battery can.

42.8 Discharging capacitors

When a capacitor has been disconnected from the supply it may still be charged and it may retain this charge for some considerable time. Thus precautions must be taken to ensure that the capacitor is automatically discharged after the supply is switched off. This is done by connecting a high value resistor across the capacitor terminals.

42.9 Inductance

Inductance is the property of a circuit whereby there is an electromotive force (e.m.f.) induced into the circuit by the change of flux linkages produced by a current change. When the e.m.f. is induced in the same circuit as that in which the current is changing, the property is called **self-inductance L**.

The **unit** of inductance is the **henry**[*], **H**:

A circuit has an inductance of one henry when an e.m.f. of one volt is induced in it by a current changing at the rate of one ampere per second.

Induced e.m.f. in a coil of N turns

$$E = -N\frac{d\Phi}{dt} \text{ volts}$$

where $d\Phi$ is the change in flux in webers, and dt is the time taken for the flux to change in seconds (i.e. $\frac{d\Phi}{dt}$ is the rate of change of flux).

Induced e.m.f. in a coil of inductance L henries

$$E = -L\frac{dI}{dt} \text{ volts}$$

where dI is the change in current in amperes and dt is the time taken for the current to change in seconds (i.e. $\frac{dI}{dt}$ is the rate of change of current).

The minus signs in each of the above two equations remind us of its direction (given by Lenz's law).

Problem 14. Determine the e.m.f. induced in a coil of 200 turns when there is a change of flux of 25 mWb linking with it in 50 ms

Induced e.m.f. $E = -N\dfrac{d\Phi}{dt} = -(200)\left(\dfrac{25 \times 10^{-3}}{50 \times 10^{-3}}\right)$

$$= -\textbf{100 volts}$$

Problem 15. A flux of 400 μWb passing through a 150-turn coil is reversed in 40 ms. Find the average e.m.f. induced

Since the flux reverses, the flux changes from $+400\,\mu$Wb to $-400\,\mu$Wb, a total change of flux of 800 μWb.

[*]Who was Henry? – **Joseph Henry** (17 December 1797–13 May 1878) was an American scientist who discovered the electromagnetic phenomenon of self-induction. To find out more about Henry go to www.routledge.com/cw/bird (For image refer to page 30)

Induced e.m.f. $E = -N\dfrac{d\Phi}{dt} = -(150)\left(\dfrac{800 \times 10^{-6}}{40 \times 10^{-3}}\right)$

$$= -\frac{150 \times 800 \times 10^3}{40 \times 10^6}$$

Hence, **the average e.m.f. induced** $E = -\textbf{3 volts}$

Problem 16. Calculate the e.m.f. induced in a coil of inductance 12 H by a current changing at the rate of 4 A/s

Induced e.m.f. $E = -L\dfrac{dI}{dt} = -(12)(4) = -\textbf{48 volts}$

Problem 17. An e.m.f. of 1.5 kV is induced in a coil when a current of 4 A collapses uniformly to zero in 8 ms. Determine the inductance of the coil

Change in current

$$dI = (4 - 0) = 4\,\text{A}, \, dt = 8\,\text{ms} = 8 \times 10^{-3}\,\text{s},$$

$$\frac{dI}{dt} = \frac{4}{8 \times 10^{-3}} = \frac{4000}{8} = 500\,\text{A/s}$$

and $\quad E = 1.5\,\text{kV} = 1500\,\text{V}$

Since $\qquad\qquad |E| = L\dfrac{dI}{dt}$

inductance $\qquad L = \dfrac{|E|}{\dfrac{dI}{dt}} = \dfrac{1500}{500} = \textbf{3 H}$

(Note that $|E|$ means the 'magnitude of E' which disregards the minus sign.)

Now try the following Practice Exercise

Practice Exercise 207   Further problems on inductance (answers on page 549)

1. Find the e.m.f. induced in a coil of 200 turns when there is a change of flux of 30 mWb linking with it in 40 ms

2. An e.m.f. of 25 V is induced in a coil of 300 turns when the flux linking with it changes by 12 mWb. Find the time, in milliseconds, in which the flux makes the change

3. An ignition coil having 10,000 turns has an e.m.f. of 8 kV induced in it. What rate of change of flux is required for this to happen?

4. A flux of 0.35 mWb passing through a 125-turn coil is reversed in 25 ms. Find the magnitude of the average e.m.f. induced

5. Calculate the e.m.f. induced in a coil of inductance 6 H by a current changing at a rate of 15 A/s

42.10 Inductors

A component called an **inductor** is used when the property of inductance is required in a circuit. The basic form of an inductor is simply a coil of wire.

Factors which affect the inductance of an inductor include:

(i) the number of turns of wire – the more turns the higher the inductance,

(ii) the cross-sectional area of the coil of wire – the greater the cross-sectional area the higher the inductance,

(iii) the presence of a magnetic core – when the coil is wound on an iron core the same current sets up a more concentrated magnetic field and the inductance is increased,

(iv) the way the turns are arranged – a short thick coil of wire has a higher inductance than a long thin one.

42.11 Practical inductors

Two examples of practical inductors are shown in Figure 42.15, and the standard electrical circuit diagram symbols for air-cored and iron-cored inductors are shown in Figure 42.16. An iron-cored inductor is often

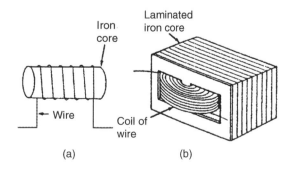

Figure 42.15

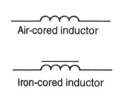

Air-cored inductor

Iron-cored inductor

Figure 42.16

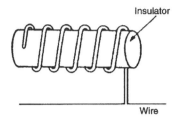

Figure 42.17

called a choke since, when used in a.c. circuits, it has a choking effect, limiting the current flowing through it. Inductance is often undesirable in a circuit. To reduce inductance to a minimum the wire may be bent back on itself, as shown in Figure 42.17, so that the magnetising effect of one conductor is neutralised by that of the adjacent conductor. The wire may be coiled around an insulator, as shown, without increasing the inductance. Standard resistors may be non-inductively wound in this manner.

42.12 Energy stored by inductors

An inductor possesses an ability to store energy. The energy stored W in the magnetic field of an inductor is given by:

$$W = \frac{1}{2}LI^2 \text{ joules}$$

Some typical small inductors are shown in Figure 42.18.

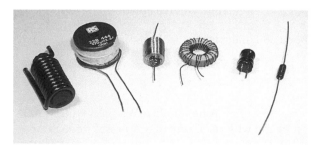

Figure 42.18

Problem 18. An 8 H inductor has a current of 3 A flowing through it. How much energy is stored in the magnetic field of the inductor?

Energy stored $W = \dfrac{1}{2}LI^2 = \dfrac{1}{2}(8)(3)^2 = \textbf{36 joules}$.

Now try the following Practice Exercise

Practice Exercise 208 Further problems on energy stored (answers on page 549)

1. An inductor of 20 H has a current of 2.5 A flowing in it. Find the energy stored in the magnetic field of the inductor

2. Calculate the value of the energy stored when a current of 30 mA is flowing in a coil of inductance 400 mH

3. The energy stored in the magnetic field of an inductor is 80 J when the current flowing in the inductor is 2 A. Calculate the inductance of the coil

42.13 Inductance of a coil

If a current changing from 0 to I amperes, produces a flux change from 0 to Φ webers, then $dI = I$ and $d\Phi = \Phi$. Then, from section 42.9,

$$\text{induced e.m.f.} \quad E = \frac{N\Phi}{t} = \frac{LI}{t}$$

from which, **inductance of coil** $L = \dfrac{N\Phi}{I}$ **henries**

Problem 19. Calculate the coil inductance when a current of 4 A in a coil of 800 turns produces a flux of 5 mWb linking with the coil

For a coil, inductance

$$L = \frac{N\Phi}{I} = \frac{(800)(5 \times 10^{-3})}{4} = \textbf{1 H}$$

Problem 20. A flux of 25 mWb links with a 1500 turn coil when a current of 3 A passes through the coil. Calculate (a) the inductance of the coil, (b) the energy stored in the magnetic field and (c) the average e.m.f. induced if the current falls to zero in 150 ms

(a) **Inductance**

$$L = \frac{N\Phi}{I} = \frac{(1500)(25 \times 10^{-3})}{3} = \textbf{12.5 H}$$

(b) **Energy stored**

$$W = \frac{1}{2}LI^2 = \frac{1}{2}(12.5)(3)^2 = \textbf{56.25 J}$$

(c) **Induced e.m.f.**

$$E = -L\frac{dI}{dt} = -(12.5)\left(\frac{3-0}{150 \times 10^{-3}}\right) = \textbf{-250 V}$$

(Alternatively,

$$E = -N\frac{d\Phi}{dt} = -(1500)\left(\frac{25 \times 10^{-3}}{150 \times 10^{-3}}\right) = \textbf{-250 V}$$

since if the current falls to zero so does the flux.)

Problem 21. When a current of 1.5 A flows in a coil the flux linking with the coil is 90 μWb. If the coil inductance is 0.60 H, calculate the number of turns of the coil

For a coil $\quad L = \dfrac{N\Phi}{I}$

thus $\quad N = \dfrac{LI}{\Phi} = \dfrac{(0.6)(1.5)}{90 \times 10^{-6}} = \textbf{10,000 turns}$.

Now try the following Practice Exercises

Practice Exercise 209 Further problems on the inductance of a coil (answers on page 549)

1. A flux of 30 mWb links with a 1200 turn coil when a current of 5 A is passing through the coil. Calculate (a) the inductance of the coil, (b) the energy stored in the magnetic field and (c) the average e.m.f. induced if the current is reduced to zero in 0.20 s

2. An e.m.f. of 2 kV is induced in a coil when a current of 5 A collapses uniformly to zero in 10 ms. Determine the inductance of the coil

3. An average e.m.f. of 60 V is induced in a coil of inductance 160 mH when a current of 7.5 A is reversed. Calculate the time taken for the current to reverse

4. A coil of 2500 turns has a flux of 10 mWb linking with it when carrying a current of 2 A. Calculate the coil inductance and the e.m.f. induced in the coil when the current collapses to zero in 20 ms

5. When a current of 2 A flows in a coil, the flux linking with the coil is 80 μWb. If the coil inductance is 0.5 H, calculate the number of turns of the coil

Practice Exercise 210   Short answer questions on capacitors and inductors

1. How can an 'electric field' be established between two parallel metal plates?

2. What is capacitance?

3. State the unit of capacitance

4. Complete the statement: Capacitance $= \dfrac{\cdots\cdots}{\cdots\cdots}$

5. Complete the statements:

   (a) 1 μF = ... F,   (b) 1 pF = ... F

6. Complete the statement: Electric field strength $E = \dfrac{\cdots\cdots}{\cdots\cdots}$

7. Complete the statement: Electric flux density $D = \dfrac{\cdots\cdots}{\cdots\cdots}$

8. Draw the electrical circuit diagram symbol for a capacitor

9. Name two practical examples where capacitance is present, although undesirable

10. The insulating material separating the plates of a capacitor is called the ......

11. 10 volts applied to a capacitor results in a charge of 5 coulombs. What is the capacitance of the capacitor?

12. Three 3 μF capacitors are connected in parallel. The equivalent capacitance is ....

13. Three 3 μF capacitors are connected in series. The equivalent capacitance is ....

14. State an advantage of series-connected capacitors

15. Name three factors upon which capacitance depends

16. What does 'relative permittivity' mean?

17. Define 'permittivity of free space'

18. What is meant by the 'dielectric strength' of a material?

19. State the formula used to determine the energy stored by a capacitor

20. Name five types of capacitor commonly used

21. Sketch a typical rolled paper capacitor

22. Explain briefly the construction of a variable air capacitor

23. State three advantages and one disadvantage of mica capacitors

24. Name two disadvantages of paper capacitors

25. Between what values of capacitance are ceramic capacitors normally available

26. What main advantages do plastic capacitors possess?

27. Explain briefly the construction of an electrolytic capacitor

28. What is the main disadvantage of electrolytic capacitors?

29. Name an important advantage of electrolytic capacitors

30. State three applications of supercapacitors

31. State five advantages and five disadvantages of supercapacitors

32. What safety precautions should be taken when a capacitor is disconnected from a supply?

33. Define inductance and name its unit

34. What factors affect the inductance of an inductor?

35. What is an inductor? Sketch a typical practical inductor

36. Explain how a standard resistor may be non-inductively wound

37. Energy W stored in the magnetic field of an inductor is given by: $W = ...$ joules

Practice Exercise 211   Multiple-choice questions on capacitors and inductance (answers on page 498)

1. The capacitance of a capacitor is the ratio

    (a) charge to p.d. between plates

    (b) p.d. between plates to plate spacing

    (c) p.d. between plates to thickness of dielectric

    (d)) p.d. between plates to charge

2. The p.d. across a 10 μF capacitor to charge it with 10 mC is:

    (a) 10 V   (b) 1 kV   (c) 1 V   (d) 100 V

3. The charge on a 10 pF capacitor when the voltage applied to it is 10 kV is

    (a) 100 μC   (b) 0.1 C

    (c) 0.1 μC   (d) 0.01 μC

4. Four 2 μF capacitors are connected in parallel. The equivalent capacitance is:

    (a) 8 μF   (b) 0.5 μF

    (c) 2 μF   (d) 6 μF

5. Four 2 μF capacitors are connected in series. The equivalent capacitance is:

    (a) 8 μF   (b) 0.5 μF

    (c) 2 μF   (d) 6 μF

6. State which of the following is false. The capacitance of a capacitor:

    (a) is proportional to the cross-sectional area of the plates

    (b) is proportional to the distance between the plates

    (c) depends on the number of plates

    (d) is proportional to the relative permittivity of the dielectric

7. Which of the following statements is false?

    (a) An air capacitor is normally a variable type

    (b) A paper capacitor generally has a shorter service life than most other types of capacitor

    (c) An electrolytic capacitor must be used only on a.c. supplies

    (d) Plastic capacitors generally operate satisfactorily under conditions of high temperature

8. The energy stored in a 10 μF capacitor when charged to 500 V is:

    (a) 1.25 mJ   (b) 0.025 μJ

    (c) 1.25 J   (d) 1.25 C

9. The capacitance of a variable air capacitor is at maximum when:

    (a) the movable plates half overlap the fixed plates

    (b) the movable plates are most widely separated from the fixed plates

    (c) both sets of plates are exactly meshed

    (d) the movable plates are closer to one side of the fixed plate than to the other

10. When a voltage of 1 kV is applied to a capacitor, the charge on the capacitor is 500 nC. The capacitance of the capacitor is:

    (a) 2×10^9 F   (b) 0.5 pF

    (c) 0.5 mF   (d) 0.5 nF

11. When a magnetic flux of 10 Wb links with a circuit of 20 turns in 2 s, the induced e.m.f. is:

    (a) 1 V   (b) 4 V

    (c) 100 V   (d) 400 V

12. A current of 10 A in a coil of 1000 turns produces a flux of 10 mWb linking with the coil. The coil inductance is:

    (a) 10^6 H   (b) 1 H

    (c) 1 μH   (d) 1 mH

13. Which of the following statements is false? The inductance of an inductor increases:

    (a) with a short, thick, coil

    (b) when wound on an iron core

    (c) as the number of turns increases

    (d) as the cross-sectional area of the coil decreases

14. Self-inductance occurs when:

(a) the current is changing

(b) the circuit is changing

(c) the flux is changing

(d) the resistance is changing

For fully worked solutions to each of the problems in Exercises 203 to 211 in this chapter, go to the website: www.routledge.com/cw/bird

Chapter 43

Electrical measuring instruments and measurements

Why it is important to understand: **Electrical measuring instruments and measurements**

Future electrical engineers need to be able to appreciate basic measurement techniques, instruments, and methods used in everyday practice. This chapter covers analogue and digital instruments, measurement errors, bridges, oscilloscopes, data acquisition, instrument controls and measurement systems. Accurate measurements are central to virtually every scientific and engineering discipline. Electrical measurements often come down to either measuring current or measuring voltage. Even if you are measuring frequency, you will be measuring the frequency of a current signal or a voltage signal and you will need to know how to measure either voltage or current. Many times you will use a digital multimeter – a DMM – to measure either voltage or current; actually, a DMM will also usually measure frequency (of a voltage signal) and resistance. The quality of a measuring instrument is assessed from its accuracy, precision, reliability, durability, and so on, all of which are related to its cost.

At the end of this chapter, you should be able to:

- recognise the importance of testing and measurements in electric circuits
- understand the advantages of electronic instruments
- understand the operation of a wattmeter
- understand the operation of an oscilloscope for d.c. and a.c. measurements
- calculate periodic time, frequency, peak to peak values from waveforms on an oscilloscope
- understand null methods of measurement for a Wheatstone bridge and d.c. potentiometer

Science and Mathematics for Engineering. 978-0-367-20475-4, © John Bird. Published by Taylor & Francis. All rights reserved.

43.1   Introduction

Tests and measurements are important in designing, evaluating, maintaining and servicing electrical circuits and equipment. In order to detect electrical quantities such as current, voltage, resistance or power, it is necessary to transform an electrical quantity or condition into a visible indication. This is done with the aid of instruments (or meters) that indicate the magnitude of quantities either by the position of a pointer moving over a graduated scale (called an analogue instrument) or in the form of a decimal number (called a digital instrument).

The digital instrument has, in the main, become the instrument of choice in recent years; in particular, computer-based instruments are rapidly replacing items of conventional test equipment, with the virtual storage test instrument, the **digital storage oscilloscope**, being the most common.

43.2   Electronic instruments

Electronic measuring instruments have advantages over instruments such as the moving-iron or moving-coil meters, in that they have a much higher input resistance (some as high as $1000\,\text{M}\Omega$) and can handle a much wider range of frequency (from d.c. up to MHz).

The **digital voltmeter** (DVM) is one which provides a digital display of the voltage being measured. Advantages of a DVM over analogue instruments include higher accuracy and resolution, no observational or parallex errors and a very high input resistance, constant on all ranges.

A **digital multimeter** is a DVM with additional circuitry which makes it capable of measuring a.c. voltage, and d.c. and a.c. current and resistance.

Instruments for a.c. measurements are generally calibrated with a sinusoidal alternating waveform to indicate root mean square (r.m.s.) values when a sinusoidal signal is applied to the instrument.

Sometimes quantities to be measured have complex waveforms, and whenever a quantity is non-sinusoidal, errors in instrument readings can occur if the instrument has been calibrated for sine waves only. Such waveform errors can be largely eliminated by using electronic instruments.

43.3   Multimeters

Digital multimeters (DMM) are now almost universally used, the **Fluke digital multimeter** being an industry leader for performance, accuracy, resolution, ruggedness, reliability and safety. These instruments measure d.c. currents and voltages, resistance and continuity, a.c. (r.m.s.) currents and voltages, temperature and much more.

43.4   Wattmeters

A **wattmeter** is an instrument for measuring electrical power in a circuit. Figure 43.1 shows typical connections of a wattmeter used for measuring power supplied to a load. The instrument has two coils:

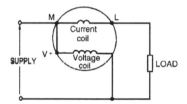

Figure 43.1

(i)   a current coil, which is connected in series with the load, like an ammeter, and

(ii)   a voltage coil, which is connected in parallel with the load, like a voltmeter.

43.5   Instrument 'loading' effect

Some measuring instruments depend for their operation on power taken from the circuit in which measurements are being made. Depending on the 'loading' effect of the instrument (i.e. the current taken to enable it to operate), the prevailing circuit conditions may change.

The resistance of voltmeters may be calculated, since each has a stated sensitivity (or 'Figure of merit'), often stated in 'kΩ per volt' of f.s.d. A voltmeter should have as high a resistance as possible (ideally infinite). In a.c. circuits the impedance of the instrument varies with frequency and thus the loading effect of the instrument can change.

Problem 1.   Calculate the power dissipated by the voltmeter and by resistor R in Figure 43.2 when (a) $R = 250\,\Omega$, (b) $R = 2\,M\Omega$. Assume that the voltmeter sensitivity (or figure of merit) is $10\,k\Omega/V$

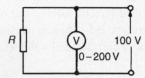

Figure 43.2

(a)   Resistance of voltmeter R_v = sensitivity × f.s.d.

Hence $R_v = (10\,k\Omega/V) \times (200\,V) = 2000\ k\Omega$
$$= 2\,M\Omega$$

Current flowing in voltmeter

$$I_v = \frac{V}{R_v} = \frac{100}{2 \times 10^6} = 50 \times 10^{-6}\,A$$

Power dissipated by voltmeter

$$= VI_v = (100)(50 \times 10^{-6}) = \mathbf{5\,mW}$$

When $R = 250\,\Omega$, current in resistor

$$I_R = \frac{V}{R} = \frac{100}{250} = \mathbf{0.4\,A}$$

Power dissipated in load resistor

$$R = VI_R = (100)(0.4) = \mathbf{40\,W}$$

Thus the power dissipated in the voltmeter is insignificant in comparison with the power dissipated in the load.

(b)   When $R = 2\,M\Omega$, current in resistor

$$I_R = \frac{V}{R} = \frac{100}{2 \times 10^6} = 50 \times 10^{-6}\,A$$

Power dissipated in load resistor R

$$= VI_R = 100 \times 50 \times 10^{-6} = \mathbf{5\,mW}$$

In this case the higher load resistance reduced the power dissipated such that the voltmeter is using as much power as the load.

Problem 2.   An ammeter has a f.s.d. of 100 mA and a resistance of 50 Ω. The ammeter is used to measure the current in a load of resistance 500 Ω when the supply voltage is 10 V. Calculate (a) the ammeter reading expected (neglecting its resistance), (b) the actual current in the circuit, (c) the power dissipated in the ammeter and (d) the power dissipated in the load

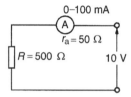

Figure 43.3

From Figure 43.3,

(a)   expected ammeter reading $= \dfrac{V}{R} = \dfrac{10}{500} = \mathbf{20\,mA}$

(b)   Actual ammeter reading $= \dfrac{V}{R + r_a} = \dfrac{10}{500 + 50}$
$$= \mathbf{18.18\,mA}$$

Thus the ammeter itself has caused the circuit conditions to change from 20 mA to 18.18 mA

(c)   Power dissipated in the ammeter

$$= I^2 r_a = (18.18 \times 10^{-3})^2 (50) = \mathbf{16.53\,mW}$$

(d)   Power dissipated in the load resistor

$$= I^2 R = (18.18 \times 10^{-3})^2 (500) = \mathbf{165.3\,mW}$$

Problem 3. (a) A current of 20 A flows through a load having a resistance of 2 Ω. Determine the power dissipated in the load. (b) A wattmeter, whose current coil has a resistance of 0.01 Ω, is connected as shown in Figure 43.4. Determine the wattmeter reading

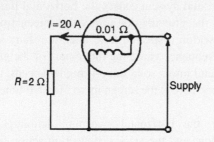

$I = 20$ A   0.01 Ω

$R = 2$ Ω   Supply

Figure 43.4

(a) Power dissipated in the load

$$P = I^2R = (20)^2(2) = \mathbf{800\ W}$$

(b) With the wattmeter connected in the circuit the total resistance R_T is $2 + 0.01 = 2.01$ Ω

The wattmeter reading is thus

$$I^2R_T = (20)^2(2.01) = \mathbf{804\ W}$$

Now try the following Practice Exercise

Practice Exercise 212   Further problems on instrument 'loading' effects (answers on page 549)

1. A 0–1 A ammeter having a resistance of 50 Ω is used to measure the current flowing in a 1 kΩ resistor when the supply voltage is 250 V. Calculate: (a) the approximate value of current (neglecting the ammeter resistance), (b) the actual current in the circuit, (c) the power dissipated in the ammeter, (d) the power dissipated in the 1 kΩ resistor

2. (a) A current of 15 A flows through a load having a resistance of 4 Ω. Determine the power dissipated in the load. (b) A wattmeter, whose current coil has a resistance of 0.02 Ω is connected (as shown in Figure 43.4) to measure the power in the load. Determine the wattmeter reading assuming the current in the load is still 15 A

43.6   The oscilloscope

The oscilloscope is basically a graph-displaying device – it draws a graph of an electrical signal. In most applications the graph shows how signals change over time. From the graph it is possible to:

- determine the time and voltage values of a signal
- calculate the frequency of an oscillating signal
- see the 'moving parts' of a circuit represented by the signal
- tell if a malfunctioning component is distorting the signal
- find out how much of a signal is d.c. or a.c.
- tell how much of the signal is noise and whether the noise is changing with time

Oscilloscopes are used by everyone from television repair technicians to physicists. They are indispensable for anyone designing or repairing electronic equipment. The usefulness of an oscilloscope is not limited to the world of electronics. With the proper transducer (i.e. a device that creates an electrical signal in response to physical stimuli, such as sound, mechanical stress, pressure, light or heat), an oscilloscope can measure any kind of phenomenon. An automobile engineer uses an oscilloscope to measure engine vibrations; a medical researcher uses an oscilloscope to measure brain waves, and so on.

Oscilloscopes are available in both analogue and digital types. An **analogue oscilloscope** works by directly applying a voltage being measured to an electron beam moving across the oscilloscope screen. The voltage deflects the beam up or down proportionally, tracing the waveform on the screen. This gives an immediate picture of the waveform.

In contrast, a **digital oscilloscope** samples the waveform and uses an analogue-to-digital converter to convert the voltage being measured into digital information. It then uses this digital information to reconstruct the waveform on the screen.

For many applications either an analogue or digital oscilloscope is appropriate. However, each type does possess some unique characteristics making it more or less suitable for specific tasks.

Analogue oscilloscopes are often preferred when it is important to display rapidly varying signals in 'real time' (i.e. as they occur).

Digital oscilloscopes allow the capture and viewing of events that happen only once. They can process the digital waveform data or send the data to a computer for processing. Also, they can store the digital waveform data for later viewing and printing. Digital storage oscilloscopes are explained in section 43.8.

43.6.1 Analogue oscilloscopes

When an oscilloscope probe is connected to a circuit, the voltage signal travels through the probe to the vertical system of the oscilloscope. Figure 43.5 gives a simple block diagram that shows how an analogue oscilloscope displays a measured signal.

Depending on how the vertical scale (volts/division control) is set, an attenuator reduces the signal voltage or an amplifier increases the signal voltage. Next, the signal travels directly to the vertical deflection plates of the cathode ray tube (CRT). Voltage applied to these deflection plates causes a glowing dot to move. (An electron beam hitting phosphor inside the CRT creates the glowing dot.) A positive voltage causes the dot to move up, while a negative voltage causes the dot to move down.

The signal also travels to the trigger system to start or trigger a 'horizontal sweep'. Horizontal sweep is a term referring to the action of the horizontal system causing the glowing dot to move across the screen. Triggering the horizontal system causes the horizontal time base to move the glowing dot across the screen from left to right within a specific time interval. Many sweeps in rapid sequence cause the movement of the glowing dot to blend into a solid line. At higher speeds, the dot may sweep across the screen up to 500,000 times each second.

Together, the horizontal sweeping action (i.e. the X direction) and the vertical deflection action (i.e. the Y direction) trace a graph of the signal on the screen. The trigger is necessary to stabilise a repeating signal. It ensures that the sweep begins at the same point of a repeating signal, resulting in a clear picture.

In conclusion, to use an analogue oscilloscope, three basic settings to accommodate an incoming signal need to be adjusted:

- the attenuation or amplification of the signal – use the volts/division control to adjust the amplitude of the signal before it is applied to the vertical deflection plates
- the time base – use the time/division control to set the amount of time per division represented horizontally across the screen

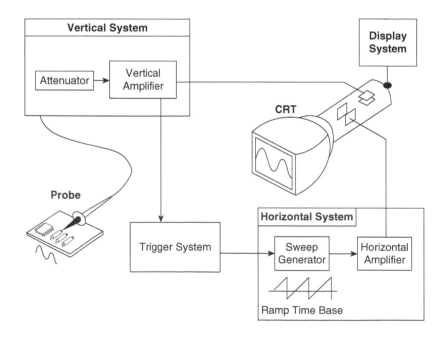

Figure 43.5

- the triggering of the oscilloscope – use the trigger level to stabilise a repeating signal, as well as triggering on a single event

Also, adjusting the focus and intensity controls enable a sharp, visible display to be created.

(i) With **direct voltage measurements**, only the Y amplifier 'volts/cm' switch on the cathode ray oscilloscope (c.r.o.) is used. With no voltage applied to the Y plates the position of the spot trace on the screen is noted. When a direct voltage is applied to the Y plates the new position of the spot trace is an indication of the magnitude of the voltage. For example, in Figure 43.6(a), with no voltage applied to the Y plates, the spot trace is in the centre of the screen (initial position) and then the spot trace moves 2.5 cm to the final position shown, on application of a d.c. voltage. With the 'volts/cm' switch on 10 volts/cm the magnitude of the direct voltage is 2.5 cm $\times$ 10 volts/cm, i.e. 25 volts.

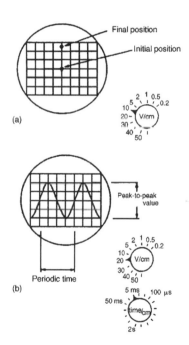

Figure 43.6

(ii) With **alternating voltage measurements**, let a sinusoidal waveform be displayed on a cathode ray oscilloscope (c.r.o.) screen as shown in Figure 43.6(b). If the time/cm switch is on, say,

5 ms/cm then the **periodic time T** of the sinewave is 5 ms/cm $\times$ 4 cm, i.e. **20 ms** or **0.02 s**.

Since frequency $f = \dfrac{1}{T}$, **frequency** $= \dfrac{1}{0.02}$
$$= \mathbf{50\,Hz}$$

If the 'volts/cm' switch is on, say, 20 volts/cm then the **amplitude** or **peak value** of the sinewave shown is 20 volts/cm $\times$ 2 cm, i.e. 40 V

Since r.m.s. voltage $= \dfrac{\text{peak voltage}}{\sqrt{2}}$

(see chapter 41),

$$\textbf{r.m.s. voltage} = \frac{40}{\sqrt{2}} = \mathbf{28.28\,volts}.$$

Double beam oscilloscopes are useful whenever two signals are to be compared simultaneously. The c.r.o. demands reasonable skill in adjustment and use. However, its greatest advantage is in observing the shape of a waveform – a feature not possessed by other measuring instruments.

43.6.2 Digital oscilloscopes

Some of the systems that make up digital oscilloscopes are the same as those in analogue oscilloscopes; however, digital oscilloscopes contain additional data processing systems – as shown in the block diagram of Figure 43.7. With the added systems, the digital oscilloscope collects data for the entire waveform and then displays it.

When a digital oscilloscope probe is attached to a circuit, the vertical system adjusts the amplitude of the signal, just as in the analogue oscilloscope. Next, the analogue-to-digital converter (ADC) in the acquisition system samples the signal at discrete points in time and converts the signals' voltage at these points to digital values called *sample points*. The horizontal systems' sample clock determines how often the ADC takes a sample. The rate at which the clock 'ticks' is called the sample rate and is measured in samples per second.

The sample points from the ADC are stored in memory as *waveform points*. More than one sample point may make up one waveform point.

Together, the waveform points make up one waveform *record*. The number of waveform points used to make a waveform record is called a *record length*. The trigger system determines the start and stop points of the record. The display receives these record points after being stored in memory.

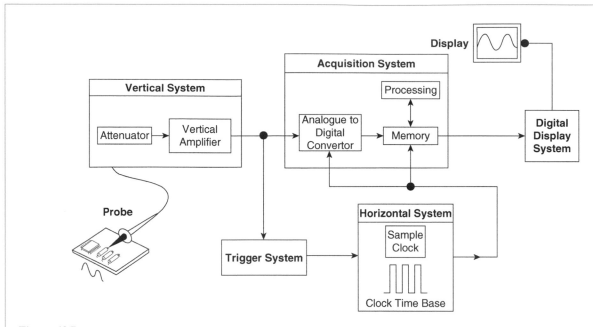

Figure 43.7

Depending on the capabilities of an oscilloscope, additional processing of the sample points may take place, enhancing the display. Pre-trigger may be available, allowing events before the trigger point to be seen.

Fundamentally, with a digital oscilloscope as with an analogue oscilloscope, there is a need to adjust vertical, horizontal, and trigger settings to take a measurement. A typical double-beam digital fluke oscilloscope is shown in Figure 43.8.

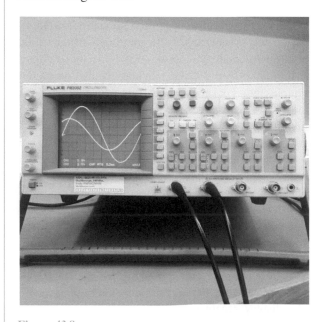

Figure 43.8

Problem 4. For the oscilloscope square voltage waveform shown in Figure 43.9 determine (a) the periodic time, (b) the frequency and (c) the peak-to-peak voltage. The 'time/cm' (or time base control) switch is on $100\,\mu s/cm$ and the 'volts/cm' (or signal amplitude control) switch is on $20\,V/cm$

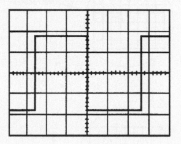

Figure 43.9

(*In Figures 43.9–43.15 assume that the squares shown are 1 cm × 1 cm.*)

(a) The width of one complete cycle is 5.2 cm.

Hence **the periodic time**

$$T = 5.2\,cm \times 100 \times 10^{-6}\,s/cm$$

$$= \mathbf{0.52\,ms}$$

(b) **Frequency** $f = \dfrac{1}{T} = \dfrac{1}{0.52 \times 10^{-3}} = \mathbf{1.92\,kHz}.$

(c) The peak-to-peak height of the display is 3.6 cm, hence the:

peak-to-peak voltage=3.6 cm×20 V/cm=**72 V**

Problem 5. For the oscilloscope display of a pulse waveform shown in Figure 43.10 the 'time/cm' switch is on 50 ms/cm and the 'volts/cm' switch is on 0.2 V/cm. Determine (a) the periodic time, (b) the frequency and (c) the magnitude of the pulse voltage

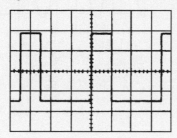

Figure 43.10

(a) The width of one complete cycle is 3.5 cm.

Hence **the periodic time**

$$T = 3.5 \text{ cm} \times 50 \text{ ms/cm} = \textbf{175 ms}$$

(b) **Frequency** $f = \dfrac{1}{T} = \dfrac{1}{0.175} = \textbf{5.71 Hz}$

(c) The height of a pulse is 3.4 cm hence

the magnitude of the pulse voltage

$$= 3.4 \text{ cm} \times 0.2 \text{ V/cm} = \textbf{0.68 V}$$

Problem 6. A sinusoidal voltage trace displayed by an oscilloscope is shown in Figure 43.11. If the 'time/cm' switch is on 500 μs/cm and the 'volts/cm' switch is on 5 V/cm, find, for the waveform, (a) the frequency, (b) the peak-to-peak voltage, (c) the amplitude and (d) the r.m.s. value

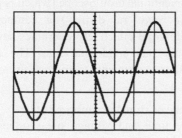

Figure 43.11

(a) The width of one complete cycle is 4 cm. Hence the periodic time, T is 4 cm × 500 μs/cm, i.e. 2 ms

Frequency $f = \dfrac{1}{T} = \dfrac{1}{2 \times 10^{-3}} = \textbf{500 Hz}$

(b) The peak-to-peak height of the waveform is 5 cm. Hence

the peak-to-peak voltage=5 cm × 5 V/cm=**25 V**

(c) **Amplitude** $= \dfrac{1}{2} \times 25 \text{ V} = \textbf{12.5 V}$

(d) The peak value of voltage is the amplitude, i.e. 12.5 V, and

r.m.s. voltage $= \dfrac{\text{peak voltage}}{\sqrt{2}} = \dfrac{12.5}{\sqrt{2}} = \textbf{8.84 V}$

Problem 7. For the double-beam oscilloscope displays shown in Figure 43.12 determine (a) their frequency, (b) their r.m.s. values and (c) their phase difference. The 'time/cm' switch is on 100 μs/cm and the 'volts/cm' switch on 2 V/cm

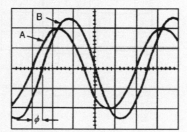

Figure 43.12

(a) The width of each complete cycle is 5 cm for both waveforms.

Hence the periodic time T of each waveform is 5 cm × 100 μs/cm, i.e. 0.5 ms.

Frequency of each waveform

$$f = \dfrac{1}{T} = \dfrac{1}{0.5 \times 10^{-3}} = \textbf{2 kHz}$$

(b) The peak value of waveform A is 2 cm × 2 V/cm = 4 V, hence **the r.m.s. value of waveform** $A = \dfrac{4}{\sqrt{2}} = \textbf{2.83 V}$

The peak value of waveform B is $2.5\,\text{cm} \times 2\,\text{V/cm} = 5\,\text{V}$, hence **the r.m.s. value of waveform**

$$B = \frac{5}{\sqrt{2}} = \mathbf{3.54\,V}$$

(c)   Since 5 cm represents 1 cycle, then 5 cm represents $360°$, i.e. 1 cm represents $\dfrac{360}{5} = 72°$

The phase angle $\phi = 0.5\,\text{cm} = 0.5\,\text{cm} \times 72°/\text{cm}$
$= 36°$

Hence, waveform A leads waveform B by $36°$

Now try the following Practice Exercise

Practice Exercise 213   Further problems on the cathode ray oscilloscope (answers on page 549)

1.   For the square voltage waveform displayed on a c.r.o. shown in Figure 43.13, find (a) its frequency, (b) its peak-to-peak voltage

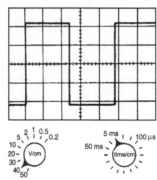

Figure 43.13

2.   For the pulse waveform shown in Figure 43.14, find (a) its frequency, (b) the magnitude of the pulse voltage

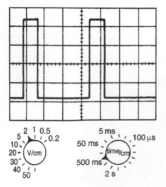

Figure 43.14

3.   For the sinusoidal waveform shown in Figure 43.15, determine (a) its frequency, (b) the peak-to-peak voltage, (c) the r.m.s. voltage

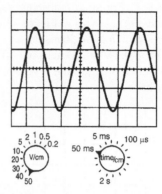

Figure 43.15

43.7   Virtual test and measuring instruments

Computer-based instruments are rapidly replacing conventional test equipment in many of today's test and measurement applications. Probably the most commonly available virtual test instrument is the digital storage oscilloscope (DSO). Because of the processing power available from the PC coupled with the mass storage capability, a computer-based virtual DSO is able to provide a variety of additional functions, such as spectrum analysis and digital display of both frequency and voltage. In addition, the ability to save waveforms and captured measurement data for future analysis or for comparison purposes can be extremely valuable, particularly where evidence of conformance with standards or specifications is required.

Unlike a conventional oscilloscope (which is primarily intended for waveform display) a computer-based virtual oscilloscope effectively combines several test instruments in one single package. The functions and available measurements from such an instrument usually include:

*   real time or stored waveform display
*   precise time and voltage measurement (using adjustable cursors)
*   digital display of voltage

- digital display of frequency and/or periodic time
- accurate measurement of phase angle
- frequency spectrum display and analysis
- data logging (stored waveform data can be exported in formats that are compatible with conventional spreadsheet packages, e.g. as .xls files)
- ability to save/print waveforms and other information in graphical format (e.g. as .jpg or .bmp files)

Virtual instruments can take various forms including:

- internal hardware in the form of a conventional PCI expansion card
- external hardware units which are connected to the PC by means of either a conventional 25-pin parallel port connector or by means of a serial USB connector

The software (and any necessary drivers) is invariably supplied on CD-ROM or can be downloaded from the manufacturer's website. Some manufacturers also supply software drivers together with sufficient accompanying documentation in order to allow users to control virtual test instruments with their own software developed using popular programming languages such as VisualBASIC or C++.

43.8 Virtual digital storage oscilloscopes

Several types of virtual DSO are currently available. These can be conveniently arranged into three different categories according to their application:

- Low-cost DSOs
- High-speed DSOs
- High-resolution DSOs

Unfortunately, there is often some confusion between the last two categories. A high-speed DSO is designed for examining waveforms that are rapidly changing. Such an instrument does not necessarily provide high-resolution measurement. Similarly, a high-resolution DSO is useful for displaying waveforms with a high degree of precision but it may not be suitable for examining fast waveforms. The difference between these two types of DSO should become clearer later on. Low-cost DSOs are primarily designed for low-frequency signals (typically signals up to around 20 kHz) and are usually able to sample their signals at rates of between 10 k and 100 k samples per second.

Resolution is usually limited to either 8-bits or 12-bits (corresponding to 256 and 4096 discrete voltage levels respectively).

High-speed DSOs are rapidly replacing CRT-based oscilloscopes. They are invariably dual-channel instruments and provide all the features associated with a conventional oscilloscope including trigger selection, time-base and voltage ranges, and an ability to operate in X–Y mode.

Additional features available with a computer-based instrument include the ability to capture transient signals (as with a conventional digital storage oscilloscope) and save waveforms for future analysis. The ability to analyse a signal in terms of its frequency spectrum is yet another feature that is only possible with a DSO (see later).

43.8.1 Upper frequency limit

The upper signal frequency limit of a DSO is determined primarily by the rate at which it can sample an incoming signal. Typical sampling rates for different types of virtual instrument are:

Type of DSO	Typical sampling rate
Low-cost DSO	20 k to 100 k per second
High-speed DSO	100 M to 1000 M per second
High-resolution DSO	20 M to 100 M per second

In order to display waveforms with reasonable accuracy it is normally suggested that the sampling rate should be *at least* twice and *preferably more* than five times the highest signal frequency. Thus, in order to display a 10 MHz signal with any degree of accuracy a sampling rate of 50 M samples per second will be required.

The 'five times rule' merits a little explanation. When sampling signals in a digital-to-analogue converter we usually apply the **Nyquist*** criterion that the sampling

---

*Who was Nyquist? – **Harry Nyquist** (7 February, 1889–4 April, 1976) was an important contributor to communication theory, with work on thermal noise, the stability of feedback amplifiers, telegraphy, facsimile, television, and the Nyquist stability criterion can now be found in all textbooks on feedback control theory. To find out more about Nyquist go to www.routledge.com/cw/bird

frequency must be at least twice the highest analogue signal frequency. Unfortunately, this no longer applies in the case of a DSO where we need to sample at an even faster rate if we are to accurately display the signal. In practice we would need a minimum of about five points within a single cycle of a sampled waveform in order to reproduce it with approximate fidelity. Hence the sampling rate should be at least five times that of highest signal frequency in order to display a waveform reasonably faithfully.

A special case exists with dual-channel DSOs. Here the sampling rate may be shared between the two channels. Thus an effective sampling rate of 20 M samples per second might equate to 10 M samples per second for *each* of the two channels. In such a case the upper frequency limit would not be 4 MHz but only a mere 2 MHz.

The approximate bandwidth required to display different types of signals with reasonable precision is given in the table:

Signal	Bandwidth required (approximate)
Low-frequency and power	d.c. to 10 kHz
Audio frequency (general)	d.c. to 20 kHz
Audio frequency (high-quality)	d.c. to 50 kHz
Square and pulse waveforms (up to 5 kHz)	d.c. to 100 kHz
Fast pulses with small rise-times	d.c. to 1 MHz
Video	d.c. to 10 MHz
Radio (LF, MF and HF)	d.c. to 50 MHz

The general rule is that, for sinusoidal signals, the bandwidth should ideally be at least double that of the highest signal frequency whilst for square wave and pulse signals, the bandwidth should be at least ten times that of the highest signal frequency.

It is worth noting that most manufacturers define the bandwidth of an instrument as the frequency at which a sine wave input signal will fall to 0.707 of its true amplitude (i.e. the -3dB point). To put this into context,

at the cut-off frequency the displayed trace will be in error by a whopping 29%!

43.8.2   Resolution

The relationship between resolution and signal accuracy (not bandwidth) is simply that the more bits used in the conversion process the more discrete voltage levels can be resolved by the DSO. The relationship is as follows:

$$x = 2^n$$

where x is the number of discrete voltage levels and n is the number of bits. Thus, each time we use an additional bit in the conversion process we double the resolution of the DSO, as shown in the table below:

Number of bits n	Number of discrete voltage levels x
8-bit	256
10-bit	1024
12-bit	4096
16-bit	65,536

43.8.3   Buffer memory capacity

A DSO stores its captured waveform samples in a buffer memory. Hence, for a given sampling rate, the size of this memory buffer will determine for how long the DSO can capture a signal before its buffer memory becomes full.

The relationship between sampling rate and buffer memory capacity is important. A DSO with a high sampling rate but small memory will only be able to use its full sampling rate on the top few time base ranges.

To put this into context, it is worth considering a simple example. Assume that we need to display 10,000 cycles of a 10 MHz square wave. This signal will occur in a time frame of 1 ms. If applying the 'five times rule' we would need a bandwidth of at least 50 MHz to display this signal accurately.

To reconstruct the square wave we would need a minimum of about five samples per cycle so a minimum sampling rate would be 5×10 MHz $= 50$ M samples per second. To capture data at the rate of 50 M samples per second for a time interval of 1 ms requires a memory that can store 50,000 samples. If each sample uses 16-bits we would require 100 kbyte of extremely fast memory.

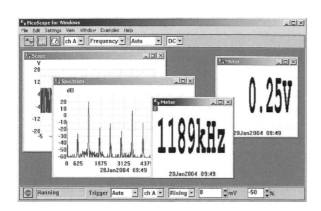

Figure 43.16

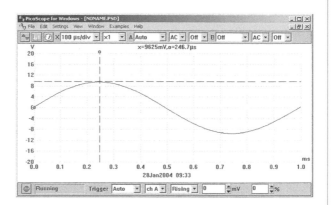

Figure 43.17

43.8.4   Accuracy

The measurement resolution or measurement accuracy of a DSO (in terms of the smallest voltage change that can be measured) depends on the actual range that is selected. So, for example, on the 1 V range an 8-bit DSO is able to detect a voltage change of one two-hundred-and-fifty-sixth of a volt or (1/256) V or about 4 mV. For most measurement applications this will prove to be perfectly adequate as it amounts to an accuracy of about 0.4% of full-scale.

Figure 43.16 depicts a PicoScope software display showing multiple windows providing conventional oscilloscope waveform display, spectrum analyser display, frequency display and voltmeter display.

Adjustable cursors make it possible to carry out extremely accurate measurements. In Figure 43.17, the peak value of the (nominal 10 V peak) waveform is measured at precisely 9625 mV (9.625 V). The time to reach the peak value (from 0 V) is measured as 246.7 μs (0.2467 ms).

The addition of a second time cursor makes it possible to measure the time accurately between two events. In Figure 43.18, event 'o' occurs 131 ns before the trigger point whilst event 'x' occurs 397 ns after the trigger point. The elapsed time between these two events is 528 ns. The two cursors can be adjusted by means of a mouse (or other pointing device) or, more accurately, using the PC's cursor keys.

43.8.5   Autoranging

Autoranging is another very useful feature that is often provided with a virtual DSO. If you regularly use a conventional oscilloscope for a variety of measurements

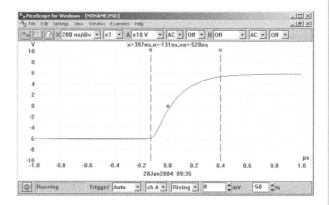

Figure 43.18

you will know only too well how many times you need to make adjustments to the vertical sensitivity of the instrument.

43.8.6   High-resolution DSOs

High-resolution DSOs are used for precision applications where it is necessary to faithfully reproduce a waveform, and also to be able to perform an accurate analysis of noise floor and harmonic content. Typical applications include small signal work and high-quality audio.

Unlike the low-cost DSO, which typically has 8-bit resolution and poor d.c. accuracy, these units are usually accurate to better than 1% and have either 12-bit or 16-bit resolution. This makes them ideal for audio, noise and vibration measurements.

The increased resolution also allows the instrument to be used as a spectrum analyser with very wide dynamic range (up to 100 dB). This feature is ideal for performing noise and distortion measurements on low-level analogue circuits.

Bandwidth alone is not enough to ensure that a DSO can accurately capture a high-frequency signal. The goal of manufacturers is to achieve a flat frequency response. This response is sometimes referred to as a Maximally Flat Envelope Delay (MFED). A frequency response of this type delivers excellent pulse fidelity with minimum overshoot, undershoot and ringing.

It is important to remember that if the input signal is not a pure sine wave it will contain a number of higher frequency harmonics. For example, a square wave will contain odd harmonics that have levels that become progressively reduced as their frequency increases. Thus, to display a 1 MHz square wave accurately you need to take into account the fact that there will be signal components present at 3 MHz, 5 MHz, 7 MHz, 9 MHz, 11 MHz, and so on.

43.8.7 Spectrum analysis

The technique of Fast Fourier Transformation (FFT) calculated with software algorithms using data captured by a virtual DSO has made it possible to produce frequency spectrum displays. Such displays can be used to investigate the harmonic content of waveforms as well as the relationship between several signals within a composite waveform.

Figure 43.19 shows the frequency spectrum of the 1 kHz sine wave signal from a low-distortion signal generator. Here the virtual DSO has been set to capture samples at a rate of 4096 per second within a frequency range of d.c. to 12.2 kHz. The display clearly shows the second harmonic (at a level of −50 dB or −70 dB relative to the

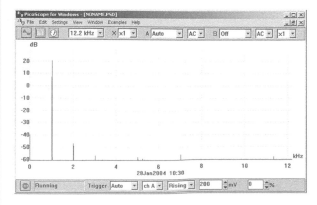

Figure 43.19

fundamental), plus further harmonics at 3 kHz, 5 kHz and 7 kHz (all of which are greater than 75 dB down on the fundamental).

Problem 8.   Figure 43.20 shows the frequency spectrum of a signal at 1184 kHz displayed by a high-speed virtual DSO. Determine (a) the harmonic relationship between the signals marked 'o' and 'x', (b) the difference in amplitude (expressed in dB) between the signals marked 'o' and 'x' and (c) the amplitude of the second harmonic relative to the fundamental signal 'o'

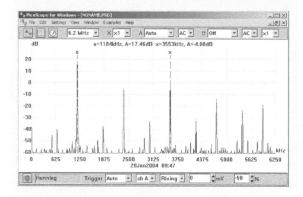

Figure 43.20

(a)   The signal 'x' is at a frequency of 3553 kHz. This is three times the frequency of the signal at 'o' which is at 1184 kHz. Thus, **'x' is the third harmonic of the signal 'o'**.

(b)   The signal at 'o' has an amplitude of +17.46 dB whilst the signal at 'x' has an amplitude of −4.08 dB. Thus, **the difference in level** = (+17.46) − (−4.08) = **21.54 dB**.

(c)   **The amplitude of the second harmonic** (shown at approximately 2270 kHz) = **−5 dB**

43.9 Null method of measurement

A **null method of measurement** is a simple, accurate and widely used method which depends on an instrument reading being adjusted to read zero current only. The method assumes:

(i)   if there is any deflection at all, then some current is flowing,

(ii) if there is no deflection, then no current flows (i.e. a null condition).

Hence it is unnecessary for a meter sensing current flow to be calibrated when used in this way. A sensitive milliammeter or microammeter with centre zero position setting is called a **galvanometer**. Two examples where the method is used are in the Wheatstone bridge (see section 43.10), and in the d.c. potentiometer (see section 43.11).

43.10 Wheatstone bridge

Figure 43.21 shows a **Wheatstone*** **bridge** circuit which compares an unknown resistance R_x with others of known values, i.e. R_1 and R_2, which have fixed values, and R_3, which is variable. R_3 is varied until zero deflection is obtained on the galvanometer G. No current then flows through the meter, $V_A = V_B$, and the bridge is said to be 'balanced'.

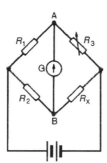

Figure 43.21

At balance,

$$R_1 R_x = R_2 R_3, \text{ i.e. } R_x = \frac{R_2 R_3}{R_1} \text{ ohms.}$$

*Who was Wheatstone? – **Sir Charles Wheatstone** (6 February 1802–19 October 1875), was an English scientist best known for his contributions in the development of the **Wheatstone bridge**, which is used to measure an unknown electrical resistance. To find out more about Wheatstone go to www.routledge.com/cw/bird (For an image refer to page 354)

Problem 9. In a Wheatstone bridge $ABCD$, a galvanometer is connected between A and C, and a battery between B and D. A resistor of unknown value is connected between A and B. When the bridge is balanced, the resistance between B and C is $100\,\Omega$, that between C and D is $10\,\Omega$ and that between D and A is $400\,\Omega$. Calculate the value of the unknown resistance

The Wheatstone bridge is shown in Figure 43.22 where R_x is the unknown resistance.

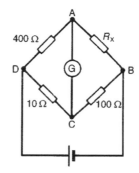

Figure 43.22

At balance, equating the products of opposite ratio arms, gives:

$$(R_x)(10) = (100)(400)$$

and

$$R_x = \frac{(100)(400)}{10} = 4000\,\Omega$$

Hence, the unknown resistance $R_x = 4\,k\Omega$

43.11 d.c. potentiometer

The **d.c. potentiometer** is a null-balance instrument used for determining e.m.f. and p.d values by comparison with a known e.m.f. or p.d. In Figure 43.23(a), using a standard cell of known e.m.f. E_1, the slider S is moved along the slide wire until balance is obtained (i.e. the galvanometer deflection is zero), shown as length l_1 The standard cell is now replaced by a cell of unknown e.m.f. E_2 (see Figure 43.23(b)) and again balance is obtained (shown as l_2).

Since $E_1 \propto l_1$ and $E_2 \propto l_2$ then

$$\frac{E_1}{E_2} = \frac{l_1}{l_2} \text{ and } E_2 = E_1\left(\frac{l_2}{l_1}\right) \text{ volts.}$$

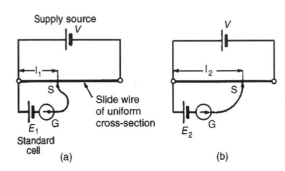

Figure 43.23

A potentiometer may be arranged as a resistive two-element potential divider in which the division ratio is adjustable to give a simple variable d.c. supply. Such devices may be constructed in the form of a resistive element carrying a sliding contact which is adjusted by a rotary or linear movement of the control knob.

Problem 10. In a d.c. potentiometer, balance is obtained at a length of 400 mm when using a standard cell of 1.0186 volts. Determine the e.m.f. of a dry cell if balance is obtained with a length of 650 mm

$E_1 = 1.0186$ V, $l_1 = 400$ mm and $l_2 = 650$ mm.

With reference to Figure 43.23,

$$\frac{E_1}{E_2} = \frac{l_1}{l_2}$$

from which, $\qquad E_2 = E_1 \left(\frac{l_2}{l_1}\right) = (1.0186)\left(\frac{650}{400}\right)$

$$= 1.655 \text{ volts}$$

Now try the following Practice Exercises

Practice Exercise 214   Further problems on the Wheatstone bridge and d.c. potentiometer (answers on page 549)

1. In a Wheatstone bridge $PQRS$, a galvanometer is connected between Q and S and a voltage source between P and R. An unknown resistor R_x is connected between P and Q. When the bridge is balanced, the resistance between Q and R is 200 Ω, that between R and S is 10 Ω and that between S and P is 150 Ω. Calculate the value of R_x

2. Balance is obtained in a d.c. potentiometer at a length of 31.2 cm when using a standard cell of 1.0186 volts. Calculate the e.m.f. of a dry cell if balance is obtained with a length of 46.7 cm

Practice Exercise 215   Short answer questions on electrical measuring instruments and measurements

1. Name two advantages of electronic measuring instruments compared with moving-coil or moving-iron instruments

2. What is a multimeter?

3. Name five quantities that an oscilloscope is capable of measuring

4. Draw a simple block diagram to show how an analogue oscilloscope displays a measured signal

5. Draw a simple block diagram to show how a digital oscilloscope displays a measured signal

6. State five functions a computer-based virtual oscilloscope is capable of

7. What is meant by a null method of measurement?

8. Sketch a Wheatstone bridge circuit used for measuring an unknown resistance in a d.c. circuit and state the balance condition

9. How may a d.c. potentiometer be used to measure p.d.s

Practice Exercise 216   Multiple-choice questions on electrical measuring instruments and measurements (answers on page 549)

A sinusoidal waveform is displayed on a c.r.o. screen. The peak-to-peak distance is 5 cm and the distance between cycles is 4 cm. The 'variable' switch is on 100 μs/cm and the 'volts/cm' switch is on 10 V/cm. In questions 1–5, select the correct answer from the following:

(a) 25 V   (b) 5 V   (c) 0.4 ms

(d) 35.4 V   (e) 4 ms   (f) 50 V

(g) 250 Hz   (h) 2.5 V   (i) 2.5 kHz

(j) 17.7 V

1.   Determine the peak-to-peak voltage

2.   Determine the periodic time of the waveform

3.   Determine the maximum value of the voltage

4.   Determine the frequency of the waveform

5.   Determine the r.m.s. value of the waveform

Figure 43.24 shows double-beam c.r.o. waveform traces. For the quantities stated

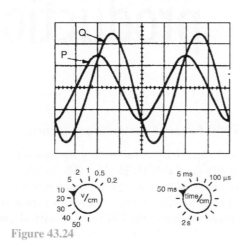

Figure 43.24

in questions 6–12, select the correct answer from the following:

(a) 30 V   (b) 0.2 s   (c) 50 V

(d) $\dfrac{15}{\sqrt{2}}$ (e) 54° leading   (f) $\dfrac{250}{\sqrt{2}}$ V

(g) 15 V   (h) 100 μs   (i) $\dfrac{50}{\sqrt{2}}$ V

(j) 250 V   (k) 10 kHz   (l) 75 V

(m) 40 μs   (n) $\dfrac{3\pi}{10}$ rads lagging   (o) $\dfrac{25}{\sqrt{2}}$ V

(p) 5 Hz   (q) $\dfrac{30}{\sqrt{2}}$ V   (r) 25 kHz

(s) $\dfrac{75}{\sqrt{2}}$ (t) $\dfrac{3\pi}{10}$ rads leading

6.   Amplitude of waveform P

7.   Peak-to-peak value of waveform Q

8.   Periodic time of both waveforms

9.   Frequency of both waveforms

10.   r.m.s. value of waveform P

11.   r.m.s. value of waveform Q

12.   Phase displacement of waveform Q relative to waveform P

13.   A potentiometer is used to:

(a) compare voltages

(b) measure power factor

(c) compare currents

(d) measure phase sequence

For fully worked solutions to each of the problems in Exercises 212 to 216 in this chapter, go to the website:

www.routledge.com/cw/bird

Chapter 44

Global climate change and the future of electricity production

Why it is important to understand: **Global climate change and the future of electricity production**

In 1831, Michael Faraday devised a machine that generated electricity from rotary motion – but it took almost 50 years for the technology to reach a commercially viable stage. In 1878, in the USA, Thomas Edison developed and sold a commercially viable replacement for gas lighting and heating using locally generated and distributed direct current electricity. The world's first public electricity supply was provided in late 1881, when the streets of the Surrey town of Godalming in the UK were lit with electric light. This system was powered from a water wheel on the River Wey, which drove a Siemens alternator that supplied a number of arc lamps within the town.

In the 1920s the use of electricity for home lighting increased and by the mid-1930s electrical appliances were standard in the homes of the better-off; after the Second World War they became common in all households. Now, some 80 years later, we have become almost totally dependent on electricity! Without electricity we have no lighting, no communication via mobile phones/smartphones, internet or computer, television, IPods or radios, no air-conditioning, no fans, no electric heating, no refrigerator or freezer, no coffee maker, no kitchen appliances, no dishwasher, no electric stoves, ovens or microwaves, no washing machines or tumble dryers, problems with drinking water if it comes from a system dependent on electrical pumps, ….. Our whole lifestyles have become totally dependent on a reliable electricity supply.

In the future, civilization will be forced to research and develop alternative energy sources. Our current rate of fossil fuel usage will lead to an energy crisis this century. In order to survive the energy crisis many companies in the energy industry are inventing new ways to extract energy from renewable sources and while the rate of development is slow, mainstream awareness and government pressures are growing. Traditional methods of generating electricity are unsustainable, and new energy sources must be found that do not produce as much carbon. The recognised need for alternative power sources is not new. Massive solar arrays are seen unveiled in vast deserts, enormous on-and-offshore wind-farms, wave-beams converting the power of our oceans, and a host of biomass solutions arrive and disappear.

Science and Mathematics for Engineering. 978-0-367-20475-4, © John Bird. Published by Taylor & Francis. All rights reserved.

One of the most important of current engineering problems centres upon the realisation that fossil fuels are unsustainable, and therefore the challenge of how to generate enough electricity using clean, renewable sources.

Global climate change is a much-discussed topic. There is increasing evidence that such change is quite rapid and there are a number of important consequences to appreciate and to act upon. For example, electric power production affects global climate, an important consideration for the future.

At the end of this chapter, you should be able to:

- appreciate, with evidence, the problems and consequences of global climate change
- briefly describe the process of generating electricity using:

    (a) coal

    (b) oil

    (c) natural gas

    (d) nuclear energy

    (e) hydro power

    (f) pumped storage

    (g) wind

    (h) tidal power

    (i) biomass

    (j) solar energy

- appreciate the advantages and disadvantages of each of the methods of generating electricity
- understand a future possibility of harnessing the power of wind, tide and sun on an 'energy island'

44.1 Introduction

In this chapter, methods of generating electricity using **coal, oil, natural gas, nuclear energy, hydro power, pumped storage, wind, tidal power, biomass** and **solar energy** are explained.

Coal, oil and gas are called '**fossil fuels**' because they have been formed from the organic remains of prehistoric plants and animals. Historically, the transition from one energy system to another, as from wood to coal or coal to oil, has proven an enormously complicated process, requiring decades to complete. In similar fashion, it will be many years before renewable forms of energy – wind, solar, tidal, geothermal, and others still in development – replace fossil fuels as the world's leading energy providers. This chapter explains some ways of generating electricity for the present and for the future.

However, before considering electricity production, some problems associated with global climate change need to be appreciated.

44.2 Global climate change

The term 'global climate change' usually refers to changes to the earth's climate brought about by a wide array of human activities. Because of predictions of a steady rise in average world-wide temperatures, global climate change is sometimes referred to as '**global warming**'. Regardless of which term is used, **different methods of electricity production can impact the earth's climate** in ways that raise extraordinary environmental issues.

There is increasing scientific evidence showing that burning fossil fuels such as coal, oil and natural gas, are altering the earth's climate. Burning fossil fuels releases carbon that has previously been locked up in coal, oil and natural gas for millions of years. The carbon in these fossil fuels is transformed into carbon dioxide, CO_2, the predominant gas contributing to the 'greenhouse effect', during the combustion process.

The greenhouse effect allows energy from the sun to pass through the earth's atmosphere and then traps some of that energy in the form of heat. This process has kept global temperatures on earth relatively stable – currently averaging 33°C (60°F) – which is liveable for human populations. Nonetheless, jumps in emissions of CO_2 and other gases, such as methane, traced to fossil fuel burning and other human endeavours, boost heat trapping processes in the atmosphere, gradually raising average world-wide temperatures.

Observations show that the surface temperature this century is as warm or warmer than any other century since at least 1400. The ten warmest years on record have all occurred since 1980. The release of vast stores of fossilised carbon threaten to raise average global temperatures at an accelerated pace.

Scientists have observed that the earth's surface warmed by approximately 1°F during the 20th century.

A 2018 report by the UN Intergovernmental Panel on Climate Change (IPCC) suggests the planet will reach the crucial threshold of 1.5°C (2.7°F) above pre-industrial levels by as early as 2030, precipitating the risk of extreme drought, wildfires, floods and food shortages for hundreds of millions of people. Avoiding going even higher will require significant action in the next few years.

If 1.5°C global warming is exceeded, then more heatwaves and hot summers, greater sea level rise, and, for many parts of the world, worse droughts and rainfall extremes will occur.

It is difficult to know precisely how quickly the earth's temperature will jump since human influences mix with natural events that may slow or accelerate these long-term trends. It is quite possible, however, to identify actions to reduce causes of climate change, thereby reducing the intense risks associated with such a hot planet. Energy-related ventures account for about 86% of all greenhouse gas emissions linked to human activities. Since power plants, and related electricity generation operations, produce around 25% of greenhouse gas emissions, reductions in this sector can play a major role in slowing global climate change.

44.3 Evidence of rapid climate change

Evidence of rapid climate change includes the following:

(a) Global temperature rise

The planet's average surface temperature has risen about 0.9°C (1.62°F) since the late 19th century, a change driven largely by increased carbon dioxide and other human-made emissions into the atmosphere. Most of the warming occurred in the last 3 or 4 decades, with the five warmest years on record taking place since 2010.

(b) Warming oceans

The oceans have absorbed much of this increased heat, with the top 700 m (about 2,300 feet) of ocean showing warming of more than 0.4°F since 1969.

(c) Shrinking ice sheets

The Greenland and Antarctic ice sheets have decreased in mass. Data from NASA's Gravity Recovery and Climate Experiment show Greenland lost an average of 281 billion tons of ice per year between 1993 and 2016, while Antarctica lost about 119 billion tons during the same time period. The rate of Antarctica ice mass loss has tripled in the last decade.

(d) Glacial retreat

Glaciers are retreating almost everywhere around the world – including in the Alps, Himalayas, Andes, Rockies, Alaska and Africa.

(e) Decreased snow cover

Satellite observations reveal that the amount of spring snow cover in the Northern Hemisphere has decreased over the past five decades and that the snow is melting earlier.

(f) Sea level rise

Global sea level rose about 3.15 cm (8 inches) in the last century. The rate in the last two decades, however, is nearly double that of the last century and is accelerating slightly every year.

(g) Declining Arctic sea ice

Both the extent and thickness of Arctic sea ice has declined rapidly over the last several decades.

(h) Extreme events

The number of record high temperature events has been increasing, while the number of record low temperature events has been decreasing, since 1950. The world has also witnessed increasing numbers of intense rainfall events.

(i) Ocean acidification

Since the beginning of the Industrial Revolution, the acidity of surface ocean waters has increased by about 30%. This increase is the result of humans emitting more carbon dioxide into the atmosphere and hence more being absorbed into the oceans. The amount of carbon dioxide absorbed by the upper layer of the oceans is increasing by about 2 billion tons per year.

44.4   Consequences of global climate change

(a) Human health impacts

Global warming poses a major threat to human health by way of increased infectious diseases. Increasing temperatures nurture the spread of disease-carrying mosquitoes and rodents. IPCC scientists project that as warmer temperatures spread north and south from the tropics and to higher elevations, malaria-carrying mosquitoes will spread with them, significantly extending the exposure of the world's people to malaria.

Extreme weather impacts

The IPCC identifies more frequent and more severe heat waves as a potential lethal effect of global warming. Some segments of the population, especially people in a weakened state of health, are vulnerable to heat stress. Though imprecise in their predictions, global weather models indicate that extreme weather events are more likely to occur from increases in global average temperatures. The ocean temperature shifts may occur more rapidly and more often, generating major changes in global weather patterns.

Coastal zones and small island flooding

As global temperatures rise, sea levels will also rise. The seawater expands as it warms. Water previously bound to mountain and polar glaciers, melts and flow into the world's seas. Much of the world's population, especially the poorer people of the world, live at or close to sea level, areas vulnerable to the lethal combination of rising sea level and increasingly severe ocean storms. The rising water table along coastlines could also encourage the release of pathogens into septic systems and waterways. More than half the world's people live within 35 miles of an ocean or sea. Areas at risk include developed coastal cities, towns and resort areas, saltwater marshes, coastal wetlands, sandy beaches, coral reefs, coral atolls, and river deltas. Sea levels have already risen 10 to 25 cm (4 to 10 inches) over the last century.

Forest devastation

Forest ecosystems evolve slowly in response to gradual natural climate cycles. The rapid pace of global climate change resulting from combustion of fossil fuels and other industrial and agricultural activities disrupts such gradual adjustments. Many tree species may be unable to survive at their present sites due to higher temperatures. Increased drought, more pests and disease attacks, and higher frequency of forest fires, are all projected to occur at spots throughout the globe. The IPCC report states that averaged over all zones, the [global] models predict that 33% of the currently forested area could be affected.

Agriculture

Agriculture depends on rainfall, which impacts how to manage crop production, the types of seeds planted, and investments in irrigation systems. Changing weather patterns associated with changing global climate patterns pose major challenges for the farmers, small and large, who feed the world's growing population. Just as forest ecosystems face the stress of loss of traditional habitat, so will the world's farming community.

44.5   How does electric power production affect the global climate?

The generation of electricity is the single largest source of CO_2 emissions. The combustion of fossil fuels such as coal is the primary source of these air emissions. Burning coal produces far more CO_2 than oil or natural gas. Reducing reliance upon coal combustion must be the cornerstone of any credible global climate change prevention plan.

Some methods of electricity production produce no or few CO_2 emissions – solar, wind, geothermal, hydropower, and nuclear systems particularly. Power plants fuelled by wood, agricultural crop wastes, livestock wastes, and methane collected from municipal landfills release CO_2 emissions but may contribute little to global climate change since they also can prevent even greater releases of both CO_2 and methane.

Biomass fuels that depend on forest resources must be evaluated carefully since the stock of forests worldwide represent a storehouse for CO_2. If forests are harvested for fuel to generate electricity, and are not replaced, global climate change could be accelerated. If electricity generators use forest or other plant stocks that are being regrown in a closed cycle of growth, combustion and regrowth, the CO_2 emissions may be offset by plant and animal growth that withdraws CO_2 from the atmosphere. Closed cycle systems such as this are carbon 'neutral'. These neutral biomass systems represent progress since they displace the fossil fuel combustion that would otherwise increase the CO_2 linked to rising temperatures.

It may be proven that methane has over 20 times the heat-trapping capacity of CO_2. Power plants that capture methane – or prevent methane releases – are therefore extremely beneficial when it comes to slowing global climate change. Methane is produced by the natural decay of organic materials underground or in other spots lacking oxygen. Municipal landfills, large piles of animal wastes, and other sites where plant wastes decay without exposure to the air, generate large volumes of methane that escape into the atmosphere. Natural gas

is simply methane produced by the decay of plant and animal matter that is captured beneath the earth's surface over millions of years. It is therefore important to stop methane production or capture and burn it so that it does not escape into the atmosphere, where it may accelerate global climate change.

44.6 Generating electrical power using coal

A coal power station turns the chemical energy in coal into electrical energy that can be used in homes and businesses.

In Figure 44.1, the coal (1) is ground to a fine powder and blown into the boiler (2), where it is burned, converting its chemical energy into heat energy. Grinding the coal into powder increases its surface area, which helps it to burn faster and hotter, producing as much heat and as little waste as possible.

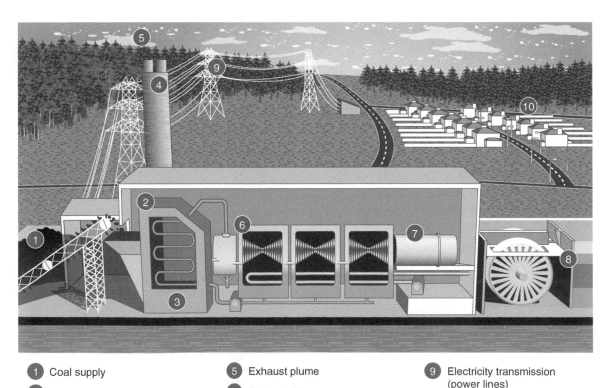

① Coal supply	⑤ Exhaust plume	⑨ Electricity transmission (power lines)
② Boiler	⑥ Steam turbine	⑩ Consumer homes and businesses
③ Ash systems	⑦ Turbine generator	
④ Exhaust stack	⑧ Water supply	

Figure 44.1

As well as heat, burning coal produces ash and exhaust gases. The ash falls to the bottom of the boiler and is removed by the ash systems (3). It is usually then sold to the building industry and used as an ingredient in various building materials, like concrete.

The gases enter the exhaust stack (4), which contains equipment that filters out any dust and ash, before venting into the atmosphere. The exhaust stacks of coal power stations are built tall so that the exhaust plume (5) can disperse before it touches the ground. This ensures that it does not affect the quality of the air around the station.

Burning the coal heats water in pipes coiled around the boiler, turning it into steam. The hot steam expands in the pipes, so when it emerges it is under high pressure. The pressure drives the steam over the blades of the steam turbine (6), causing it to spin, converting the heat energy released in the boiler into mechanical energy.

A shaft connects the steam turbine to the turbine generator (7), so when the turbine spins, so does the generator. The generator uses an electromagnetic field to convert this mechanical energy into electrical energy (as described in chapter 40).

After passing through the turbine, the steam comes into contact with pipes full of cold water (8); in coastal stations this water is pumped straight from the sea. The cold pipes cool the steam so that it condenses back into water. It is then piped back to the boiler, where it can be heated up again, and turned into steam to keep the turbine turning.

Finally, a transformer converts the electrical energy from the generator to a high voltage. The national grid uses high voltages to transmit electricity efficiently through the power lines (9) to the homes and businesses that need it (10). Here, other transformers reduce the voltage back down to a usable level.

Coal has been a reliable source of energy for many decades; however coal is considered to produce the highest amount of carbon emissions of any form of electricity generation. A new technology called Carbon Capture and Storage (CCS) is being developed to remove up to 90% carbon dioxide from power station emissions and store it underground. CCS will be an important way to cut emissions and meet international low carbon targets.

Advantages of coal

Coal is easily combustible, and burns at low temperatures, making coal-fired boilers cheaper and simpler than many others. It is widely and easily distributed all over the world, is comparatively inexpensive to buy on the open market due to large reserves and easy accessibility, has good availability for much of the world (i.e. coal is found in many more places than other fossil fuels) and is mainly simple to mine, making it by far the least expensive fossil fuel to actually obtain. It is economically possible to build a wide variety of sizes of generation plants, and a fossil-fuelled power station can be built almost anywhere, so long as you can get large quantities of fuel to it – most coal-fired power stations have dedicated rail links to supply the coal.

Disadvantages of coal

Coal is non-renewable and fast depleting, has the lowest energy density of any fossil fuel, i.e. it produces the least energy per ton of fuel, and has the lowest energy density per unit volume, meaning that the amount of energy generated per cubic metre is lower than any other fossil fuel. Coal has high transportation costs due to the bulk of coal; coal dust is an extreme explosion hazard, so transportation and storage must take special precautions to mitigate this danger. Storage costs are high, especially if required to have enough stock for a few years to assure power production availability. Burning fossil fuels releases carbon dioxide, a powerful greenhouse gas, that had been stored in the earth for millions of years – contributing to global warming – and coal leaves behind harmful by-products upon combustion (both airborne and in solid-waste form), thereby causing a lot of pollution (air pollution due to burning coal is much worse than any other form of power generation, and very expensive 'scrubbers' must be installed to remove a significant amount of it; even then, a non-trivial amount escapes into the air). Mining of coal leads to irreversible damage to the adjoining environment, coal will eventually run out, cannot be recycled, and prices for all fossil fuels are rising, especially if the real cost of their carbon is included.

44.7 Generating electrical power using oil

An oil power station turns the chemical energy in oil into electrical energy that can be used in homes and businesses.

In Figure 44.2, the oil (1) is piped into the boiler (2), where it is burned, converting its chemical energy into heat energy. This heats water in pipes coiled around the

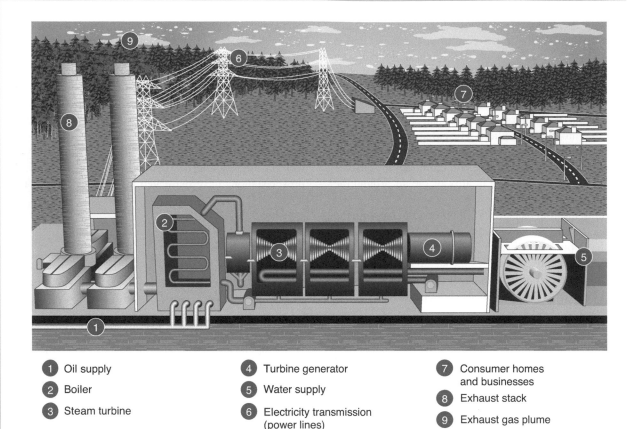

1. Oil supply
2. Boiler
3. Steam turbine
4. Turbine generator
5. Water supply
6. Electricity transmission (power lines)
7. Consumer homes and businesses
8. Exhaust stack
9. Exhaust gas plume

Figure 44.2

boiler, turning it into steam. The hot steam expands in the narrow pipes, so when it emerges it is under high pressure.

The pressure drives the steam over the blades of the steam turbine (3), causing it to spin, converting the heat energy released in the boiler into mechanical energy. A shaft connects the steam turbine to the turbine generator (4), so when the turbine spins, so does the generator. The generator uses an electromagnetic field to convert this mechanical energy into electrical energy.

After passing through the turbine, the steam comes into contact with pipes full of cold water (5). The cold pipes cool the steam so that it condenses back into water. It is then piped back to the boiler, where it can be heated up again, and turned into steam again to keep the turbine turning.

Finally, a transformer converts the electrical energy from the generator to a high voltage. The national grid uses high voltages to transmit electricity efficiently through the power lines (6) to the homes and businesses that need it (7). Here, other transformers reduce the voltage back down to a usable level.

As well as heat, burning oil produces exhaust gases. These are piped from the boiler to the exhaust stack (8), which contains equipment that filters out any particles, before venting into the atmosphere. The stack is built tall so that the exhaust gas plume (9) can disperse before it touches the ground. This ensures that it does not affect the quality of the air around the station.

Advantages of oil

Oil has a high energy density (i.e. a small amount of oil can produce a large amount of energy), has easy availability and infrastructure for transport, is easy to use, is crucial for a wide variety of industries, is relatively easy to produce and refine, is a constant power source and is highly reliable.

Disadvantages of oil

Oil produces greenhouse gas emissions (GHG), can cause pollution of water and earth, emits harmful substances like sulphur dioxide, carbon monoxide and acid rain, and can lead to production of very harmful and toxic materials during refining (plastic being one of the most harmful of substances).

44.8 Generating electrical power using natural gas

A gas power station turns the chemical energy in natural gas into electrical energy that can be used in homes and businesses.

In Figure 44.3, natural gas (1) is pumped into the gas turbine (2), where it is mixed with air (3) and burned, converting its chemical energy into heat energy. As well as heat, burning natural gas produces a mixture of gases called the combustion gas. The heat makes the combustion gas expand. In the enclosed gas turbine, this causes a build-up of pressure.

The pressure drives the combustion gas over the blades of the gas turbine, causing it to spin, converting some of the heat energy into mechanical energy. A shaft connects the gas turbine to the gas turbine generator (4), so when the turbine spins, the generator does too. The generator uses an electromagnetic field to convert this mechanical energy into electrical energy.

After passing through the gas turbine, the still-hot combustion gas is piped to the heat recovery steam generator (5). Here it is used to heat pipes full of water, turning the water to steam, before escaping through the exhaust stack (6). Natural gas burns very cleanly, but the stack is still built tall so that the exhaust gas plume (7) can disperse before it touches the ground. This ensures that it does not affect the quality of the air around the station.

The hot steam expands in the pipes, so when it emerges it is under high pressure. These high-pressure steam jets spin the steam turbine (8), just like the combustion gas spins the gas turbine. The steam turbine is connected by a shaft to the steam turbine generator (9), which converts the turbine's mechanical energy into electrical energy.

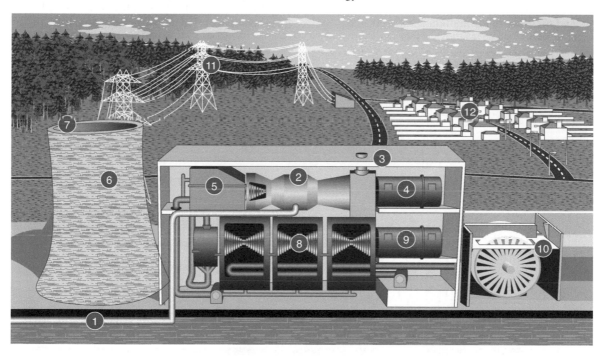

1	Gas line in	**6**	Stack for exhaust gases	**10**	Cooling water supply
2	Gas turbine	**7**	Exhaust gas plume	**11**	Electricity transmission (power lines)
3	Air in	**8**	Steam turbine	**12**	Consumer homes and businesses
4	Gas turbine generator	**9**	Steam turbine generator		
5	Heat recovery steam generator				

Figure 44.3

After passing through the turbine, the steam comes into contact with pipes full of cold water (10). In coastal stations this water is pumped straight from the sea. The cold pipes cool the steam so that it condenses back into water. It is then piped back to the heat recovery steam generator to be re-used.

Finally, a transformer converts the electrical energy from the generator to a high voltage. The national grid uses high voltages to transmit electricity efficiently through the power lines (11) to the homes and businesses that need it (12). Here, other transformers reduce the voltage back down to a usable level.

Advantages of natural gas

Gas is used to produce electricity, is less harmful than coal or oil, is easy to store and transport, has much residential use, is used for vehicle fuel, burns cleaner without leaving any smell, ash or smoke, is an instant energy, ideal in kitchens, has much industrial use for producing hydrogen, ammonia for fertilizers and some paints and plastics, and is abundant and versatile.

Disadvantages of natural gas

Gas is toxic and flammable, can damage the environment, is non-renewable and is expensive to install.

44.9 Generating electrical power using nuclear energy

A nuclear power station turns the nuclear energy in uranium atoms into electrical energy that can be used in homes and businesses.

In Figure 44.4, the reactor vessel (1) is a tough steel capsule that houses the fuel rods – sealed metal cylinders containing pellets of uranium oxide. When a neutron – a neutrally charged subatomic particle – hits a uranium atom, the atom sometimes splits, releasing two or three more neutrons. This process converts the nuclear energy that binds the atom together into heat energy.

The fuel assemblies are arranged in such a way that when atoms in the fuel split, the neutrons they release are likely to hit other atoms and make them split as well. This chain reaction produces large quantities of heat.

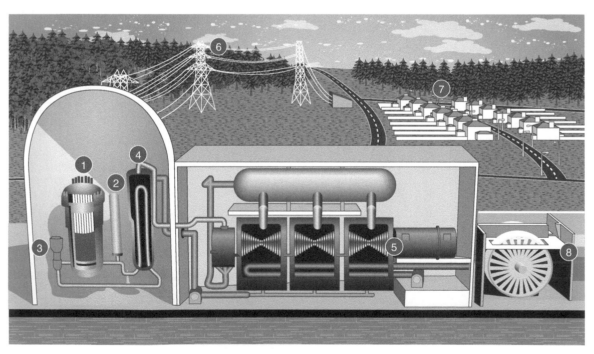

① Reactor vessel	④ Steam generator	⑦ Consumer homes and businesses
② Pressuriser	⑤ Turbine generator and turbines	⑧ Cooling via sea water
③ Reactor coolant pump	⑥ Electricity transmission (power lines)	

Figure 44.4

Water flows through the reactor vessel, where the chain reaction heats it to around 300°C. The water needs to stay in liquid form for the power station to work, so the pressuriser (2) subjects it to around 155 times atmospheric pressure, which stops it boiling.

The reactor coolant pump (3) circulates the hot pressurised water from the reactor vessel to the steam generator (4). Here, the water flows through thousands of looped pipes before circulating back to the reactor vessel. A second stream of water flows through the steam generator, around the outside of the pipes. This water is under much less pressure, so the heat from the pipes boils it into steam.

The steam then passes through a series of turbines (5), causing them to spin, converting the heat energy produced in the reactor into mechanical energy. A shaft connects the turbines to a generator, so when the turbines spin, so does the generator. The generator uses an electromagnetic field to convert this mechanical energy into electrical energy.

A transformer converts the electrical energy from the generator to a high voltage. The national grid uses high voltages to transmit electricity efficiently through the power lines (6) to the homes and businesses that need it (7). Here, other transformers reduce the voltage back down to a usable level.

After passing through the turbines, the steam comes into contact with pipes full of cold water pumped in from the sea (8). The cold pipes cool the steam so that it condenses back into water. It is then piped back to the steam generator, where it can be heated up again, and turned into steam again to keep the turbines turning.

Advantages of nuclear energy

Nuclear energy is reliable, has low fuel costs, low electricity costs, no greenhouse gas emissions/air pollution, has a high load factor and huge potential.

Disadvantages of nuclear energy

Disadvantages of nuclear energy includes the fear of nuclear and radiation accidents, the problems of nuclear waste disposal, the low level of radioactivity from normal operations, the fear of nuclear proliferation, high capital investment, cost overruns and long gestation time, the many regulations for nuclear energy power plants, and fuel danger – uranium is limited to only a few countries and suppliers.

44.10 Generating electrical power using hydro power

A hydroelectric power station converts the kinetic – or moving – energy in flowing or falling water into electrical energy that can be used in homes and businesses. Hydroelectric power can be generated on a small scale with a 'run-of-river' installation, which uses naturally flowing river water to turn one or more turbines, or on a large scale with a hydroelectric dam.

A hydroelectric dam straddles a river, blocking the water's progress downstream. With reference to Figure 44.5, water collects on the upstream side of the dam, forming an artificial lake known as a reservoir (1). Damming the river converts the water's kinetic energy into potential energy: the reservoir becomes a sort of battery, storing energy that can be released a little at a time. As well as being a source of energy, some reservoirs are used as boating lakes or drinking water supplies.

The reservoir's potential energy is converted back into kinetic energy by opening underwater gates, or intakes (2), in the dam. When an intake opens, the immense weight of the reservoir forces water through a channel called the penstock (3) towards a turbine. The water rushes past the turbine, hitting its blades and causing it to spin, converting some of the water's kinetic energy into mechanical energy. The water then finally flows out of the dam and continues its journey downstream.

A shaft connects the turbine to a generator (4), so when the turbine spins, so does the generator. The generator uses an electromagnetic field to convert this mechanical energy into electrical energy.

As long as there is plenty of water in the reservoir, a hydroelectric dam can respond quickly to changes in demand for electricity. Opening and closing the intakes directly controls the amount of water flowing through the penstock, which determines the amount of electricity the dam is generating.

The turbine and generator are located in the dam's power house (5), which also houses a transformer. The transformer converts the electrical energy from the generator to a high voltage. The national grid uses high voltages to transmit electricity efficiently through the power lines (6) to the homes and businesses that need it (7). Here, other transformers reduce the voltage back down to a usable level.

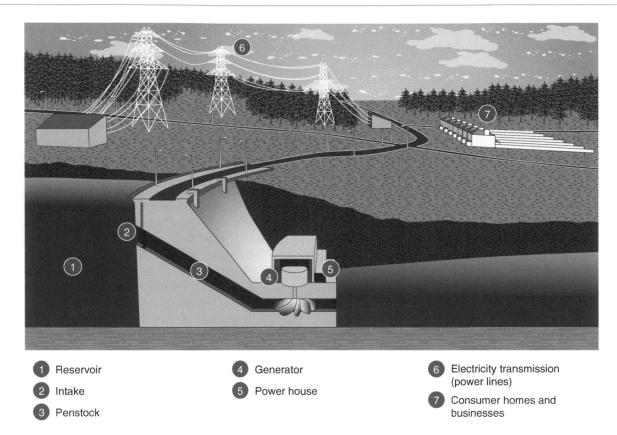

1 Reservoir
2 Intake
3 Penstock
4 Generator
5 Power house
6 Electricity transmission (power lines)
7 Consumer homes and businesses

Figure 44.5

Advantages of hydro power

With hydro power there are no fuel costs, low operating costs and little maintenance, low electricity costs, no greenhouse gas emissions/air pollution, energy storage possibilities, small size hydro plants possible, reliability, a high load factor and long life.

Disadvantages of hydro power

Disadvantages of hydro power includes environmental, dislocation and tribal rights difficulties, wildlife and fish being affected, the possibility of earthquake vulnerability, siltation, dam failure due to poor construction or terrorism, the fact that plants cannot be built anywhere, and long gestation times.

44.11 Generating electrical power using pumped storage

Pumped storage reservoirs provide a place to store energy until it's needed. There are fluctuations in demand for electricity throughout the day. For example

when a popular TV programme finishes, many people put the kettle on, causing a peak in demand for electrical power.

When electricity is suddenly demanded, a way is needed of producing power which can go from producing no power to full power immediately, and keep generating power for half an hour or so until other power stations can catch up with the demand for energy. This is why pumped storage reservoirs are so useful.

A pumped storage plant has two separate reservoirs, an upper and a lower one. When electricity is in low demand, for example at night, water is pumped into the upper reservoir.

When there is a sudden demand for power, giant taps known as the head gates are opened. This allows water from the upper reservoir to flow through pipes, powering a turbine, into the lower reservoir.

The movement of the turbine turns a generator which creates electricity. The electricity is created in the generator by using powerful magnets and coils of wire. When the coils are spun quickly inside the magnets, they produce electricity.

Water exiting from the pipe flows into the lower reservoir rather than re-entering a river and flowing downstream. At night, the water in the lower reservoir can be pumped back up into the upper reservoir to be used again.

In terms of how pumped storage and dammed water generate electricity, the methods are the same. The difference is that in pumped storage, the water is continually reused, whereas in hydroelectric dams, the water which generates electricity continues flowing downriver after use.

Pumped storage in the UK: Most pumped storage plants are located in Scotland, except the largest of all, Dinorwig, which is in North Wales. Dinorwig, built in 1984, produces 1728 MW – which is enough electricity to power nearly 7 million desktop computers. Dinorwig has the fastest 'response time' of any pumped storage plant in the world – it can provide 1320 MW in 12 seconds.

Advantages of pumped storage

Pumped storage provides a way to generate electricity instantly and quickly, no pollution or waste is created, and there is little effect on the landscape, as typically pumped storage plants are made from existing lakes in mountains.

Disadvantages of pumped storage

Pumped storage facilities are expensive to build, and once the pumped storage plant is used, it cannot be used again until the water is pumped back to the upper reservoir.

44.12 Generating electrical power using wind

Wind turbines use the wind's kinetic energy to generate electrical energy that can be used in homes and businesses. Individual wind turbines can be used to generate electricity on a small scale – to power a single home, for example. A large number of wind turbines grouped together, sometimes known as a wind farm or wind park, can generate electricity on a much larger scale.

A wind turbine works like a high-tech version of an old-fashioned windmill. The wind blows on the angled blades of the rotor, causing it to spin, converting some of the wind's kinetic energy into mechanical energy. Sensors in the turbine detect how strongly the wind is blowing and from which direction. The rotor automatically turns to face the wind, and automatically brakes in dangerously high winds to protect the turbine from damage.

With reference to Figure 44.6, a shaft and gearbox connect the rotor to a generator (1), so when the rotor spins, so does the generator. The generator uses an electromagnetic field to convert this mechanical energy into electrical energy.

The electrical energy from the generator is transmitted along cables to a substation (2). Here, the electrical energy generated by all the turbines in the wind farm is combined and converted to a high voltage. The national grid uses high voltages to transmit electricity efficiently through the power lines (3) to the homes and businesses that need it (4). Here, other transformers reduce the voltage back down to a usable level.

Advantages of wind energy

Wind energy has no pollution and global warming effects, low costs, a large industrial base, no fuel costs, and offshore advantages.

Disadvantages of wind energy

Wind energy has low persistent noise, can cause a loss of scenery, requires land usage and is intermittent in nature.

44.13 Generating electrical power using tidal power

The tide moves a huge amount of water twice each day, and harnessing it could provide a great deal of energy – for example, around 20% of Britain's needs. Although the energy supply is reliable and plentiful, converting it into useful electrical power is not easy.

There are eight main sites around Britain where tidal power stations could usefully be built, including the Severn, Dee, Solway and Humber estuaries. Only around 20 sites in the world have been identified as possible tidal power stations.

Tidal energy is produced through the use of tidal energy generators, as shown in Figure 44.7. These large underwater turbines are placed in areas with high tidal movements, and are designed to capture the kinetic motion of the ebbing and surging of ocean tides in order to produce electricity. Tidal power has great potential for future power and electricity generation because of the massive size of the oceans.

Advantages of tidal power

Tidal power is renewable, non-polluting and carbon negative, predictable, needs no fuel, has low costs, long life, high energy density and high load factor.

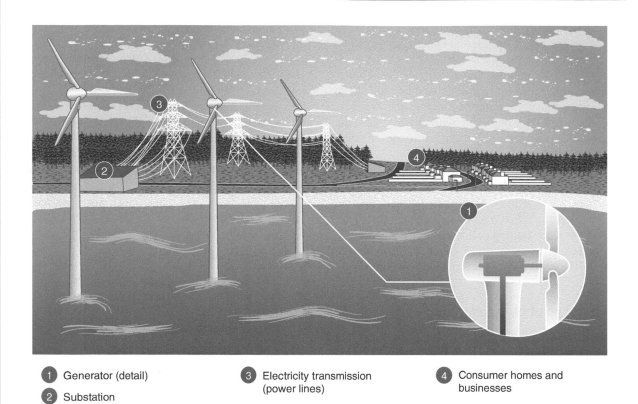

1. Generator (detail)
2. Substation
3. Electricity transmission (power lines)
4. Consumer homes and businesses

Figure 44.6

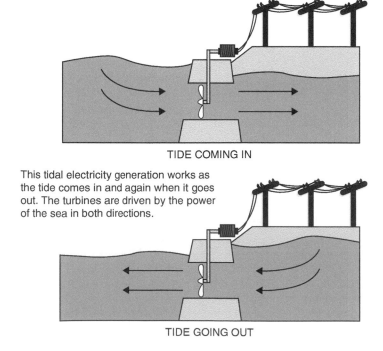

TIDE COMING IN

This tidal electricity generation works as the tide comes in and again when it goes out. The turbines are driven by the power of the sea in both directions.

TIDE GOING OUT

Figure 44.7

Disadvantages of tidal power

Tidal power has high initial capital investment, limited locations, detrimental effects on marine life, immature technology, long gestation time, some difficulties in transmission of tidal electricity, and weather effects that can damage tidal power equipment.

44.14 Generating electrical power using biomass

Biomass is fuel that is developed from organic materials, a renewable and sustainable source of energy used to create electricity or other forms of power, as stated on page 397.

There are several methods to convert biomass into electricity.

One way is to simply burn biomass directly, heat water to steam and send it through a steam turbine, which then generates electricity. The second way requires gasification of biomass. A biomass gasifier takes dry biomass, such as agriculture waste, and with the absence of oxygen and high temperatures produces synthesis gas (CO + H2), also known as pyrolysis of biomass. The gasification process turns wet biomass, such as food waste and manure, into methane (CH4) in a digestion tank. Both methane and synthesis gas (syngas) can be used in a gas engine or a gas turbine for electricity production. A third way to produce electricity from gasified biomass is by using fuel cells. If biogas/bio-syngas is available with high enough purity fuel cells can be used to produce bio-electricity. However, the fuel cells break down quickly if the gas in any way contains impurities. This technology is not yet commercial.

Biofuels, like ethanol, biodiesel and bio-oil, can be also be used for power production in most types of power generators built for gasoline or diesel.

The generation of electricity from biomass results not only in electricity, but also a lot of heat. A traditional gas engine will have an efficiency of 30–35%. Gas turbines and steam turbines will end up at around 50%.

It is common to also use the heat from these processes, thus increasing the overall energy efficiency. There are many plants that combine heat and power production from biomass – the biogas plant cost increases, but the long-term savings that come with better energy efficiency is in most cases worth it.

Advantages of biomass energy

Biomass energy is carbon neutral, uses waste efficiently, is a continuous source of power, can use a large variety of feedstock, has a low capital investment, can be built in remote areas and on a small scale, reduces methane, is easily available and is a low cost resource.

Disadvantages of biomass energy

With biomass energy, pollution can occur where poor technology is used, continuous feedstock is needed for efficiency, good management of biomass plants are required, it has limited potential compared to other forms of energy like solar, hydro etc., and biomass plants are unpopular if constructed near homes.

44.15 Generating electrical power using solar energy

Solar panels turn energy from the sun's rays directly into useful energy that can be used in homes and businesses. There are two main types: solar thermal and photovoltaic, or PV. Solar thermal panels use the sun's energy to heat water that can be used in washing and heating. PV panels use the photovoltaic effect to turn the sun's energy directly into electricity, which can supplement or replace a building's usual supply.

A PV panel is made up of a semiconducting material, usually silicon-based, sandwiched between two electrical contacts. Referring to Figure 44.8, to generate as much electricity as possible, PV panels need to spend as much time as possible in direct sunlight (1a). A sloping, south-facing roof is the ideal place to mount a solar panel.

A sheet of glass (1b) protects the semiconductor sandwich from hail, grit blown by the wind and wildlife. The semiconductor is also coated in an antireflective substance (1c), which makes sure that it absorbs the sunlight it needs instead of scattering it uselessly away. When sunlight strikes the panel and is absorbed, it knocks loose electrons from some of the atoms that make up the semiconductor (1d). The semiconductor is positively charged on one side and negatively charged on the other side, which encourages all these loose electrons to travel in the same direction, creating an electric current. The contacts (1e and 1f) capture this current (1g) in an electrical circuit.

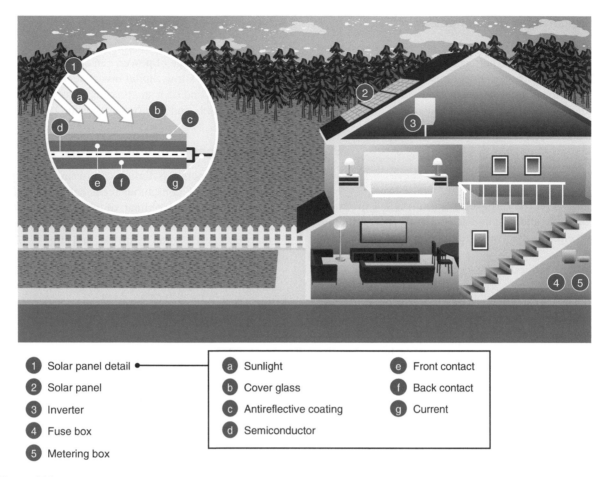

1 Solar panel detail
2 Solar panel
3 Inverter
4 Fuse box
5 Metering box

a Sunlight
b Cover glass
c Antireflective coating
d Semiconductor

e Front contact
f Back contact
g Current

Figure 44.8

The electricity PV panels (2) generate direct current (d.c.). Before it can be used in homes and businesses, it has to be changed into alternating current (a.c.) electricity using an inverter (3). The inverted current then travels from the inverter to the building's fuse box (4) and from there to the appliances that need it.

PV systems installed in homes and businesses can include a dedicated metering box (5) that measures how much electricity the panels are generating. As an incentive to generate renewable energy, energy suppliers pay the system's owner a fixed rate for every unit of electricity it generates – plus a bonus for units the owner doesn't use, because these can help supply the national grid. Installing a PV system is not cheap, but this deal can help the owner to earn back the cost more quickly – and potentially even make a profit one day.

Advantages of solar power

Solar power is environmentally friendly, has declining costs, no fuel, low maintenance, no pollution, has almost unlimited potential, has the advantage that installations can be any size, and installation can be quick.

Disadvantages of solar power

Solar power has higher initial costs than fossil energy forms, is intermittent in nature and has high capital investment.

44.16 Harnessing the power of wind, tide and sun on an 'energy island' – a future possibility?

Man-made 'energy islands' anchored, say, in the North Sea and the English Channel could help the world meet increasing power demands – and tackle the problem of expanding population by providing homes for people. The artificial structures – the brainchild of mechanical engineer Professor Carl Ross, of the University of Portsmouth – would produce energy by harnessing the power of the wind, tide and sun, and could be towed far out to sea to avoid complaints of noise and unsightliness.

A particular problem, especially in smaller overcrowded countries, such as those found in Europe and Asia, is the NIMBY syndrome – that is, the 'Not In My Back Yard' reaction of people. Complaints that using these renewable methods of producing energy takes up valuable land space, is unsightly and causes noise are common. Putting these three renewable energy-producing forms on a floating island, all these negative points can be avoided. Two-thirds of the Earth's surface is covered by water, so dry land is not being 'lost' – in fact, the oceans are being colonised. Professor Ross feels that once this technology is matured, humans can even start living on the floating island to help counter over-population.

In 2011 the United Nations announced that the global population had reached seven billion, and the momentum of a growing population does not show any signs of slowing down.

The floating islands would be attached to the sea bed by tubular pillars with vacuum chambers in their bases, similar to offshore drilling rigs. The islands would support wind turbines and solar panels on their upper surface, while underneath, tidal turbines would harness the power of the oceans.

Professor Ross proposes that the floating energy islands be deployed in locations including the deep sea region of the North Sea, the west coast of Scotland and the opening of the English Channel.

Each one would supply power for some 119,500 homes. The islands come with a hefty price tag – an estimated £1.7 billion – but Professor Ross believes the initial outlay would be recovered through household energy bills after 11 years. However, this estimate is based mostly on wind farms situated on land, and not at sea. For wind farms at sea, the wind is about twice that on dry land. Moreover the kinetic energy produced on the wind farm is proportional to the mass of the wind, times its velocity squared; but the mass of the wind hitting the wind turbine is proportional to its velocity; thus the kinetic energy produced on the wind turbine at sea is proportional to the velocity cubed of the wind hitting it, and NOT on its velocity squared! This means that a wind turbine at sea is likely to produce about 8 times the kinetic energy than that produced on dry land, so that instead of the wind farm paying for itself in about 11 years it will pay for itself in less than 2 years! This revelation is likely to favourably change the economic argument in favour of wind turbines at sea.

[For more, see *The Journal of Ocean Technology* Vol 10, No 4, 2015, pages 45 to 52, 'Floating Energy Islands – producing multiple forms of renewable energy' by Carl T F Ross and Tien Y Sien]

Now try the following Practice Exercise

Practice Exercise 217   Short answer questions on ways of generating electricity

1. State five examples of evidence of rapid climate change.

2. State four examples of consequences of global climate change.

3. How does electricity production affect the global climate?

4. What is meant by 'fossil fuels'?

5. Briefly explain with a diagram how electricity is generated using coal.

6. State five advantages and five disadvantages of coal.

7. Briefly explain with a diagram how electricity is generated using oil.

8. State four advantages and three disadvantages of oil.

9. Briefly explain with a diagram how electricity is generated using natural gas.

10. State four advantages and three disadvantages of natural gas.

11. Briefly explain with a diagram how electricity is generated using nuclear energy.

12. State four advantages and four disadvantages of nuclear energy.

13. Briefly explain with a diagram how electricity is generated using hydro power.

14. State five advantages and four disadvantages of hydro power.

15. Briefly explain with a diagram how electricity is generated using pumped storage.

16. State three advantages and two disadvantages of pumped storage.

17. Briefly explain with a diagram how electricity is generated using wind.

18. State five advantages and three disadvantages of wind.

19. Briefly explain with a diagram how electricity is generated using tidal power.

20. State five advantages and five disadvantages of tidal power.

21. What is biomass?

22. Briefly explain three ways to convert biomass to electricity.

23. State five advantages and four disadvantages of biomass.

24. Briefly explain with a diagram how electricity is generated using solar energy.

25. State five advantages and three disadvantages of coal.

For fully worked solutions to each of the problems in Exercises 217 to in this chapter, go to the website: www.routledge.com/cw/bird

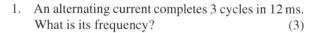

Revision Test 15: Alternating voltages and currents, capacitors and inductors, measurements and electricity production

This assignment covers the material contained in chapters 41 to 44. *The marks for each question are shown in brackets at the end of each question.*

1. An alternating current completes 3 cycles in 12 ms. What is its frequency? (3)

2. For the periodic waveform shown in Figure RT15.1 determine: (a) the frequency, (b) the average value over half a cycle (use 8 intervals for a half cycle), (c) the r.m.s. value, (d) the form factor, (e) the peak factor. (12)

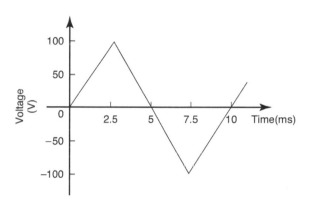

Figure RT15.1

3. A sinusoidal voltage has a mean value of 3.0 V. Determine its maximum and r.m.s. values. (4)

4. An e.m.f. of 2.5 kV is induced in a coil when a current of 2 A collapses to zero in 5 ms. Calculate the inductance of the coil. (4)

5. A flux of 15 mWb links with a 2000-turn coil when a current of 2 A passes through the coil. Calculate (a) the inductance of the coil, (b) the energy stored in the magnetic field, (c) the average e.m.f. induced if the current falls to zero in 300 ms. (6)

6. A sinusoidal voltage trace displayed on an oscilloscope is shown in Figure RT15.2; the 'time/cm' switch is on 50 ms/cm and the 'volts/cm' switch is on 2 V/cm. Determine for the waveform (a) the frequency, (b) the peak-to-peak voltage, (c) the amplitude, (d) the r.m.s. value. (8)

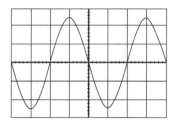

Figure RT15.2

7. An ammeter has a full scale defection. of 200 mA and a resistance of 40 Ω. The ammeter is used to measure the current in a load of resistance 400 Ω when the supply voltage is 20 V. Calculate (a) the ammeter reading expected (neglecting its resistance), (b) the actual current in the circuit, (c) the power dissipated in the ammeter, (d) the power dissipated in the load. (8)

8. In a Wheatstone bridge *PQRS*, a galvanometer is connected between *P* and *R*, and a battery between *Q* and *S*. A resistor of unknown value is connected between *P* and *Q*. When the bridge is balanced, the resistance between *Q* and *R* is 150 Ω, that between *R* and *S* is 25 Ω, and that between *S* and *P* is 500 Ω. Calculate the value of the unknown resistance. (3)

9. There is evidence of global climate changing rapidly. Name five examples and briefly describe each. (10)

10. Name and briefly describe four examples of the consequences of global climate change. (12)

11. Explain briefly how electricity production affects the global climate? (8)

12. (a) Briefly explain how fossil fuels are used to generate electricity.

    (b) State four advantages of coal, oil and natural gas.

    (c) State four disadvantages of coal, oil and natural gas. (17)

13. List five advantages and five disadvantages of using nuclear power to generate electricity. (10)

14. (a) State five renewable methods used to generate electricity.

    (b) State for each renewable method listed in part (a) two advantages and two disadvantages.

    (25)

For lecturers/instructors/teachers, fully worked solutions to each of the problems in Revision Test 15, together with a full marking scheme, are available at the website:

www.routledge.com/cw/bird

Section IV

Engineering systems

Chapter 45

Introduction to engineering systems

Why it is important to understand: **Introduction to engineering systems**

Common types of systems used in engineering include electromechanical, communications, fluidic, electrochemical, information, control and transport systems, and many more besides. Knowledge of system diagrams and system control – analogue, digital, sequential and combination control systems – are considered and briefly explained in this chapter. Open-loop, closed-loop, on/off, hysteresis and proportional control methods are also considered. System response, negative and positive feedback and evaluation of system response is also briefly explored. Whatever area of engineering is undertaken, knowledge of engineering systems will be important.

At the end of this chapter, you should be able to:

- define the component parts of a basic system
- name types of engineering systems
- define a transducer
- sketch system diagrams
- explain analogue and digital control systems
- explain control methods
- appreciate system response
- distinguish between negative and positive feedback
- describe how system response is evaluated

Science and Mathematics for Engineering. 978-0-367-20475-4, © John Bird. Published by Taylor & Francis. All rights reserved.

45.1 Introduction

The study of engineering systems is a considerable topic in itself and whole books have been written on the subject. This chapter offers a brief overview of the subject area, sufficient to form a basis for further study.

45.2 Systems

Systems comprise a number of elements, components or sub-systems that are connected together in a particular way to perform a desired function. The individual elements of an engineering system interact together to satisfy a particular functional requirement.

A basic system has (i) a function or purpose, (ii) inputs (for example, raw materials, energy and control values) and outputs (for example, finished products, processed materials, converted energy), (iii) a boundary and (iv) a number of smaller linked components or elements.

45.3 Types of systems

Some of the most common types of system used in engineering are:

(i) **Electromechanical systems** – for example, a vehicle electrical system comprising battery, starter motor, ignition coil, contact breaker and distributor.

(ii) **Communications systems** – for example, a local area network comprising file server, coaxial cable, network adaptors, several computers and a laser printer.

(iii) **Fluidic systems** – for example, a vehicle braking system comprising foot-operated lever, master cylinder, slave cylinder, piping and fluid reservoir.

(iv) **Electrochemical systems** – for example, a cell that uses gas as a fuel to produce electricity, pure water and heat.

(v) **Information systems** – for example, a computerised airport flight arrival system.

(vi) **Control systems** – for example, a micro-computer-based controller that regulates the flow and temperature of material used in a die casting process.

(vii) **Transport systems** – for example, an overhead conveyor for transporting gravel from a quarry to a nearby processing site.

45.4 Transducers

A **transducer** is a device that responds to an input signal in one form of energy and gives an output signal bearing a relationship to the input signal but in a different form of energy. For example, a **tyre pressure gauge** is a transducer, converting tyre pressure into a reading on a scale. Similarly, a photocell converts light into a voltage, and a motor converts electric current into shaft velocity.

45.5 System diagrams

A simple practical **processing system** is shown in Figure 45.1 comprising an input, an output and an energy source, together with unwanted inputs and outputs that may, or may not, be present.

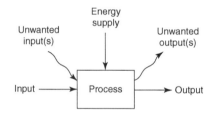

Figure 45.1

A typical **public address system** is shown in Figure 45.2. A microphone is used as an input transducer to collect acoustic energy in the form of sound pressure waves and convert this into electrical energy in the form of small voltages and currents. The signal from the microphone is then amplified by means of an electronic circuit containing transistors and/or integrated circuits before it is applied to a loudspeaker. This output transducer converts the electrical energy supplied to it back into acoustic energy.

A **sub-system** is a part of a system that performs an identified function within the whole system; the amplifier in Figure 45.2 is an example of a sub-system.

A **component** or **element** is usually the simplest part of a system that has a specific and well-defined function – for example, the microphone in Figure 45.2. The illustrations in Figures 45.1 and 45.2 are called

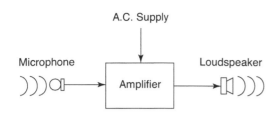

Figure 45.2

block diagrams, and engineering systems, which can often be quite complicated, can be better understood when broken down in this way. It is not always necessary to know precisely what is inside each sub-system in order to know how the whole system functions.

As another example of an engineering system, Figure 45.3 illustrates a **temperature control system** containing a heat source – a gas boiler, a fuel controller – an electrical solenoid valve, a thermostat and a source of electrical energy. The system of Figure 45.3 can be shown in block diagram form as in Figure 45.4; the thermostat compares the actual room temperature with the desired temperature and switches the heating on or off.

There are many types of engineering systems and all such systems may be represented by block diagrams.

45.6 System control

The behaviour of most systems is subject to variations in input (supply) and in output (demand). The behaviour of a system may also change in response to variations in the characteristics of the components that make up the system. In practice, it is desirable to include some means of regulating the output of the system so that it remains relatively immune to these three forms of variation. System control involves maintaining the system output at the desired level regardless of any disturbances that may affect (i) the input quantities, (ii) any unwanted variations in systems components or (iii) the level of demand (or 'loading') on the output.

Different control methods are appropriate to different types of system. The overall control strategy can be based on analogue or digital techniques and may be classed as either sequential or combinational.

45.6.1 Analogue control

Analogue control involves the use of signals and quantities that are continuously variable. Within analogue control systems, signals are represented by voltages and currents that can take any value between two set limits. Figure 45.5(a) shows how the output of a typical analogue system varies with time.

Analogue control systems are invariably based on the use of **operational amplifiers** (see Figure 45.6). These devices are capable of performing mathematical operations such as addition, subtraction, multiplication, division, integration and differentiation.

45.6.2 Digital control

Digital control involves the use of signals and quantities that vary in discrete steps. Values that fall between two adjacent steps must take one or other value. Figure 45.5(b) shows how the output of a typical digital system varies with time.

Digital control systems are usually based on logic devices (such as AND-gates, OR-gates, NAND-gates and NOR-gates), or microprocessor-based computer systems.

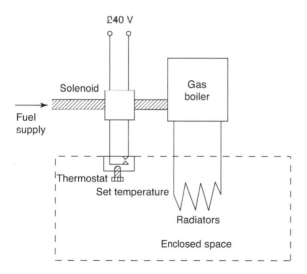

Figure 45.3

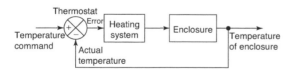

Figure 45.4

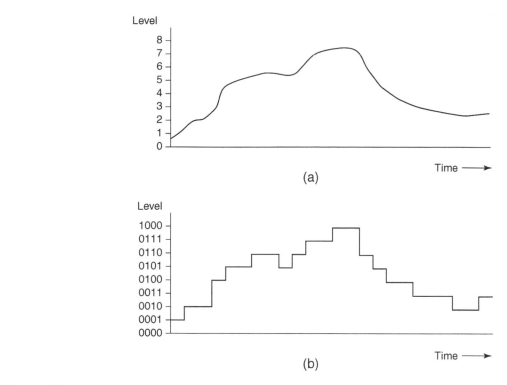

Figure 45.5

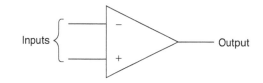

Figure 45.6

45.6.3 Sequential control systems

Many systems are required to perform a series of operations in a set order. For example, **the ignition system of a gas boiler** may require the following sequence of operations: (i) operator's start button is pressed, (ii) fan motor operates, (iii) delay 60 seconds, (iv) open gas supply valve, (v) ignition operates for 2 seconds, (vi) if ignition fails, close gas supply valve, delay 60 seconds, then stop fan motor, (vii) if ignition succeeds, boiler will continue to operate until either the stop switch operates or flame fails.

The components of simple sequential systems often include timers, relays, counters and so on; however, digital logic and microprocessor-based controllers are used in more complex systems.

45.6.4 Combinational control systems

Combinational control systems take several inputs and perform comparisons on a continuous basis. In effect, everything happens at the same time – there are no delays or predetermined sequences that would be associated with sequential controllers. An **aircraft instrument landing system (ILS)** for example, makes continuous comparisons of an aircraft's position relative to the ILS radio beam. Where any deviation is detected, appropriate correction is applied to the aircraft's flight controls.

45.7 Control methods

45.7.1 Open-loop control

In a system that employs open-loop control, the value of the input variable is set to a given value in the expectation that the output will reach the desired value. In such a system there is no automatic comparison of the actual output value with the desired output value in order to compensate for any differences. A simple example of an open-loop control method is the manual adjustment of the regulator that controls the flow of gas to a burner on the hob of a **gas cooker**. This adjustment is carried out in the expectation that food will be raised to the correct temperature in a given time and without burning. Other than the occasional watchful eye of the chef, there is no means of automatically regulating the gas flow in response to the actual temperature of the food.

Another example of an open-loop control system is that of an **electric fan heater** as shown in the block diagram of Figure 45.7.

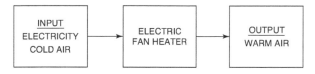

Figure 45.7

45.7.2   Closed-loop control

From the above, open-loop control has some significant disadvantages. What is required is some means of closing the loop in order to make a continuous automatic comparison of the actual value of the output compared with the setting of the input control variable. In the above example, the chef actually closes the loop on an intermittent basis. In effect, the cooker relies on human intervention in order to ensure consistency of the food produced.

All practical engineering systems make use of closed-loop control. In some cases, the loop might be closed by a human operator who determines the deviation between the desired and actual output. In most cases, however, the action of the system is made fully automatic and no human intervention is necessary other than initially setting the desired value of the output. The principle of a closed-loop control system is illustrated in Figure 45.8.

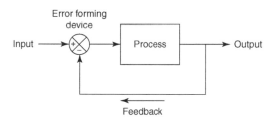

Figure 45.8

Reasons for making a system fully automatic include:

(i)   Some systems use a very large number of input variables and it may be difficult or impossible for a human operator to keep track of them all.

(ii)   Some processes are extremely complex and there may be significant interaction between the input variables.

(iii) Some systems may have to respond very quickly to changes in variables.

(iv)   Some systems require a very high degree of precision.

The analogue closed-loop control system shown in Figure 45.9 provides **speed control for the d.c. motor M**. The actual motor speed is sensed by means of a small d.c. tachogenerator G coupled to the output shaft. The voltage produced by the tachogenerator is compared with that produced at the slider of a potentiometer R which is used to set the desired speed. The comparison of the two voltages (i.e. that of the tachogenerator with that corresponding to the set point), is performed by an operational amplifier connected as a comparator. The output of the comparator stage is applied to a power amplifier that supplies current to the d.c. motor. Energy is derived from a mains-powered d.c. power supply comprising transformer, rectifier and smoothing circuits.

45.7.3   On/off control

On/off control is the simplest form of control and it merely involves turning the output on and off repeatedly in order to achieve the required level of output variable. Since the output of the system is either fully on or fully off at any particular instant of time (i.e. there is no half-way state), this type of control is sometimes referred to as 'discontinuous'.

The most common example of an on/off control system is that of a simple **domestic room heater**. A variable thermostat is used to determine the set point (SP) temperature value. When the actual temperature is below the SP value, the heater is switched on (i.e. electrical energy is applied to the heating element). Eventually, the room temperature will exceed the SP value and, at this point, the heater is switched off. Later, the temperature falls due to heat loss, the room temperature will once again fall below the SP value and the heater will once again be switched on (see Figure 45.10). In normal operation, the process of switching on and off will continue indefinitely.

45.7.4   Hysteresis

In practice, a small amount of hysteresis is built into most on/off control systems. In the previous example, this hysteresis is designed to prevent the heater switching on and off too rapidly which, in turn, would result in early failure of the switch contacts fitted to the thermostat. Note, however, that the presence of

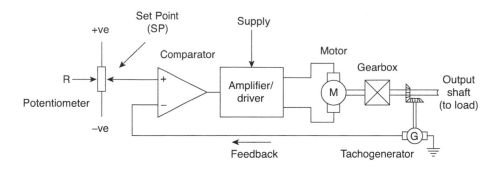

Figure 45.9

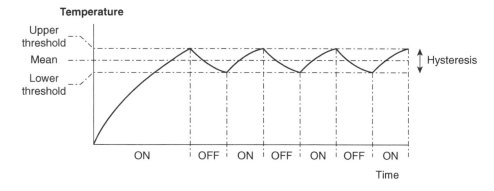

Figure 45.10

hysteresis means that, at any instant of time, the actual temperature will always be somewhere between the upper and lower threshold values (see Figure 45.10).

45.7.5 Proportional control

On/off control is crude but can be effective in many simple applications. A better method is to vary the amount of correction applied to the system according to the size of the deviation from the set point (SP) value. As the difference between the actual value and the desired output becomes less, the amount of correction becomes correspondingly smaller. A simple example of this type of control is in the control of the water level in a **water tank**.

45.8 System response

In a perfect system, the output value C will respond instantaneously to a change in the input S. There will be no delay when changing from one value to another and no time required for the output to 'settle' to its final value.

The ideal state is shown in Figure 45.11(b). In practice, real-world systems take time to reach their final state. Indeed, a very sudden change in output may, in some cases, be undesirable. Furthermore, inertia is present in many systems.

Consider the case of the **motor speed control system** shown in Figure 45.12, where the output shaft is connected to a substantial flywheel. The flywheel effectively limits the acceleration of the motor speed when the set point (SP) is increased. Furthermore, as the output speed reaches the desired value, the inertia present will keep the speed increasing despite the reduction in voltage C applied to the motor. Thus the output shaft speed overshoots the desired value before eventually falling back to the required value. Increasing the gain present in the system will have the effect of increasing the acceleration but this, in turn, will also produce a correspondingly greater value of overshoot. Conversely, decreasing the gain will reduce the overshoot but at the expense of slowing down the response. The actual response of the system represents a compromise between speed and an acceptable value of overshoot. Figure 45.11(c) shows the typical response of a system to the step input shown in Figure 45.11(a).

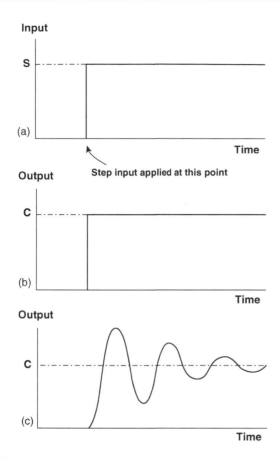

Figure 45.11

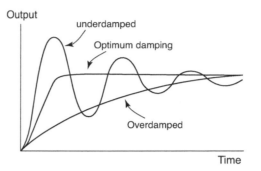

Figure 45.13

prevents overshoot. When a system is '**underdamped**', some overshoot is still present. Conversely, an '**overdamped**' system may take a significantly greater time to respond to a sudden change in input. These damping states are illustrated in Figure 45.13.

45.9 Negative and positive feedback

Most systems use negative feedback in order to precisely control the operational parameters of the system and to maintain the output over a wide variation of the internal parameters of the system. In the case of an amplifier, for example, negative feedback can be used not only to stabilise the gain but also to reduce distortion and improve bandwidth.

The amount of feedback determines the overall (or closed-loop) gain. Because this form of feedback has the effect of reducing the overall gain of the circuit, this form of feedback is known as '**negative feedback**'. An alternative form of feedback, where the output is fed back in such a way as to reinforce the input (rather than to subtract from it) is known as '**positive feedback**'.

45.8.1 Second-order response

The graph shown in Figure 45.11(c) is known as a 'second-order' response. This response has two basic components, an exponential growth curve and a damped oscillation. The oscillatory component can be reduced (or eliminated) by artificially slowing down the response of the system. This is known as '**damping**'. The optimum value of damping is that which *just*

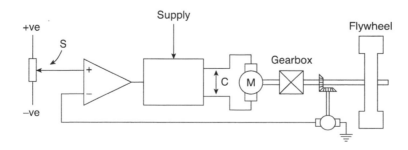

Figure 45.12

45.10 Evaluation of system response

Depending on the equipment available, the evaluation of systems can be carried out practically. This can involve **measurement** of (i) accuracy, (ii) repeatability, (iii) stability, (iv) overshoot, (v) settling time after disturbances, (v) sensitivity and (vi) rate of response. When evaluating systems, it is **important to recognise** that (i) all variables are dependent on time, (ii) velocity is the rate of change of distance, (iii) current is the rate of change of charge and (iv) flow is the rate of change of quantity.

Sources of error also exist, for example (i) dimensional tolerance, (ii) component tolerance, (iii) calibration tolerance, (iv) instrument accuracy, (v) instrument resolution and (vi) human observation.

Now try the following Practice Exercises

Practice Exercise 218   Short answer questions on engineering systems

1. Define a system

2. State, and briefly explain, four common types of engineering systems, drawing the system block diagram for each

3. Define a transducer and give four examples

4. State three methods of system control

5. Briefly explain: (a) open-loop control, (b) closed-loop control

6. State three reasons for making a system fully automatic

7. Define (a) negative feedback, (b) positive feedback

8. Explain the terms: (a) underdamped, (b) overdamped

9. State four measurements that may be involved in evaluating a system response

10. State four possible sources of error that may exist when evaluating system response

Practice Exercise 219   Multiple-choice questions on engineering systems (answers on page 549)

1. A networked computer database containing details of railway timetables is an example of:

    (a) an electromechanical system

    (b) a transport system

    (c) a computer system

    (d) an information system

2. An anti-lock braking system fitted to an articulated lorry is an example of:

    (a) an electromechanical system

    (b) a fluidic system

    (c) a logic system

    (d) an information system

3. Which one of the following system components is a device used to store electrical energy?

    (a) a flywheel   (b) an actuator
    (c) a battery   (d) a transformer

4. Which one of the following is *not* a system output device?

    (a) a linear actuator   (b) a stepper motor
    (c) a relay   (d) a tachogenerator

5. Which one of the following is *not* a system input device?

    (a) a microswitch   (b) a relay
    (c) a variable resistor   (d) a rotary switch

6. Which one of the following is *not* an advantage of using negative feedback?

    (a) the overall gain produced by the system is increased

    (b) the system is less prone to variations in the characteristics of individual components

    (c) the performance of the system is more predictable

    (d) the stability of the system is increased

7. A computerised airport flight arrival system is an example of:

    (a) an electromechanical system

    (b) an information system

    (c) a chemical system

    (d) a fluidic system

8. An oil refinery is an example of:

    (a) an electromechanical system

    (b) a fluidic system

    (c) a chemical system

    (d) an information system

For fully worked solutions to each of the problems in Exercises 218 to 219 in this chapter, go to the website:
www.routledge.com/cw/bird

List of formulae for Science and Mathematics for Engineering

Mathematics formulae

Laws of indices

$$a^m \times a^n = a^{m+n} \qquad \frac{a^m}{a^n} = a^{m-n} \qquad (a^m)^n = a^{mn}$$

$$a^{\frac{m}{n}} = \sqrt[n]{a^m} \qquad a^{-n} = \frac{1}{a^n} \qquad a^0 = 1$$

Areas of plane figures

Rectangle

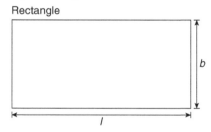

$$\text{Area} = l \times b$$

Parallelogram

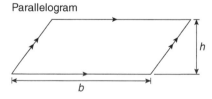

$$\text{Area} = b \times h$$

Trapezium

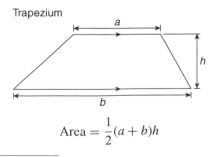

$$\text{Area} = \frac{1}{2}(a+b)h$$

Triangle

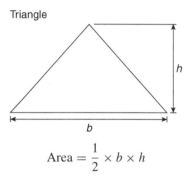

$$\text{Area} = \frac{1}{2} \times b \times h$$

Circle

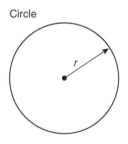

$$\text{Area} = \pi r^2$$

$$\text{Circumference} = 2\pi r$$

Radian measure: 2π radians $= 360$ degrees

Sector of a circle

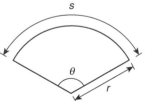

$$\text{arc length } s = \frac{\theta^\circ}{360}(2\pi r) = r\theta \qquad (\theta \text{ in rad})$$

$$\text{area} = \frac{\theta^\circ}{360}(\pi r^2) = \frac{1}{2}r^2\theta \qquad (\theta \text{ in rad})$$

Science and Mathematics for Engineering. 978-0-367-20475-4, © John Bird. Published by Taylor & Francis. All rights reserved.

Volumes and surface areas of regular solids

Cone

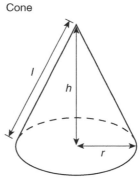

Rectangular prism (or cuboid)

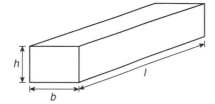

$$\text{Volume} = l \times b \times h$$

$$\text{Surface area} = 2(bh + hl + lb)$$

$$\text{Volume} = \frac{1}{3}\pi r^2 h$$

$$\text{Curved surface area} = \pi rl$$

$$\text{Total surface area} = \pi rl + \pi r^2$$

Cylinder

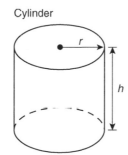

Sphere

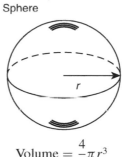

$$\text{Volume} = \pi r^2 h$$

$$\text{Total surface area} = 2\pi rh + 2\pi r^2$$

$$\text{Volume} = \frac{4}{3}\pi r^3$$

$$\text{Surface area} = 4\pi r^2$$

Right-angled triangles

Theorem of Pythagoras $b^2 = a^2 + c^2$

Pyramid

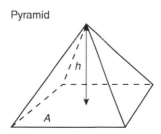

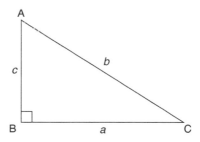

If area of base $= A$ and

perpendicular height $= h$ then:

$$\text{Volume} = \frac{1}{3} \times A \times h$$

Total surface area

= sum of areas of triangles forming sides

+ area of base

Trigonometric ratios

$$\sin C = \frac{c}{b}$$

$$\cos C = \frac{a}{b}$$

$$\tan C = \frac{c}{a}$$

Non right-angled triangles

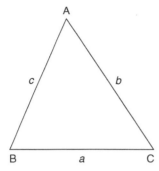

Sine rule $\dfrac{a}{\sin A} = \dfrac{b}{\sin B} = \dfrac{c}{\sin C}$

Cosine rule $a^2 = b^2 + c^2 - 2bc\cos A$

Area of any triangle

(i) $\dfrac{1}{2} \times \text{base} \times \text{perpendicular height}$

(ii) $\dfrac{1}{2}ab\sin C$ or $\dfrac{1}{2}ac\sin B$ or $\dfrac{1}{2}bc\sin A$

Graphs

Equation of a straight line $y = mx + c$

Definition of a logarithm:

If $y = a^x$ then $x = \log_a y$

Laws of logarithms:

$$\log(A \times B) = \log A + \log B$$

$$\log\left(\dfrac{A}{B}\right) = \log A - \log B$$

$$\log A^n = n \times \log A$$

Length in metric units:

$$1\,\text{m} = 100\,\text{cm} = 1000\,\text{mm}$$

Areas in metric units:

$1\,\text{m}^2 = 10^4\,\text{cm}^2$ $1\,\text{cm}^2\ 10^{-4}\text{m}^2$

$1\,\text{m}^2 = 10^6\,\text{mm}^2$ $1\,\text{mm}^2 = 10^{-6}\,\text{m}^2$

$1\,\text{cm}^2 = 10^2\,\text{mm}^2$ $1\,\text{mm}^2 = 10^{-2}\,\text{cm}^2$

Volumes in metric units:

$1\,\text{m}^3 = 10^6\,\text{cm}^2$ $1\,\text{cm}^3\ 10^{-6}\text{m}^3$

$1\,\text{litre} = 1000\,\text{cm}^3$

$1\,\text{m}^3 = 10^9\,\text{mm}^3$ $1\,\text{mm}^3 = 10^{-9}\,\text{m}^3$

$1\,\text{cm}^3 = 10^3\,\text{mm}^3$ $1\,\text{mm}^3 = 10^{-3}\,\text{cm}^3$

Mechanical formulae

Formula	Formula symbols	Units
Density $= \dfrac{\text{mass}}{\text{volume}}$	$\rho = \dfrac{m}{V}$	kg/m^3
Average velocity $= \dfrac{\text{distance travelled}}{\text{time taken}}$	$v = \dfrac{s}{t}$	m/s
Acceleration $= \dfrac{\text{change in velocity}}{\text{time taken}}$	$a = \dfrac{v - u}{t}$	m/s^2
Force $=$ mass $\times$ acceleration	$F = ma$	N
Weight $=$ mass $\times$ gravitational field	$W = mg$	N
Centripetal acceleration	$a = \dfrac{v^2}{r}$	m/s^2
Centripetal force	$F = \dfrac{mv^2}{r}$	N
Work done $=$ force $\times$ distance moved	$W = Fs$	J
Efficiency $= \dfrac{\text{useful output energy}}{\text{input energy}}$		
Power $= \dfrac{\text{energy used (or work done)}}{\text{time taken}} =$ force $\times$ velocity	$P = \dfrac{E}{t} = Fv$	W
Potential energy $=$ weight $\times$ change in height	$E_p = mgh$	J
Kinetic energy $= \dfrac{1}{2} \times$ mass $\times$ (speed)^2	$E_k = \dfrac{1}{2}mv^2$	J
Moment $=$ force $\times$ perpendicular distance	$M = Fd$	N m
Angular velocity	$\omega = \dfrac{\theta}{t} = 2\pi n$	rad/s
Linear velocity	$v = \omega r$	m/s
Relationships between initial velocity u, final velocity v, displacement s, time t and constant acceleration a	$\begin{cases} s = ut + \dfrac{1}{2}at^2 \\ v^2 = u^2 + 2as \end{cases}$	m (m/s)^2
Relationships between initial angular velocity ω_1, final angular velocity ω_2, angle θ, time t and angular acceleration α	$\begin{cases} \theta = \omega_1 t + \dfrac{1}{2}\alpha t^2 \\ \omega_2^2 = \omega_1^2 + 2\alpha\theta \end{cases}$	rad (rad/s)^2
Frictional force $=$ coefficient of friction $\times$ normal force	$F = \mu N$	N
Force ratio $= \dfrac{\text{load}}{\text{effort}}$		

Formula	Formula symbols	Units
Movement ratio $=\dfrac{\text{distance moved by effort}}{\text{distance moved by load}}$		
Efficiency $=\dfrac{\text{force ratio}}{\text{movement ratio}}$		
Stress $=\dfrac{\text{applied force}}{\text{cross-sectional area}}$	$\sigma = \dfrac{F}{A}$	Pa
Strain $=\dfrac{\text{change in length}}{\text{original length}}$	$\varepsilon = \dfrac{x}{L}$	
Young's modulus of elasticity $=\dfrac{\text{stress}}{\text{strain}}$	$E = \dfrac{\sigma}{\varepsilon}$	Pa
Stiffness $=\dfrac{\text{force}}{\text{extension}}$		N/m
Momentum = mass × velocity		kg m/s
Impulse = applied force × time = change in momentum		kg m/s
Torque = force × perpendicular distance	$T = Fd$	N m
Power = torque × angular velocity	$P = T\omega = 2\pi nT$	W
Torque = moment of inertia × angular acceleration	$T = I\alpha$	N m
Pressure $=\dfrac{\text{force}}{\text{area}}$	$p = \dfrac{F}{A}$	Pa
Pressure = density × gravitational acceleration × height	$p = \rho g h$	Pa
1 bar $= 10^5\,$Pa		
Absolute pressure = gauge pressure + atmospheric pressure		
Quantity of heat energy = mass × specific heat capacity × change in temperature	$Q = mc(t_2 - t_1)$	J
Kelvin temperature = degrees Celsius + 273		
New length = original length + expansion	$L_2 = L_1[1 + \alpha(t_2 - t_1)]$	m
New surface area = original surface area + increase in area	$A_2 = A_1[1 + \beta(t_2 - t_1)]$	m^2
New volume = original volume + increase in volume	$V_2 = V_1[1 + \gamma(t_2 - t_1)]$	m^3
Characteristic gas equation	$\dfrac{p_1 V_1}{T_1} = \dfrac{p_2 V_2}{T_2} = k$ $pV = mRT$	

Electrical formulae

Formula	Formula symbols	Units
Charge = current × time	$Q = It$	C
Resistance = $\dfrac{\text{potential difference}}{\text{current}}$	$R = \dfrac{V}{I}$	Ω
Electrical power = potential difference × current	$P = VI = I^2R = \dfrac{V^2}{R}$	W
Terminal p.d. = source e.m.f. − (current)(resistance)	$V = E - Ir$	V
Resistance = $\dfrac{\text{resistivity} \times \text{length of conductor}}{\text{cross-sectional area}}$	$R = \dfrac{\rho l}{A}$	Ω
Total resistance of resistors in series	$R = R_1 + R_2 + \cdots$	Ω
Total resistance of resistors in parallel	$\dfrac{1}{R} = \dfrac{1}{R_1} + \dfrac{1}{R_2} + \cdots$	
Magnetic flux density = $\dfrac{\text{magnetic flux}}{\text{area}}$	$B - \dfrac{\Phi}{A}$	T
Force on conductor = flux density × current × length of conductor	$F = BIl$	N
Force on a charge = charge × velocity × flux density	$F = QvB$	N
Induced e.m.f. = flux density × current × conductor velocity	$E = Blv$	V
Induced e.m.f. = number of coil turns × rate of change of flux	$E = -N\dfrac{d\Phi}{dt}$	V
Induced e.m.f. = inductance × rate of change of current	$E = -L\dfrac{dI}{dt}$	V
Inductance = $\dfrac{\text{number of coil turns} \times \text{flux}}{\text{current}}$	$L = \dfrac{N\Phi}{I}$	H
Mutually induced e.m.f	$E_2 = -M\dfrac{dI_1}{dt}$	V
For an ideal transformer	$\dfrac{V_1}{V_2} = \dfrac{N_1}{N_2} = \dfrac{I_2}{I_1}$	
Electric field strength = $\dfrac{\text{p.d. across dielectric}}{\text{thickness of dielectric}}$	$E = \dfrac{V}{d}$	V/m
Electric flux density = $\dfrac{\text{charge}}{\text{area}}$	$D = \dfrac{Q}{A}$	C/m^2

Formula	Formula symbols	Units
Charge = capacitance × potential difference	$Q = C \times V$	C
Parallel plate capacitor	$C = \dfrac{\varepsilon_0 \varepsilon_r A(n-1)}{d}$	F
Total capacitance of capacitors in series:	$\dfrac{1}{C} = \dfrac{1}{C_1} + \dfrac{1}{C_2} + \cdots$	
Total capacitance of capacitors in parallel:	$C = C_1 + C_2 + \cdots$	F
Energy stored in capacitor	$W = \dfrac{1}{2}CV^2$	J
Wheatstone bridge	$R_x = \dfrac{R_2 R_3}{R_1}$	Ω
Potentiometer	$E_2 = E_1 \left(\dfrac{l_2}{l_1}\right)$	V
Periodic time	$T = \dfrac{1}{f}$	s
r.m.s. current	$I = \sqrt{\dfrac{i_1^2 + i_2^2 + \cdots + i_n^2}{n}}$	A

For a sine wave

Average or mean value	$I_{AV} = \dfrac{2}{\pi} I_m$	
r.m.s. value	$I = \dfrac{1}{\sqrt{2}} I_m$	

$$\text{Form factor} = \frac{\text{r.m.s.}}{\text{average}}$$

$$\text{Peak factor} = \frac{\text{maximum}}{\text{r.m.s.}}$$

These formulae are available for downloading at the website:
www.routledge.com/cw/bird

Answers to practice exercises

Chapter 1

Exercise 1 (Page 5)

1. 19 kg **2.** 479 mm **3.** −66 **4.** £565
5. −225 **6.** −2136 **7.** £107,701 **8.** −4
9. 1487 **10.** −70872 **11.** $15,333

Exercise 2 (Page 7)

1. (a) 468 (b) 868 **2.** (a) £1827 (b) £4158
3. (a) 8613 kg (b) 584 kg **4.** (a) 351 mm (b) 924 mm
5. (a) 48 m (b) 89 m **6.** (a) 259 (b) 56
7. (a) 8067 (b) 3347 **8.** 18 kg
9. 90 cm **10.** 29

Exercise 3 (Page 8)

1. (a) 4 (b) 24 **2.** (a) 12 (b) 360
3. (a) 10 (b) 350 **4.** (a) 90 (b) 2700
5. (a) 2 (b) 210 **6.** (a) 3 (b) 180
7. (a) 5 (b) 210 **8.** (a) 15 (b) 6300

Exercise 4 (Page 9)

1. 59 **2.** 14 **3.** 88 **4.** 5
5. 33 **6.** 22 **7.** 68 **8.** 5

Chapter 2

Exercise 5 (Page 12)

1. $2\frac{1}{7}$ **2.** $\frac{22}{9}$ **3.** $\frac{4}{11}$ **4.** $\frac{3}{7}, \frac{4}{9}, \frac{1}{2}, \frac{3}{5}, \frac{5}{8}$
5. $\frac{11}{15}$ **6.** $\frac{17}{30}$ **7.** $\frac{9}{10}$ **8.** $\frac{3}{16}$
9. $\frac{43}{77}$ **10.** $\frac{47}{63}$ **11.** $1\frac{1}{15}$ **12.** $\frac{4}{27}$
13. $8\frac{51}{52}$ **14.** $1\frac{9}{40}$

Exercise 6 (Page 14)

1. $\frac{8}{35}$ **2.** $\frac{6}{11}$ **3.** $\frac{5}{12}$ **4.** $\frac{3}{28}$
5. $\frac{3}{5}$ **6.** $\frac{1}{13}$ **7.** $1\frac{1}{2}$ **8.** $\frac{8}{15}$
9. $2\frac{2}{5}$ **10.** $3\frac{3}{4}$ **11.** $\frac{12}{23}$ **12.** $\frac{3}{4}$
13. $\frac{1}{9}$ **14.** 13 **15.** 400 litres **16.** 2880 litres

Exercise 7 (Page 15)

1. $2\frac{1}{18}$ **2.** $-\frac{1}{9}$ **3.** $1\frac{1}{6}$ **4.** $4\frac{3}{4}$
5. $\frac{13}{20}$ **6.** $\frac{7}{15}$ **7.** $4\frac{19}{20}$ **8.** 2

Science and Mathematics for Engineering. 978-0-367-20475-4, © John Bird. Published by Taylor & Francis. All rights reserved.

Exercise 8 (Page 16)

1. 36:1   **2.** 3.5:1 or 7:2   **3.** 47:3   **4.** 96 cm, 240 cm

5. $5\frac{1}{4}$ hours or 5 hours 15 minutes

6. £3680, £1840, £920

7. 12 cm

8. 8960 bolts

Exercise 9 (Page 17)

1. 14.18   **2.** 2.785   **3.** 65.38
4. 43.27   **5.** 1.297   **6.** 0.000528

Exercise 10 (Page 17)

1. 80.3   **2.** 329.3   **3.** 72.54   **4.** −124.83   **5.** 295.3

Exercise 11 (Page 19)

1. 4.998	**2.** 47.544	**3.** 456.9
4. 434.82	**5.** 626.1	**6.** 0.444
7. 0.62963	**8.** 1.563	**9.** 13.84

10. (a) 24.81   (b) 24.812
11. (a) 0.00639   (b) 0.0064
12. (a) 8.$\dot{4}$ (b) 62.$\dot{6}$

Exercise 12 (Page 20)

1. 0.32%	**2.** 173.4%	**3.** 5.7%
4. 0.20	**5.** 0.0125	**6.** 68.75%

7. (a) 21.2%   (b) 79.2%   (c) 169%
8. (b) 52.9   (d) 54.5   (c) 55.6   (a) 57.1
9. $\frac{5}{16}$ **10.** $\frac{9}{16}$ **11.** 21.8 kg
12. 9.72 m   **13.** (a) 496.4 t (b) 8.657 g (c) 20 73 s
14. (a) 14%   (b) 15.$\dot{6}$%   (c) 5.36%
15. 37.49%   **16.** 17%   **17.** 38.7%
18. 2.7%   **19.** 2.5%   **20.** 779 Ω to 861 Ω
21. (a)(i) 544 Ω (ii)816 Ω (b)(i) 44.65 KΩ (ii) 49.35 kΩ
22. 2592 rev/min

Chapter 3

Exercise 13 (Page 25)

1. 27   **2.** 128   **3.** 100,000   **4.** 96
5. ± 5   **6.** 100   **7.** 1   **8.** 64

Exercise 14 (Page 28)

1. 128	**2.** 3^9	**3.** 16
4. $\frac{1}{9}$	**5.** 1	**6.** 8
7. 100	**8.** 1000	**9.** $\frac{1}{100}$ or 0.01
10. 5	**11.** 7^6	**12.** $3^6 = 729$
13. 3^{-5} or $\frac{1}{3^5}$ or $\frac{1}{243}$	**14.** 49	**15.** $\frac{1}{2}$ or 0.5
16. 1	**17.** $\frac{1}{3 \times 5^2}$	**18.** $\frac{1}{7^3 \times 3^7}$
19. $\frac{3^2}{2^5}$	**20.** 9	**21.** 3
22. $\frac{1}{2}$	**23.** $+\frac{2}{3}$	

Exercise 15 (Page 30)

1. cubic metres, m^3	**2.** farad
3. square metres, m^2	**4.** metres per second, m/s
5. kilogram per cubic metre, kg/m^3	**6.** joule
7. coulomb	**8.** watt
9. volt	**10.** mass
11. electrical resistance	**12.** frequency
13. acceleration	**14.** electric current
15. inductance	**16.** length
17. temperature	**18.** angular velocity
19. $\times 10^9$	**20.** m, $\times 10^{-3}$
21. $\times 10^{-12}$	**22.** M, $\times 10^6$

Exercise 16 (Page 32)

1. (a) 7.39×10 (b) 2.84×10 (c) 1.9762×10^2
2. (a) 2.748×10^3 (b) 3.317×10^4 (c) 2.74218×10^5
3. (a) 2.401×10^{-1} (b) 1.74×10^{-2} (c) 9.23×10^{-3}
4. (a) 1.7023×10^3 (b) 1.004×10 (c) 1.09×10^{-2}
5. (a) 5×10^{-1} (b) 1.1875×10 (c) 1.306×10^2
   (d) 3.125×10^{-2}
6. (a) 1010   (b) 932.7   (c) 54,100   (d) 7
7. (a) 0.0389   (b) 0.6741   (c) 0.008
8. (a) 1.35×10^2 (b) 1.1×10^5
9. (a) 2×10^2 (b) 1.5×10^{-3}
10. (a) $2.71 \times 10^3 \, \text{kg m}^{-3}$ (b) 4.4×10^{-1}
    (c) $3.7673 \times 10^2 \, \Omega$ (d) 5.11×10^{-1} MeV
    (e) $9.57897 \times 10^7 \text{C kg}^{-1}$ (f) $2.241 \times 10^{-2} \text{ m}^3\text{mol}^{-1}$

Exercise 17 (Page 33)

1. $60 \, \text{kPa}$ 2. $150 \, \mu \text{W}$ or   3. $50 \, \text{MV}$
                      $0.15 \, \text{mW}$
4. $55 \, \text{nF}$ 5. $100 \, \text{kW}$ 6. $0.54 \, \text{mA}$ or
                      $540 \, \mu \text{A}$
7. $1.5 \, \text{M}\Omega$ 8. $22.5 \, \text{mV}$ 9. $35 \, \text{GHz}$
10. $15 \, \text{pF}$ 11. $17 \, \mu \text{A}$ 12. $46.2 \, \text{k}\Omega$
13. $3 \, \mu \text{A}$ 14. $2.025 \, \text{MHz}$ 15. $62.50 \, \text{m}$
16. $0.0346 \, \text{kg}$ 17. 13.5×10^{-3} 18. 4×10^3
19. $3.8 \times 10^5 \, \text{km}$ 20. $0.053 \, \text{nm}$ 21. $5.6 \, \text{MPa}$
22. 4.3×10^{-3} m or 4.3 mm

Exercise 18 (page 34)

1. 2450 mm   2. 167.5 cm   3. 658 mm
4. 25.4 m   5. 56.32 m   6. 4356 mm
7. 87.5 cm   8. (a) 4650 mm (b) 4.65 m
9. (a) 504 cm (b) 5.04 m
10. 148.5 mm to 151.5 mm

Exercise 19 (page 35)

1. $8 \times 10^4 \, \text{cm}^2$ 2. $240 \times 10^{-4} \, \text{m}^2$ or $0.024 \, \text{m}^2$
3. $3.6 \times 10^6 \, \text{mm}^2$
4. $350 \times 10^{-6} \text{m}^2$ or $0.35 \times 10^{-3} \, \text{m}^2$
5. 5000 mm^2 6.   2.5 cm^2
7. (a) 288000 mm^2 (b) 2880 cm^2 (c) 0.288 m^2

Exercise 20 (page 37)

1. $2.5 \times 10^6 \, \text{cm}^3$ 2. $400 \times 10^{-6} \, \text{m}^3$
3. $0.87 \times 10^9 \, \text{mm}^3$ or $870 \times 10^6 \, \text{mm}^3$
4. $2.4 \, \text{m}^3$
5. $1500 \times 10^{-9} \, \text{m}^3$ or $1.5 \times 10^{-6} \, \text{m}^3$
6. $400 \times 10^{-3} \, \text{cm}^3$ or $0.4 \, \text{cm}^3$
7. $6.4 \times 10^3 \, \text{mm}^3$
8. $7500 \times 10^{-3} \, \text{cm}^3$ or $7.5 \, \text{cm}^3$
9. (a) $0.18 \, \text{m}^3$ (b) $180000 \, \text{cm}^3$ (c) 180 litres

Exercise 21 (page 40)

1. 18.74 in   2. 82.28 in   3. 37.95 yd
4. 18.36 mile   5. 41.66 cm   6. 1.98 m
7. 14.36 m   8. 5.91 km
9. (a) 22,745 yd (b) 20.81 km   10. 9.688 in^2
11. 18.18 yd^2 12. 30.89 acres   13. 21.89 mile^2
14. 41 cm^2 15. 28.67 m^2 16. 10117 m^2
17. 55.17 km^2 18. 12.24 in^3 19. 7.572 ft^3
20. 17.59 yd^3 21. 31.7 fluid pints 22. 35.23 cm^3
23. 4.956 m^3 24. 3.667 litre   25. 47.32 litre
26. 34.59 oz   27. 121.3 lb
28. 4.409 short tons   29. 220.6 g
30. 1.630 kg   31. (a) 280 lb (b) 127.2 kg
32. 131°F   33. 75 °C

Chapter 4

Exercise 22 (Page 44)

1. 53.832   2. 1.0944   3. 50.330
4. 12.25   5. 1.296×10^{-3} 6. 2.4430
7. 30.96   8. 0.0549   9. 0.571
10. 40   11. 137.9   12. 14.96
13. 19.4481   14. 515.36×10^{-6} 15. 1.0871
16. 52.70   17. 11.122   18. 0.8307
19. 2.571   20. 1.256   21. 1.068
22. 3.5×10^6 23. 37.5×10^3
24. 4.2×10^{-6} 25. 18.32×10^6

Exercise 23 (Page 45)

1. $\dfrac{13}{14}$ 2. 4.458   3. $\dfrac{1}{21}$ 4. $3\dfrac{1}{3}$
5. 0.0776   6. 0.2719   7. 0.4424   8. 0.0321
9. 0.4232   10. 0.1502   11. -0.6992 12. 5.8452
13. 4.995   14. 25.72   15. 591.0   16. 17.90
17. 3.520   18. 0.3770

Exercise 24 (Page 47)

1. $C = 52.78\,\text{mm}$ **2.** $R = 37.5$ **3.** $159\,\text{m/s}$

4. $5.02\,\text{mm}$ **5.** $0.144\,\text{J}$ **6.** 224.5

7. $14{,}230\,\text{kg/m}^3$ **8.** $281.1\,\text{m/s}$ **9.** $2.526\,\Omega$

10. $508.1\,\text{W}$ **11.** $V = 2.61\,\text{V}$ **12.** $F = 854.5$

13. $E = 3.96\,\text{J}$ **14.** $I = 12.77\,\text{A}$ **15.** $s = 17.25\,\text{m}$

16. $A = 7.184\,\text{cm}^2$

Chapter 5

Exercise 25 (Page 52)

1. $-3a$ **2.** $a + 2b + 4c$ **3.** $6ab - 3a$

4. $6x - 5y + 8z$ **5.** $4x + 2y + 2$ **6.** $3a + 5b$

7. $-2a - b + 2c$ **8.** p^2q^3r **9.** $8a^2$

10. $6q^2$ **11.** 46 **12.** $-\dfrac{1}{2}$

13. 6 **14.** $\dfrac{1}{7y}$

15. $3a^2 + 2ab - b^2$ **16.** $\dfrac{1}{3b}$

Exercise 26 (Page 54)

1. z^8 **2.** a^8 **3.** n^3

4. b^{11} **5.** b^{-3} **6.** c^4

7. m^4 **8.** x^{-3} or $\dfrac{1}{x^3}$ **9.** x^{12}

10. y^{-6} or $\dfrac{1}{y^6}$ **11.** t^8 **12.** c^{14}

13. a^{-9} or $\dfrac{1}{a^9}$ **14.** b^{-12} or $\dfrac{1}{b^{12}}$ **15.** b^{10}

16. s^{-9} **17.** $a^2b^{-2}c$ or $\dfrac{a^2c}{b^2}$, 9 **18.** $a^{-4}b^5c^{11}$

19. $a^{11/6}b^{1/3}c^{-3/2}$ or $\dfrac{\sqrt[6]{a^{11}}\,\sqrt[3]{b}}{\sqrt{c^3}}$

Exercise 27 (Page 55)

1. $x^2 + 5x + 6$ **2.** $2x^2 + 9x + 4$

3. $4x^2 + 12x + 9$ **4.** $2j^2 + 2j - 12$

5. $4x^2 + 22x + 30$ **6.** $2pqr + p^2q^2 + r^2$

7. $x^2 + 12x + 36$ **8.** $25x^2 + 30x + 9$

9. $4x^2 - 24x + 36$ **10.** $4x^2 - 9$

11. $3ab - 6a^2$ **12.** $2x^2 - 2xy$

13. $2a^2 - 3ab - 5b^2$ **14.** $13p - 7q$

15. $7x - y - 4z$ **16.** $4a^2 - 25b^2$

17. $x^2 - 4xy + 4y^2$ **18.** 0

19. $4 - a$ **20.** $4ab - 8a^2$

Exercise 28 (Page 56)

1. $2(x + 2)$ **2.** $2x(y - 4z)$ **3.** $p(b + 2c)$

4. $2x(1 + 2y)$ **5.** $4d(d - 3f^5)$ **6.** $4x(1 + 2x)$

7. $2q(q + 4n)$ **8.** $r(s + p + t)$ **9.** $x(1 + 3x + 5x^2)$

10. $bc(a + b^2)$ **11.** $3xy(xy^3 - 5y + 6)$

12. $2pq^2(2p^2 - 5q)$ **13.** $7ab(3ab - 4)$

14. $2xy(y + 3x + 4x^2)$ **15.** $2xy(x - 2y^2 + 4x^2y^3)$

Exercise 29 (Page 57)

1. $2x + 8x^2$ **2.** $12y^2 - 3y$ **3.** $4b - 15b^2$

4. $4 + 3a$ **5.** $\dfrac{3}{2} - 4x$ **6.** 1

7. $10y^2 - 3y + \dfrac{1}{4}$ **8.** $9x^2 + \dfrac{1}{3} - 4x$

9. $6a^2 + 5a - \dfrac{1}{7}$ **10.** $-15t$

Chapter 6

Exercise 30 (Page 60)

1. 1 **2.** 2 **3.** 6 **4.** -4 **5.** 2

6. 2 **7.** $\dfrac{1}{2}$ **8.** 0 **9.** 3 **10.** 2

11. -10 **12.** 6 **13.** -2 **14.** 2.5 **15.** 2

Exercise 31 (Page 62)

1. 5 **2.** -2 **3.** $-4\dfrac{1}{2}$ **4.** 2

5. 15 **6.** -4 **7.** $5\dfrac{1}{3}$ **8.** 2

9. 13 **10.** -11 **11.** -6 **12.** 9

13. $6\dfrac{1}{4}$ **14.** 3 **15.** 10 **16.** ± 12

Exercise 32 (Page 63)

1. 10^{-7} **2.** $8\,\text{m/s}^2$ **3.** 3.472
4. (a) $1.8\,\Omega$ (b) $30\,\Omega$ **5.** $800\,\Omega$ **6.** $176\,\text{MPa}$

Exercise 33 (Page 64)

1. 0.004 **2.** 30 **3.** $45°\text{C}$ **4.** 50
5. $12\,\text{m}, 8\,\text{m}$ **6.** $3.5\,\text{N}$

Chapter 7

Exercise 34 (Page 68)

1. $d = c - e - a - b$ **2.** $x = \dfrac{y}{7}$ **3.** $v = \dfrac{c}{p}$

4. $a = \dfrac{v-u}{t}$ **5.** $y = \dfrac{1}{3}(t-x)$ **6.** $r = \dfrac{c}{2\pi}$

7. $x = \dfrac{y-c}{m}$ **8.** $T = \dfrac{I}{PR}$ **9.** $L = \dfrac{X_L}{2\pi f}$

10. $c = \dfrac{Q}{m\Delta T}$ **11.** $R = \dfrac{F}{I}$ **12.** $x = a(y-3)$

13. $C = \dfrac{5}{9}(F-32)$ **14.** $R = \dfrac{pV}{mT}$

Exercise 35 (Page 69)

1. $r = \dfrac{S-a}{S}$ or $r = 1 - \dfrac{a}{S}$

2. $x = \dfrac{d}{\lambda}(y+\lambda)$ or $x = d + \dfrac{yd}{\lambda}$

3. $f = \dfrac{3F-AL}{3}$ or $f = F - \dfrac{AL}{3}$

4. $D = \dfrac{A\,B^2}{5Cy}$ **5.** $t = \dfrac{R-R_0}{R_0\alpha}$

6. $R_2 = \dfrac{RR_1}{R_1 - R}$

7. $R = \dfrac{E-e-Ir}{I}$ or $R = \dfrac{E-e}{I} - r$

8. $b = \sqrt{\left(\dfrac{y}{4ac^2}\right)}$ **9.** $V_2 = \dfrac{P_1 V_1 T_2}{P_2 T_1}$

10. $L = \dfrac{t^2 g}{4\pi^2}$ **11.** $u = \sqrt{v^2 - 2as}$

12. $a = N^2 y - x$ **13.** $v = \sqrt{\dfrac{2L}{\rho ac}}$

14. $T_2 = \dfrac{P_2 V_2 T_1}{P_1 V_1}$

Exercise 36 (Page 70)

1. $a = \sqrt{\left(\dfrac{xy}{m-n}\right)}$ **2.** $R = \sqrt[4]{\left(\dfrac{M}{\pi} + r^4\right)}$

3. $r = \dfrac{3(x+y)}{(1-x-y)}$ **4.** $L = \dfrac{mrCR}{\mu - m}$

5. $b = \dfrac{c}{\sqrt{1-a^2}}$ **6.** $r = \sqrt{\left(\dfrac{x-y}{x+y}\right)}$

7. $v = \dfrac{uf}{u-f}, 30$ **8.** $t_2 = t_1 + \dfrac{Q}{mc}, 55$

9. $v = \sqrt{\left(\dfrac{2dgh}{0.03L}\right)}, 0.965$ **10.** $l = \dfrac{8S^2}{3d} + d, 2.725$

11. $64\,\text{mm}$ **12.** $w = \dfrac{2R-F}{L}; 3\,\text{kN/m}$

13. $t_2 = t_1 - \dfrac{Qd}{kA}$ **14.** $r = \dfrac{v}{\omega}\left(1 - \dfrac{s}{100}\right)$

15. $F = EI\left(\dfrac{n\pi}{L}\right)^2; 13.61\,\text{MN}$ **16.** $r = \sqrt[4]{\left(\dfrac{8\eta\ell V}{\pi p}\right)}$

Chapter 8

Exercise 37 (Page 74)

1. $x = 4$, $y = 2$ **2.** $x = 3$, $y = 4$
3. $x = 2$, $y = 1.5$ **4.** $x = 4$, $y = 1$
5. $x = 3$, $y = 2$ **6.** $a = 2$, $b = 3$
7. $a = 5$, $b = 2$ **8.** $x = 1$, $y = 1$
9. $s = 2$, $t = 3$ **10.** $x = 3$, $y = -2$
11. $m = 2.5$, $n = 0.5$ **12.** $a = 6$, $b = -1$

Exercise 38 (Page 76)

1. $p = -1$, $q = -2$ **2.** $x = 4$, $y = 6$
3. $a = 2$, $b = 3$ **4.** $s = 4$, $t = -1$
5. $x = 3$, $y = 4$ **6.** $u = 12$, $v = 2$

Exercise 39 (Page 77)

1. $a = 0.2$, $b = 4$ **2.** $I_1 = 6.47$, $I_2 = 4.62$
3. $u = 12$, $a = 4$, $v = 26$ **4.** $m = -0.5$, $c = 3$
5. $a = 12$, $b = 0.40$ **6.** $F_1 = 1.5$, $F_2 = -4.5$
7. $R_1 = 5.7\,\text{kN}, R_2 = 6.3\,\text{kN}$

Chapter 9

Exercise 40 (page 80)

1. log 6 **2.** log 15 **3.** log 2 **4.** log 3

5. log 12 **6.** log 500 **7.** log 2

8. log 243 or log 3^5 or 5 log 3 **9.** log 16 or log 2^4 or 4 log 2

10. log 64 or log 2^6 or 6 log 2 **11.** $x = 2.5$ **12.** $t = 8$
13. $b = 2$ **14.** $x = 2$ **15.** $a = 6$

Exercise 41 (page 81)

1. 1.690 **2.** 3.170 **3.** 6.058 **4.** 2.251 **5.** 2.542

6. −0.3272 **7.** 316.2 **8.** 0.057 m^3 **9.** 5.69

Exercise 42 (page 83)

1. (a) 0.1653 (b) 0.4584 (c) 22030

2. (a) 5.0988 (b) 0.064037 (c) 40.446

3. (a) 4.55848 (b) 2.40444 (c) 8.05124

4. (a) 48.04106 (b) 4.07482 (c) −0.08286

5. 2.739 **6.** 120.7 m

Exercise 43 (page 84)

1. 3.95, 2.05 **2.** 1.65, −1.30

3. (a) 28 cm^3 (b) 116 min

4. (a) 70°C (b) 5 minutes

Exercise 44 (page 86)

1. (a) 0.55547 (b) 0.91374 (c) 8.8941

2. (a) 2.2293 (b) −0.33154 (c) 0.13087

3. −0.4904 **4.** −0.5822 **5.** 2.197 **6.** 816.2

7. 11.02 **8.** 1.522 **9.** 1.485 **10.** 1.962

11. 4.901 **12.** 500 **13.** 992 m/s **14.** 348.5 Pa

Exercise 45 (page 88)

1. (a) 150°C (b) 100.5°C **2.** 99.21 kPa

3. (a) 29.32 volts (b) 71.31 × 10^{-6}s

4. (a) 1.993 m (b) 2.293 m

5. (a) 50°C (b) 55.45 s **6.** 30.37 N

7. (a) 3.04 A (b) 1.46 s **8.** 2.45 mol/cm^3

9. (a) 7.07 A (b) 0.966 s

10 (a) 100% (b) 67.03% (c) 1.83% **11.** 2.45 mA

12. 142 ms **13.** 99.752%

14. 20 min 38 s

Chapter 10

Exercise 46 (Page 96)

1. (a) Horizontal axis, 1 cm = 4 V (or 1 cm = 5 V);
   vertical axis, 1 cm = 10 Ω

 (b) Horizontal axis, 1 cm = 5 m;
   vertical axis, 1 cm = 0.1 V

 (c) Horizontal axis, 1 cm = 10 N;
   vertical axis, 1 cm = 0.2 mm

2. (a) −1 (b) −8 (c) −1.5 (d) 5
3. 14.5 **4.** (a) −1.1 (b) −1.4
5. The 1010 rev/min reading should be 1070 rev/min;
 (a) 1000 rev/min (b) 167 V

Exercise 47 (Page 100)

1. Missing values: −0.75, 0.25, 0.75, 1.75, 2.25, 2.75
 Gradient = $\dfrac{1}{2}$

2. (a) 4, −2 (b) −1, 0 (c) −3, −4 (d) 0, 4
3. (a) 6, −3 (b) −2, 4 (c) 3, 0 (d) 0, 7
4. (a) $\dfrac{3}{5}$ (b) −4 (c) $-1\dfrac{5}{6}$
5. (a) and (c), (b) and (e)
6. (2, 1)
7. (a) 89 cm (b) 11 N (c) 2.4
 (d) $l = 2.4 W + 48$

Exercise 48 (Page 103)

1. (a) 40°C   (b) 128 Ω
2. (a) 850 rev/min   (b) 77.5 V
3. (a) 0.25   (b) 12 N   (c) $F = 0.25L + 12$
   (d) 89.5 N   (e) 592 N   (f) 212 N
4. (a) 22.5 m/s   (b) 6.5 s   (c) $v = 0.7t + 15.5$
5. $m = 26.8L$
6. (a) $\dfrac{1}{5}$ (b) 6   (c) $E = \dfrac{1}{5}L + 6$
   (d) 12 N   (e) 65 N
7. $a = 0.85$, $b = 12$, 254.3 kPa, 275.5 kPa, 280 K

Chapter 11

Exercise 49 (Page 107)

1. 24 m   2. 9.54 mm   3. 20.81 cm   4. 7.21 m
5. 11.18 cm   6. 24.11 mm   7. 20.81 km
8. 3.35 m, 10 cm   9. 132.7 nautical miles

Exercise 50 (Page 109)

1. $\sin Z = \dfrac{9}{41}$, $\cos Z = \dfrac{40}{41}$, $\tan X = \dfrac{40}{9}$, $\cos X = \dfrac{9}{41}$
2. $\sin A = \dfrac{3}{5}$, $\cos A = \dfrac{4}{5}$, $\tan A = \dfrac{3}{4}$, $\sin B = \dfrac{4}{5}$,
   $\cos B = \dfrac{3}{5}$, $\tan B = \dfrac{4}{3}$
3. $\sin A = \dfrac{8}{17}$, $\tan A = \dfrac{8}{15}$
4. (a) $\dfrac{15}{17}$ (b) $\dfrac{15}{17}$ (c) $\dfrac{8}{15}$
5. (a) $\sin \theta = \dfrac{7}{25}$ (b) $\cos \theta = \dfrac{24}{25}$
6. 9.434

Exercise 51 (Page 111)

1. 2.7550   2. 4.846   3. 36.52
4. (a) −0.1010 (b) 0.5865   5. 42.33°   6. 15.25°
7. 73.78°   8. 7° 56′   9. 31°22′   10. 41°54′
11. 29.05°   12. 20°21′   13. (a) 40° (b) 6.79 m

Exercise 52 (Page 113)

1. (a) 12.22   (b) 5.619   (c) 14.87
   (d) 8.349   (e) 5.595   (f) 5.275
2. (a) $AC = 5.831$ cm, $\angle A = 59.04°$, $\angle C = 30.96°$
   (b) $DE = 6.928$ cm, $\angle D = 30°$, $\angle F = 60°$
   (c) $\angle J = 62°$, $HJ = 5.634$ cm, $GH = 10.60$ cm
   (d) $\angle L = 63°$, $LM = 6.810$ cm, $KM = 13.37$ cm
   (e) $\angle N = 26°$, $ON = 9.125$ cm, $NP = 8.201$ cm
   (f) $\angle S = 49°$, $RS = 4.346$ cm, $QS = 6.625$ cm
3. 6.54 m   4. 9.40 mm   5. 36.15 m

Exercise 53 (Page 118)

1. $C = 83°$, $a = 14.1$ mm, $c = 28.9$ mm, area $= 189$ mm^2
2. $A = 52°2′$, $c = 7.568$ cm, $a = 7.152$ cm, area $= 25.65$ cm^2
3. $D = 19°48′$, $E = 134°12′$, $e = 36.0$ cm, area $= 134$ cm^2
4. $E = 49°0′$, $F = 26°38′$, $f = 15.09$ mm, area $= 185.6$ mm^2
5. $p = 13.2$ cm, $Q = 47.35°$, $R = 78.65°$, area $= 77.7$ cm^2
6. $p = 6.127$ m, $Q = 30.83°$, $R = 44.17°$, area $= 6.938$ m^2
7. $X = 83.33°$, $Y = 52.62°$, $Z = 44.05°$, area $= 27.8$ cm^2
8. $X = 29.77°$, $Y = 53.50°$, $Z = 96.73°$, area $= 355$ mm^2

Exercise 54 (Page 118)

1. 193 km   2. (a) 11.4 m (b) 17.55°
3. 163.4 m   4. $BF = 3.9$ m, $EB = 4.0$ m
5. 6.35 m, 5.37 m   6. 32.48 A, 14.31°
7. $x = 69.3$ mm, $y = 142$ mm
8. 13.66 mm

Chapter 12

Exercise 55 (Page 126)

1. $p = 105°$, $q = 35°$ 2. $r = 142°$, $s = 95°$
3. $t = 146°$

Exercise 56 (Page 130)

1. (i) rhombus   (a) $14\,cm^2$ (b) $16\,cm$
  (ii) parallelogram (a) $180\,mm^2$ (b) $80\,mm$
 (iii) rectangle   (a) $3600\,mm^2$ (b) $300\,mm$
 (iv) trapezium   (a) $190\,cm^2$ (b) $62.91\,cm$
2. $35.7\,cm^2$ 3. (a) $80\,m$ (b) $170\,m$
4. $27.2\,cm^2$ 5. $18\,cm$ 6. $1200\,mm$
7. (a) $29\,cm^2$ (b) $650\,mm^2$
8. $560\,m^2$ 9. $3.4\,cm$ 10. $6750\,mm^2$
11. $43.30\,cm^2$ 12. 32

Exercise 57 (Page 131)

1. $482\,m^2$ 2. (a) $50.27\,cm^2$ (b) $706.9\,mm^2$
                                     (c) $3183\,mm^2$
3. $2513\,mm^2$ 4. (a) $20.19\,mm$ (b) $63.41\,mm$
5. (a) $53.01\,cm^2$ (b) $129.9\,mm^2$
6. $5773\,mm^2$ 7. $1.89\,m^2$

Exercise 58 (Page 134)

1. $1932\,mm^2$ 2. $1624\,mm^2$
3. (a) $0.918\,ha$ (b) $456\,m$

Exercise 59 (Page 135)

1. $80\,ha$ 2. $80\,m^2$ 3. $3.14\,ha$

Chapter 13

Exercise 60 (Page 138)

1. $45.24\,cm$ 2. $259.5\,mm$ 3. $2.629\,cm$ 4. $47.68\,cm$
5. $38.73\,cm$ 6. $12{,}730\,km$ 7. $97.13\,mm$

Exercise 61 (Page 139)

1. (a) $\dfrac{\pi}{6}$ (b) $\dfrac{5\pi}{12}$ (c) $\dfrac{5\pi}{4}$
2. (a) 0.838 (b) 1.481 (c) 4.054
3. (a) $210°$ (b) $80°$ (c) $105°$
4. (a) $0°43'$ (b) $154°8'$ (c) $414°53'$
5. $104.7\,rad/s$

Exercise 62 (Page 141)

1. $113\,cm^2$ 2. $2376\,mm^2$ 3. $1790\,mm^2$ 4. $802\,mm^2$
5. $1709\,mm^2$ 6. $1269\,m^2$ 7. $1548\,m^2$
8. (a) $106.0\,cm$ (b) $783.9\,cm^2$ 9. $21.46\,m^2$
10. $17.80\,cm$, $74.07\,cm^2$
11. (a) $59.86\,mm$ (b) $197.8\,mm$ 12. $26.2\,cm$
13. $82.5°$ 14. 748
15. (a) $0.698\,rad$ (b) $804.2\,m^2$ 16. $10.47\,m^2$
17. (a) $396\,mm^2$ (b) 42.24% 18. $701.8\,mm$
19. $7.74\,mm$

Chapter 14

Exercise 63 (Page 147)

1. $1.2\,m^3$ 2. $5\,cm^3$ 3. $8\,cm^3$
4. (a) $3840\,mm^3$ (b) $1792\,mm^2$
5. $972\,litres$ 6. $15\,cm^3$, $135\,g$ 7. $500\,litres$
8. $1.44\,m^3$ 9. (a) $35.3\,cm^3$ (b) $61.3\,cm^2$
10. (a) $2400\,cm^3$ (b) $2460\,cm^2$ 11. $37.04\,m$
12. $8796\,cm^3$ 13. $4.709\,cm$,   14. $2.99\,cm$
                             $153.9\,cm^2$
15. $28{,}060\,cm^3$, $1.099\,m^2$
16. $8.22\,m$ by $8.22\,m$
17. $4\,cm$ 18. $4.08\,m^3$

Exercise 64 (Page 150)

1. $201.1\,cm^3$, $159.0\,cm^2$ 2. $7.68\,cm^3$, $25.81\,cm^2$
3. $113.1\,cm^3$, $113.1\,cm^2$ 4. $3\,cm$ 5. $2681\,mm^3$
6. (a) $268{,}083\,mm^3$ or $268.083\,cm^3$
   (b) $20{,}106\,mm^2$ or $201.06\,cm^2$
7. $8.53\,cm$
8. (a) $512 \times 10^6\,km^2$ (b) $1.09 \times 10^{12}\,km^3$
9. 664 10. $92\,m^3$, $92{,}000\,litres$

Exercise 65 (Page 154)

1. $5890\,mm^2$ or $58.90\,cm^2$
2. (a) $56.55\,cm^3$ (b) $84.82\,cm^2$
3. $13.57\,kg$ 4. $29.32\,cm^3$ 5. $393.4\,m^2$
6. (i) (a) $670\,cm^3$ (b) $523\,cm^2$
   (ii) (a) $180\,cm^3$ (b) $154\,cm^2$
  (iii) (a) $56.5\,cm^3$ (b) $84.8\,cm^2$
  (iv) (a) $10.4\,cm^3$ (b) $32.0\,cm^2$
   (v) (a) $96.0\,cm^3$ (b) $146\,cm^2$
  (vi) (a) $86.5\,cm^3$ (b) $142\,cm^2$
(vii) (a) $805\,cm^3$ (b) $539\,cm^2$

7. (a) 17.9 cm (b) 38.0 cm **8.** 10.3 m^3, 25.5 m^2

9. 6560 litres **10.** 657.1 cm^3, 1027 cm^2

11. (a) 1458 litres (b) 9.77 m^2 (c) £140.45

Exercise 66 (Page 156)

1. 8:125 **2.** 137.2 g

Exercise 67 (Page 160)

1. (b) **2.** (d) **3.** (b) **4.** (b) **5.** (a) **6.** (c) **7.** (c)
8. (d) **9.** (b) **10.** (c) **11.** (a) **12.** (d) **13.** (a) **14.** (b)
15. (a) **16.** (d) **17.** (a) **18.** (b) **19.** (d) **20.** (d) **21.** (a)
22. (a) **23.** (a) **24.** (a) **25.** (a) **26.** (d) **27.** (b) **28.** (c)
29. (d) **30.** (b) **31.** (a) **32.** (c) **33.** (b) **34.** (b) **35.** (c)
36. (c) **37.** (b) **38.** (d) **39.** (c) **40.** (c) **41.** (a) **42.** (a)
43. (c) **44.** (b) **45.** (d) **46.** (b) **47.** (d) **48.** (b) **49.** (d)
50. (b) **51.** (a) **52.** (a) **53.** (b) **54.** (a) **55.** (a) **56.** (d)
57. (d) **58.** (c) **59.** (b) **60.** (b) **61.** (c) **62.** (c) **63.** (c)
64. (d) **65.** (a) **66.** (a) **67.** (c) **68.** (b) **69.** (b) **70.** (d)
71. (d) **72.** (a) **73.** (d)

Chapter 15

Exercise 68 (Page 171)

1. (a) 0.052 m (b) 0.02 m^2 (c) 0.01 m^3
2. (a) 12.5 m^2 (b) 12.5 × 10^6 mm^2
3. (a) 37.5 m^3 (b) 37.5 × 10^9 mm^3
4. (a) 0.0063 m^3 (b) 6300 cm^3 (c) 6.3 × 10^6 mm^3

Exercise 69 (Page 173)

1. 11,400 kg/m^3 **2.** 1.5 kg **3.** 20 litres **4.** 8.9
5. 8000 kg/m^3 **6.** (a) 800 kg/m^3
　　　　　　　　　　　　(b) 0.0025 m^3 or 2500 cm^3

Exercise 70 (Page 173)

Answers found within the text of the chapter, pages 169 to 173.

Exercise 71 (Page 173)

1. (c) **2.** (d) **3.** (b) **4.** (a) **5.** (b)
6. (c) **7.** (b) **8.** (b) **9.** (c) **10.** (a)

Chapter 16

Exercise 72 (Page 179)

1. (a) mixture (b) compound (c) mixture (d) element
2. 1.2 kg **3.** See page 176 **4.** See page 178

Exercise 73 (Page 179)

Answers found within the text of the chapter, pages 175 to 179.

Exercise 74 (Page 179)

1. (d) **2.** (c) **3.** (a) **4.** (b)
5. (a) **6.** (b) **7.** (b) **8.** (c)

Chapter 17

Exercise 75 (Page 182)

1. (a) 72 km/h (b) 20 m/s **2.** 3.6 km
3. 3 days 19 hours

Exercise 76 (Page 184)

1. 0 to A, 30 km/h; A to B, 40 km/h; B to C, 10 km/h;
0 to C, 24 km/h; A to C, 20 km/h
2. 4 m/s, 2.86 m/s, 3.33 m/s, 2.86 m/s, 3.2 m/s
3. 119 km **4.** 333.33 m/s or 1200 km/h
5. 2.6 × 10^{13} km

Exercise 77 (Page 185)

1. 12.5 km **2.** 4 m/s

Exercise 78 (Page 186)

Answers found within the text of the chapter, pages 181 to 186.

Exercise 79 (Page 187)

1. (c) **2.** (g) **3.** (d) **4.** (c) **5.** (e) **6.** (b) **7.** (a)
8. (b) **9.** (d) **10.** (a) **11.** (a) **12.** (b) **13.** (c) **14.** (d)

Chapter 18

Exercise 80 (Page 191)

1. 50 s **2.** 9.26×10^{-4} m/s^2
3. (a) 0.0231 m/s^2 (b) 0 (c) -0.370 m/s^2
4. A to B, -2 m/s^2, B to C, 2.5 m/s^2, C to D, -1.5 m/s^2

Exercise 81 (Page 193)

1. 12.25 m/s **2.** 10.2 s **3.** 7.6 m/s
4. 39.6 km/h **5.** 2 min 5 s **6.** 1.35 m/s, -0.45 m/s^2

Exercise 82 (Page 193)

Answers found within the text of the chapter, pages 189 to 193.

Exercise 83 (Page 193)

1. (b) **2.** (a) **3.** (c) **4.** (e) **5.** (i)
6. (g) **7.** (d) **8.** (c) **9.** (a) **10.** (d)

Chapter 19

Exercise 84 (Page 198)

1. 873 N **2.** 1.87 kN, 6.55 s
3. 0.560 m/s^2 **4.** 16.46 kN
5. (a) 11.62 kN (b) 19.62 kN (c) 24.62 kN

Exercise 85 (Page 200)

1. 988 N, 34.86 km/h **2.** 0.3 m/s^2 **3.** 10 kg

Exercise 86 (Page 200)

Answers found within the text of the chapter, pages 195 to 200.

Exercise 87 (Page 200)

1. (c) **2.** (b) **3.** (a) **4.** (d) **5.** (a) **6.** (b)
7. (b) **8.** (a) **9.** (a) **10.** (d) **11.** (d) **12.** (c)

Chapter 20

Exercise 88 (Page 207)

1. 4.0 kN in the direction of the forces
2. 68 N in the direction of the 538 N force
3. 36.8 N at 20° **4.** 12.6 N at 79°
5. 3.4 kN at -8°

Exercise 89 (Page 208)

1. 2.6 N at -20° **2.** 15.7 N at -172°
3. 26.7 N at -82° **4.** 0.86 kN at -101°
5. 15.5 N at -152°

Exercise 90 (Page 208)

1. 11.52 kN at 105° **2.** 36.84 N at 19.83°
3. 3.38 kN at -8.19° **4.** 15.69 N at -171.67°
5. 0.86 kN at -100.94°

Exercise 91 (Page 210)

1. 3.1 N at -45° to force A
2. 53.5 kN at 37° to force 1 (i.e. 117° to the horizontal)
3. 72 kN at -37° to force P
4. 43.2 N, 39° East of North

Exercise 92 (Page 212)

1. 5.86 N, 8.06 N **2.** 9.96 kN, 7.77 kN
3. 100 N force at 148° to the 60 N force, 90 N force at 277° to the 60 N force

Exercise 93 (Page 215)

1. 10.08 N, 20.67 N   **2.** 9.09 N at 93.92°
3. 3.06 N at −45.40° to force *A*
4. 53.50 kN at 37.27° to force 1
    (i.e. 117.27° to the horizontal)
5. 18.82 kN at 210.03° to the horizontal
6. 34.96 N at −10.23° to the horizontal

Exercise 94 (Page 215)

Answers found within the text of the chapter, pages 203 to 215.

12. 15 N acting horizontally to the right
13. 10 N at 230°   **15.** 5 N at 45°

Exercise 95 (Page 216)

1. (b) **2.** (a)   **3.** (b)   **4.** (d)   **5.** (b)   **6.** (c)   **7.** (b)
8. (b) **9.** (c) **10.** (d) **11.** (c) **12.** (d) **13.** (d) **14.** (a)

Chapter 21

Exercise 96 (Page 222)

1. 75 kJ **2.** 20 kJ   **3.** 50 kJ
4. 1.6 J **5.** (a) 2.5 J   (b) 2.1 J   **6.** 6.15 kJ

Exercise 97 (Page 224)

1. 75%   **2.** 1.2 kJ   **3.** 4 m   **4.** 10 kJ

Exercise 98 (Page 227)

1. 600 kJ   **2.** 4 kW
3. (a) 500 N (b) 625 W   **4.** (a) 10.8 MJ (b) 12 kW
5. 160 m   **6.** (a) 72 MJ   (b) 100 MJ
7. 13.5 kW   **8.** 480 W
9. 100 kW   **10.** 90 W

Exercise 99 (Page 231)

1. 8.31 m   **2.** 196.2 J
3. 90 kJ   **4.** 176.6 kJ, 176.6 kJ
5. 200 kJ, 2039 m   **6.** 8 kJ, 14.0 m/s
7. 4.85 m/s (a) 2.83 m/s (b) 14.70 kN

Exercise 100 (Page 231)

Answers found within the text of the chapter, pages 218 to 231.

Exercise 101 (Page 232)

1. (b)   **2.** (c)   **3.** (c)   **4.** (a)   **5.** (d)
6. (c)   **7.** (a)   **8.** (d)   **9.** (c) **10.** (b)
11. (b) **12.** (a) **13.** (d) **14.** (a) **15.** (d)

Chapter 22

Exercise 102 (Page 235)

1. 4.5 N m   **2.** 200 mm   **3.** 150 N

Exercise 103 (Page 237)

1. 50 mm, 3.8 kN   **2.** 560 N
3. $R_A = R_B = 25$ N   **4.** 5 N, 25 mm

Exercise 104 (Page 240)

1. 2.0 kN, 24 mm **2.** 1.0 kN, 64 mm   **3.** 80 m
4. 36 kN   **5.** 50 N, 150 N
6. (a) $R_1 = 3$ kN, $R_2 = 12$ kN   (b) 15.5 kN

Exercise 105 (Page 241)

Answers found within the text of the chapter, pages 234 to 240.

Exercise 106 (Page 241)

1. (a) **2.** (c) **3.** (a) **4.** (d)   **5.** (a)
6. (d) **7.** (c) **8.** (a) **9.** (d) **10.** (c)

Chapter 23

Exercise 107 (Page 246)

1. $\omega = 90$ rad/s, $v = 16.2$ m/s
2. $\omega = 40$ rad/s, $v = 10$ m/s

Exercise 108 (Page 247)

1. 275 rad/s, $8250/\pi$ rev/min
2. 0.4π rad/s^2, 0.1π m/s^2

Exercise 109 (Page 249)

1. 2.693 s 2. 222.8 rad/s^2 3. 187.5 revolutions
4. (a) 160 rad/s (b) 4 rad/s^2 (c) $12,000/\pi$ rev

Exercise 110 (Page 251)

1. 83.5 km/h at $71.6°$ to the vertical
2. 4 min 55 s, $60°$
3. 22.79 km/h in a direction E $9.78°$ N

Exercise 111 (Page 251)

Answers found within the text of the chapter, pages 244 to 251.

Exercise 112 (Page 251)

1. (b) 2. (c) 3. (a) 4. (c) 5. (a) 6. (d) 7. (c)
8. (b) 9. (d) 10. (c) 11. (b) 12. (d) 13. (a)

Chapter 24

Exercise 113 (Page 255)

1. 1250 N 2. 0.24 3. (a) 675 N (b) 652.5 N

Exercise 114 (Page 256)

Answers found within the text of the chapter, pages 253 to 256.

Exercise 115 (Page 256)

1. (c) 2. (c) 3. (f) 4. (e) 5. (i)
6. (c) 7. (h) 8. (b) 9. (d) 10. (a)

Chapter 25

Exercise 116 (Page 260)

1. (a) 3.3 (b) 11 (c) 30% 2. 8
3. (a) 5 (b) 500 mm
4. (a) $F_e = 0.04F_l + 170$ (b) 25 (c) 83.3%
5. 410 N, 48%

Exercise 117 (Page 262)

1. 60% 2. 1.5 kN, 93.75%

Exercise 118 (Page 263)

1. 9.425 kN 2. 16.91 N

Exercise 119 (Page 265)

1. 6, 10 rev/s 2. 6, 4.32 3. 250, 3.6

Exercise 120 (Page 266)

1. 1.25 m 2. 250 N 3. 333.3 N, $1/3$

Exercise 121 (Page 266)

Answers found within the text of the chapter, pages 258 to 266.

Exercise 122 (Page 267)

1. (b) 2. (f) 3. (c) 4. (d) 5. (b) 6. (a)
7. (b) 8. (d) 9. (c) 10. (d) 11. (d) 12. (b)

Chapter 26

Exercise 123 (Page 273)

1. 250 MPa   **2.** 12.78 mm   **3.** 69.44 MPa
4. 2.356 kN   **5.** 2.06 mm   **6.** 38.2 MPa

Exercise 124 (Page 275)

1. 2.25 mm   **2.** (a) 0.00002 (b) 0.002%
3. 0.00025, 0.025%
4. (a) 212.2 MPa (b) 0.002 or 0.20%
5. (a) 25 kPa (b) 0.04 or 4%

Exercise 125 (Page 278)

1. 800 N   **2.** 10 mm
3. (a) 43.75 kN (b) 1.2 mm
4. (a) 19 kN (b) 0.057 mm
5. 132.6 GPa   **6.** (a) 27.5 mm (b) 68.2 GPa
7. 0.10%

Exercise 126 (Page 279)

Answers found within the text of the chapter, pages 270 to 279.

Exercise 127 (Page 280)

1. (c) **2.** (c) **3.** (a) **4.** (b) **5.** (c) **6.** (c) **7.** (b)
8. (d) **9.** (b) **10.** (c) **11.** (f) **12.** (h) **13.** (d)

Chapter 27

Exercise 128 (Page 285)

1. 250 kg m/s   **2.** 2.4 kg m/s   **3.** 64 kg
4. 22,500 kg m/s **5.** 90 km/h   **6.** 2 m/s
7. 10.71 m/s   **8.** (a) 26.25 m/s (b) 11.25 m/s

Exercise 129 (Page 287)

1. 8 kg m/s   **2.** 1.5 m/s   **3.** 1 kN
4. (a) 200 kg m/s (b) 8 kN   **5.** 314.5 kN
6. (a) 1.5 kg m/s (b) 150 N   **7.** 3600 kg m/s

Exercise 130 (Page 287)

Answers found within the text of the chapter, pages 282 to 287.

Exercise 131 (Page 287)

1. (d) **2.** (b) **3.** (f) **4.** (c) **5.** (a) **6.** (c)
7. (a) **8.** (g) **9.** (f) **10.** (f) **11.** (b) **12.** (e)

Chapter 28

Exercise 132 (Page 291)

1. 70 N m **2.** 160 mm

Exercise 133 (Page 293)

1. 339.3 kJ   **2.** 523.8 rev   **3.** 45 N m
4. 12.66 N m **5.** 13.75 rad/s **6.** 16.76 MW
7. 27.52 kW **8.** (a) 3.33 N m   (b) 540 kJ

Exercise 134 (Page 295)

1. 272.4 N m   **2.** 30 kg m^2
3. 502.65 kJ, 1934 rev/min
4. 1.885 kJ, 1209.5 rev/min
5. 6.283 kN m   **6.** 16.36 J
7. (a) 2.857 × 10^{-3} kg m^2 (b) 24.48 J

Exercise 135 (Page 297)

1. 6.05 kW   **2.** 70.37%, 280 mm   **3.** 57.03%
4. 404.2 N, 300 rev/min   **5.** 545.5 W
6. 4.905 MJ, 1 min 22 s

Exercise 136 (Page 298)

Answers found within the text of the chapter, pages 290 to 298.

Exercise 137 (Page 298)

1. (d) **2.** (b) **3.** (c) **4.** (a) **5.** (c) **6.** (d)
7. (a) **8.** (b) **9.** (c) **10.** (d) **11.** (a) **12.** (c)

Chapter 29

Exercise 138 (Page 301)

1. 28 kPa **2.** 4.5 kN **3.** 56 kPa

Exercise 139 (Page 303)

1. 343 kPa **2.** (a) 303 Pa (b) 707 Pa **3.** 1400 kg/m^3

Exercise 140 (Page 303)

1. 100 kPa **2.** 134 kPa **3.** 97.9 kPa **4.** 743 mm

Exercise 141 (Page 305)

1. 1.215 kN **2.** 15.25 N **3.** 0.03222 m^3
4. 148.6 N **5.** 2.47 kN **6.** 4.186 tonne/m^3, 4.186
7. 159 mm **8.** 1.551 kN

Exercise 142 (Page 311)

Answers found within the text of the chapter, pages 300 to 311.

Exercise 143 (Page 311)

1. (b) **2.** (d) **3.** (a) **4.** (a) **5.** (c) **6.** (d) **7.** (b)
8. (c) **9.** (c) **10.** (d) **11.** (d) **12.** (d) **13.** (c) **14.** (b)
15. (c) **16.** (a) **17.** (b) **18.** (f) **19.** (a) **20.** (b) **21.** (c)

Chapter 30

Exercise 144 (Page 316)

1. (a) 324 K (b) 195 K (c) 456 K
2. (a) 34°C (b) −36°C (c) 142°C

Exercise 145 (Page 317)

1. 2.1 MJ **2.** 390 kJ **3.** 5 kg **4.** 130 J/kg °C **5.** 65°C

Exercise 146 (Page 319)

1. Similar to Figure 30.1, page 318 – see section 30.5, page 318

Exercise 147 (Page 320)

1. 8.375 MJ **2.** 18.08 MJ **3.** 3.98 MJ **4.** 12.14 MJ

Exercise 148 (Page 324)

1. (a) 6.06 MW (b) 11 minutes 13 seconds

2. 35.4% **3.** 31 minutes 43 seconds

4. (a) 2.62 MJ/kg (b) 54.58%

5. 45.39% **6.** (a) 2.60 MJ/kg (b) 78.14%

Exercise 149 (Page 325)

Answers found within the text of the chapter, pages 314 to 325.

Exercise 150 (Page 325)

1. (d) **2.** (b) **3.** (a) **4.** (c) **5.** (b)
6. (b) **7.** (b) **8.** (a) **9.** (c) **10.** (b)
11. (d) **12.** (c) **13.** (d)

Chapter 31

Exercise 151 (Page 331)

1. 50.0928 m   **2.** 6.45×10^{-6} K^{-1}
3. 20.4 mm   **4.** (a) 0.945 mm   (b) 0.045%
5. 0.45 mm   **6.** 47.26°C
7. 6.75 mm   **8.** 74 K

Exercise 152 (Page 334)

1. 2.584 mm^2 **2.** 358.8 K
3. (a) 0.372 mm   (b) 280.5 mm^2 (c) 4207 mm^3
4. 29.7 mm^3 **5.** 2.012 litres
6. 50.4 litres   **7.** 0.03025 litres

Exercise 153 (Page 335)

Answers found within the text of the chapter, pages 328 to 334.

Exercise 154 (Page 335)

1. (b)   **2.** (c)   **3.** (a)   **4.** (d)   **5.** (b)
6. (c)   **7.** (c)   **8.** (a)   **9.** (c)   **10.** (b)

Chapter 32

Exercise 155 (Page 339)

1. 0.3 m^3 **2.** 125 kPa   **3.** 400 kPa   **4.** 24 kN
5. 2000 mm of mercury

Exercise 156 (Page 340)

1. 1.02 m^3 **2.** −57°C   **3.** 54.5°C
4. 28.5 litres

Exercise 157 (Page 342)

1. 141 kPa   **2.** 720°C

Exercise 158 (Page 343)

1. 180 kPa

Exercise 159 (Page 344)

1. 0.363 m^3 **2.** 2.30 m^3 **3.** 34.0 litres
4. 1.24 kg   **5.** 1.24 kg   **6.** 300 J/(kg K)
7. 4160 J/(kg K)   **8.** 291°C   **9.** 177°C
10. 0.181 kg   **11.** 184 J/(kg K)

Exercise 160 (Page 346)

1. 1.24 MPa   **2.** 0.60 m^3, 0.72 m^3
3. (a) 57.4 kPa, 142.6 kPa   (b) 1.94 kg
4. (a) 0.0068 m^3 (b) 0.050 kg
5. (a) 2.33 kg   (b) 1.83 m^3

Exercise 161 (Page 347)

Answers found within the text of the chapter, pages 337 to 346.

Exercise 162 (Page 347)

1. (a)   **2.** (d)   **3.** (b)   **4.** (b)   **5.** (c)
6. (d)   **7.** (b)   **8.** (c)   **9.** (c)   **10.** (b)

Chapter 33

Exercise 163 (Page 353)

1. 1270°C

Exercise 164 (Page 355)

1. 203°C

Exercise 165 (Page 358)

Answers found within the text of the chapter, pages 349 to 358.

Exercise 166 (Page 358)

1. (c) **2.** (b) **3.** (d) **4.** (b) **5.** (i) **6.** (a)
7. (e) **8.** (d) **9.** (e) or (f) **10.** (k) **11.** (b) **12.** (g)

Chapter 34

Exercise 167 (Page 366)

1. 5 s **2.** 3600 C **3.** 13 min 20 s

Exercise 168 (Page 368)

1. 7 Ω **2.** 0.25 A **3.** 2 mΩ, 5 mΩ **4.** 30 V

Exercise 169 (Page 371)

1. 0.4 A, 100 W **2.** 20 Ω, 2.88 kW, 57.6 kWh
3. 0.8 W **4.** 9.5 W **5.** 2.5 V

6. (a) 0.5 W (b) 1.33 W (c) 40 W
7. 8 kW, 20 A **8.** 5 kW **9.** £26.70
10. 3 kW, 90 kWh, £12.15

Exercise 170 (Page 372)

1. 3 A, 5 A

Exercise 171 (Page 372)

Answers found within the text of the chapter, pages 363 to 372.

Exercise 172 (Page 373)

1. (d) **2.** (a) **3.** (c) **4.** (b) **5.** (d) **6.** (d)
7. (b) **8.** (c) **9.** (a) **10.** (a) **11.** (c) **12.** (c)
13. (b) **14.** (a) **15.** (c) **16.** (b) **17.** (d) **18.** (d)

Chapter 35

Exercise 173 (Page 378)

1. (a) 8.75 Ω (b) 5 m **2.** (a) 5 Ω (b) 0.625 mm^2
3. 0.32 Ω **4.** 0.8 Ω
5. 1.5 mm^2 **6.** 0.026 $\mu\Omega$m
7. 0.216 Ω

Exercise 174 (Page 380)

1. 69 Ω **2.** 24.69 Ω **3.** 488 Ω **4.** 26.4 Ω
5. 70°C **6.** 64.8 Ω **7.** 5.95 Ω

Exercise 175 (Page 382)

1. 68 kΩ $\pm$2% **2.** 4.7 Ω $\pm$20%
3. 690 Ω $\pm$5%
4. Green-brown-orange-red
5. brown-black-green-silver **6.** 1.8 MΩ to 2.2 MΩ
7. 39.6 kΩ to 40.4 kΩ
8. (a) 0.22 Ω $\pm$2% (b) 4.7 kΩ $\pm$1%
9. 100KJ **10.** 6M8M

Exercise 176 (Page 382)

Answers found within the text of the chapter, pages 375 to 382.

Exercise 177 (Page 382)

1. (c) **2.** (d) **3.** (b) **4.** (d) **5.** (d)
6. (c) **7.** (b) **8.** (a) **9.** (d)

Chapter 36

Exercise 178 (Page 389)

1. (a) 18 V, 2.88 Ω (b) 1.5 V, 0.02 Ω
2. (a) 2.17 V (b) 1.6 V (c) 0.7 V
3. 0.25 Ω **4.** 18 V, 1.8 Ω **5.** (a) 1 A (b) 21 V
6. (i)(a) 6 V (b) 2 V (ii)(a) 4 Ω (b) 0.25 Ω
7. 0.04 Ω, 51.2 V

Exercise 179 (Page 400)

Answers found within the text of the chapter, pages 384 to 400.

Exercise 180 (Page 401)

1. (d) **2.** (a) **3.** (b) **4.** (c) **5.** (b)
6. (d) **7.** (d) **8.** (b) **9.** (c) **10.** (d)
11. (c) **12.** (a) **13.** (c)

Chapter 37

Exercise 181 (Page 406)

1. (a) 22 V (b) 11 Ω (c) 2.5 Ω, 3.5 Ω, 5 Ω
2. 10 V, 0.5 A, 20 Ω, 10 Ω, 6 Ω
3. 4 A, 2.5 Ω **4.** 45 V   **5.** (a) 1.2 Ω (b) 12 V
6. 6.77 Ω **7.** (a) 4 Ω (b) 48 V

Exercise 182 (Page 412)

1. (a) 3 Ω (b) 3 A (c) 2.25 A, 0.75 A
2. 2.5 A, 2.5 Ω
3. (a)(i) 5 Ω (ii) 60 kΩ (iii) 28 Ω (iv) 6.3 kΩ
   (b)(i) 1.2 Ω (ii) 13.33 kΩ (iii) 2.29 Ω (iv) 461.54 Ω
4. 8 Ω **5.** 27.5 Ω **6.** 2.5 Ω, 6 A
7. (a) 1.6 A (b) 6 Ω **8.** (a) 30 V (b) 42 V
9. I_1 = 5 A, I_2 = 2.5 A, I_3 = 1.67 A, I_4 = 0.83 A, I_5 = 3 A,
   I_6 = 2 A, V_1 = 20 V, V_2 = 5 V, V_3 = 6 V
10. 1.8 A

Exercise 183 (Page 414)

1. 400 Ω **2.** (a) 70 V (b) 210 V

Exercise 184 (Page 414)

Answers found within the text of the chapter, pages 403 to 414.

Exercise 185 (Page 415)

1. (a)   **2.** (c)   **3.** (c)   **4.** (c)   **5.** (a)
6. (d)   **7.** (b)   **8.** (c)   **9.** (d)   **10.** (d)

Chapter 38

Exercise 186 (Page 422)

1. I_3 = 2 A, I_4 = −1 A, I_6 = 3 A
2. (a) I_1 = 4 A, I_2 = −1 A, I_3 = 13 A
   (b) I_1 = 40 A, I_2 = 60 A, I_3 = 120 A,
      I_4 = 100 A, I_5 = −80 A

3. 2.162 A, 42.09 W
4. 2.716 A, 7.407 V, 3.951 V
5. (a) 60.38 mA (b) 15.10 mA
   (c) 45.28 mA (d) 34.20 mW
6. I_1 = 1.26 A, I_2 = 0.74 A, I_3 = 0.16 A,
   I_4 = 1.42 A, I_5 = 0.58 A

Exercise 187 (Page 423)

Answers found within the text of the chapter, pages 418 to 422.

Exercise 188 (Page 423)

1. (a) **2.** (d) **3.** (c) **4.** (b) **5.** (c)

Chapter 39

Exercise 189 (Page 431)

1. 1.5 T   **2.** 2.7 mWb   **3.** 32 cm   **4.** 4 cm by 4 cm

Exercise 190 (Page 434)

1. 21.0 N, 14.8 N   **2.** 4.0 A   **3.** 0.80 T   **4.** 0.582 N
5. (a) 14.2 mm (b) towards the viewer

Exercise 191 (Page 435)

1. 8×10^{-19} N   **2.** 10^6 m/s

Exercise 192 (Page 435)

Answers found within the text of the chapter, pages 424 to 435.

Exercise 193 (Page 435)

1. (d) **2.** (d) **3.** (a) **4.** (a) **5.** (b) **6.** (b)
7. (d) **8.** (c) **9.** (d) **10.** (a) **11.** (c) **12.** (c)

Chapter 40

Exercise 194 (Page 442)

1. 0.135 V   **2.** 25 m/s   **3.** (a) 0 (b) 0.16 A
4. 1 T, 0.25 N   **5.** 1.56 mV
6. (a) 48 V (b) 33.9 V (c) 24 V
7. (a) 10.21 V (b) 0.408 N

Exercise 195 (Page 444)

1. 4.5 V   **2.** 1.6 mH   **3.** 250 V
4. (a) −180 V (b) 5.4 mWb

Exercise 196 (Page 446)

1. 96   **2.** 990 V   **3.** 16 V
4. 3000 turns   **5.** 12 V, 60 A   **6.** 50 A
7. 16 V, 45 A   **8.** 225 V, 3:2

Exercise 197 (Page 446)

Answers found within the text of the chapter, pages 438 to 446.

Exercise 198 (Page 447)

1. (c)   **2.** (b)   **3.** (c)   **4.** (a)   **5.** (d)
6. (a)   **7.** (b)   **8.** (c)   **9.** (d)   **10.** (a)
11. (b)   **12.** (a)   **13.** (d)   **14.** (b) and (c)

Chapter 41

Exercise 199 (Page 453)

1. (a) 0.4 s (b) 10 ms (c) 25 μs
2. (a) 200 Hz (b) 20 kHz (c) 5 Hz
3. 800 Hz

Exercise 200 (Page 456)

1. (a) 50 Hz (b) 5.5 A, 3.1 A (c) 2.8 A (d) 4.0 A
2. (a) (i) 100 Hz (ii) 2.50 A (iii) 2.87 A
     (iv) 1.15 (v) 1.74
     (b) (i) 250 Hz (ii) 20 V (iii) 20 V (iv) 1.0 (v) 1.0
     (c) (i) 125 Hz (ii) 18 A (iii) 19.56 A
       (iv) 1.09 (v) 1.23
     (d) (i) 250 Hz (ii) 25 V (iii) 50 V (iv) 2.0 (v) 2.0
3. (a) 150 V (b) 170 V
4. 212.1 V
5. 282.9 V, 180.2 V
6. 84.8 V, 76.4 V
7. 23.55 A, 16.65 A

Exercise 201 (Page 459)

Answers found within the text of the chapter, pages 450 to 459.

Exercise 202 (Page 459)

1. (b)   **2.** (c)   **3.** (d)   **4.** (d)   **5.** (a)
6. (d)   **7.** (c)   **8.** (b)   **9.** (c)   **10.** (b)

Chapter 42

Exercise 203 (Page 465)

1. 2.5 mC   **2.** 2 kV   **3.** 2.5 μF
4. 1.25 ms   **5.** 2 kV   **6.** 750 kV/m
7. 312.5 μC/m², 50 kV/m

Exercise 204 (Page 466)

1. 885 pF   **2.** 0.885 mm   **3.** 7   **4.** 2.97 mm   **5.** 200 pF

Exercise 205 (Page 467)

1. (a) $8\,\mu F$ (b) $1.5\,\mu F$
2. $15\,\mu F$
3. (a) $0.06\,\mu F$ (b) $0.25\,\mu F$
4. $2.4\,\mu F$, $2.4\,\mu F$
5. (a) $1.2\,\mu F$ (b) $100\,V$
6. $4.2\,\mu F$ each

Exercise 206 (Page 468)

1. (a) $0.02\,\mu F$ (b) $0.4\,mJ$ 2. $20\,J$ 3. $550\,V$
4. (a) $0.04\,mm$ (b) $361.6\,cm^2$ (c) $0.02\,J$ (d) $1\,kW$

Exercise 207 (Page 472)

1. $-150\,V$ 2. $144\,ms$ 3. $0.8\,Wb/s$
4. $3.5\,V$ 5. $-90\,V$

Exercise 208 (Page 474)

1. $62.5\,J$ 2. $0.18\,mJ$ 3. $40\,H$

Exercise 209 (Page 474)

1. (a) $7.2\,H$ (b) $90\,J$ (c) $180\,V$
2. $4\,H$ 3. $40\,ms$
4. $12.5\,H$, $1.25\,kV$ 5. $12,500$

Exercise 210 (Page 475)

Answers found within the text of the chapter, pages 461 to 474.

Exercise 211 (Page 476)

1. (a) 2. (b) 3. (c) 4. (a) 5. (b)
6. (b) 7. (c) 8. (c) 9. (c) 10. (d)
11. (c) 12. (b) 13. (d) 14. (a)

Chapter 43

Exercise 212 (Page 481)

1. (a) $0.250\,A$ (b) $0.238\,A$ (c) $2.832\,W$ (d) $56.64\,W$
2. (a) $900\,W$ (b) $904.5\,W$

Exercise 213 (Page 486)

1. (a) $41.7\,Hz$ (b) $176\,V$
2. (a) $0.56\,Hz$ (b) $8.4\,V$
3. (a) $7.14\,Hz$ (b) $220\,V$ (c) $77.78\,V$

Exercise 214 (Page 492)

1. $3\,k\Omega$ 2. $1.525\,V$

Exercise 215 (Page 492)

Answers found within the text of the chapter, pages 478 to 492.

Exercise 216 (Page 492)

1. (f)   2. (c)   3. (a)   4. (i)   5. (j)
6. (g)   7. (c)   8. (b)   9. (p)   10. (d)
11. (o)   12. (n)   13. (a)

Chapter 44

Exercise 217 (page 510)

Answers found within the text of the chapter, pages 494 to 510.

Chapter 45

Exercise 218 (Page 522)

Answers found within the text of the chapter, pages 515 to 522.

Exercise 219 (Page 522)

1. (d)   2. (a)   3. (c)   4. (d)
5. (b)   6. (a)   7. (b)   8. (c)

Glossary of terms

Acceleration: The amount by which the velocity of an object increases in a certain time.

Acceleration of free fall: The acceleration experienced by bodies falling freely in the Earth's gravitational field. It varies from place to place around the globe, but is assigned a standard value of 9.80665m/s^2, called 'g'. Ignoring air resistance, the acceleration does not vary with the size or shape of the falling body. The value of 'g' on the equator $\approx 9.78 \text{ m/s}^2$ is less than its value at the poles, where $g \approx 9.83 \text{ m/s}^2$.

Acid: A chemical compound containing hydrogen that can be replaced by a metal or other positive ion to form a salt. Acids dissociate in water to yield aqueous hydrogen ions, thus acting as proton donors; the solutions are corrosive, have a sour taste, give a red colour to indicators, and have a pH below 7. Strong acids, such as sulphuric acid, are fully dissociated into ions whereas weak acids, such as ethanoic acid, are only partly dissociated.

Algebra: The branch of mathematics dealing with the study of equations that are written using numbers and alphabetic symbols, which themselves represent quantities to be determined.

Alkali: A soluble base that reacts with an acid to form a salt and water. A solution of an alkali has a pH greater than 7 and turns litmus dye blue. Alkali solutions are used as cleaning materials.

Alloy: A combination of two or more metals.

Ampere: The SI unit of electric current. The symbol is A.

Angular acceleration: The rate of change of angular velocity.

Angular momentum: The product of the moment of inertia I and the angular velocity ω of an object.

Angular velocity: The rate of change of an object's angular position relative to a fixed point.

Anode: The positive electrode of an electrolytic cell.

Archimedes' principle: A body immersed in a fluid is pushed up by a force equal to the weight of the displaced fluid.

Atmospheric pressure: The downward force exerted by the atmosphere because of its weight, (gravitational attraction to the Earth), measured by barometers, and usually expressed in units of millibars. Standard atmospheric pressure at sea level is 1013.25 mb.

Atomic number: The number of protons in the nucleus of an atom of an element, which is equal to the number of electrons moving around that nucleus.

Bar: Unit of pressure – the pressure created by a column of mercury 75.006 cm high at 0°C, or about 33.45 feet of water at 4°C. It is equal to 10^5 pascal. Standard atmospheric pressure (at sea level) is 1.01325 bar, or 1013.25 mb.

Barometer: An instrument for measuring atmospheric pressure. There are two main types – the mercury barometer, and the aneroid barometer.

Base: In chemistry, any compound that accepts protons. A base will neutralise an acid to form salt and water.

Battery: A collection of electrochemical cells that convert chemical energy into electrical energy.

Boyle's law: The volume of a gas at constant temperature is inversely proportional to the pressure. This means that as pressure increases, the volume of a gas decreases.

Calorie: A unit of heat. A calorie is the amount of heat required to raise 1 g of water by 1°C between the temperatures of 14.5°C and 15.5°C. The SI system uses the joule (1 calorie = 4.184 joules) instead of the calorie. 1000 gram calories = 3.968 Btu (British thermal unit). 1 J = 1 N m.

Science and Mathematics for Engineering. 978-0-367-20475-4, © John Bird. Published by Taylor & Francis. All rights reserved.

Capacitance: The ability of an electric circuit to store charge. Capacitance, symbol C, is measured in farads.

Capacitor: An electrical circuit component that has capacitance. Having at least two metal plates, it is used principally in alternating current circuits.

Cathode: The negative electrode of an electrolytic cell.

Celsius: The temperature scale based on the freezing point of water (0°C) and the boiling point of water (100°C). The interval between these points is divided into 100 degrees. The scale was devised by Anders Celsius.

Centre of gravity: Point at which the weight of a body can be considered to be concentrated and around which its weight is evenly balanced. In a uniform gravitational field, the centre of gravity is the same as the centre of mass.

Centripetal force: In circular or curved motion, the force acting on an object that keeps it moving in a circular path. For example, if an object attached to a rope is swung in a circular motion above a person's head, the centripetal force acting on the object is the tension in the rope. Similarly, the centripetal force acting on the Earth as it orbits the sun is gravity. In accordance with Newton's laws, the reaction to this can be regarded as a centrifugal force, equal in magnitude and opposite in direction.

Change of state: The change that takes place when matter turns from one physical phase (gas, liquid or solid) into another.

Charles' law: The volume of a gas at constant pressure is directly proportional to its absolute temperature.

Coefficient of friction: The number characterising the force necessary to slide or roll one material along the surface of another. If an object has a weight N and the coefficient of friction is μ, then the force F necessary to move it without acceleration along a level surface is $F = \mu N$. The coefficient of static friction determines the force necessary to initiate movement; the coefficient of kinetic friction determines the force necessary to maintain movement. Kinetic friction is usually smaller than static friction.

Coefficient of linear expansion: The fractional increase in length per unit temperature rise.

Coefficient of superficial expansion: The fractional increase in area per unit temperature rise.

Coefficient of cubic expansion: The fractional increase in volume per unit temperature rise.

Compound: A substance formed by chemical combination of two or more elements that cannot be separated by physical means.

Conduction, thermal: The transfer of heat from a hot region of a body to a cold region.

Conductor: A substance or object that allows easy passage of free electrons, thereby allowing heat energy or charged particles to flow easily. Conductors have a low resistance.

Convection: The transfer of heat by flow of currents within fluids due to kinetic theory.

Conservation of energy, law of: States that energy cannot be created or destroyed.

Cosine: In trigonometry, is the ratio of the length of the side adjacent to an acute angle to the length of the hypotenuse in a right-angled triangle.

Coulomb: Is the SI unit of electric charge. It is defined as the charge carried by a current of 1 ampere in 1 second.

Couple: Two equal and opposite parallel forces, which do not act in the same line. The forces produce a turning effect or torque.

Dalton's law: The pressure exerted by each gas in a mixture of gases does not depend on the pressures of the other gases, provided no chemical reaction occurs. The total pressure of such a mixture is therefore the sum of the partial pressures exerted by each gas (as if it were alone in the same volume as the mixture occupies).

Density: The ratio of mass to volume for a given substance expressed in SI units as kilograms per cubic metre. The symbol for density is ρ (Greek rho).

Dielectric: Non-conducting material such as an electrical insulator that separates two conductors in a capacitor.

Dielectric strength: The maximum field that a dielectric can withstand without breaking down by becoming ionised.

Ductility: Ability of metals and some other materials to be stretched without being weakened.

Dynamics: The branch of mechanics that deals with objects in motion. Its two main branches are kinematics, which studies motion without regards to its cause, and kinetics, which also takes into account forces that cause motion.

Efficiency: The work a machine does (output) divided by the amount of work put in (input), usually expressed as a percentage. For simple machines, efficiency can be defined as the force ratio (mechanical advantage) divided by the distance ratio (velocity ratio).

Elasticity: Capability of a material to recover its size and shape after deformation by stress. When an external force (stress) is applied, the material develops strain (a change in dimension). If a material passes its elastic limit, it will not return to its original shape.

Electric circuit: System of electric conductors, appliances or electronic components connected together so that they form a continuously conducting path for an electric current.

Electric current: A movement of electrons along a conductor.

Electric field: The region around an electric charge in which any particle experiences a force.

Electrolysis: The chemical reaction caused by passing a direct current through an electrolyte.

Electron: A subatomic particle with a negative elementary electric charge.

Electromagnet: A magnet constructed from a soft iron core around which is wound a coil of insulating wire.

Element: A substance that cannot be split into simpler substances by chemical means.

Energy: The capacity for doing work; it is measured in joules.

Equilibrium: A stable state in which forces acting on a particle or object negate each other, resulting in no net force.

Expansion: A change in the size of an object with change in temperature. Most substances expand on heating, although there are exceptions – water expands when it cools from 4°C to its freezing point at 0°C.

Fahrenheit: The temperature scale based on the freezing point of water (32°F) and the boiling point of water (212°F). The interval between these points is divided into 180 equal parts. Although replaced by the Celsius scale, the Fahrenheit scale is still sometimes used for non-scientific measurements.

Farad: The SI unit of capacitance; the symbol is F. The farad is name after Michael Faraday.

Fluid: Any substance that is able to flow. Of the three common states of matter, gas and liquid are considered fluid, while solid is not.

Force: A push, a pull or a turn. A force acting on an object may (i) balance an equal but opposite force or a combination of forces to maintain the object in equilibrium (so that it does not move), (2) change the state of motion of the object (in magnitude or direction), or (3) change the shape or state of the object. The unit of force is the newton.

Force ratio: The factor by which a simple machine multiplies an applied force. It is the ratio of the load (output force) to the effort (input force).

Free fall: The state of motion of an unsupported body in a gravitational field.

Freezing point: The temperature at which a substance changes phase (or state) from liquid to solid. The freezing point for most substances increases as pressure increases. The reverse process, from solid to liquid, is melting; melting point is the same as freezing point.

Friction: The resistance encountered when surfaces in contact slide or roll against each other, or when a fluid (liquid or gas) flows along a surface. Friction is directly proportional to the force pressing the surfaces together and the surface roughness. Before the movement begins, it is opposed by static friction up to a maximum 'limiting friction' and then slipping occurs.

Fulcrum: Point about which a lever pivots.

Fuse: A safety device to protect against overloading. Fuses are commonly strips of easily melting metal placed in series in an electric circuit such that, when overloaded, the fuse melts, breaking the circuit and preventing damage to the rest of the system.

Galvanometer: An instrument for detecting, comparing or measuring electric current.

Gear wheel: Is usually toothed, attached to a rotating shaft. The teeth of one gear engage those of another to transmit and modify rotary motion and torque. The smaller member of a pair of gears is called a pinion. If the pinion is on the driving shaft, speed is reduced and turning force increased. If the larger gear is on the driving shaft, speed is increased and turning force reduced. A screw-type driving gear, called a worm, gives the driven gear a greatly reduced speed.

Gravity: The gravitational force of attraction at the surface of a planet or other celestial body. The Earth's gravity produces an acceleration of around 9.8m/s^2 for any unsupported body.

Heat: A form of energy associated with the constant vibration of atoms and molecules.

Hertz: The SI unit of frequency named after Heinrich Hertz. 1 Hz is equivalent to 1 cycle per second.

Hooke's law: Within the limit of proportionality, the extension of a material is proportional to the applied force. Approximately, it is the relationship between stress and strain in an elastic material when it is stretched. The law states that the stress (force per unit area) is proportional to the strain (a change in dimensions). The law, which holds only approximately and over a limited range, was discovered in 1676 by Robert Hooke.

Horsepower: Unit indicating the rate at which work is done. The electrical equivalent of one horsepower is 746 watts.

Hydraulics: The physical science and technology of the behaviour of fluids in both static and dynamic states. It deals with practical applications of fluid in motion and devices for its utilisation and control.

Hydrostatics: The branch of mechanics that deals with liquids at rest. Its practical applications are mainly in water engineering and in the design of such equipment as hydraulic presses, rams, lifts and vehicle braking and control systems.

Ideal gas laws: The law relating pressure, temperature and volume of an ideal (perfect) gas pV = mRT, where R is the gas constant. The law implies that at constant temperature T, the product of pressure p and volume V is constant (Boyle's law), and at constant pressure, the volume is proportional to the temperature (Charles' law).

Impedance: In an electrical circuit, a measure of the opposition to the passage of a current when a voltage is applied.

Imperial system: The units of measurement developed in the UK. Formerly known as the fps system, which is an abbreviation for the 'foot-pound-second system' of units.

Inductance: The property of an electric circuit or component that produces an electromotive force following a change in the current. The SI unit of inductance is the henry, H.

Inertia: The property possessed by all matter that is a measure of the way an object resists changes to its state of motion.

Ion: An atom or molecule in which the total number of electrons is not equal to the total number of protons, giving the atom a net positive or negative electrical charge.

Joule: The SI unit of energy. One joule is the work done by a force of one newton acting over a distance of one metre. The symbol is J, where 1 J = 1 N m.

Kelvin: The SI unit of temperature. The Kelvin temperature scale has a zero point at absolute zero and degree intervals (kelvins), the same size as degrees Celsius. The freezing point of water occurs at 273K (0°C) and the boiling point at 373 K (100°C).

Kinetic energy: Energy that an object possesses because it is in motion. It is the energy given to an object to set it in motion. On impact, it is converted into other forms of energy such as strain, heat, sound and light.

Latent heat: The heat absorbed or given out by a substance as it changes its phase (of matter) at constant temperature – from a solid to a liquid state or from a liquid to a gas.

Latent heat of fusion: The heat necessary to transform ice into water at constant temperature.

Latent heat of vaporisation: The heat necessary to transform water into steam at constant temperature.

Lever: A simple machine used to multiply the force applied to an object, usually to raise a heavy load. A lever consists of a rod and a point (fulcrum) about which the rod pivots. In a crowbar, for example, the applied force (effort) and the object to be moved (load) are on opposite sides of the fulcrum, with the point of application of the effort farther from it. The lever multiplies the force applied by the ratio of the two distances.

Machine: A device that modifies or transmits a force in order to do useful work. In a simple machine, a force (effort) opposes a larger force (load). The ratio of the load (output force) to the effort (input force) is the machine's force ratio, formerly called mechanical advantage. The ratio of the distance moved by the load to the distance moved by the effort is the distance or movement ratio, formerly known as the velocity ratio. The ratio of the work done by the machine to that put into it is the efficiency, usually expressed as a percentage.

Magnetic field: The region surrounding a magnet, or a conductor through which a current is flowing, in which magnetic effects, such as the deflection of a compass needle, can be detected.

Magnetic flux: The measure of the strength and extent of a magnetic field. The unit of magnetic flux is the weber, Wb.

Malleability: Property of materials (or other substances) that can be permanently shaped by hammering or rolling without breaking. In some cases, it is increased by raising temperature.

Manometer: A device for measuring pressure.

Mechanical advantage: The factor by which a simple machine multiplies an applied force. It is the ratio of the load (output force) to the effort (input force).

Mechanics: The branch of physics concerned with the behaviour of matter under the influence of forces. It may be divided into solid mechanics and fluid mechanics. Another classification is as statics, the study of matter at rest, and dynamics, the study of matter in motion.

Metric system: The decimal system of weights and measures based on a unit of length called the metre and a unit of mass called the kilogram. Devised by the French in 1791, the metric system is used internationally (SI units) and has been adopted for general use by most Western countries, although the imperial system is still commonly used in the USA and for certain measurements in Britain.

Moment of inertia: For a rotating object. The sum of the products formed by multiplying the elements of mass of the rotating object by the squares of their distances from the axis of the rotation. Finding this distribution of mass is important when determining the force needed to make the object rotate.

Momentum: The product of the mass and linear velocity of an object. One of the fundamental laws of physics is the principle that the total momentum of any system of objects is conserved (remains constant) at all times, even during and after collisions.

Motion, laws of: Three laws proposed by Isaac Newton form the basis of the classical study of motion and force. According to the **first law**, a body resists changes in its state of motion – a body at rest tends to remain at rest unless acted upon by an external force, and a body in motion tends to remain in motion at the same velocity unless acted on by an external force. This property is known as inertia. The **second law** states that the change in velocity of a body as a result of a force is directly proportional to the force and inversely proportional to the mass of the body. According to the

third law, to every action there is an equal and opposite reaction.

Nautical mile: The unit used to measure distances at sea, it is defined as the length of one minute of arc of the Earth's circumference. The international nautical mile is equal to 1852 m (6076.04 feet), but in the UK it is defined as 6080 feet (1853.18 m). A speed of one nautical mile per hour is called a knot, a term used both at sea and in flying.

Newton: The SI unit of force with the symbol N. One newton is the force that gives a mass of one kilogram an acceleration of one metre per second per second. One kilogram weighs 9.807 N.

Ohm: Is the SI unit of electrical resistance, named after Geog Simon Ohm.

Ohm's law: A statement that the amount of steady current through a material is proportional to the voltage across the material.

Parallelogram of vectors: A method of calculating the sum of two vector quantities. The direction and size of the vectors is determined by trigonometry or scale drawing. The vectors are represented by two adjacent sides of a parallelogram and the sum is the diagonal through their point of intersection.

Pascal: The SI unit of pressure with the symbol Pa. It is equal to a pressure of 1 newton per square metre.

Permittivity: A measure of the extent to which a dielectric can resist the flow of charge.

Potential difference: The difference in electric potential between two points in a circuit or electric field, usually expressed in volts.

Potential energy: An object's ability to do work because of a change in the object's position or shape.

Power: The rate of doing work or of producing or consuming energy. The unit of power is the watt, W, where $1 \text{ W} = 1 \text{ N m/s}$.

Pressure: The force on an object's surface divided by the area of the surface. The SI unit is the pascal (symbol Pa), which is 1 newton per square metre. In meteorology, the millibar, which equals 100 pascals, is commonly used. $1 \text{ bar} = 10^5 \text{Pa} = 14.5 \text{ psi}$.

Principle of moments: A law that states that the moments of two bodies balanced about a central pivot or fulcrum are equal (the moment of a body being the product of its mass and its distance from the pivot).

Pulley: A simple machine used to multiply force or to change the direction of its application. A simple pulley consists of a wheel, often with a groove, attached to a fixed structure. Compound pulleys consist of two or more such wheels, some movable, that allow a person to raise objects much heavier than he or she could lift unaided.

Pyrometer: A thermometer for use at extremely high temperatures, well above the ranges of ordinary thermometers.

Radian: The angle formed by the intersection of two radii at the centre of a circle, when the length of the arc cut off by the radii is equal to one radius in length. Thus, the radian is a unit of angle equal to $57.296°$, and there are 2π radians in $360°$.

Radiation: The transmission of energy by subatomic particles of electromagnetic waves.

Refrigeration: The process by which the temperature in a refrigerator is lowered. In a domestic refrigerator, a refrigerant gas, such as ammonia or chlorofluorocarbon (CFC) is alternately compressed and expanded. The gas is first compressed by a pump, causing it to warm up. It is then cooled in a condenser where it liquefies. It is then passed into an evaporator where it expands and boils, absorbing heat from its surroundings and thus cooling the refrigerator. It is then passed through the pump again to be compressed.

Relative density: The ratio of the density of one substance to that of a reference substance (usually water) at the same temperature and pressure. Formerly called specific capacity.

Scalar: A quantity that only has magnitude; mass, energy and speed are examples of scalars.

Screw: A variant of a simple machine, the inclined plane. It is an inclined plane cut around a cone, usually of metal, in a helical spiral. When force is exerted radially on the screw, for example by a screwdriver or the lever of a screw-jack, the screw advances to an extent determined by its pitch (the distance between crests of its thread).

Shearing force: The force tending to cause deformation of a material by slipping along a plane parallel to the imposed stress.

SI units: The Systeme International d'Unites – the internationally agreed system of units, derived from the MKS system (metre, kilogram and second). The seven basic units are: the metre (m), kilogram (kg), second (s), ampere (A), kelvin (K), mole (mol), and candela (cd).

Specific heat capacity: The heat necessary to raise the temperature of 1 kg of a substance by 1 K. It is measured in J/kg K.

Statics: The study of matter at rest. In statics, the forces on an object are balanced and the object is said to be in equilibrium; static equilibrium may be stable, unstable or neutral.

STP: An abbreviation of standard temperature and pressure.

Strain: Change in dimensions of an object subjected to stress. Linear strain is the ratio of the change in length of a bar to its original length. Shearing strain describes the change in shape of an object whose opposite faces are pushed in different directions. Hooke's law for elastic materials states that strain is approximately proportional to stress up to the material's limit of proportionality.

Stress: Force per unit area applied to an object. Tensile stress stretches an object, compressive stress squeezes it, and shearing stress deforms it sideways. In a fluid, no shearing stress is possible because the fluid slips sideways, so all fluid stresses are pressures.

Temperature: A measure of the hotness or coldness of an object.

Tensile strength: The resistance that a material offers to tensile stress. It is defined as the smallest tensile stress required to break the body.

Tesla: The SI unit of magnetic flux density with symbol T.

Thermistor: A type of semiconductor whose resistance sharply decreases with increasing temperature. At $20°C$ the resistance may be of the order of a thousand ohms and at $100°C$ it may be only ten ohms. Thermistors are used to measure temperature and to compensate for temperature changes in other parts of the circuit.

Thermocouple: A thermometer made from two wires of different metals joined at one end, with the other two ends maintained at constant temperature. The junction between the wires is placed in the substance whose temperature is to be measured. An e.m.f. is generated which can be measured and which, in turn, is a measure of temperature.

Thermopile: A device used to measure radiant heat, consisting of several thermocouples connected together in series. Alternate junctions are blackened for absorbing radiant heat, the other junctions are shielded from the radiation. The e.m.f. generated by the temperature difference between the junctions can be measured. From this, the temperature of the blackened junctions can be calculated, and thus the intensity of the radiation measured.

Torque: Turning effect of a force. An example is a turbine that produces a torque on its rotating shaft to turn a generator. The unit of measurement is newton metre (N m).

Torsion: The strain in material that is subjected to a twisting force. In a rod or shaft, such as an engine drive shaft, the torsion angle of twist is inversely proportional to the fourth power of the rod diameter multiplied by the shear modulus (a constant) of the material. Torsion bars are used in the spring mechanism of some car suspensions.

Triangle of vectors: A triangle, the sides of which represent the magnitude and direction of three vectors about a point that are in the same plane and are in equilibrium. A triangle of vectors is often used to represent forces or velocities. If the magnitude and direction of two forces are known, then two sides of the triangle can be drawn. Using scale drawing or trigonometry, the magnitude and direction of the third force can be calculated.

Vacuum flask: A container for keeping things (usually liquids) hot or cold. A vacuum flask is made with double, silvered glass walls separated by a near vacuum. The vacuum prevents heat transfer by conduction and convection, and the silvering on the glass minimizes heat transfer by radiation.

Vaporisation: The conversion of a liquid or solid into its vapour, such as water into steam.

Vector: A quantity that has both magnitude (size) and direction; velocity, acceleration and force are examples of vectors.

Velocity: The rate of motion of a body in a certain direction.

Velocity ratio: In a simple machine. The distance moved by the point of application of the effort (input force) divided by the distance moved by the point of application of the load (output force).

Watt: The SI unit of power, named after the Scottish engineer James Watt. A machine consuming one joule of energy per second has a power output of one watt. 1 W = 1 J/s = 1 N m/s. One horsepower corresponds to 746 watts.

Weber: The SI unit of magnetic flux, with symbol Wb.

Weight: The force of attraction on a body due to gravity. A body's weight is the product of its mass and the gravitational field strength at that point. Mass remains constant, but weight depends on the object's position on the Earth's surface, decreasing with increasing altitude.

Work: The energy transferred in moving the point of application of a force. It equals the magnitude of the force multiplied by the distance moved in the direction of the force.

Young's modulus: Ratio of the stress exerted on a body to the longitudinal strain produced.

Index

BOOKS TO HELP YOU
AT EVERY STEP OF YOUR
ENGINEERING EDUCATION

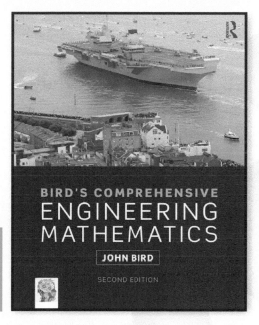

2nd Edition
June 2018
978-0-8153-7814-3

Comprehensively covers the key mathematics for real-life engineering.

BIRD'S COMPREHENSIVE
ENGINEERING MATHEMATICS
JOHN BIRD
SECOND EDITION

8th Edition
May 2017
978-1-138-67359-5

Mathematical theories and interactive problems for students who require more than the basics.

SI UNITS USED
EIGHTH EDITION
ENGINEERING MATHEMATICS
JOHN BIRD

6th Edition
March 2017
978-1-138-67352-6

Introduces electrical and electronic principles and technology through examples rather than theory.

SI UNITS USED
SIXTH EDITION
ELECTRICAL AND ELECTRONIC PRINCIPLES AND TECHNOLOGY
JOHN BIRD

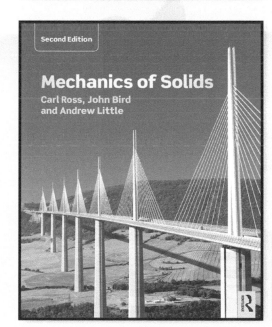

Second Edition

Mechanics of Solids

Carl Ross, John Bird and Andrew Little

2nd Edition
January 2016
978-1-138-90467-5

Essential introduction to the behaviour of solid materials and their properties for first year undergrads.

Routledge
Taylor & Francis Group

Routledge... think about it
www.routledge.com/furthereducation